Geophysical Monograph 29
Maurice Ewing Volume 5

Climate Processes and Climate Sensitivity

Edited by

James E. Hansen
Taro Takahashi

American Geophysical Union
Washington, D.C.
1984

Climate Processes and Climate Sensitivity

Library of Congress Cataloging in Publication Data

Maurice Ewing Symposium (4th : 1982 : Palisades, N.Y.)
Climate processes and climate sensitivity.

(Maurice Ewing series ; 5) (Geophysical monograph, ISSN 0065-8448 ; v. 29)
"Based on papers presented at the Fourth Biennial Maurice Ewing Symposium held at Palisades, New York, October 25-27, 1982"
Includes bibliographies.
1. Climatology—Congresses. 2. Geophysics—Congresses. 3. Ocean-atmosphere interaction--Congresses. I. Hansen, James E. (James Edward), 1941- . II. Takahashi, Taro. III. American Geophysical Union. IV. Title. V. Series. VI. Series: Geophysical monograph ; v. 29.

QC980.M38 1982 551.5 84-6256
ISBN 0-87590-404-1

2000 Florida Avenue, NW
Washington, D.C. 20009

Printed in the United States of America

CONTENTS

OCEAN CHEMISTRY

Preface

The objective of this book is to contribute to the understanding of the sensitivity of the climate system by bringing together the work of modelers and others involved in the analysis of observational data. The approach considers climate variations and climate processes (focusing on climate feedback processes) on a broad range of time scales.

We have organized the book into conventional areas such as the atmosphere, ocean and cryosphere, but these are bound together by the feedback processes; thus the papers in different sections overlap extensively. If the book contributes a small amount toward a better understanding of the role which feedbacks play in coupling different parts of the climate system and in determining climate sensitivity, it will have served its purpose well.

We have included, following this preface, an introduction by E. E. David, Jr., President of the EXXON Research and Engineering Company. David speaks of another feedback process, the interactions between progressive understanding of the earth's climate and possible societal adaptation, particularly with regard to carbon dioxide climate effects. His thoughtful discourse should be of interest to those who study the climate system as well as to planners. There is a wide range of opinion as to when greenhouse climate effects will become important, but no doubt that eventually, as man-made climate effects become apparent, scientists will be pressed to provide the best possible data and understanding of the climate system.

This monograph discusses climate processes and climate sensitivity and is the result of a symposium held at the Lamont-Doherty Geophysical Observatory in Palisades, New York. The meeting was the fourth biennial Maurice Ewing symosium in honor of the founder and first director of the Observatory. The subject, differing from the topics in solid earth geophysics of the previous Ewing symposia, was chosen deliberately to emphasize the breadth of both Professor Ewing's research interests and the scientific contributions of the Observatory.

The symposium was made possible by a generous grant from the EXXON Research and Engineering Company, and was held October 25-27, 1982. Additional support was provided by the National Climate Program Office of the National Oceanographic and Atmospheric Administration, and by the Carbon Dioxide Research Division of the Department of Energy. We thank Wallace Broecker of L-DGO, David Rind and William Rossow of Goddard Institute of Space Studies, Hank C. Hayworth of EXXON for advice on the program, Margaret Swan for helping with the logistics and Andrea Calarco and Christy Bohol for handling and typing most of the manuscripts.

James Hansen
Taro Takahashi

INVENTING THE FUTURE: ENERGY AND THE CO_2 "GREENHOUSE" EFFECT

E. E. David Jr.

President, Exxon Research and Engineering Company
Remarks at the Fourth Annual Ewing Symposium, Tenafly, NJ

Dennis Gabor, A winner of the Nobel Prize for Physics, once remarked that man cannot predict the future, but he can invent it. The point is that while we do not know with certainty how things will turn out, our own actions can play a powerful role in shaping the future. Naturally, Gabor had in mind the power of science and technology, and the model includes that of correction or feedback.

It is an important: Man does not have the gift of prophecy. Any manager or government planner would err seriously by masterminding a plan based unalterably on some vision of the future, without provision for mid-course correction. It is also a comforting thought. With man's notorious inability to create reliable predictions about such matters as elections, stock markets, energy supply and demand, and, of course, the weather, it is a great consolation to feel that we can still retain some control of the future.

As you may know, Exxon is a hundred years old this year; we have a long corporate memory of the very profound social and economic transformations that our business activities have helped bring about, and of how we and society have had to adapt further in response. That includes the at least temporary respite given to the whales through substituting kerosene lighting fuel for their rendered blubber; as well as the revolutionary changes wrought by the automobile and other machinery powered by liquid hydrocarbon fuels. The primary factors guiding such developments were technology and economic markets, though political systems also played their role.

But faith in technologies, markets, and correcting feedback mechanisms is less than satisfying for a situation such as the one you are studying at this year's Ewing Symposium. The critical problem is that the environmental impacts of the CO_2 buildup may be so long delayed. A look at the theory of feedback systems shows that where there is such a long delay the system breaks down unless there is anticipation built into the loop. The question then becomes how to anticipate the future sufficiently far in advance to prepare for it.

One answer is to invent the future in another way--through a system of contingency planning based on an assessment of a number of futures. As Harvey Brooks has noted, scenarios have limited use if they are merely surprise free projections of current trends; instead, they must somehow take into account those clouds on the horizon no bigger than a man's hand that can turn out to be dominant influences in twenty years. Inadequate scenario-making explains the poor performance of most social research to date--which so often gives the sense of too little too late, whether the topic is toxic waste, frost belt and sun belt, or the shift from manufacturing to information society. The key is to undertake research that will tend to be independent of future events, or, rather, <u>relevant</u> across a broad spectrum of scenarios.

This is not easy to do, but some of Exxon's own research and development strategy is aimed in that direction. And Exxon is not the only company with this attitude. That is why we and others in the petroleum industry have taken a strong interest in the issue of the greenhouse effect and your work. It is why we have participated in several initiatives to promote your research; it is why we are pleased to contribute to the holding of this symposium and to participate in it. And it is why we have begun our own modest research effort in the field, motivated also by the belief that perhaps the only way to understand a field is to do research in it. You have seen some of the results in a paper delivered yesterday afternoon. We are also in the process of evaluating the data on CO_2 concentrations collected over two years by an Exxon tanker plying between the Gulf of Mexico and the Gulf of Arabia.

Organization

Few people doubt that the world has entered an energy transition away from dependence upon fossil fuels and toward some mix of renewable resources that will not pose problems of CO_2 accumulation. The question is how do we get from here to there while preserving the health of our political, economic and environmental support systems. What I will do in the remainder of this talk is indicate how the world may invent a successful energy future, using the sort of corrective feedback system I have described. My perspective is of course an Exxon perspective, reflecting our own assumptions about the economic and social paths societies will prefer. And since fossil fuels, and liquid chemical fuels, are really the heart of the energy and the CO_2 problem, I will focus on those.

My plan of attack is, first, to consider the implications of recent energy developments. Then I will describe some of the key assumptions that are guiding Exxon's own R&D planning and which, I think, we have in common with many other actors in the scene. Finally, I will go on to mention some of the technical possibilities that may present themselves well beyond our usual twenty-year outlook period, that is, fifty years or more into the future.

While I am far from certain about the details, I think you'll find that I'm generally upbeat about the chances of coming through this most adventurous of all human experiments with the ecosystem.

Recent Energy History

It is ironic that the biggest uncertainties about the CO_2 buildup are not in predicting what the climate will do, but in predicting what people will do. The scientific community is apparently reaching some consensus about the general mechanisms of the greenhouse effect. It is considerably less agreed on how much fossil fuels mankind will burn; how fast economies will grow; what energy technologies societies will foster and when; and so how fast the buildup will occur.

But we do know about the recent past and the present. In the aftermath of the energy price increase of the past decade, consumers have reacted to the price feedback mechanism very much as classic economic theory would predict. They have sharply reduced their energy consumption and, in particular, their consumption of oil. They have substituted other fuels like coal and nuclear for petroleum, although more coal use does <u>increase</u> CO_2 emissions. Consumers have also conserved by turning to more energy efficient technologies, including smaller cars in the U.S. And they have done without.

It is difficult to disentangle the effects of conservation from the effects of recession. According to a recent report from the International Energy Agency, they are about equal. We think conservation effects are larger, but regardless, energy consumers have certainly broken the lockstep relationship between economic activity and energy consumption that seemed to prevail for a quarter century following World War II. For example, according to the International Energy Agency, it now takes 16 percent less energy and 26 percent less oil to produce 1 percent more of output in the non-communist industrialized countries than in 1973.

This development carries great significance for the CO_2 buildup. Consumers and technologists have been inventing and applying a wealth of methods to extract more work from less energy. For example, as one of our own biggest energy customers, we at Exxon have stepped up the efficiency of our refineries by twenty percent since 1973. Because refining is so energy-intensive, the energy savings, and the corresponding reductions of CO_2 emissions, have been very large indeed. Last year the savings amounted to the equivalent of some 28 million barrels of oil--equal to the production from a world-scale, 50,000-barrel-per-day synthetic fuels plant. On top of that, we have set the goal of doubling our refining efficiency by the year 2000, and we think the goal is realistic.

How far will the conservation trend go? It is too early to say for sure, but we think the implications apply very far into the future. And how far will the energy mix tend to favor fuels, such as coal, that produce large amounts of CO_2, rather than fuels with high ratios of hydrogen to carbon, such as gasoline and methane? To some extent the answer to that question depends upon our ability to come up with a source of low cost hydrogen based on non-fossil energy--a point I'll return to later.

Fossil Fuel Outlook: Key Assumptions

In assessing alternative futures, I would offer three assumptions in the form of predictions about the use of energy and fossil fuels.

First, nearly all societies will continue to give primacy to economic growth. The human desire to improve material conditions burns as bright as ever, if not brighter. As we have seen most recently in Poland, governments that fail to deliver at least a convincing promise of growth suffer dire consequences as a rule. With the overall world population expected to double over the next 50 years, economies and energy use will have to grow at a good clip just to hold per capita incomes even. Naturally, the pressures for growth will be greatest in the developing world, where populations are growing fastest.

A second assumption, one that follows from the first, is that in pursuit of growth most societies will prefer least-cost energy alternatives. I say this with the recognition that at least a few developing countries will prefer options that utilize local resources in order to conserve foreign exchange or use local labor, no matter what the cost. An example is Brazil's resort to alcohol fuels extracted from its sugar cane. However, such exceptions will not materially alter the world future.

The third assumption is that societies will continue to prefer the efficiencies of fossil-based liquid fuels in transportation uses. Because conventional petroleum resources will not suffice to meet the demand, a major industry will begin to grow around the turn of the century to produce synthetic fuels from oil sands, oil shale, and coal.

Despite the trend toward electricity, the electric vehicle will have trouble making significant inroads in transportation markets over the next twenty years. One problem is storage, which is partly a problem of energy density. Today's lead-acid batteries store about 1/300th the energy of a like weight of gasoline. We can improve on that; in fact, Exxon is in the middle of developing a zinc-bromine battery with two to three times the capacity of conventional lead-acid batteries. Another problem is the cost of batteries. They are expensive, mainly because of the cost of raw materials and typically short life cycles. Incidentally, we expect that load leveling, rather than the electric car, will be one of the earliest applications of our new battery. However, we would certainly not rule out the electric car one day--perhaps initially in the form of hybrid vehicles powered by batteries in tandem with small gasoline or diesel engines.

Another alternative features electric guideway systems in which vehicles use batteries on the feeder roads and electrically induced power along the main arteries. But the capital costs of such a system would be immense--making it a viable option only for much richer societies than we can foresee.

Granted, liquid fuels-like all chemical fuels-have their share of problems. In burning they may synthesize some unfriendly substances--such as PNA's, NO_x, SO_x and CO_2. Still, there are also well-known problems with producing electricity through non-chemical means, such as nuclear power. Solar voltaics overcome many of these drawbacks, but the inherent problems of the duty cycle and storage make me skeptical that solar voltaics will penetrate a large fraction of the electricity market in the near future, except in remote applications.

But to reiterate my main theme, such assumptions only act as a guide in determining where R&D managers can most usefully concentrate resources for inventing the future, subject to correction and further feedback. In any case we are not up against fatal, malthusian limits to growth. On the distant horizon, we may discern a peaking of petroleum production; because for more than a decade the world has been consuming pertroleum faster than the industry has been replacing it. But remaining non-petroleum fossil fuel resources are immense. As an example, in 1980 oil and gas production accounted for nearly 70 percent of the world's production of fossil energy. But oil and gas reserves account for only a little over 11 percent of the world's estimated total recoverable fossil energy resources.

As a practical matter, you would surely agree that the world economy is committed to using fossil resources for some time to come. The massiveness of the energy system in place simply forbids immediate displacement of one fuel or energy source by another. Historical market studies going back to wood and coal confirm this idea, suggesting that a new energy source requires about 50 years to achieve just half the total energy market.

What are Exxon's projections for fossil fuel use? Over the twenty years encompassed by our normal outlook we estimate the fossil fuel use will grow at the equivalent of about two percent per year. Much of this growth will occur in the developing countries, as they modernize their economies.

Beyond our normal twenty-year outlook period, we recently attempted a forecast of the CO_2 build-up. We assumed different growth rates at different times, but with an average growth rate in fossil fuel use of about one percent a year starting today, our estimate is that the doubling of atmospheric CO_2 levels might occur sometime late in the 21th century. That includes the impacts of synfuels industry. Assuming the greenhouse effect occurs, rising CO_2 concentrations might begin to induce climatic changes around the middle of the 21th century.

Manufacturing synthetic fuels will produce more CO_2 than conventional petroleum fuels, but the impact of substituting synthetics for depleting petroleum supplies will be relatively small. If, in our estimate, we back out synfuels and replace them with conventional petroleum fuels, the difference in CO_2 emissions would only add about five years to the doubling time. This is a highly conservative estimate, because it assumes that industry in the 21st century will continue using today's "Dinosaur" technologies for manufacturing synfuels, with no increase in the efficiency of these highly energy-intensive processes. And it takes no notice of the trends we are already seeing today in this budding information age. As John Pierce, the inventor of satellite communcations, likes to say, soon we may be traveling for pleasure but communicating to work. Such developments could eventually go very far in reducing the energy intensity and CO_2 emissions of advanced economies.

Exxon's Response in Science and Technology

The real point of these extrapolations is to get an understanding of how soon the problem may become serious enough to require action. And the lesson is that, while the issue is clearly important, we can still afford further research on the problem. And the world will have time to accumulate the material and scientific resources required to contend with the problem.

The same point is emphasized in the energy study published last year by the International Institute for Applied System Analysis, or IIASA. The study involved some 150 top scientists at one time or another and represents one of the most comprehensive assessments of the outlook for the next 50 critical years of what may well be in absolute terms the world's period of greatest population growth.

The IIASA study concludes that to make a successful transition from fossil fuels to an energy system based on renewable resources, the world economy must expand its productive powers. It must expand in all dimensions, but, most importantly, in the new knowledge and human skill that enlarge the technological base. For such knowledge and skill more than brute capital, is what enables societies in this age to use the same or even fewer resources to produce more.

The IIASA strategy for inventing that future resembles the one I have suggested: a strategy first, of gradual transition from clean, high quality resources--natural gas and oil--to dirtier unconventional fossil resources. The study also takes note of the CO_2 issue, recommending that society incorporate sufficient non-fossil options in the energy supply system so as to allow expansion of that base, if necessary, as the effects of carbon dioxide become better quantifiable through further research.

That means pursuing research leads in technologies that may not seem attractive by the fashionable standards of financial analysis. In a recent landmark article, Professors Hayes and Abernathy of the Harvard Business School warn strongly against such financially biased practices in American Industry; trying to outguess the economics of untried and untested technological approaches can be the death of an industry, and I might add, of a society, too. Some of the tools of this trade--for example, discounted cash flow analysis--are completely unrealistic. Sometimes they are called the Astrophysics of a non-existent universe.

As I have already suggested, Exxon's own R&D philosophy dictates searching for a diversified mix of short- and long-range technological options. I have already alluded to our efforts to boost the energy efficiency of our refineries--a highly immediate and apparent need to management. This need is apparent even though our R&D in some areas may not pay out for years--for example, in advanced separation systems that do not employ normal heat distillation techniques. Another of our major thrusts is in developing more versatile technologies for converting crude residuums to light transportation fuels. The need stems from an evident shift of demand in that direction and from the reduced quality of the average crude oil today. Exxon has begun deployment of an innovation in this area called FLEXICOKING, a processing "garbage can" that can convert virtually any heavy crude or residuum into transportation fuels and fuel gas.

As industry moves down to lower quality resources, there is synergism between such "resid" conversion technologies and our efforts to develop improved synthetics technologies. With the exception of established synthetics operations in South Africa and Canada, falling crude prices and escalating project costs have nipped the synthetic fuels industry in the bud. Many synthetics technologies have turned out to be far more expensive than anyone thought. So price feedback has told us that we must use R&D to bring capital and operating costs down through developing synthetics technologies adaptable to local conditions, resources, and markets. In the process, as I suggested earlier, we will certainly succeed in increasing their efficiency and so reducing CO_2 emissions. In the crucial conversion step, many of today's synthetics technologies operate at efficiences in the range of 60 percent. By the year 2000 we see possibilities for stepping up those efficiencies to nearly 80 percent. And this is not a fundamental limit.

Exxon is working on a very wide variety of synthetics options, including advanced shale retorting and direct coal liquefaction; a catalytic process for producing methane directly from coal; the generation of CO and Hydrogen, or synthesis gas, from coal, lignite, or remotely located natural gas; and the conversion of synthesis gas to fuels and chemicals. On the non-fossil fuel side, Exxon has for many years been doing R&D aimed at improving the fabrication of nuclear fuel elements; and we have been one of several companies in the race to produce cheaper solar voltaic cells made from amorphous silicon.

These efforts suggest primarily the early stages of the transition. For the later stages, some interesting options are beginning to present themselves. A prime difficulty with synthetics resources is their high carbon content. Chemically, that means low ratio of hydrogen to carbon. While the ratio is about four to one in natural gas and 1.8 in crude oil, it is only about 1.5 in oil sands bitumen or raw shale oil, and less than one in coal. In simple terms, a result is that producing these fuels means generating larger amounts of CO_2 than to produce comparable fuels from petroleum. Synfuels require more processing to manufacture and hence more processing heat generated by burning part of the resource.

Prompted by concerns about CO_2 emissions, among other things, some people have suggested a hydrogen economy, a fuel cycle based on hydrogen generated from water not using heat generated by fossil fuels. Perhaps there are ways to generate cheap hydrogen which could then feed directly into a synthetics process. One possible method would be to use thermochemical processes to split water, with advanced solar collectors or nuclear reactors supplying the process heat. The IIASA study notes that in manufacturing coal synthetics such a scheme would cut CO_2 emissions by one-fourth to one-third, compared to the usual coal conversion technologies envisioned. If they could generate hydrogen cheaply, such technologies would also cut overall costs sharply. For example, in the Exxon donor solvent coal liquefaction process, hydrogen accounts for well over a third of the total cost of producing coal liquids.

Summary and Conclusion

To sum up, the world's best hope for inventing an acceptable energy transition is one that favors multiple technical approaches subject to correction--feedback from markets, societies, and politics, and scientific feedback about external costs to health and the environment. This approach is not easy, or comforting to the uninitiated, because there is no overall neat and understandable plan. But prophecies leading to masterminded solutions that commit a society unalterably to a single course are likely to be dangerous and futile. A good sign is that, without any central plan, the world economy has already adopted conservation technologies that are reducing the rate of CO_2 buildup.

In shaping strategies for energy research and development, we must recognize that, generally, societies will give primacy to economic growth, to least-cost energy alternatives, and, in most transportation uses, to liquid fuels. Fortunately, these conditions give science and engineering a lot of room to maneuver. It appears we still have time to generate the wealth and knowledge we will need to invent the transition to a stable energy system.

I hope I do not appear too sanguine about the collective wisdom of our species. History bears ample testimony to the human capacity for grievous folly, as well as achievement and excellence. Clearly, there is vast opportunity for conflict. For example, it is more than a little disconcerting the the few maps showing the likely effects of global warming seem to reveal the two superpowers losing much of the rainfall, with the rest of the world seemingly benefitting. An acceptable future may require a degree of international cooperation that has eluded our grasp to date. An exception is of course science itself and in particular climatology, which even by the standards of science has been distinguished by a remarkable degree of interdisciplinary and international cooperation. As the world continues to grapple with the profound issues posed by the CO_2 buildup, it could seek few better models of international cooperation than what have already achieved.

FEEDBACKS BETWEEN DYNAMICAL HEAT FLUXES AND TEMPERATURE STRUCTURE IN THE ATMOSPHERE

Peter H. Stone

Department of Meteorology and Physical Oceanography
Massachusetts Institute of Technology

Abstract. Observational studies of meridional and vertical dynamical fluxes of heat and temperature structure in the atmosphere indicate that the primary dynamical modes affecting the temperature structure are large-scale eddies (mainly transient eddies) and moist convection. Theoretical parameterizations of these processes suitable for heat balance models are reviewed and the observational evidence supporting them is summarized. Some of the implied feedback mechanisms have been investigated with simple climate models and can have a sizeable impact on model sensitivity. Other dynamical feedbacks are omitted from current one-dimensional and two-dimensional models and have not yet been assessed. In particular, the impact on climate sensitivity of feedbacks associated with large-scale vertical eddy heat fluxes and β-effects need to be determined. It is advocated that simulations of seasonal changes in vertical and meridional dynamical fluxes and temperature structure be documented more thoroughly for general circulation models, as an aid in validating their simulations of dynamical feedbacks. The simulations of these feedbacks can be seriously compromised in experiments which use annual mean insolation or omit ocean transports.

Introduction

The fundamental drive for atmospheric motions is differential solar heating and the resulting gradients of entropy and temperature. The motions themselves transport heat and modify these gradients and consequently play a fundamental role in determining the climate of the atmosphere-earth-ocean system. Similarly, the feedbacks between the dynamical heat transports and temperature structure play a crucial role in determining how sensitive the climate system is to external change.

Much of what is known about the interactions between atmospheric motions and temperature structure is implicit in the modern theory of the general circulation of the atmosphere (see Lorenz, 1967, for a good summary of the modern theory and its development). However, the immediate stimulus for the development of this theory came from attempts to understand the momentum balance of the atmosphere, rather than its heat balance, and consequently atmospheric dynamicists did not turn their attention to the development of analytic theories of how atmospheric motions and temperature structure interact until the last dozen years or so. The stimulus for these more recent studies has come not only from an increased interest in climate problems, but also from an increased interest in other planetary atmospheres (e.g., see Stone, 1972, and Leovy, 1973). The problem of how atmospheric motions and temperature structure interact is fundamental to understanding not only climate sensitivity, but also the differences between planetary atmospheres.

There is still much that is not known about these interactions. In this paper we will focus on those areas where progress has been made -- in particular, we will focus on the interactions of meridional and vertical dynamical heat fluxes with meridional and vertical temperature structure on the planetary scale. We will not discuss zonal fluxes or interactions between dynamical fluxes and zonal structure -- these are potentially quite important, but they are not yet well understood. We will summarize and discuss both observational and theoretical studies of the interactions between the dynamics and the temperature structure; and also applications of climate models which tell us something about how these interactions affect climate sensitivity.

Symbols

a	radius of the earth
f	coriolis parameter, $2\,\Omega \sin \phi$
g	acceleration of gravity
H	density scale height, RT/g
N	Brunt-Vaisala frequency, $(\alpha g \frac{\partial \overline{\theta}}{\partial z})^{1/2}$

Table 1. Components of the Annual Mean Meridional Flux in the Northern Hemisphere

Component	Latitude where component is a maximum	Maximum flux	Flux at 35°
Ocean	20°	34 ± 11	23 ± 8
Atmosphere	45°	32 ± 2	25 ± 2
Atmospheric zonal mean motions	10°	9	-5
Atmospheric eddies	50°	37	30
Atmospheric eddy sensible heat flux	50°	28	17
Atmospheric eddy latent heat flux	30°	15	13
Total Atmosphere-ocean flux	35°	48	48

Units: 10^{14} Watts

Souces: Oort (1971), Vonder Haar and Oort (1973).

q_s saturated specific humidity

R perfect gas constant

T temperature

u zonal wind

z height above the surface

C_p specific heat at constant pressure

L_v latent heat of vaporization of water

α thermal expansion coefficient

β $\frac{1}{a}\frac{df}{d\phi} = \frac{2\Omega}{a}\cos\phi$

ϕ latitude

ρ density

θ potential temperature

Ω angular rotation rate of the earth

K eddy mixing coefficient

$(\overline{\ })$ mean value of ( )

Meridional Fluxes and Meridional Temperature Structure

Components of the Meridional Flux

Table 1 summarizes observational results from Oort (1971) and Vonder Haar and Oort (1973) showing the contributions of various components of the dynamical flux to the annual mean meridional flux in the Northern Hemisphere. Both the maximum values and the values at 35°N, where the total flux is a maximum, are shown. Vonder Haar and Oort's estimates of the probable errors in their estimates of the ocean and atmosphere fluxes are also included. The ocean and atmosphere are equally important contributors, with the ocean carrying most of the flux in low latitudes and the atmosphere carrying most of it in middle and high latitudes.

In this paper we are only concerned with the flux carried by the atmosphere. However, it is clear that some understanding of the interaction between the ocean flux and temperature structure will also be necessary in order to understand climate and climate sensitivity. We do note that there is evidence that this interaction is very indirect -- the seasonal changes in the ocean flux do not correlate with the seasonal changes in the meridional temperature gradient (Stone and Miller, 1980). This is perhaps not surprising since surface wind stresses play a major role in driving ocean currents (Rooth, 1983). Thus changes in the ocean fluxes may involve changes in the momentum balance of the atmosphere as well as changes in the heat balance of the climate system.

The atmospheric flux may be divided into a component due to the zonal mean motion and a component due to eddies. Table 1 shows that the eddies are the dominant mechanism for transporting heat in the atmosphere. The eddies are primarily a middle and high latitude phenomenon, and in those latitudes their transports greatly exceed the transports by the zonal mean motions. Thus the non-interaction theorem for low dissipation systems, in which these two components sum to zero, does not apply to the atmosphere.

In low latitudes most of the atmospheric transport is accomplished by the zonal mean motions. However, the atmospheric transport in low latitudes is much less than the ocean transport. Thus an understanding of the former, without an understanding of the latter, is not very useful for furthering our understanding of climate. Consequently, we will not discuss the low-latitude transports by the zonal mean motions in the atmosphere. We note however that a good foundation for understanding this transport has been established by Schneider (1977) and Held and Hou (1980).

Atmospheric eddies carry heat in two primary forms, sensible heat and latent heat (the transports of potential energy and kinetic energy are negligible -- Oort, 1971). As shown in Table 1 both forms are important. Since the total atmospheric flux can be approximated by the sum of the eddy sensible and latent heat fluxes, we will concentrate our discussion of the meridional fluxes in the atmosphere on these two components.

Stationary Eddies vs. Transient Eddies

It is conventional to divide atmospheric eddies into transient and stationary components. About three quarters of the meridional eddy heat transport is associated with the transient eddies and one quarter with the stationary eddies (Oort, 1971). The stationary eddies arise from the stationary components of the eddy forcing, i.e., from topography, land-ocean contrasts in surface heating, etc. Transient eddies would exist even in the absence of such stationary forcing because of the baroclinic instability of the atmosphere (e.g., see Lorenz, 1967).

However, the distinction between transient and stationary eddies may be artificial as far as the meridional heat transport is concerned. There is considerable evidence that there is a very strong negative feedback between the stationary and transient components of the eddy heat fluxes and that the two components should be considered to be a joint response to differential heating. The evidence is a follows:

<u>Experiments with General Circulation Models (GCM's)</u>. Manabe and Terpstra (1974) performed an experiment with the Geophysical Fluid Dynamics Laboratory (GFDL) GCM, in which all mountains were removed from the model. Thus in the experiment the forcing for stationary eddies in the northern hemisphere was greatly reduced, and the stationary eddy meridional heat transport was reduced by 75%. However, at the same time the transient eddy meridional heat transport increased by 70%, so that the total eddy transport was only decreased by 25%.

<u>Correlations with changes in the temperature gradients</u>. Stone and Miller (1980) correlated seasonal changes in the meridional eddy heat transports and the surface meridional temperature gradient. The results are shown in Fig. 1. The correlation for the total eddy flux is much larger than for the two individual components, and the correlations of the two components vary in a complementary way. This indicates that it is the total eddy flux rather than the individual components which is responding to changes in the temperature gradient.

<u>Eddy energy cycles</u>. The Lorenz energy cycles associated with the transient eddies and the stationary eddies are qualitatively similar. In particular the stationary eddies only transport a significant amount of heat in winter (Oort, 1971) and Holopainen (1970) has shown that the energy cycle associated with the stationary eddies in winter includes the following features:

i) zonal mean available potential energy is the main source of eddy available potential energy;

ii) eddy available potential energy is the main source of eddy kinetic energy;

iii) dissipation is the main sink of eddy kinetic energy, i.e., the non-interaction theorem does not apply.

The same features are characteristic of the transient eddy energy cycle (Oort and Peixoto, 1974) and they suggest that baroclinic instability is at work in both kinds of eddies. Yao (1980), using a greatly simplified process model has

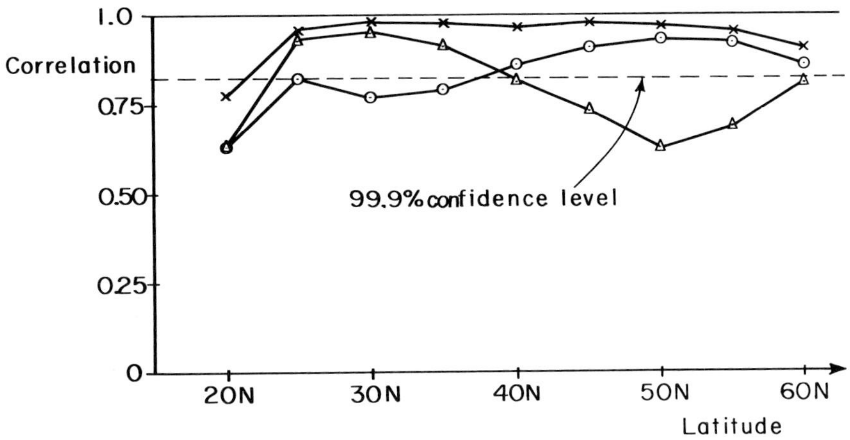

Fig. 1. Correlations between surface meridional temperature gradient and the meridional flux of sensible heat by all eddies (x), by transient eddies alone (Δ), and by stationary eddies alone (0) as a function of latitude. From Stone and Miller (1980).

demonstrated that, in a system which is sufficiently strongly unstable, baroclinically, the two types of eddies can in fact behave in a similar fashion.

All the theories that have been developed for how eddy fluxes interact with the temperature structure are in fact based on the hypothesis that heat-transporting eddies are generated by baroclinic instability. In light of the above evidence it is plausible that such theories will work well even when stationary forcing is important. In any case the transient eddies are more readily identified with baroclinic instability and they are the main contributors to the total eddy meridional heat flux.

Eddy Sensible Heat Flux

The most sophisticated parameterization for the meridional eddy sensible heat flux developed to date, i.e., the only one that includes β-effects, is that of Branscome (1980, 1983). He used approximate analytic solutions of Green's (1960) model of baroclinic instability to derive a general expression for the characteristic vertical structure of the most unstable baroclinic wave. This general expression allowed him to combine the parameterizations of Stone (1972) and Held (1978) which are applicable to deep waves ($\beta \to 0$) and shallow waves ($\beta \to \infty$), respectively. All of these parameterizations are based on mixing length theory, with the meridional mixing length taken to be the radius of deformation, and the eddy velocity amplitude taken to be equal to that of the zonal mean flow. The former choice has been shown to be appropriate for baroclinic waves by Killworth's (1980) analysis of the baroclinic instability problem with the effects of meridional shear included. The latter choice can be derived by extrapolating the results of analyses of finite amplitude baroclinic instabilities (e.g., Pedlosky, 1970 and 1979) to conditions far from neutral stability.

Branscome's general result for the meridional eddy flux of sensible heat per unit area is

$$F_{MESH} = 0.6\,\frac{\rho\, C_p\, N\, d^2}{\alpha\, g}\left(\frac{\partial \bar{u}}{\partial z}\right)^2 e^{-z/d} \qquad (1)$$

where

$$f\,\frac{\partial \bar{u}}{\partial z} = -\,\frac{\alpha\, g}{a}\,\frac{\partial \bar{T}}{\partial \phi} \qquad (2)$$

$$d \cong \frac{h\, H}{h + H} \qquad (3)$$

and

$$h = \frac{f^2 \frac{\partial \bar{u}}{\partial z}}{\beta\, N^2} \qquad (4)$$

The dependence of the flux on the temperature structure, i.e., on $\partial\bar{T}/\partial\phi$ and $\partial\bar{\theta}/\partial z$ (which enters through N), is rather complicated, but can be written simply in the two extreme cases studied by Stone and Held. In these cases we obtain for the vertical integral of the flux:

$\beta \to 0$ $(d \to H)$:

$$\int_0^\infty F_{MESH}\, dz \;\alpha\; \left(\frac{\partial \bar{\theta}}{\partial z}\right)^{1/2} \left(\frac{\partial \bar{T}}{\partial \phi}\right)^2 , \qquad (5)$$

$\beta \to \infty$ $(d \to h \to 0)$:

$$\int_0^\infty F_{MESH}\, dz \;\alpha\; \left(\frac{\partial \bar{\theta}}{\partial z}\right)^{-5/2} \left(\frac{\partial \bar{T}}{\partial \phi}\right)^5 . \qquad (6)$$

The flux increases monotonically as $\partial\bar{T}/\partial\phi$ increases, so there is a negative feedback between the flux and the temperature gradient. As the static stability increases, the flux increases to a maximum value and then decreases.

Branscome (1983) has shown that the above parameterization gives realistic magnitudes for the eddy sensible heat flux and has a qualitatively correct height and latitude dependence. The best check on how the flux depends on the temperature structure comes from observations of seasonal changes. In seasonal changes there is very little change in the static stability in low and middle latitudes, so one can use the observed changes to infer the dependence of the flux on just the temperature gradient. Stone and Miller (1980) assumed a functional dependence of the form

$$\int_0^\infty F_{MESH}\, dz \;\alpha\; \left(\frac{\partial \overline{T}}{\partial \phi}\right)^n \qquad (7)$$

and derived values of n from the seasonal changes. In high latitudes $\beta \to 0$, $h \to \infty$, and $d \to H$, so one would expect from the parameterization that Eq. (5) would apply. In low latitudes $f \to 0$, $h \to 0$, and $d \to h$, so one would expect that Eq. (6) would apply. Stone and Miller's results for the exponent, n, are shown in Fig. 2 and are in fact consistent with these expectations. A regime corresponding to deep waves appears to apply in mid-latitudes (Eq. 5), while a regime corresponding to eddies of intermediate height (with a sensitivity intermediate to Eqs. 5 and 6) appears to apply around 30°N.

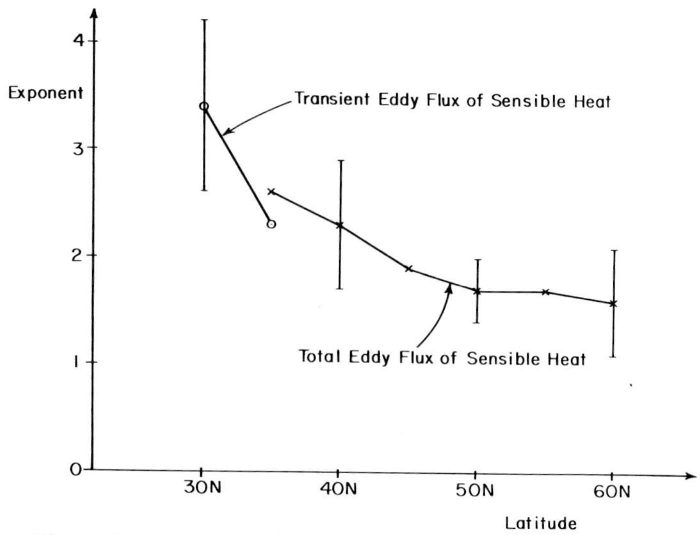

Fig. 2. Exponent in Eq. (7) which gives the best fit between the integrated eddy sensible heat flux and the surface meridional temperature gradient at various latitudes. The transient eddy flux closely approximates the total eddy flux at 30°N and 35°N. From Stone and Miller (1980).

Eddy Latent Heat Flux

A simple and effective way for calculating eddy latent heat fluxes has been suggested by Leovy (1973), based on the following approximations:

i) fluctuations in relative humidity are small compared to fluctuations in specific humidity

ii) fluctuations in temperature are small compared to mean temperatures

iii) the relative humidity is approximately 100%.

The first two approximations are quite good for typical baroclinic waves, and the last is reasonable for at least the lowest layers of the troposphere, where most of the latent heat fluxes are concentrated. The first and third approximations work because of the presence of the oceans over much of the earth's surface, so that, in effect, there is an infinite moisture source at the surface which keeps the atmosphere nearly saturated in its lowest layers.

With the above approximations the meridional eddy flux of latent heat per unit area can be related to the eddy flux of sensible heat per unit area as follows (Mullan, 1978):

$$F_{MELH} = \frac{L_v}{C_p}\frac{\partial q_s}{\partial T} F_{MESH} \qquad (8)$$

This implies that the eddy latent heat flux reinforces the same feedbacks present in the eddy sensible heat flux. In addition, it interacts strongly with the temperature field because the saturated specific humidity, q_s, is very sensitive to temperature. According to the Clausius-Clapeyron equation, for typical temperatures,

$$q_s \;\alpha\; e^{-\frac{5400°K}{T}} \; . \qquad (9)$$

Therefore small increases in temperature lead to relatively large increases in the latent heat flux. This is turn increases the moisture convergence, precipitation, etc. Consequently, the heat flux and the hydrological cycle will be rather sensitive to the temperature of the system. This sensitivity has been demonstrated in experiments with the GFDL GCM (Manabe and Wetherald, 1980; Manabe and Stouffer, 1980). The former study also illustrated the feedback between eddy sensible heat flux and temperature gradient: an increase in CO_2 caused an increase in temperature and eddy latent heat flux, and this in turn caused a decrease in meridional temperature gradient and eddy sensible heat flux (see Eqs. 5 and 6).

Mullan (1978) used observations to test the validity of the above parameterization. A comparison of the observed and calculated moisture fluxes is shown in Figs. 3 and 4, for January and July, respectively. The main error in the parameterization occurs in lower latitudes, particularly in summer at 20°N to 35°N. The error is due to the parameterization's not being able to represent transports associated with monsoonal

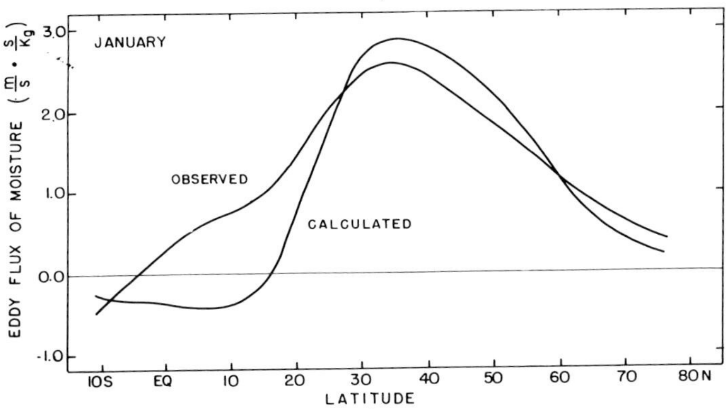

Fig. 3. Comparison of the observed meridional eddy flux of moisture and that calculated from Leovy's parameterization for January. From Mullan (1979).

circulations. In these circulations the fluctuations in relative humidities are not small compared to fluctuations in specific humidity. However, the parameterization does quite well at higher latitudes where most of the latent heat transport is due to baroclinic waves.

Role of Meridional Fluxes in Climate Sensitivity

The one-dimensional interaction of the meridional dynamical fluxes and the meridional temperature structure can be examined in the meridional heat balance models developed by Budyko (1969) and Sellers (1969). In this type of model it is conventional to represent the total meridional flux in the atmosphere-ocean system by an equation of the following form:

$$F = -K\left(\frac{\partial T}{\partial \phi}\right)^n \quad . \tag{10}$$

Most commonly n is chosen to be unity, corresponding to a simple diffusive parameterization. As we discussed above, for just the atmospheric component of the flux, values $n \geq 2$ would be appropriate. A purely empirical fit of the seasonal changes in the total atmosphere-ocean flux to the seasonal changes in

$$\frac{\partial T}{\partial \phi}$$

yields a value of n = 1.8 (Stone and Miller, 1980). The value of n is in effect a measure of the strength of the feedback between the dynamical flux and the temperature gradient. The feedback is negative as long as n > 0, and n = 0 corresponds to no feedback, i.e., the flux is immutable.

The impact of this feedback on climate sensitivity is illustrated in Fig. 5. x_s is the sine of the latitude of the edge of the polar ice cap -- i.e., x_s = 1 corresponds to no polar ice cap and x_s = 0 corresponds to an ice-covered earth. S is the solar constant normalized by its current value. The Figure shows how the size of the ice-cap varies as the solar constant changes, for different values of n, i.e., for different strengths of the dynamical feedback. The curves were calculated using a two-mode version of the Budyko-Sellers model (North, 1975), and with parameterizations derived by Wang and Stone (1980). In particular, the parameterization of the thermal emissions to space was a "fixed cloud temperature" (FCT) parameterization, and the planetary albedo parameterization allowed for the contribution to planetary albedo in high latitudes due to large solar zenith angles (Lian and Cess, 1977). The values of K were chosen in each case to position the edge of the ice-cap at x_s = 0.95 (ϕ = 72°) for the current value of the solar constant.

Fig. 5 shows that the sensitivity of the climate depends crucially on the strength of the dynamical feedback when the polar ice cap is small (x_s > 0.8). With no feedback (n = 0), decreasing the solar constant requires a smaller polar ice cap for equilibrium, i.e., the current climate would be unstable without any dynamical feedback. With a moderate negative feedback (n = 1), by contrast, the current climate is rather stable. Increasing the negative feedback to the limit where the temperature gradient is fixed ($n \to \infty$) results in only a slight further stabilization compared to the diffusive (n = 1) case.

On the other hand, the ice-covered-earth instability is not sensitive to the strength of the dynamical feedback, as was noted by North (1975). This instability sets in when the ice cap reaches the point where a further increase in its size would require a larger solar constant for equilibrium. In this model this occurs when the solar constant falls by about 13 1/2% regardless of the value of n. However, as shown by Held and Suarez (1974), the existence of the dynamical flux is still necessary for the existence of the ice-covered-earth instability, i.e., as K decreases, the instability sets in at

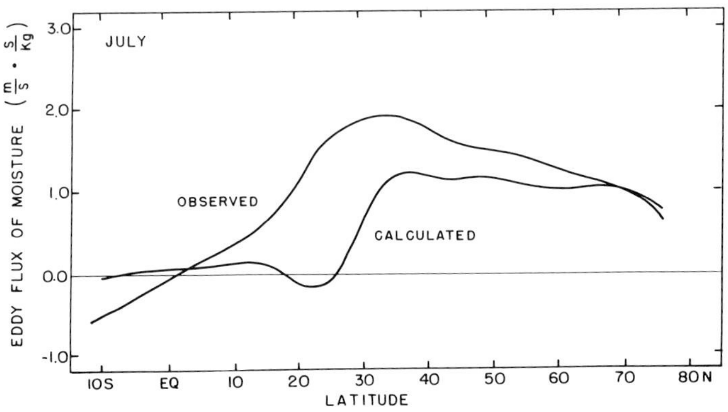

Fig. 4. Same as Fig. 3, but for July. From Mullan (1979).

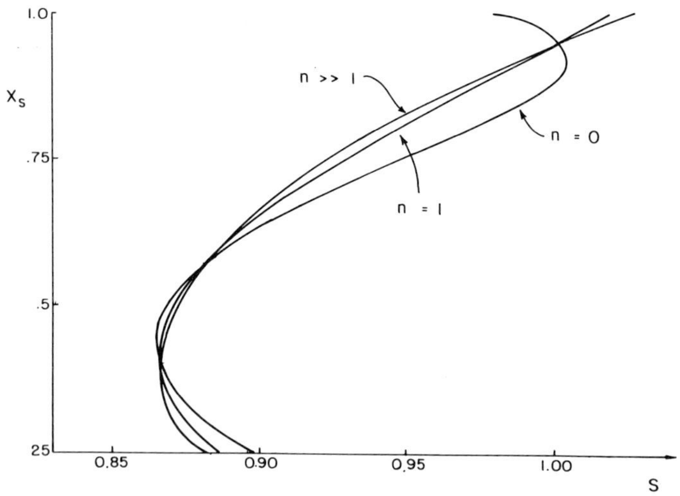

Fig. 5. Position of the edge of the polar ice cap vs. solar constant, for various strengths of the dynamical feedback.

lower and lower latitudes, and finally disappears when K = 0.

The impact of the dependence of the latent heat fluxes on temperature has been studied by Gal-Chen and Schneider (1976). They examined the sensitivity of the climate to changes in the solar constant, using a Budyko-Sellers type of model. In particular, they looked at the effect when the eddy coefficient, K, is modified to include an exponential term like that in Eq. (9), so as to simulate the temperature dependence of the Clausius-Clapeyron equation in that portion of the total flux attributable to the eddy latent heat flux. They found that the global mean temperature changes induced by changes in the solar constant were reduced by about 10% when the Clausius-Clapeyron effect was included -- i.e., the climate was stabilized slightly with respect to changes in the solar constant.

Vertical Fluxes and Vertical Temperature Structure

Components of the Vertical Flux

Table 2 summarizes Hantel's (1976) results for the hemispheric mean contributions to the total vertical flux in the Northern Hemisphere. Hantel's analysis of the hemispheric imbalances indicates that the probable error in these fluxes is about 5 W/m^2 in winter and 2 W/m^2 in summer. The contribution by the zonal mean motions is negligible, and on average the radiative flux is closely balanced by the eddy flux. Since the eddy flux was not measured directly, but was calculated as a residual, Hantel's analysis does not reveal how this eddy flux is broken up into contributions by large or small scale eddies, transient or stationary eddies, sensible or latent heat. However, Oort and Rasmusson (1971) have calculated vertical fluxes due to the large-scale stationary eddies. From their results we calculate that the flux across 500 mb in the Northern Hemisphere due to stationary eddies is ~2 W/m^2 in winter and ~1/W m^2 in summer. Therefore the stationary eddy vertical flux is also negligible.

The above results imply that the important dynamical fluxes in the vertical are those due to large-scale transient eddies and to small-scale convection. Because the atmosphere is nearly saturated, small-scale convection occurs primarily in the form of moist convection (except in a thin boundary layer adjacent to the surface and below

Table 2. Components of the Vertical Flux in the Northern Hemisphere at 500 mb

Component	Winter*	Summer*	Average
Radiation	-4	-112	-58
Zonal mean motions	+2	-2	0
Eddies	+51	+60	+56
Total	+49	-53	-2

Units: W/m^2

Source: Hantel (1976)

*Winter is the months of December, January, and February; summer is the months of June, July, and August.

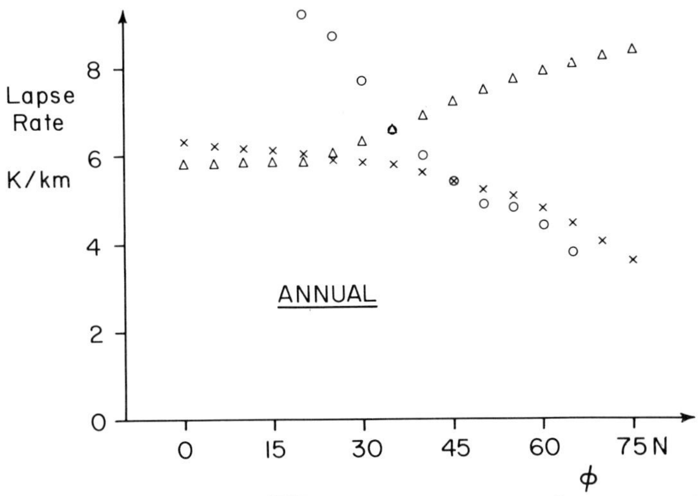

Fig. 6. Tropospheric mean values of observed lapse rates (x), moist adiabatic lapse rates (Δ), and the critical lapse rate for large-scale baroclinic instability (0) vs. latitude for annual mean conditions. From Stone and Carlson (1979).

the condensation level). The relative importance of these two mechanisms can be inferred from a study by Stone and Carlson (1979). They compared observed tropospheric lapse rates with critical values of the lapse rate associated with the onset of moist convection and of unstable baroclinic waves extending throughout the troposphere. The former is the moist adiabatic lapse rate and the latter is given by h = H.

The results are illustrated in Fig. 6. We see that the atmosphere is stable to deep baroclinic disturbances in low latitudes but marginally unstable in middle and high latitudes. By contrast, it is marginally unstable to moist convection in low latitudes but stable in middle and high latitudes. Therefore, the atmosphere appears to fall into two regimes: a low-latitude regime where lapse rates are controlled primarily by small-scale moist convection, and a mid- to high-latitude regime where lapse rates are controlled primarily by deep baroclinic disturbances. Consequently we will focus our discussion of vertical dynamical fluxes on these two mechanisms.

Moist Convection

The most commonly used method for simulating the effect of moist convection on temperature structure is the moist adiabatic adjustment. In its simplest form this method merely replaces lapse rates that exceed the moist adiabatic lapse rate by the moist adiabatic lapse rate itself. The assumption implicit in using this method is that moist convection, when it occurs, is so efficient that actual lapse rates will not appreciably exceed the moist adiabatic lapse rate. This is equivalent to assuming a very strong negative feedback between the vertical flux generated by moist convection and the vertical lapse rate when the lapse rate exceeds the critical lapse rate, and no feedback when it is less than the critical lapse rate. These assumptions appear to be justified by the observational evidence, showing that the lapse rate may be considerably less than the moist adiabatic lapse rate, but cannot appreciably exceed it (see Fig. 6). Another implicit assumption is that the atmosphere is nearly saturated, so that the moist adiabatic lapse rate rather than the dry adiabatic lapse rate is the appropriate critical lapse rate.

In spite of their lack of sophistication, moist adiabatic adjustment schemes are commonly used, even in GCM's. More sophisticated schemes (e.g., Arakawa and Schubert, 1974) do not give appreciably different results, at least as far as the vertical temperature structure is concerned.

Large-Scale Eddy Fluxes

The most sophisticated parameterization for the vertical eddy flux of sensible heat is again Branscome's based on the theory of baroclinic instability (Branscome, 1980, 1983). The vertical flux per unit area is

$$F_{VESH} = 0.6\,\frac{\rho\, C_p\, f\, d^2}{\alpha g N}\left(\frac{\partial \bar{u}}{\partial z}\right)^3\left(\frac{z}{d} - \frac{z^2}{4d^2}\right)e^{-z/d} \quad . \quad (11)$$

The dependence on the temperature structure is again illustrated most easily by looking at the asymptotic limits:

$\beta \to 0\ (d \to H)$

$$F_{VESH} \propto \left(\frac{\partial \bar{\theta}}{\partial z}\right)^{-1/2}\left(\frac{\partial \bar{T}}{\partial \phi}\right)^3 \quad , \qquad (12)$$

$\beta \to \infty\ (d \to h \to 0)$

$$F_{VESH} \propto \left(\frac{\partial \bar{\theta}}{\partial z}\right)^{-5/2}\left(\frac{\partial \bar{T}}{\partial \phi}\right)^5 \quad . \qquad (13)$$

Therefore, the vertical flux increases monotonically as the temperature gradient increases, and decreases monotonically as the static stability increases (a negative feedback).

The critical lapse rate for the baroclinic instability of deep waves, plotted in Fig. 6 and given by h = H, corresponds to

$$d = \frac{H}{2} \quad .$$

Therefore, the critical lapse rate separates two regimes in which the eddy flux has the two asymptotic behaviors given above. For lapse rates less than the critical lapse rate,

$$d < \frac{H}{2} \quad ,$$

the flux is carried mainly by shallow baroclinic waves, and the negative feedback between the eddy flux and vertical temperature structure is relatively weak -- i.e., it is easy to realize lapse rates less than the critical lapse rate. On the other hand, when the lapse rate is greater than the critical lapse rate,

$$d > \frac{H}{2} \quad ,$$

the flux is carried mainly by deep baroclinic waves, and the negative feedback is much stronger, i.e., it is more difficult to realize lapse rates greater than the critical lapse rate. This is at least qualitatively consistent with the observational evidence, indicating that the lapse rate never appreciably exceeds the critical lapse rate for baroclinic instability (see Fig. 6).

Leovy's (1973) suggestion can also be used to parameterize the vertical eddy flux of latent heat. One obtains

$$F_{VELH} = \frac{L_V}{C_p} \frac{\partial q_s}{\partial T} F_{VESH} \quad . \qquad (14)$$

Once again the eddy latent heat flux reinforces the feedbacks in the eddy sensible heat flux, and the latent heat flux is sensitive to the temperature.

We can obtain one observational check on the above formulas by using them to calculate the total large-scale vertical eddy flux. If we evaluate the parameters at 40°N in January, using Oort and Rasmusson's (1971) analysis, we find that the vertical flux peaks at about 650 mb, with values

$$F_{VESH} \cong 27 \text{ W/m}^2$$

$$F_{VELH} \cong 22 \text{ W/m}^2$$

The sum of these agrees very well with Hantel's winter value of the eddy flux required for heat balance (see Table 2). Similar estimates for July are much smaller, but Stone and Carlson's (1979) study of the lapse rates indicates that the large-scale eddies will only dominate in winter, and the moist convection will make an important contribution to the vertical flux in summer even in middle and high latitudes.

Role of Vertical fluxes in Climate Sensitivity

The one-dimensional interaction between the vertical dynamical fluxes and the vertical temperature structure can in principle be studied in a radiative-convective model for the global mean temperature structure. The prototype of this kind of model was introduced by Manabe and Strickler (1964). In their version the effect of vertical dynamical fluxes is parameterized by imposing a "convective" adjustment, such that the lapse rate is not allowed to exceed a critical value of 6.5°K/km -- i.e., the lapse rate is first determined from the condition of radiative equilibrium, and is then modified by replacing lapse rates which exceed the critical lapse rate by the critical lapse rate itself. This parameterization is still frequently employed.

A convective adjustment parameterization of the above kind contains two implicit assumptions. One is that there is a very strong negative feedback between the total dynamical flux and the difference between the lapse rate and its critical value when the difference is positive, but no feedback when it is negative. As we discussed above, this assumption appears to be appropriate for the important dynamical processes operating in the atmosphere. Somewhat surprisingly, the impact of such a strong negative feedback on climate sensitivity has not been evaluated. This could be done in principle simply by calculating the vertical dynamical flux necessary to maintain the critical lapse rate under current conditions, and then keeping this flux fixed as the climate is varied. One would expect the elimination of the dynamical feedback to destabilize the climate, but to what extent is not clear.

The second implicit assumption is that the critical lapse rate is fixed. Although seasonal changes in lapse rates are indeed small (about 5% for the Northern Hemisphere mean tropospheric lapse rate) the lapse rate does change with latitude (see Fig. 6) and one would expect the two critical lapse rates to be changeable on physical grounds. In particular, the moist adiabatic lapse rate, through its dependence on q_s, has a strong temperature dependence (see Eq. 9), and the critical lapse rate associated with baroclinic instability, given by h = H, is proportional to the meridional temperature gradient (cf. Eqs. 2 and 4). As discussed by Stone and Carlson, the small seasonal changes in lapse rates in mid-latitudes appear to be caused by compensating seasonal changes in the activity of large-scale eddies and moist convection, and one would not expect such compensation in an arbitrary climate change.

The effect of the changeability of the moist-adiabatic lapse rate on climate sensitivity in a global radiative-convective model has been investigated by Wang et al. (1981) and Hansen et al. (1981). They found that the sensitivity of the global mean surface temperature to changes in CO_2 concentration or solar constant was decreased by about 30% compared to the sensitivity when the critical lapse rate was fixed. Use of more sophisticated parameterizations of moist convection in radiative-convective models gives very similar results (Lindzen et al., 1982). On the other hand, the impact of changes in the vertical fluxes or critical lapse rate associated with baroclinic instability cannot be evaluated in radiative-convective models because of their implicit neglect of meridional temperature structure. We note that climate change experiments with GCM's do show changes in global mean lapse rates (Manabe and Stouffer, 1980; Hansen et al., 1983b).

Fluxes and Temperature Structure in More than One Dimension

Limitations of One-Dimensional Models

Dynamical fluxes of heat are in general three-dimensional vectors. One-dimensional heat balance models can eliminate the explicit appearance of two of the components by averaging the thermodynamic equation over two dimensions. However, the remaining component of the flux usually depends on the three-dimensional temperature structure, and therefore an implicit dependence on events in the other two dimensions remains. Such multi-dimensional interactions are necessarily excluded from one-dimensional models.

It is clear from our discussion in the preceding section that this exclusion is potentially a very serious limitation on the ability of radiative-convective models to simulate climate change. There we pointed out that these models are unable to simulate changes in the critical lapse rate associated with baroclinic instability, because it depends on the meridional temperature gradient. This same limitation appears in a somewhat different form in Branscome's parameterizations for the meridional and vertical fluxes of sensible heat associated with baroclinic instability. For example, consider the vertical flux in the regime $d \to H$, and consider what happens if a climate change induces an increase in the static stability. Such an increase will cause a decrease in the vertical eddy flux of sensible heat (see Eq. 12). This is a direct negative feedback which could in principle be included in global radiative-convective models. However, the increase in static stability will also cause an increase in the meridional eddy flux of sensible heat (see Eq. 5), and this in turn will cause a decrease in the magnitude of the meridional temperature gradient and the magnitude of the vertical eddy flux of sensible heat (see Eq. 12). This indirect change represents an additional negative feedback which is excluded a priori from global radiative-convective models. We also note that in the regime $d \to 0$ this excluded feedback is positive rather than negative.

The same kind of limitation applies to the Budyko-Sellers type of model for the meridional temperature structure. For example, consider what happens to the meridional flux in Branscome's parameterization in the regime $d \to 0$. If a climate change induces an increase in the meridional temperature gradient, the direct effect will be an increase in the meridional flux (see Eq. 6), i.e., a negative feedback. The indirect effect will be an increase in the vertical flux (see Eq. 13), which will lead to an increase in the static stability and a decrease in the meridional flux (see Eq. 6), i.e., a positive feedback. This latter feedback is excluded a priori from the Budyko-Sellers type of model.

Two-Dimensional Models

Zonally averaged models which parameterize eddy fluxes and deal explicitly with the vertical meridional structure can in principle include two-dimensional interactions such as those described above. A number of these models have been developed (e.g., see Saltzman, 1978; Taylor, 1980). However, none of them have included β-effects. Without these effects it is not possible to simulate accurately many of the feedbacks described above, and in fact one cannot even simulate the observed, vertical, latitudinal, and seasonal structure of the eddy fluxes (Branscome, 1983). With Branscome's improved parameterization, which does include β-effects, these models can be improved greatly. Consequently, the potential of the two-dimensional models for studying climate problems needs to be re-examined.

General Circulation Models

In principle, three-dimensional general circulation models (GCM's) are free from all of the above difficulties. Not only do they have the capacity for simulating accurately all of the feedbacks we have discussed, but they also have the capacity to simulate other potentially important feedbacks, such as the interactions of the fluxes with zonal structure. However, before GCM's are applied to climate problems, it is important to verify that they are exploiting this capacity correctly. It is particularly important to verify that GCM's can simulate accurately changes in the dynamical fluxes -- otherwise one can have no confidence in their ability to simulate dynamical feedbacks.

The one well-documented climate change that can be used for such a verification is the seasonal change. In particular, model simulated changes in the meridional and vertical dynamical fluxes can be compared with the observational analyses of Oort (1971) and Hantel (1976), respectively. Very few such comparisons have been documented. The only documented comparisons for the total meridional flux (sensible heat plus latent heat plus potential energy) are from simulations with the GCM's developed at the Goddard Institute for Space Studies (GISS). The old GISS fine-resolution GCM underestimates the total meridional flux in the atmosphere by about 15% in both January and July (Stone et al., 1977). The new GISS medium-resolution GCM overestimates the same flux by about 45% in January and 20% in July (Hansen et al., 1983a). No comparisons at all have been made between seasonal changes in simulated vertical fluxes and observations. It is not at all clear that GCM's are yet exploiting their inherent capacity to simulate dynamical feedbacks better than one- and two-dimensional models can.

A particular problem that GCM's suffer from in climate applications in their enormous computational requirements. These requirements often force compromises that seriously limit the models' ability to simulate dynamical feedbacks. For example, some experiments omit the seasonal cycle and run the models with annual mean insolation (e.g., Manabe and Wetherald, 1975 and 1980). The very strong nonlinearity in the dynamical feedbacks (as well as in other feedbacks) raises the possibility that climate sensitivity will be changed significantly by the omission of the seasonal cycle.

Another example is the omission of horizontal transports by ocean currents. This omission tends to increase meridional temperature gradients, which in turn increases the atmospheric meridional transports (see Eqs. 5 and 6) -- i.e., the atmospheric transports pick up much of the slack caused by the omission of the oceanic transports. The net effect is that dynamical feedbacks associated with ocean transports are replaced by the dynamical feedbacks associated with atmospheric transports. Consider, for example, the sensitivity of the hydrological cycle to climate changes found in experiments with the GFDL GCM which omit the oceanic transport (Manabe and Wetherald, 1980; Manabe and Stouffer, 1980). This sensitivity is unrealistically large because the model's meridional latent heat transport for current conditions is as much as two times the observed value.

Summary

Our review of the observational and theoretical evidence concerning dynamical fluxes in the atmosphere and their interaction with temperature structure can be summarized as follows:

1) Both the atmosphere and the oceans make important contributions to the meridional heat transport, with the former dominating in middle and high latitudes and the latter in low latitudes in the Northern Hemisphere.

2) The atmospheric meridional transport of heat is dominated by the eddy fluxes of sensible and latent heat.

3) Strong negative feedback betwen the meridional fluxes by stationary and transient eddies blurs the destinction between these two components of the meridional flux.

4) The meridional eddy flux of sensible heat supplies a negative feedback to the meridional temperature gradient which is stronger in low latitudes than in high latitudes.

5) The meridional and vertical eddy fluxes of latent heat reinforce the feedbacks in the eddy fluxes of sensible heat, and in addition, are very sensitive to the temperature.

6) The negative feedback in the meridional fluxes has a strong stabilizing effect on the current climate.

7) The vertical dynamical fluxes in the atmosphere are dominated by small-scale moist convection and large-scale transient eddies.

8) Both these vertical fluxes supply a negative feedback to the static stability.

9) "Critical" lapse rates are in principle changeable, and this changeability can have a substantial impact on climate sensitivity.

10) Large-scale eddy fluxes depend on both the meridional and vertical temperature structure. Therefore, one-dimensional models automatically exclude some feedbacks, and these same feedbacks have not yet been included accurately in two-dimensional zonal mean models.

11) GCM's need to have their ability to simulate dynamical feedbacks validated by comparison of model-generated and observed seasonal changes in vertical and meridional fluxes and temperature structure.

12) Compromises forced on GCM's by their computational requirements, such as the use of annual mean insolation or the omission of ocean transports, can seriously compromise their simulation of the feedbacks in atmospheric dynamical fluxes.

References

Arakawa, A., and W. H. Schubert, Interaction of a cumulus cloud ensemble with the large-scale environment, Part I, J. Atmos. Sci., 31, 674-701, 1974.

Branscome, L. E., Scales and structures of baroclinic waves and their influence on climatic states, Ph. D. Thesis, Massachusetts Institute of Technology, 201 pp, 1980.

Branscome, L. E., A parameterization of transient eddy heat flux on a beta-plane, submitted to J. Atmos. Sci., 1983.

Budyko, M. I., The effect of solar radiation variations on the climate of the earth, Tellus, 21, 661-619, 1969.

Gal-Chen, T., and S. H. Schneider, Energy balance climate modeling: Comparison of radiative and dynamic feedback mechanisms, Tellus, 28, 108-121, 1976.

Green, J. S. A., A problem in baroclinic stability, Quart. J. Roy. Met. Soc., 86, 237-251, 1960.

Hansen, J., D. Johnson, A. Lacis, S. Lebedeff, P.

Hansen, J., D. Johnson, A. Lacis, S. Lebedeff, P. Lee, D. Rind, and G. Russell, Climate impact of increasing atmospheric carbon dioxide, Science, 213, 957-966, 1981.

Hansen, J., G. Russell, D. Rind, P. Stone, A. Lacis, S. Lebedeff, R. Ruedy, and L. Travis, Efficient three-dimensional global models for climate studies: Models I and II, Mon. Wea. Rev., 111, 609-662, 1983a.

Hansen et al., In Ewing Symposium volume, 1983b.

Hantel, M., On the vertical eddy transports in the northern atmosphere 1, vertical eddy heat transport for summer and winter, J. Geophys. Res., 81, 1577-1588, 1976.

Held, I. M., The vertical scale of an unstable

baroclinic wave and its importance for eddy heat flux parameterizations, J. Atmos. Sci., 35, 572-576, 1978.

Held, I. M., and A. Y. Hou, Nonlinear axially symmetric circulations in a nearly inviscid atmosphere, J. Atmos. Sci., 37, 515-553, 1980.

Held, I. M., and M. J. Suarez, Simple albedo feedback models of the icecaps, Tellus, 26, 613-627, 1974.

Holopainen, E. O., An observational study of the energy balance of the stationary disturbances in the atmosphere, Quart. J. Roy. Met. Soc., 96, 626-644, 1970.

Killworth, P. D., Barotropic and baroclinic instability in rotating stratified fluids, Dyn. Atmos. Oceans, 4, 143-184, 1980.

Leovy, C. B., Exchange of water vapor between the atmosphere and surface of Mars, Icarus, 18, 120-125, 1973.

Lian, M. S., and R. D. Cess, Energy balance climate models: a reappraisal of ice-albedo feedback, J. Atmos. Sci., 34, 1058-1062, 1977.

Lindzen, R. S., A. Y. Hou, and B. F. Farrell, The role of convective model choice in calculating the climate impact of doubling CO_2, J. Atmos. Sci., 39, 1189-1205, 1982.

Lorenz, E. N., The Nature and Theory of the General Circulation of the Atmosphere, World Meteorological Organization, WMO No. 218. TP. 115, 161 pp, 1967.

Manabe, S., and R. J. Stouffer, Sensitivity of a global climate model to an increase of CO_2 concentration in the atmosphere, J. Geophys. Res., 85, 5529-5554, 1980.

Manabe, S., and R. Strickler, Thermal equilibrium of the atmosphere with a convective adjustment, J. Atmos. Sci., 21, 361-385, 1964.

Manabe, S., and T. B. Terpstra, The effects of mountains on the general circulation of the atmosphere as identified by numerical experiments, J. Atmos. Sci., 31, 3-42, 1974.

Manabe, S., and R. T. Wetherald, The effects of doubling the CO_2 concentration on the climate of a general circulation model, J. Atmos. Sci., 32, 3-15, 1975.

Manabe, S., and R. T. Wetherald, On the distribution of climate change resulting from an increase in CO_2 content of the atmosphere, J. Atmos. Sci., 37, 99-118, 1980.

Mullan, A. B., A mechanistic model for midlatitude mean temperature structure, Ph. D. Thesis, Massachusetts Institute of Technology, 248 pp, 1979.

North, G. R., Theory of energy balance climate models, J. Atmos. Sci., 32, 2033-2043, 1975.

Oort, A. H., The observed annual cycle in the meridional transport of atmospheric energy, J. Atmos. Sci., 28, 325-339, 1971.

Oort, A. H., and J. P. Peixoto, The annual cycle of the energetics of the atmosphere on a planetary scale, J. Geophys. Res., 79, 2705-2719, 1974.

Oort, A. H., and E. M. Rasmusson, Atmospheric Circulation Statistics, NOAA Prof. Paper 5, 1971.

Pedlosky, J., Finite amplitude baroclinic waves, J. Atmos. Sci., 27, 15-30, 1970.

Pedlosky, J., Finite amplitude baroclinic waves in a continuous model of the atmosphere, J. Atmos. Sci., 36, 1908-1924, 1979.

Rooth, C., in Ewing Symposium volume, 1983.

Saltzman, B., A survey of statistical-dynamical models of the terrestrial climate, Adv. in Geophysics, 20, (Academic Press, N. Y.), 183-304, 1978.

Schneider, E. K., Axially symmetric steady-state models of the basic state for instability and climate studies, Part II: nonlinear calculations, J. Atmos. Sci., 34, 280-297, 1977.

Sellers, W. D., A global climatic model based on the energy balance of the earth-atmosphere system, J. Appl. Meteor., 8, 392-400, 1969.

Stone, P. H., A simplified radiative-dynamical model for the static stability of rotating atmospheres, J. Atmos. Sci., 29, 405-418, 1972.

Stone, P. H., and J. H. Carlson, Atmospheric lapse rate regimes and their parameterization, J. Atmos. Sci., 36, 415-423, 1979.

Stone, P. H., and D. A. Miller, Empirical relations between seasonal changes in meridional temperature gradients and meridional fluxes of heat, J. Atmos. Sci., 37, 1708-1721, 1980.

Stone, P. H., S. Chow, and W. J. Quirk, The July climate and a comparison of the January and July climates simulated by the GISS general circulation model, Mon. Wea. Rev., 105, 170-194, 1977.

Taylor, K. E., The roles of mean meridional motions and large scale eddies in zonally averaged circulations, J. Atmos. Sci., 37, 1-19, 1980.

Vonder Haar, T., and A. H. Oort, New estimate of annual poleward energy transport by Northern Hemisphere oceans, J. Phys. Oceanogr., 3, 169-172, 1973.

Wang, W.-C., W. B. Rossow, M.-S. Yao, and M. Wolfson, Climate sensitivity of a one-dimensional radiative-convective model with cloud feedback, J. Atmos. Sci., 38, 1167-1178, 1981.

Wang, W.-C., and P. H. Stone, Effect of ice-albedo feedback on global sensitivity in a one-dimensional radiative-convective climate model, J. Atmos. Sci., 37, 545-552, 1980.

Yao, M.-S., Maintenance of quasi-stationary waves in a 2-level quasi-geostrophic spectral model with topography, J. Atmos. Sci., 37, 29-43, 1980.

ON THE ROLE OF GLOBAL-SCALE WAVES IN ICE-ALBEDO AND VEGETATION-ALBEDO FEEDBACK

Dennis L. Hartmann

Department of Atmospheric Sciences and Quaternary Research Center
University of Washington, Seattle, Washington 98195

Abstract. Quasi-stationary planetary waves are the major contributor to interannual variability of local climate. These waves can interact with climate feedback mechanisms to enhance the sensitivity of the Earth's climate. In high latitudes planetary waves can encourage ice-albedo feedback over the continents. In the tropics planetary-scale circulations are associated with the movement of precipitation between land and ocean. The movement of precipitation between land and ocean gives rise to planetary albedo changes through vegetation-albedo feedback and through a new, more direct mechanism involving the contrast in clear-sky albedo between land and ocean areas. It is demonstrated that interannual and intraseasonal variability in mid-latitude planetary waves can cause tropical precipitation to shift between land and ocean. The importance of the location of tropical cloudiness for the radiation budget and the ability of the internal dynamics of the atmosphere to move cloudiness combine to produce a potentially important feedback mechanism.

Introduction

In attempting to explain past climate fluctuations and predict future responses to anthropogenic perturbations to the Earth's climatic system, it is useful to consider the sensitivity of this system to quite general forcing and to try to identify mechanisms within the climatic system which control sensitivity. Estimates of the sensitivity of the Earth's climate can be obtained from observation or inference of present and past climate fluctuations, if the perturbing influences responsible for these fluctuations can be identified quantitatively. An example of such an approach is found in the orbital parameter theory for the succession of Quaternary ice ages. The history of global ice volume and sea-surface temperature can be inferred from isotopic and paleontological evidence. One can then assume that some part of this history is "explained" by orbital parameter variations, and thence arrive at an estimate of what the sensitivity of the climate to this specific kind of forcing must be (Imbrie and Imbrie, 1980; Held, 1982). Such quantitative estimates of climate sensitivity provide important constraints against which climate models can be tested.

Another estimate of the sensitivity of the Earth's climate can be obtained by studying the interannual variability of climate. Because of the much different times scales involved, interannual variability is not an exact analogy for climate change. We can study interannual variability in much greater detail than past climate changes, however, and this detail may provide some clues about the mechanisms of climate change. Most of this interannual variability results from shifts in the position and intensity of quasi-stationary waves of planetary scale rather than from global mean changes. The quasi-stationary waves are associated with orography and surface thermal contrasts. Because the dynamics of these waves are dominated by advection, and because the thermal and orographic forcing are dependent on the flow field and hence the waves themselves, the quasi-stationary waves structure tends to be highly variable even on seasonal time scales. It is this variability which is responsible for the the interannual variability of the climate of particular regions. Interannual variability of the climate of particular localities is of significant amplitude and has social and economic importance. Furthermore, there appears to be no external forcing mechanism of measurable amplitude with which to associate it. Because of this combination of large signal and small forcing, an estimate of the sensitivity of the climate on this time scale must necessarily be very high.

An important question is whether this interannual variability, that is associated with small global mean surface temperature variations, has any bearing on long-term global climate change. It is one thesis of this paper that the interannual variability of quasi-stationary planetary wave structure can play a central role in determining climate sensitivity and change. In support of this thesis it can be said that the dynamical constraints which produce the structure of the modes of interannual variability must also operate in controlling the structure of climate change, so that the structure of interannual variability may reveal something about the mechanisms of climate change. Interannual variability, since it requires no external forcing, reveals the structures which require the least external forcing. If these structures are such that interannual variability interacts with the feedback mechanisms which lead to global climate change, then these same structures are probably important in climate change. It will also be argued that quasi-stationary waves play an important role in determining climate sensitivity by controlling continent-ocean temperature contrasts in middle latitudes and continent-ocean precipitation contrasts in the tropics and subtropics.

In the following sections the potential role of quasi-stationary waves in the albedo feedback sensitivity mechanism in high and low latitudes will first be reviewed. Then the structure of present-day variations in quasi-stationary wave variations will be examined using observed data from the mid-latitudes and the tropics. It will be observed that this structure suggests links between high latitude albedo feedback and tropical albedo feedback.

Albedo Feedback in High and Low Latitudes

Ice-albedo feedback has long been recognized as a potentially strong contributor to climate sensitivity (e.g. Koppen and Wegener, 1924). Its importance arises from the comparatively high albedo of snow and ice and the inference of rather more extensive snow and ice coverage during the last glacial age. During the last glacial maximum perennial icecover extended as far south as 40N over North America, sea ice extended farther equatorward than at present in both hemispheres, and mountain glaciers in many areas expanded.

Quite recently it has become known that vegetation-albedo feedback may have played an important role in increasing the global mean albedo during the last glacial maximum. In the tropics and subtropics the continents were much drier during the glacial maximum than at present and had the higher surface albedos characteristic of deserts or semi-arid grasslands (e.g. Williams, 1975; Climap, 1976; Street and Grove, 1979; Sarnthein, 1978). The drier and brighter continents during the ice age are believed to have contributed to the maintenance of colder climate, and tropical aridity may play an important role in sensitivity. In addition there is evidence that tropical continental moisture and vegetation were significantly enhanced during the climatic optimum ~6000BP, which could have contributed to warming via decreased continental surface albedos. In contrast to ice-albedo feedback, which occurs primarily in high latitudes, vegetation-albedo feedback is most important in the tropics and subtropics.

Estimates of the magnitude of the surface albedo changes between glacial and interglacial conditions depend sensitively on reconstructions of past vegetation patterns and their associated surface albedos. Cess (1978) estimated that the planetary albedo changes associated with shifts in vegetation patterns in the tropics between glacial and present conditions could contribute nearly as much to sensitivity as ice in high latitudes. Dickinson (1983) reevaluated the vegetation-albedo feedback mechanism using the surface albedo data of Kukla and Robinson (1980) and concluded that the vegetation-albedo effect was small. Street-Perrott and Roberts (1983) have shown evidence that African lake levels not only decreased during the last glacial maximum at ~18000BP, but also for short intervals around 10200BP and 7400BP. These episodes of tropical continental aridity are important because they provide evidence that the changes in tropical surface hydrology are not solely a result of ice age cooling or sea level changes, but rather that tropical hydrology possesses an inherent sensitivity. This sensitivity is also apparent in modern processes (e.g. Glantz, 1977). Because of this sensitivity, changes in tropical hydrology may be an active agent in the earliest stages of climate change, and not merely a response to climate changes produced by other means.

Planetary Waves and Ice-Albedo Feedback

During the last glacial maximum the latitude of the perennial icecover varied greatly with longitude. It has been pointed out by Hartmann and Short (1979) that the north-south meandering of the perennial iceline causes the albedo enhancing effect of the ice sheets to be greater than they would be if the ice were confined to a zonally symmetric polar cap of the same area. This is because when the ice protrudes equatorward over the continents it extends into a region which is more strongly insolated and where the snow-free planetary albedo is smaller. Thus the albedo increase induced by snowcover at lower latitudes is more effective in reducing absorbed solar radiation. The distorted shape of the iceline results from the different thermal and dynamic properties of the continents and the oceans. The land areas have essentially no heat capacity compared to the oceans so that it is much easier for ice to form in fall and winter and eventually to persist through a complete annual cycle (e.g. North et al., 1983). The orography of the continents can also contribute to the maintenance of ice, either through the effect of elevation or by forcing perturbations of the flow field (planetary waves) which advect cold air equatorward over the continental surface. Linear theory shows that orography in mid-latitudes tends to produce equatorward flow over the elevated topography and cold air near the surface over and slightly downstream of the mountain (e.g. Hendon and Hartmann 1982, Fig. 6). Thus for a continent such as North America the purely dynamical effect of the mountains is to produce an import of cold air from higher latitudes over the continent. In this way the mountains help to encourage the maintenance of perennial snowcover. This role for the mountains probably becomes more important during an ice age since the topography is then augmented by the continental ice sheets, and the winds blowing over the continents are greater because of the enhanced north-south temperature gradient. The change in the strength of the zonal winds would be particularly strong during the summer months, for which we expect the change in temperature gradient between present and ice age conditions to be greatest. The expected change in summertime planetary wave amplitude and structure may be especially important for maintaining the continental snowcover during the critical summer months. Thus planetary waves may play an essential role in mid-latitude albedo feedback and thereby contribute substantially to climate sensitivity.

The contrast between present conditions and paleoclimatic reconstructions of glaciations suggests that during the onset of a glaciation the continents cool more than the oceans, thus increasing the land-ocean temperature contrast. The contact of the colder continental air with the relatively warm ocean to the east (downstream) of the continent can be expected to increase the heating of the atmosphere by the ocean in this region. Under present conditions, most of the heating of the atmosphere by the oceans takes place downstream of the continents (Bunker, 1976) and the heating rate is greatest in the lower troposphere (Lau, 1979). Linear steady-state calculations show that such low-level heating in the mid-latitudes produces cold air near the surface upstream of the heat source (e.g. Hendon and Hartmann, 1982, Fig. 5). Thus the expected increase in air-ocean temperature contrast downstream of North America during an ice age would cause a heating increase which would support colder than zonal average surface temperatures over the continent. The dynamical responses to both heating and orography support cold surface temperature over the central North American continent and may provide a positive feedback to climate change. These dynamical effects must be added to the purely thermodynamic effects of land-sea contrast in assessing the importance of zonal asymmetries for climate sensitivity.

Planetary Waves and Vegetation-Albedo Feedback

Changes in the large-scale circulation in the tropics must be associated with the changes in continental hydrology and vegetation which take place between glacial and interglacial periods. A mechanism relating vegetation-albedo feedback and local large-scale circulations has been presented by Charney (1975). Increased surface albedo and drying of the surface favor large scale subsidence which further promotes drying of the air column. This reasoning is consistent with the ice age simulation of Manabe and Hahn (1977). They found that when the increased continental albedos in the tropics during the last ice age were imposed on a general circulation model, it caused a decrease in the continental precipitation relative to the oceanic precipitation. Because albedo increases promote drier continents, the dry, bright continents are somewhat self-sustaining or, in other words, tropical continental hydrology is "sensitive". It then becomes of interest to ask whether changes in tropical hydrology are a response to climate change or are in fact active agents of change.

The differences in the large scale circulation in the tropics between glacial conditions and present conditions and between hypsithermal conditions and present conditions are indicated schematically in Figure 1. These changes are consistent with the numerical simulation of Manabe and Hahn (1977) and with the available paleoclimatic evidence. During the glacial age relatively more precipitation over the ocean and an accompanying change in large-scale circulation produces more rising over the ocean and sinking over the land. The sinking over the land and low level land to sea airflow help to maintain a relatively dry continental surface. Conversely, during an interglacial the precipitation over the land is greater than at present, there is more of a tendency for rising motion over the land, and the low-level flow carries moist air from sea to land.

Cloud-Surface Albedo Feedback

Figure 1 suggests two mechanisms whereby the movement of convection relative to the continents in the tropics can produce increased sensitivity

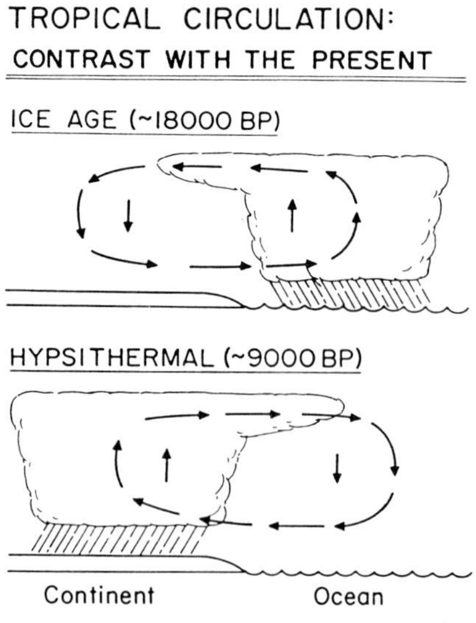

Fig. 1. Schematic drawing of the contrast in the tropical east-west circulations between present, ice-age, and hypsithermal conditions.

Table 1. Zonal-Mean Albedo Change Possible because of Zonal Cloud Movement

ϕ	A_L	A_C[1]	α_L[2]	α_C[1]	α_O[2]	α_{OB}[3]	α_{max}	α_{min}	$\Delta\alpha$	%
90.00	0.00	0.50	0.59	0.59	0.59	0.59	0.59	0.59	0.00	0.00
80.00	0.10	0.58	0.57	0.57	0.57	0.57	0.57	0.57	0.00	0.00
70.00	0.35	0.62	0.50	0.55	0.50	0.50	0.53	0.53	0.00	0.00
60.00	0.65	0.64	0.25	0.50	0.15	0.43	0.41	0.37	0.04	3.30
50.00	0.55	0.61	0.25	0.46	0.13	0.38	0.38	0.33	0.05	6.63
40.00	0.47	0.52	0.25	0.43	0.10	0.33	0.34	0.27	0.07	13.73
30.00	0.40	0.44	0.25	0.40	0.10	0.29	0.29	0.23	0.06	14.86
20.00	0.34	0.42	0.30	0.38	0.10	0.26	0.29	0.22	0.07	19.93
10.00	0.27	0.48	0.20	0.36	0.10	0.25	0.25	0.22	0.03	8.74
0.00	0.25	0.51	0.20	0.35	0.10	0.25	0.25	0.23	0.02	8.37
-10.00	0.25	0.48	0.20	0.36	0.10	0.24	0.25	0.22	0.03	8.10
-20.00	0.25	0.47	0.20	0.38	0.10	0.24	0.26	0.23	0.03	7.33
-30.00	0.15	0.50	0.20	0.40	0.10	0.27	0.26	0.25	0.01	3.71
-40.00	0.05	0.60	0.20	0.43	0.10	0.33	0.30	0.30	0.00	0.97
-50.00	0.00	0.72	1.00	0.46	0.11	0.39	0.36	0.36	0.00	0.00
-60.00	0.00	0.78	1.00	0.50	0.12	0.47	0.42	0.42	0.00	0.00
-70.00	0.40	0.61	0.60	0.55	0.40	0.55	0.57	0.49	0.08	4.34
-80.00	1.00	0.52	0.61	0.61	0.45	0.61	0.61	0.61	0.00	0.00
-90.00	1.00	0.45	0.62	0.62	0.50	0.62	0.62	0.62	0.00	0.00

[1] Cloud fractions and albedos are from Cess (1976)
[2] Clear-sky albedos are from Ellis (1978)
[3] Observed albedos are from Ellis and Vonder Haar (1976)

of the climate. The first of these is the vegetation-albedo feedback which has already been mentioned. If decreased continental precipitation is somehow associated with cold climates, then the increased surface albedo associated with this decreased continental precipitation will lead to further cooling. The second mechanism is the direct cloud-surface albedo effect. Since the continental surface albedos are generally higher than the surface albedo of the ocean, there will be a global mean albedo change when the clouds and precipitation move from the land to the ocean regardless of whether there is any surface albedo change. When the clouds are over the land area and expose more of the ocean surface to sunlight the global mean albedo must decrease. When the clouds are over the ocean, an observer in space sees a relatively brighter planet composed of clouds and land surface. The potential magnitude of this direct cloud-surface albedo effect can be estimated from available data on zonal-mean quantities.

Table 1 shows, for individual latitudes ϕ, the fractional area coverage of land A_L, the average clear-sky albedo over land α_L, the average albedo of cloudy sky α_c, the average clear-sky albedo over ocean α_o, the observed zonal-mean albedo α_{ob}. Using these data it is possible to calculate the maximum and minimum albedos which can be attained by a simple east-west shift of the cloudiness relative to the land areas. If the clear-sky albedo over land is greater than the clear-sky albedo over ocean, then the highest albedos are obtained when the clouds cover the ocean area to the maximum possible extent.

$$\alpha_{max} = \begin{cases} A_c\alpha_c + A_L\alpha_L + (1-A_c-A_L)\alpha_o; & A_c<1-A_L \\ A_c\,\alpha_c + (1-A_c)\,\alpha_L\,; & A_c>1-A_L \end{cases} \quad (1)$$

The minimum albedo is attained when the clouds cover the land.

$$\alpha_{min} = \begin{cases} A_c\alpha_c + (1-A_c)\alpha_o\,; & A_c>A_L \\ A_c\alpha_c + (A_L-A_c)\alpha_L + (1-A_L)\alpha_c; & A_c<A_L \end{cases} \quad (2)$$

The cloudy sky albedos in column 5 of Table 1 were inferred for zonal mean conditions. In applying equations (1) and (2) we assume that the cloudy sky albedo does not depend on the albedo of the underlying surface, and that maritime and continental clouds have the same albedos. In truth, some of the albedo contrast of the underlying surface is transmitted through the cloud to the top of the atmopshere by multiple scattering. This means that the numbers in column 10 of Table 1 will be systematic overestimates of the actual top of the atmosphere albedo contrast. In order to properly account for this effect we would need separate estimates of the albedo of cloudy skies over land and ocean, which are not currently available. It appears, however, that the error associated with the assumption that the cloudy sky albedo is independent of surface conditions is probably small and can justifiably be ignored at the current level of approximation.

Support for this assertion can be found in radiative transfer calculations for clouds which are not exceptionally thin. Results of such calculations have been provided by S.G. Warren (private communication). As an example we consider the contrast of the planetary albedo when a cloud is placed over water and over new snow. Over these surfaces we introduce a cloud with a liquid water content of 230 g m^{-2} and a visible optical depth of 40, which we consider typical. These scenes are illuminated by the sun at a zenith angle of 53 degrees (cosine of solar zenith angle 0.6). The clear sky albedos over water and new snow are 0.108 and 0.671, respectively. The calculated planetary albedos when the cloud is placed over these surfaces are 0.661 and 0.741. The clear sky surface albedos for the water and snow are 0.074 and 0.836. Therefore a change in clear sky surface albedo of 0.762 results in a change in the planetary albedo of the cloudy sky of 0.084, so that only 11% of the surface albedo contrast is observed at the top of the atmosphere. Such a small difference will not affect our general conclusions. It should be noted, however, that for very thin clouds the surface albedo is, of course, more important and that the incorporation of more realistic cloud geometry cloud also alter the conclusion regarding the importance of surface albedo for the planetary albedo above clouds.

The eighth, ninth and tenth columns of Table 1 show the estimates for α_{max} and α_{min} and the difference between them. When these are weighted according to annual mean insolation and averaged over the area of the globe they yield minimum and maximum global mean albedos of 0.274 and 0.306 and a difference of 0.032. The difference between these maximum and minimum possible albedos is quite large. Hartmann and Short (1979) estimated a global mean albedo change due to surface albedo changes between present and glacial conditions, due to both ice and vegetation changes, of only slightly greater than half this value. If a global mean planetary albedo change of 0.032 is imposed on the simplest global mean energy balance climate model with the outgoing terrestrial radiation linearly dependent on the surface temperature with the coefficient 2.0 Watts per meter-squared per degree Kelvin, then a global mean surface temperature change of 5.4K is implied. This is a large as the global mean temperature change between ice age and present conditions. Although the 0.032 upper bound on the albedo change due simply to moving the clouds relative to the land is probably not attained in practice, the potential is large enough that even a fraction of this potential would be important for climate sensitivity. It is also important to note that the direct cloud-surface albedo effect acts in the same sense as the vegetation-albedo feedback and would tend to augment it.

Dickinson (1983) has pointed out that the surface albedos of vegetated land used by Cess (1978) and Hartmann and Short (1979) were unrealistically low and that vegetation-surface albedo feedback may not be a strong contributor to climate sensitivity. Increasing the surface albedo of tropical forest and grassland, however, increases the potential importance of the direct cloud-surface albedo feedback, since the contrast between the clear-sky albedos of land and ocean is thereby increased. The net effect of tropical circulation and hydrology on climate sensitivity may thus remain important, even though surface albedo change over the continents is not as powerful a feedback mechanism as previously believed.

The last column in Table 1 gives the percent contributions by each 10 degree latitude belt to the 0.032 maximum albedo change. It is notable that the largest contributions come from the northern subtropics where there is a large expanse of relatively high albedo land which is well insolated. Although the basic data in Table 1 have uncertainties in them and are taken from different data sources, it is interesting to note that the currently observed albedos in the Northern Hemisphere are close to the maximum values we can obtain by shifting the clouds in the east-west direction. It is suggested then that most of the potential albedo feedback by the direct cloud-surface albedo mechanism in this region is for producing a warming through the movement of precipitation over the dry land areas. Most of the potential vegetation-albedo feedback is also in the direction of warming, since most of the land in the northern subtropics is already dry and bright.

Observed Relations between Mid-Latitude Waves and Tropical Convection

We have tried to show that large-scale zonally asymmetric circulations play an important role in enhancing ice-albedo feedback in high latitudes and combined vegetation-albedo feedback and cloud-surface albedo feedback in the tropics. In this section we would like to investigate the possibility that large-scale quasi-stationary waves may also perform the important task of coupling ice-albedo feedback in high latitudes to vegetation-albedo feedback in low latitudes. Paleoclimatic data suggest that high latitude continental glaciations tend to occur simultaneously with aridity of the tropical continents. If the coupling between cold mid-latitude continents and dry low-latitude continents also occurs on much shorter time scales such as the interannual or intraseasonal time scales, then it can be argued that the two types of albedo feedback are really part of a single, very powerful sensitivity mechanism.

A prerequisite for such a coupling to exist is

that there be a relation between mid-latitude planetary wave structure and tropical wave structure and precipitation. Secondarily there should be a suggestion that the relation is of the proper sign so that cold mid-latitude continents tend to occur simultaneously with drier than normal tropical continents. In order to examine the relation between mid-latitude planetary waves and tropical convection, we compare 500mb height analyses with simultaneous measurements of outgoing infrared radiation. In the tropics variations of outgoing IR are a good measure of cloudiness and, when averaged over space or time, of precipitation and large-scale vertical motion (Griffith et al., 1978; Richards and Arkin, 1981). For the results shown here the data are averaged into five day means in order to remove some of the high frequency variability. We will consider only the winter season of December, January and February. The data cover the period from June 1974 through May 1981 and the mean annual cycle for this period was removed before the data analysis described here were performed. Thus the variations considered include interannual variability and low-frequency intraseasonal variability but exclude the annual cycle. The geopotential height data for the 500mb surface are the NMC Northern Hemisphere analyses and the IR data are a 5 x 5 degree subset of the 2.5 x 2.5 degree latitude-longitude analysis of the NOAA Scanning Radiometer data (Gruber and Winston, 1978). Some characteristics of the means and standard deviations of the IR data are given in Hartmann and Short (1980) and the substantial interannual variability of the seasonal means of IR data are described in Liebmann and Hartmann (1982).

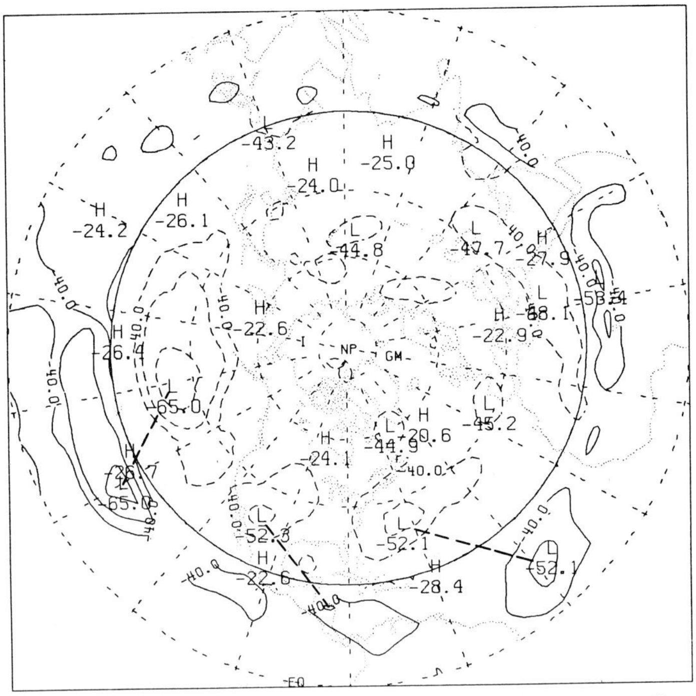

Fig. 2. Contours of the maximum negative correlations between 5-day means of the 500 mb height field at any point north of 20N with the IR at any point between the equator and 20N for the Dec-Feb season. The smallest contour plotted is 40 percent, and the contour interval is 10 percent.

The Minimum Correlation Map

To begin this study we would like to perform an objective search for relationships between extratropical height anomalies and tropical IR anomalies. As a means of conducting this search we calculate the correlation between the 500mb height at every point poleward at 20N and the outgoing IR at every point between the equator and 20N. The largest negative correlations obtained after trying every possible combination of points are contoured in Figure 2. There are three prominent pairs of centers of maximum negative correlation in the Western Hemisphere. There are three centers at about 35N and 160W, 100W and 60W where the height field is strongly negatively correlated with the outgoing IR at between 10N and 15N and 140W, 85W and 40W, respectively. These centers are connected by heavy dashed lines in Figure 2. The largest correlation coefficient of -0.65 occurs for the pair of centers over the Pacific Ocean. The interpretation of this pair is that when the geopotential height in the vicinity of 35N-160W is low (high) the outgoing IR near 15N-140W tends to be high (low). High IR corresponds to less than normal cloud and precipitation and low IR to more cloud and precipitation than normal. Since the zonal component of the group velocity of quasi-stationary Rossby waves is normally eastward (e.g. Hoskins and Karoly, 1981) and the heating occurs to the southwest of the center of the height anomaly, one must interpret the mid-latitude height field as driving the tropical precipitation change rather than the other way around, for the three pairs of centers emphasized in Figure 2.

If one assumes that every other five day average is an independent datum then the nearly 7 years of winter season data used to construct Figure 2 yield a required minimum correlation coefficient of 0.3 to be significantly different from zero at the 99 percent confidence level. Figure 2 thus indicates quite clearly that there are significant relationships between the 500 mb height in mid-latitudes and tropical convection. This fact in itself is sufficient to indicate that there is some degree of coupling between mid-latitude quasi-stationary waves and tropical convection and possibly also between ice-albedo feedback and vegetation-albedo feedback. An examination of the structure of the anomalies associated with the correlations in Figure 2 will further illustrate this coupling. In addition, the consistency of the structures obtained with what we expect a priori based purely upon dynamical reasoning gives a measure of confidence in the significance of the results which could not be gained from statistical tests alone.

Composites of Extreme Events

In order to determine the characteristic structure in the mid-latitude height and the tropical convection anomalies associated with the correlations in Figure 2 we construct composites of the extreme positive and negative events. To provide a specific example that illustrates a connection between mid-latitude planetary wave structure and tropical convection we will consider the correlation pair over the Atlantic Ocean which is between the 500mb height at 35N-65W (actually 33N-63W) and the outgoing IR at 10N-35W. Figure 3a shows the average 500mb height anomaly for the 18 pentads (5-day means) when the anomaly at 35N-65W is positive and larger than one standard deviation. Figure 3b shows the average anomaly for the 16 pentads when the height is negative and larger than one standard deviation. The difference between positive and negative anomaly composites is shown in Figure 3c. The significance of the difference between the high and low anomalies can be estimated using a statistic of the form

$$Z = \frac{\overline{x_1} - \overline{x_2} - \mu}{\left(\frac{\sigma_1^2}{N_1} + \frac{\sigma_2^2}{N_2}\right)^{1/2}} = \frac{\overline{x_1} - \overline{x_2}}{\sigma\left(\frac{1}{N_1} + \frac{1}{N_2}\right)^{1/2}} \qquad (2)$$

where x_1 and x_2 are the two means, σ_1^2 and σ_2^2 are the true variances of the populations from which the samples were taken and N_1 and N_2 are the sizes of the two samples. To arrive at the final form on the right we assume that the variances of the two populations are equal to the variance of the winter season for the entire seven-year data set, $\sigma_1 = \sigma_2 = \sigma$. The null hypothesis is then that the true means of the two samples are equal ($\mu = 0$). In order to reject this null hypothesis at the 95 percent level of certainty the value of Z must exceed 1.96. In Figure 3d we contour the values of the Z-statistic at intervals of 0.62, beginning at 1.96. The second contour level of 2.58 corresponds formally to the 99 percent significance level. The Z-statistic can be used as a guide in assessing the significance of particular features in the anomaly difference map. It also draws out tropical features that are not apparent in the height anomaly contour map because of the small amplitude of the height anomlies in the tropics.

Figures 3 shows the structure of the hemispheric height field assoicated with high and low anomalies at 35N-65W. The dominant feature is a dipole pattern with an anomaly of opposite sign centered roughly over Hudson Bay associated with anomalies over the Western Atlantic at 35N. The Z-statistic draws out a subtropical anomaly off the west coast of North America and also a tropical feature to the south of the reference anomaly at 35N-65W. This tropical feature will be reflected in large anomalies in tropical cloudiness associated with the reference anomaly. It contributes to the large correlation exhibited in Figure 2 between the height field at 35N-65W and the IR field at 10N-35W.

The IR anomaly maps for the same conditions as in Figure 3 are shown in Figure 4. The anomalies over Mexico and the North Atlantic Ocean are associated with the height anomalies just to the west of each of those locations. The large IR anomaly over the tropical Atlantic Ocean is associated with the correlation in Figure 2 and the tropical height anomaly suggested in Figure 3d. Comparison of the high and low anomaly maps together with the Z-statistic suggest that there may be a consistent relationship between precipitation over tropical South America and geopotential height at 35N-65W. The phase of the relationship is such that when the height at 35N-65W is higher than normal there is a greater than normal precipitation over the tropical North Atlantic Ocean and less than normal precipitation over South America. The South American IR anomalies are significant over the upper Amazon Basin and to the east of the Brazilian Highlands, if _a priori_ tests are used. Additional confidence in the significance of these anomalies is gained by noting that the anomalies occur in both the positive and negative composites in Figures 4a and 4b.

The 500mb height anomaly pattern in Figure 3 associated with extreme values in the 5-day means for the point 35N-65W is similar to the Western Atlantic pattern described by Wallace and Gutzler (1981). This pattern was obtained by searching for horizontal correlation patterns in monthly mean data. The pattern shown in Figure 3 can thus be expected to appear on longer time scales than the 5-day means we have used. This is also suggested by Table 2 which shows the distribution of the occurrences of the extreme high and low anomalies in the height field. There are several occurrences of runs of anomalies of one sign, suggesting some low frequency components in the time variation of this pattern. These low frequencies increase the potential importance of these anomaly patterns for climate sensitivity, since it is probable that they can contribute to variance in seasonal means.

The magnitude of the IR anomalies are quite large. The anomalies in Figure 3c over the Americas and the Atlantic Ocean are nearly as large as the standard deviation of daily values in those regions (see Hartmann and Short, 1980). The contrast of the mean IR over the tropical North Atlantic Ocean (~250 W/m^{-2}) with that over the dry zone in the tropical Southern Atlantic Ocean (~280 W/m^{-2}) is about 30 W/m^{-2} (e.g. Liebmann and Hartmann, 1982). The anomaly difference for the tropical North Atlantic region shown in Figure 3c is 25 W/m^{-2}. Therefore the contrast in IR at 10N-35W associated with the height variations at 35N-65W suggests that the

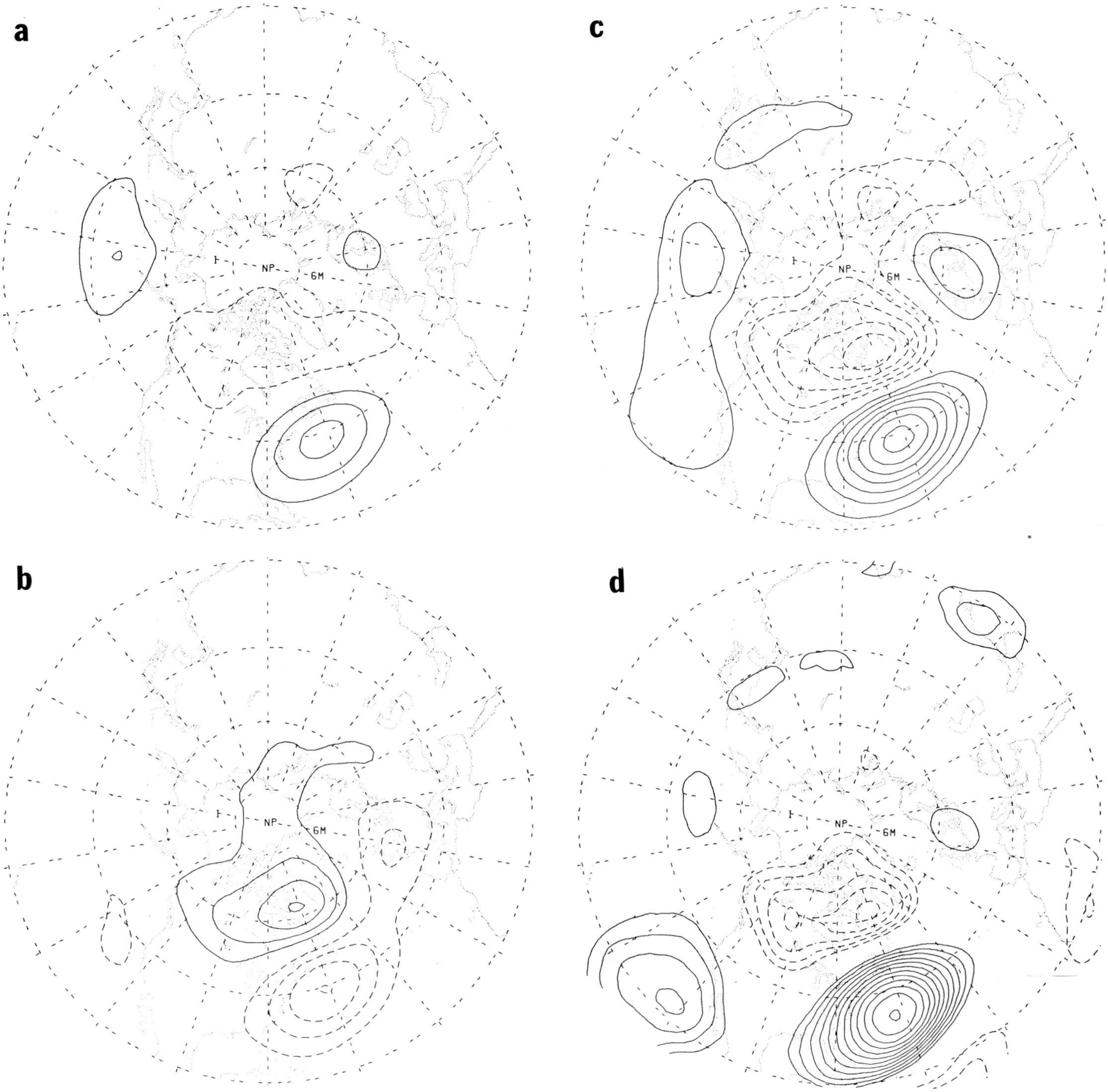

Fig. 3. Polar sterographic contour maps of (a) the height field anomalies when the height at 35N-65W is more than one standard deviation above the mean, (b) below the mean, (c) high minus low, (d) Z-statistic for high minus low. Contour interval of height is 40 m and of Z-statistic is 0.62.

precipitation is varying between nearly zero and nearly twice its mean value. Similarly inferred relative changes in precipitation over South America are only slightly smaller. The IR anomalies thus represent significant low-frequency variations in tropical precipitation.

Conclusion

In this paper we have reviewed the ice-albedo and vegetation-albedo feedback mechanisms and emphasized the role of the dynamics of planetary-scale zonally asymmetric motions in them.

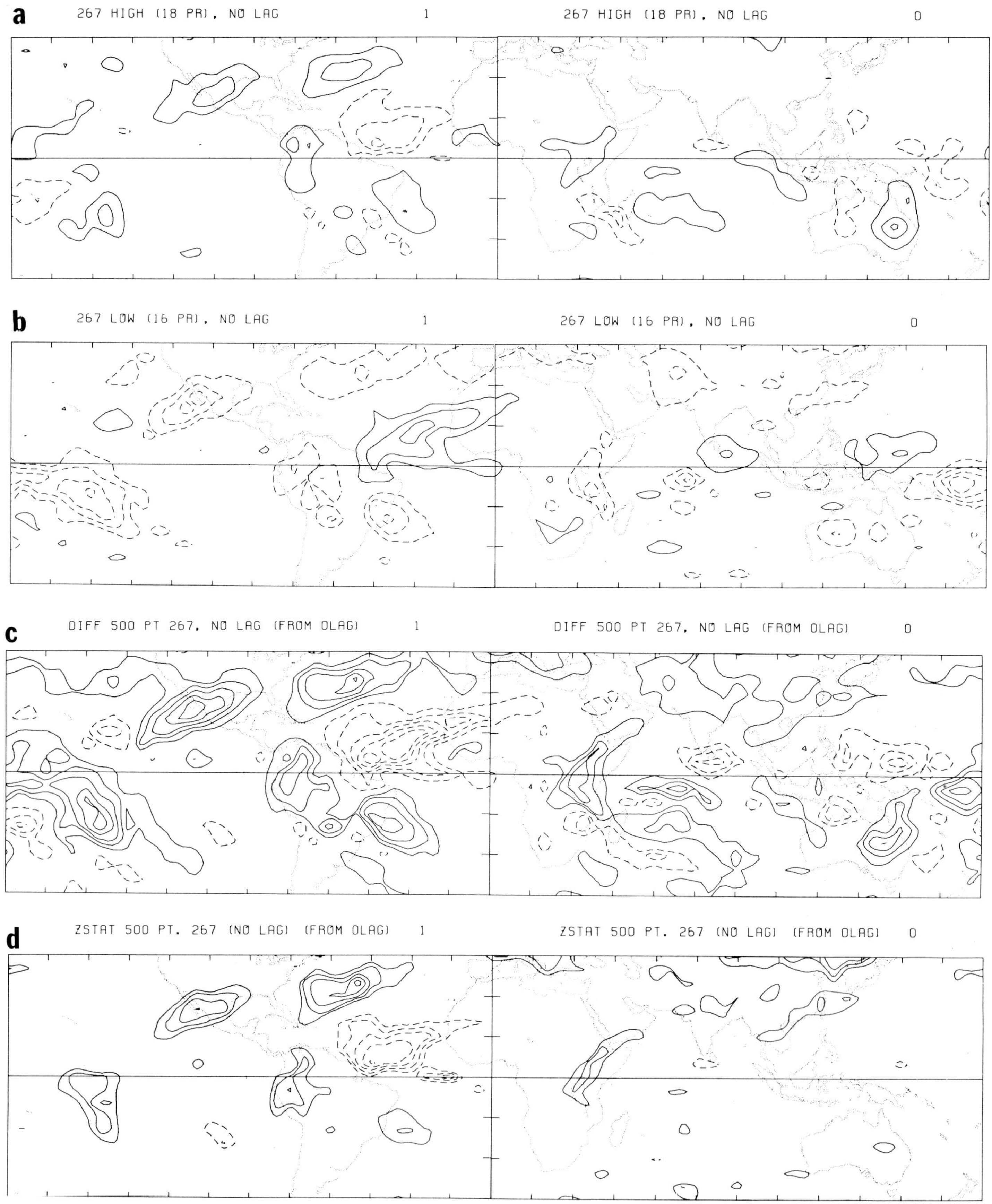

Fig. 4. Same as Figure 3 except a cylindrical equidistant plot of the outgoing IR anomalies. Contour of outgoing IR anomalies is 5 W/m^{-2}.

Table 2. Occurrence of extreme high (+) and low (-) anomalies of the 500mb height field at 36N-65N. Columns are for the 18 pentads during the Dec-Feb season, rows correspond to the years indicated.

	1	2	3	4	5	6	7	8	9	10	11	12	13	14	15	16	17	18
74-75									+		+	+					+	+
75-76																+	+	+
76-77	+					-	-			-	-							+
77-78	+									+			-		-		-	
78-79							+					-	-		-			+
79-80		+				-			+		-		-					
80-81										-	-			+	+			-

Planery albedo increase associated with the equatorward migration of perennial snowcover over the continents is an important contributor to climate sensitivity. The maintenance of colder than zonal-mean temperatures near the surface over the continents is aided by the dynamical response of the atmosphere to continental orography and air-sea heat exchange downstream of the continent. Both mountains on the western side of a continent and warm waters east of a continent in middle latitudes give rise to a linear planetary wave response which advects cold air equatorward over the continent. The resulting low temperatures near the surface over the continent promote the formation and maintenance of snowcover at relatively low latitudes. Since enhanced continental elevation and enhanced air-sea temperature contrast both contribute to increased forcing of cold air advection over a continent, the dynamical response of planetary waves can be a positive feedback mechanism for climate change.

In low latitudes the shifting of precipitation between land and ocean areas can lead to planetary albedo changes. This albedo change can result both from surface hydrological changes affecting surface albedo through vegetation and soil changes, and from a direct cloud albedo, surface albedo effect. Clouds over the ocean screen a lower albedo surface from sunlight than clouds over land. By moving the current cloudcover in an east-west direction between land and ocean the global mean planetary albedo can be varied by 0.032. This albedo change is sufficient to produce an approximately 5K global mean temperature change, without any other feedbacks. The largest potential for direct cloud albedo, surface albedo sensitivity is in the direction of warming.

With the help of observed height and outgoing IR data it was illustrated that mid-latitude planetary waves and tropical precipitation are coupled to an important degree. The structure of the anomalies associated with the regions of greatest correlation between mid-latitude height and tropical IR suggest that it is more likely that mid-latitude disturbances propagate into the tropics and modulate convection than that tropical heating anomalies excite mid-latitude planetary waves as has so often been suggested.

In the one specific example shown a height anomaly pattern resembling the Western Atlantic pattern found in correlations of monthly mean data was associated with a variation in outgoing IR over the tropical Atlantic Ocean and South America. The phase of the pattern was such that a low 500mb height over Hudson Bay occurred simultaneously with high heights at 35N over the western Atlantic Ocean, low heights in the tropics, high precipitation over the tropical Atlantic and low precipitation over much of South America. Although the application of this result to climate change would be highly conjectural, the pattern does suggest that an association between cold North American climate at high latitudes and drier than normal conditions over South America is a possibility, and suggests a mechanism whereby this association may result. The mechanism is the propagation of planetary waves from the mid-latitudes to the tropics where they can influence precipitation and the tropical circulation. Low frequency variability in mid-latitudes propagating into the tropics also provides a mechanism for moving precipitation between land and ocean areas within the tropics, which can produce a planetary albedo change.

The role of low-frequency, planetary-scale waves in climate sensitivity may be an important one and should be thoroughly studied. Modern and paleoclimatic data and a broad range of modeling efforts can be brought to bear on this question.

Acknowledgments. Brant Liebmann performed the analysis of the 500mb height and outgoing IR data and is continuing a more complete study of the connections between mid-latitude waves and tropical convection including their time dependence. This work war supported by the Climate Dynamics Programs, Climate Dynamics Research Section, Atmospheric Sciences Division, National Science Foundation under Grant ATM-81-06099.

References

Bunker, A.F., Computations of surface energy flux and annual air-sea interaction cycles of the North Atlantic Ocean, Mon. Wea. Rev., 104, 1122-1140, 1976.

Cess, R.D., Climate change: An appraisal of atmospheric feedback mechanisms employing zonal climatology, J. Atmos. Sci., 33, 1831-1843, 1976.

Cess, R.D., Biosphere-albedo feedback and climate modeling, J. Atmos. Sci., 35, 1765-1768, 1978.

Charney, J.G., Dynamics of deserts and drought in the Sahel, Quart. J. Roy. Meteor. Soc., 101, 193-202, 1975.

Dickinson, R.E., Effects of tropical deforestation on climate. In Blowing in the Wind: Deforestation and Long-Range Implications, Pub. No. 14, Studies in Third World Societies, Dept. of Anthropology, College of William and Mary, Williamsburg, VA, 1980.

Dickinson, R.E., Vegetation/albedo feedback, this volume, 1983.

Ellis, J.S., Cloudiness, the Planetary Radiation Budget, and Climate., Ph.D. thesis, Colorado State University, 129 pp., 1978.

Ellis, J.S. and T.H., Vonder Haar, Zonal Averaged Earth Radiation Budget Measurements from Satellites for Climate studies, Atmos. Sci. Paper No. 240, Colorado State University, 50 pp, 1976.

Glantz, M.H., Desertification, Westview Press, Boulder, Colorado, 346 pp., 1977.

Griffith, G.C., W.L. Woodley, P.G. Grude, D.W. Martin, J. Stout, and D.N. Sikdar, Rain estimation from geosynchronous satellite imagery - visible and infrared studies, Mon. Wea. Rev., 106, 1153-1171, 1978.

Gruber, A., and J.S. Winston, Earth-atmosphere radiative heating based on NOAA scanning radiometer measurements, Bull. Amer. Meteor. Soc., 59, 1570-1573, 1978.

Hartmann, D.L. and D.A. Short, On the role of zonal asymmetries in climate change, J. Atmos. Sci., 36 519-528, 1979.

Hartmann, D.L. and D.A. Short, On the use of Earth radiation budget statistics for studies of clouds and climate, J. Atmos. Sci., 37, 1233-1250, 1980.

Held, I.M., Climate models and the astronomical theory of ice ages, Icarus, 50, 449-461, 1982.

Hendon, H.H. and D. L. Hartmann, Stationary waves on a sphere: Sensitivity to thermal feedback, J. Atmos. Sci., 39, 1906-1920, 1982.

Hoskins, B.J. and D. Karoly, The steady linear response of a spherical atmosphere to thermal and orographic forcing, J. Atmos. Sci., 38, 1179-1196, 1981.

Imbrie, J. and J.Z. Imbrie, Modeling the climatic response to orbital variations, Science, 207, 943-953, 1980.

Koppen, W. and A. Wegener, Die Klimate der Geologischen Vorzeit, Verlag Gebruder Borntraeger, Berlin, 256 pp, 1924.

Kukla, G. and D. Robinson, Annual cycle of surface albedo, Mon. Wea. Rev., 108, 56-68, 1980.

Lau, N.-C., The observed structure of tropospheric stationary waves and local balances of vorticity and heat, J. Atmos. Sci., 36, 996-1016, 1979.

Liebmann, B. and D.L. Hartmann, Interannual variations in outgoing IR associated with tropical circulation changes during 1974-78, J. Atmos. Sci., 39, 1153-1162, 1982.

Livingston, D.A., Speculations on the climate history of mankind, Amer. Sci., 59, 332-337, 1971.

Manabe, S. and D.G. Hahn, Simulations of the tropical climate of an ice age. J. Geophys. Res., 82, 3889-3911, 1977.

North, G.R., J.G. Mengel and D.A. Short, A simple energy balance model resolving the seasons and the continents, submitted to J. Geophys. Res., 1983.

Richards, F. and P. Arkin, On the relationship between satellite - observed cloud cover and precipitation, Mon. Wea. Rev., 109, 1081-1093, 1981.

Sarnthein, M., Sand deserts during glacial maximum and climatic optimum, Nature, 272, 43-46, 1978.

Street, F.A. and A.T. Grove, Environmental and climatic implications of late Quaternary lake - level fluctuations in Africa, Nature, 261, 385-390, 1976.

Street, F.A. and A.T. Grove, Global maps of lake-level fluctuations, since 30,000 BP, Quat. Res., 12, 83-118, 1979.

Street-Perrott, F.A. and N. Roberts, Fluctuations in closed basin lakes as an indicator of past atmospheric circulation patterns, Variations in the Global Water Budget, D. Reidel, Dordrecht, in press 1983.

Wallace, J.M. and D.S. Gutzler, Teleconnections in the geopotential height field during Northern Hemisphere winter, Mon. Wea. Rev., 109, 784-812, 1981.

Williams, M.A., Jr., Late Pleistocene tropical aridity sychronous in both hemispheres?, Nature, 253, 617-618, 1975.

THE OCEAN'S TRANSIENT RESPONSE TO GLOBAL SURFACE TEMPERATURE ANOMALIES

K. Bryan and F.G. Komro

Geophysical Fluid Dynamics Laboratory/NOAA, Princeton, New Jersey 08540

C. Rooth

Rosenstiel School of Oceanography, University of Miami, Miami, FL 33149

Abstract. Transient tracers are not perfect analogues for the downward penetration of a heat anomaly associated with climate change. Buoyancy effects associated with a temperature anomaly can significantly alter the stratification and thermohaline circulation. To investigate these effects a three-dimensional model of the world ocean is perturbed by spatially uniform surface anomalies, and the response calculated over a 50 year period. The penetration depth for a negative temperature anomaly of 0.5°C is 25% greater than that for a positive anomaly. The penetration depth is found to be approximately one half the pycnocline depth after 10 years, and one pycnocline depth after 40 years. Both a simple two box model, and a box diffusion model provide a reasonable fit to the globally averaged results of the more general three-dimensional model.

Introduction

Measurements of atmospheric carbon dioxide taken at Mauna Loa Observatory and other locations indicate a monotonic increase since 1955 (Keeling and Bacastow, 1977). The effect of the increase of atmospheric carbon dioxide, and other radiatively active gases produced by man's activities is receiving increased attention. The oceans are believed to be a critical factor in both the rate at which carbon dioxide builds up in the atmosphere, and as a moderating influence on any climate response which may ensue. The effectiveness of the ocean as a sink of carbon dioxide has been the subject of many studies. A widely referenced report on the carbon dioxide/climate question (Climate Research Board, NAS, 1979) suggests that the heat capacity of the ocean may delay a full climate response to increasing carbon dioxide by "several decades". A more precise prediction of the delaying effect of the ocean which will allow for a varying rate of atmospheric carbon dioxide increase is urgently needed. Data on ocean currents and water mass properties provide a source of information for building ocean models. Transient tracer data for tritium, bomb-produced C_{14} and freon would appear to be valuable analogues for the vertical penetration of heat anomalies. On the basis of the tracer data an interesting one-dimensional diffusion model (Oeschger et al., 1975) has been constructed. In this study we have attempted to go one step further and use a fully three-dimensional, dynamic model of the ocean to study the penetration of a heat anomaly. The model has the obvious advantage of being free of the heavy dependence on empirical evidence, and allows some insight on the actual mechanisms involved in transferring substances introduced at the ocean surface to lower levels. It also permits the study of the effect of buoyancy perturbations associated with a heat anomaly, an effect which is absent in the case of purely passive tracers which have been used to calibrate the available one-dimensional models.

Some experience exists in the use of similar three-dimensional models for the study of the ocean's geochemical tracers. Holland (1971) has used a similar model to study the steady state distribution of tracers like C_{14} in the ocean. Sarmiento (1982) has studied the tritium entry into the ocean by the same method. Schneider and Thompson (1981) studied the transient response to increasing atmospheric carbon dioxide of a highly simplified coupled ocean-atmosphere model, in which the ocean is approximated by a static mixed layer increasing in depth from equator to pole. A more detailed atmospheric and ocean model was used in a similar study by Bryan et al. (1982). A conclusion of the latter study was that a slow increase in atmospheric carbon dioxide (doubling time greater than 30 years) will produce a climatic response which has nearly the same latitudinal distribution as the equilibrium response obtained in previous climate modeling studies by Manabe and Wetherald (1980) and others. That is; the response is rather small in the surface temperature of the tropics, but increases to large values at high latitudes. The coupled model of Bryan et al. (1982) showed that after a sudden "switch on", four-fold increase of atmospheric carbon dioxide, 50% of the final equilibrium response was achieved at all latitudes in air temperature after 10 years. Only after 20 years is a 50% response achieved in sea surface temperature at all latitudes. To obtain the actual delaying effect of the ocean requires a careful interpretation of these results for actual scenarios of atmospheric carbon dioxide increase.

The quantitative aspects of the response of the coupled model of Bryan et al. (1982) naturally depend on the details of the ocean model. In the present study of a more general model of the ocean is examined which takes into account the actual geometry of the World Ocean. On the other hand, the complications due to the coupling of the ocean model with an atmospheric model are absent. In the present model, the initial conditions are simply equilibrium solutions found by Bryan and Lewis (1979) for an ocean with prescribed temperature, salinity and wind stress at the surface. Seasonal effects are only included in one special case. A uniform anomaly is imposed by modifying the surface boundary condition, and a numerical integration of the model is made over a 50 year period to study the response. Matched integrations with, and without, the perturbed boundary conditions are compared to eliminate any possible transient departures from the initial conditions. The behavior of temperature anomalies is compared with tracers and a special integration is made to examine seasonal effects.

Model

Details of the model are given in Bryan and Lewis (1979) and the numerical method is given in Bryan (1969) and Semtner (1974). Let $\underline{v}$ be the horizontal velocity vector, and ∇ be the horizontal gradient operators.

The governing equations are:

$$\frac{d}{dt}\underline{v} + f\,\underline{k} \times \underline{v} + \frac{1}{\rho_o}\nabla p = A_{VM}\,\partial_z\partial_z\,\underline{v} + A_{HM}\,(\nabla^2\,\underline{v} + \ldots) \quad (1)$$

$$\rho g + \partial_z\, p = 0 \quad (2)$$

$$\nabla\cdot\underline{v} + \partial_z w = 0 \quad (3)$$

$$\frac{d}{dt}\theta = \partial_z\,(A_{VH}\,\partial_z\,\theta) + A_{HH}\,\nabla^2\,\theta \quad (4)$$

$$\frac{ds}{dt} = \partial_z\,(A_{VH}\,\partial_z\,S) + A_{HH}\,\nabla^2\,S \quad (5)$$

and the equation of state may be written

$$\rho = F(\theta,\ S,\ p) \quad . \quad (6)$$

Thus, θ,S and ρ represent potential temperature, salinity and density, respectively. p is pressure and F(θ, S, p) is the nonlinear equation of state of sea water. f is the Coriolis force, w is the vertical velocity and $\underline{k}$ is a unit vector in the vertical direction. The left hand side of equations (1), (4) and (5) represents the explicitly resolved fields. The right hand side represents the closure term representing unresolved motions including convection. In general, linear closure is used, and the mixing coefficients are chosen to be the minimum values for the resolution of the numerical grid (Bryan, Manabe and Pacanowski, 1975). Several terms which are only important in polar regions are omitted from the right side of (1). Definitions and numerical values of coefficients are given in Table 1. Convection is included by allowing vertical mixing to be infinite for statically unstable cases.

Temperature and salinity at the surface are not directly specified. Instead surface heating Q and net water flux, P - E are specified in the following way,

$$Q = \mu\rho c_p\,(\theta^* - \theta)\,\Delta Z \quad (7)$$

$$P\text{-}E = -\,\mu(S^* - S)\,\Delta Z / S_o \quad (8)$$

ΔZ is the thickness of the surface layer, and S_o is a reference surface salinity (S_o = 35 ‰). Based on a study of the surface heat balance of the ocean Haney (1971) determined that μ^{-1} = 20 days is a reasonable value over a wide range of latitudes. Our time constant is slightly larger (i.e., 30 days). The formulation in (7) and (8) allows the buoyancy flux to be determined locally by the model. It will tend to be more vigorous, when θ^* is lowered than when it is increased because a negative anomaly is favorable for convective overturning.

The geometry of the world ocean model is shown in Figure 1. Wind stress data was taken from tables by Hellerman (1968) and the surface reference values, θ^* and S^* from a digitized data base from an atlas published by Levitus (1982). Salinity reference values specified at the surface along the Antarctic Continent are increased to simulate the ice formation taking place there. Thus S^* is chosen to represent the salinity of water formed in winter along the continental shelves, rather than the low salinity water normally observed at the surface. Since this is only done in areas adjacent to Antarctica, it has no direct effect on water formation at other latitudes. The initial balanced state is obtained by the "relaxation" method given in the Appendix of Bryan and Lewis (1979). Short time scales are stretched out by distorting the speeds of internal gravity waves, Kelvin waves and short Rossby waves. Long time scales associated with diffusion in the deep abyssal water are shortened by reducing the local heat capacity. The result is an iterative procedure in which time-steps are limited only by the Courant-Friedrichs-Lewy

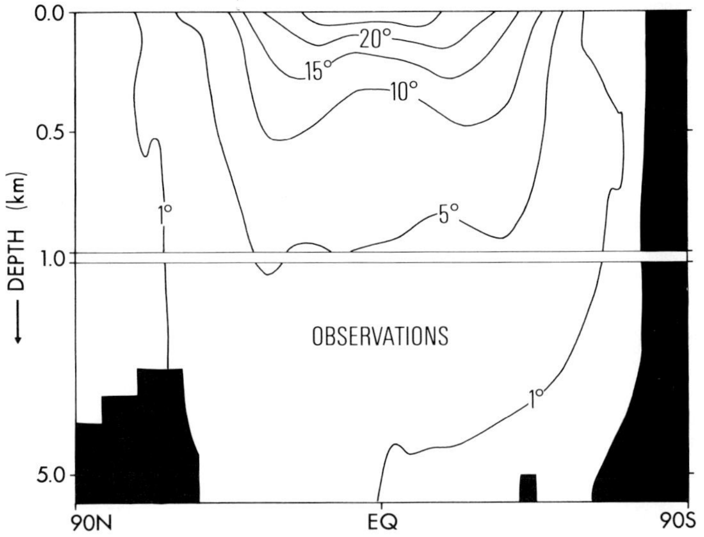

Fig. 2. The observed field of zonally averaged potential temperature from Levitus (1982).

tion and includes seasonal variations. The mid-level counterclock-wise circulation in the Northern Hemisphere shown in Figure 4a is due to the thermohaline circulation of the North Atlantic. In the model it only has an intensity of 10 Mt/s (Mt/s=10^9 kg/s), about 50% of estimates made by Roemmich (1980) and Bryden and Hall (1980) at 25°N. The deep water formation of the Southern Hemisphere is also about 10 Mt/s, except in the close proximity of Antarctica where it reaches 20 Mt/s. As pointed out by Bryan and Lewis (1979) the Ekman flow produced by the strong westerlies of the Southern Ocean produce a powerful meridional gyre which reaches downward quite deeply in the weakly stratified waters of the Southern Ocean.

A key advantage of the present model is that it allows a feedback of the thermal anomaly on the ocean circulation. Figures 4b and 4c indicate that to first approximation the anomalies cause a simple strengthening or weakening of the existing circulation shown in Figure 4a. The ± 0.5°C anomalies cause change of about 1-3 Mt/s or about 10-30% of the undisturbed circulation. Closer inspection reveals some interesting departures from this simple picture. The negative anomaly intensifies the meridional circulation more than the positive anomaly weakens it, an obvious effect of a nonlinearity. In the case of Figure 4c a very intense overturning is indicated at 50°S at a depth of 1 kilometer. Analysis of horizontal patterns of temperature and velocity (not shown) indicated that the intense overturning is mainly localized in the southern Pacific off the coast of South America.

The fact that the anomalous temperature pattern is bowed up at the equator for the +0.5°C case, but not for the -0.5°C case appears to be due to the differences in total upwelling in the equatorial zone. Although the patterns are similar, the meridional circulation is clearly stronger in the negative anomaly case, and weaker in the positive anomaly case, thus destroying the mirror symmetry between the two cases. Stronger upwelling produces a cooling, allowing the negative anomaly to propagate downward effectively.

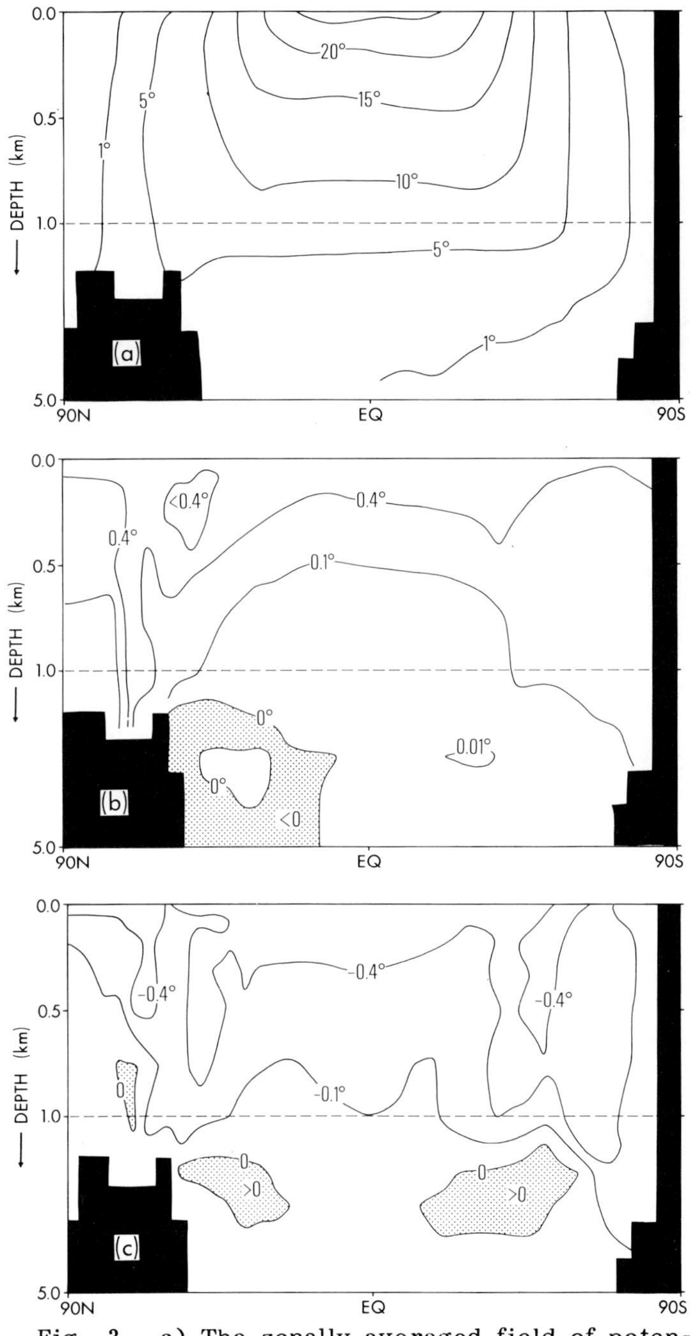

Fig. 3. a) The zonally averaged field of potential temperature for the model initial conditions, b) the difference in zonally averaged potential temperature 50 years after "switch on" for the + 0.5°C anomaly. c) Same as (b) for the - 0.5°C anomaly.

Table 1. Important constants used in a global ocean model by Bryan and Lewis (1979). Where upper and lower values are indicated, a smooth analytical function of depth connects values in the upper thermocline with those indicated for deep water. The vertical viscosity, A_{VM}, has a uniform value of $10^2 m^2/s$.

Name	Symbol	Low resolution (m^2s^{-1})	High resolution (m^2s^{-1})
E-W spacing	Δ	$\pi/32$	$\pi/64$
N-S spacing	$\Delta\phi$	$\pi/38$	$\pi/76$
Lateral diffusion coefficient of heat	A_{HH}(upper)	$2.5 \times 10^3 m^2s^{-1}$	$10^3 m^2s^{-1}$
and salt	A_{HH}(lower)	0.5×10^3	0.5×10^3
Lateral viscosity	A_{HM}	8.0×10^5	10^5
Vertical diffusion coefficient of heat	A_{VH}(upper)	0.3×10^{-4}	0.3×10^{-4}
and salt	A_{VH}(lower)	1.3×10^{-4}	1.3×10^{-4}

condition (see Richtmyer and Morton, 1967) in the temperature and salinity equations. The method provides enormous economy over a straightforward numerical integration to equilibrium of the full model (1) - (6).

In the 50 year numerical integrations made in the present study the distortion of very long time scales was dropped in order to preserve the correct time scales of thermal response at all depths. The purely dynamical distortion introduced by slowing internal waves was retained, however, to economize the numerical integration.

Transient Temperature Fields

The observed field of zonally averaged potential temperature for the World Ocean is shown in Figure 2. The main features of the thermocline are a strong front in the Southern Ocean associated with the Circumpolar Current, and a tendency for the isotherms to bow upward at the equator. The thermocline computed in Case I of Bryan and Lewis (1979) is shown in Figure 3a. This pattern represents the initial condition for the transient anomaly numerical experiments. Note that the thermocline tends to be deeper than observed, and lacks the shallow thermocline at the equator. The volume enclosed between any pair of isotherms will obviously be larger in the model than in the real ocean. This fact should be kept in mind in the interpretation of the results.

The penetration of the zonally averaged anomaly after 50 years is shown in Figures 3b and 3c. The solution of a numerical integration is subtracted from a control run without an anomaly at the end of 50 years after "switch on". Figure 3b shows the effect of a positive anomaly and Figure 3c the effect of a negative anomaly. In both cases there is greatest penetration at high latitudes. The weak stratification of the polar areas obviously provides more efficient downward pathways for the anomaly. At high latitudes the predominate mechanism for vertical heat transport is convection, which is extremely sensitive to small changes in surface heating and cooling.

A very obvious difference between the +0.5°C and the -0.5°C anomaly cases is in the temperature structure at the surface. In the positive anomaly case (Figure 3b), the penetration is relatively weak at the equator, while it is much greater in the negative anomaly case. A clue explaining this difference lies in the meridional circulation shown in Figure 4. The initial circulation is shown in Figure 4a. The pattern is not identical to that given by Bryan and Lewis (1979) Case III, which has higher spatial resolu-

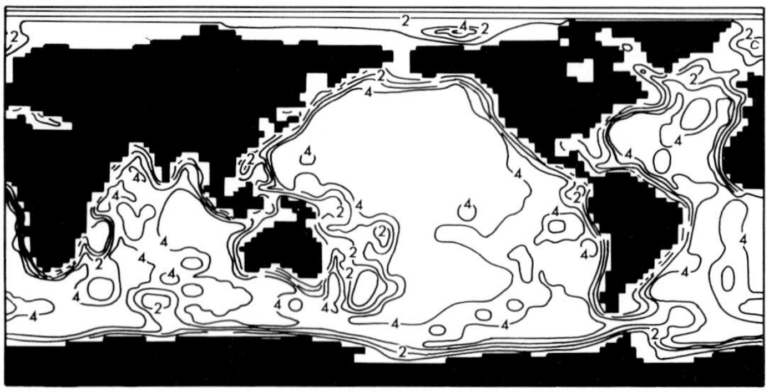

Fig. 1. The geometry of the Bryan and Lewis (1979) model. Depth is given in km.

Comparison With Simple Globally Averaged Models

Any globally averaged model must necessarily contain a highly simplified representation of the vertical transfer processes in the ocean. All spatial variations in input and vertical pathways must be averaged together. The surface flux condition (7) and (8) allows a spatially non-uniform heat input, depending on the local difference between θ^* and S^* in the surface layer. Thus, the numerical model used in this study takes into account any spatial correlation or lack of correlation which may exist between surface sources of an anomaly and favored downward pathways into the thermocline. Since the globally averaged models, however, have been widely used to interpret data, it is of considerable interest to also analyze our more general three-dimensional model in the same manner. This allows us to compare the coefficients of simple models determined by empirical fits to data, with the corresponding values derived from our numerical results.

Let us first consider a very simple case in which the ocean is represented by a number of uniformly mixed reservoirs, which are all in contact with the surface. Each reservoir contains a fraction, V_n, of the total ocean value, V_o. There is also a characteristic mixing time, t_n, for each reservoir. Let T_n' be the departure of temperature from its initial equilibrium value. The change in temperature in each reservoir is governed by the equation.

$$\frac{dT_n'}{dt} = \frac{1}{t_n} (T^* - T_n') \tag{9}$$

T^* is the imposed surface temperature perturbation. The total rate of heating of the entire ocean is then governed by the equation

$$Q = c_p \rho V_o T^* \sum_n V_n t_n^{-1} \exp(-t/t_n) \quad . \tag{10}$$

We can compare the heating of the model with this expression. The mixed layer will dominate the heat uptake initially. In the three-dimensional numerical model the mixed layer corresponds to the top layer which is 50 m thick. After the mixed layer reaches "saturation" deeper layers will dominate. If the time scales of various reservoirs are well separated, i.e., $V_{n+1}t_n << V_n t_{n+1}$, then the uptake of heat at any one time will be dominated by at most two of them, eliminating much of the difficulty in separating exponential contributions.

Box models have been criticized (Oeschger et al., 1975) as a tool for handling phenomena with different time scales. For example, a box model fitted to one set of data involving vertical transfer on time scale, may be useless in describing vertical transfer at another scale. Although this point is certainly valid in a formal sense, it is interesting to see in Fig. 5a that two exponentials roughly describe the three dimensional numerical heat input. The initial adjustment involving the mixed layer indicates a time scale of about 5 years. The second time scale is about 100 years. A more extended integration would undoubtedly indicate the existence of still longer time scales, but for the range of the scales from 1-50 years, a two box model provides a reasonable fit.

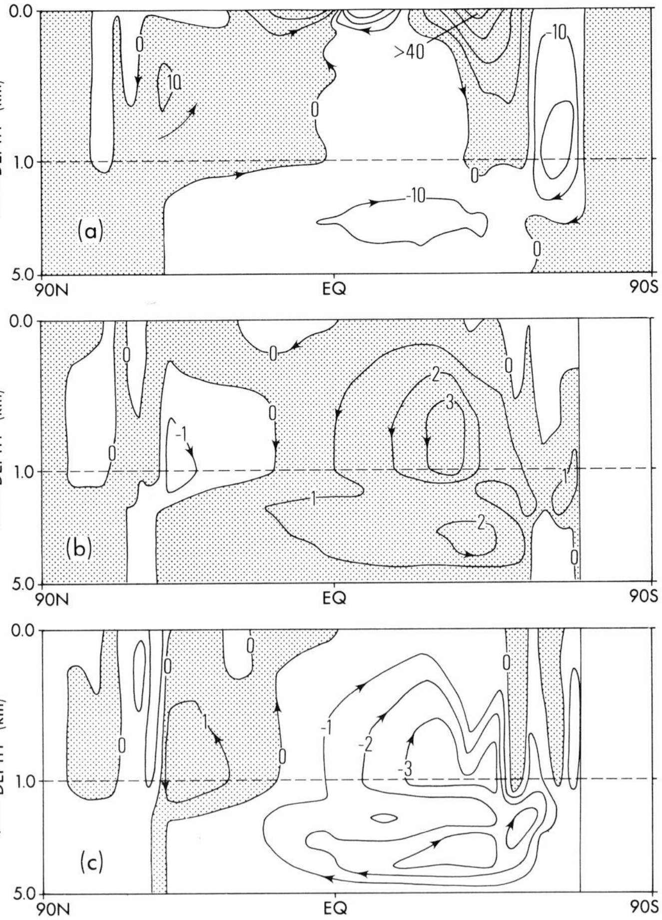

Fig. 4. The meridional circulation of the model in units of Mt/s. a) Initial condition, b) Difference after 50 years between the + 0.5°C case and control run, c) Difference between the - 0.5°C case and control run,

As an improvement on box models Oeschger et al. (1975) suggested the "box" diffusion model. In this model the global ocean is divided into a surface mixed layer above a diffusive slab. The original model assumes a finite depth for the diffusive slab but this is an unimportant refinement for time scales of less than a few centuries. For our purposes we have assumed that the diffusive slab has infinite depth. The governing equations are,

$$\partial_t T_M' = \frac{1}{t_M} (T^* - T_M') - \frac{\kappa}{h} \partial_z T' \Big|_{z=-h} \tag{11}$$

$$\partial_t T_D' = \kappa \partial_z \partial_z T_D \tag{12}$$

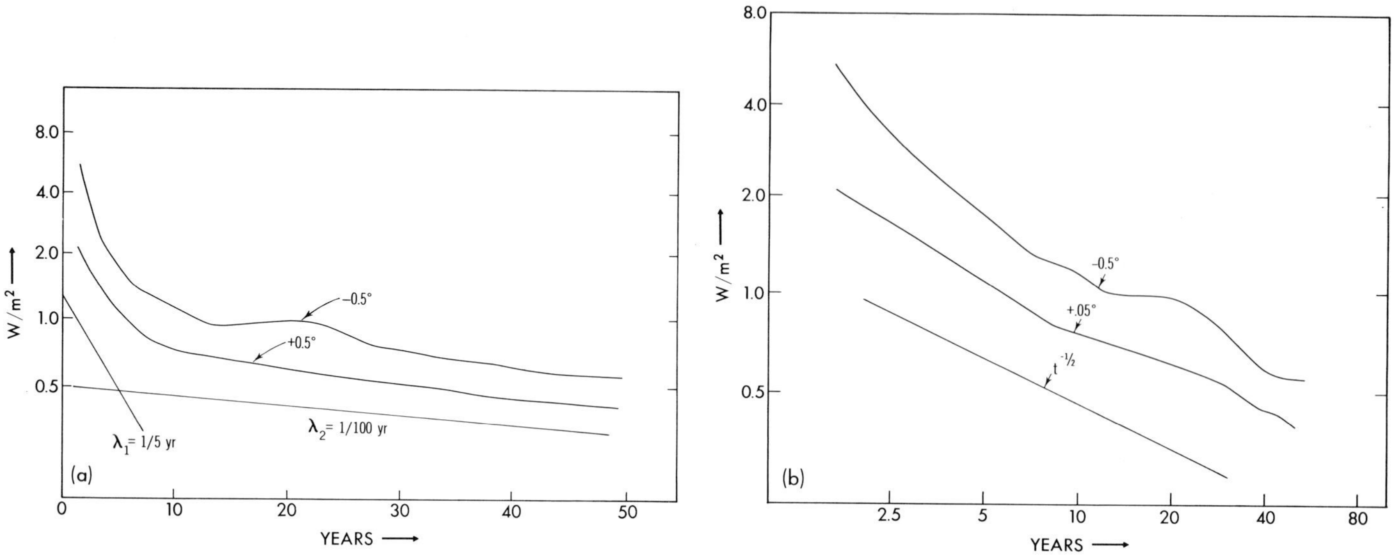

Fig. 5. Heat uptake or cooling of the model after "switch on" a) semi-log coordinates for comparison with a multi-box model, b) log-log coordinates for comparison with a diffusive model.

where T_M' is the temperature of the mixed layer, and T_D' is the temperature of the diffusive slab; h is the depth of the mixed layer, and κ is the coefficient of diffusion. Boundary conditions are

$$T_M' = T_D' \qquad z = -h$$

$$T_D' = 0 \qquad z = -\infty \tag{13}$$

For the case of a simple "switch on" of an anomaly,

$$T^* = \begin{cases} 0 & t < 0 \\ 1 & t > 0 \end{cases} \tag{14}$$

The analytic solution is a rather complicated expression involving error functions. In practice it was found to be more convenient to use finite difference solutions of (11) and (12) using 10 meter intervals.

For the asymptotic range $t >> t_m$, the box-diffusion model can be approximated by the diffusive slab alone with a specified temperature at the upper boundary. In that case,

$$T_M' = T^* \left[1 - \text{erf}\,[(z + h)^2/4\,\kappa t]^{\frac{1}{2}}\right] \tag{15}$$

and the heating rate is,

$$Q = \rho c_p T^* \,(\kappa/\pi t)^{1/2} \tag{16}$$

where it assumed that the mixed layer is already in equilibrium. To compare the heating rate of the present three-dimensional model with the asymptotic solution of the box-diffusion model, Q is shown in a log-log plot in Figure 5b. Note that the heating curve roughly corresponds to $t^{-1/2}$. From the results shown in Figure 5a and 5b it appears that the three-dimensional model is fitted reasonably well by both a simple two-box model, and the slab diffusion model.

Another way to represent the results of model is to consider the penetration depth,

$$d_p = \frac{1}{\rho c_p T_M'} \int_0^t Q\, d \tag{17}$$

where Q is defined in (7). Since the fit to a purely diffusive one-dimensional model is quite good, the penetration depths are multiplied by a factor of $(10\ \text{yr}/t)^{1/2}$ and replotted in Figure 6a. This plot includes an experiment for a passive tracer, and one in which seasonal variations in the surface boundary conditions are included. The passive tracer case behaves in almost exactly the same way as the positive anomaly. The small differences in penetration depth are probably not significant. From this result we conclude that buoyancy effects are not important in the +0.5°C anomaly case. The anomaly is still sufficiently small for a linear regime to exist so that feedback on the dynamics is unimportant. The significantly greater penetration for the negative anomaly is an indicator of the importance of convection. It was thought that winter convection associated with seasonally varying boundary conditions would cause a much greater penetration. Seasonal variations do produce a significantly faster initial penetration, which is also brought out in Sarmiento's (1982) study of tritium in the North Atlantic. There is very little difference at longer times, however, and results become nearly identical beyond 20 years after

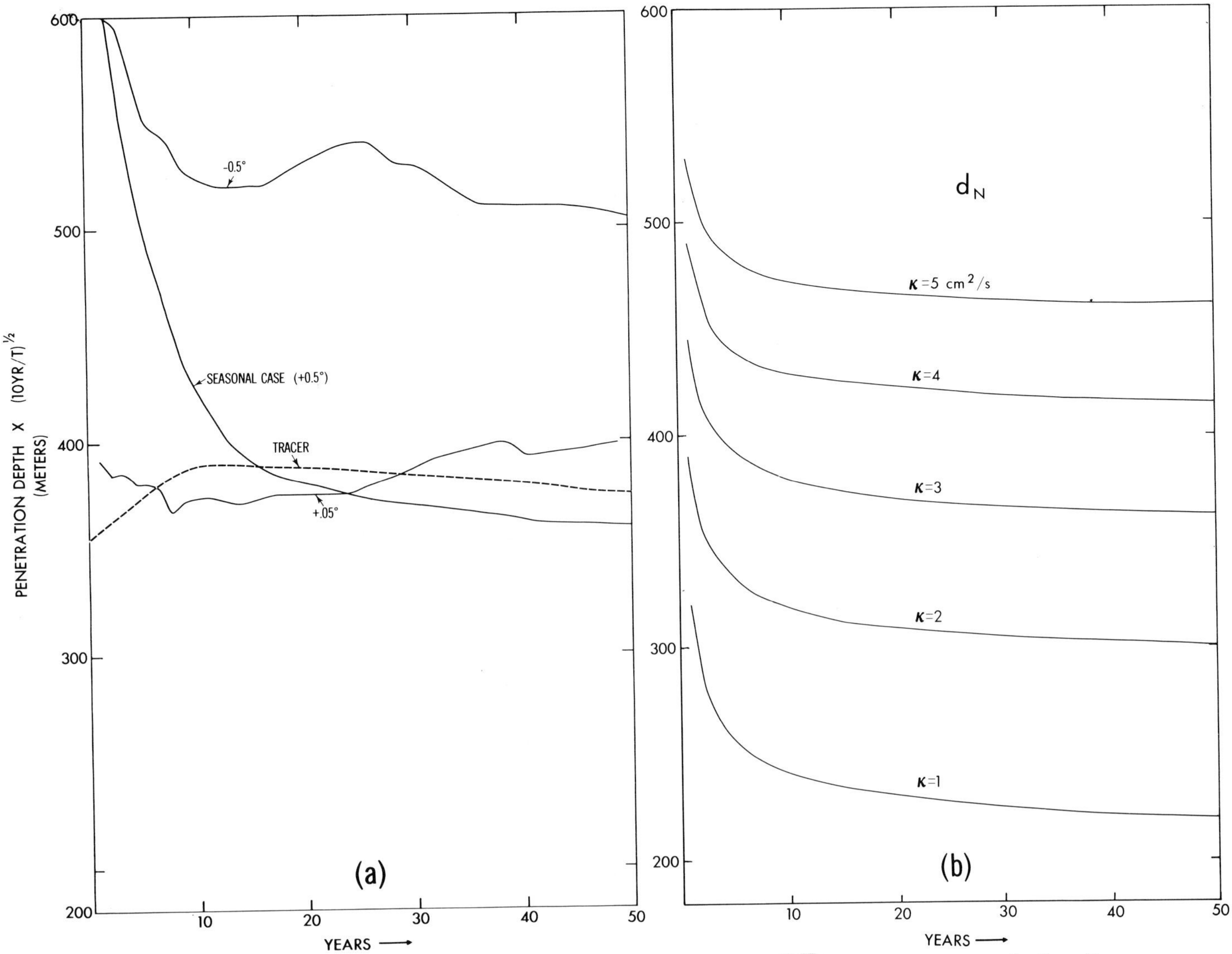

Fig. 6. The penetration depth multiplied by $(10\ yr/t)^{1/2}$. a) Results of the three-dimensional model, b) the one-dimensional model of Oeschger et al. (1975) for a mixed layer depth of 50 m.

"switch on." The initial condition for the seasonal case is Case II of Bryan and Lewis (1979),

For purposes of comparison the corresponding penetration depths are shown for the box-diffusion model with a 50 m mixed layer in Figure 6b. Separate curves are shown for different values of vertical diffusion. Based on various sources of tracer data Oeschger et al. (1975) found a best fit for a mixed layer depth of 70 m, and a diffusion of 1.5 cm^2/s. Based on GEOSECS tritium data Broecker et al. (1978) has estimated a vertical diffusion for the World Ocean of about 2.0 cm^2/s. The model suggests higher values, about 3 cm^2/s for the positive anomaly and tracer case, and over 5 cm^2/s for the unstable case. Sarmiento (1982) has found values this high in modeling the tritium data in the North Atlantic. Since the North Atlantic is known to be a region of very strong overturning, however, such high values would not be expected to be typical for the World Ocean.

Comparing Fig. 2 and Fig. 3a, we can see a possible source of bias in the model. The scale depth of the thermocline in the Bryan and Lewis (1979) World Ocean is almost twice that of the real ocean. This implies that the volume enclosed between any two isopycnal surfaces is much greater than observed, and consequently has a greater capacity to absorb tracers originating at the surface. A rough way to allow for this difference between model and prototype is to scale the results by the depth of the thermocline. To do this we define a thermocline scale depth based on the density structure. Let σ_0' be the globally averaged potential density minus the value of potential density in deep water. Let σ_{0s}' be the surface value of the same quantity. A scale depth of the thermocline may then be

Table 2. Pycnocline scale depth based on potential density and the penetration depth 10 years after "switch on." Normalized penetration depths are divided by the pycnocline scale depth.

	Observed Levitus (1982)	Bryan et al. (1982)	Present study
Pycnocline scale depth	389	445	746
Penetration depth at 10 yr	-	270	400
Normalized penetration depth at 10 yr	-	.60	.53

defined as,

$$d_s = \frac{1}{\sigma_{os}'} \int_{-H}^{0} \sigma_o' dz \qquad (18)$$

where H is total depth of the ocean. Values of d_s for the real ocean based on data by Levitus (1982), and d_s for the model are shown in Table 2. Note that the pycnocline depth of the World Ocean model is almost twice that observed. This implies the total volume between adjacent isopycnal surfaces is much larger in the model, providing a greater capacity to take up a surface heat anomaly. To remove this possible source of bias the results of Fig. 6 are rescaled. In Fig. 7a the response of the model is scaled by the pycnocline scale depth of the model. In Fig. 7b the one-dimensional results are scaled by the pycnocline scale depth observed. In comparison with estimates of 1.5-2.0 cm^2/s of effective diffusion based on tritium data (Broecker, 1981) the best fit of the model is closer to 1.0 cm^2/s for a tracer or positive heating anomaly. While Fig. 6 suggests that the World Ocean model is mixing downward more actively than the tracer data indicate, Figure 7 gives the reverse impression.

Summary

In this model study we attempted to answer two questions. The first question revolves around the analogue between a tracer introduced at the ocean surface and surface heat anomalies. Is a heat anomaly of moderate amplitude carried downward into the ocean in the same manner as a passive tracer which doesn't effect the buoyancy field? The question is an important one, since goechemical field data exists on the downward movement of passive tracers, but none for large scale heat anomalies. The second question concerns the appropriateness of one-dimensional models for calculating downward transfer of heat anomalies. Can these models, and in particular the box-diffusion model, be applied with the same parameters to a variety of situations?

The basic tool for this study is a numerical model developed by Bryan and Lewis (1979) to study the origin of water masses in the World Ocean. The model includes the large scale dynamics of the ocean circulation, allowing for the feedback on the circulation associated with a positive or negative temperature anomaly. For small anomalies (± 0.5°C) it was found that the tracer analogue works quite well in the stable case for a positive temperature perturbation. Convection destroys the symmetry between the two cases, producing a penetration depth 50% larger in the case of the cold anomaly.

A comparison showed that global averages of the three-dimensional vertical penetration could be fitted almost equally well to a very simple two-box model, and to the box-diffusion of Oeschger et al. (1975). The penetration depth for temperature anomalies indicate a vertical mixing larger than obtained by Oeschger et al. (1975) and Broecker et al. (1978) from geochemical data. To allow for the fact that the model thermocline is deeper than that of the real ocean, and therefore provides larger reservoirs between pairs of isopycnal surfaces, the results are scaled by the thermocline depth. Results are shown in Table 2. The thermocline depth is shown for the observed ocean, the present model, and a joint ocean-atmosphere model with simplified geometry by Bryan et al. (1982). The thermocline depth of the joint ocean-atmosphere model is in fair agreement with observations while the present model is almost twice as large. In the Bryan et al. (1982) case the positive anomaly is caused by a "switch on" of a four-fold increase of atmospheric carbon dioxide. A comparison of the penetration depth calculated for the joint ocean-atmosphere model, and the present study indicates a large discrepancy in the penetration depth of a positive heat anomaly. Scaling the penetration depths by the total thermocline depth, however, brings the results of the two model studies into reasonable agreement.

The present study does indicate a diffusive analogue for the globally averaged penetration of a tracer and explains the time dependent behavior of a more complete three-dimensional model

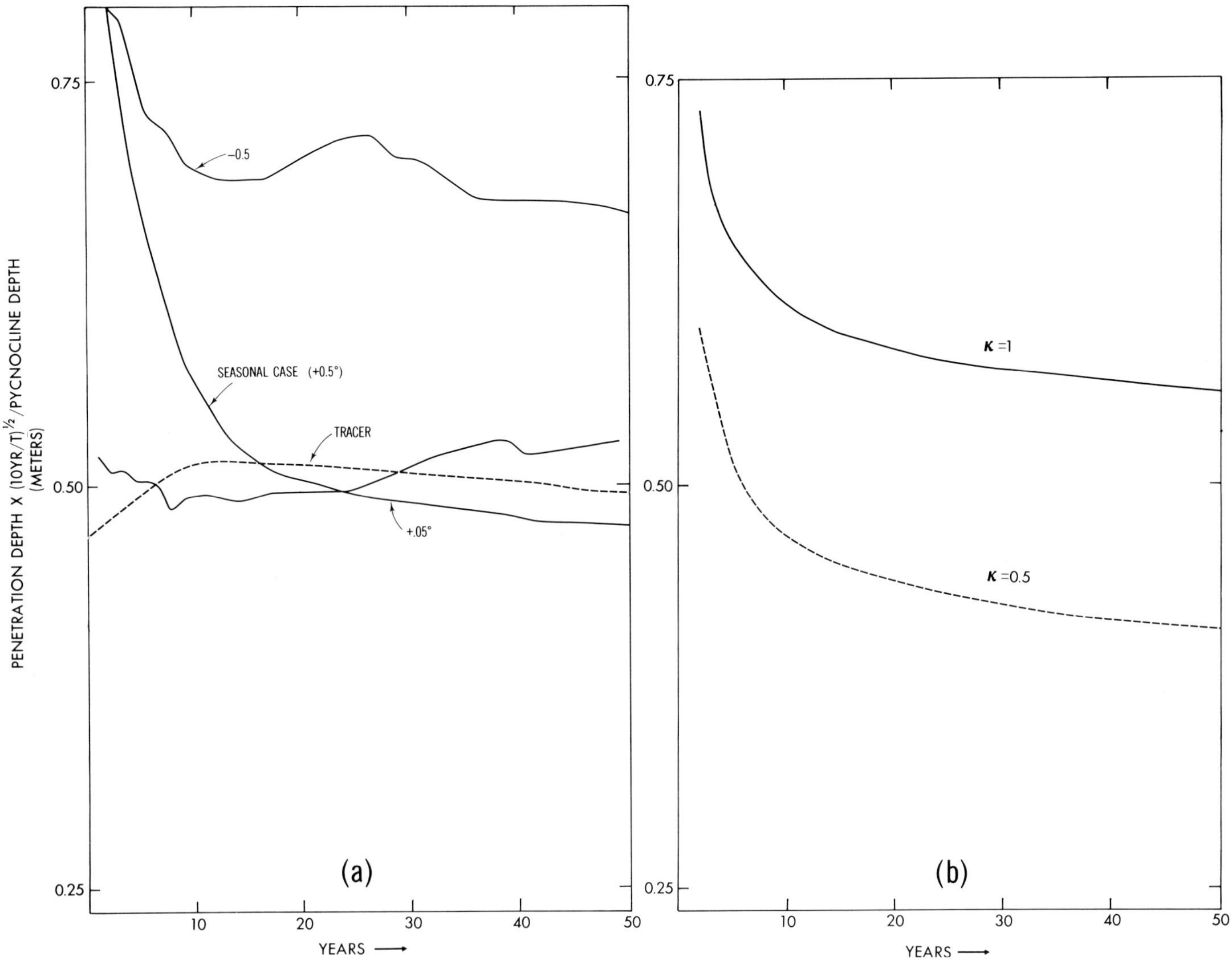

Fig. 7. Same as Fig. 7 with the penetration depth sealed by the pycnocline depth of the model in (a) and the pycnocline depth of real ocean in (b).

reasonably well. This study is limited to only one type of time dependent forcing. Future studies could explore the relationship of one-dimensional and three-dimensional models in more detail by considering other types of surface forcing, such as a "switch on, switch off" event that would simulate the transient tracers produced by the major bomb tests of the late fifties and early sixties. The penetration depth is only a first moment of the vertical distribution. It is of some interest to examine higher order moments of the vertical distribution of heat anomalies and tracers through a comparison of observations, one-dimensional and three-dimensional models.

References

Broecker, W.S., T.-H. Peng and M. Stuiver, An estimate of the upwelling rate in the equatorial Atlantic based on the distribution of bomb radiocarbon, J. Geophys. Res., 83, 6179-6186, 1978.

Broecker, W.S., Geochemical tracers and ocean circulation in Evolution of Physical Oceanography, edited by B.A. Warren and C. Wunsch, pp. 434-460, MIT Press, Cambridge, MA 1981.

Bryan, K., A numerical method for the study of the circulation of the World Ocean, J. Comput. Phys., 4, 347-376, 1969.

Bryan, K. and L.J. Lewis, A water mass model of the World Ocean, J. Geophys. Res., 84, 2503-2517, 1979.

Bryan, K., F.G. Komro, S. Manabe and M.J. Spelman, Transient climate response to increasing atmospheric carbon dioxide, Science, 215, 56-58, 1982.

Bryan, K., S. Manabe and R. Pacanowski, A global ocean-atmosphere climate model. Part III. The ocean circulation, J. Phys. Oceanog., 5, 30-46, 1975.

Bryden, H.L. and M.M. Hall, Heat transport by ocean currents across 25°N latitude in the Atlantic, Science, 207, 884-886, 1980.

Climate Research Board, Carbon dioxide and climate: A scientific assessment. 22 pp, National Academy of Sciences, Washington, D.C. 1979.

Haney, R.L., Surface thermal boundary condition for ocean circulation models, J. Phys. Oceanogr., 1, 241-248, 1971.

Hellerman, S., An updated estimate of wind stress on the World Ocean, Mon. Weather Rev., 96, 63-74, 1968.

Holland, W.R., Ocean tracer distributions - a preliminary numerical experiment, Tellus, 23, 371-372, 1971.

Keeling, C.D. and R.B. Bacastow, Energy and Climate, p.72, National Academy of Sciences, Washington, D.C., 1977.

Levitus, S. Climatological Atlas of the World Ocean, NOAA Prof. Paper, No. 13, U.S. Gov't Printing Office, Washington, D.C., 1982.

Manabe, S. and R.T. Wetherald, On the distribution of climate change resulting from an increase in carbon dioxide content of the atmosphere, J. Atmos. Sci., 37, 99-118, 1980.

Oeschger, H., U. Siegenthaler, U. Schotterer and A. Gugelman, A box diffusion model to study the carbon dioxide exchange in nature, Tellus, 27, 168-192, 1975.

Richtmyer, R.D. and K.W. Morton, Difference Methods for Initial Value Problems, 399 pp, Interscience, New York, 1967.

Roemmich, D., Estimation of meridional heat flux in the North Atlantic by inverse methods, J. Phys. Oceanogr., 10, 1972-1983, 1980.

Sarmiento, J.L., A simulation of bomb-tritium entry into the Atlantic Ocean, (submitted to J. Phys. Oceanogr., 1982).

Schneider, S.H. and S.L. Thompson, Atmospheric CO_2 and climate: Importance of the transient response, J. Geophys. Res., 86, 3135-3147, 1981.

Semtner, A.J., An oceanic general circulation model with bottom topography. Numerical simulation of weather and climate, Tech. Rept., No. 9, 99 pp, Univ. of California, Los Angeles, 1974.

A RECENT Θ-S SHIFT IN THE DEEP WATER OF THE NORTHERN NORTH ATLANTIC

James H. Swift

Marine Life Research Group, Scripps Institution of Oceanography, La Jolla, CA

Abstract. North Atlantic ocean data from the Transient Tracers in the Ocean North Atlantic Study (1981) show that a deep water Θ-S shift has occurred since GEOSECS (1972). For a given density, the deep water north of 50°N is now colder by about 0.15°C and less saline by about 0.02×10^{-3}. Other hydrographic properties such as dissolved oxygen have also changed, but the configuration of the deep density fields is nearly unchanged. The deep changes were observed in at least three distinct deep water masses. Although the shifts are nearly universal north of 50°N, they have not yet spread significantly south of the Grand Banks. The reduced deep water salinities are caused by a freshening since the mid-1970's of the upper layers in the deep water mass formation regions. This upper-layer freshening may in turn be related to shifts in the large-scale atmospheric forcing over the northern North Atlantic and Norwegian-Greenland Sea.

Introduction

Until recently there has been virtually no modern-day evidence of any large-scale alteration in deep ocean salinity. Of course, we have not been accumulating high-precision deep salinity data long, but nonetheless the concept of a "steady-state ocean" has been broadened by tradition to include the deep temperature and salinity fields. In the North Atlantic, the local invariance of these deep water properties has led to the tracing of elements of the circulation via a salinity "anomaly" calculated as a local salinity difference at some reference temperature from a large-scale mean Θ-S correlation (e.g., cf. Worthington and Metcalf, 1961). Small variations in deep water salinity have been observed in the western North Atlantic, particularly on the often-run Woods Hole to Bermuda section (e.g., cf. Barrett, 1969), but these have been so near the noise level that they have not yet received wider attention.

The GEOSECS (Geochemical Ocean Sections) program provided sparse, large-scale coverage of the North Atlantic in 1972. While the Θ-S correlations for the deep GEOSECS samples were similar to those from past expeditions, the GEOSECS stations did show new features, notably a deep tongue of tritium-enriched water extending southward from Denmark Strait (Ostlund et al., 1976). The presence of the deep tritium signal showed that the bottom layers in the northwest Altantic clearly responded on a time scale of 10 years or less to the injection of tritium into the atmosphere by thermonuclear testing. Swift et al. (1980) noted from other hydrographic evidence that the deep northwest Atlantic was potentially sensitive to perturbations generated at the winter sea surface north of Iceland, though they found no evidence for such changes in their data base (other than the tritium signal observed during GEOSECS), which extended from 1958-73 in the waters south of Denmark Strait.

Indeed, in general the best location to examine the deep ocean for evidence of recent changes is the northern North Atlantic: it is close to the deep water mass formation regions, and has been surveyed by expeditions 10-20 years apart.

The apparent sensitivity of the deep northwest Atlantic on 10-year (or less) time scales added immediate interest to the data gathered in 1981 - nine years after GEOSECS - by the Transient Tracers in the Ocean North Atlantic Study (TTO/NAS). A total of 250 high-quality hydrographic/geochemical stations were occupied north of 17°N (cf. Figure 1), a region covered by only 43 stations during GEOSECS. The TTO track provided some basin-scale sections, thus giving adequate coverage and facilitating comparison with other data from different years. Early reports in the summer and fall of 1981 from TTO/NAS indicated detectable changes in deep ocean properties in the northern North Atlantic during the nine years following GEOSECS. Brewer et al. (1983) have outlined the essentials of the new observations: a shift toward colder, fresher deep water in the northern North Atlantic, and a larger freshening in the upper layers there. Here I will describe in greater detail the nature and extent of the hydrographic shifts observed by TTO/NAS, with special emphasis upon the deepest, densest layers, and relate these results to the sensitivity of the deep North Atlantic to certain perturbations. This paper has thus been designed as a companion to that by Brewer et al. (1983), adding details to their basic exposition.

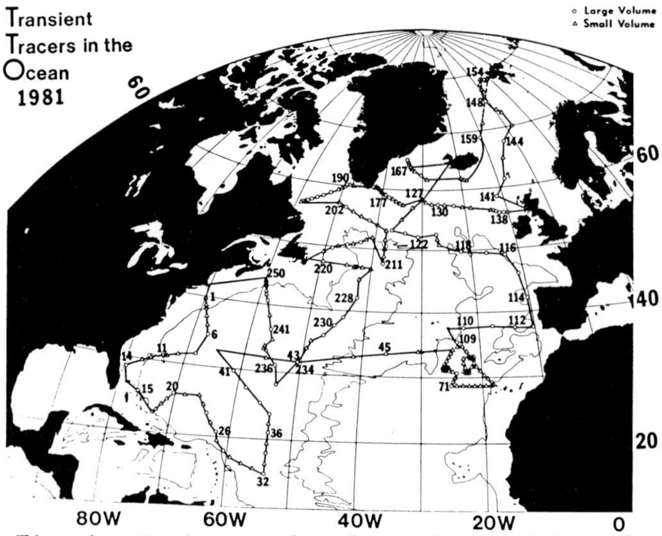

Fig. 1. Cruise track and station positions for the Transient Tracers in the Ocean North Atlantic Study (TTO/NAS), April-October 1981.

The 1981 Observations vis-a-vis Those From Earlier Expeditions

Property maps such as those shown by Mantyla and Reid (1983), Talley and McCartney (1982), and many others, have been made from file data from various cruises in different years. This is possible because data from different years show nearly the same deep water properties at any given location. For example, the potential temperature-salinity correlations in Figure 2 include four stations from 1962-72 near the same location northeast of the Grand Banks. These stations show remarkable similarity in the densest layers from cruise-to-cruise or year-to-year. The variability at potential temperatures above 3°C in Figure 2 is due to inter-annual fluctuations in the Θ-S characteristics of newly formed Labrador Sea Water (LSW; e.g., cf. Clarke and Gascard, 1983, or Talley and McCartney, 1982). The relative salinity maximum underneath (derived from the Iceland-Scotland overflow from the Norwegian Sea which forms Iceland-Scotland Overflow Water, or ISOW; e.g., cf. Swift, 1984) and the lower salinity bottom layer (derived from the Denmark Strait overflow) show no such fluctuations in the 1962-72 data. But at the same location off the Grand Banks, the TTO data in Figure 2 show marked differences from previous years. In 1981 all the deep water was colder and less saline, for a given density, than in the period 1962-72. Because the data sets have been checked carefully for errors, these deep water differences are real. Each of the three principal deep water layers here is substantially altered in 1981. This was by no means an isolated occurrence: such findings were nearly universal in the northern North Atlantic.

One straightforward method to provide a more general comparison of the 1981 data with earlier work is by plotting sections from past years reoccupied in 1981. The northernmost pair shown in Figure 3 extends from Greenland to the Rockall Plateau, cutting across ISOW in the northeast Atlantic, the northward extension of this layer into the Irminger Sea from the Charlie-Gibbs Fracture Zone, and the outflow from Denmark Strait. Above these two deepest layers are LSW derivatives which have spread east and north. (See Swift, 1984, for a discussion of deep water circulation in the northern North Atlantic.) One of the sections of the northern pair is from a February 1967 Hudson cruise (Grant, 1968) and the other is from July-August 1981 TTO data.

A convenient way to search out property differences is to display the data versus a density parameter. Figure 4 shows salinity and oxygen for the two cruises plotted against sigma-2, which is the density in kg m^{-3} referred to 2000 db (20 M Pa), less 1000 kg m^{-3}. These sections have Greenland on the left and the Rockall Plateau on the right. The apparent bottom contour shows the density of the bottom water, not its depth. The sections illustrate only the denser, deeper layers and do not extend to the surface except for a small area near Greenland denser than 36.7 in the 1967 data, which was gathered in winter.

In nearly every layer, salinities are lower in 1981 than in 1967. Since the data are displayed versus density, and the densities of the deep layers have not changed much, every layer is colder, too. The ISOW layer along the bottom of the east flank of the mid-ocean ridge in the northeast Atlantic had a salinity above 35.00 x 10^{-3} in 1967 but only slightly above 34.98 x 10^{-3}

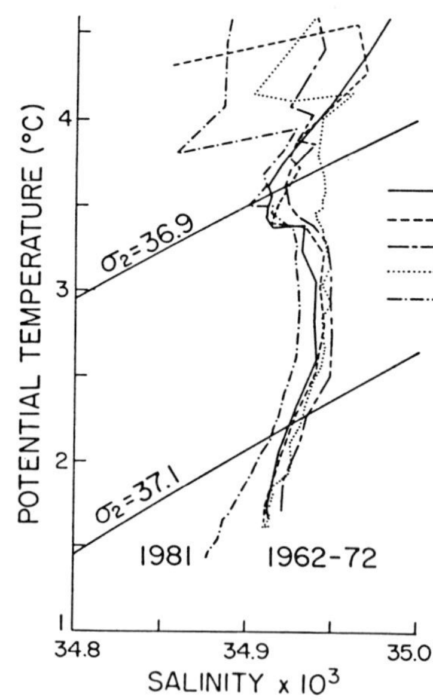

Fig. 2. Potential temperature - salinity correlations below 4.5°C for four stations from 1962-72 (Erika Dan sta. 340, 4/62, Atlantis sta. 362, 4/64, Hudson sta. 95, 4/66, GOESECS sta. 3, 7/72) and from TTO/NAS (sta. 214, 9/81). Location shown in Figure 3.

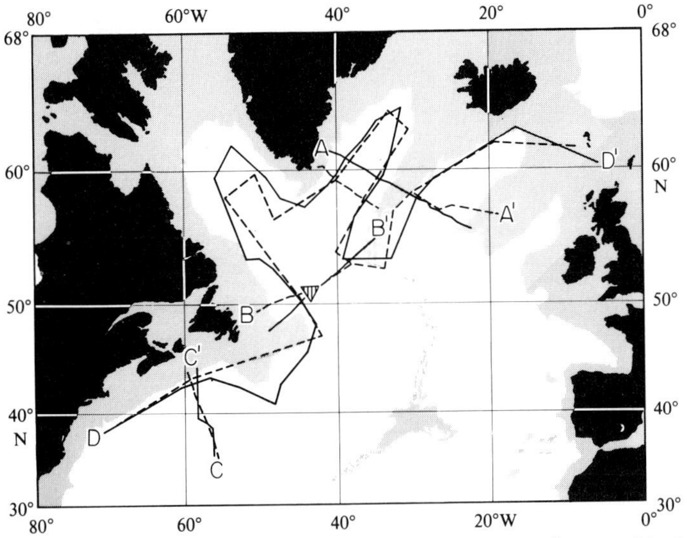

Fig. 3. Location of sections and stations cited in the text. The triangle symbol encloses the locations of the stations in Figure 2. The northernmost pair of sections oriented across bathymetric contours is shown in Figure 4, the section pair extending northeast from Newfoundland is shown in Figure 5, and the section pair extending south from Nova Scotia is shown in Figure 6. The long, winding pair of sections is shown in Figure 7. The TTO/NAS sections are shown by dashed lines, and sections from earlier cruises are shown by solid lines. Shaded areas are shallower than 2000 m.

in 1981. The ISOW is about the same amount fresher in 1981 in the northwest Atlantic, too. The outflow from Denmark Strait ($\sigma_2 > 37.13$) is slightly less saline in 1981, although the freshening there amounts to only about 0.01 x 10^{-3}. The LSW extension in the northeast Atlantic ($36.83 < \sigma_2 < 36.93$) is about 0.02 x 10^{-3} less saline in 1981, and that in the western half of the sections is about 0.1 - 0.2 x 10^{-3} fresher.

The outflow from Denmark Strait, ISOW, and LSW are principal northern hemisphere components of the lower deep water of the World Ocean - and they all have become colder (by an average of about 0.15°C) and fresher (by an average of about 0.02 x 10^{-3}) in the interval between 1972 and 1981. These are the largest changes ever observed in this deep domain.

There are few instances where the deep layers in Figure 4 are not fresher in 1981. Brewer et al. (1983) showed that in the density range $36.9 < \sigma_2 < 37.0$ (ca. $3.0 < T < 3.5$°C) the TTO salinities across 59°N were often close to or even slightly greater than the earlier values. The salinity plots in Figure 4 show that the similarity extends across the northeast Atlantic (in the σ_2 range 36.9 - 36.95) and into the eastern half of the Irminger Sea, but that the western Irminger Sea is, again, about 0.02 x 10^{-3} fresher in 1981. As Brewer et al. have stated, surface climatic conditions are inappropriate for the formation of the denser, classical form of LSW, and new, fresher, and less dense waters are formed, leaving the older LSW relatively undisturbed. The balance in the σ_2 range 36.9 - 37.0 between LSW and ISOW derivatives has therefore been shifted toward the relatively saline ISOW, and so the waters in this density range have not freshened in recent years, despite the freshening of the ISOW core.

The deep oxygen fields were also different during 1981. Oxygen versus density sections for the 1967 and 1981 data sets are shown in Figure 4b. The bottom layer in the western half - that is, the Denmark Strait overflow - has about the same oxygen content in both years. The upper part of the LSW layer there is better oxygenated in 1981, a sign of recent renewal. But the lower, denser LSW, that between 36.8 and 36.9 (σ_2), is lower in oxygen in 1981. These densities might not have been renewed as recently or vigorously, in accord with the salinity/density evidence discussed by Brewer et al. (1983).

In the northeast Altantic, all the layers denser than 36.9 (σ_2) are lower in oxygen in 1981. Since the evidence in 1981 was that the properties of Norwegian Sea water supplying the Faeroe Bank Channel outflow were unchanged, the decrease in ISOW oxygens - downstream from the sill - must have been caused by a decrease in dissolved oxygen in the water which mixes with the Norwegian Sea outflow to form ISOW. It may also be possible that in recent years the rate of supply of new water to the deep northeast Altantic has been slower, and the balance of ventilation and consumption has slightly shifted, lowering in situ oxygen concentrations.

Similar results are obtained for comparisons of TTO data with those from other cruises, or in other regions of the northern North Atlantic. For example, the section pair extending northeast from the grand Banks in Figure 3 compares data from the 1962 Erika Dan cruise (Worthington and Wright, 1970) with the TTO data. In Figure 5, Newfoundland is on the left, and the mid-ocean ridge is on the right. Again, salinity and oxygen are displayed versus a density parameter (σ_2). Because the Erika Dan data in Figure 5 extend farther northeast than the TTO stations, the comparison cannot be extended quite to the ridge crest. The cruise-to-cruise salinity changes in Figure 5a are virtually identical to those in Figure 4a. One difference is that the salinity decrease in the bottom layer is stronger in Figure 5a than in Figure 4a. Since closer to its source the Denmark Strait overflow changes were relatively small (e.g., Figure 4a), the bottom changes northeast of the Grand Banks may be caused more by an increased downstream presence of relatively pure Denmark Strait water rather than a change in its properties at the source.

At σ_2 values greater than 36.9, the pattern of dissolved oxygen similarities and differences is the same in Figures 4b and 5b. The only incon-

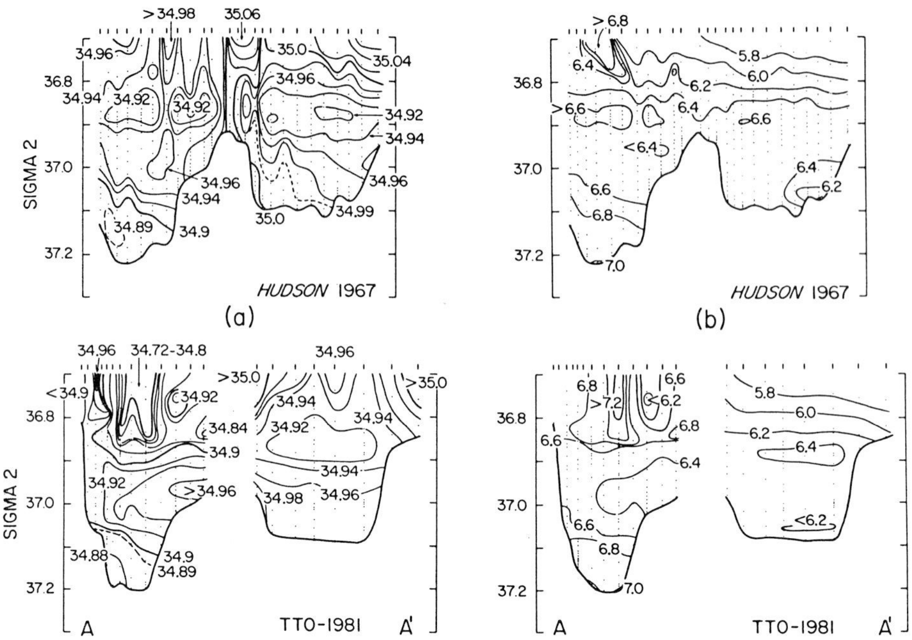

Fig. 4. Sections of a) salinity (x 10^3), and b) dissolved oxygen (x 10^3) versus σ_2 from Greenland (left) to the Rockall Plateau (right), drawn from data from a February 1967 Hudson cruise (upper sections) and from TTO/NAS cruises in mid-1981 (lower sections). Salinities are contoured only for S $\geq$ 34.8 x 10^{-3}; contour interval 0.02 x 10^{-3}. Dissolved oxygen contour interval is 0.2 x 10^{-3}. (Dissolved oxygen units, which have been expressed previously as milliliters liter^{-1}, or 10^{-3} volume volume^{-1}, are here written 10^{-3}.) Location shown in Figure 3. Note that the TTO sections are a composite of two shorter sections not run concurrently, hence the break over the mid-ocean ridge.

sistency this brings out is that because the near-bottom oxygens are about the same in both years in both the Irminger Sea (left side of Figure 4b) and northeast of the Grand Banks (Figure 5b),

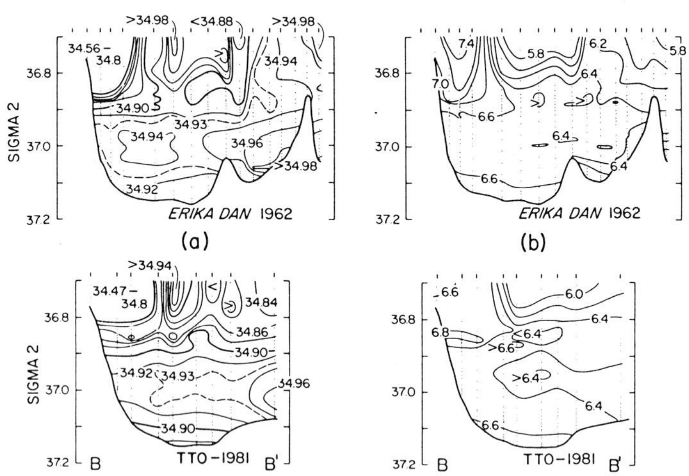

Fig. 5. Sections of a) salinity (x 10^3), and b) dissolved oxygen (x 10^3) versus σ_2 from Newfoundland (left to the mid-ocean ridge (right), drawn from February 1962 Erika Dan data (upper sections) and from TTO/NAS cruise in September 1981 (lower sections). Contouring as in Figure 4. Location shown in Figure 3.

there is no evidence in the dissolved oxygens for the increased presence in 1981 of relatively high oxygen Denmark Strait water in Figure 5b which might be expected from the bottom salinities in Figure 5a. (The oxygen data are notoriously difficult to interpret here, however, because there are only relatively small variations about high values.)

At σ_2 values less than 36.9 in Figure 5b, the LSW layer near the Grand Banks is significantly less well oxygenated in 1981, as if it had not been so recently or vigorously renewed, although most of the water further northeast was better oxygenated during TTO. In these layers, the oxygen saturation ratio may be a more convenient indicator.

Deep changes in salinity and temperature such as those observed in the northern North Atlantic are much less prominent, if they have occurred at all, south of the Grand Banks. The cross-section comparison in Figure 6 is drawn from Gulf Stream '60 data (Fuglister, 1964) and TTO data (location in Figure 3). While there are deep differences, they are much more subtle south of the Grand Banks than in the north. It seems from this figure that the deep Θ-S shift from the north has not propagated strongly to the south, at least in 1981.

The cross sections show details of the changes, but it is easier to visualize the overall

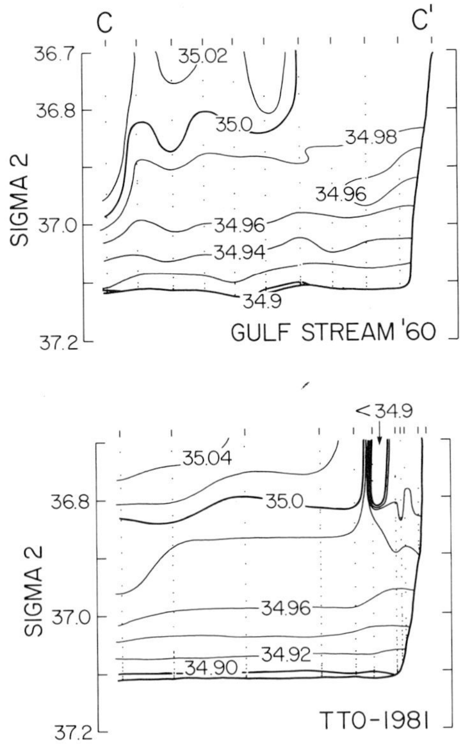

Fig. 6. Sections of salinity (x 10^3) versus σ_2 south from Nova Scotia (right), drawn from April 1960 Gulf Stream '60 data (upper section) and from a TTO/NAS cruise in October 1981 (lower section). Contouring as in Figure 4, except contoured only for $S \geq 34.9 \times 10^3$. Location shown in Figure 3.

effects by looking along the path of propagation of the densest layers. A long, winding path, from the Faeroe Bank Channel, along the mid-ocean ridge, through the Charlie-Gibbs Fracture Zone, around the northern Irminger Sea (where it picks up the Denmark Strait overflow), around the Labrador Sea, and finally around the Grand Banks, can be used for this purpose (e.g., cf. Swift, 1984). The pair of long sections shown in Figure 3 are from file data from 1960-72, and from TTO (1981) data. Using sections such as these, all the deep water masses of the region may be examined at once.

The sections in Figure 7 show salinity versus density, not depth, and, as before, only for the relatively dense deep water masses. A more detailed discussion of the historical data section (upper section) can be found in Swift (1984), though it should be noted that it includes some winter data, and thus the sea surface density appears in the Labrador Sea region where LSW is formed. The propagation of the Norwegian-Greenland Sea overflow layers is more easily seen in Figure 7 than on the cross-oriented sections: the ISOW layer (ca. σ_2 = 37.0), which is formed by mixing of relatively fresh Norwegian Sea water with much more saline water near the Faeroe Bank Channel sill, and the lower-salinity Denmark Strait overflow, which retains its high densities well away from the sill, and thus forms the bottom water in the northwest Atlantic. Of course, the newer section is based upon fewer stations, so does not show the same degree of detail and variability.

The larger of the salinity shifts in Figure 7 in the 1981 data (and the implied temperature shifts) are relatively easy to identify, although many of the details take time to distinguish. Freshening of about 0.04 to 0.10 x 10^{-3} was observed in the deep water in 1981 in the LSW near σ_2 = 36.85 around the southern tip of Greenland and into the north/central Labrador Sea. This was the largest deep water salinity change, at least in regions away from frontal zones. The LSW change has spread to the northeast Atlantic and around the Grand Banks to the waters off the New England continental shelf, as seen by the freshening of about 0.03 at σ_2 = 36.8 - 36.85 in these regions. As mentioned earlier, the LSW-ISOW transition layer immediately underneath LSW, in the σ_2 range 36.9 - 36.95, registered very little change in the interval between the two data sets. Figure 7 shows that the similarity extends from the northeast Atlantic to the central Labrador Sea. Again, this appears to be the result of a failure of more recent winter convection in the Labrador Sea to produce LSW with maximum densities as great as "classical" LSW.

Underneath the transition layer, the core of the ISOW layer in the northern North Atlantic (ca. σ_2 = 37.0) was, however, fresher in 1981, by about 0.005 - 0.025 x 10^{-3}, depending on location. Careful comparison of the sections in Figure 7 shows that the freshening tends to be greater in the northwest Atlantic than in the northeast Atlantic, probably due to further freshening of the ISOW layer by mixing with the now-freshened outflow from Denmark Strait. The TTO stations in Figure 7 were chosen both to duplicate the path of the historical data section and to yield maximum salinities, especially in the ISOW layer, so as to under-estimate, if anything, the magnitude of the deep freshening. In the northeast Atlantic, the bottom densities at the selected TTO stations were sometimes considerably less than at the historical counterparts (e.g., Figure 7). Farther east of the section path, away from the mid-ocean ridge, the TTO stations did exhibit bottom densities similar to the earlier values, though the bottom water was about 0.02 - 0.03 x 10^{-3} fresher there in 1981. In general, the northeast Atlantic water with σ_2 > 37.0 was 0.03 x 10^{-3} fresher in 1981. As mentioned earlier, since the properties of the Norwegian Sea outflow water in the Faeroe Bank Channel were unchanged in 1981, the bottom water freshening in the northeast Atlantic must be derived from a freshening of the Atlantic water at about 500-700 m which mixes with the dense outflow (e.g., cf. Steele et al., 1962).

In the northern Irminger Sea the dense, relatively low salinity outflow from Denmark Strait

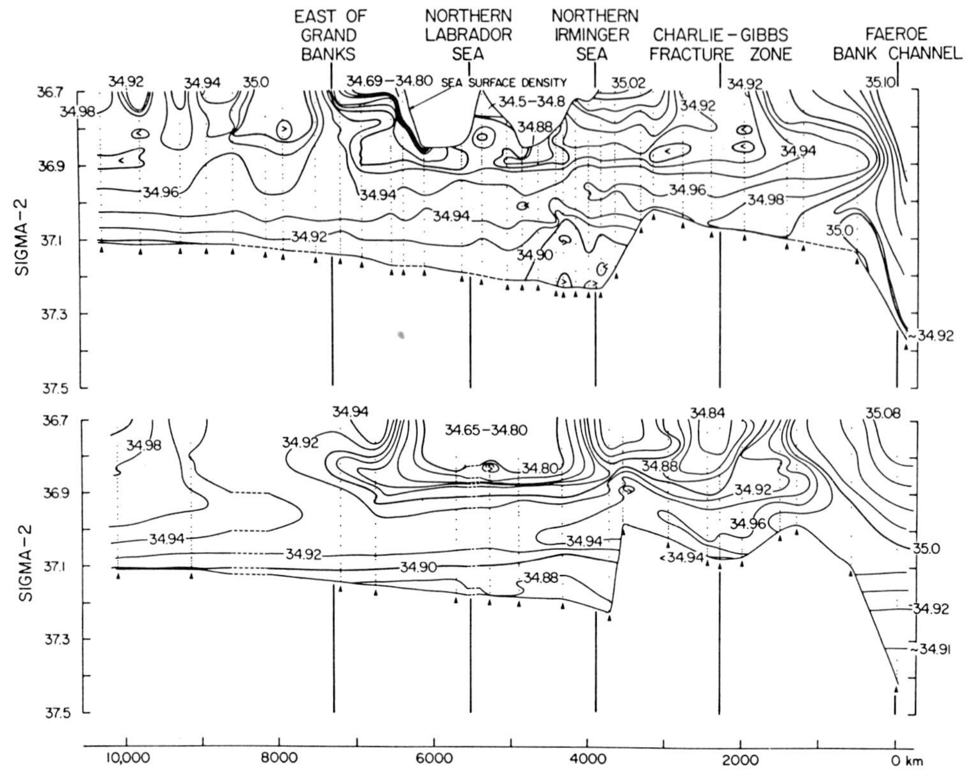

Fig. 7. Sections of salinity (x 10^3) versus σ_2 on a long, winding path from the Faeroe Bank Channel to the waters off the U.S. east coast (location shown in Figure 3). The upper section is drawn from a composite of historical data (Chain 1960 and 1972, Crawford 1960 and 1965, Erika Dan 1962, Atlantis 1964, Hudson 1967, and Knorr 1972 and 1976) and the lower section is drawn from April-October 1981 TTO/NAS data. The only post-1972 data in the upper section is a single Knorr 1976 station in the Faeroe-Shetland Channel. Contouring as in Figure 4.

(σ_2 > 37.1) is fresher in 1981 by about 0.005 - 0.02 x10^{-3}. Downstream, in the northern Labrador Sea and northeast of the Grand Banks, the bottom layer is about 0.02 - 0.035 x 10^{-3} fresher. Again, this is evidence for a greater downstream presence of relatively unaltered overflow water in 1981 than in the earlier data (here mostly Erika Dan data from 1962).

On the whole, the deep waters in the northern North Atlantic are nearly 0.02 x 10^{-3} less saline and 0.15°C colder than before. The changes evident in Figure 7 occur in nearly all regions of the northern North Atlantic, throughout virtually all of the deep water column. In 1981, the deepest changes had not propagated strongly southward, around the Grand Banks, except for a freshening of about 0.005 - 0.01 x 10^{-3} for σ_2 > 37.04 at the first TTO station southwest of the Grand Banks in Figure 7. Instead, the regime north of ca. 50°N seems only loosely coupled to that to the south, since by 1981 the LSW and overflow-derived changes had spread nearly uniformly throughout the northern deep water (e.g., cf. Figures 4, 5).

Discussion

The changes in Figures 4, 5 and 7 can indeed arise in under 10 years from advection of the water from the water mass formation areas, even in the ISOW layer off the Grand Banks: a fluid parcel moving at only 5 cm/sec would reach the Grand Banks area of the section 4-1/2 years after leaving the Faeroe Bank Channel. An alternative viewpoint is simply that these deep water property changes indicate that the time scale for spreading and recirculation, perhaps initially strongest along the boundary then later into the interior of the basins, is under 10 years for the northern North Atlantic.

The deep changes must be related to lower salinities of the surface layers in the water mass formation regions. There is evidence over the last 5-10 years of just such a reduction in salinities of the upper layers of much of northern North Atlantic, including the Labrador and Norwegian-Greenland seas. For example, Figure 8 illustrates the year-to-year changes in mean salinity of the upper waters in the Faeroe-Shetland Channel. This example (from Martin, 1979) is

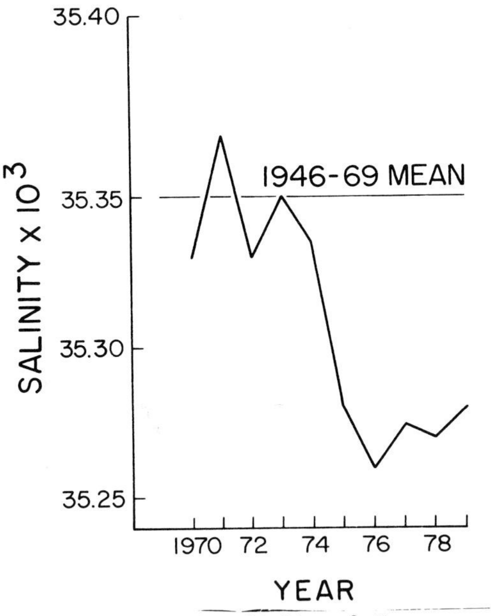

Fig. 8. Mean annual salinity of the upper 200 m of North Atlantic water in the Faeroe-Shetland Channel (from Martin, 1979).

particularly interesting because this water is not only coming from a broad region of the northeast Atlantic, but is going into the Norwegian-Greenland Sea, where it is the major supplier of salt. It is this upper-layer northeast Atlantic water which mixes with the Faeroe Bank Channel overflow, now reduced in salinity (e.g., cf. Figure 7) and in the north the Norwegian Current waters have been freshened by just this amount (Solonitsyna, 1979). Lazier (1980) has described a freshening of waters in the Labrador Sea from 1964-72, and other papers showing similar upper layer salinity changes have been published by Taylor and Stephens (1980), Ellett (1976, 1982), Solonitsyna (1979), and Smed et al. (1982).

Brewer et al. (1983) have noted that surface salinities were lower than 1955-65 values almost everywhere along the TTO track north of 50°N. Comparison calculations of average salinity in the upper 200 m at a number of locations (more or less along the track of the sections in Figure 7) indicate a recent net upper layer freshening of about $0.1 - 0.2 \times 10^{-3}$.

The lower surface salinities propagate into the deep water because in the water mass formation areas the downward penetrated of the cooled water in winter depends upon its density, not its temperature or salinity alone. Thus, other factors equal, the net result of an admixture of lower salinity surface water is to produce colder, less saline water for a given density. A consistent shift of this type could produce the deep changes seen in the 1981 data. One limiting factor is that if in any particular winter the upper layers are too stably stratified by low salinity water, the upper layers may freeze before they convect into the denser layers. But the isopycnals corresponding to the Denmark Strait overflow continue to outcrop in winter north of Iceland (various personal observations and communications), and the central Labrador Sea does not freeze.

Although the deep temperature, salinity, and dissolved oxygen fields have recently changed, this has occurred without major restructuring of the deep density fields, such as those which might accompany a change in the patterns of the deep field of motion. For the most part, the comparison cross-section of density parameters vs. depth for the data in Figures 4-6 are so nearly identical that meaningful differences, if any, are not apparent. Hence, such sections are not shown here. One exception is that a careful overlay of the density sections extending northeast from the Grand Banks (e.d., cf. Figure 5) shows that the slopes of the deep isopycnals ($\sigma_2 \geq 36.9$) were slightly greater in 1962 than in 1981. Deep isopycnals at the same depth in both years 400 km northeast of the boundary current off the Grand Banks separated about 100 m as they rose 1000 m to the boundary current, with the 1962 data rising higher. A similar separation was observed on this comparison section from the center to the northeast, near the mid-ocean ridge. This suggests that the deep cyclonic circulation in the Labrador Sea might have been weaker in 1981 than 1962. If such a relaxation occurred throughout the water column, for example in response to a change in wind forcing, a surface salinity reduction could be achieved by the resulting greater penetration of the low-salinity surface boundary currents into the deep water mass formation region nearer the central Labrador Sea.

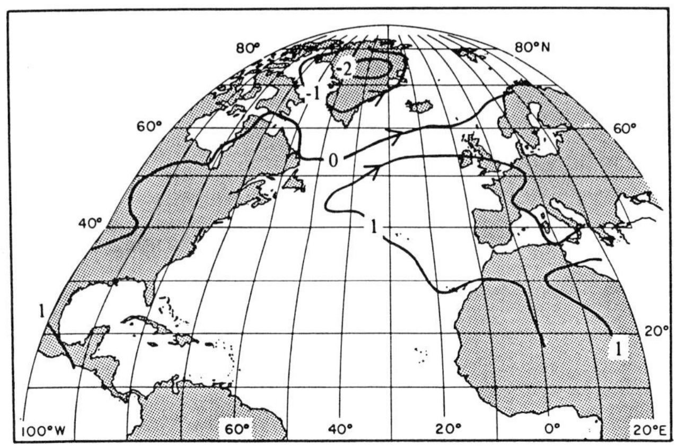

Fig. 9. Differences of long-term mean annual sea level pressures (mb), average of 1973 to 1981 minus average of 1955 to 1972. Contour interval as 1 mb, or approximately one standard deviation. Contoured from computations on a 5° grid, data from the Scripps Institution of Oceanography Climate Research Group.

The driving agent behind such larger-scale changes must be large-scale itself. Brewer et al. (1983) discuss several appropriate mechanisms. For example, although the firm evidence for this is not yet in hand (beyond the discussion by Brewer et al.), one possibility is that occasional shifts in wind forcing are now allowing a greater proportion of low salinity surface water, such as that from the East Greenland Current, to penetrate the areas of water mass formation than has been the case in the 1950's and 1960's. One promising avenue for future work is illustrated in Figure 9, which shows the difference of average sea level pressures in the period 1973-81 minus those from 1955-72, the years most of the previous high-quality deep water hydrographic data were gathered. The implied difference in average geostrophic winds in Figure 9 is consistent with a shift in the surface Ekman flow towards a greater transport of low salinity water away from the coast of Greenland. Similar shifts persist southeast from Greenland to approximately 45°N, suggesting that the relatively saline waters of the North Atlantic Drift may have been swept more strongly southward, partially blocking the supply of saline surface waters to the northeast Atlantic, and hence to the Labrador and Norwegian seas. Small circulation changes such as these may affect the local balance between fresh and salty sources, producing a moderate reduction in surface salinities in the deep water mass formation regions. Future examination of TTO data of an ocean tracer such as tritium, which in the 1970's was relatively abundant in the East Greenland Current (e.g., cf. Dorsey and Peterson, 1976), may indicate whether the low surface salinities in the water mass formation regions arise from such a shift or from other causes, such as increased precipitation.

The overall effect of the present change is difficult to assess, if indeed it is of any consequence whatsoever to the thermohaline circulation of the World Ocean. Reid and Lynn (1971) have shown sections of potential temperature and salinity versus a density parameter, starting from the Norwegian-Greenland Sea and Denmark Strait, and proceeding along the western side of the Atlantic Ocean, across the southern Indian Ocean, and into the Pacific Ocean. The sections show the contribution to the deep World Ocean of the cold, saline tongue from the northern North Atlantic. This tongue originates at the sea surface and carries with it sea surface signals, such as anthropogenic CO_2 or other passive substances. It is clear that mixing along the propagation path of the tongue would reduce the size of a density-preserving change of 0.15°C and 0.02×10^{-3} introduced in the deep waters at only the northern source. But if the change should persist, since the waters above the tongue are warmer and fresher, while those below are colder and fresher, there could be a small change in the gradients. This might cause the tongue to erode slightly differently than now, but probably would not make much of a dent in the global thermohaline circulation.

What we can learn from this small shift is more of the maintenance of the structure of the deep ocean, and perhaps the ties to large-scale atmospheric forcing of the deep thermohaline circulation. This area near the water mass formation regions is responding on less than 10-year time scales, but what we are looking at is a piece of the "100-year" circulation of the deep ocean.

The upper layer freshening in the northern North Atlantic has already been called the "great salinity anomaly of the 1970's" (cf. Dickson and Blindheim, 1983, in preparation). The fact that it occurred in a region in relatively intimate communication with the deep ocean has provided oceanographers with a new tool with which to study the deep circulation.

Acknowledgements. The TTO Steering Committee permitted publication of these results. The TTO field program was carried out by the officers and crew of the R/V Knorr, Woods Hole Oceanographic Institution, and by the staff of the Physical and Chemical Oceanographic Data Facility of Scripps Institution of Oceanography under the leadership of R.T. Williams. I am particularly grateful for the assistance provided by S. Anderson, S. McBride, J. Washington, and the Scripps Marine Life Research Group Illustration Facility. Funding for this analysis was provided by the Office of Naval Research under Contract N00014-80-C-0440.

References

Barrett, J.R., Salinity changes in the western North Atlantic, Deep-Sea Res., 16(Suppl.), 7-16, 1969.

Brewer, P.G., W.S. Broecker, W.J. Jenkins, P.B. Rhines, C.G. Rooth, J.H. Swift, T. Takahashi, and R.T. Williams, A climatic freshening of the deep North Atlantic (north of 50°N) over the past 20 years, Science, 222, 1237-1239, 1983.

Clarke, R.A., and J.-C. Gascard, The Formation of Labrador Sea Water. Part I: Large-Scale Processes, J. Phys. Oceanogr., 13, 1764-1778, 1983.

Dickson, R., and J. Blindheim, On the abnormal hydrographic conditions in the European arctic during the 1970's, in preparation, 1983.

Dorsey, H.G., and W.H. Peterson, Tritium in the Arctic ocean and East Greenland Current, Earth Planet. Sci. Lett., 32, 342-350, 1976.

Ellett, D.J., Surface temperature and salinity anomalies in the Rockall Channel, 1948-1976, Annales Biologiques, 33, 27-28, 1976.

Ellett, D.J., Long-term water-mass changes to the west of Britain, in Time Series of Ocean Measurements, World Climate Report WCP21, 245-254, 1982.

Fuglister, F.C., Gulf Stream '60, Prog. Oceanogr., 1, 265-373, 1964.

Grant, A.B., Atlas of Oceanographic Sections,

1965-1967. Atlantic Oceanographic Laboratory, Bedford Institute of Oceanography, Dartmouth, N.S., Canada. Report A.O.L. 68-5, 80 pp., (Unpublished document).

Lazier, J.R.N., Oceanographic conditions at Ocean Weather Ship Bravo, 1964-1974, Atmosphere-Ocean, 18, 227-238, 1980.

Mantyla, A.W., and J.L. Reid, On the abyssal circulation of the World Ocean, Deep-Sea Res., 30, 805-833, 1983.

Martin, J.H.A., Hydrographic conditions in the Faeroe-Shetland Channel in 1970-1979, Annales Biologiques, 36, 55-56, 1979.

Ostlund, H.G., H.G. Dorsey, and R. Brescher, GEOSECS Atlantic Radiocarbon and Tritium Results (Miami), Tritium Laboratory Data Report #5, University of Miami, 14 pp. plus tables and graphs, 1976. (Unpublished document).

Reid, J.L., and R.J. Lynn, On the influence of the Norwegian-Greenland and Weddell seas upon the bottom waters of the Indian and Pacific oceans, Deep-Sea Res., 18, 1063-1088, 1971.

Smed, J., J. Meincke, and D.J. Ellett, Time series of oceanographic measurements in the ICES area, in Time Series of Ocean Measurements, World Climate Programme Report WCP21, 225-244, 1982.

Solonitsyna, L.R., Hydrographic conditions in the Norwegian Sea in 1979, Annales Biologiques, 36, 51-53, 1979.

Steele, J.H., J.R. Barrett, and L.V. Worthington, Deep currents south of Iceland, Deep-Sea Res., 9, 465-474, 1962.

Swift, J.H., The circulation of the Denmark Strait and Iceland-Scotland overflow waters in the North Atlantic, Deep-Sea Res., in press, 1984.

Swift, J.H., K. Aagaard, and S.-A. Malmberg, The contribution of the Denmark Strait overflow to the deep North Atlantic, Deep-Sea Res., 27, 29-42, 1980.

Talley, L.D., and M.S. McCartney, Distribution and circulation of Labrador Sea Water, J. Phys. Oceanogr., 12, 1189-1205, 1982.

Taylor, A.H., and J.A. Stephens, Seasonal and year-to-year variations in surface salinity at the nine North-Atlantic Ocean Weather Stations, Oceanol. Acta, 3, 421-430, 1980.

Worthington, L.V., and W.G. Metcalf, The relationship between potential temperature and salinity in deep Atlantic water, Rapp. P.-v. Cons. perm. int. Explor. Mer., 149, 122-128, 1961.

Worthington, L.V. and W. R. Wright, North Atlantic Ocean Atlas of Potential Temperature and Salinity in the Deep Water, Including Temperature, Salinity, and Oxygen Profiles from the Erika Dan Cruise of 1962, Woods Hole Oceanographic Institution Atlas Series, 2, 58 plates, 1970.

VARIATION OF MONSOONAL UPWELLING: A RESPONSE TO CHANGING SOLAR RADIATION

Warren L. Prell

Department of Geological Sciences, Brown University, Providence, RI 02912-1846

Abstract. In the western Arabian Sea, the distribution of planktonic foraminifera is related to coastal upwelling and therefore to the low-level summer monsoon winds which force the upwelling. Records of faunal variation over the past 30 kyr reveal that the Indian Ocean summer monsoon has been both stronger (9 kyr) and weaker (18 kyr). These paleoceanographic reconstructions and analogies to general circulation model simulations suggest that changes in the distribution of seasonal solar radiation and surface albedo are responsible for the changes in monsoon intensity. This study illustrates that the Milankovitch Hypothesis of climate change, which changes the seasonal distribution of solar radiation as a function of orbital variations, applies to low-latitude as well as high-latitude climate changes.

Introduction

Paleoceanographic records can be used to infer the relative importance of various climate processes in the past. However, such inferences must be based on a clear strategy that relates climatic processes with paleoceanographic data. First, one must select a part of the climate system that has a well-known forcing that causes a specific and strong oceanic response. Next, the oceanic response must have a geologic record that can be calibrated to modern environmental conditions. If these criteria can be met, then a time series of the paleoceanic index may be used to interpret the relative importance of past climatic processes.

The Indian Ocean summer monsoon is a distinct climatic phenomenon associated with a specific upwelling response that, in turn, is recorded in deep-sea sediments. This geologic time series of monsoonal upwelling can be used to address the following questions. How has the Indian Ocean summer monsoon behaved over the past 30,000 years? What climatic mechanisms and processes are consistent with past variations of the monsoon? Do changes in seasonal solar radiation affect low-latitude climate, and in particular, monsoonal upwelling.

To answer these questions, I have studies the spatial and temporal distribution of planktic foraminifers in the Arabian Sea. The patterns of faunal variation have been related to patterns of sea surface temperature and coastal upwelling and therefore to the windfield. These data and modern relationships are used to reconstruct the relative intensity of the Indian Ocean monsoon over the past 30 kyr. This study concentrates on the changing direction and intensity of low-level winds and their effect on coastal upwelling in the western Arabian sea.

Models for the Intensity of Indian Ocean Monsoon Circulation Patterns

The Indian Ocean summer monsoon is a global phenomenon as well as an intense seasonal variation of climate on the regional scale. The characteristics and causes of the Indian Monsoon circulation have been the subject of numerous books (for example, Ramage, 1971; Miller and Keshavamurthy, 1968), papers, and intensive large-scale studies (Garp, 1981). The synoptic patterns and atmospheric dynamics of the Southwest Indian Monsoon are extremely complex and only their large-scale time-averaged characteristics will be examined here.

Past variations in the strength of the Southwest Monsoon may be interpreted to reflect changes in large-scale boundary conditions of the climate. Figure 1 illustrates the major components of the Southwest Monsoon in conceptual fashion. In this simplified view of the Southwest Monsoon, the major boundary conditions are the Indian Ocean SST, the size and elevation of the Asian continent, the surface albedo of the Asian continent, and the seasonal distribution of solar radiation. The major response to these boundary conditions is the near-surface heating of Asia during the summer. This summer heating, which is due to the low heat capacity of the continent, creates the low pressure zone over Asia relative to the Indian Ocean, which has relatively constant SST and sea-level atmospheric pressure. This pressure gradient between the Asian continent

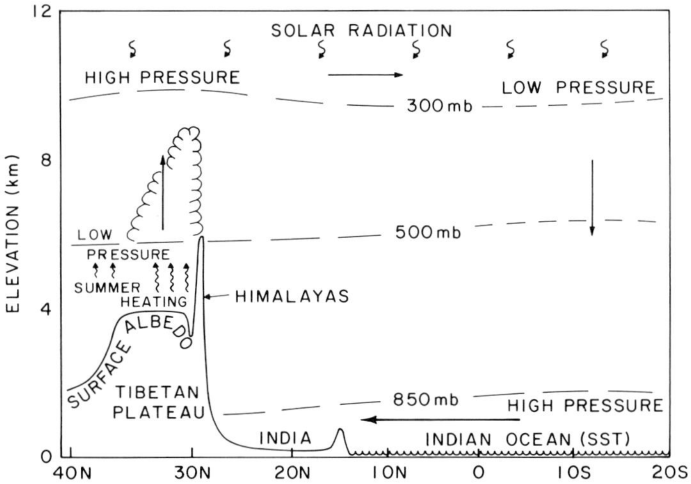

Fig. 1. Conceptional illustration of the boundary conditions and climate processes controlling the Southwest Indian Monsoon circulation. The elevations of isobar surfaces are somewhat exaggerated to show the gradients.

and the Indian Ocean reverses seasonally and is the primary cause of the low level Southwest Monsoon winds.

The relative importance of these various boundary conditions can be inferred from synoptic studies and general circulation model (GCM) experiments. In particular, GCM results have provided insight into the importance of topography, albedo, and solar radiation on the intensity of the Southwest Monsoon. Hahn and Manabe (1975) used the GFDL GCM to show that the great elevation of the Tibetan Plateau in central Asia was responsible, in part, for the great intensity of the Southwest Monsoon. Simulations with no mountains gave a much weaker monsoonal circulation. Hence, the topography of the Tibetan Plateau in central Asia is thought to be an important boundary condition for the Southwest Monsoon. This continental elevation is considered constant over 10^3 years but may not be constant over intervals of 10^5 to 10^6 years.

The effect of surface albedo over Asia on the monsoon was demonstrated by Manabe and Hahn (1977). Their simulation of the last glacial maximum produced a weaker monsoonal circulation, which they attributed to the effect to higher albedo over Asia. An experiment using modern terrestrial boundary conditions and glacial SSTs (CLIMAP, 1976) generated a monsoon circulation with an intensity similar to modern conditions. Hence, Manabe and Hahn concluded that albedo changes were more important than SST changes in controlling the intensity of the monsoonal circulation. Importantly, the solar radiation during the summer at 18,000 YBP is very similar to modern conditions.

Recent experiments by Kutzbach (1981) and Kutzbach and Otto-Belisner (1982) have demonstrated the importance of changes in seasonal radiation patterns. They simulated the climate at 9,000 YBP by utilizing modern boundary conditions and imposing the solar radiation pattern of 9,000 YBP. Their results showed an intensified monsoonal circulation, with low level winds over the Arabian Sea increased by almost 50%. This set of simulations clearly indicates that simple changes in the distribution of seasonal solar radiation can cause major fluctuations of the Southwest Monsoon. In light of these observations and our simple model of the monsoon

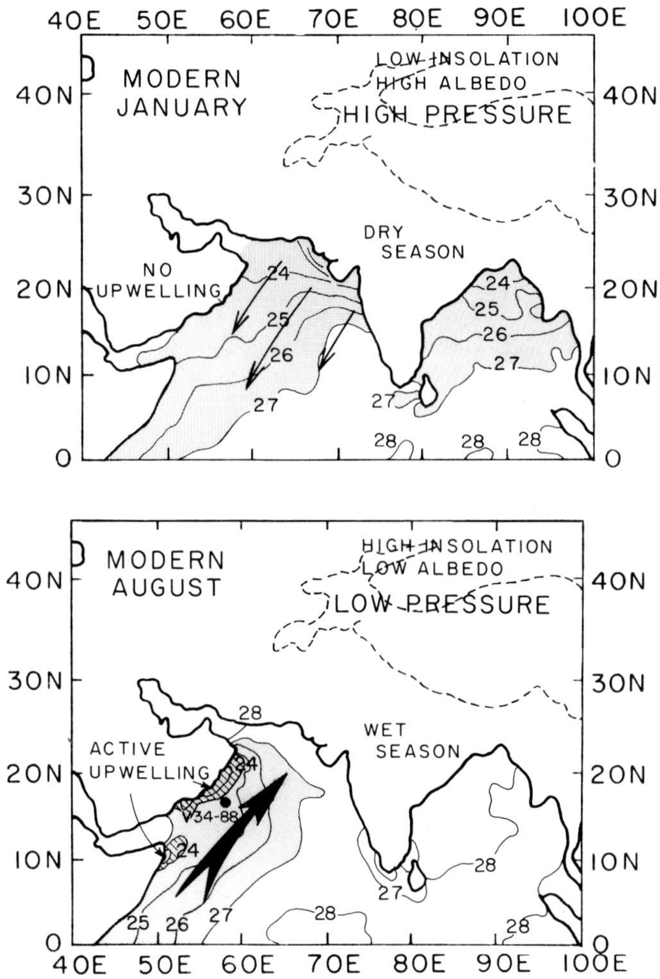

Fig. 2. Top: Boundary conditions and circulation for the modern winter monsoon. The dashed line over Asia indicates elevations greater than 3000 meters. The shaded area shows the distribution of SSTs less than 27°C. The circulation arrows indicate moderate wind speeds. Bottom: Boundary conditions and circulation during the modern summer monsoon. Cold water (shaded area is less than 27°C) is concentrated along the western Arabian Sea and the upwelling (diagonal lines show SSts less than 24°C) is active off arabia. The bold arrow represents high wind speeds and intense circulation.

(Figure 1), the geologic data will be interpreted in terms of changes in Indian Ocean SST, surface albedo, and solar radiation.

Atmospheric Circulation of the Modern Monsoon

The near-surface winds of the Arabian Sea are characterized by a complete reversal on a seasonal basis. The winter Northeast Monsoon is characterized by low seasonal insolation over Asia and relatively high albedo due to seasonal snow cover. These boundary conditions produce a high pressure cell in the low-level atmosphere over Asia which results in a northeasterly wind flow over the Arabian Sea (Figure 2a). During the winter monsoon (December, January, and February), winds are from the northeast off Asia with relatively low velocities.

The summer Southwest Monsoon is characterized by high seasonal insolation over Asia and relatively low albedo (Figure 2b). These conditions create a low-level, low-pressure cell centered at about 30°N over Asia. During this season, the southern Indian Ocean acts as a high pressure cell and the resulting wind flow is from the southwest across the Arabian Sea (see Figure 2b). During the summer monsoon (June, July, and August), the winds are from the southwest at extremely high velocities. Long-term climate records (sixty years) show that the resultant winds over much of the western Arabian Sea average >10 ms^{-1} during this season (Hastenrath and Lamb, 1979). This dramatic reversal and increased intensity of the winds is also associated with major changes in cloudiness, heat transport, and precipitation.

Oceanic Circulation of the Modern Monsoon

The response of the Arabian Sea to the atmospheric forcing is most clearly observed in the sea-surface temperature (SST) patterns. During the winter monsoon, the isotherms of the Arabian Sea are roughly trend from east-west to northeast-southwest with lower SSTs (23°C) to the north and higher SSTs (27°C) to the south (Figure 2a, Hastenrath and Lamb, 1979). Much of the Arabian Sea and Bay of Bengal are less than 27°C due to lowered insolation and northerly wind flow. The cool waters off Africa represent southward transport in the winter Somali Current.

This relatively simple pattern completely changes during the summer monsoon when the isotherms trend roughly northeast-southwest parallel to the coasts of Africa and Arabia (Figure 2b). During the summer, a time of expected high temperatures at this latitude, long-term average SSTs as low as 22°C to 23°C are observed along the coast of Africa and Arabia and grade to 27°C to 28°C in the central Arabian Sea (Hastenrath and Lamb, 1979). The reason for this abrupt change in the SST pattern is the coastal upwelling along the coast of

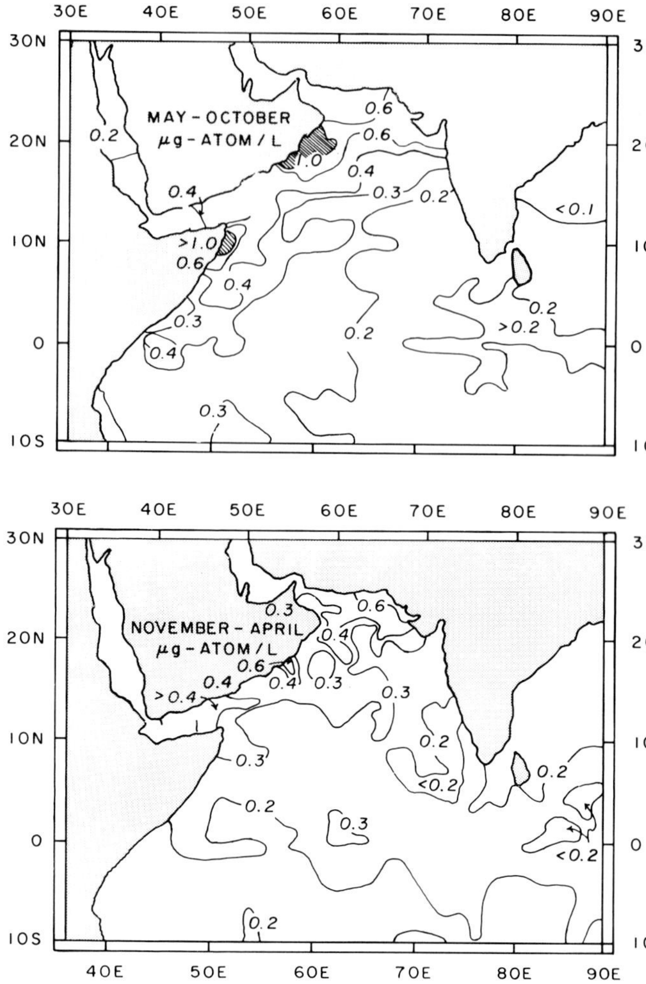

Fig. 3. Phosphorus content of the surface waters (g-atom per liter) for the summer half-year (May to October) and the winter half-year (November to April). Redrawn from Wyrtki, 1971.

Africa and Arabia. This upwelling brings cooler waters to the surface and distorts the normal seasonal pattern (Duing and Leetma, 1980; Prell and Streeter, 1982). During the Southwest Monsoon interval, about 66% of the variance in local SST anomalies along the coast of Arabia can be explained by simple Ekman transport (Prell and Streeter, 1982). Hence, the monsoonal winds and coastal SST are closely linked during the summer.

The effect of coastal upwelling is also observed in various chemical and biotic measures. For example, the concentration of nutrients (Wyrtki, 1971) and chlorophyll (Krey and Babenerd, 1976) in the surface waters show intermediate values during the winter half-year whereas they show high values along the coast of Africa and Arabia during the summer half-year (Figure 3). These nutrient concentrations negatively covary with the SST to provide a geographic pattern for the variation of coastal

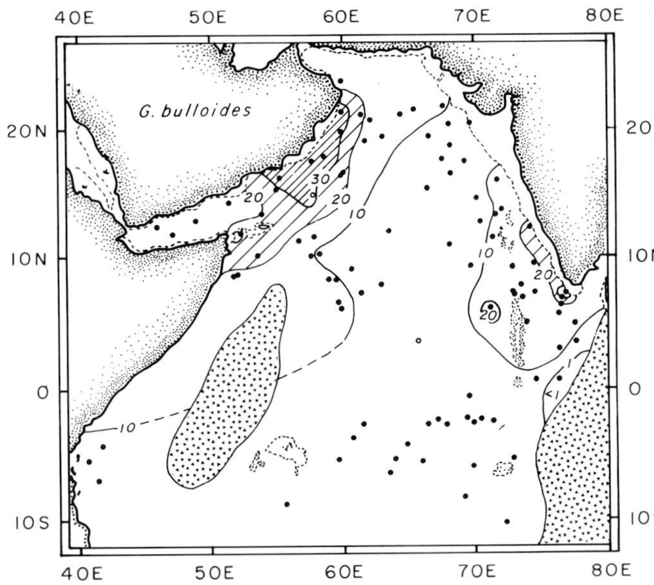

Fig. 4. The percent of G. bulloides in surface sediments of the Arabian Sea. Note that G. bulloides maximum abundance coincides with low SST (Figure 2b) and high nutrient content (Figure 3) of the surface waters. Samples in the dotted areas contain less than 10 forams/g and have not been used to construct species biogeography.

upwelling and its associated effects. These data show that monsoonal upwelling is a strong and spatially restricted oceanographic response that can be used as a measure of monsoonal wind intensity.

Plankton Patterns

The plankton of the Indian Ocean, specifically the planktic foraminifers, respond to the SST and chemical gradients caused by monsoonal upwelling. Eventually, the foraminifers are deposited in the underlying sediments to preserve a record of changes in monsoonal upwelling. The response of the planktic foraminifers can be determined by mapping their distribution in the surficial sea-floor sediments and comparing their abundance and size gradients to various upwelling-associated gradients such as SST.

Among the various planktic foraminifers in the Indian Ocean, one species in particular, Globigerina bulloides, seems particularly well attuned to the environments caused by upwelling. A map of the relative abundance of G. bulloides reveals that it is most abundant along the coast of Arabia and decreases in abundance toward the central Arabian Sea (Figure 4). The abundance of G. bulloides off southwest India is associated with another upwelling system (i.e. lower SSTs), which is probably remotely forced. This spatial pattern is similar to that of the summer SST (Figure 2b) and phosphate content (Figure 3). A lack of suitable samples prevents us from documenting a similar pattern off the coast of Somalia. The relative abundance of G. bulloides versus the sea-surface temperatures during August, a measure of the upwelling gradient, gives a correlation coefficient of -0.91 (Figure 5). Although some scatter exists in this core-top data set, the high degree of correlation clearly documents that G. bulloides is associated with the monsoonal upwelling pattern.

I have used the abundance of a single species, G. bulloides rather than the somewhat standard factor analysis assemblages and SST estimates for several reasons. First, in the Northern Indian Ocean, the major factor assemblage associated with upwelling contains significant abundances of both G. bulloides and Globigerinita glutinata (Cullen, 1980). G. bulloides is clearly associated with low-latitude upwelling zones, such as off North Africa (Ganssen and Sarnthein, 1983) and in the Cariaco Trench, Caribbean Sea (Rogl and Bolli, 1973 and references therein). However, no clear association with environmental data or upwelling exist for the species G. glutinata. Second, the calibration of paleotemperature equations for the Northern Indian Ocean is complex and in need of further study. In ocean-wide equations such as FI-2 (Hutson and Prell, 1980), G. bulloides is a sub-polar species and is associated with SSTs much less than in the Arabian Sea upwelling zone. Preliminary attempts at local calibrations (Cullen, 1980) have produced noisy estimates with low multiple correlation

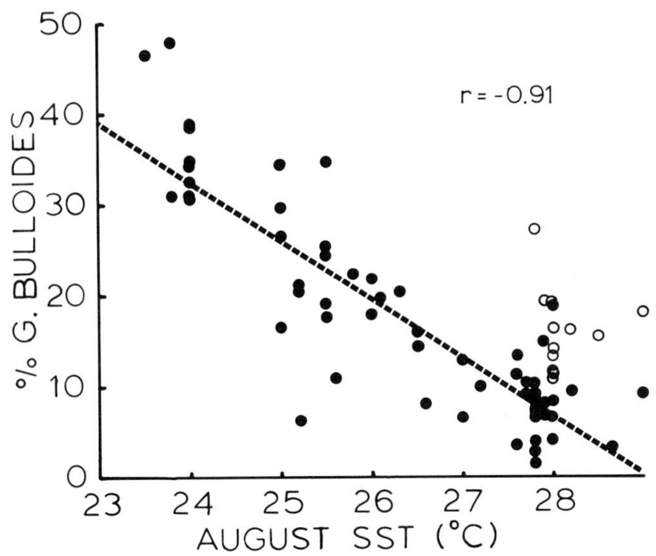

Fig. 5. The abundance of G. bulloides in the surface sediments versus August SST (°C). Filled circles represent open Arabian Sea samples and open circles represent Chargos-Lacadive Sea samples. The correlation of SST and G. bulloides in open Arabian Sea samples is -0.91.

coefficients for the Southwest Monsoon season. In addition, the definition of seasonality is complicated in the Arabian Sea. In light of these possible sources of misinterpretation, I have elected to use a single species index, which is imperfect, but at this time probably the best index of upwelling.

A subset of samples also reveals that the average size of G. bulloides increases with cooler temperatures or toward the upwelling maxima. Hence, G. bulloides is largest where it is most abundant, another indication that the upwelling zone is its optimum environment. These abundance data relate the distributon of planktic foraminifers in the deep-sea sediments to the environmental patterns of monsoonal upwelling, and in turn, to the intensity of monsoonal winds. Hence, the abundance of G. bulloides will be used as an indication of the intensity of monsoonal upwelling. Next, we examine the geologic record of this index of monsoonal upwelling.

Records of Monsoonal Upwelling

The relative abundance of G. bulloides has been measured in the late Quaternary sections of nine Arabian Sea cores (Figure 6). The two transects of cores represent conditions proximal to the coastal upwelling zone (A-A′) and conditions in the eastern Arabian Sea and the central Arabian Sea (B-B′) distal to the upwelling. Examination of the abundance data and the oxygen isotope stratigraphy in each core (Figure 6) reveals several distinct patterns.

The oxygen isotope stratigraphy reveals that all records extend from the late Holocene at core-top through Termination I (isotope stage 1/2 boundary), include isotope stage 2 (the last glacial maximum), and most extend into isotope stage 3. Termination I is characterized by a rapid $\delta^{18}O$ depletion and has little secondary structure in the low accumulation rate cores but displays more structure in cores with high accumulation rates. Core MD77-203 exhibits a two-stage termination (1a and 1b) similar to that observed in the Atlantic Ocean (Duplessy et al., 1981). Here, we are interested in the G. bulloides abundance relative to the isotope stratigraphy.

The first major faunal pattern is that G. bulloides decreases in abundance away from the coast (Figure 6). Maximum values range from over 40% near the coast (MD77-203 and RC9-161) to less than about 10% in the eastern and central Arabian Sea (for example, V34-101 and V19-188). This gradient persists throughout the 30,000-year record but is significantly lower during certain intervals.

A second major pattern is the relative abundance peak of G. bulloides which occurs on Termination I. This acme can be correlated in cores near the upwelling zone (dashed line in Figure 6) and is best observed near the coast (MD77-203). The abundance peak decreases in magnitude offshore over several hundred miles and does not occur in the eastern or central Arabian Sea. The age of the abundance peak datum is about 9000 years on the basis of ^{14}C-dated $\delta^{18}O$ stratigraphy (Duplessy et al., 1981; Prell et al., 1980) and limited ^{14}C dates of V34-88.

A third major pattern is the low relative abundance of G. bulloides at the base of Termination I (correlated by the solid line in Figure 6). At this time, all cores proximal to the upwelling have about 20% G. bulloides with no apparent gradient within the proximal cores. The age of this minimum in G. bulloides abundance is estimated to be <18,000 YBP. Hence, G. bulloides was less abundant at the last glacial maximum (18,000 YBP) and, by inference, monsoonal upwelling was weaker (Prell et al., 1980). Alternatively, G. bulloides was more abundant and monsoonal upwelling more intense during the glacial-interglacial transition. Here, I note that the abundance peak is not coincident with so-called hypsithermal (approximately 6,000 YBP) but clearly falls on the transition between glacial and interglacial modes. Examination of cores from the central Arabian Sea (Figure 6) reveals only small variations in the abundance of G. bulloides. Hence, monsoonal upwelling variations are restricted to the western Arabian Sea, as expected for a coastal upwelling system.

These data provide us with a rough time series of how monsoonal upwelling has varied over the past 30,000 years. Of particular interest is the inference of weaker than present monsoonal upwelling during the last glacial maximum and stronger than present monsoonal upwelling during the glacial-interglacial transition. Next, I examine what climatic mechanisms or processes may have changed to account for the observed variation in monsoonal upwelling.

Past Monsoons

The geologic data reveal that the Southwest Monsoonal upwelling has been both weaker and stronger. The interpretation of these geologic data lead us to ask what boundary conditions have changed. Among the boundary conditions previously discussed, the seasonal distribution of solar radiation and the surface albedo are the most likely to have shown major changes over this time interval.

The albedo of the Asian continent definitely changes on both seasonal and long time scales (10^3 years). The most obvious albedo changes are due to the spatial extent and persistence of seasonal snow cover, which is likely linked to solar radiation patterns. Albedo changes are also caused by the succession of vegetational patterns as the region passes from glacial to interglacial climates (CLIMAP, 1976, 1981).

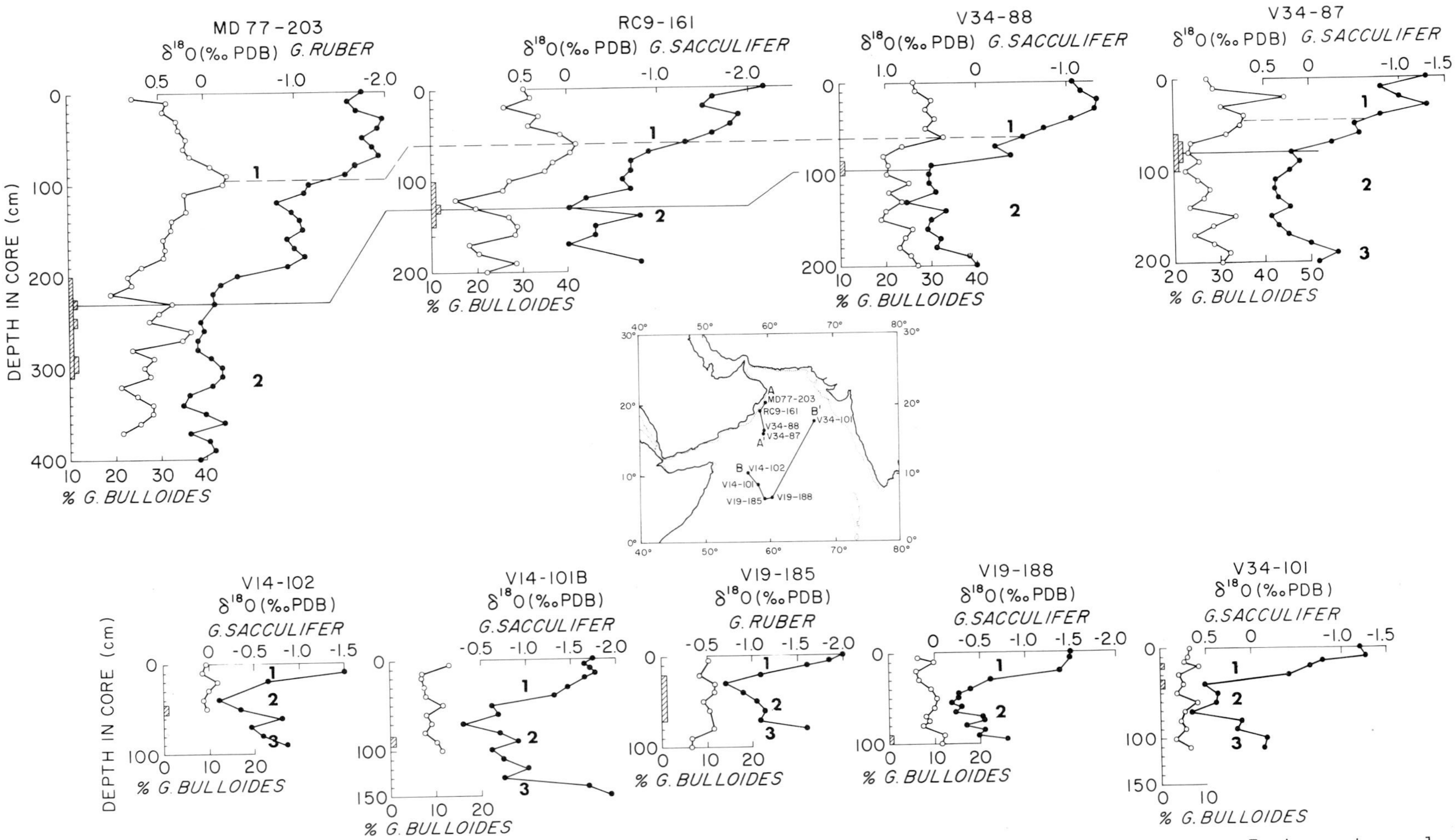

Fig. 6. Location, oxygen isotope stratigraphy, and percent G. bulloides for nine Arabian Sea cores. Isotope stages 1, 2 and 3 are shown in each core. Transect A-A' is proximal to coastal upwelling off Arabia, whereas transect B-B' is distal to the upwelling. The dashed line in transect A-A' correlates the G. bulloides maxima and the solid line correlates the G. bulloides minima and a brief occurrence of Globorotalia truncatulinoides (shown by bars on the depth axis). Isotopic data are from Prell et al., 1980; J. C. Duplessy (MD77-203) and this paper. Open circles denote the abundance of G. bulloides, and filled circles the oxygen isotope data.

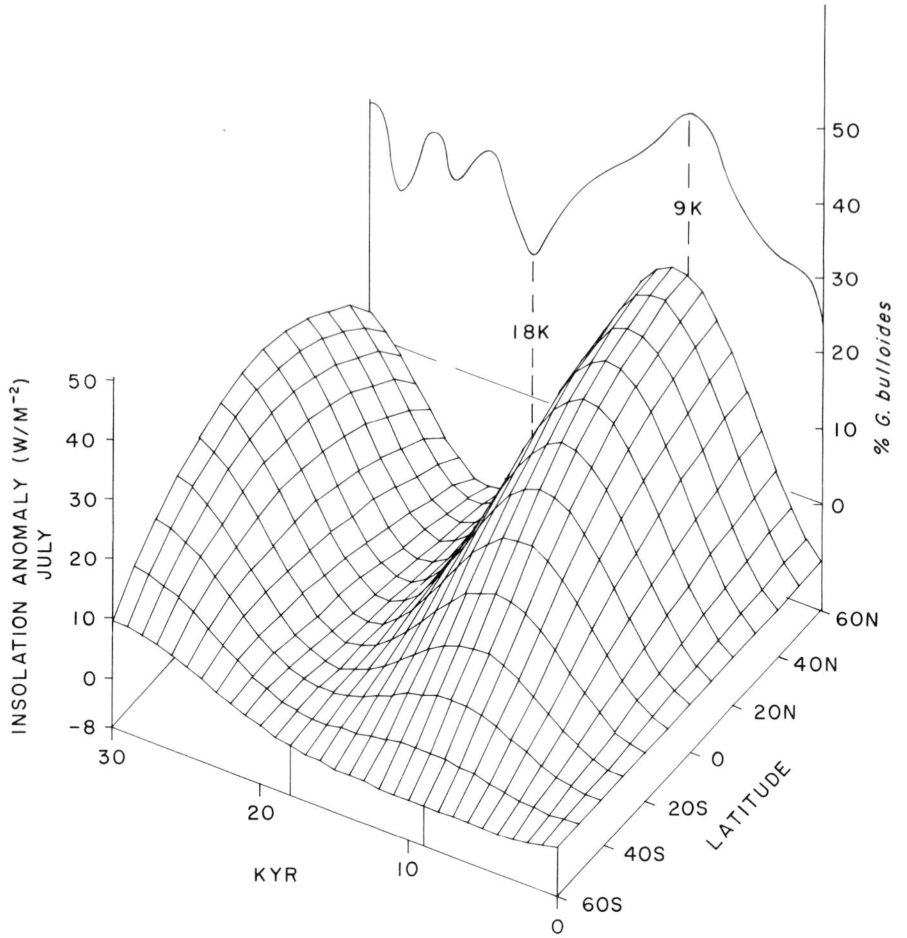

Fig. 7. Variation of solar insulation anomaly for July (past insulation-modern insulation, W/M^2) as a function of time (KYR=10^3yr) and latitude. Insulation values are calculated from Berger, 1979. A generalized curve of relative percent of G. bulloides, the upwelling index, is also shown on the same time scale.

Unfortunately, these vegetation and snow-cover patterns, and hence the resultant albedo, are difficult to quantify and map for past intervals. However, major changes in the distribution of Himalayan glaciers (Ahmad and Mayewski, in press) and vegetation patterns suggest that significant changes in albedo have occurred over Asia during the past 30,000 years.

The seasonal distribution of solar radiation has also exhibited significant changes over the past 30,000 years. The pattern of change is related to the shape of the Earth's orbit around the sun and has been calculated by numerous astronomers (for example, Milankovitch, 1941; Vernekar, 1972; and Berger, 1979). As an illustration of how solar radiation has changed with time and latitude, an insolation anomaly (past insolation - modern insolation) for July was calculated over the past 30,000 years for latitudes from 60°S to 60°N (Figure 7). This time-space diagram of the July insolation anomaly was selected because summer insolation should be the dominant driving force of the Southwest Monsoon. This diagram reveals several important features:

1. A maximum insolation anomaly occurs in the Northern Hemisphere at approximately 11,000 years B.P.
2. The insolation anomaly is minimal at 18,000 YBP and is quite similar to the modern distribution. Hence, during July the glacial maximum and modern solar radiation patterns are the same.
3. The insolation anomalies are strongly related to latitude.
4. The anomaly patterns for winter, not shown here, are essentially the reverse of the summer patterns. Hence, the annual anomaly is approximately zero.

These data show that the pattern of solar insolation during July undergoes major variations, which are latitude-dependent, over the past 30,000 years. Note also that the pattern of G. bulloides variation (Figure 7),

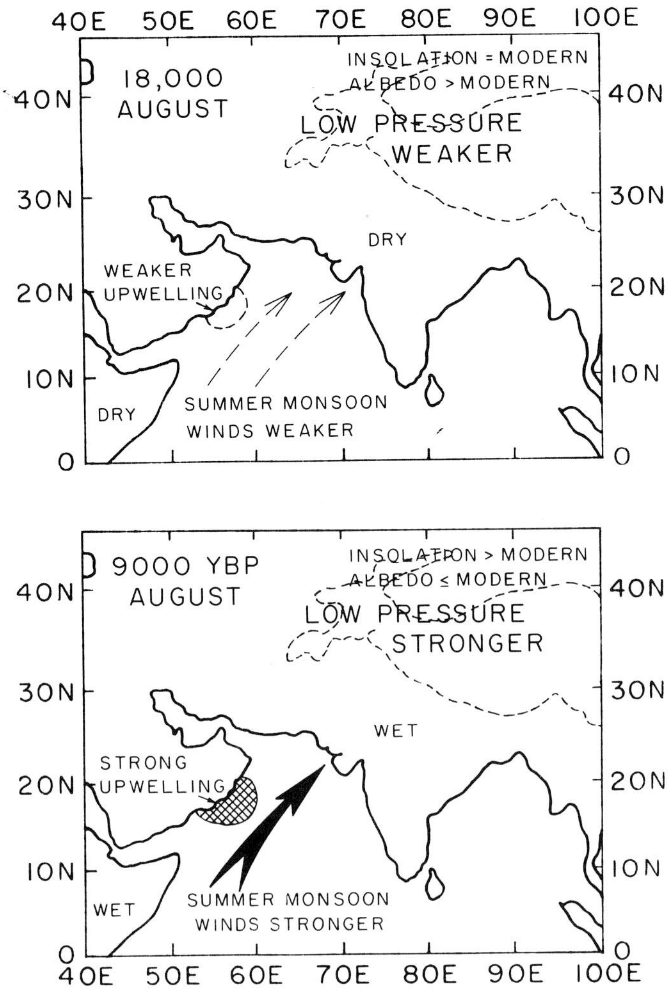

Fig. 8. Top: The inferred boundary conditions and circulation of the summer monsoon during the last glacial maximum (18,000 YBP) summer monsoon. Dashed arrows indicate weak monsoonal winds. Bottom: The inferred boundary conditions and circulation for the summer monsoon at 9000 YBP (the deglacial monsoon). The bold arrow indicates stronger monsoonal sinds and the checked area off Arabia indicates stronger than modern upwelling.

the index to monsoonal upwelling, is similar to that of the solar insolation anomaly pattern in the Northern Hemisphere.

The Glacial Maximum Monsoon (18,000 YBP)

Examination of the geologic and insolation data for the last glacial maximum about 18,000 years ago reveals that the insolation anomaly for July was about the same as today's but that monsoonal upwelling was weaker (Figure 7). Albedo was higher (CLIMAP, 1976, 1981). These observations suggest that the higher albedo led to less effective heating of the Tibetan Plateau and Asian continent and therefore the resultant summer low-pressure cell was weaker than in the modern climate. This weaker low-pressure cell over Asia created a lower pressure gradient from the central Indian Ocean to the Asian continent and hence less intense low-level monsoonal winds. Thus, the weak upwelling off Arabia is interpreted to reflect this decreased monsoonal intensity (Figure 8a). The weaker monsoonal upwelling is consistent with low lake levels throughout the region (Street and Grove, 1979) and enhanced precipitation in the equatorial zone (Duplessy, 1982). This interpretation is also consistent with GCM simulations of the last glacial maximum (Gates, 1976; Manabe and Hahn, 1977), which indicate a weaker monsoonal circulation.

The Deglacial Monsoon (9000 YBP)

Examination of the geologic and insolation data at 9000 YBP reveals that the amount of solar radiation is greater during the northern latitude summers and monsoonal upwelling is stronger (Figure 7). Note, however, that the maximum of upwelling does not exactly coincide with the maximum solar radiation in the Northern Hemisphere. Albedo is thought to be about the same as modern values. These observations, along with the results of Kutzbach (1981) and Kutzbach and Otto-Bliesner (1982), suggest that the higher summer insolation increased the heating of the Tibetan Plateau and Asian continent resulting in a stronger low-pressure cell. This enhanced low-pressure cell increases the pressure gradient between the Indian Ocean and the continental mass and hence causes the stronger monsoonal winds and increased upwelling (Figure 8b). The observation of increased upwelling is consistent with high lake levels in the surrounding areas (Street and Grove, 1979) and the transport of pollen off Africa into the Arabian Sea by southwesterly winds (van Campo et al., 1982).

Summary

In the modern climate, the Southwest Indian Monsoonal winds cause distinct areas of coastal upwelling along Arabia. This coastal upwelling is characterized by low sea-surface temperatures and high concentrations of nutrients. These upwelling conditions support distinct planktic faunal assemblages which are preserved in deep-sea sediments to record past variations in the monsoonal upwelling. Examination of this geologic record reveals that, compared to modern conditions, the Southwest Monsoonal upwelling has been both stronger and weaker during the past 30,000 years. This clear mechanistic linkage of the geologic record to the strength of monsoonal winds allows one to interpret the intensity of the Southwest Monsoon and to infer the boundary conditions which are consistent with the geologic record.

Examination of the large-scale boundary conditions which control the intensity of monsoonal circulation suggest that changes in the distribution of seasonal solar radiation and surface albedo are responsible for the changes in the monsoon intensity. Changes in solar radiation, possibly modified by albedo feedbacks, are translated into climate change through their effect on the heating of the Asian continent and the creation of the pressure gradient between Asia and the Indian Ocean. Since the distribution of solar radiation is controlled by the variations of the Earth's orbit, the changing intensity of the Southwest Monsoon illustrates how small orbital variations may be translated into significant climate changes. These data suggest that the Milankovitch hypothesis applies to low-latitude climates, especially the monsoonal circulation, as well as high-latitude climate changes.

Acknowledgments. This research was supported by the Division of Atmospheric Sciences, National Science Foundation under grant ATM82-13296 to Brown University (W. L. Prell).

References

Ahmad, N., and Paul A. Mayewski, ____, Himalayan glaciation: an unresolved issue. (ms.)

Berger, A. L., 1979, Insolation signatures of Quaternary climatic changes: Il Nuovo Cimento, v. 2C, no. 1, p. 63-87.

CLIMAP Project Members, 1981, Seasonal reconstructions of the earth's surface at the last glacial maximum: Geol. Soc. Amer. Map and Chart MC-36.

CLIMAP Project Members, 1976, The surface of the ice-age Earth: Science, v. 191, p. 1131-1137.

Cullen, James L., 1980, Microfossil evidence for changing salinity patterns in the Bay of Bengal over the last 20 000 years: Palaeogeogr., Palaeoclimatol., Palaeoecol., v. 35: p. 315-356.

Duing, Walter, and Ants Leetma, 1980, Arabian Sea cooling: a preliminary heat budget: J. Phys. Oceanog., v. 10, no. 2, p. 307-312.

Duplessy, J.-C., G. Delibrias, J. L. Turon, C. Pujol, and J. Duprat, 1981, Deglacial warming of the northeastern Atlantic Ocean: correlation with the paleoclimatic evolution of the European continent: Palaeogeogr., Palaeoclimatol., Palaeoecol., v.35, p. 121-144.

Duplessy, J.-C., 1982, Glacial to interglacial contrasts in the northern Indian Ocean revealed by oxygen isotopes: Nature, v. 295, p. 494-498.

Ganssen, Gerald, and Sarnthein, Michael, 1983, Stable-isotope composition of foraminifers: The surface and bottom water record of coastal upwelling: E. Suess and J. Thiede (eds.): Coastal Upwelling: Its Sediment Record

Gates, L. W., 1976, The numerical simulation of ice-age climate with a global general circulation model: J. Atmospheric Sci., v. 33, p. 184-1873.

Hahn, Douglas G., and Syukuro Manabe, 1975, The role of mountains in the south Asian monsoon circulation: J. Atmospheric Sci., v. 32, no. 8, p. 1515-1541.

Hastenrath, Stefan, and Peter J. Lamb, 1979, Climatic Atlas of the Indian Ocean: Univ. of Wisconsin Press, Madison, 97 charts, 19 p.

Hutson, William H., and Prell, Warren L., 1980, A paleoecological transfer function, FI-2, for Indian Ocean planktonic foraminifera: J. Paleon., v. 54, no. 2, p. 381-399.

Joint Scientific Committee for World Climate Research Programme and Global Atmospheric Research Programme, 1981, International Conference on the Scientific Results of the Monsoon Experiment: International Council of Scientific Unions, World Meteorological Organization Information Note No. 4, 164 p.

Krey, J., and B. Babenerd, 1976, Phytoplankton Production: Atlas of the International Indian Ocean Expedition: Intergovernmental Oceanographic Commission, UNESCO, 70 p.

Kutzbach, J. E., and B. L. Otto-Bliesner, 1982, The sensitivity of the African-Asian monsoonal climate to orbital parameter changes for 9000 years B.P. in a low-resolution general climate model: J. Atmospheric Sci., v. 39, no. 6, p. 1177-1188.

Kutzbach, J., 1981, Monsoon climate of the early Holocene: climate experiment using the Earth's orbital parameters for 9,000 years ago: Science, v. 214, p. 59-61.

Manabe, S., and D. G. Hahn, 1977, Simulation of the tropical climate of an Ice Age: J. Geophys. Res., v. 82, p. 3889-3911.

Milankovitch, M. M., 1941, Canon of Insolation and the Ice-Age Problem: Beograd, Koninglich Serbische Akademie. English translation by the Israel Program for Scientific Translation and published for the U. S. Department of Commerce and the National Science Foundation, Washington, D. C., 484 p.

Miller, Forrest R., and R. N. Keshavamurthy, 1968, Structure of an Arabian Sea Summer Monsoon System: Internat. Indian Ocean Expedition Meteor. Monogr. No. 1, East-West Center Press, Honolulu, 9 p.

Prell, Warren L., and Harold F. Streeter, 1982, Temporal and spatial patterns of monsoonal upwelling along Arabia: a modern analogue for the interpretation of Quaternary SST anomalies: J. Mar. Research, v. 40, no. 1, p. 143-155.

Prell, W. L., W. H. Hutson, D. F. Williams, A. W. H. Be, K. Geitzenauer, and B. Molfino, 1980, Surface circulation of the Indian Ocean during the last glacial maximum, approximately 18,000 yr B.P.: Quat. Res., v. 14, p. 309-336.

Ramage, C. S., 1971, Monsoon Meteorology: Internat. Geophys. Series No. 15, Academic Press, New York, 296 p.

Rogl, Fred, and Bolli, Hans M., 1973, Holocene to Pleistocene planktonic foraminifera of Leg 15, Site 147 (Cariaco Basin, Caribbean Sea) and their climatic interpretation: Initial Reports of the Deep Sea Drilling Project, v. XV, p. 553-615.

Street, F. Alayne, and A. T. Grove, 1979, Global maps of lake-level fluctuations since 30,000 yr B.P.: Quat. Res., v. 12, p. 83-118.

Van Campo, E., J.-C. Duplessy, and M. Rossignol-Strick, 1982, Climatic conditions deduced from a 150-kyr oxygen isotope-pollen record from the Arabian Sea: Nature, v. 296, p. 56-59.

Vernekar, A. D., 1972, Long-Period Global Variations of Incoming Radiation: Meteorol. Monogr. No. 34, Amer. Meteor. Soc., 21 p.

Wyrtki, Klaus, 1971, Oceanographic Atlas of the International Ocean Expedition: National Science Foundation, Washington, D. C., 531 p.

MODELING EVAPOTRANSPIRATION FOR THREE-DIMENSIONAL GLOBAL CLIMATE MODELS

Robert E. Dickinson

National Center for Atmospheric Research, Boulder, Colorado 80307

Abstract. A parameterization is developed for the calculation of evapotranspiration in three-dimensional atmospheric models. It distinguishes separately between evaporation from the ground and evapotranspiration from plant foliage. Soil water is stored in an active layer of 1 m depth and a 10 cm surface layer is separately distinguished. The evaporation from this soil is parameterized using a high resolution multilayer model for comparison.

This parameterized evaporation from the soil is defined by either the potential evaporation rate or by the maximum rate at which water can diffuse to the surface, depending on which rate is smaller. The maximum rate is obtained empirically in terms of various soil-hydraulic parameters.

The evapotranspiration from plants occurs either as the evaporation of water stored on the surface of the foliage or as the transpiration of water extracted by roots from the soil. The flux of water from the outer surface of foliage to the atmosphere above the canopy is determined by the decrease in water vapor concentration from the foliage surface to the overlying atmosphere and by the resistance of the foliage molecular boundary layers and the bulk aerodynamic resistance of the canopy. Transpired water encounters an additional stomatal resistance in passing from the inside to the outside of leaves. The foliage temperature and saturation vapor pressure are calculated from a model of the plant canopy energy balance. Soil moisture determines the maximum rate at which roots can extract water from the soil, and if the transpiration demand exceeds this maximum rate, stomatal closure occurs until the demand matches the root supply.

A parameterization of land evapotranspiration at the level of detail described in this paper may be required to obtain a realistic diurnal cycle of surface temperature and evapotranspiration for use in mesoscale or global climate models. However, application of the developed procedures is made difficult by the extreme complexity and small-scale detail of surface processes, the lack of adequate data sets for surface parameters, and the need for satisfactory parameterizations of other components of GCMs such as their rainfall and planetary boundary layer treatments. Because of these difficulties, the development of validated models of land surface processes first requires detailed sensitivity studies to establish what further data sets are most urgently required and what model improvement should be given highest priority.

Introduction

Past GCMs (global circulation models) have used extremely simple parameterizations of surface hydrological processes. Many model versions have avoided considering surface hydrology entirely by prescribing surface Bowen ratios. Where surface hydrology was included, the algorithms that have been generally used can be summarized as follows: Rainfall is assumed to fill the soil with water until the soil is at its field capacity for water. At that point, all further water is put into runoff. This "bucket model" for runoff was first used in global circulation models (GCMs) at the Geophysical Fluid Dynamics Laboratory (Manabe, 1969), where a soil water capacity of 15 cm has been assumed. Evapotranspiration is assumed to occur at its potential value until soil moisture drops below some critical value. For lower values of soil moisture, the evapotranspiration is assumed to be proportional to its potential value, with the proportionality factor beta, referred to as the evaporation efficiency and equal to the ratio of the model soil moisture to the critical value. This parameterization for evaporation was earlier developed by climatologists as a simple means to relate daily evapotranspiration to daily potential evapotranspiration using surface climatic data to define the latter (e.g., Thornthwaite and Mather, 1955; Budyko, 1974).

Many global climate models use diurnally averaged solar heating. With such an assumption, the models are perhaps further justified in using very simple surface parameterizations since, in reality, surface physical processes are so dominated by the diurnal cycle of solar heating that they cannot be modeled realistically in any case. However, for some purposes such as for the realistic modeling of moist convection over land

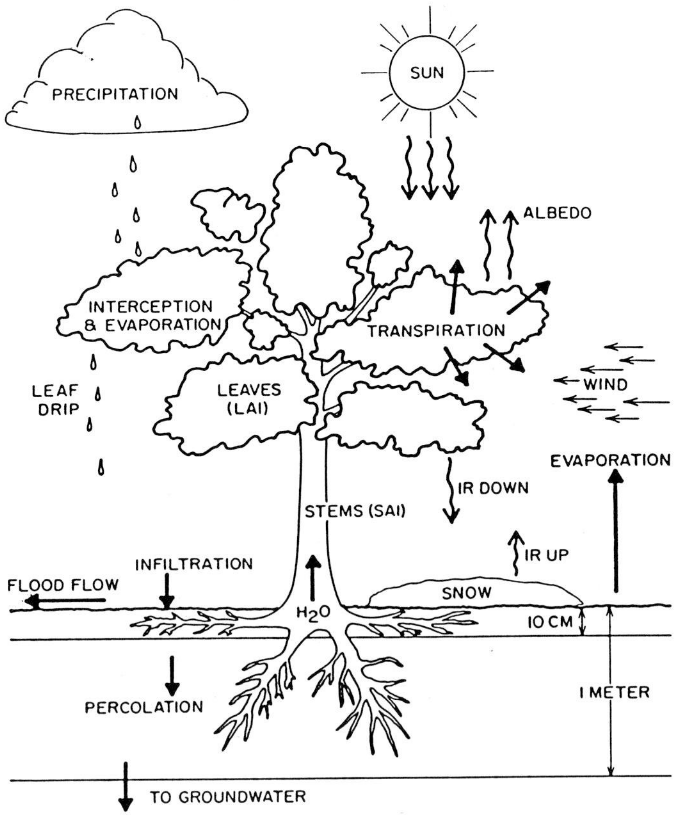

Fig. 1. Schematic of processes involved in surface physics parameterization.

or of land surface climate itself, inclusion of a diurnal variation of solar radiation at the surface and a realistic description of latent and sensible heat surface fluxes becomes essential.

In order to model adequately the diurnal course of evapotranspiration, it is necessary to parameterize the movement of water through soil and plants. The movement of water through vegetation is an especially crucial question since, on the average, more than half the total flow of water from land surfaces to the atmosphere travels through plants (Molz, 1981), and the water supply available to plants is of considerable practical importance. In the past it was believed that, with plentiful soil moisture, evapotranspiration was dominated by meteorological factors independent of the nature of the soil or underlying vegetation (Penman, 1948; Priestly and Taylor, 1972) and that plants acted simply as wicks to move water from the soil to the air. It is, however, now realized that this concept is approximately correct only for short vegetation. In forests, the evaporation from wet leaves can be much faster than from a surface of water exposed to the same meteorological conditions (Murphy and Knoerr, 1975; Stewart, 1977). Furthermore, transpiration from a dry canopy can be significantly less than that from a surface of water (Shuttleworth and Calder, 1979; Thom and Oliver, 1977).

The developments reported in this paper are motivated by three considerations: first, a number of recent GCM modeling studies have shown considerable sensitivity of their simulated climatologies, especially their rainfall and surface temperature, to drastic changes in the formulation of model land evapotranspiration (e.g., Charney et al., 1977; Kurbatkin et al., 1979; Rind, 1982; Shukla and Mintz, 1982) indicating a need for realistic modeling of land evapotranspiration; second, as reviewed above, there have been considerable advances in the last decade in understanding surface micrometeorology in the presence of vegetation, and it is desirable where possible to incorporate these advances in global models; and, third, many of the practical questions now being asked climate modelers, such as what climate change might result from tropical deforestation (e.g., as reviewed by Dickinson, 1980) depend on a detailed knowledge of the effects of land vegetation on surface hydrology.

The intent of this paper is to describe parameterizations that have been developed for surface hydrological processes in the presence of vegetation. These parameterizations are intended to be simple enough to be of practical use in GCMs which have a diurnal cycle, but to include the essential surface physical processes determining the interactions between the land surface and the atmosphere. They have been sufficiently tested in GCMs at NCAR to demonstrate their practicability. However, detailed sensitivity studies have not yet been carried out to establish what are their most crucial parameters or what improvements should be expected over use of simpler methods. This paper concentrates on describing these parameterizations and their rationalization rather than on model output.

The essential physical processes of vegetated land are summarized in Fig. 1. Precipitation incident on the surface is, in part, intercepted by the vegetation foliage; some of the intercepted precipitation is reevaporated and some drops off to reach the ground along with the non-intercepted precipitation. Part of the water incident on the ground infiltrates into the soil, and the rest travels along the surface or near-surface to enter streams. The water in the soil passes downward and may travel below the active surface layer tapped by roots into groundwater reservoirs which also eventually supply streamflow. At the same time, plants extract water from the soil through their roots and move it to the atmosphere by transpiration through their leaves. The energy balance of a plant canopy is an important factor in determining its evapotranspiration. Furthermore, shading of the ground by the canopy can drastically alter energy balance processes at the ground and so alter ground evaporation and snow melting.

The following sections describe further suitable parameterizations for these processes. One of the biggest conceptual difficulties in parameterizing land surface processes is the extremely varied subgrid-scale mosaic of vegetation and soil types (e.g., Warrick and Nielsen, 1980), and the importance of subgrid-scale variability of rainfall intensities (Eagleson, 1978; Freeze, 1980). I have not yet found convincing procedures to allow for these complexities in global models using existing data bases; my intent is first to understand better the role of hypothetical spatially uniform surface processes before exploring treatments of subgrid-scale structure.

Concept of Water Supply and Demand

The concept used in past GCMs that the ratio of actual to potential evaporation depends only on soil water content is without a physical basis. It can be established empirically only for diurnal mean evaporation; the beta parameter can vary strongly with the diurnal cycle even for fixed soil moisture. The flow of water through soils and plants generally involves the diffusion of water in its liquid or gaseous phase (e.g., Hillel, 1980). The sum of these diffusion processes cannot always supply water at a rate proportional to the maximum demand for water at the soil or leaf surface, that is, the potential evapotranspiration. Rather, only for small-enough actual transpiration will the water supply satisfy the potential evapotranspiration demand. On the other hand, for large-enough actual transpiration, the water supply occurs at a maximum rate determined by rate-limiting diffusion processes in the soil or plant roots (e.g., Phillip, 1957, 1966; Gardiner and Hillel, 1962; Cowan, 1965; Van Bavel and Hillel, 1976).

Thus, over the course of a diurnal cycle, evapotranspiration can change from the rate of potential evapotranspiration in the morning to some maximum rate later in the day that is less than but otherwise independent of the instantaneous rate of potential evapotranspiration. The level of soil moisture determines the demand rate at which occurs the switchover from demand-limited to supply-limited evapotranspiration; hence, if potential evapotranspiration does not vary too much from day to day, the level of soil moisture may correlate well with the ratio of diurnal averages of the actual to potential evapotranspiration (Federer, 1982).

The next section develops a simple but physically based model for soil evaporation, and the subsequent section, a model for evapotranspiration from the surfaces of plant foliage. Both soil and plant canopy models use the concept of matching the supply and demand of water to be evaporated or transpired (as also advocated for plant canopy models in Federer, 1982); the models require specification of some information regarding soil hydraulic parameters and vegetative cover. Earlier versions of these parameterizations and further details as to their implementation with NCAR's fourth-order grid-point GCM are given in Dickinson et al. (1981).

Soil Water Movement

Vertical movement of water through soil, in principle, depends on soil temperature, structure and moisture. However, neglecting the effects of temperature coupling to water diffusion has only a small effect on calculated evaporation rates (as indicated by numerical studies of the question by Milly and Eagleson (1982)). With the coupling to temperature gradients neglected, water upward flow F through soil at some depth z is given by:

$$F = K(z) \frac{\partial \Phi(z)}{\partial z} \qquad (1)$$

where K = hydraulic conductivity of the soil, and $\Phi = z - \Phi_w$ is the soil water suction (negative potential) in units of length, Φ_w = the matric potential which is determined by surface tension and other forces binding water to the soil, and z = the negative of the gravitational potential or equivalently, the soil depth. The terms K and Φ_w for a given soil can be expressed as functions of soil water which are not necessarily unique due to hysteresis effects, but are here assumed to be so. Besides the dependence of K and Φ_w on soil water, it is necessary to specify their values, depending on soil type, at some reference soil moisture level. Eq. (1) combined with the equation of continuity of soil water then becomes a diffusion equation for soil water that may be solved by standard numerical procedures.

I have programmed such a solution for use in developing simple soil water parameterizations. The detailed model uses simple finite differences, 100 layers, and extends to a depth of 10 m. To ensure adequate resolution for large gradients near the surface, the grid is stretched, with half the layers in the top 10 cm and three-quarters in the top 1 m. The numerical scheme was developed primarily to obtain adequately accurate solutions of the soil water diffusion equation. It uses a time step of 100 sec and implicit time differencing. Gravitational drainage was assumed through the lower boundary. Several boundary conditions were considered at the top, as discussed later.

For numerical calculations over a wide range of soil properties, it is convenient to use analytic relationships between Φ, K, and soil water s, defined as the ratio of soil water volume to the volume of voids in the soil, i.e., the total volume available to the water. In particular, it is convenient to assume that

$$\Phi_w = - \Phi_0 \, s^{-B} \qquad (2)$$

$$K = K_0 \, s^{2B+3} \quad ,$$

where B is a nondimensional parameter, and Φ_0 and K_0 are the values of Φ and K at saturation. These expressions have been found applicable to a wide variety of soils (Clapp and Hornberger, 1978) and can be inferred from a mechanistic model of soil water diffusion (e.g., Campbell, 1974).

My objective has been to find simple parameterized expressions for soil evaporation involving a surface soil layer of nominally 10 cm depth and the total soil 'active' layer of nominally 1 m depth. The parameters used for approximating the soil evaporation are s_1 = the average soil water s in the surface layer, and s_0 = the average soil water s in the total soil active layer. I have not yet attempted accurate modeling of the infiltration of water into the soil; infiltration processes over the area of a GCM grid square appear to be dominated by subgrid-scale surface and rainfall structure, so if subgrid-scale surface is neglected, it appears to me that tuning is necessary to obtain the detailed partition between infiltration and surface runoff in agreement with observation. Milly and Eagleson (1982) have developed an accurate parameterization for inflitration also suitable for determining evaporation and have examined the impact of subgrid-scale surface and rainfall structure on infiltration.

Eqs. (2) become invalid as $s \rightarrow 1$ or $s \rightarrow 0$. The first of these limits is not important for considering evaporation. However, the surface soil layers easily dry to the point where diffusion is controlled by vapor rather than liquid movement. With vapor diffusion, the water diffusivity in the soil changes much more slowly with water content than implied by Eqs. (2) (Jackson et al., 1975). This limit is simply accounted for in the numerical model by not allowing the diffusivity = $2B\Phi K/s$ to drop below some limit D_{min} set equal to 10^{-9} m^2 sec^{-1}. The parameter K_0 ranges from about 10^{-4} m sec^{-1} or more for sandy soils tc about 10^{-6} m sec^{-1} or less for clay-like soils and B is typically about 5, but is sometimes smaller for sandy soils and usually larger for clay-like soils (Clapp and Hornberger, 1978). The parameter $\Phi 0$ does not seem to vary systematically with soil type except for being somewhat smaller for sandy soils and is nominally assumed to be 0.5 m. Solution of the multi-layer soil water model shows that sandy soils can have diffusion-limited maximum evaporation rates of about 2×10^{-7} m sec^{-1} (2 cm day^{-1}) or greater, whereas clay-like soils can have maximum evaporation rates of about 10^{-8} m sec^{-1} (0.1 cm day^{-1}) or smaller. The objective here is to develop a parameterization that can encompass this range of possibilities and provides the decrease of evapotranspiration with increasing soil dryness (approximately proportional to $t^{-1/2}$, where t is the number of days since drying began).

To develop a parameterization empirically, an initial water distribution near field capacity and constant with depth is assumed. A diurnally varying evaporation demand E_D is applied as a top boundary condition as obtained from

$$E_D = \rho_a C_D V(q_g - q_a) \quad .$$

It is assumed as a typical value that $\rho_a C_D$ V = 0.01 kg m^{-2} sec^{-1}. The individual terms are ρ_a = air density, C_D = surface bulk transfer coefficient, and V = mean wind near the surface. The air-water mixing ratio q_a is prescribed from an assumed air temperature and relative humidity. The ground surface mixing ratio is found from the saturation value q_g^{SAT} at a prescribed temperature of the ground T_g, using the thermodynamic expression (Phillip, 1957)

$$q_g = q_g^{SAT}(T_g) \exp(-g\Phi(0)/(R_w T_g)) \ ,$$

where R_w = water vapor gas constant, g = gravity. The evaporation over one time step is limited by the requirement that soil moisture at the surface stay positive. However, if this condition was frequently applied, the time step was reduced. In preliminary calculations, q_g in the definition of evaporation demand was assumed to be at its saturated value so that E_D = PE, the potential evaporation, i.e., the maximum evaporative demand. Even more simply, a diurnally varying PE was prescribed. The parameterizations inferred using any of the above assumptions about evaporation demand for top boundary conditions were not significantly different. The actual evaporation at rates less than the potential value is limited by the maximum rate at which water can diffuse upward to a dry surface, and so does not depend on the potential value.

After an integration begins from the prescribed initial moisture distribution, the evaporation obtained as described above is initially given by the diurnal cycle of PE. After several days, its mid-day values become diffusion-limited, with the maximum evaporation rate depending on the soil water content and other hydrological parameters. A good fit to the evaporation rate F_q is obtained from

$$F_q = \text{minimum}\ (PE,\ F_{qm}) \qquad (3)$$

where F_{qm} is given by

$$F_{qm} = C_k\ D\ s_0/(Z_0 Z_1)^{1/2} \qquad (4a)$$

where Z_0 is the depth of the soil active layer (between 0.5 and 2 m), Z_1 is the depth of the surface soil layer (restricted to be between 1 and 20 cm thickness), and where D is an average soil diffusivity defined by

$$D = 1.02\ D_{max}\ s_1^{B}\ s_0^{2}\ (s_0/s_1)^{Bf} \qquad (4b)$$

using

$$D_{max} = B\Phi_0 K_0/\rho_{wm} \quad , \qquad (4c)$$

where ρ_{wm} is the fraction of saturated soil filled by water, nominally 0.5, and K_0 and Φ_0 the maximum conductivity and soil suction of Eqs. (2),

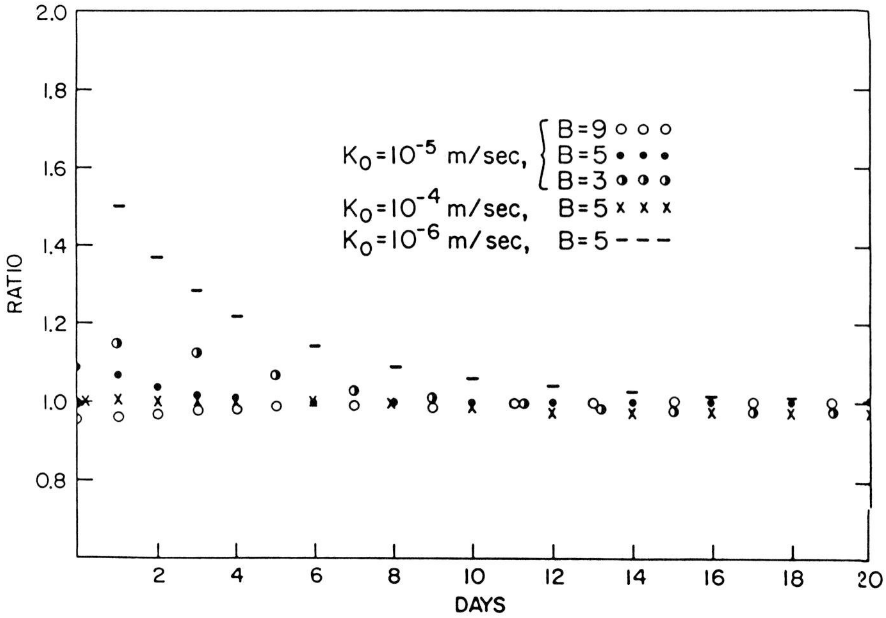

Fig. 2. Ratio of daily averaged exact to daily averaged approximate soil evaporation for three soils with: loam-like hydraulic conductivities ($K_0 = 10^{-5}$ m sec^{-1}), a sandy soil ($K_0 = 10^{-4}$ m sec^{-1}), and a clay-like soil $K_0 = 10^{-6}$ m sec^{-1}).

and defining

$$B_f = 5.5 - 0.8\ B[1+0.1(B-4)\ LOG_{10}(K_0/K_R)] \quad (4d)$$

with $K_R = 10^{-5}$ m/sec. The term C_k in Eq. (4a) is

$$C_k = (1 + 1550\ D_m/D_{max})\ \left[\frac{B - 3.7 + 5/B}{B + 5}\right] \quad (4e)$$

Eq. (3) was tested for values of Φ_0 from 0.1 to 1.0 m sec^{-1}. The nominal value of D_{min} is 10^{-9} m^2 sec^{-1} and the expression was tested for values of D_{min} from 10^{-8} m^2 sec^{-1} to 10^{-10} m^2 sec^{-1}. The general structure of the terms (4a,b,c,d,e) is indicated from dimensional analysis and physical reasoning, but their detailed structure was inferred from trial and error numerical integrations.

Fig. 2 shows examples of the ratio of exact (i.e., high resolution model) to approximate (i.e., two-layer model) soil evaporation averaged over daytime hours for 20 days. In all cases the soil was assumed to have initial soil water concentrations of s = 0.6 (60% of saturation). After a few days of adjustment time, the approximate evaporation is generally within several percent of the exact. Similar accuracy is achieved for all values of B between 1 and 12 and K_0 between 10^{-3} m sec^{-1} and 5×10^{-8} m sec^{-1} with the same initial conditions.

The error in the daily average of the parameterized evaporation is rather insensitive to the imposed daytime potential evapotranspiration. However, for maximum rates of daytime potential evaporation, the instantaneous parameterized evaporation rates may be significantly in error. When parameterized rates are less than potential evaporation, they are nearly flat across the course of the day. However, the exact evaporation develops a morning pulse of evaporation of moisture that had been diffused upward the previous night, which may exceed the parameterized rates by 50% or more within the first two hours after sunrise. In the afternoon the exact evaporation rates then drop below the parameterized rates by 10-20% or so.

Tests were made of the parameterized evaporation with different soil moisture, e.g., s = 0.4 and s = 0.5 as initial wet soils and s = 0.2 as an initial dry soil. The approximate expression usually overestimated the actual evaporation for wet soils and underestimated if for dry soils. However, for light to medium texture soils, i.e., $K_0 > 10^{-5}$ m sec^{-1}, the error was acceptable (< 5% for the wet case, < 50% for the dry case). The actual rate of evaporation for the dry case is an order of magnitude smaller than that of the wet case so these errors are small compared to the range of evaporation that has been parameterized. For $K_0 = 10^{-6}$ m sec^{-1}, the daily mean evaporation was overestimated by about 20% for the wet case and underestimated by a factor of 4 for the dry case. This factor of 4 error is largely a result of the soil diffusivity reaching its lower limit throughout the thick layer near the surface and is reduced to a 40% overestimate if the minimum diffusivity D_{min} is reduced by an order of magnitude. Errors similar to those described above were obtained if only the soil below 0.5 m was initially wet or dry. It is believed that, in practice, the largest source of error in applying the parameterization will come from difficulties in specifying the soil hydraulic properties due to their vertical and horizontal variability.

During the process of evaporation in response to daytime heating, the soil moisture distribution approaches a self-similar profile. For such a profile, the water loss from the top layer is proportional to the water loss from the total active layer, with a proportionality factor that is just the ratio of the water contents within these two layers, i.e., $(s_1 Z_1)/(s_0 Z_0)$. In that case, the down-gradient rate of transfer of water from the lower soil layer to the surface layer, denoted T_w, is given by

$$T_w = T_{w_0} = C_k\, D\, (s_0 - \varepsilon\, s_1)/(Z_0 Z_1)^{1/2} \quad (5a)$$

where $\varepsilon = Z_1/Z_0$ is the ratio of the depth of the upper layer to the depth of the total active soil layer. Although (5a) with $\varepsilon = Z_1/Z_0$ works well for the assumed regime of daily evaporation, it gives unrealistic results during infiltration when s_1 exceeds s_0. With infiltration it is more realistic to assume that the downward transfer by diffusion is proportional to $(s_0 - s_1)$. To interpolate between these two limits, I assume

$$\varepsilon = \frac{Z_1/Z_0}{\text{maximum of } \left[Z_1/Z_0 \text{ or } (1-s_1/s_0)\div(1-\hat{s}_1/\hat{s}_0)\right]} \quad (5b)$$

where $\hat{s}_1/\hat{s}_0$ is the ratio of s_1 to s_0 during evaporation with a self-similar profile. For standard parameters $B = 5$, $K_0 = 10^{-5}$ m/s, $Z_1 = 0.1$ m and $Z_0 = 1$ m, $s_1/s_0 = 0.8$. An additional gravitational drainage term is also added so that the model water transfer between layers is evaluated from

$$T_w = T_{w_0} - (s_1/s_0)^{B+1.5}\, R_g \quad (5c)$$

where $R_g = K_0\, s_0^{2B+3}$ is the rate of gravitational drainage of soil water from the active soil layer to underlying ground water. The geometric mean of s_1 and s_0 has been assumed for the value of s at the interface between the upper and lower layer. The possibility of a gradient-driven upward water flow from an underlying saturated zone is neglected. With use of Eq. (5a,b,c) water in the surface layer is generally transferred downward as rapidly as generated by rainfall of GCM grid square intensity and for soils other than heavy clay. Hence, only infrequently does water pond on the surface due to saturation of the surface layer. It is necessary to assume additional surface runoff near saturation, tuned to get the observed surface runoff (globally about half the total runoff), or to assume a subgrid-scale variation of rainfall intensity.

Water Flow Through Plants

Parameters Required to Model Plant Water Budgets

The following parameters are used here for including the effects of vegetation in GCMs: (a) the fraction of a model grid square covered (i.e., shaded) by vegetation (denoted σ_f), (b) the surface area of transpiring vegetation per unit vegetated area, referred to as the leaf area index LAI. (The common definition of projected area is used for the LAI, corresponding to half the total area for flat leaves and the total area divided by π for cylindrical needles.) (c) the surface area of nontranspiring vegetation, referred to as the stem area index SAI, and defined in the same manner as the LAI. The SAI parameter includes not only stems but also standing dead leaves. Fallen vegetative material in contact with the ground could also be included as part of the SAI, but if important perhaps should be parameterized as an additional moisture holding layer. The total foliage or leaf stem area index (LSAI) is defined as the sum, LSAI = LAI + SAI.

Vegetation Storage of Intercepted Precipitation and Dew

When it rains, the surfaces of vegetation become covered with a film of water before drip through and stem flow carries water to the ground. This water can then reevaporate to the air, but at the same time transpiration is suppressed from the surface of wet green leaves. Similarly, the formation of nighttime dew can keep foliage cool in the morning and likewise suppress transpiration. Typical values for reevaporation of intercepted rainfall are in the range of 5 to 50% of rainfall, depending primarily on rainfall intensity and duration (e.g., Rutter, 1975). The suppression of transpiration by wet leaves has been less studied but may also be significant.

Snowfall is also intercepted by foliage and frost formation on foliage commonly occurs. Intercepted snow and frost are of somewhat less significance for the water budget than are intercepted rain and dew because of lower evapotranspiration rates at low temperatures, but they can significantly modify canopy albedos. The parameterization described here assumes vegetation storage of solid water is the same as liquid water and ignores blow off. In practice, more water can be stored on leaves in the form of snow than of liquid. On the other hand, snow is susceptible to blowing off.

The maximum water storage per unit projected area of plant material is taken to be 0.02 cm. The water stored on foliage per unit land surface area denoted W_f is calculated from the sum of the incident precipitation P and the transpiration E_{tr} minus the water flux E_f from the plant surface, i.e.

$$\frac{\partial W_f}{\partial t} = [\sigma_f P - E_f + E_{tr}] \tag{6}$$

If $W_f > W_{fmax} = 0.02$ cm × σ_f × LSAI, W_f is set equal to W_{fmax} and the excess leaf moisture is added to the precipitation on the soil, either as water or snow. The numerical factor in W_{fmax} is inferred from the range of values for total canopy storage quoted in Rutter (1975), i.e., 0.4 to 2.0 mm. With typical values of $\sigma_f = 0.8$ and LSAI = 6, W_{fmax} = 1 mm. More general drip formulas are discussed by Massman (1980).

Leaf Evapotranspiration

I first discuss the evaporation from wet foliage. The water flux from dry foliage follows similar considerations, but in addition, the resistance to water flux by stomata needs consideration. The water on wet foliage (leaves and stems) evaporates according to

$$E_f^{WET} = A_f \rho_a r_{\ell a}^{-1} (q_f^{SAT} - q_{af}) \tag{7}$$

where A_f is the projected area of the wet surface, which for the whole wet canopy is σ_f × LSAI, q_f^{SAT} is the saturation water vapor mixing ratio at temperature of the foliage T_f, q_{af} is the water vapor mixing ratio of air within the canopy, and $r_{\ell a}$ is the aerodynamic resistance to moisture and heat flux of the foliage molecular boundary layer per unit foliage projected area. Eq. (7), if negative, gives the rate of accumulation of dew on foliage whether it is wet or not.

I assume for the conductance for heat and vapor flux from leaves an expression

$$r_{\ell a}^{-1} = C_f \times (U_{af}/D_f)^{1/2} \tag{8a}$$

where $C_f = 0.01$ m sec$^{-1/2}$, D_f is the characteristic dimension of the leaves in the direction of wind flow, and U_{af} is the magnitude of the wind velocity incident on the leaves. Eq. (8a) corresponds to models of laminar boundary layers past flat plates (as discussed in Gates (1980)). Eq. (7) gives the evaporation at each surface within the canopy; however a different definition of D_f in (8a) would be needed to apply it to stems and needles. It is also assumed that Eq. (7) can be applied to the canopy as a whole.

The exact form for the foliage conductance and the most appropriate procedure to match fluxes inside the canopy to the air above the canopy is somewhat controversial, in part because of the scarcity of data. Brutsaert (1979) has developed an analytic solution for the transport of a tracer into a canopy, derived from an eddy mixing model and the assumption that the value of the tracer is held fixed on the leaves. His theory can be used to infer a reasonable definition of U_{af} in Eq. (8a). He finds that the effective leaf conductance per unit leaf area $r_{\ell a}^{-1}$ is proportional to $C_t u_*$ where C_t is a bulk transfer coefficient for the foliage elements and u_* is the friction velocity ($u_* = C_D^{1/2} V_a$, where C_D and V_a are the drag coefficient and mean wind above the canopy evaluated at some level within the surface mixed layer). The proportionality factor is determined by the additional resistance due to the eddy mixing and is typically about one-half. In his model, the wind at the top interface of the canopy is 2.5 u_*. Brutsaert questions the correctness of the laminar foliage boundary layer model adopted here and in fitting one data set for artificial grass finds C_t to vary with u_* to the minus one-quarter power.

The largest departures of foliage temperature and humidity from the air temperature and humidity are likely to occur in the upper part of the canopy, because radiative heating and cooling are there largest. With such a radiatively driven profile of temperature and humidity variation, the rates of heat and moisture transfer from the canopy could be larger than inferred from Brutsaert's theory. Thus it is not unreasonable to assume the above mentioned proportionality factor to be unity for heat and moisture, as implied by the canopy energy and moisture balance relationships assumed in the next section. For $u_* = 0.5$ m sec^{-1} and $D_f = 0.04$ m, Eq. (8a) agrees numerically with the value of $C_t u_*$ inferred by Brutsaert, provided

$$U_{af} = u_* \tag{8b}$$

The assumption that $U_{af} = u_*$ in general is of adequate accuracy for the present model, considering the uncertainty in the other aspects of the foliage conductance.

Two further parameters are needed to evaluate Eq. (7), i.e., the foliage temperature T_f and the water mixing ratio q_{af} for air within the canopy. These are obtained from energy and mass balance requirements detailed in the next section.

The flux E_f from foliage that is only partly wet, with wetted fraction denoted f_w, now follows from Eq. (7) as

$$E_f = r'' E_f^{WET} \tag{9a}$$

where

$$r'' = \left[1 - \delta(E_f^{WET}) \left(1 - f_w - f_d \left(\frac{r_{\ell a}}{r_{\ell a} + r_s}\right)\right)\right], \tag{9b}$$

where r_s = 'stomatal' resistance to be discussed further below, $f_d = (1 - f_w)$ × LAI/LSAI is the fraction of canopy surface which is transpiring and where δ is a step function, and is one for positive argument, and zero for negative argument. It has been assumed that the fraction of wetted area over nontranspiring foliage is the same as for transpiring foliage. Both fractions are calculated from the ratio of foliage water W_f to its maximum value, using the formula proposed by Deardorff (1978). The general framework used here for partially wetted vegetation is also that proposed by Deardorff.

Transpiration E_{tr} occurs only from dry leaf surfaces and is only outward, i.e.,

$$E_{tr} = \delta(E_f^{WET})\, f_d\left(\frac{r_{\ell a}}{r_{\ell a} + r_s}\right) E_f^{WET} \qquad (10)$$

The above formulation neglects possible temperature differences between wet and dry surfaces, as well as the effects of several other possible inhomogeneities. For example, many plants, such as hardwood trees, have leaf stomata that occur only on the underside of leaves, whereas leaf water may be stored largely on the upper surface of the leaves.

Stomatal Resistance

The term stomatal resistance, as used here, refers to the total mechanical resistance encountered by diffusion from inside a leaf to outside. This term is sometimes referred to as leaf resistance to distinguish it from the resistance due to the stomata alone. Water vapor inside leaves is maintained at or very near its saturated value, for otherwise the mesophyllic cells of the leaf would desiccate and the leaf wilt. The stomata are pores which when open are the main conduits for transpired water. Hence, the net resistance r_s to water passing from the inside to the outside of the leaf depends largely on the size, distribution, and degree of opening of these stomata. However, some water diffusion also occurs through leaf cuticles, which can be the primary route for transpiration when the stomata are closed. In general, the opening of the stomata and hence r_s changes with various environmental parameters, including inability of roots to supply adequately the transpiration demand.

The stomatal resistance factor is taken to be

$$r_s = r_{smin} \times R_\ell \times S_\ell \times M_\ell \;. \qquad (11)$$

The factors on the right hand side are discussed, e.g., by Hinkley et al. (1978); R_ℓ gives the dependence of r_s on solar radiation. Studies have been made of the dependence of r_s on visible fluxes (c.f., e.g., Hinckley et al., 1978, Fig. 3; Watts, 1977; Figs. 10-11; Denmead and Millar, 1976, Fig. 1). These studies show that a minimum stomatal resistance is reached from between 10% of full sun (approximately true for the trees studied) and full sun as indicated by Denmead for wheat. Full sun corresponds to about 500 W m^{-2} of incident visible solar radiation (λ < 0.7 μm). The factor R_ℓ varies between about 1 for overhead sun and r_{smax}/r_{smin} for nighttime, where r_{smax} is the cuticular resistance of the leaves. It is assumed to be of the form

$$R_\ell = \frac{1 + f}{f + r_{smin}/r_{smax}} \qquad (12)$$

where $f = F_v/F_{vc}$ with r_{smax} = 5000 sec/m, F_v = flux of visible solar radiation, and F_{vc} the visible solar flux for which R_ℓ is about double its minimum value. Present model values are F_{vc} = 30 W m^{-2} for trees, and F_{vc} = 100 W m^{-2} for grasslands and crops. The dependence of stomatal resistance on vapor pressure deficit is not included because it appears to be highly variable from species to species (mechanistic models for the dependence of stomatal resistance on vapor pressure deficit have been reviewed by Farquhar, 1978). Hence, r_{smin}, which takes different values for different kinds of vegetative cover, is intended to apply for average vapor pressure deficit conditions.

Eq. (12) is averaged over the different leaves in the canopy which receive different amounts of radiation. The canopy is divided into an upper and lower fraction and it is assumed that 0.75 of the radiation is absorbed by the upper canopy and 0.25 is absorbed by the lower canopy, and that the leaf-stem area index of upper (LSAIU) and lower (LSAIL) canopy are each one-half of the total leaf-stem area index, i.e., LSAIU = LSAIL = 0.5 LSAI, and hence

$$F_{vu} = 0.75\, F_{vi}/\mathrm{LSAIU} \;,$$

$$F_{vl} = 0.25\, F_{vi}/\mathrm{LSAIL} \;,$$

where F_{vi} is the visible radiation incident at the top of the canopy. Eq. (12) is applied using F_{vu} and then F_{vl} for F_v and the resulting conductivities averaged to get R_ℓ in the model. Singh and Szeicz (1980) have described a similar division of canopy into sunlight and shaded parts in determining stomatal resistance.

The seasonal temperature factor S_ℓ depends on foliage temperature T_f and varies from 1 at summer conditions to r_{smax}/r_{smin} for $T_f \leq$ 273K. The moisture factor M_ℓ depends on the soil moisture and the ability of plant roots to take water readily from the soil for a given level of root moisture. Initially, M_ℓ = 1, and if the plants' transpiration exceeds a maximum value, depending on soil moisture as described below, M_ℓ is increased so that the transpiration is maintained at the maximum value. Finally, if r_s as given by Eq. (11) exceeds r_{smax}, it is set to r_{smax}.

Root Resistance

The transpiration rate calculated from Eq. (10) must be consistent with the maximum transpiration the vegetation can sustain as defined below. If E_{tr} is redetermined so that $E_{tr} = E_{trmx}$, the maximum transpiration available from the plant. The determination of this maximum is based on a simplification of the SPAC (soil-plant-atmosphere model) of Federer (1979). For recent reviews of root-water models, see Hillel (1980) and Molz (1981). The plant water uptake in each soil layer is limited by the difference between soil and the leaf potential divided by an effective resistance. This effective resistance depends on the total length of root per unit area and the internal plant resistance per unit root length. When the soil is dry enough, the diffusion of water from the soil to the roots also contributes to this resistance. Lumping various terms together yields the following formula

$$E_{trmx} = T_{r_0} \sum_i R_{ti} (1 - WLT_i) \qquad (13)$$

where T_{r_0} is the maximum transpiration that can be sustained. The sum is over soil layers, each designated by subscript i, R_{ti} is the fraction of roots in a given soil layer, and WLT_i is a soil dryness (or plant wilting) factor. I have taken $T_{r_0} = (2 \times 10^{-7}$ m/sec) $\times \sigma_f \times$ SEASB, where SEASB is a factor that is one during the growing season and drops to zero when the soil is frozen. For typical vegetation $\sigma_f \simeq 0.8$, and near field capacity, $WLT_i \simeq 0.05$ so that typically for well-watered growing season vegetation, $T_{r_0} \simeq 1.5 \times 10^{-7}$ m sec^{-1}, corresponding to a maximum latent energy flux of about 380 W m^{-2}. Observed latent fluxes from unwetted foliage under intense solar radiation rarely exceed 400 W m^{-2}. Federer (1979), modeling data for birch and maple forests, cf. his Fig. 1, finds maximum transpiration of about 1.5×10^{-7} m/s, although with sufficient root capacity, the transpiration could have exceeded 2.0×10^{-7} m/s.

The negative of the leaf potential can be approximated by the maximum value it takes before leaf desiccation. The term WLT then is $(\Phi - \Phi_0) \div (\Phi_{max} - \Phi_0)$ where Φ is the soil water suction and Φ_{max} is the negative of the minimum value of potential of leaves before desiccation. Hence, for soil suction of the form in Eqs. (2)

$$WLT_i = \frac{s_i^{-B} - 1}{s_w^{-B} - 1} \qquad (14)$$

where s_i is the soil water of the i'th layer, and s_w the soil water for which transpiration goes to zero. The current model value is $s_w = 0.125$. Normally, soils of clay-like texture would have larger values and sandy textured soils lower values. The factor WLT_i ranges from 0 at saturation to 1 at the "permanent wilting point".

The ratio of maximum water extraction given by Eq. (14) for a given soil layer to the maximum water extraction for the total active soil layer multiplied by the fraction of roots in the given layer gives the fraction of total transpiration that is taken from the given layer. With a nominal two-layer soil structure and the first 0.1 of the active soil layer designated the upper layer, the density of roots in the upper layer is assumed to be five times the total root density, or equivalently, 50% of the roots are then in the upper layer.

Energy Balance of Plant Canopy and Soil

The air within the canopy has negligible heat capacity and so heat flux from the foliage H_f and from the ground H_g must be balanced by heat flux to the atmosphere H_a, i.e.,

$$H_a = H_f + H_g \qquad (15)$$

where flux to the atmosphere is given by

$$H_a = \rho_a C_p C_D V_a (T_{af} - T_a) \qquad (16)$$

where C_p is the specific heat of air, V_a is the magnitude of atmospheric wind above the canopy, and C_D is the aerodynamic bulk transfer coefficient between canopy air and atmosphere above, assumed to be the same for heat and moisture as for momentum. To follow more closely the analysis of Brutsaert (1979), the canopy bulk transfer coefficients for heat and water should be referred to the difference between atmospheric values and the values of temperature and water vapor at the top of the canopy, by multiplying C_D by $(1.0 - \delta)^{-1}$, using $\delta = \ln [(h-d)/z_0] / \ln [(z-d)/z_0]$ and where h is the canopy height, d the zero displacement level, z_0 the momentum roughness length, and z the height level above the canopy where C_D is defined. For $z >> h$ this correction factor can be approximated by 1.

The heat flux from the foliage is given by

$$H_f = \sigma_f \, LSAI \, r_{\ell a}^{-1} \rho_a C_p (T_f - T_{af}) \, . \qquad (17)$$

The flux from the ground is assumed to be

$$H_g = \rho_a C_p C_D \left((1-\sigma_f) V_a + \sigma_f U_{af} \right) \left(T_g - T_{af} \right) \qquad (18)$$

Eqs. (15) to (18) are solved for T_{af}, i.e.,

$$T_{af} = (c_A T_a + c_F T_f + c_G T_g) \div (c_A + c_F + c_G) \qquad (19)$$

where

$$c_A = C_D V_a \, , \qquad (20a)$$

$$c_F = \sigma_f \, LSAI \, r_{\ell a}^{-1} \qquad (20b)$$

$$c_G = C_D \left((1 - \sigma_f) V_a + \sigma_f U_{af} \right) \qquad (20c)$$

are conductances for heat flux, to the atmosphere above the canopy, and from foliage, and the ground, respectively. Similarly, the canopy air is assumed to have zero capacity for water vapor storage so that the flux of water from the canopy air E_a balances the flux from the foliage E_f and the flux from the ground E_g, i.e.,

$$E_a = E_f + E_g \quad , \tag{21}$$

where E_f was defined earlier, cf. Eq. (9a,b)

$$E_a = \rho_a \, c_A \, (q_{af} - q_a) \tag{22}$$

$$E_g = \rho_a \, c_G \, f_g \, (q_g - q_{af}) \quad , \tag{23}$$

where q_g is the saturated ground water vapor concentration and f_g is a wetness factor, defined as the ratio of actual to potential ground evaporation as obtained from the soil moisture parameterization. Eq. (21) is solved for q_{af} using the definitions of E_a, E_f, E_g, i.e.,

$$q_{af} = (c_A q_a + c_V q_f + c_G f_g q_g) \div (c_A + c_V + f_g c_G), \tag{24}$$

where $c_V = r'' \, c_F$ is the average conductance of foliage to water vapor flux.

The net energy flux to the ground over a model grid square is calculated as follows. Ground albedo is prescribed separately for visible and near-infrared solar fuxes and, in general, depends on water content of the surface soil layer and on the solar zenith angle. As in Deardorff (1978), net radiative flux incident on the ground is taken to be $(1 - \sigma_f)$ of that which would exist in the absence of vegetation plus σ_f times net black body emission between the ground and the foliage at temperature T_f. The ground temperature is calculated using Deardorff's force restore method. The subsoil is assumed to be an energy conserving reservoir which by exchanges of heat with the surface soil relaxes to surface ground temperature at a rate of $(5 \text{ day})^{-1}$. Soil conductivity and heat capacity are calculated from the model soil moisture, and prescribed quartz and organic soil fractions, using simple fits to the theory of deVries (1963; 1975). Modifications applied to the energy balance calculations for snow cover include partial covering of the vegetation by the snow which reduces the effective vegetative cover, appropriate modifications of the ground albedo, and of the ground conductivity and heat capacity as used in the force restore method. Freezing and thawing of the ground are included in the energy budgets for surface and subsurface soil (Dickinson et al., 1981).

The final budget equation required to obtain the transpiration from vegetation is an equation of foliage energy conservation, i.e.,

$$R_n(T_f) = L \, E_f(T_f) + H_f(T_f) \quad , \tag{25a}$$

where R_n is the net radiation absorbed by the foliage (Dickinson et al., 1981). Eq. (25) is solved for foliage temperature T_f by Newton-Raphson iteration. In the model formulation, the emissivity of the ground and foliage is assumed to be unity, the foliage is assumed to have zero heat capacity and photosynthetic and respiratory energy transformations are neglected.

Let $\hat{R}_n$ be net radiation that would be absorbed by the foliage if the foliage was at the temperature of the ground, i.e.,

$$R_n = R_n + 2\sigma_f \sigma \, (T_f^4 - T_g^4) \quad , \tag{25b}$$

where σ = the Stefan-Boltzman constant. The term $(T_f^4 - T_g^4)$ is approximated by $4T_g^3(T_f - T_g)$. Let the superscript N denote quantities evaluated on the N'th iteration. The foliage temperature at the next iteration is then obtained from

$$T_f^{N+1} = \frac{\hat{R}_n + \sigma_f \, (8\sigma T_g^4) + c_H \, (c_A T_a + f_g c_G T_g) - L_V \hat{E}_f^{N+1}}{\left[\sigma_f \, (8\sigma T_g^3) + c_H \, (c_A + f_g \, c_G) + L_V \dfrac{\partial E_f^N}{\partial T_f} \right]} \tag{26a}$$

where L_V is the latent heat of vaporization, and where

$$\hat{E}_f^{N+1} = E_f^N - \frac{\partial E_f^N}{\partial T_f} T_f^N \tag{26b}$$

$$c_H = c_F/(c_A + c_F + f_g \, c_G) \quad . \tag{26c}$$

Alternatively, if fluxes to the ground are negligible. Eqs. (15), (16), (17), (21), (22), (24) and (25) can be used to eliminate all the temperatures and obtain the evapotranspiration E_a from a Monteith combination formula

$$E_a = \frac{L_V^{-1} \, R_n + B_e \, \rho_a \, \Delta q/(r_A + r_F)}{1 + (r_A + r_V) \, B_e/(r_A + r_F)} \tag{27}$$

where Δq = the difference between saturated and ambient water vapor mixing ratios above the canopy, B_e is the equilibrium Bowen ratio = $C_p/(L_V \, \partial q^{SAT}/\partial T)$ (e.g., Priestly and Taylor, 1972), and r_A, r_F and r_V are, respectively, the resistances to bring air into the canopy, and to move heat and water vapor to the foliage, i.e.,

$$r_A = c_A^{-1} = (C_D V_a)^{-1} \tag{28}$$

$$r_F = c_F^{-1} = (\sigma_f \, \text{LSAI})^{-1} \, r_{\ell a} \tag{29}$$

$$r_V = c_V^{-1} = \frac{(r_{\ell a} + r_s)}{(\sigma_f \, \text{LSAI}) \, (f_d + f_w \, (1 + r_s/r_{\ell a}))} \tag{30}$$

Eq. (27) can also be applied to rate of dew formation ($E_a < 0$) provided r_V in Eq. (30) is equated to $r_F = r_{\ell a}/(\sigma_f LSAI)$. Eq. (27) was first derived by Thom (1972) for dry leaves and by Shuttleworth (1977) for partially wetted leaves (with different notation). Its dry form is apparently equivalent to what Sinclair et al. (1976) used as their "big leaf model." They showed that their "big leaf model" gave rates of evapotranspiration and photosynthesis in excellent agreement with a high resolution multilayer model for a corn canopy. The Sinclair et al., study thus provides some validation for the present parameterization as a reasonable description of evapotranspiration for a completely vegetated area. For partial vegetation, the model could be improved by distinguishing between the temperature of shaded and unshaded ground. However, this addition would require specifying the diurnal course of shading of the ground, and hence more detail with regard to canopy structure than now necessary with T_g presumed to be an average of shaded and unshaded ground.

It also should be noted that the assumption used above that wet and dry surfaces within a canopy have the same temperature is somewhat unrealistic. This assumption is referred to as the "single canopy limit" by Shuttleworth (1978) who suggests the "double canopy limit" of an independent calculation of the wet and dry surface temperatures could be more realistic.

Another somewhat questionable parameter is the "foliage conductance or resistance" term defined by Eqs. (8a,b), (29). Besides the evidence previously mentioned that $u_* r_F$ should depend more weakly on u_* than $\sim u_*^{1/2}$, there appears to be little evidence one way or another for the dependence on canopy surface area, the LSAI term. An alternative expression for $u_* r_F$ is

$$u_* r_F = k^{-1} \ln(z_0/z_{H,V}) \qquad (31)$$

where $k = 0.4$ is the von Karman constant, and $z_{H,V}$ is the temperature or moisture roughness length. Garrett (1978) reports observational values of z_0/z_H to lie in the range 12 ± 6 for savannahlike vegetation with 25% tall trees, and that previous values for forests were near 3. The present representation for $u_* r_F$ as obtained from Eqs. (8a,b) with $u_* = 0.5$ m/sec, $D_f = 0.04$ m, becomes

$$u_* r_F = 14/(\sigma_f LSAI) \qquad (32)$$

Eq. (32) equated to Eq. (31) for sparse vegetation ($\sigma_f LSAI = 2.$) gives $z_0/z_H = 16$, and for more dense vegetation ($\sigma_f LSAI = 5.$), $z_0/z_H = 3.$, so that for typical eddy momentum flux to the surface and foliage length scales, Eq. (29) gives a foliage resistance to sensible heat fluxes comparable to that observed.

Finally, it should be noted that the connection between canopy vapor resistance as defined in Eq. (30) and the stomatal or leaf surface resistance term r_s is only qualitatively supported by detailed observational studies.

Discussion and Conclusions

This paper reports parameterizations for surface hydrological processes in global climate models. The essential concept of both soil evaporation and plant transpiration is that fluxes are given either by the demand by atmospheric processes or by the maximum supply rate provided by the soil or vegetation. The demand rate is obtained through modeling of surface energy balance, water vapor concentrations, temperatures, and winds. Refinements in the parameterizations allow the model demand to adjust to the supply when water flow is diffusion-limited. Soil rainfall infiltration and surface runoff are but crudely treated and could readily be improved, but probably without much impact on other aspects of the surface energy balance processes. The question of subgrid scale rainfall distribution also needs to be considered.

Use of these parameterizations together with a diurnal cycle of solar radiation should lead to a physically more realistic description of surface energy balance processes and evapotranspiration than now incorporated in GCMs. Improvements in GCM performance should be most notable for near ground and ground climate parameters and might not be achieved unless the model also includes high quality simulations of vertical transport in the planetary boundary layer, atmospheric rainfall processes, cloudiness and incidence of solar and longwave radiation at the surface.

Application of the parameterizations described in this paper requires prescription of soil hydraulic parameters and properties of vegetation over model grid areas. Most surface processes as well as convective rainfall have much smaller spatial structure than the grid scale of global models. Further improvements in the parameterization of land evapotranspiration in global models may require recognition of other statistical properties than just average values of surface parameters. However, at present it is difficult to prescribe even the needed grid square average parameters due to lack of suitable global data sets for land surface properties. The present parameterizations are already more complex than most GCM modelers would be comfortable with. Yet they are incredibly oversimplified compared to the actual processes that occur. Also, it is not clear that they can lead to any significant improvements in GCM simulations unless other aspects of model simulations, such as cloudiness, rainfall, and planetary boundary layer dynamics, are improved. How can further progress be made?

At this point, it is crucial that thorough sensitivity studies be carried out to determine the relative importance for climate simulations of the parameters included in the land surface parameterizations such as described in this paper. It is likely that model results will turn out to be insensitive to the precise values of at least some of the soil and vegetation parameters. On the other hand, the processes and parameters on which model results most sensitively depend are prime candidates for further improvements in the parameterizations and for emphasis in collection of future data sets. Achieving these future improvements will require collaboration with other disciplines that have traditionally been more concerned with surface processes such as evapotranspiration than have global climate modelers.

This paper has discussed only the parameterization of fluxes of heat and moisture from vegetated surfaces. However, the same general framework applies to the dry deposition of various trace gases and aerosols of interest. In particular, the flux to vegetated surfaces of ozone, sulfur dioxide, and nitrogen dioxide, depend in part on the stomatal resistance of green foliage (e.g., Wesley and Hicks, 1977; Wesley et al., 1982). Thus the parameterizations of this paper provide a suitable framework for the land surface sink component of global tropospheric chemistry models. Plant uptake of carbon dioxide also occurs through stomata and can be modeled with the use of a photosynthesis parameterization for describing the CO_2 concentrations within the leaf cavity (e.g., Gates, 1980).

In conclusion, it is expected that inclusion of a realistic description of land surface processes, such as attempted in this paper, should significantly enhance the credibility of GCMs in their application to questions of climate change due to such external perturbations as increasing concentrations of atmospheric carbon dioxide, and especially to questions of the impact of land surface changes. However, better GCM simulations may also require improvements in other aspects of model formulation. The question as to what further improvements are needed in a land evapotranspiration parameterization including better data sets can now best be answered by sensitivity studies with parameterizations of the level of detail presented in this paper and using atmospheric models of varying complexity and dimensions.

Acknowledgments. The author is grateful for the many comments and suggestions given on the manuscript by his colleagues, including R. Chervin, B. Hanson, P. Kennedy, A. H. Sellers, M. Wilson, M. Verstraete, and an anonymous reviewer.

Symbols

A_f projected area of wet surface used in Eq. (7).

B soil parameter defining change with soil water of soil water potential and hydraulic conductivity, cf. Eqs. (2).

B_e equilibrium Bowen ratio.

B_f nondimensional parameter for soil water diffusion parameterization as defined by Eq. (4d).

$c_{A,F,V}$ various bulk conductances as defined in text, Eqs. (28)-(30).

C_D aerodynamic bulk transfer coefficient between canopy air or unvegetated soil and atmosphere above.

C_f factor in Eq. (8a) equal to 0.01 m sec^{-1}.

C_G effective resistance of ground to flux of heat and moisture.

C_k nondimensional parameter for soil water diffusion parameterization as defined by Eq. (4e).

C_p specific heat of air per unit mass.

C_t bulk transfer coefficient for the foliage elements as used by Brutsaert, (1979).

d zero displacement level for surface mixed layer wind profile.

D average soil diffusivity as defined by Eq. (4b).

D_f characteristic dimension of foliage elements (leaves) in direction of wind flow.

D_{min} minimum value of soil diffusivity.

D_{max} maximum value of soil diffusivity = $B\Phi K/\rho_{wm}$.

E_D evaporation demand as determined by the air-ground humidity difference.

$E_{a,f,g}$ water flux per unit land area; f refers to origin at foliage, g at ground, and a to the total flux; a superscript N refers to the flux at the Nth iteration for leaf temperature.

E_f^{WET} water flux per unit land area that would occur if the foliage were totally wet.

E_{tr} transpiration rate per unit land area.

E_{trmx} maximum transpiration as allowed by root resistance per unit land area.

f ratio of F_v to F_{vc}.

f_d fraction of foliage that is green and unwetted.

f_w fraction of wet foliage.

F_{qm} maximum soil evaporation as limited by diffusion.

F_v flux of visible solar radiation incident on leaf surfaces; added to the v, i indicates flux at the top of the canopy, u flux on leaves in the upper canopy and l flux on leaves in the lower canopy.

F_{vc} flux of visible solar radiation at which stomatal resistance is double its minimum value.

g gravity.

h height of canopy.

$H_{a,f,g}$ heat flux per unit land area to the atmosphere; f refers to origin at foliage, g at ground, a to the total flux.

k von Karman constant.

K(z) soil hydraulic conductivity.

K_0 soil hydraulic conductivity at saturation.

LAI projected area of green leaf per unit surface area.

LSAI total projected area of foliage per unit surface area.

L_v latent heat of vaporization.

M_ℓ factor giving the dependence of stomatal resistance on root resistance.

PE potential evaporation, i.e., evaporation that would occur from a saturated ground as the same temperature as present.

q_a water vapor mixing ratio at reference level above unvegetated soil or above canopy.

q_{af} water vapor mixing ratio of canopy air.

q_g saturated water vapor concentration immediately adjacent to ground.

r" ratio of water vapor flux from foliage to the flux if the surface were wet at the same temperature, cf. Eq. (9b).

r_A resistance to bring air into canopy.

r_F $r_{\ell a}/(\sigma_f \mathrm{LSAI})$ bulk foliage resistance.

$r_{\ell a}$ resistance to heat or moisture transfer through the laminar boundary layer at the foliage surface.

r_s resistance to transfer of water vapor from the inside to the outside of leaves.

$r_{s\,max}$ the maximum value of r_s as implied by the resistance of the leaf cuticles.

$r_{s\,min}$ the minimum value of r_s occurring under average water vapor conditions and otherwise optimum conditions for transpiration.

r_v average resistance for transfer of water vapor from foliage due to stomatal and aerodynamic resistance of the foliage surface.

R_ℓ factor giving the dependence of stomatal resistance on solar radiation.

$\hat{R}_n$ net radiation absorbed by the foliage.
R_n net radiation that would be absorbed by the foliage if the foliage was at the temperature of the ground, cf. Eq. (25b).

R_{ti} fraction of roots in the i'th soil layer.

R_w water vapor gas constant.

s ratio of soil water volume to soil water volume at saturation; subscript i denotes value for i'th layer.

s_0 average value of s in soil surface layer (nominally the top 0.1 m of soil).

s_1 average value of s in the total active soil layer (nominally the top 1 m of soil).

s_w value of s at which transpiration goes to zero.

S_ℓ factor giving the dependence of stomatal resistance on season.

T_a temperature of the air above the canopy.

T_{af} temperature of the canopy air.

T_f foliage temperature; superscript N refers to its value on the N'th iteration.

T_g temperature of the ground at the surface.

T_{r0} maximum transpiration that can be sustained.

T_w rate of transfer of water by diffusion to the upper soil layer from the lower soil layer; subscript 0 denotes the rate in the absence of gravity.

u_* friction velocity $C_D^{1/2} V_a$.

U_{af} canopy air mean wind speed (set equal to u_*).

V, V_a magnitude of atmospheric wind at reference level above unvegetated soil or above canopy.

W_f leaf water storage per unit land area.

W_{fmax} maximum leaf water storage per unit land area.

WLT_i soil dryness (plant wilting factor).

z soil depth or atmospheric reference level at which C_D is defined.

z_0 aerodynamic roughness length for momentum flux.

$z_{H,V}$ roughness length for sensible or latent heat fluxes.

Z_0 depth of total active soil layer, varying from 0.5 to 2. m in thickness.

Z_1 depth of surface soil layer, varying from 1 to 20 cm in thickness.

$\delta()$ Step function, is one for positive argument and zero otherwise.

ρ_a density of air at reference level.

ρ_{wm} nondimensional water density of saturated soil equal to the fraction of voids in the soil and nominally taken to be 0.5.

σ_f fraction of grid square shaded by vegetation.

Φ_{max} negative of the minimum value of leaf potential achieved before desiccation.

$\Phi(z)$ soil water suction (negative potential).

Φ_w soil matric potential.

Φ_0 soil water suction for saturated soil.

References

Brutsaert, W., Heat and mass transfer to and from surfaces with dense vegetation or similar permeable roughness. Boundary Layer Meteorology, 16, 365-388, 1979.

Budyko, M. I., Climate and Life, Academic Press, New York, 508 pp., 1974.

Campbell, G. S., A simple method for determining unsaturated conductivity from moisture retention data, Soil Science, 117, 311-314, 1974.

Charney, J., W. Quirk, S. Chow and J. Kornfield, A comparative study of the effects of albedo change on drought in semi-arid regions, J. Atmos. Sci., 34, 1366-1385, 1977.

Clapp, R. B. and G. M. Hornberger, Empirical equations for some soil hydraulic properties, Water Res. Res., 14, 601-604, 1978.

Cowan, I. R., Transport of water in the soil-plant-atmosphere system, J. Appl. Ecol., 2, 221-229, 1965.

Deardorff, J., Efficient prediction of ground temperature and moisture with inclusion of a layer of vegetation, J. Geophys. Res., 83, 1889-1903, 1978.

Denmead, O. T. and B. D. Millar, Field studies of the conductance of wheat leaves and transpiration, Agron. J., 68, 307-311, 1976.

de Vries, D. A., Thermal properties of soils, In Physics of Plant Environment, North Holland Publishing Co., 594 pp., 1963.

de Vries, D. A., Heat transfer in soils, In Heat and Mass Transfer in the Biosphere, Part I., D.A. de Vries and N. H. Afgan, Eds., Scripts Book Co., 594 pp., 1975.

Dickinson, R. E., J. Jaeger, W. M. Washington and R. Wolski, Boundary subroutine for the NCAR global climate model, NCAR Technical Note 173+1A, 75 pp., 1981.

Eagleson, P. S., Climate, soil, and vegetation, The distribution of annual precipitation derived from observed storm sequences, Water Resour. Res., 14, 713-721, 1978.

Farquhar, G. D., Feedforward responses of stomata to humidity, Ann. J. Plant Physiol., 5, 787-800, 1978.

Federer, C. A., A soil-plant atmosphere model for transpiration and availability of soil water, Water Resour. Res., 15, 555-562, 1979.

Federer, C. A., Transpirational supply and demand: Plant, soil, and atmospheric effects evaluated by simulation, Water Resour. Res., 18, 355-362, 1982.

Freeze, R. A., A stochastic-conceptual analysis of rainfall-runoff processes on a hillslope, Water Resour. Res., 16, 391-408, 1978.

Gardner, W. R. and D. I. Hillel, The relation of external evaporative conditions to the drying of soils, J. Geophys. Res., 67, 4317-4325, 1962.

Garratt, Transfer characteristics for a heterogeneous surface of large aerodynamic roughness, Quart. J. Roy. Met. Soc., 104, 491-502, 1978.

Gates, D. M., Biophysical Ecology, Springer Verlag, New York, 611 pp., 1980.

Hillel, D., Applications of Soil Physics, Academic Press, New York, 385 pp., 1980.

Hinckley, T. M., J. P. Lassoie and S. W. Running, 1978: Temporal and spatial variations in the water status for forest trees, For. Sci. Monogr., 20, Society of American Foresters, 72 pp., 1978.

Jackson, R. D., B. A. Kimball, R. J. Reginato, S. B. Idso and F. S. Nakayama, Heat and water transfer in a nocturnal soil environment, In Heat and Mass Transfer in the Biosphere, Part I,

Transfer processes in the plant environment, D. A. deVries and N. H. Afgan, Eds., Scripts Book Co., 594 pp., 1975.

Kurbatkin, G. P., S. Manabe and D. G. Hahn, The moisture content of the continents and the intensity of summer monsoon circulation, Meteorologiya i Gidrologiya, 11, 5-11, translated as Soviet Meteorology and Hydrology, 11, 1-6, 1979.

Manabe, S., Climate and the ocean circulation: I, The atmospheric circulation and the hydrology of the earth's surface, Mon. Wea. Rev., 97, 739-744, 1969.

Massman, W. J., Water storage on forest foliage: a general model, Water Resour. Res., 16, 210-216, 1980.

Milly, P. C. D. and P. S. Eagleson, Parameterization of moisture and heat fluxes across the land surface for use in atmospheric general circulation models, MIT Report No. 279, 226 pp., 1982.

Molz, F. J., Models of water transport in the soil-plant system: a review, Water Resour. Res., 17, 1245-1260, 1981.

Murphy, C. E. Jr. and K. R. Knoerr, The evaporation of intercepted rainfall from a forest stand: An analysis by simulation, Water Resour. Res., 11, 273-280, 1975.

Penman, Natural evaporation from open water, bare soil, and grass, Proc. Roy. Soc. London, A193, 120-145, 1948.

Phillip, J. R., Evaporation, moisture and heat fields in the soil, J. Meteorol., 14, 354-366, 1957.

Phillip, J. R., Plant water relations: some physical aspects, Ann. Rev. Plant Physiol., 17, 245-268, 1966.

Priestly, H. B. and K. J. Taylor, On the assessment of surface heat flux and evaporation using large-scale parameters, Mon. Wea. Rev., 100, 81-92, 1972.

Rind, D., The influence of ground moisture conditions in North America on summer climate as modeled in the GISS GCM, Mon. Wea. Rev., 110, 1487-1493, 1982.

Rutter, A. J., The hydrological cycle in vegetation, In "Vegetation and the Atmosphere, I," J. L. Monteith, ed., 111-154, Academic Press, 1975.

Shukla, J. and Y. Mintz, Influence of land-surface evapotranspiration on the earth's climate, Science, 215, 1498-1500, 1982.

Shuttleworth, W. J., A simplified one-dimensional theoretical description of the vegetation-atmosphere interaction, Bound. Layer Meteor., 14, 3-27, 1978.

Shuttleworth, W. J. and I. R. Calder, Has the Priestly-Taylor equation any relevance to forest evaporation? J. Appl. Met., 18, 639-646, 1979.

Sinclair, T. R., C. E. Murphy, Jr. and K. R. Knoerr, Development and evaluation of simplified models for simulation, canopy photosynthesis and transpiration, J. Appl. Ecol., 13, 813-829, 1976.

Singh, B. and G. Szeicz, Predicting the canopy resistance of a mixed hardwood forest, Agric. Meteor., 21, 49-58, 1980.

Stewart, J. B., Evaporation from the wet canopy of a pine forest, Water Resour. Res., 13, 915-921, 1977.

Thom, A. S., Momentum, mass and heat exchange of vegetation, Quart. J. Roy. Meteor. Soc., 98, 124-134, 1972.

Thom, A. S. and H. R. Oliver, On Penman's equation for estimating regional evaporation, Quart. J. Roy. Meteor. Soc., 103, 345-357, 1977.

Thornwaite, C. W. and J. R. Mather, The water balance, Publications in Climatology, Vol. VIII, No. 1, Drexel Institute of Technology Laboratory of Climatology, Centerton, NJ, 1955.

Van Bavel, C. H. M. and D. I. Hillel, Calculating potential and actual evaporation from a bare soil surface by simulation of concurrent flow of water and heat, Agric. Met., 17, 453-476, 1976.

Warrick, A. W. and D. R. Nielsen, Spatial variability of soil physical properties in the field, In Hillel, 1980, loc. cit., 319-343, 1980.

Watts, W. R., Field studies of stomatal conductance, Environmental Effects in Crop Physiology, J. J. Landsberg and C. V. Cutting, Eds., Academic Press, New York, 388 pp.

Wesley, M. L. and B. B. Hicks, Some factors that affect the deposition rates of sulfur dioxide and similar gases on vegetation, J. Air. Poll. Contr. Assoc., 27, 1110-1116, 1977.

Wesley, J. L., J. A. Eastman, D. H. Stedman and E. D. Yalvac, An eddy-correlation measurement of NO_2 flux to vegetation and comparison to O_3 flux, Atmos. Environ., 16, 815-820, 1982.

THE INFLUENCE OF VEGETATION ON THE HYDROLOGIC CYCLE IN A GLOBAL CLIMATE MODEL

D. Rind

NASA Goddard Space Flight Center, Institute for Space Studies
New York, N. Y. 10025

Abstract. The influence of plants on climate has many different aspects. Plants affect the surface albedo, the water holding capacity of the soil-vegetation system, the transport of water from deeper levels within the ground, the roughness of the surface, and the influence of snow cover on albedo. The first three of these characteristics are investigated in a global climate model (GCM) by removing each influence individually over all land grid points. Each experiment was run for two years and compared with a control run of five years. The results indicate that the effect on surface albedo and on water holding capacity are of similar magnitude in the model but maximize in somewhat different regions. The albedo effect is most important in areas with plentiful moisture associated with evaporation off ocean surfaces and in seasons when large-scale dynamics or convection controls rain production. The water holding capability influence dominates where local evaporation is essential for precipitation. The transport of moisture within the ground affects the phase of water loss in most areas. These experiments emphasize the importance of vegetation in the global climate system. They also indicate the importance of including full seasonal cycles in the estimation of climate change associated with vegetation removal.

Introduction

The ability to model the current climate and potential climate changes is intimately connected to our understanding of the hydrologic cycle. We must understand how to parameterize the evaporation of water from the ocean and land, how to advect it through both large scale motions and convection, how to condense it to form clouds and precipitation, how to reevaporate the precipitation on the way down if necessary, and how to distribute the precipitation at the surface between runoff and accumulation at various levels within the ground. Many of these processes are currently uncertain; in particular, the process of evaporation is modeled with parameterizations of turbulence, convection with subgrid-scale schemes related to large-scale variables, condensation and cloud formation with simple saturation schemes, and runoff and soil moisture relationships based on little data. The modeled hydrologic system is validated chiefly by comparison with observations of precipitation, which are themselves somewhat uncertain, and which can be reproduced in a general sense by accurate simulations of processes largely independent of the hydrologic system (e.g. correct low pressure/high pressure location simulation).

The importance of the correct reproduction of the processes enumerated above becomes clearer when one looks at the regional climate scale. In certain areas at certain times clouds and precipitation result more from local processes such as evaporation and convection than from larger scale advective effects (for example, in summer, or in light wind situations). Here the questions of how water is stored and transported in the soil, and how surface heating produces convection and subsequent condensation can determine how much precipitation is modeled, what the surface temperature is, how much cloud cover occurs, etc. The generalized schemes that are used in models are unlikely to be equally applicable to all areas, which means that the ability to simulate climate and climate changes will vary from area to area. It is necessary to include in models all the characteristics of a particular area which may impact these processes, and allow that area to differ from others with different physical characteristics. One of the chief characteristics that varies from area to area and impacts hydrologic processes is the vegetation.

The vegetation effect is felt in several ways. (1) It influences the surface albedo, and thus the stability of the lower atmosphere and its convective processes. This is the property emphasized by Charney (1975, 1977) in his study of the effect of overgrazing on the Sahel region. The albedo associated with vegetation is on the order of 20% lower than that of deserts, which means land in vegetated areas can absorb more radiation and, from this effect alone, increase in temperature relative to nonvegetated areas. If the ground level heating is sufficient, it may initiate convection and precipitation which would not have occurred otherwise. (2) Vegetation can affect the albedo in snow covered areas by its masking characteristics. When snow falls in regions of trees the canopy provides a cover which masks the snow and presents a significantly darker appearance than does snow in uncovered areas. The reduced albedo provides for additional heating which can more easily melt the snow, producing a positive feedback loop, especially in regions of snow cover extending towards lower latitudes where sunshine is more abundant. (3) Vegetation influences the water holding capacity of a region, directly by incorporating water during the growing season and by catching from 5 to 20% or more of incident rainfall on leaves which can then be evaporated (Dickinson, 1980; Lockwood and Sellers, 1982) and indirectly by attracting moisture in the vicinity of the roots. It also stabilizes the soil in an area, decreased vegetation allows the wind to erode the soil cover, as in the dust bowl epic of the 1930s, thus decreasing the soil depth and the water holding capacity of the ground itself. In addition vegetation increases the rate of infiltration of water into the soil, retarding runoff, further increasing the soil moisture. (4) Vegetation influences the amount of water available for evaporation by transporting water from deeper levels to the surface and then to the atmosphere through the leaf stomata in the process known as transpiration. Diffusion upward from deeper levels through the soil by itself is a much slower process. Typically, evaporation from bare soil is negligible after the top 30 cm are dry, while the roots of forests can extract water down to several meters. (5) Vegetation influences the surface roughness of an area-by increasing turbulence it helps establish the small scale eddies responsible for surface fluxes of heat and moisture away from the ground.

To properly simulate the effect of vegetation on climate, it is necessary to account for these various processes in the model. Table 1 lists the parameters associated with different vegetation types which have been incorporated into a global climate model, model II described by Hansen et al. (1983). The distribution of the vegetation types over the earth has been compiled by Matthews (1983), on a 1° resolution basis, and the proportion of each type in the 8° x 10° grid determined for the model. It is this model which is used in the subsequent experiments.

The purpose of this study is to assess the influence of vegetation on the hydrologic cycle. As noted above, the vegetation has numerous effects. In order to elucidate the character of each, it was decided to remove the vegetation influence of the most important effects one at a time. The masking depth effect on snow cover, and the influence on surface roughness, while important in some regions, appear to have a minor effect on the hydrologic cycle in the model in general and will thus not be discussed further. The remaining influences, on water holding capacity, vertical diffusion of water from deeper levels, and ground albedo are the subject of the following study.

Model

The ground hydrology scheme used in the standard version of the model is programmed as follows. There are two layers in the soil which have a water holding capacity (f_i) dependent upon the vegetation in the grid box, as in Table 1 (see also Fig. 1). The wetness (W_i) of each layer is defined as the ratio of the actual water in the layer (w_i) to the water holding capacity, viz.,

$$W_i = w_i/f_i \tag{1}$$

where $0 < W_i < 1$. In the first layer the wetness is changed by the excess of precipitation rate (P_r) over evaporation and runoff rates (E,R), and by diffusion with the second layer, as

$$\frac{\partial W_1}{\partial t} = \frac{P_r - E - R}{f_1} + \frac{W_2 - W_1}{\tau} \tag{2}$$

where τ is the time constant for diffusion of moisture between layers. The wetness of the lower layer is determined by

$$\frac{\partial W_2}{\partial t} = \frac{f_1}{f_2} \left(\frac{W_1 - W_2}{\tau} \right) \quad . \tag{3}$$

The water holding capacity is determined by area weighting the values associated with each vegetation type over the grid box. The values listed in Table 1 are used to represent the total water holding capacity of the soil-vegetation system in which the vegetation types is found (not just the water held by that vegetation).

Precipitation is calculated by the model as the supersaturation amount and evaporation is a function of the wetness of the ground (W_1) and potential evaporation based on a drag law parameterization

$$E = W_1 E_p \tag{4}$$

$$E_p = \rho C_q V_s (q_g - q_s) \tag{5}$$

with C_q a humidity transfer coefficient, V_s the surface wind speed, q the specific humidity and subscripts s and g referring to surface air and ground; q_g is the saturated value of the air at the ground temperature. C_q is a function of stability and wind speed; for further details see Hansen et al. (1983).

Runoff is parameterized as a function of precipitation rate and ground wetness, as

$$R = W_1 P_r/2 \tag{6}$$

with additional runoff as necessary to keep $W_1 < 1$.

Table 1. Vegetation characteristics in the general circulation model.

		Desert	Tundra	Grass	Shrub	Woodland	Deciduous	Evergreen	Rainforest
Visual Albedo	Winter	0.35	0.07	0.09	0.09	0.08	0.10	0.07	0.06
	Spring	0.35	0.06	0.10	0.10	0.07	0.05	0.07	0.06
	Summer	0.35	0.08	0.09	0.14	0.08	0.06	0.08	0.06
	Autumn	0.35	0.08	0.09	0.11	0.06	0.05	0.06	0.06
Near-IR Albedo	Winter	0.35	0.20	0.27	0.27	0.23	0.30	0.20	0.18
	Spring	0.35	0.21	0.35	0.30	0.24	0.22	0.20	0.18
	Summer	0.35	0.30	0.36	0.42	0.30	0.29	0.25	0.18
	Autumn	0.35	0.25	0.31	0.33	0.20	0.22	0.18	0.18
Field Capacity ($g\ m^{-1}$)	Layer 1	10	30	30	30	30	30	30	200
	Layer 2	10	200	200	300	300	450	450	450
Masking Depth (m)		0.1	0.2	0.2	0.5	2	5	10	25
Roughness Length (m)		0.005	0.01	0.01	0.018	0.32	1	1	2

The model produces realistic values of precipitation (see Fig. 2), evaporation and runoff -again see Hansen et al. (1983) for more details and comparisons.

The upward diffusion time constant from the lower to upper layer is parameterized by assuming that it is essentially the result of evapotranspiration processes during the growing season. Thus in May-August north of 30°N, six months later south of 30°S, and all year at latitudes in between, upward diffusion is assumed to be infinitely fast wherever there is vegetation. At other times, and in the desert portions of the globe, the upward diffusion is set to zero. The downward diffusion time is set to one day everywhere.

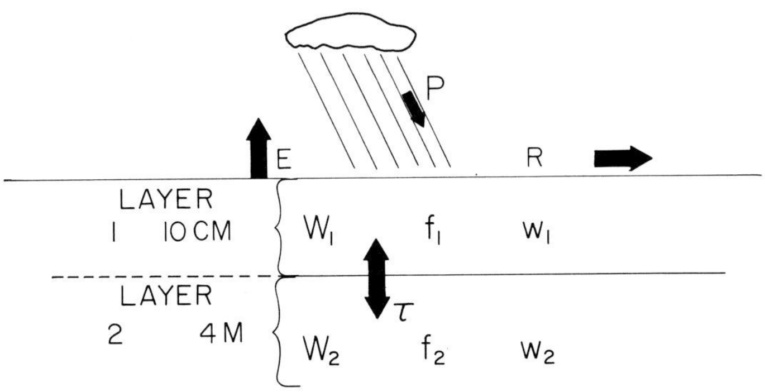

Fig. 1. Schematic of model hydrology; see text for details.

The ground albedo of the model is determined by its vegetative cover, as in Table 1, in addition to values associated with snow, sea ice and ocean. See Hansen et al. (1983) for further details. The value for deserts shown in the table refers to completely dry conditions. In general,

$$A_g = A_{gd} \left(1 - \frac{W_1}{2} \right) \tag{7}$$

where A_g is the desert ground albedo, A_{gd} is the dry ground value and W_1 the wetness of the first layer. Typical wetness values in desert regions average about 5% so the albedo under current conditions is generally close to the dry ground value.

The version of the model used for these stu-

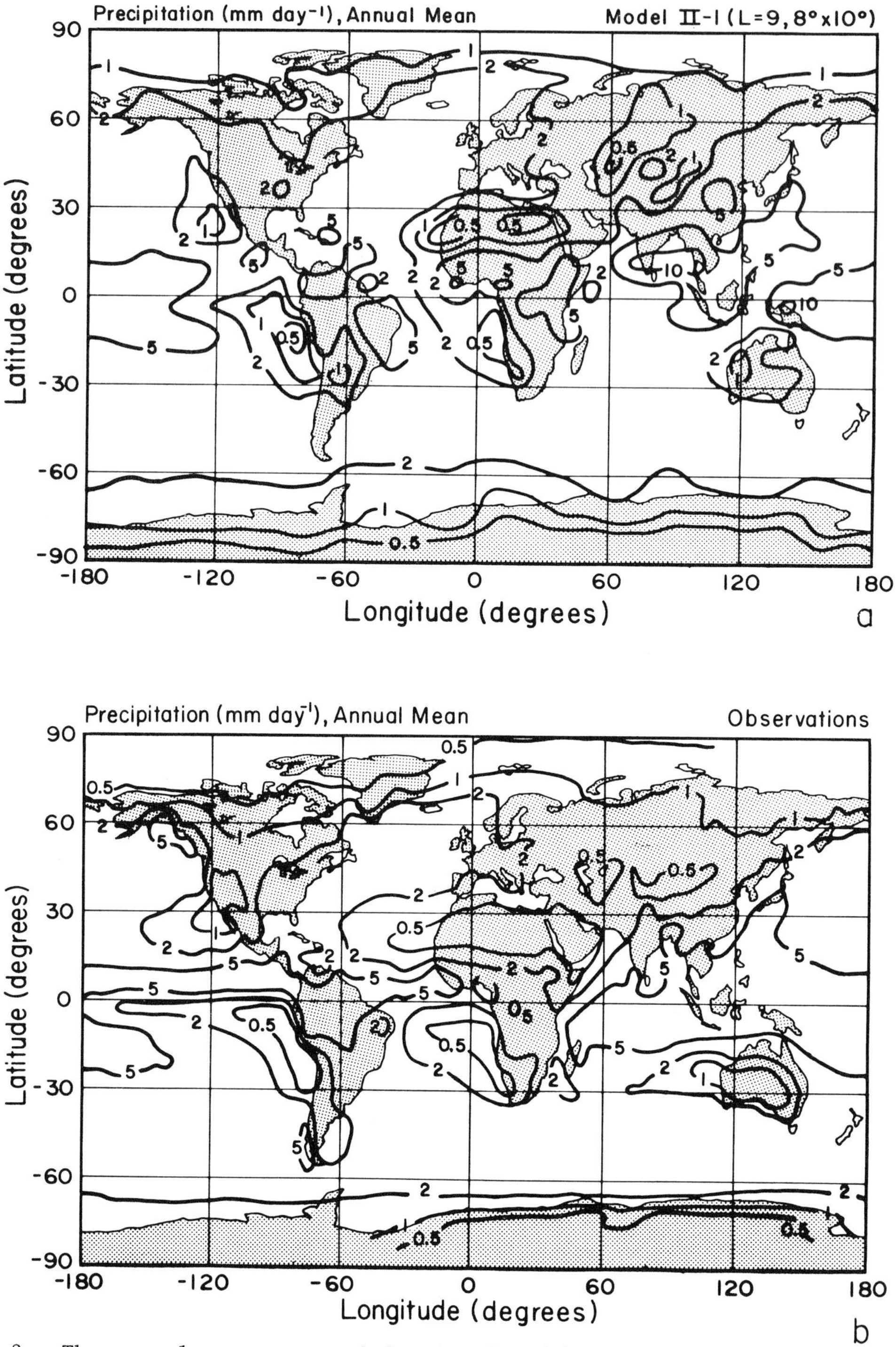

Fig. 2. The annual average precipitation for (a) five years of the control run; (b) observations from Schutz and Gates (1971).

dies has an 8° x 10° resolution, nine layers between the ground and 10 mb, realistic topography, radiation schemes, cloud cover generation and specified sea surface temperatures and ocean ice cover. The specification of sea surface temperature limits the response of the model for long term climate effects - this will be discussed more in the next section.

Experiments

Three different experiments were performed to isolate the impact of the different vegetation influences on the hydrologic cycle. Each experiment was run for two years starting from January 1, and the results to be discussed are from the second year. Comparison was made with a control run which included all the vegetation effects and was run for five years, from which means and standard deviations of various quantities were calculated. The experiments were:

(1) eliminate the instantaneous upward diffusion of water vapor from the second ground layer associated with transpiration; instead, replace it with upward diffusion associated with movement through the soil with a five day time constant;

(2) remove the water holding capacity associated with the different vegetation types in Table 1 and replace it with the value associated with deserts;

(3) replace the ground albedo associated with the different vegetation types (Table 1) with the value for deserts.

Each of these changes was incorporated over all the land areas to determine which regions were most sensitive, although there is no guarantee that the response in a particular region results from the changes incorporated in that region alone. The five day diffusion time constant in experiment one for water diffusion upward in soils may actually be too quick; nevertheless, it was picked as representing a considerable retardation in the speed with which water from the second layer would be made available to the first. A larger time constant will amplify the effect.

The water holding capacity, altered in the second experiment, is really a function of soil type in addition to vegetation. The assumption here is that with the removal of all vegetation, erosion processes would eventually reduce the soil depth, and alter its capacity as well. Alternately one could consider the effect as occurring due to enhanced runoff from the soil-vegetation system. Vegetation limits runoff by intercepting precipitation and allowing evaporation to occur from the canopy surface, thus increasing the water holding capacity of the region. Nevertheless, adopting the values associated with deserts represents a drastic assumption which will exaggerate any effect. The same can be said of replacing the albedo values with desert values; soils have a range of albedos, varying from less than 10% to greater than 40% depending on both the organic and inorganic content. Without vegetation the soils will lose their organic topsoil cover and increase in albedo, in general. The final albedo will depend on the characteristics of the lower soil horizons and may differ considerably from the 35% value used for deserts. The emphasis here, then, is not on "realistic" climate perturbations, as is obvious from the removal of vegetation everywhere, but rather on experiments which highlight the vegetations' influence.

All three experiments were run for two years with the model which specifies the sea surface temperatures at their monthly climatic values. This means that the sea surface temperatures are not allowed to change, which limits the total response of the model. Were the ocean temperatures allowed to adjust, and the model run long enough, it is conceivable that the results presented below would be altered somewhat. It is thought, however, that the experiments as conducted do fulfill the aim of delineating the differing vegetation influences on the hydrologic cycle.

Results

Global

The effects of vegetation absence in the three experiments differed depending upon which of its characteristics was omitted. The experiment which reduced the water capacity of the ground (experiment 2 - the "capacity" experiment) reduced the ability of the land portion of the model to evaporate water. Annually-averaged evaporation for the globe decreased by 6%, and over land by 33% (both percentage changes are relative to the control run); this can be compared to interannual standard deviations of 0.1% and 1.3% respectively from the five years of the control run. (A difference of three standard deviations is significant at the 90% confidence level using the 2-sided t-test). The decreased evaporation was associated with an increase in surface air temperature, as excess heat had to be removed by sensible heat fluxes. The surface air temperature over land warmed by 1.5°C, despite the fixed sea surface temperatures.

In contrast, the experiment which replaced the vegetation albedos with those for desert (experiment 3 - the "albedo" experiment) tended to cool the globe - surface air temperatures over land were about 1°C cooler, and the global temperatures for the entire atmosphere were about 0.5°C cooler, a factor of ten greater than the interannual standard deviation. (And were the sea surface temperatures allowed to adjust, the change would likely have been several times greater). This cooling also helped to promote reduced evaporation, 1% globally, and 14% over land, due to the reduced energy available for evaporation. Thus on the global average reduced evaporation resulted from both effects, for different reasons.

The experiment which altered the time constant for the transfer of water from the second to the first ground layer during the growing season (experiment 1 - the "diffusion" experiment) had smaller effects on evaporation on the annual and global average - evaporation over land decreased by 5% relative to the control. However, the altered diffusion produced a change in phase of the evaporation as will be discussed further in the next section.

The annually averaged precipitation as a func-

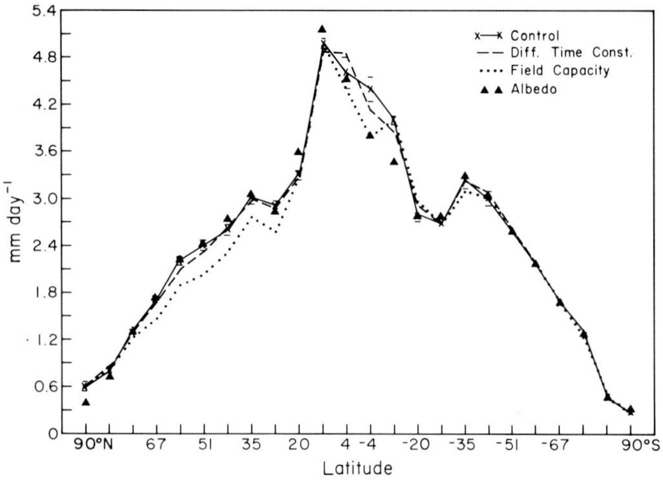

Fig. 3. Annually averaged precipitation at each latitude for the five year control run and the second year of each experiment. In this and subsequent figures the bars indicate one standard deviation above and below the values for the control run, determined from the five year integration.

tion of latitude for the three experiments is shown in Fig. 3, as are the values and standard deviations for the control run. This result includes precipitation over the ocean as well, which was not directly affected by the changes, to indicate the impact of the vegetation effects on the global rainfall. There are two regions of the globe where the latitudinal average of precipitation appears to have significantly decreased: at mid-latitudes in the northern hemisphere, and at low latitudes, especially in the southern hemisphere. The first region has a high percentage of land (approximately 53%) as opposed to water. The second region is dominated by tropical rainforest and undergoes the largest shift in values when the vegetation is replaced by desert (see Table 1; in the case of albedo the change is augmented by the large amount of sunshine received at these latitudes). These zones will be examined in more detail below.

Latitudinal Zones

Figs. 4a-c present the precipitation, evaporation, and their difference as a function of month averaged for all land grid points between 40-55°N. This is one of the regions shown in Fig. 3 which has a significant reduction in annually averaged precipitation; it is also an area which encompasses a major portion of the North American and Siberian "bread basket" regions. Results for the control run show that maximum precipitation occurs in the months of May through July (Fig. 4a), in agreement with observations (see for example, Berry et al., 1973). Evaporation is also a maximum in these months (Fig. 4b), and the difference between precipitation and evaporation (Fig. 4c) indicates that there

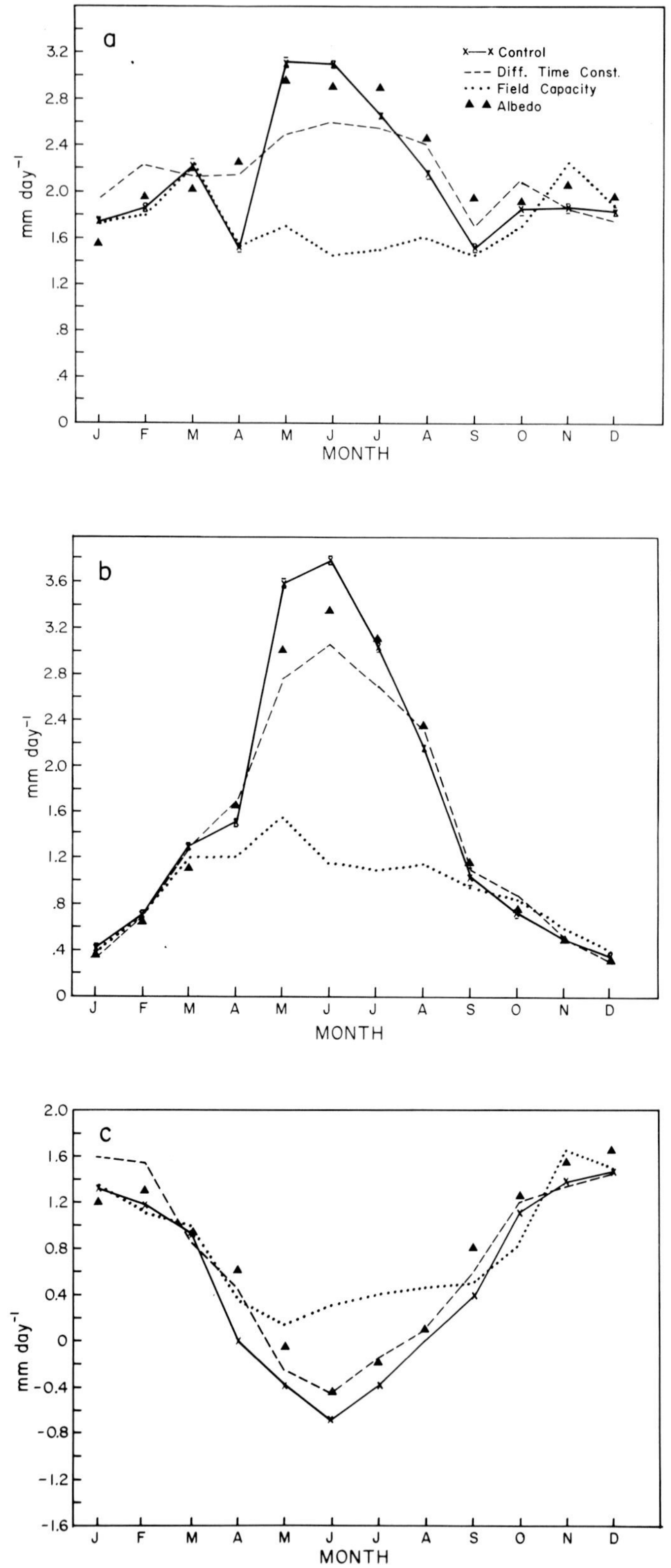

Fig. 4. Results from the five year control run and the second year of the different experiments at 40-55°N over land for each month for: (a) precipitation; (b) evaporation; (c) precipitation - evaporation.

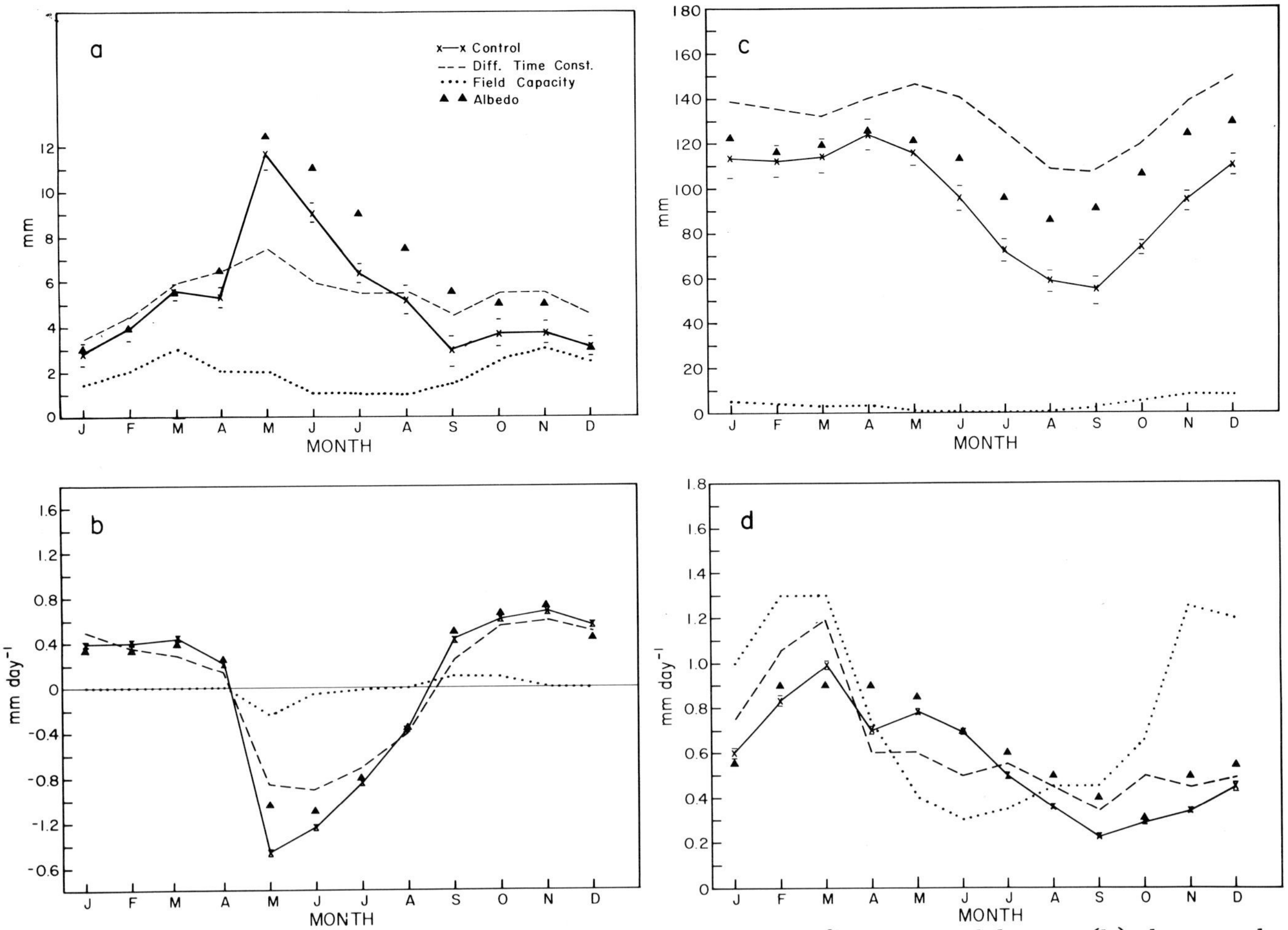

Fig. 5. As in Fig. 4 for: (a) soil moisture in the first ground layer; (b) downward diffusion between the first and second ground layers; (c) soil moisture in the second ground layer; (d) runoff.

is a net loss of moisture from the ground during the spring and summer.

Figs. 5a-d show the ground hydrological parameters as a function of month for the same region. The liquid water in the first ground layer (Fig. 5a) maximizes in May, due to a combination of ice and snow melting and the excess of precipitation over evaporation during the winter (Fig. 4c). Starting in May, the excess of evaporation over precipitation causes a decrease in the upper layer soil moisture, and causes there to be a net flux of water up from the second layer of the ground (Fig. 5b). This is also the time when the beginning of the growing season allows transpiration to bring water up from the second layer in the model. The liquid water in the second ground layer (Fig. 5c) decreases during the summer due to this demand. The combination of reduced precipitation and minimum ground wetness in September result in the runoff being lowest in the month (Fig. 5d). The runoff maximizes in late winter when there is substantial precipitation and a maximum amount of water plus ice in the two ground layers.

How were these results altered by removing the individual vegetation influences? Beginning with Fig. 4 we note that the diffusion experiment reduced the evaporation for the first part of the summer, and, while precipitation was reduced as well, the deficit of water was less negative during these months than in the control run. By August precipitation and evaporation were both slightly higher than in the control run. Fig. 5a shows that the ground water in the first layer was less than that of the control run for the first part of the summer, but was slightly higher by August. With the five day time constant for drawing water from the second ground layer into the first replacing the infinitely fast upward diffusion associated with vegetation, the water was not drawn upward fast enough to provide for as wet an upper layer, and hence as much evaporation and precipitation, as occurred in the control run during the first part of the summer. By the second half of the summer, however, the situation reversed, as the excess water remaining in the second ground layer (Fig. 5c) was sufficient to allow the first ground layer to be even

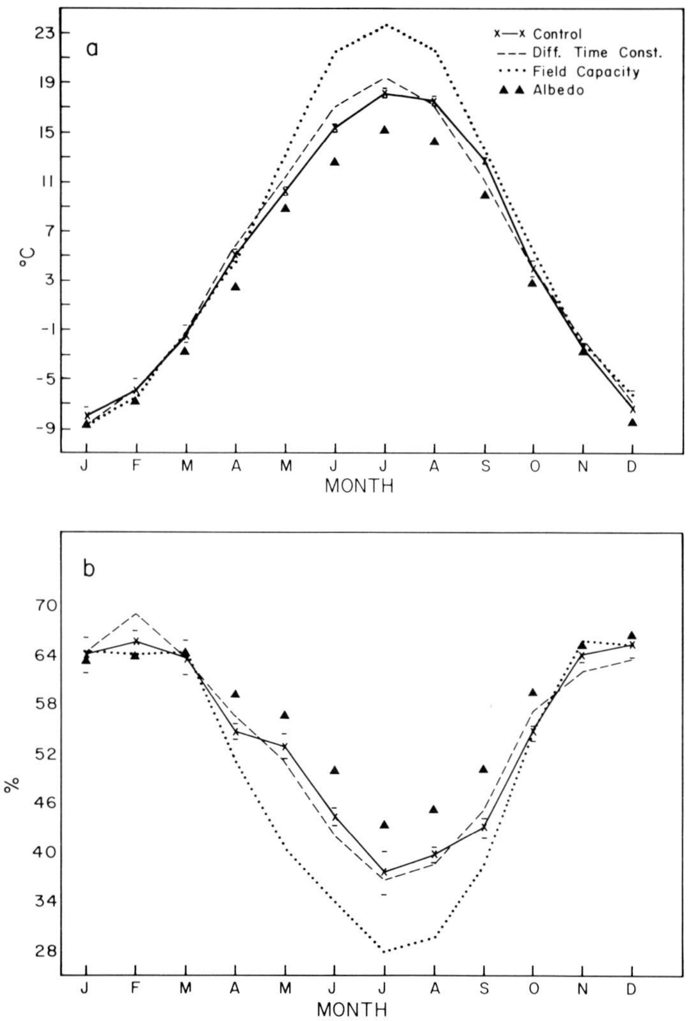

Fig. 6. As in Fig. 4 for: (a) surface air temperature; (b) cloud cover.

wetter than that in the control run, despite the slower diffusion rate still prevailing. The result thus was to change the phase of the water loss, stretching it out over the entire summer while reducing the early season maximum.

Fig. 6a shows the surface temperature as a function of month for this region. Relative to the control run, the diffusion experiment has warmer temperatures during the first half of the summer, when evaporation is reduced, and cooler temperatures later on, when evaporation is augmented. With the reduced hydrologic cycle, cloud cover was slightly lower during the summer (Fig. 6b) which, including the effects of both short wave and long wave radiation, resulted in an excess of 1.5 W/m^2 radiation hitting the ground in July compared to the control run. In contrast, the evaporative flux decreased by 8 W/m^2.

The results associated with reducing the water holding capacity were much more extreme. As can be seen in Fig. 4, both the precipitation and evaporation were strongly reduced during the summer, and while there is an excess of precipitation over evaporation during the summer months (Fig. 4c), ground water levels are forced to remain quite low (Figs. 5a,c). As shown in Fig. 5d, one effect of the reduced holding capacity is to produce large runoff values during the winter, when precipitation is not affected. This emphasizes that it is during the summer when precipitation is derived from local moisture sources in the ground; in winter traveling cyclones advect moisture from the ocean.

Fig. 6a shows that the result of the lower ground water was to increase the surface temperatures strongly throughout the summer in this experiment. This effect was mainly due to the change in evaporative heat flux, by 52 W/m^2 relative to the control run. The cloud cover decrease had a smaller effect, for while the planetary albedo dropped from 34% to 26%, thermal radiation escaped more easily, so the net radiation at the surface amounted to only a few W/m^2 higher than in the control run.

The albedo experiment also resulted in reduced evaporation and precipitation during May and June, as shown in Fig. 4, although for very different reasons. It did not result from reduced ground water - as can be seen in Figs. 5a and 5c, where ground water is consistently higher than in the control run. Rather it is associated directly with the reduced surface temperatures, shown in Fig. 6a. The ground albedo in this region went from 18% in the control run to 31% in this experiment; the planetary albedo went from 33% to 40%, with the slight cloud increase also contributing. Including the slight reduction in long wave cooling, the net radiation at the ground decreased by 30 W/m^2 in June, 2/3 of which was reflected in a drop of the sensible heat flux from the surface, with the remaining third associated with reduced evaporation. By July and August the excess ground water available apparently overcame the reduced surface energy effect, as precipitation and evaporation increased relative to the control run, even though the surface temperature remained significantly cooler (and net radiation significantly reduced). This latter result emphasizes that the vegetative albedo effect will not necessarily result in reduced precipitation as in the experiment of Charney (1975). It is necessary to include a full seasonal cycle to gauge how the soil moisture will vary during the year. The result here was somewhat similar to the result for the diffusion experiment - by keeping the ground wet later into the summer, it eventually resulted in increased precipitation.

The other latitude region which experienced a significant precipitation decrease was centered around 4°S. Figs. 7a-c shows the variation of precipitation, evaporation and the difference as a function of month, for the land grid points between 0 and 8°S. The seasonal variation in precipitation is associated with the movement of the intertropical convergence zone, with minimum values during Northern Hemisphere summer when the zone has moved significantly north of the equator. Precipitation exceeds evaporation for this region over land for all months except May, and occasionally June. Only a little more than 20% of this latitude zone is occupied by land, and the ocean provides moisture for land rainfall. In addition the large scale circulation converges moisture into the region during December through March.

Figs. 8a-d show the ground hydrological variables at this latitude. The ground water peaks at about the time of the precipitation peak, as does the runoff. Moisture is drawn up from the second layer essentially only during the two months when precipitation does not significantly exceed evaporation. This can be contrasted to the situation at 40-55°N over land where upward transport of water was required throughout the summer (Fig. 5b).

The variation of surface temperature and cloud cover is shown in Figs. 9a,b. Cloud cover is a minimum in June along with the precipitation. The temperature shows a slight maximum late in the Southern Hemisphere winter, associated with less loss of heat by evaporation. The evaporation appears to decrease largely because of the reduction in ground moisture shown in Figs. 8a,c. At neighboring grid points over the ocean, evaporation is higher in August than in January, as surface wind speeds are greater and the atmosphere in the layer above the ground is drier.

The results for the diffusion experiment show that the influence of the second ground layer on evaporation at this latitude occurs mostly during the months of August through October, when a decrease occurs in this experiment (Fig. 7b) along with an increase in surface temperature (Fig. 9a). Precipitation itself appears to be affected only during July (Fig. 7a) and the ground water in the first layer is replenished by there being less runoff in August (Fig. 8). The major effect of this experiment is to increase the soil moisture in the second layer in most months (Fig. 8c). In contrast with the situation at higher latitudes, the climate here is not that dependent on the phase of the water withdrawn from the second soil layer due to the availability of water from other sources, as discussed for the control run.

The capacity experiment allows us to clarify how much of the precipitation results from evaporation of water from the ground. Reducing the ability of both ground layers to hold water does reduce the precipitation, especially in the months of April through October, when there is no large-scale dynamic convergence of moisture. The evaporation is reduced much more, however, so there is an excess of precipitation minus evaporation throughout the year (Fig. 7c). Again, the ability of the model to evaporate water from the nearby ocean undoubtedly influences this result. With the decreased evaporation, and decrease in cloudiness (Fig. 9b), the surface temperature

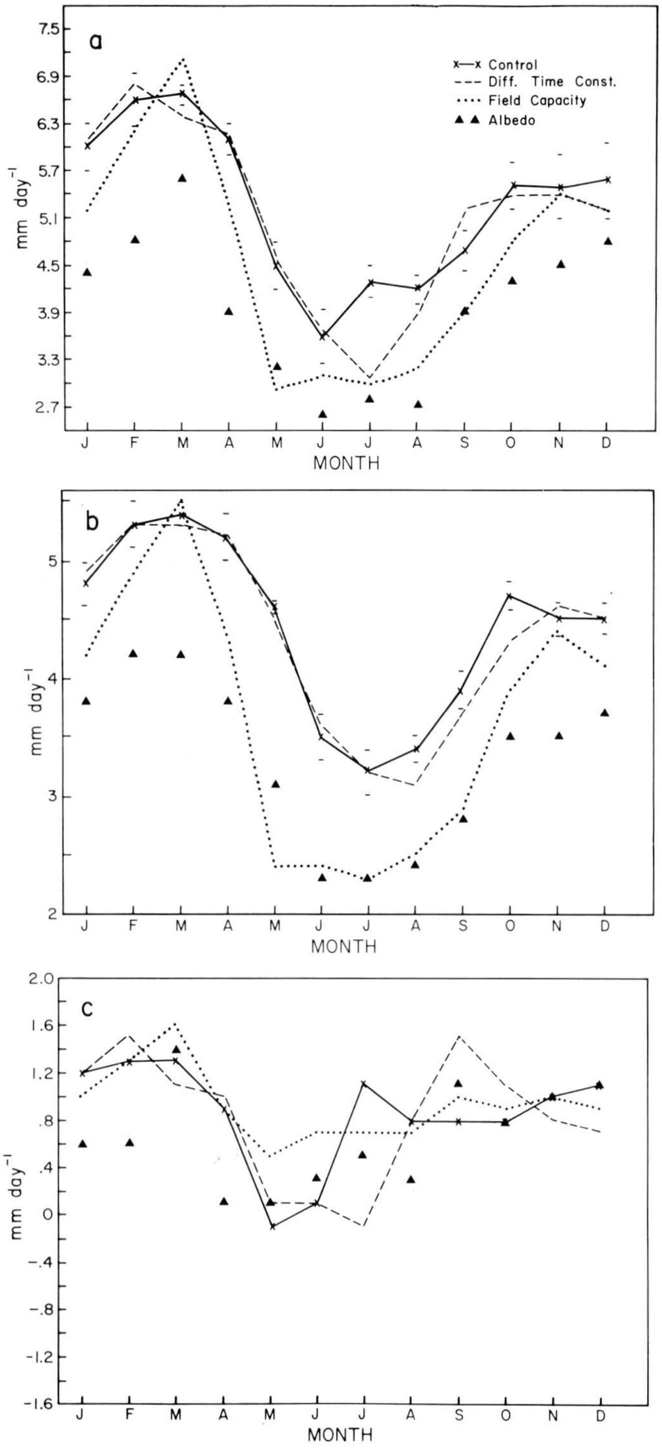

Fig. 7. Results from the five year control run and the second year of the different experiments at 4°S over land for each month for: (a) precipitation; (b) evaporation; (c) precipitation minus evaporation.

shows a significant increase from May through September. The cooler surface air temperatures during the other months are caused by cooler ground temperatures during these months - these

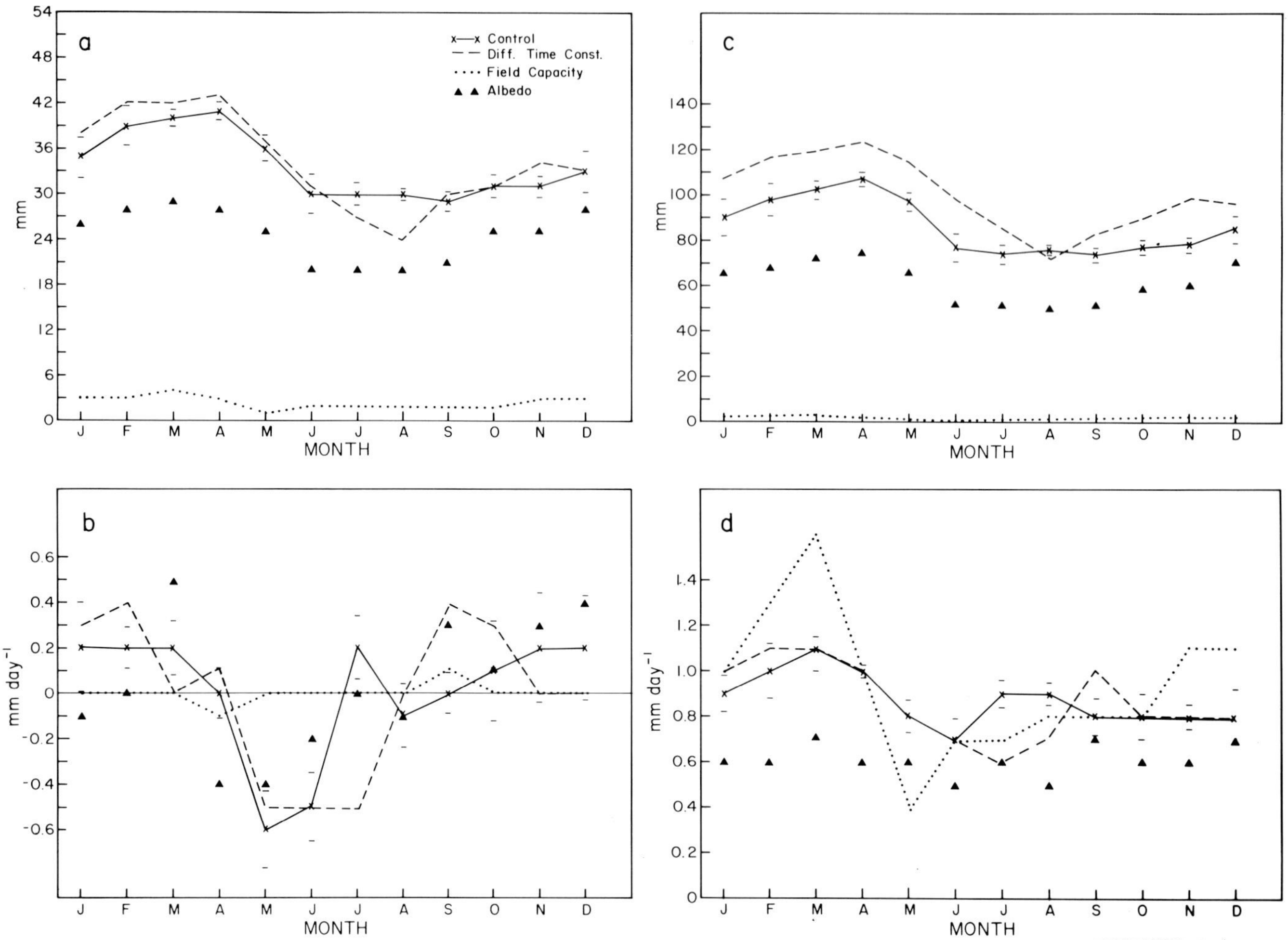

Fig. 8. As in Fig. 7 for: (a) soil moisture in the first ground layer; (b) downward diffusion from the first to second ground layer; (c) soil moisture in the second layer; (d) runoff.

in turn result from a changed thermal conductivity, which is a function of ground moisture, so the reduced ground moisture alters the phase of the ground temperature variations.

Results for the albedo experiment show that precipitation and evaporation are decreased, as was the case for the capacity experiment. Unlike the capacity experiment, however, the albedo experiment produced significant decreases in all seasons, including the months of January-March, when, as noted before, the precipitation resulted mainly from dynamical processes. Fig. 10 shows the tropospheric vertical velocity averaged for the grid points between 10°N and 18°S over land for the different months. The control run, diffusion and capacity experiments produce similar results, except that the capacity experiment has slightly greater upward velocities during June and July due to the increased surface heating associated with the experiment. In contrast, the albedo experiment produces relatively more subsidence in all months, with the greatest differences occurring between October and March when, as concluded from the other experiments, dynamical factors were largely responsible for the precipitation. The reduced precipitation during these months resulted from large-scale changes in the dynamics, with less vertical motion over land in the tropics, and a weaker Hadley circulation south of the equator in January (by a factor of two).

This effect of the albedo experiment at low latitudes can be contrasted to the result at higher northern latitudes discussed earlier-there the mean circulation is normally weak during the months most affected (e.g. July), and there was no significant change in the mean circulation. The precipitation decrease in that month in the albedo experiment was associated with the reduced energy available for local evaporation and subsequent precipitation, along with a reduced tendency for low level heating to incite moist convection, again a local effect. This influence does appear to a slight degree at these low latitudes as well - Fig. 9a shows the surface temperature is slightly cooler, and evaporation has decreased (although not as much as precipitation in January). For the months of

May-September, the decreased ground water is likely to have been of greater significance in reducing the precipitation.

The dominant effects on the hydrologic cycle from these experiments in the two latitude regions are summarized in Table 2. In general the result of removing the effects of vegetation was to lower the annual precipitation in these two regions (except for the albedo experiment at 40-55°N), but the seasonal variation and physical cause of the change differed in the two latitude zones.

Geographical Results

The differences between the annually averaged precipitation of the diffusion, capacity and albedo experiments and that of the control run are shown in Figs. 11-13. For the diffusion experiment, the

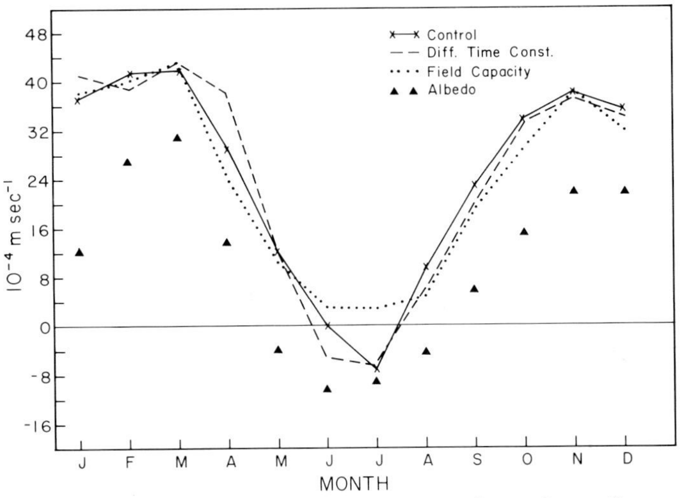

Fig. 10. Vertical velocity as a function of month for the five year control run and the experiments as a function of month between 10°N and 18°S over land.

absolute values of the difference is, in general, relatively small (Fig. 11a). The significance of the changes can be gauged by comparing the difference to the standard deviation of the annual average as determined from the five year control run (Fig. 11b). Significant increase occurs in the western and northern United States, as well as in several ocean areas, while decreases occur around India and in and around Australia. As noted in the previous section, whether the diffusion experiment increases or decreases precipitation averaged over the year depends on how much the local precipitation depends upon evaporation, on whether this evaporation is hindered by the slower upward diffusion of water from the second ground layer, or whether it is aided as the slower loss of water during the year results in more soil moisture being available for evaporation.

The results for the capacity experiment, and their significance, appear in Figs. 12a,b. Significant decreases occur at high latitudes over land, and in some other areas including the Amazon basin and parts of the Sahara desert and Sahel regions. Increases occur on coastlines and over the oceans, except for increase over Australia. This result shows which areas in the model depend upon locally stored ground water for their precipitation; however, it is not the only effect, as indicated by the decrease of precipitation in African desert regions whose capacity was changed very little in this experiment. The tendency for there to be compensatory increases near or over the ocean areas where water is available is associated with the land heating up and moist convective instability being realized in these locations.

Fig. 13 shows the precipitation difference resulting from the albedo experiment, as well as its significance. At low latitudes the precipitation decreases over land, arising in conjunction

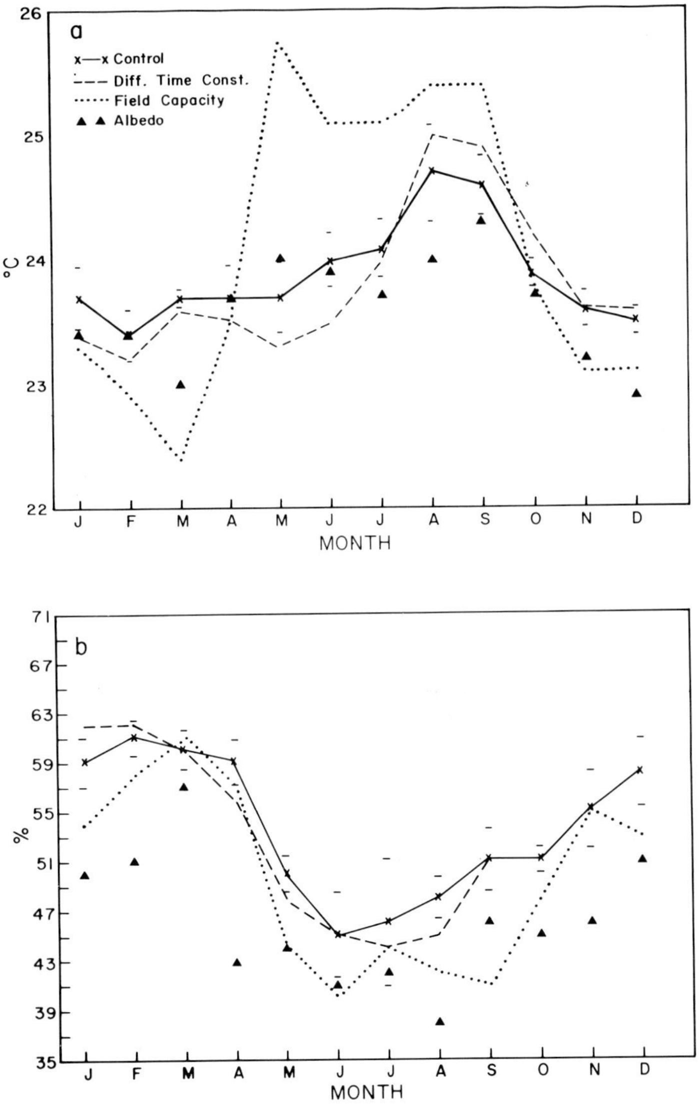

Fig. 9. As in Fig. 7 for: (a) surface air temperature; (b) cloud cover.

Table 2. Summary of results in the two latitude zones.

EXPERIMENT	LATITUDE	EFFECT OF HYDROLOGIC CYCLE
DIFFUSION	40-55°N	Reduced water loss, evap., precip. in early summer
		Increased water loss, evap., precip. in late summer
		Little effect in winter
CAPACITY	40-55°N	Strong reduction in evap., precip. in spring, summer
		Little effect in winter
ALBEDO	40-55°N	Reduced evap., precip. in spring; more ground water led to slightly more evap., precip. in summer
		Little effect in winter
DIFFUSION	4°S	Slightly reduced soil moisture, evap., precip. in winter
		Little effect in summer
CAPACITY	4°S	Reduced evap., precip. in winter
		Less effect in summer
ALBEDO	4°S	Reduced soil moisture, evap., precip. all year

with the greater atmospheric subsidence, can be seen to be accompanied by precipitation increases over the ocean. Vertical velocities over the ocean showed greater ascent, especially in the region just south of India. The increase in albedo associated with the removal of vegetation over the African continent has stimulated a local Walker-type east-west circulation cell with rising motion centered at 75°E and sinking centered at 20°E. A weaker effect of the same nature is visible extending from the Pacific to South America.

At higher latitudes the albedo effect on precipitation is mixed but generally results in increased precipitation, as the reduced surface temperatures lead to reduced evaporation, increased ground moisture, and increased precipitation, especially during the summer (cf. Figs.4-6).

Several areas are deserving of special attention. The original Charney experiment concentrated on the Sahel region, located at about 17°N from 15°W to 35°E. As can be seen in Table 3, which shows the change in annually averaged precipitation for various regions compared to the control run (and also the standard deviation of rainfall in the region from the five year control run), the yearly precipitation was reduced in the Sahel region in all three experiments, although the change in the diffusion experiment was less than one standard deviation. The mid U.S. and southern Canada were both very sensitive to the capacity and albedo changes, with large decreases in precipitation in both areas from both effects. Areas in the USSR were in general more affected by the capacity experiment than the other runs. The rainforest regions were strongly affected by both the capacity and albedo experiment, but the Amazon rainforest belt was relatively more affected by changing the surface albedo, which affected the large-scale dynamics, while the African rainforest was also strongly affected by capacity changes, responsible for more locally generated rainfall. This indicates that in the model rain in the Amazon is derived less from local (on land) sources than is rain in equatorial Africa. Table 3 and Fig. 12 show that the middle of the North American and European continents are also strongly dependent on soil moisture derived precipitation in the model.

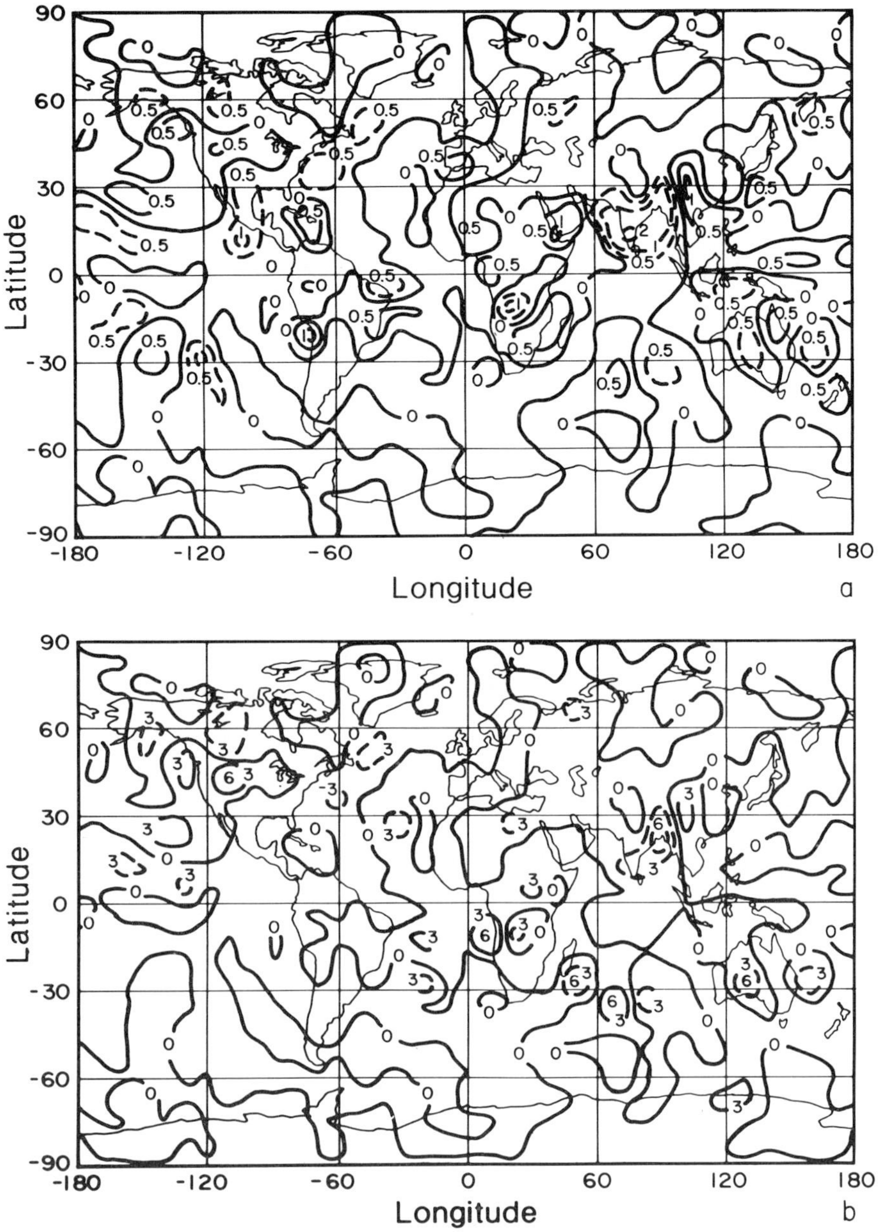

Fig. 11. (a) Change in annually averaged precipitation the diffusion experiment minus control run. In this and subsequent figures, dashed lines indicate negative values. (b) Change in precipitation divided by the standard deviation of precipitation from the control run. Only the contours for zero, three and six times the standard deviation are shown.

Discussion

As noted earlier, the effect of increased albedo on rainfall was originally emphasized by Charney (1975) in connection with the Sahel region. Subsequent experiments of this nature have been performed by Charney et al. (1977), Chervin (1979), Rowntree (1982) and Sud and Fennessy (1982). Using different general circulation models, each work altered the albedo in different areas, generally including the Sahel area; one difference among the experiments which is thought to be significant relates to the way soil moisture is calculated, whether it is fixed at wet or dry values, or interactive. This difference has even produced opposing results in certain regions (cf. Sud and Fennessy 1982). In general these experiments were run for the month of July. The results from these experiments are unanimous in finding decreased precipitation in the Sahel region when its albedo is increased, with values ranging from 1-4 mm/day. The results are partly caused by increased subsidence associated with surface cooling due to decreased

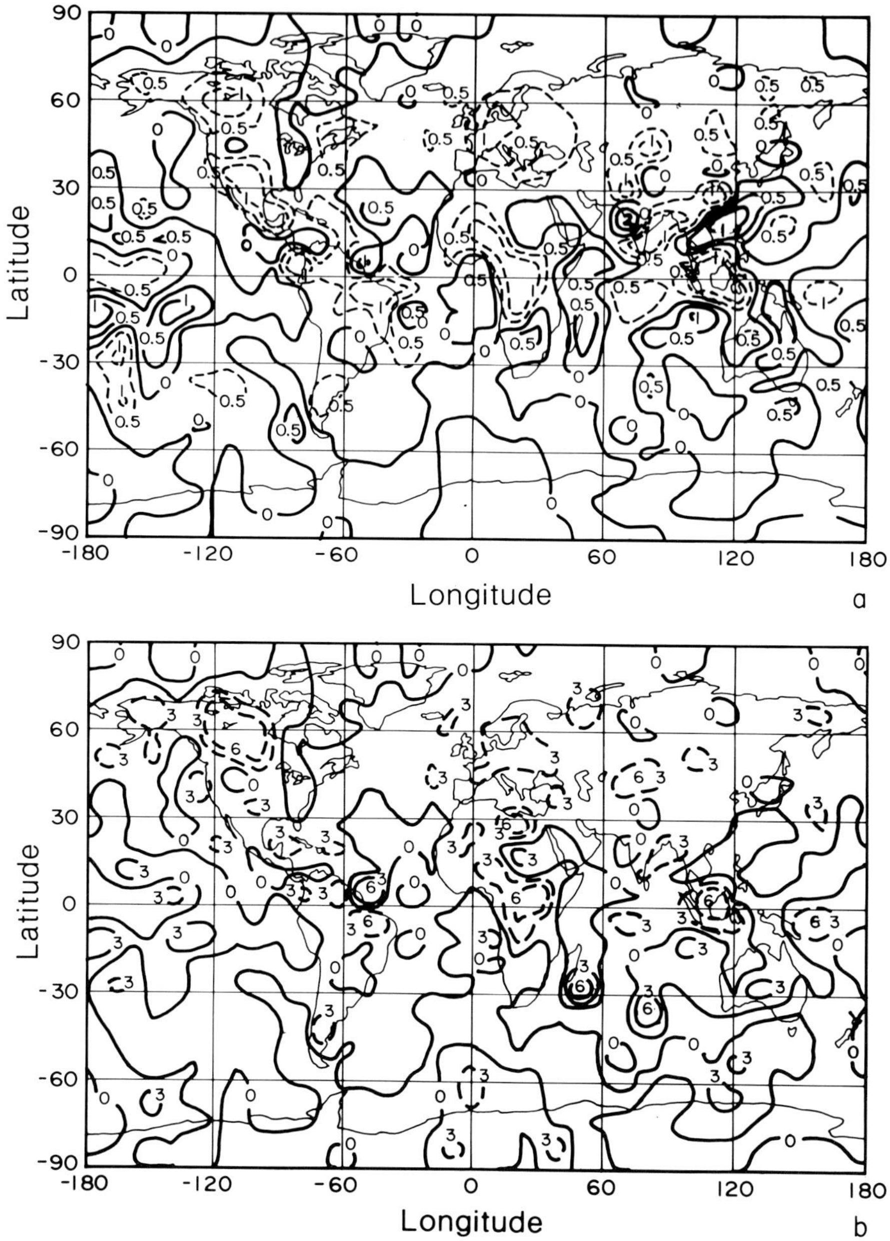

Fig. 12. (a) As in Fig. 11a for the capacity experiment, (b) As in Fig. 11b for the capacity experiment.

short wave, or short wave plus long wave effects; in the models with interactive ground moisture, reduced values, associated with the initial precipitation changes, amplified the effect.

The results from Table 3 indicate that the annual average rainfall in the Sahel decreased in this albedo experiment as well; however, as can be seen in Fig. 14, in summer the rainfall over the Sahel actually increased. There are two important differences between this experiment and most of the others. One is that the reduction in surface albedo occurred everywhere, not only in the Sahel area. While the surface albedo reduction was 13% in the Sahel, it was 17% in the African rainforest. As shown in Fig. 14, it is equatorial Africa which suffers the large decrease in precipitation. Note that the land/ocean albedo contrast exists in this experiment, as in those by other experimenters, only in this case it results in a land/ocean temperature contrast that is not focused but occurs for the whole continent. Rowntree (1982) also noted that regions of precipitation increase and decrease over land occur when surface albedo was altered over a wide area. The second difference is that this experiment was run for two years and thus incorporated several seasonal cycles. As noted in the previous section the results for any month are dependent on how the hydrologic cycle, and soil moisture in particular, are altered in previous

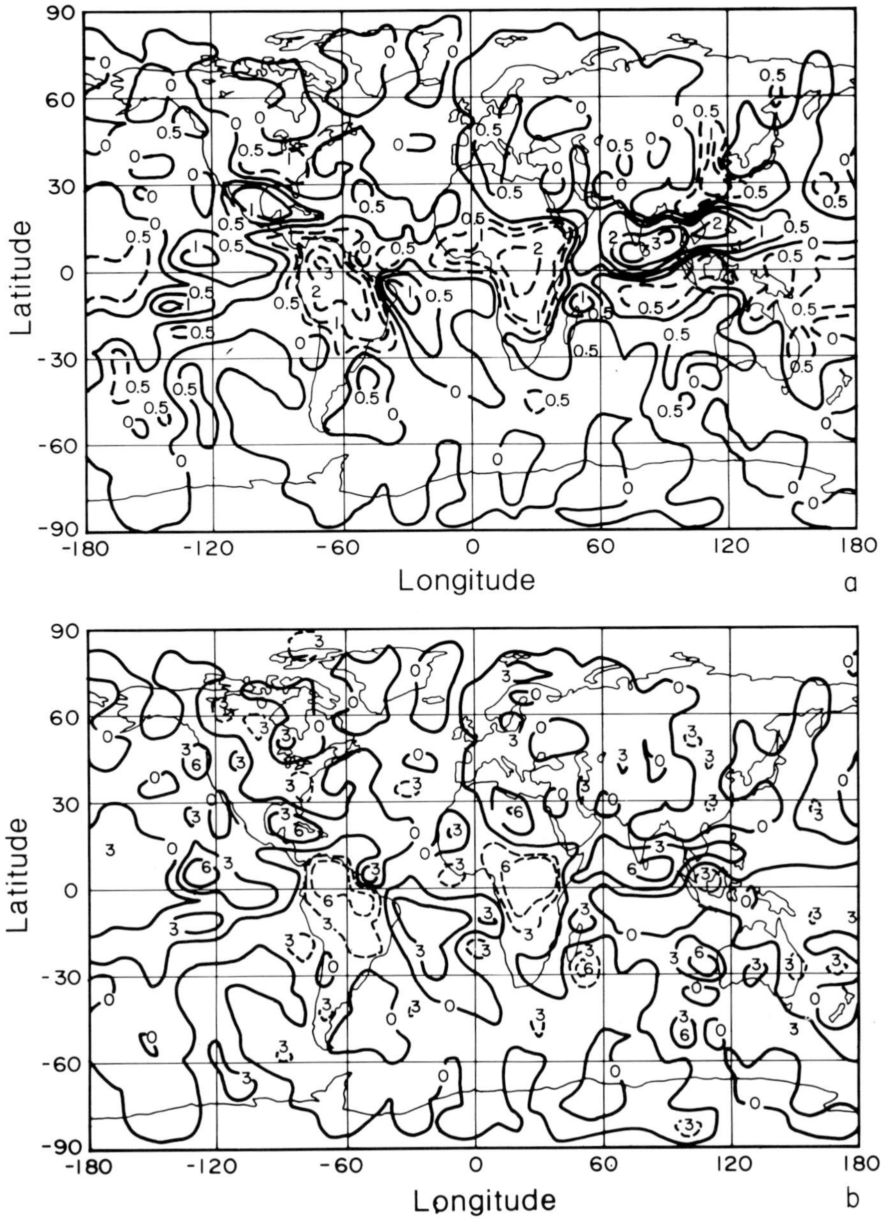

Fig. 13. (a) As in Fig. 11a for the albedo experiment; (b) As in Fig. 11b for the albedo experiment.

months. The albedo experiment limits evaporation during spring, inceasing the available soil moisture in summer.

The impact of soil moisture on rainfall has been studied by various modelers (ua. Manabe, 1975; Charney et al., 1977; Walker and Rowntree, 1977; Mintz, 1981; Rowntree, 1982; Rind, 1982). Results are consistent in showing that reduced soil moisture produces reduced rainfall in July in continental areas, especially in those regions without a nearby oceanic water vapor source. The effect may last for several months or even longer. As shown in Fig. 15 the capacity experiment produces similar changes although results are far from uniform. This figure emphasizes which regions in the months of June through August depend upon moisture evaporated from the ground (not necessarily locally, although this is most likely) for their rainfall, while Fig. 14 indicates which regions depend upon instability or large-scale circulations associated with vegetation albedos to produce rainfall during these months in the model. Except for a few areas such as Bengaladesh, southeast China, and the central portions of the United States, the regions do not overlap. Thus in addition to indicating the effect of vegetation on climate, these experiments provide an indication of the origin and cause of rainfall in the model in the different geographical locations.

Table 3. Change in annual average precipitation (mm day^{-1}) between the experiments and the control run for different geographic regions. Also shown is the standard deviation of the control run.

REGION	DIFFUSION	CAPACITY	ALBEDO	STAND DEV
WESTERN U.S.	.06	-.72	.27	.33
MID U.S.	-.16	-.75	-.72	.11
EASTERN U.S.	-.12	.09	-.71	.63
SOUTHERN CANADA	-.78	-1.08	-.47	.07
MID EUROPE	-.19	-1.13	.35	.12
NORTHERN RUSSIA	-.13	-.31	.31	.16
WESTERN SIBERIA	-.38	-.23	.05	.06
SIBERIAN PLATEAU	.03	-.22	-.04	.06
SOUTHERN CHINA	.46	-.51	-.69	.30
CHINA DESERT	-.27	-.48	-.36	.18
INDIAN DESERT	-.09	-.59	.39	.50
AUSTRALIAN DESERT	-.33	.15	.43	.22
NORTHERN SAHARA	-.14	-.26	.22	.04
SOUTHERN SAHARA	-.02	-.27	.06	.21
AFRICAN SAHEL	-.13	-.50	-.50	.23
AFRICAN RAINFOREST	.22	-1.35	-2.33	.22
AMAZON RAINFOREST	-.07	-.60	-1.88	.21

Figure 16 shows the change in precipitation in the albedo experiment in the months of December through February. As discussed in Section 4b, changes in the albedo instituted changes in the large-scale dynamics, with more subsidence in the ITCZ area - this produced the large deficits extending south of the equator in South America and Africa. Note that six months later (Fig. 14) the regions of maximum deficits had moved north of the equator, with the ITCZ. At higher latitudes, in Northern Hemisphere winter, the changed albedo in general resulted in slightly greater precipitation.

The changes for these months produced by the capacity experiments are shown in Fig. 17. The largest effects of reduced evaporation from the ground occur over land in the Southern Hemisphere, near the southern tip of South America, in equatorial Africa and in southeast Australia. Compari son with Fig. 15 shows that the changes are less extensive in December through February presumably because the summer hemisphere, which is the one primarily affected by the change, is the Southern Hemisphere which has a greater availability of oceanic water vapor than does the Northern Hemi-sphere.

The region around India deserves special comment. All of these results indicate large changes, both over India itself and in the ocean regions nearby. Lowering the surface albedo reduced the movement of the ITCZ north during the monsoon months (Fig. 14), with increased precipitation over neighboring ocean regions. This appears to result partially from changes in the local circulation contrasting the Indian land mass and the nearby waters, and partially due to large scale circulation changes, as noted earlier, with an east-west Walker circulation extending from

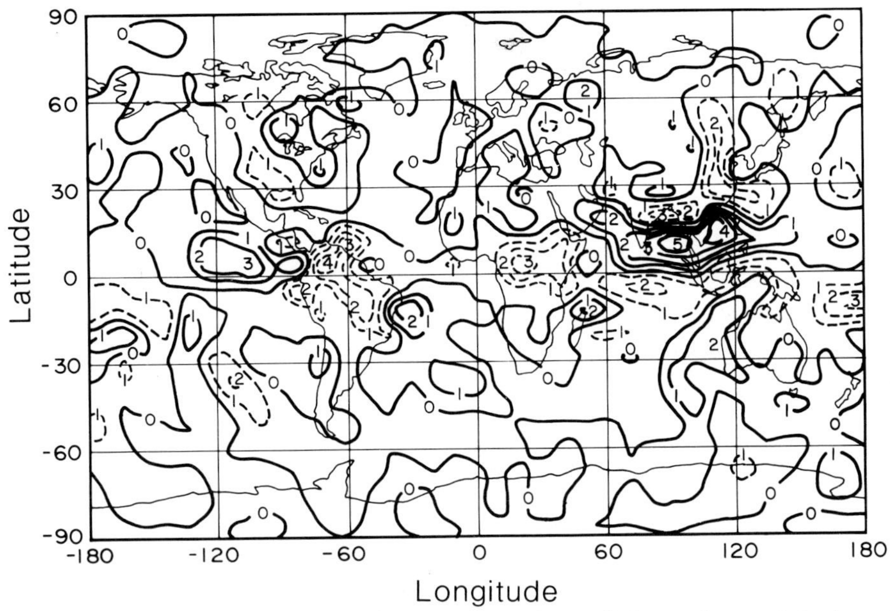

Fig. 14. Difference in precipitation for the months of June-August between the albedo experiment and the control.

the ocean waters south of India to the African continent. As can be seen in Fig. 16, in winter the whole pattern moves further south, along with the ITCZ. The capacity experiments also resulted in decreased precipitation over India, with the effect moving slightly northeastward from winter to summer with the monsoon movement. The sensitivity of this region to both soil moisture content and albedo in the model indicates the close interaction between these quantities and variations in monsoon intensity and duration.

Conclusions

The aim of these experiments was to examine the influence of vegetation on the hydrologic cycle in the GCM. The vegetation influence on the water holding capacity of the ground affects rainfall that is derived from local evaporation; the vegetation influence on the surface albedo affects rainfall associated with convection, either locally or in conjunction with large-scale circulations. These two effects often tend to be real-

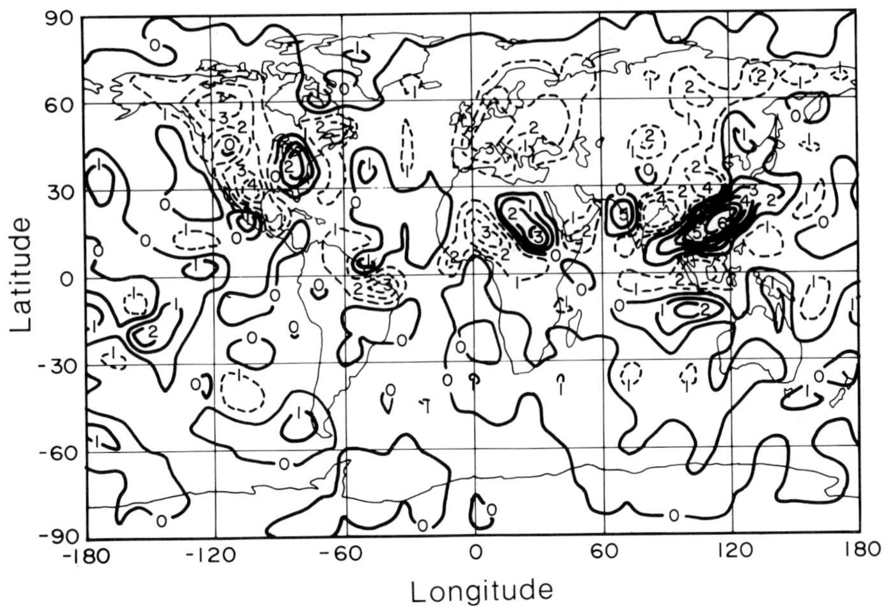

Fig. 15. Same as Fig. 14 for the capacity experiment.

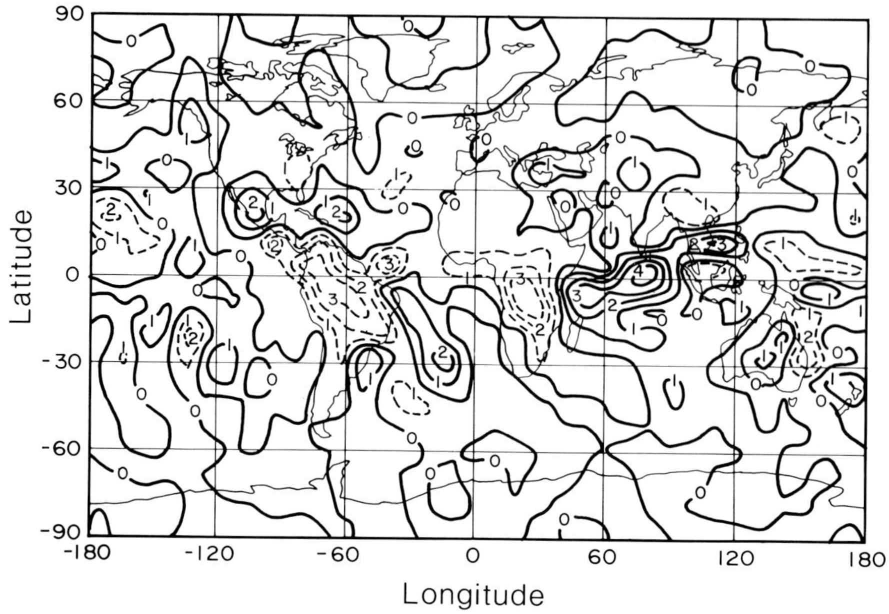

Fig. 16. Difference in precipitation for the months of December-February between the albedo experiment and the control.

ized in different geographical regions and in different seasons. The diffusion experiment altered the rate at which water is made available from deeper levels, and its main effect was a stretching out of the time of soil moisture loss, which affected the seasonal phase of precipitation.

These experiments were not designed to predict climate changes associated with minor vegetation alterations; the results are more applicable to large climate perturbations, such as those associated with ice ages, or to significantly different biological eras, such as those occurring before the appearance of land plants some 400 million years ago. In addition, a more sophisticated treatment of the effect of vegetation would include additional parameterizations, such as the ability of forests to limit evaporative losses by closing leaf stomata during times of water stress, and the effect of increased dust at high levels in the atmosphere due to soil erosion, as well as using parameters associated with the actual soil

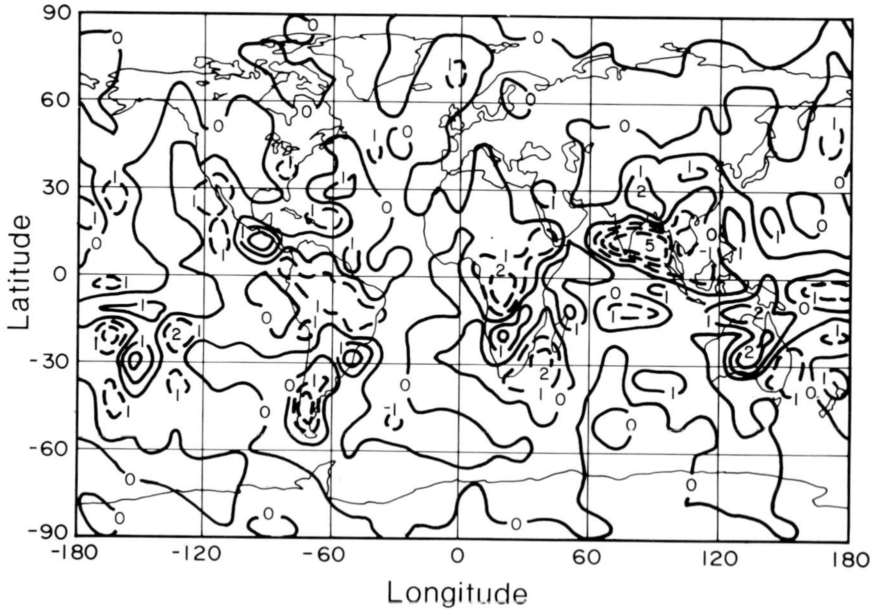

Fig. 17. Same as Fig. 16 for the capacity experiment.

in each area. The treatment here has examined the gross effects of vegetative influence, which appear to be highly significant. This emphasizes the necessity for incorporating vegetation influence as realistically as possible into global climate models.

Acknowledgments. Thanks go to R. Suozzo and J. Lerner for computer assistance. This work was supported by the National Aeronautics and Space Administration Climate Program Office managed by Dr. R. Schiffer.

References

Berry, F., E. Bollay and N. Beers (ed.), Handbook of meteorology, McGraw-Hill Book Company, 1068 pp. 1973.

Charney, J. G., Dynamics of deserts and drought in the Sahel, Quart. J. Roy. Met. Soc., 101, 193-202, 1975.

Charney, J. G., W. Quirk, S.-H. Chow and J. Kornfield, A comparative study of the effects of albedo change on drought in semi-arid regions, J. Atm. Sci., 34, 1366-1385, 1977.

Chervin, R. M., Response of the NCAR general circulation model to changed land surface albedo, Report of the JOC Study Conference on Climate Models: Performance, intercomparison and sensitivity studies, Volume 1, 563-581, 1979.

Dickinson, R. E., Effects of tropical deforestation on climate, Blowing in the Wind: Deforestation and Long-Range Implications. Publication 14, Studies in Third World Societies, Dept. of Anthropology, College of William and Mary, Williamsburg, VA, 411-441, 1980.

Hansen, J. E., G. Russell, D. Rind, P. Stone, A. Lacis, S. Lebedeff, R. Ruedy, L. Travis, Efficient three dimensional global models for climate studies: models I and II, Mon. Wea. Rev., 111, 609-662, 1983.

Lockwood, J. and P. Sellers, Comparisons of interception loss from tropical and temperate vegetation canopies, J. App. Meteor., 21, 1405-1412, 1982.

Manabe, S., A study of the interaction between the hydrological cycle and climate using a mathematical model of the atmosphere, Summary of presentation at Meeting on Weather-Food Interactions, Massachusetts Institute of Technology, May 4-11, 1975.

Matthews, E., Global vegetation and land use: new high resolution data bases for climate studies, J. Clim. and App. Meteor., 22, 474-487, 1983.

Mintz, Y., The influence of soil moisture on rainfall and circulation: a review of simulation experiments, Paper presented at the GARP Study Conference on Land Surface Processes in Atmospheric General Circulation Models, Greenbelt, Md., January 5-10, 1981.

Rind, D., The influence of ground moisture conditions in North America on summer climate as modeled in the GISS GCM, Mon. Wea. Rev., 110, 1487-1494, 1982.

Rowntree, P. R., Numerical experiments relevant to the Sahel drought, paper presented at the COSPAR Sahel Workshop, Paris, January 27-28, 1982.

Schutz, C. and W. Gates, Global climatic data for surface, 800mb, 400mb, The Rand Corporation, R-915-ARPA, R-915/1-ARPA, R-1029-ARPA, 1971.

Sud, Y. and M. Fennessy, A study of the influence of surface albedo on July circulation in semi-arid regions using the GLAS GCM, J. Clim., 2, 105-125, 1982.

Walker, J. M. and P. R. Rowntree, The effect of soil moisture on circulation and rainfall in tropical model, Quart. J. R. Met. Soc., 103, 29-46, 1982.

CLOUD/CLIMATE SENSITIVITY EXPERIMENTS

John O. Roads, Geoffrey K. Vallis and Lorraine Remer

Climate Research Group, Scripps Institution of Oceanography
University of California, San Diego, A-024, La Jolla, California 92093

Abstract. A study of the relationships between large-scale cloud fields and large-scale circulation patterns is presented. The basic tool is a multi-level numerical model comprising conservation equations for temperature, water vapor and cloud water and appropriate parameterizations for evaporation, condensation, precipitation and radiative feedbacks. Incorporating an equation for cloud water in a large-scale model, which is somewhat novel, allows the formation and advection of clouds to be treated explicitly. The model is run on a two-dimensional, vertical-horizontal grid with constant winds. It is shown that cloud cover increases with decreased eddy vertical velocity, decreased horizontal advection, decreased atmospheric temperature, increased surface temperature, and decreased precipitation efficiency. The cloud field is found to be well correlated with the relative humidity field except at the highest levels. When radiative feedbacks are incorporated and the temperature increased by increasing CO_2 content, cloud amounts decrease at upper-levels or equivalently cloud top height falls. This reduces the temperature response, especially at upper levels, compared with an experiment in which cloud cover is fixed.

Introduction

The compensating features of the cloud greenhouse and the cloud albedo effects, allied to possible changes in cloud height, distribution, and characteristics complicate our understanding of the response of the climate to external stimuli, such as increased CO_2 levels. The problem is twofold. First it is necessary to understand the radiative effects of cloud cover. This has been discussed by Ohring and Clapp (1980), Cess and Ramanathan (1978), Schneider (1972) and many others. The second, and less well understood, problem lies in understanding the causes of cloud variability, or alternatively predicting cloud response in a given situation. Predicting cloud response does not simply involve predicting total cloud cover. For example, Schneider has demonstrated the importance of the cloud top height. If cloud top height increases then the effective emitting temperature decreases and the greenhouse effect is enhanced. On the other hand, if total cloud area is simply increased then it is likely that the albedo effect will outweigh the greenhouse effect (e.g., Hartmann and Short, 1980). However, Cess (1976) and Cess et al. (1982) argue that the competing cloud effects may compensate each other very closely at all latitudes and seasons. Presumably, this does not mean the effects at individual longitudes will also be closely compensated (Shukla and Sud, 1981).

Much has been written about the physical characteristics affecting cloud formation and associated precipitation (for a comprehensive account see Ludlam, 1980). The task facing climatologists is not only to understand where and why clouds of various shapes, sizes and characteristics form but to parameterize this in terms of the large scale. Although simple parameterizations of cloud cover, such as setting it a function of relative humidity (first suggested by Smagorinsky, 1960; see also Slingo, 1980) or allowing cloud where condensation occurs, appear at first glance to give climatologically acceptable results in GCM's, high clouds are poorly treated and hence even qualitative uncertainty in the radiative feedbacks must arise. Furthermore, it seems that the parameterizations must be tuned to the particular GCM used, clearly an unsatisfactory procedure.

At the risk of oversimplifying cloud physics, let us assume that clouds occur primarily where parcels of air become supersaturated with respect to a plane surface. If the rate at which supersaturation occurs is sufficiently large to produce sufficiently large amounts of cloud water, then coalescence and accretion of various cloud drops produce still larger drops which eventually fall to the surface. Transport processes will extend clouds into regions (both space and time) where cloudwater evaporates. While the above notions are partially understood in specific cases, the associated large-scale processes that will produce regions of supersaturation and subsaturation are numerous and not well understood. A number of workers, including Manabe and Wetherald (1975), Roads (1978a), Schneider et al. (1978), and

Wetherald and Manabe (1980) have shown that middle tropospheric clouds tend to decrease with increasing temperature even though the intensity of the hydrologic cycle (precipitation and evaporation) increased. The explanation is that with a more intense hydrologic cycle, the vertical velocity variance tends to increase, resulting in more precipitation in regions of ascent and more drying in regions of descent. In regions of descent where cloud cover was already present or under conditions where horizontal advection of dry air is important, the drying effects have more effect on the cloud cover than the moistening effects and hence cloudiness decreases with increasing temperature.

There are certainly other factors influencing the cloud fields. Roads (1978b), for example, has shown that under certain conditions zonal mean cloud cover can be decreased by increasing the mean downward zonal velocity, by increasing the diabatic heating, by increasing the static stability and by decreasing the scaled evaporation, q_γ. Scaled evaporation is the ratio of the surface saturated humidity, q_s, to an atmosphere saturated humidity, q_a. If temperature alone increases, but lapse rate remains the same, the scaled evaporation decreases, since $\partial/\partial T(q_\gamma) < 0$ for $T_s > T_a$.

In a GCM experiment, Wetherald and Manabe (1980) found that low level fractional cloud coverage actually increased with increasing temperature in high latitudes. (Similar results were also found by Potter et al., (1981)). This occurred over a region in which surface albedo decreased rapidly, due to the poleward retreat of the surface ice sheet, causing surface temperatures to rise noticeably - a rise of 19K was found at latitude 80° in Wetherald and Manabe's model due to a 6% rise of solar constant. Atmospheric temperatures did not rise as rapidly, because of the moderating influences of cross-latitude heat exchange. Vallis (1982) and Roads and Vallis (1983) explained this using relatively simple zonally averaged models. As the ice sheet retreated, the lower atmosphere (in high latitudes) was destabilized and an increase in scaled surface evaporation ensued which caused cloud cover to increase. In lower latitudes however, where surface albedo stayed fixed, Roads and Vallis found that the lapse rate was stabilized and scaled evaporation fell. Acting alone this would cause cloud cover to fall; however this was compensated for by the effect of an increase in the net radiative cooling required to balance the increased precipitation.

Several mechanisms have thus been identified that affect the large-scale cloud and relative humidity fields: (1) eddy vertical velocity, (2) horizontal advection, (3) low-level temperature inversion, (4) radiative effects. How important these mechanisms are for models with more verisimilitude to nature is still unknown. For example, one of the chief simplifications (excepting the GCM work) has been the restriction of clouds to one-level, so prohibiting understanding of the vertical structure of the cloud and relative humidity fields. Since the vertical extent of cloud fields is crucial in determining the relative importance of the solar and infrared radiation components of the radiation field, this simplification is ultimately unacceptable. Nor does any wholly satisfactory scheme currently exist to relate cloud cover and cloud optical properties to the dynamical fields (e.g., relative humidity fields) predicted by a GCM.

The study to be described in this paper addresses the problem of the vertical cloud structure using a two-dimensional model with variations in the zonal and vertical directions. The problem of relating the clouds to the dynamical and thermodynamical fields is addressed by including an explicit equation for cloud water. The model is written in such a manner that the condensation and radiation algorithms could easily be added to a more comprehensive general circulation model. Some of the suggested hypotheses regarding cloud variability could then be tested further.

Model

The model comprises conservation equations for temperature, water vapor, cloud water and parameterizations for the appropriate sources and sinks. The continuous form of the equations may be written:

Thermodynamic

$$\frac{\partial\theta}{\partial t} + \nabla\cdot\underline{v}\theta + \frac{\partial}{\partial p}\omega\theta = g\frac{\partial}{\partial p}F_\theta + \{\frac{L}{C_p}\dot{C}_\theta\}p^{-\kappa} + K\nabla^2\theta \quad (1)$$

Water vapor

$$\frac{\partial q}{\partial t} + \nabla\cdot\underline{v}q + \frac{\partial}{\partial p}\omega q = g\frac{\partial}{\partial p}F_q + \dot{C}_q + K\nabla^2 q \quad (2)$$

Cloud water

$$\frac{\partial c}{\partial t} + \nabla\cdot\underline{v}c + \frac{\partial}{\partial p}\omega c = g\frac{\partial}{\partial p}F_c + \dot{C}_c + K\nabla^2 c \quad (3)$$

Precipitation (steady state form)

$$-F_l = \int_0^p \dot{C}_l\frac{dp}{g} \quad (4)$$

where

θ potential temperature $= Tp^{-\kappa}$

T temperature

$\underline{v}$ horizontal velocities

ω pressure velocity

g gravity

C_p heat capacity at constant pressure

L latent heat of condensation and sublimation

q water vapor mixing ratio

c cloud water mixing ratio

t time

$\nabla\cdot$ horizontal divergence operator

p pressure

$p^{-\kappa}$ $\left(\frac{100 \text{ k Pa}}{p}\right)^{\kappa}$

F_θ thermodynamic eddy + radiative fluxes

F_q water vapor eddy fluxes

F_c cloud water eddy fluxes

F_l precipitation fluxes

$\dot{C}$ conversion terms for water substance

K horizontal diffusion coefficient = 10^4 m^2 s^{-1}

R perfect gas constant

κ R/C_p

Numerics

The continuous equations are finite differenced in the horizontal and vertical. The vertical grid consists of points (9) equi-distant in pressure between 1000 and 100 mb. Fictitious boundary points below 1000 and above 100 mb, are used to define values (via boundary conditions) at 1000 and 100 mb. These boundary conditions at 100 mb are

$$\frac{\partial T}{\partial p} = 0$$

$$\frac{\partial q}{\partial p} = 0$$

$$\frac{\partial c}{\partial p} = 0$$

$$\omega = 0$$

and at 1000 mb

$$k\frac{\partial \theta}{\partial p} = k_s(\theta_g - \theta)$$

$$k\frac{\partial q}{\partial p} = \beta k_s(q_g - q)$$

$$k\frac{\partial c}{\partial p} = -k_s c$$

$$\omega = 0$$

where β, k and k_s are empirical coefficients and will be explained later.

Sixteen equally spaced gridpoints are used to represent variations in the zonal direction, and periodic boundary conditions are imposed. In the north-south direction the model currently assumes no variation.

The equations are stepped forward in time using a leapfrog and periodic restart scheme for the dynamical terms and a forward step for the parameterizations. It was found necessary to filter the temperature and moisture in the horizontal to remove 2 gridpoint waves. The condensation routines are then called in order to maintain convective stability and remove any supersaturation.

The velocity fields are preset and remain constant throughout the integration. In addition to a constant zonal wind of constant shear, a divergent wind is also included such that with the pressure velocity the mass continuity equation is satisfied. That is

$$\nabla\cdot\underline{v} = -\frac{\partial \omega}{\partial p}$$

and ω=0 at the top and bottom is imposed. The imposition of two-dimensionality and fixed velocity fields are the primary simplifications of this model. The paper thus becomes a study of the relationship of cloud structure to a given dynamical field. Complete verisimilitude is sacrificed in the interest of isolating particular mechanisms. Clearly the dynamical effects and feedbacks of cloud systems on the large-scale must be neglected (e.g., Moncrieff and Green, 1972 Stevens and Lindzen, 1978).

Vertical Eddy Fluxes and Diabatic Heating

The surface fluxes of heat, moisture and cloud-water are described by simple bulk aerodynamic laws

$$F_\theta = k_s(\theta_g - \theta)$$

$$F_q = \beta k_s(q_g^* - q)$$

$$F_c = k_s(-c)$$

where β, which is set to unity here, is related to soil moisture. Further up in the atmosphere the diffusive fluxes are

$$F_\theta = k\frac{\partial \theta}{\partial p}$$

$$F_q = k\frac{\partial q}{\partial p}$$

$$F_c = k\frac{\partial c}{\partial p}$$

The boundary condition used at the surface is that these two fluxes are equal.

The diabatic heating due to radiation is initially parameterized by a Newtonian cooling law

$$g\frac{\partial}{\partial p}F_\theta(\text{rad}) = (\theta_e - \theta)/10^6\text{s} \quad .$$

More elaborate radiative parameterizations (i.e., Lacis and Hansen (1976) for solar and Rodgers (1967) for infrared) will be incorporated in Section 4.

Water Conversions and Convection

The cloud water conversion and precipitation parameterizations are essentially based on simplifications of physically based schemes. A comprehensive treatment of the more complex theory is given by Kessler (1969). Here we assume that elaborate microphysical and eddy transport schemes are unnecessary and that the transports of water vapor and temperature are accomplished by large-scale advection or can be adequately parameterized by a convective adjustment scheme.

The conversion terms, $\dot{C}$, are complex nonlinear terms that describe the conversion between water vapor and liquid water or ice and the conversions between different number densities and size distributions. The conversion in this model is simplified by the perhaps rather arbitrary distinction between water/ice droplets of very small diameter and negligible fall velocity (cloud water) and water/ice droplets of large diameter and a substantial fall velocity (precipitation).

Conversion of cloud water to precipitation. The following processs are thought to be the dominant contributors to the conversion of cloud water to precipitation. Initially, small cloud droplets self-collect to form larger cloud droplets and small precipitation particles in a process referred to as auto-conversion by Kessler. The process seems to have a cubic dependence upon cloud water. The so-called Bergeron process also contributes to the formation of small precipitation particles by augmenting ice particles at the expense of liquid water in addition to sublimating water vapor. Presumably this process decreases the auto-conversion time scales. Once small precipitation particles are formed, they fall through the cloud and further collect cloud droplets. This is referred to as accretion and is dependent upon the amount of cloud water and precipitable water as well as their respective number densities and size distributions. In the limit of large accretion, cloud water is probably converted to precipitation as the 4.5 power of cloud water.

It would thus seem that precipitation is converted to cloud water in a highly nonlinear fashion (see also Sunqvist, 1981). This insures that clouds have a fairly uniform mixing ratio (compared to, say, the mixing ratio for water vapor) that precipitation falls from only dense clouds, and that drastic changes in the circulation parameters are only likely to give small changes in the cloud water. For this study we feel it is adequate to use a simple conversion rate, namely

$$\dot{C}_{cl} = ac^3 \qquad (5)$$

where $a = 10^4\ s^{-1}$. If the precipitation rate is typically $1.9\text{x}10^{-5}$ kg m^{-2} s^{-1} then the average cloud water mixing ratio is $5.7\text{x}10^{-5}$ kg/kg. If clouds exist in only 50% of the atmosphere then the average cloud water content in the clouds is $7.2\text{x}10^{-5}$ and the average cloud water content (cloudy + noncloudy regions) is $3.6\text{x}10^{-5}$. These estimates of cloud water content are all within Sasamori's (1975) estimates. The muliplying factor and exponent will be tested later to determine the sensitivity of the model to the cloud physics parameterization. As will be shown, the parameterization is crucial for determining the characteristics of the cloudwater but negligible for determining the cloud fractional cover and large-scale vapor fields, because the amount of cloudwater in the atmosphere is negligible compared to the amount of water vapor.

Because of the highly nonlinear rate at which cloud water is converted to precipitation and the desire to maintain a fairly large time-step sufficient only for large-scale dynamical stability, it is assumed that the process can be described as

$$\dot{C}_{cl} = -\int_t^{t+\Delta t} ac^3 dt$$

or

$$\dot{C}_{cl} = \frac{c}{\Delta t}\left(1 - \frac{1}{(1+2\Delta tac^2)^{1/2}}\right) \qquad (6)$$

Conversion of precipitation and cloud water to water vapor. Once precipitation is formed it falls through the cloud and into the free atmosphere. If the precipitation is falling in a region of unsaturated air then the precipitation is assumed to evaporate completely or to the point where the air is just saturated. That is, the following set of equations are solved

$$C_p\delta T + L\delta q = 0$$

$$q + \delta q = q^*(T,p) + \frac{\partial q^*}{\partial T}\delta T$$

$$\delta q - \Delta t\,\frac{g\,\delta F_l}{\Delta p} = 0$$

where δ denotes the differential changes and an * denotes saturation values (saturation is over water for temperatures above freezing and is over ice for temperatures below freezing). The conversion of precipitation to water vapor is

$$\dot{C}_{lq} = \frac{(q^{*}-q)}{\Delta t(1+\frac{L\delta q^{*}}{C_p \delta T})} .$$

If this results in $-(F_1 + \delta F_1) < 0$ then the precipitation is completely evaporated without saturating the atmosphere and the conversion is

$$\dot{C}_{lq} = -\frac{g}{\Delta p} F_1 .$$

Cloud water in unsaturated conditions is treated in a similar manner to give C_{cq}.

Conversion of water vapor to cloud water. The conversion of water vapor to cloud water requires simply that supersaturation is reduced to a saturation value instantaneously; hence for supersaturation conditions the following set of equations is solved

$$C_p \delta T + L\delta q = 0$$

$$q + \delta q = q^{*}(T,p) + \frac{\partial q^{*}}{\partial T} \delta T$$

or

$$\dot{C}_{qc} = \frac{(q-q^{*})}{\Delta t(1+\frac{L}{C_p}\frac{\partial q^{*}}{\partial T})}$$

Convective condensation. Ostensibly the conversion terms have now been parameterized for the model, but one process remains that will alter the condensation process, namely moist convection. As stated by Manabe et al. (1965) and Sarachik (1980) our knowledge of this process is insufficient to warrant a very detailed parameterization of moist convection. Further, Lindzen et al. (1982) found that sensitivity experiments with a moist-adiabatic adjustment yielded similar results to experiments with an alternative cumulus cloud model. Thus, we employ a moist convection adjustment scheme, which operates when the vertical gradient of moist static energy is negative. In the adjustment, it is assumed that turbulent processes occur which equilibrate the moist static energy between the levels, maintain saturated conditions, conserve the total energy, and equilibrate the cloud water. Momentum could also be equilibrated in a similar fashion. The equations for two contiguous levels (1 denotes the upper level and 2 the lower level) are:

$$C_p \delta T_2 + C_p \delta T_1 + L\delta q_2 + L\delta q_1 = 0$$

$$\Delta h + \frac{C_p}{2}(p_1^{\kappa} + p_2^{\kappa})\left\{\frac{\delta T_2}{p_1^{\kappa}} - \frac{\delta T_1}{p_2^{\kappa}}\right\} + L(\delta q_2 - \delta q_1) = 0$$

$$q_2 + \delta q_2 = q_2^{*} + \frac{\partial q_2^{*}}{\partial T} \delta T_2$$

$$q_1 + \delta q_1 = q_1^{*} + \frac{\partial q_1^{*}}{\partial T} \delta T_1$$

$$\delta c_1 + \delta c_2 = -(\delta_{q1} + \delta_{q2})$$

$$c_1 + \delta c_1 = c_2 + \delta c_2$$

where the difference in moist static energy between two levels has been defined as

$$\Delta h = \frac{C_p}{2}(p_1^{\kappa} + p_2^{\kappa})(\theta_2 - \theta_1) + L(q_2 - q_1) .$$

The convective process occurs whenever $\Delta h > 0$ and $(c_1 + \delta c_1 + c_2 + \delta c_2) \geq 0$. The process in the model begins by testing and solving contiguous levels from the bottom of the atmosphere and moving upward. Although more than two levels could be solved simultaneously, idealized tests showed that iteration of the above two-level process produced the same results as multi-level systems; hence, for simplicity, only the two-level system is used. Also, since the time stepping was sufficiently small (approximately 1 hour), iterations at each time step produced the same steady state results as no iterations. However, if the condensation routines are called infrequently, it becomes necessary to iterate. For rapidly changing dynamical conditions it may also be necessary to iterate.

To summarize the above processes, then, the model initially tests for moist static instability. If instability exists the temperature and moisture content of two levels are adjusted to equalize the moist static energy and convert excess water vapor to cloud water. The convection is assumed to equalize cloud water mixing ratio in the vertical. This is repeated through the depth of the atmosphere. Cloud water is then advected, and converted to precipitation according as (6). The precipitation then falls, either to reach the surface or to be evaporated in non-saturated layers. If no convection occurs, we may still have local cloud water vapor to cloud water conversion, and the precipitation process follows in precisely the same fashion.

Having calculated the new stable state, it becomes necessary to define the condensation and the accompanying turbulent transport. These are given by

$$g\frac{\partial F_{T1}}{\partial p} = \frac{\delta T_1}{\Delta t}$$

$$g\frac{\partial F_{T2}}{\partial p} = \frac{\partial T_2}{\Delta t} - \frac{L}{C_p}\dot{C}_{qc}$$

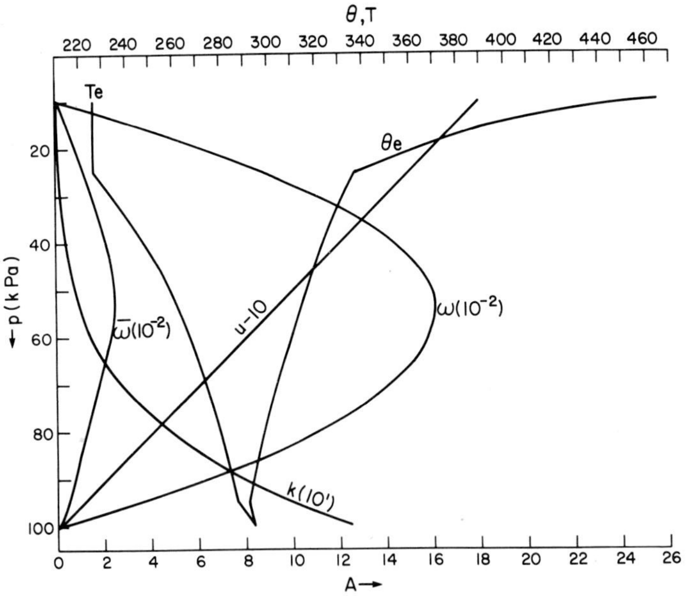

Fig. 1. Vertical structure of various control parameters. A denotes a general component for either ω, u-10, k, ω and the multiplying factor is denoted on the figure. The temperature scale for T_e, θ_e is given at the top of the figure. All values are in mks units.

$$g\frac{\partial Fq_1}{\partial p} = \frac{\delta q_1}{\Delta t}$$

$$g\frac{\partial Fq_2}{\partial p} = \frac{\delta q_2}{\Delta t} + \dot{C}_{qc}$$

$$g\frac{\partial Fc_1}{\partial p} = \frac{\delta c_1}{\Delta t}$$

$$g\frac{\partial Fc_2}{\partial p} = \frac{\delta c_2}{\Delta t} - \dot{C}_{qc}$$

where

$$\dot{C}_{qc} = -\frac{(\delta q_1+\delta q_2)}{\Delta t}$$

Dry convection also occurs infrequently whenever

$$\Delta s = \frac{C_p}{2}(p_1^\kappa + p_2^\kappa)(\theta_2 - \theta_1) > 0 \quad .$$

As in the moist convective process, q and c are assumed to be well mixed while conserving the dry static energy.

Having found the various functions, the equation (1,2,3) for the time rates of changes due to the conversion and convective terms and the precipitation equation (4) become

$$\frac{\partial\theta}{\partial t} = g\frac{\partial}{\partial p}F_\theta + \left\{\frac{L}{C_p}(\dot{C}_{qc} - \dot{C}_{cq} - \dot{C}_{lq})\right\}p^{-\kappa}$$

$$\frac{\partial q}{\partial t} = g\frac{\partial F_q}{\partial p} - \dot{C}_{qc} + \dot{C}_{cq} + \dot{C}_{lq}$$

$$\frac{\partial c}{\partial t} = g\frac{\partial F_c}{\partial p} - \dot{C}_{qc} + \dot{C}_{cq} + \dot{C}_{cl}$$

$$-F_l = \int_0^p (\dot{C}_{cl} - \dot{C}_{lq})dp/g$$

Experiments (Newtonian Cooling)

The experiments with a simple Newtonian cooling scheme are designed to illustrate the effects of changing the horizontal advection, vertical velocity and surface temperature, and to study the effects of incorporating an explicit cloud water equation. The velocity fields consist of a mean zonal wind, plus a zonally varying vertical velocity wave. The velocity fields are not meant specifically to reproduce a particular dynamical situation, but to allow the effects of horizontal and vertical motion to be studied in a simple framework.

Figure 1 shows the vertical profiles of the various constants for the control run. The vertical velocity is a maximum ($1.6\text{x}10^{-1}$ kg m^{-1} s^{-2}) at 550 mb and decreases to zero at 1000 and 100 mb. The vertical mixing parameter k decreases from $12.6\text{x}10^{-1}$ at 1000 mb to $16.5\text{x}10^{-3}$ (kg^2 m^{-3} s^{-3}) at 100 mb. The zonally averaged wind increases from 10 at 1000 mb to 28 (m s^{-1}) at 100 mb. Since the divergence is assumed to be zero, the zonal wind also includes a component that varies in x. This component is zero at 550 mb and increases linearly to a wave with an amplitude of $4.46ms^{-1}$ at 1000 and 100 mb. If the pressure velocity varies as a cosine wave, then u varies as a sine wave with positive amplitude below 550 mb and negative amplitude above.

Because of the constant value of the winds, steady state solutions for the equations are obtained. We will describe in detail the results for the control experiment and then results from other experiments motivated in part by the simple one-level cloud system of Roads (1978b).

Figure 2 shows a steady-state solution as a function of longitude for the control parameters. Here the periodic domain extends from west to east with a distance of $5\text{x}10^3$ km and grid points every 312.5 km. Essentially wavenumber 6 phenomena are studied if these scales are appropriate for midlatitude circulation systems. The vertical velocity wave is shown in the upper part of the figure to have maximum descending motion on the edges of the domain and maximum ascending motion in the center of the domain. The precipitation at the surface is correlated with the vertical motion field although precipitation is certainly narrower, occuring significantly over less than 1/3 of the domain. Note also that the precipitation occurs almost strictly on the leeward side of the vertical velocity wave. The characteristic features suggest that for the chosen parameters, horizontal advection of moist air is extremely important for the precipitation field.

Various measures of the total cloud cover are also shown in Figure 2. Of course, the most natural criterion to use is simply the presence of cloud water, although clearly this cannot be used in GCMs which do not predict cloud water. The narrowest measure is the area of positive condensation which also corresponds with the area of precipitation at the surface. If the presence of cloud water or a relative humidity criterion (H = 1) is used then the cloud boundary extends a further distance on the windward side. Smaller relative humidity criterions naturally give larger coverages still.

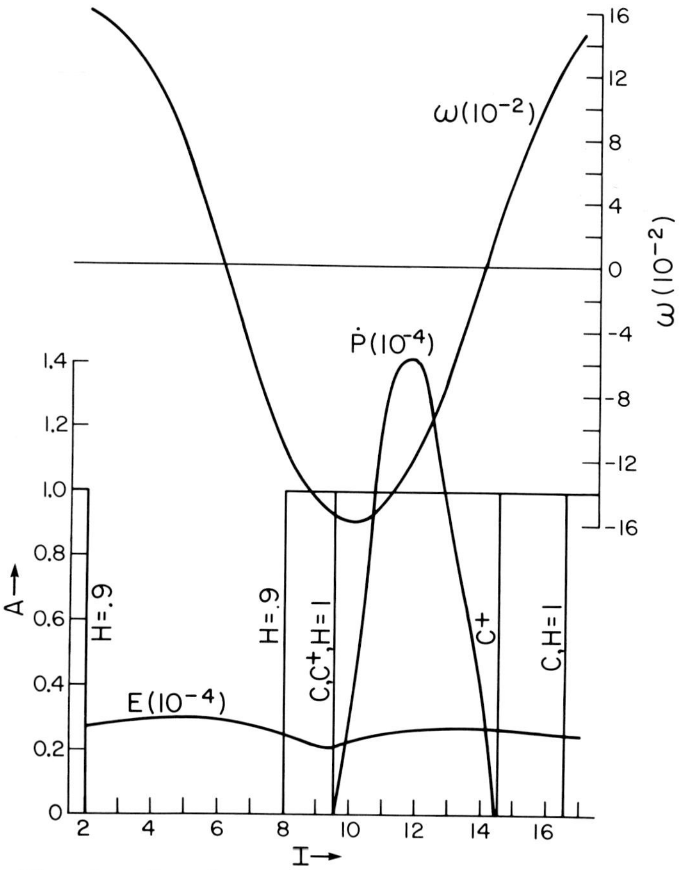

Fig. 2. Zonal structure of various components as a function of I, the zonal gridpoint running from east to west. ω denotes the eddy pressure velocity, P the precipitation rate, and E the evaporation rate in kg m^{-2} s^{-1}. Various measures of the total cloud cover: C, cloudwater > 0; C+, condensation > 0; and relative humidity H ⩾ 1; H ⩾ .9 are also given.

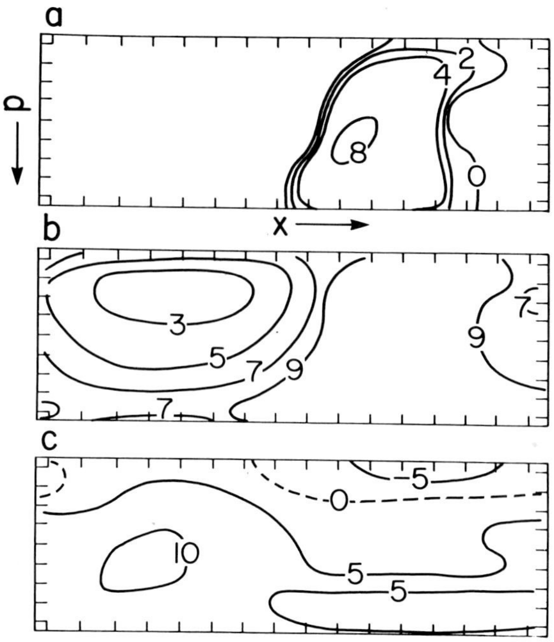

Fig. 3. a) The two-dimensional structure of the cloudwater for the control case. The tick marks denote the gridpoints of the two-dimensional x, p grid with gridpoints every 10 kPa in p and every 312.5 km in x. Contours are every, 0, 4, 8 x 10^{-5} kg/kg. b) Same as 3a except for relative humidity. Contours cover the range 30% to 90%. c) Same as 3a except for deviations of potential temperature field from the radiative equilibrium values. Contours are every 5 degrees; positive values indicate Newtonian cooling and negative values indicate Newtonian heating.

The two-dimensional structure of the cloud, relative humidity and temperature fields are shown in Figure 3a,b,c. First note that the cloud has an eastward tilt with altitude indicating the dry air advection occurs on the windward side and cloud water advection occurs on the leeward side. The relative humidity field gives a good indication of this structure and also shows that a minimum in relative humidity exists below the tropopause. The temperature field is that which would be expected for a forced circulation with the strongest positive deviations eastward of regions of descent and the strongest negative deviations in regions eastward of the ascending regions. The deviations are substantial with a maximum eastward temperature difference of about 15°C indicating differences in heating of 1.5×10^{-5} K s^{-1}. Of interest also is that the deviations are negative in the region of the cloud top indicating that the parameterized heating rate is positive in this region.

Various experiments were run to see the contributions and sensitivity of various mechanisms. Figure 4 shows the response when the mean zonal wind is removed. In this case the cloud cover decreases at the lowest level and substantially increases at the upper level. The

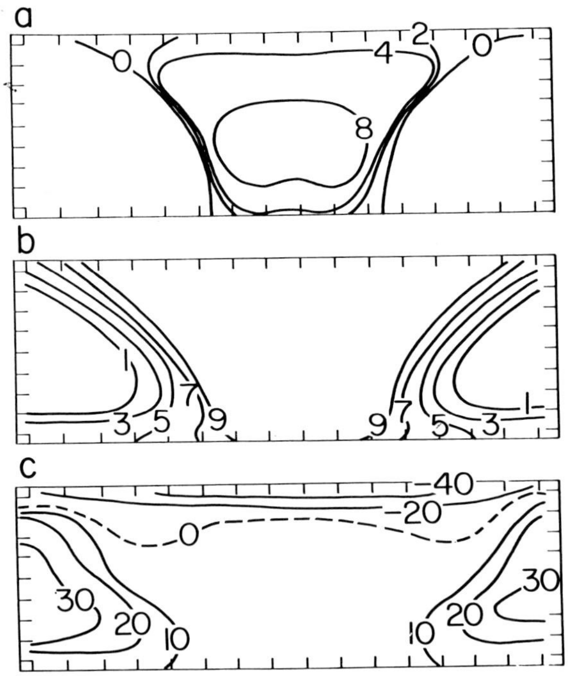

Fig. 4. Same as 3 except u = 0.

relative humidity becomes less in the descending regions and the atmosphere becomes relatively warmer in the descending regions. Not much temperature change occurs in regions of ascent except at the top of the cloud where strong negative temperature deviations occur indicating strong radiative heating. Due to this drastic change in circulation pattern, the change in the zonally averaged relative humidity does not correspond to the change in cloud coverage. For example, in the middle levels fractional cloud cover increased while the zonally averaged relative humidity decreased.

Another experiment in which the vertical velocity amplitude was decreased an order of magnitude yielded cloud coverage at all longitudes and levels except for the uppermost level and part of the next uppermost level. Although the decrease in vertical velocity yielded cloud coverage and evaporation everywhere, the precipitation is much less for this case as may be seen in Figure 5. Note also that the precipitation is strongest in the center of the rising motion and much weaker in regions of descent, decreasing to zero in regions of strong descent. Thus the cloud and precipitation characteristics are strongly dependent upon the ability of the atmosphere to dry itself out in addition to being able to produce clouds.

Another mechanism that is important for latitudinal differences in cloudiness is the mean zonal vertical velocity. Although its inclusion is inconsistent because meridional divergence is not allowed, a term of the form

$$\omega \frac{\partial A}{\partial p}$$

was added to the flux

$$\nabla \cdot \underline{v}A + \frac{\partial}{\partial p} \omega A \quad ,$$

where A is temperature or mixing ratio. ω had the same vertical profile as the eddy vertical velocity term (see Fig. 1) but a much smaller amplitude maximum of 2.5×10^{-2} kg m^{-1} s^{-3}. Because the continuity equation is no longer satisfied, precipitation was much less than evaporation. However, a steady state was reached and is shown in Fig. 6. Note that a low-level cloud of small areal coverage is formed in a very dry and hot atmosphere.

When the surface temperature is varied, large changes in the cloud structure result. Fig. 7a,b show the response in the cloud and relative humidity fields for an increase of surface temperature of 10K and Fig. 7c,d show the response in the cloud and relative humidity fields when the surface temperature is decreased by 10K. Note that clouds strongly increase for a more unstable boundary layer almost everywhere. Also of interest was that the clouds were almost completely formed by moist convection for the unstable boundary case and completely by local supersaturation condensation for the stable boundary case. As discussed by Vallis (1982) and Roads and Vallis (1983) these cases have applications to models in which the ice line is allowed to change, because of large changes in low-level static stability and cloudiness in regions of the ice line.

It was pointed out above that a major dif-

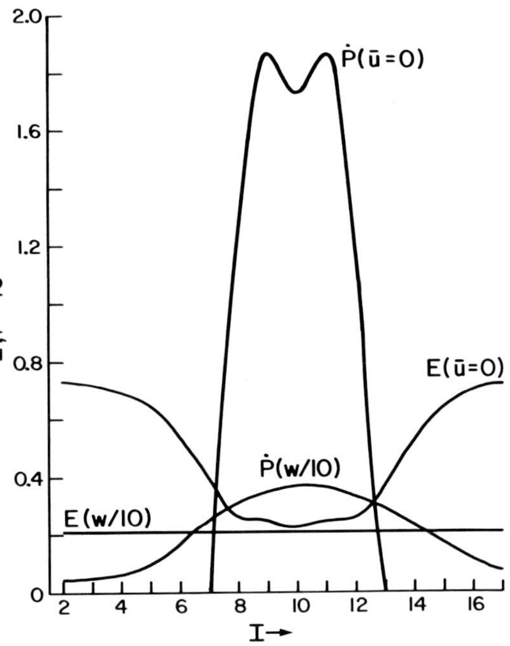

Fig. 5. Zonal structure of the precipitation and evaporation for the cases u = 0 and ω = .1ω (control).

ference between this model and others is the explicit inclusion of cloud water. If all cloud water is precipitated out as soon as it forms, then essentially the model reduces to the conventional types. A comparison of cases with and without cloud water and its concomitant advection is useful. Thus three of the above described experiments were rerun with immediate precipitation of cloudwater. For these cases it is shown in Figure 8 that the relative humidity fields are more or less the same with or without cloud water except at the top of the atmosphere where cirrus cloudiness is dependent upon cloud water being present. Because of the large influences that cirrus cloudiness has on the radiation field it would seem to be better to retain the cloud-water parameterizations. However, good estimates of cloud cover in the lower regions are obtained by equating cloud cover to values of relative humidity fields near unity. At the other extreme Figure 9 shows the response when $\dot{C}_{cl} = 10^2 c^3$ (as opposed to $10^4 c^3$ see eq. 5). The cloud water content increases by a factor of 5 with the largest areal change occurring at the upper levels where cloud water can now extend over much more extensive regions because of the importance of advection. The change in the relative humidity and temperature field are negligible however at lower levels. Thus, from the two previous experiments, we conclude that cloud characteristics and variations are determined by the circulation parameters that result in saturation and that modifications to these occur chiefly at upper levels where cirrus clouds detrain from the main cloud. The precise form of the cloud microphysical constants appears unimportant.

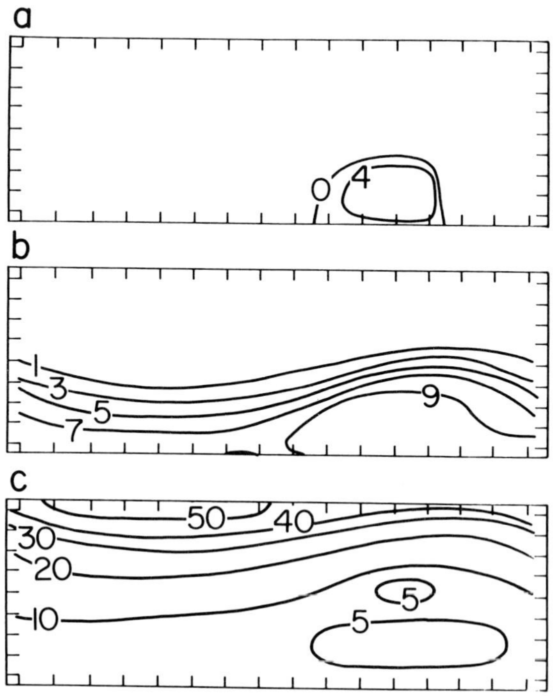

Fig. 6. Same as 3 except $\omega = 2.5\text{x}10^{-2}$.

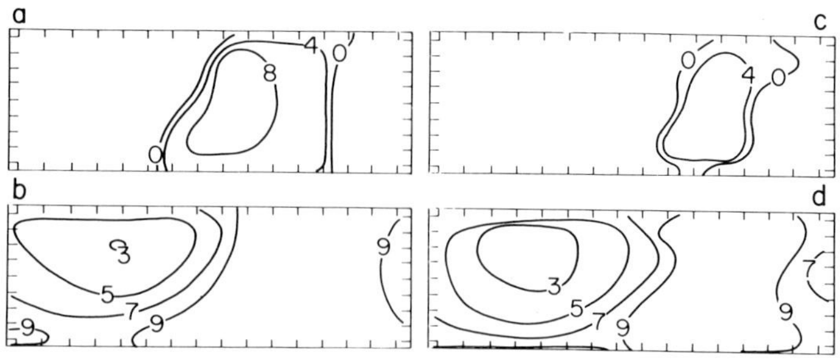

Fig. 7. a) Same as 3a except $T_g = T_g$ (control +10; b) Same as 3b except $T_g = T_g$ (control) + 10; c) Same as 3a except $T_g = T_g$ (control) - 10; d) Same as 3b except $T_g = T_g$ (control) - 10.

The change in the Newtonian heating profile produced some interesting effects. Figure 10 shows the response when the equilibrium temperature was either decreased by 10° everywhere, including the surface, or increased by 10°, including the surface, everywhere. Note that the cloud fields remain about the same in the lower to upper troposphere but that the run with the increased temperature resulted in decreased upper-level cloudiness.

Somewhat different responses occur when the cloud water is totally removed at each time step, Fig. 10c and 10f, because then the amount of upper level cloudiness ($H \geqslant 1$) is actually decreased at the upper levels (except for the topmost level) with decreasing temperature. So if cloudwater is included, the upper levels tend to detrain cloudwater over a longer distance. Presumably this occurs because at the higher temperatures the relative humidity deficit is sufficient to evaporate the cloud water but a lower temperatures the cloudwater (which is maintained at a fairly uniform level by the nonlinear precipitation processes) is much larger than the relative humidity deficit and can travel over much larger distances before finally being evaporated in the driest regions. Thus, if cloudwater is retained in the model, upper level cloudiness will increase with decreasing temperature whereas if cloud water is not retained, upper level cloudiness will increase with increasing temperature.

Cloud Radiation Feedbacks

Schemes for calculating the infrared and solar radiation fluxes were incorporated in order to allow a more self-consistent model. The schemes are fully interactive in that the water vapor and cloud water content predicted by the model are used in the radiative calculations.

Infrared Radiation

The infrared scheme is based on the emissivity approximation (see Rodgers, 1967), in which the integration over all spectral wavelengths is incorporated into simple transmission formulas.

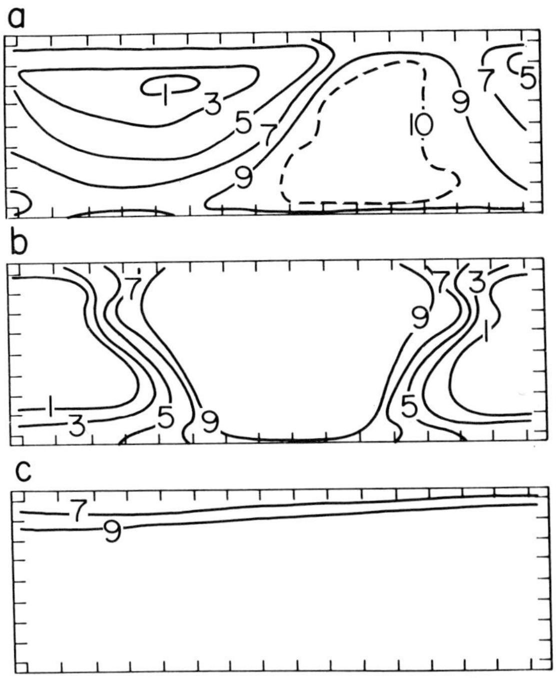

Fig. 8. a) Same as 3b except cloudwater is totally precipitated; b) Same as 4b except cloudwater is totally precipitated; c) Relative humidity field for case where $\omega/10$ is used and cloudwater totally precipitated.

In terms of transmissivity the equations are

$$F\uparrow(p) = \int_{\tau(p_s,p)}^{1} B(p')\,d\tau(p,p') + B_g\,\tau(p_s,p)$$

$$F\downarrow(p) = \int_{1}^{\tau(p,0)} B(p')\,d\tau(p,p')$$

where

$F\uparrow(p)$	the upward flux
$F\downarrow(p)$	the downward flux
$\tau(p,p')$	transmissivity between p and p'
$B(p)$	$\sigma T^4(p)$
$T(p)$	absolute temperature
σ	Stefan-Boltzmann constant
B_g	σT_g^4
T_g	surface temperature

The transmissivity for water vapor is taken from Rodgers. Carbon dioxide is accounted for by assuming that the total transmissivity is the product of the transmissivity of H_2O and CO_2 where

$$\tau(CO_2) = .926 - .02\ \ln(\Delta u)$$

$$\Delta u = 120\ \frac{\Delta p^2}{p_s^2}$$

where p_s is the surface pressure and Δp^2 is the pressure squared differential (using a value of 120 implies a CO_2 content of approximately 300 ppm). Thus Δu is a pressure weighted path length. In a similar manner, the transmissivity in the presence of cloud is the product of cloud, H_2O and CO_2 transmissivities. Currently the cloud transmissivity is taken to be

$$\tau = 1 - u/a;\ u < a$$

$$\tau = 0;\ u > a$$

where

$$du = c\ \frac{dp}{g}$$

$$a = 3x10^{-2}\ kg\ m^{-2}\quad .$$

The dependence of τ on u produces nearly black clouds everywhere except at the cirrus levels. For example, at the uppermost level the typical c is on the order of $1x10^{-5}$ kg/kg and the typical τ is on the order 2/3.

Solar Radiation

The solar radiation scheme uses the multiple scattering method described by Lacis and Hansen (1974) and Somerville et al. (1974). Simplified two stream approximations and absorption coef-

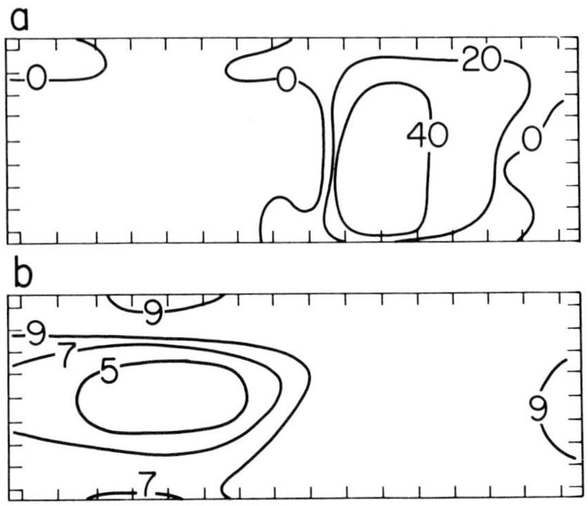

Fig. 9. Same as 3 except $C_{cl} = c^{3x10^2}$

ficient approximated by a sum of 6 gray absorbing intervals are used in partially cloudy skies. In an entirely clear sky and above the highest clouds the absorption is calculated using an integrated absorption coefficient. In this study single layer optical depths are assumed to be given by

$$\tau = c \frac{\Delta p}{g} 10^2 \quad .$$

The typical optical depth in the lower layers is then around 8 for a pressure interval of 100 mb and a cloudwater mixing ratio of 8×10^{-5} kg/kg. In the upper-most layers the optical depth is about 1 for a pressure interval of 100 mb and a cloudwater mixing ratio of 1×10^{-5} kg/kg.

As a typical cosine zenith angle we use $\mu = .225$ which is appropriate for the annually and diurnally averaged zenith angle at 45° latitude. The solar constant is set at 1372 W m^2.

Surface Temperature

The surface temperature equation can be written in the form

$$C_g \frac{\partial T_g}{\partial t} = F_T\downarrow + S\downarrow - \sigma T_g^4 - LE - S_H$$

where

$F_T\downarrow$ downcoming infrared radiation

$S\downarrow$ net downward solar flux

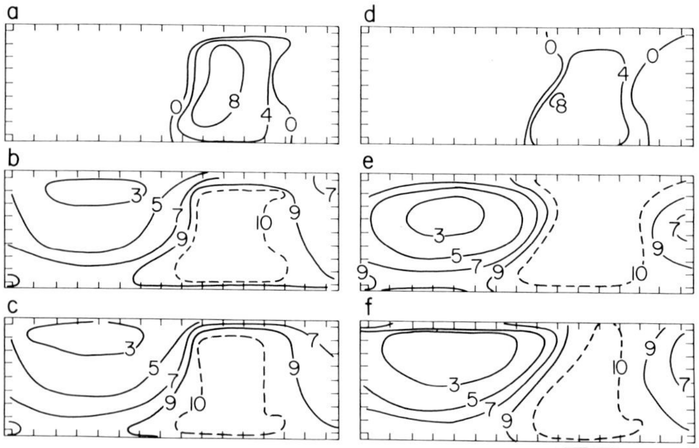

Fig. 10. a) Same as 3a except $\theta_e = \theta_e$ (control) + 10 and $T_g = T_g$ (control) + 10; b) Same as 3b except $\theta_e = \theta_e$ (control) + 10 and $T_g = T_g$ (control) + 10; c) Same as 10b except for experiment in which all cloudwater is precipitated; d) Same as 3a except $\theta_e = \theta_e$ (control) - 10 and $T_g = T_g$ (control) - 10; e) Same as 3b except $\theta_c = \theta_e$ (control) - 10 and $T_g = T_g$ (control) - 10; f) Same as 10e except for experiment in which all cloud water is precipitated.

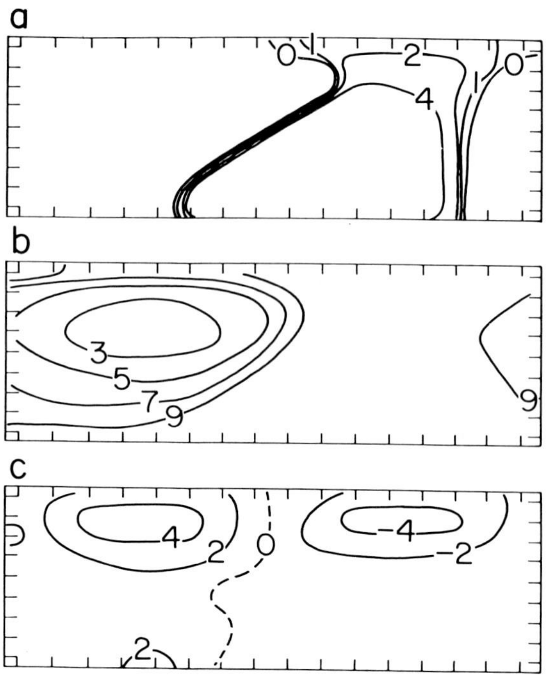

Fig. 11. Control run for model with cloud radiative feedbacks: a) Cloudwater with contours at 0, 1, 2, 4, 8 x 10^{-5} kg/kg; b) Relative humidity field; c) Potential temperature field minus the zonal mean temperature.

E evaporation flux

S_H sensible heat flux

C_g surface heat capacity

At present a very small surface heat capacity is used ($C_g = 3 \times 10^3$ J m^{-2}) so that the temperature is determined mainly as a flux balance between the outgoing evaporation, sensible heat, and infrared flux and incoming solar and infrared radiation fluxes. The equation is time differenced implicitly so that no instability develops.

Experiments

The equilibrium profiles of cloud water, relative humidity and temperature are given in Figure 11. The major differences between the previous cases is that the lower boundary tends to be more unstable which produces cloud at the lower levels everywhere the motion is upward. The upper level cloud field is similar however.

The radiation cooling rates, $\partial\theta/\partial t$, are shown next in Figure 12. Immediately noticeable is that in the atmosphere infrared radiation is dominant especially near the cloud boundaries. Above the lower level clouds, cooling rates of up to 15°/day can occur whereas below the upper level clouds heating rates up to 6°/day can occur. In contrast is the surface flux balance shown in

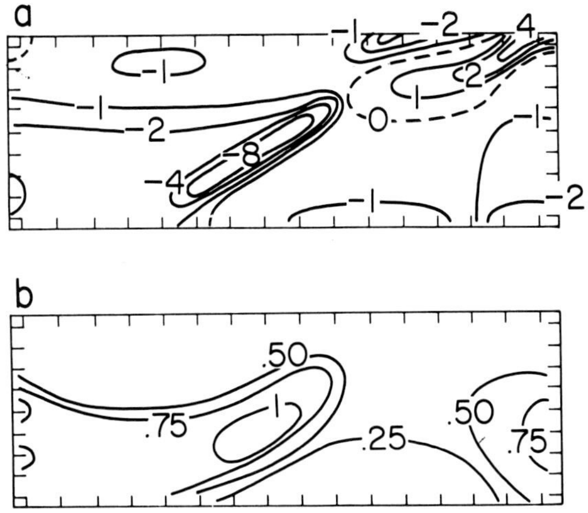

Fig. 12. Control run for model with cloud radiative feedbacks: a) Infrared heating rates, $\partial\Theta/\partial t$, with contours every 4, 2, 1, 0, -1, -2, -4, -8 x 10^{-5} K^{-1}; b) Solar heating rates, $\partial\Theta/\partial t$, with contours every .25, .5, .75, 1 x 10^{-5} K s^{-1}.

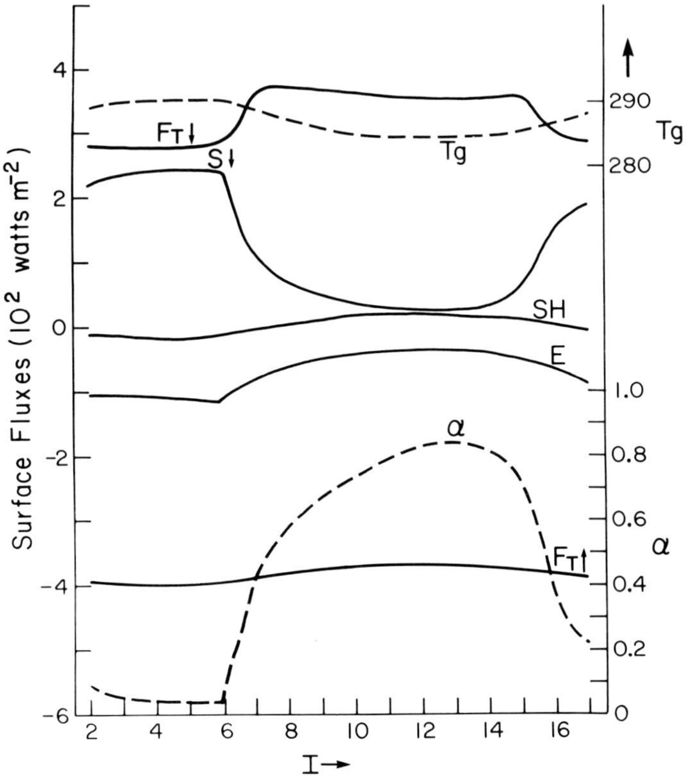

Fig. 13. Surface fluxes in watts m^{-2}. Positive values increase surface temperature and negative values decrease surface temperature. $F_T\downarrow$, S, SH, E, $F_T\uparrow$ refer to downward thermal radiation, downward solar radiation, sensible heat, evaporation, and upward thermal radiation. Also shown are the surface temperature T_g and planetary albedo α.

Figure 13, here it is the solar flux that varies most strongly from east to west with substantially less radiation below the clouds. This solar radiation variation is balanced by an increase in downward radiation, a decrease in upward radiation and decreased sensible and latent heat fluxes to the atmosphere.

Figure 14 shows the response when the cloud radiation feedbacks are modified. Here the model was only integrated for 30 days and although the solution had not yet reached a steady state the solution was in statistically stationary state and was converging slowly toward equilibrium with a slowly increasing exponential damping rate of > 30 days. Note that in the absence of cloud radiation feedbacks the temperature increases almost everywhere except at the upper levels. In the presence of infrared radiation feedbacks alone the atmosphere is still warmer, demonstrating the cloud greenhouse effect. Solar radiation, however, dominates the total cloud radiation feedback and this model atmosphere tends to be colder because of clouds. These results are similar to results found with the GLAS GCM (Herman et al., 1980).

Finally, we show a calculation for the response of the model atmosphere to a quadrupling of CO_2. This amount of CO_2 increase is well within current projections (Bolin et al., 1981) which show that by the end of the 23rd century the atmosphere could contain 4 to 8 times present amounts. Two experiments were run to test the effect of the clouds on the solution. In the fixed cloud, FC, experiment, the clouds which interacted with the radiation field were the clouds found from the control run. In the variable cloud experiment the clouds were allowed to change and interact with the radiation field. The time variation of the two runs for an arbitrary surface temperature is shown in Figure 15. Note that the variable cloud experiment produces a shorter convergence time than the one with fixed clouds, indicating a negative feedback.

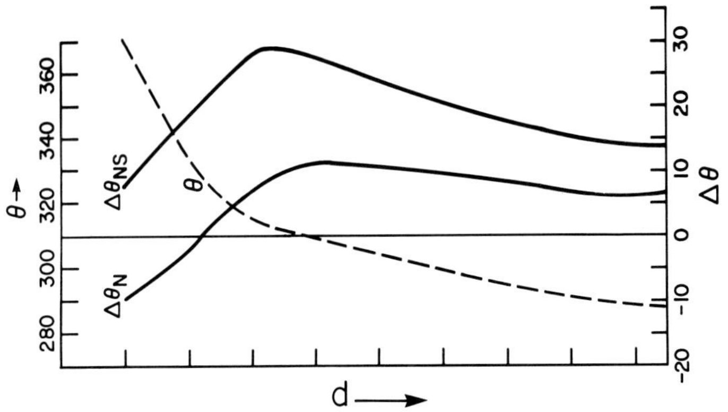

Fig. 14. Change in temperature after 30 days for the experiment with no cloud radiation feedbacks, $\Delta\Theta_N$, and no solar-cloud radiation feedbacks, $\Delta\Theta_{NS}$. The latter experiment includes infrared feedbacks.

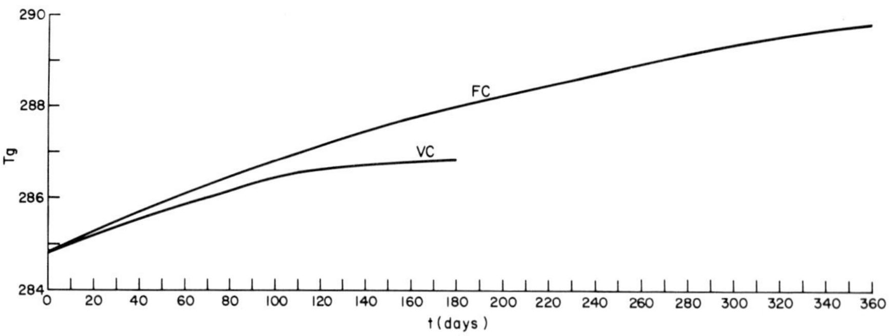

Fig. 15. Fixed cloud, FC, and variable cloud, VC, temperature changes due to CO_2 quadrupling, as a function of time. Both experiments were initialized from the end of the control run.

The change in the cloud field is shown in Figure 16. Note that the upper level cloudiness tends to decrease, and cloud cover at the eastern and western side of the original system also tends to fall in spite of the general overall increase in cloud water in the center of the cloud. In fact, the one positive region on the left edge of the cloud was substantial, increasing by 7×10^{-5} and strongly increasing the infrared cooling in this region. As shown in Figure 17, this change in cloudwater decreases the albedo of the atmosphere indicating that the quadrupling of CO_2 causes slightly more absorption of solar radiation for the variable cloud experiment than the fixed cloud experiment. However, because of the increased water vapor content the albedos are reduced in both cases.

Figure 18 shows the zonally averaged potential temperature response for a CO_2 quadrupling for the fixed and variable cloud experiments. In the variable cloud experiments the quadrupling produced a potential temperature change of about 2K at the surface increasing to about 4K at 50kPa and then decreasing in the upper troposphere with an overall increase in potential temperature. In the lower to mid troposphere the fixed cloud experiments showed similar albeit larger responses and in the upper troposphere the responses are completely opposite with the fixed cloud experiments showing a very strong increase in potential temperature.

In the fixed cloud experiments the vertical temperature structure tended to change according to changes in the moist adiabatic lapse rate. However, in the variable cloud experiments temperature responded similarly up to about 40kPa but then the temperature change decreased rapidly. Presumably, the decrease in upper level cloudiness has a strong influence on the results of the variable cloud experiments: reduction in upper level cloudiness allows radiation from lower in the atmosphere to more easily escape to space.

This experiment shows that assumptions about cloudiness can have a large influence upon the sensitivity of the upper troposphere, and to a smaller degree the lower tropospheric response. In this respect we are reminded of the experiments of Weare and Snell (1974) which also showed a reduced sensitivity to CO_2 because of their cloud model assumptions (see also Charlock, 1982; Paltridge, 1980).

Conclusions

This study has been concerned with the relationship of cloud fields with various features of atmospheric circulation patterns. In particular

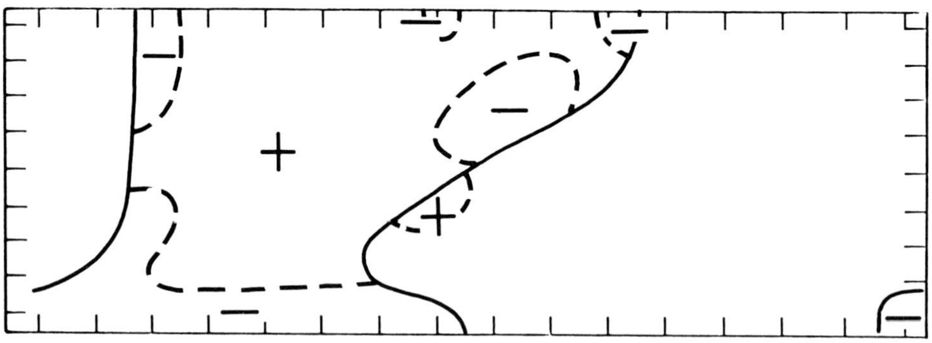

Fig. 16. Differences in cloudwater field for VC CO_2 experiment minus control. Only plus and minus values are indicated because of the large variance of cloud water concentrations. In the negative region at the top of the cloud the maximum cloud water change is -1.1×10^{-5} kg/kg.

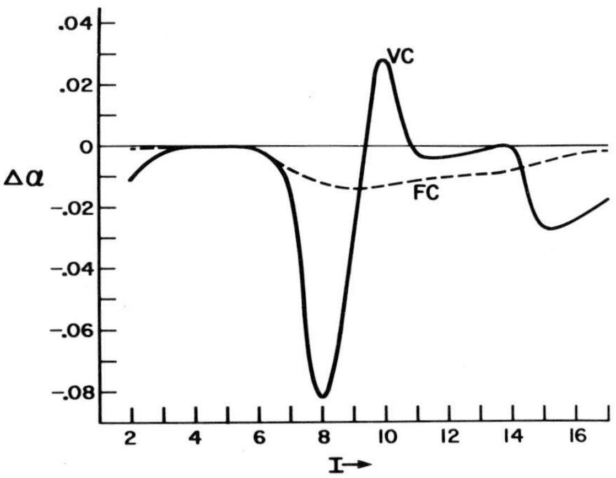

Fig. 17. Change in planetary albedo for sensitivity experiments minus the control run. With the exception of one small area for the variable cloud experiment, the albedo change is negative indicating that model atmosphere absorbs more solar radiation with increased CO_2.

we have examined the effects of explicitly including cloud water and the effects on cloud cover of increasing temperature with increased atmospheric carbon dioxide. We have also considered the effects of variations in horizontal and vertical advection.

The effect of including cloud water is clearly seen only at the uppermost levels. At mid and lower levels, cloud cover is very well correlated with the relative humidity and condensation fields. Only at upper levels, where advection of cloud water over large distances can occur, does its explicit inclusion make a qualitative difference. Further, the cloud cover is not strongly dependent on the particular microphysical parameterizations used, although the cloud water content (and hence the albedo) of the atmosphere is. When the temperature is altered, cloud cover generally responds in a similar fashion as relative humidity, except at upper levels. With cloudwater, the upper-level cloud cover tends to decrease with increased temperature because enhanced evaporation of cloud water prevents extensive advection. However, with no cloudwater, relative humidity increased with increasing temperature at the upper levels.

The incorporation of a fully self-consistent radiative model did not essentially alter these conclusions. The cloud field that developed was similar to the ones which used Newtonian cooling although it was slightly larger at the lower levels. It was shown that infrared radiation was most important for determining the atmospheric temperature structure whereas solar radiation was most important for determining surface variations. If the cloud radiation feedbacks were excluded then the model atmosphere warmed slightly albeit much less than if only the solar radiative feedbacks were excluded. This indicates that the infrared and solar cloud feedbacks were almost compensatory but that the solar radiation feedback was slightly stronger.

On the other hand, in a sensitivity experiment in which the CO_2 was quadrupled the dominant feedback was the infrared component. Increasing CO_2 levels warmed the atmosphere and generally decreased upper level cloud cover, but increased cloud water content slightly at mid levels. Mid level areal cloud cover was little changed. The reduction in upper level cloud cover caused a reduction in the cloud-greenhouse effect (larger than any albedo effect) and reduced the total warming, compared with the case in which cloud cover was fixed. It would thus seem that the explicit incorporation of cloud water is necessary to predict correctly the response to CO_2 changes. In this model, clouds act to reduce the sensitivity of the model response to CO_2 perturbations.

The relationship of the cloud cover to the

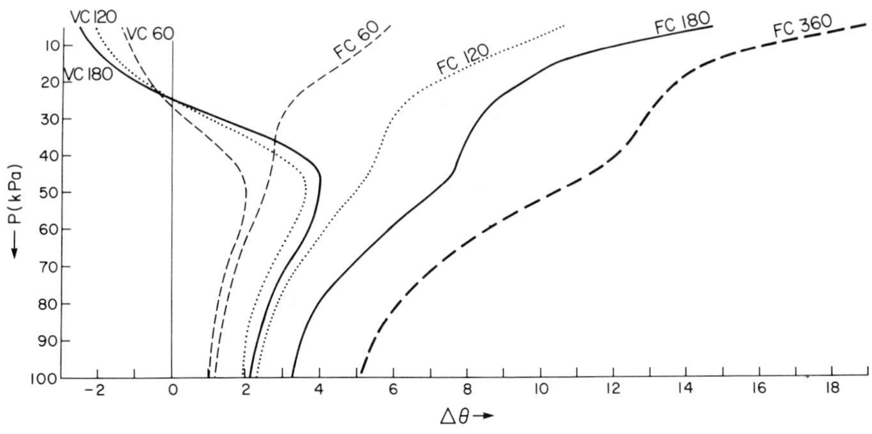

Fig. 18. Vertical structure of the potential temperature change (perturbed minus control) induced by CO_2 quadrupling for the fixed cloud experiments at 60, 120, 180, 360 days (FC60, FC120, FC180, FC360) and the variable cloud experiments at 60, 120, 180 days (VC60, VC120, VC180).

dynamical fields are generally consistent with the previous pictures obtained using one-level systems (e.g., Roads, 1978b). Thus, areal cloud cover is increased by decreasing eddy vertical velocity, increasing the mean upward vertical velocity, decreasing the zonal advection.

Finally we mention again some of the limitations of this model. The main simplifications lie in the assumption of two-dimensionality and in not allowing the large-scale cloud systems to change the winds in response to temperature changes. Thus the increased temperature - increased intensity of hydrology cycle - increased vertical velocity - reduced cloud cover feedback was not included in the CO_2 experiments. Further, the stratosphere was assumed to be in isothermal layer over an active troposphere. The relaxation of such assumptions within the framework of a simple model will be an important next step in understanding the cloud and climate response to changes in external forcing.

Acknowledgments. This research was funded in part by Calspace. J. O. Roads was also supported by NASA grant NAG5-105. Helpful comments were given by R. Somerville and C. Gautier of Scripps and G. Diak of Wisconsin. The manuscript was expertly text-edited by G. Johnston and the figures expertly drafted by F. Crowe. Helpful comments by the reviewers are also appreciated.

References

Bolin, G., A. Bjorkstrom, C. D. Keeling, R. Bacastow, U. Siegenthaler, In Carbon Cycle Modeling, Edited by Bolin, John Wiley and Sons, 1-28, 1981.

Cess, R. D., Climate change: An appraisal of atmospheric feedback mechanisms employing zonal climatology, J. Atmos. Sci., 33, 1831-1843, 1976.

Cess, R. D. and V. Ramanathan, Averaging of infrared cloud opacities for climate modeling, J. Atmos. Sci., 35, 919-922, 1978.

Cess, R. D., B. P. Bruglel and M. S. Liam, Low latitude cloudiness and climate feedback. Comparative estimates from satellite data, J. Atmos. Sci., 39, 53-59, 1982.

Charlock, T. P., Cloud optical feedback and climate stability in a radiative convective model, Tellus, 34, 245-254, 1982.

Hartmann, D. L. and D. A. Short, On the use of earth budget statistics for studies of clouds and climate, J. Atmos. Sci., 37, 1233-1250, 1980.

Herman, G. F., M. C. Wu and W. Johnson, The effect of clouds on the earth's solar and infrared radiation budgets, J. Atmos. Sci., 37, 1251-1261, 1980.

Kessler, F., On the distribution and continuity of water substance in atmospheric circulations, Meteorological Monographs, 10, Amer. Met. Soc., 84 pp, 1969.

Lacis, A. A. and J. E. Hansen, A parameterization for the absorption of solar radiation in the earth's atmosphere, J. Atmos. Sci., 31, 118-133, 1974.

Lindzen, R. S., A. Y. Hou and B. F. Farrell, The role of convective model choice in calculating the climate impact of doubling CO_2, J. Atmos. Sci., 39, 1189-1205, 1982.

Ludlam, F. H., Clouds and Storms, Pennsylvania State Univ. Press, 405 pp, 1980.

Manabe, S. and R. T. Wetherald, The effects of doubling the CO_2 concentration on the climate of a general circulation model, J. Atmos. Sci., 32, 3-15, 1975.

Manabe, S., J. Smagorinsky and R. F. Strickler, Simulated climatology of a general circulation model with a hydrologic cycle, Mon. Wea. Rev., 93, 769-798, 1965.

Moncrieff, M. W. and J. S. A. Green, The propagation and transfer properties of steady convective overturning in shear, Q. J. Roy. Met. Soc., 98, 336-357, 1972.

Ohring, G. and P. Clapp, The effect of changes in the cloud amount on the net radiation at the top of the atmosphere, J. Atmos. Sci., 37, 447-454, 1980.

Paltridge, G. W., Cloud radiation feedback to climate, Q. J. Roy. Met. Soc., 106, 895-899, 1980.

Potter, G. L., H. W. Ellsaesser, M. C. MacCracken and C. S. Mitchell, Climate change and cloud feedback: The possible radiative effects of latitudinal redistribution, J. Atmos. Sci., 38, 489-493, 1981.

Roads, J. O., Numerical experiments on the climatic sensitivity of an atmospheric hydrologic cycle, J. Atmos. Sci., 35, 753-773, 1978a.

Roads, J. O., Relationships among fractional cloud coverage, relative humidity and condensation in a simple wave model, J. Atmos. Sci., 35, 1450-1462, 1978b.

Roads, J. O. and G. K. Vallis, An energy balance climate model with cloud feedbacks, Tellus, (in press), 1983.

Rodgers, C. D., The use of emissivity in atmospheric radiation calculations, Q. J. Roy. Met. Soc., 93, 43-54, 1967.

Sarachik, E., Cloud generation in climate models. In Clouds in Climate: Modeling and Satellite observational Studies, Report of Workshop held at NASA Goddard Institute for Space Studies, New York, NY, Oct 29-31, 1980.

Sasamori, T., A statistical model for stationary atmospheric cloudiness, liquid water content and rate of precipitation, Mon. Wea. Rev., 103, 1037-1049, 1975.

Schneider, S. H., Cloudiness as a global climate feedback mechanism. The effects on the radiation balance and surface temperature of variations in cloudiness, J. Atmos. Sci., 29, 1413-1422, 1972.

Schneider, S. H., W. M. Washington and R. M. Chervin, Cloudiness as a climate feedback mechanism: Effects on cloud amounts of prescribed global and regional surface temperature changes in the NCAR GCM, J. Atmos. Sci., 35, 2207-2221, 1978.

Shukla, J. and Y. Sud, Effect of cloud-radiation feedback on the climate of a general circulation model, J. Atmos. Sci., 38, 2337-2353, 1981.

Smagorinsky, J., On the dynamical prediction of large scale condensation by numerical methods, Geophys. Monogr., 5, Am. Geophys. Union, 71-78, 1960.

Somerville, R.C.J., P.H. Stone, M. Halem, J.E. Hansen, J. S. Hogan, L. M. Druyan, G. Russell, A. A. Lacis, W. J. Quirk and J. Tenenbaum, The GISS model of the global atmosphere, J. Atmos. Sci., 31, 84-117, 1974.

Sundqvist, H., Prediction of stratiform clouds: Results from a 5-day forecast with a global model, Tellus, 33, 242-253, 1981.

Stevens, D. E. and R. S. Lindzen, Tropical wave - CISK with a moisture budget and cumulus friction, J. Atmos. Sci., 35, 940-961, 1978.

Vallis, G. K., A statistical - dynamical climate model with a simple hydrology cycle, Tellus, 34, 211-227, 1982.

Wetherald, R. T. and S. Manabe, Cloud cover and climate sensitivity, J. Atmos. Sci., 37, 1485-1510, 1980.

ENERGY BALANCE MODELS INCORPORATING EVAPORATIVE BUFFERING OF EQUATORIAL THERMAL RESPONSE

Brian P. Flannery and Andrew J. Callegari

Corporate Research Science Laboratories, Exxon Research and Engineering Co.
Linden, New Jersey 07036

Martin I. Hoffert

Department of Applied Science, New York University
New York, New York 10003

Abstract. We describe two non-standard energy balance models which include effects of latent heat on climate sensitivity, and apply them to study sensitivity to uniform variations in insolation and changes in the concentration of atmospheric CO_2. The Tropical Equatorial Constraint (TEC) model incorporates the proposal of Newell and Dopplick (1979) that evaporative losses from tropical sea surface strongly limit thermal response. The two phase model includes an approximate treatment of energy transfer both as sensible heat and as latent heat of water vapor.

We compare the mathematical and physical assumptions underlying each model, and contrast their solutions with results from a standard model in which the diffusion coefficient remains constant with forcing. Both the TEC and two phase model produce stronger thermal response in polar regions than does the standard model. The TEC model shows that the implication of Newell and Dopplick (1979), that equatorial buffering will limit climate response to increasing CO_2, is not justified when its global consequences are considered. The two phase model closely reproduces results for temperature change with latitude found by Manabe and Wetherald (1980) using a general circulation model. Polar amplification found in the two phase model is attributable to the behavior of a temperature dependent effective diffusion coefficient that increases with warming.

1. Introduction

In previous papers we developed two types of energy balance models, (EBMs): the TEC (for Tropical Equatorial Constraint), and two phase model, that incorporate different aspects of the possible influence of latent heat on climate sensitivity. The TEC model developed in Hoffert et al. (1982, hereafter referred to as Paper I) explores the consequences of the proposal of Newell and Dopplick (1979, hereafter ND) that evaporative losses from sea surface strongly limit thermal response in the tropics. That idea was incorporated as an explicit boundary condition on equatorial temperature as a function of solar constant. Here we extend the TEC model to the case of forcing by changes in the concentration of atmospheric CO_2. Flannery (1982, hereafter referred to as Paper II) developed a two phase EBM that includes the effect of energy transport in the form of latent energy of water vapor and the thermal energy of air.

Here, in section 2, we describe the physical basis and mathematical form of a standard EBM (see North et al., 1981) and of the TEC and two phase models, and in section 3 we systematically compare the behavior of these models for sensitivity to solar and CO_2 forcing. In section 4, we compare a few special cases with corresponding cases calculated using a general circulation model (GCM) (Manabe and Wetherald 1980, hereafter referred to as MW). We will especially focus on the question of the latitudinal distribution of temperature change with forcing. We discuss and summarize our results in section 5.

2. Energy Balance Models

The Standard EMB

We consider annual mean, steady state, latitudinally resolved, symmetric, EBMs which prescribe the distribution of surface temperature $T(x)$ as a function of latitude $\Theta = \sin^{-1}(x)$. The standard model equation, as discussed in North et al. (1981), describes conservation of energy on a sphere subject to solar heating $S(x)$ attenuated by

TABLE 1. Standard EBM Coefficients, Tuned to Current Climate

INSOLATION	$Q_o = 334\ W\ m^{-2}$	$s_o = 1.246$	$s_2 = 0.738$
IR RADIATION	$A = 205\ W\ m^{-2}$	$B = 2.23\ W\ m^{-2}\ °C^{-1}$	
ABSORPTION		$a_o = 0.785$	$a_2 = 0.263$
		$b_o = 0.380$	$T_s = -10°C$

fractional absorption $a(x,T)$, longwave infrared cooling $I[T,F(CO_2)]$, and diffusive energy transport characterized by diffusion coefficient D

$$-D \frac{d}{dx} (1-x^2) \frac{dT}{dx} = a(x,x_s)\ S(x) - I[T,F(CO_2)] . \quad (1)$$

For a symmetric model, boundary conditions are: at the pole

$$x=1: \frac{dT}{dx} = \frac{a(x,x_s)\ S(x) - I[T,F(CO_2)]}{2D} , \quad (2a)$$

at the equator

$$x = 0: \frac{dT}{dx} = 0 . \quad (2b)$$

Functionally, insolation includes a scale factor Q, which is 1/4 the solar constant, and geometrical projection factors

$$S(x) = Q(s_o - s_2x^2) . \quad (3)$$

To study sensitivity to solar forcing, we vary Q from its present value Q_o. We represent longwave radiation as a linear function of surface temperature, plus a term dependent on the atmospheric concentration of CO_2,

$$I[T,F(CO_2)] = A + BT + C \ln[F(CO_2)] , \quad (4)$$

where $F(CO_2)$ is the concentration of atmospheric CO_2 relative to its current value. The temperature dependence has been discussed by Warren and Schneider (1979). Satellite observations constrain A and B in order of magnitude, but they require tuning to match a prescribed temperature distribution. Following Hoffert et al. (1980), we include a logarithmic dependence on the atmospheric concentration of CO_2 relative to its present value, and we adjust C to agree with consensus values for global mean temperature rise from CO_2 doubling, $\Delta \langle T\rangle \approx 2.5$ K. Our representation of albedo comes from Hartmann and Short (1979); it includes the important non-linear effect of ice-albedo feedback when temperature falls below a critical value T_s for the formation of ice or snow cover

$$a(x,T) = a_o - a_2x^2, \text{ for } T > T_s , \quad (5)$$

$$a(x,T) = b_o, \text{ for } T \leqslant T_s . \quad (5b)$$

Table 1 lists values for the coefficients adopted here. They have been adjusted to match a current climate with T(eq) = 29°C, and an ice line at $x_s = 0.947$ ($\Theta = 71°$) which also results in T(pole) = -15°C, and a global mean temperature $\langle T\rangle = 14.33°C$. These parameterizations and coefficient values are in close agreement with values found and discussed in North et al. (1981). We refer to the model defined by these equations as the standard model. In contrast to the TEC and two phase models, the principal distinguishing features of the standard model are that (1) transport is simply proportional to the temperature gradient and (2) the diffusion coefficient D does not vary when Q or $F(CO_2)$ change.

Fortunately, the standard model can be analyzed exactly by representing the temperature and net source term as expansions in Legendre polynomials

$$a(x,x_s)S(x) = \sum_{L=0}^{\infty} \xi_L P_L(x), \quad T(x) = \sum_{L=0}^{\infty} T_L P_L(x). \quad (6)$$

The coefficients of the source term depend only on the location of the ice-line

$$\xi_L(x_s) = \int_{-1}^{1} a(x,x_s)S(x)P_L(x)dx . \quad (7)$$

The coefficients T_L are given by

$$T_o = (\xi_o - A)/B;\ T_L = \xi_L/[L(L+1)D + B], \quad (8)$$

for $L \geqslant 1$.

As discussed in North et al. (1980) a consistent solution for T(x) and Q can be found by iterating until $T(x_s) = T_s$. Note that the coefficient T_o is also the global mean temperature $\langle T\rangle$. Because the radiative loss term is linear with T, and diffusion only redistributes energy, losses at the mean temperature exactly compensate mean net insolation ξ_o. This is simply a restatement of global energy balance.

The TEC EBM

ND proposed that an evaporative constraint on heating of tropical seas would limit global warming by increases in atmospheric CO_2 to values an order of magnitude lower than previously accepted model predictions, i.e., 0.25 K compared with 2-3 K. They based their idea on seasonal and historical observations, and on quantitative analysis of surface energy balance between sea and air, identifying the strong cooling effect of evaporative losses as the controlling feature. Their proposal became the subject of several subsequent studies since it contradicted consensus ideas for global warming from CO_2. GCMs, such as those of Manabe and Wetherald (1975), do include the supposed limiting evaporative effects of ND, but do not find such constrained temperature rise: so other effects must operate that overcome the evaporative limit. The discrepancy between ND and GCM results is hardly surprising because the ND analysis neglects important contributing effects in the highly coupled climate system that are included in the GCM.

The discrepancy has been resolved by attributing the GCM rise to additional feedback from water vapor emission, calculated fully self-consistently in the numerical models (see Ramanathan, 1981; Newell and Dopplick, 1981; Kandel, 1981). However, some aspects of the problem remain controversial: ND point out that the tropical temperature and concentration of water vapor in the Manabe and Wetherald (1975) model for current climate significantly exceed observed values; Kandell notes that analyses differ substantially in the magnitude of terms in the surface energy balance; and results of line-by-line radiative transfer calculations by Wiscombe quoted in Paper I indicate that water vapor feedback is less pronounced under equatorial conditions than in mid-latitudes characteristic of global analyses--this is because water vapor bands are nearly saturated under conditions of high absolute humidity in the tropics (see also Kiehl and Ramanathan, 1982). Limited response of tropical sea surface temperature also receives support from studies of paleoclimate, e.g. Frakes (1979, page 178); Schopf (1980, page 142).

The TEC model developed in Paper I explicitly examines the consequences of the effect proposed by ND. We fixed the response of the equatorial temperature to changes in solar insolation, Q/Q_o, by imposing a boundary condition

$$T_{eq}\ (°C) = 29 + 17.3 \ln(Q/Q_o) \quad . \qquad (9)$$

This form is based on analysis of surface energy balance closely resembling that of ND. Since the standard model described by Eqs. (1-5) cannot satisfy Eq. (9), except for $Q/Q_o = 1$, an additional degree of freedom was provided by allowing D to vary with Q -- mathematically D(Q) becomes an adjustable parameter. (We continued to keep D independent of latitude). However, physically plausible solutions do not exist for arbitrary values of Q. In Paper I, we showed that D becomes negative outside the limits $0.80 \leq Q/Q_o \leq 1.15$. To maintain equatorial temperatures higher than in the standard model as Q decreases so must D, until when D = 0 the model reaches local radiative equilibrium at $Q/Q_o = 0.80$. Similarly, as Q rises so does D to maintain temperatures lower than in the standard model. Ultimately, D approaches infinity, and the model becomes isothermal at $Q/Q_o = 1.15$. Thus, ND's hypothesis can only be incorporated into a standard EBM by forcing the meridional transport, as parameterized by the diffusion coefficient D, to be a function of tropical conditions.

Paper I compared the TEC and standard EBM results for changes produced by varying insolation. Relative to the standard model, the TEC constraint buffers equatorial temperature change, but it amplifies response in polar regions. The result is readily explained by global energy balance, using the analytic results in Eqs. (6-8). With increasing Q, T must rise, but the rise at the pole will be enhanced if the equatorial temperature is constrained. Consequently, imposing such a limit strongly alters the latitudinal distribution of temperature change.

Here we also apply the TEC model to variations in the concentration of atmospheric CO_2. A straight forward extension of the simplified surface energy balance analysis of Paper I to include CO_2 radiative effects does not change Eq. (9), i.e., the equatorial temperature does not change at all with rising CO_2. This occurs because we neglect variations in IR feedback under the assumption that water vapor feedback dominates, but is small, since the water vapor bands are already near saturation under tropical conditions. Clearly, such strong anchoring overestimates the buffering effect, but not materially. If mean global warming of 2 K occurs, then, for the rest of the model, it hardly matters whether equatorial temperature changes by 0 K or 0.2 K. For the same reasons as for solar forcing, the TEC model shows strong amplification of thermal response at the poles for CO_2 forcing.

The Two Phase Model

In Paper II we described an EBM formulation that includes both sensible heat of air and latent energy of water vapor as agents of diffusive energy transfer. In the two phase model, the fundamental equation

$$-D'\frac{d}{dx}(1-x^2)\frac{d(1+\Psi_L)T}{dx} = a(x,x_s)S(x) - I[T,F(CO_2)], \qquad (10)$$

closely resembles the standard equation, but includes the energy associated with water vapor through the term Ψ_L, which is the ratio of latent to sensible energy,

$$\Psi_L = \frac{r_a q_{sat}(T) L(T)}{c_p T} \quad , \qquad (11)$$

where L(t) is the latent energy of water vapor, r_a is the relative humidity, $q_{sat}(T)$ is the saturation vapor pressure and c_p is the specific heat of air at constant pressure. This expression can be obtained from physical arguments that identify the diffusive flux as arising from the exchange of parcels of air between adjacent latitudinal segments of the atmosphere with different temperature and humidity, as described in Paper II. The principal physical dependence introduced by adding latent heat to the model is the exponential variation of saturation vapor pressure with temperature given by the Clausius-Clapeyron equation

$$\ln(q_{sat}) = \text{const} - L/(RT) \qquad (12)$$

where R is the gas constant of water vapor. $L/R \approx 5300$ K at 300 K.

The two phase model requires a prognostic equation for the distribution of water vapor. Paper II adopted a simple but reasonably accurate assumption for relative humidity that has the advantage of allowing one to define a useful diagnostic concept, the effective diffusion coefficient $D_{eff}(T)$. With reasonable accuracy, the mean distribution of water vapor in the atmosphere can be represented by assuming constant humidity $r_a = 0.8$, independent of latitude, and results, such as those in MW, suggest that relative humidity tends to remain constant during climate change. In that case, Ψ_L is only a function of temperature, and the latitudinal gradient of Ψ_L can be converted into an expression proportional to the temperature gradient. This results in an expression that can be interpreted as an effective diffusion coefficient $D_{eff}(T)$ characterizing the diffusive energy flux

$$F \propto -D' \frac{d(1+\Psi_L)T}{dr} = -D' \left[1+\Psi_L \frac{L}{RT}\right] \frac{dT}{dr} \qquad (13a)$$

$$D_{eff} = D' \left[1+\Psi_L \frac{L}{RT}\right] \qquad (13b)$$

In Paper II, the two phase model was tuned to reproduce T(eq) = 29°C, $x_s = 0.947$, as in the standard and TEC models discussed above. This required adjustment of both D', from 0.623 to 0.317 W m^{-2} K^{-1}, and the radiative loss coefficient B, from 2.23 to 1.95 W m^{-2} (deg C)$^{-1}$, because the models are slightly different in the distribution T(x). The fiducial two phase model gives a mean temperature $\langle T \rangle$ = 16.55°C, and polar temperature -17.55°C, compared with 14.33°C and -14.69°C, respectively, in the standard and TEC models.

In the sensitivity studies of Paper II, D' is held constant, as is D in the standard model, but the effective diffusion coefficient varies strongly with latitude and forcing, through its dependence on temperature. In the fiducial model of current climate obtained with the two phase model, $D_{eff}(T)$ varies from D/2 at the pole to 2D at the equator. Also, as forcing produces a warmer climate $D_{eff}(T)$ rises at all latitudes, but most rapidly in the tropics. This behavior causes temperature changes to be relatively larger at the pole than at the equator. The actual latitudinal distribution of warming with latitude agrees very well with the GCM calculations of MW for Q/Q_o = (1.02, 1.04), and $F(CO_2)$ = (2,4).

3. Results for Solar and CO_2 Forcing

We investigated sensitivity to solar and CO_2 forcing for the standard, TEC, and two phase EBMs using numerical solutions to the model equations for values of scaled insolation in the range $0.87 \le Q/Q_o \le 1.15$, and $0.5 \le F(CO_2) \le 4.0$. For all models, we used identical numerical techniques employing relaxation on a finite difference grid to solve the two point boundary value problem. This method works readily for the highly non-linear two phase model, and allows us to incorporate the adjustment of the diffusion coefficient in the TEC model in a direct fashion. In Table 2, we list the explicit differences associated with the tuned models, parameterizations, and boundary conditions in the three EBMs.

Figure 1 illustrates the major differences in sensitivity to solar forcing. We only examine those solutions closely related to the "current climate" solution, and we do not show results appropriate to the well-known alternate branches of the standard EBM formulation: the ice-covered earth branch, or the unstable branch for which $dx_s/dQ < 0$ (Held and Suarez 1974). If sensitivity is measured by the retreat of ice with increasing insolation, then the TEC model is most sensitive, with $dx_s/dQ(\text{TEC}) = 4dx_s/dQ$ (standard), and the two phase model intermediate, but much closer in response to the standard model. Figure 1(b) shows the monotonic increase of the diffusion coefficient with insolation in the TEC model. While the global mean temperature change with insolation is nearly identical in all the models, the response with latitude varies considerably, as shown in Figures 1(c) and 1(d). The TEC model constraint produces a small equatorial variation, but, since the mean temperature rise is not greatly altered, this forces large polar amplification. Again, the two phase model exhibits polar amplification and equatorial buffering intermediate between the standard and TEC models.

Polar amplification can readily be understood in terms of global energy balance and the analysis of results using Legendre expansions, as in Eqs. (6-8). Radiative losses at the global mean temperature exactly compensate for net warming at the global mean insolation rate, which only depends on the location of the ice-line. When the

TABLE 2. Comparison Between Standard, TEC and Two Phase Models

Fiducial Model	x_s	T(eq)	T(pole)	<T>
Standard	0.947	29.00 °C	-14.69 °C	14.33 °C
TEC	0.947	29.00 °C	-14.69 °C	14.33 °C
Two Phase	0.947	29.00 °C	-17.55 °C	16.54 °C

Fiducial Parameters	Diffusion	B(in I = A + BT)
Standard	0.623 W m^{-2} K^{-1}	2.23 W m^{-2} $°C^{-1}$
TEC	0.623	2.23
Two Phase	0.317	1.95

Forcing Constraints	T(eq)	Diffusion
Standard	Free	Fixed
TEC	Prescribed	Adjusts
Two Phase	Free	Fixed

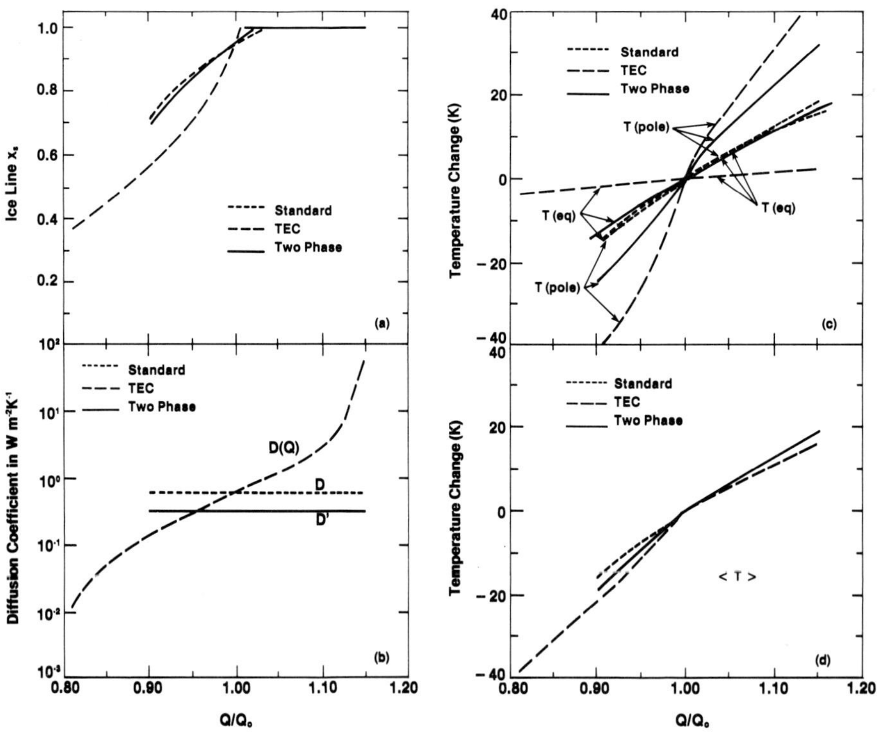

Fig. 1. Comparison of changes induced by scaled changes in insolation Q/Q_0 derived from three types of EBM: (a) location of the ice-line, (b) diffusion coefficient, (c) change in equatorial and polar temperature, and (d) global mean temperature.

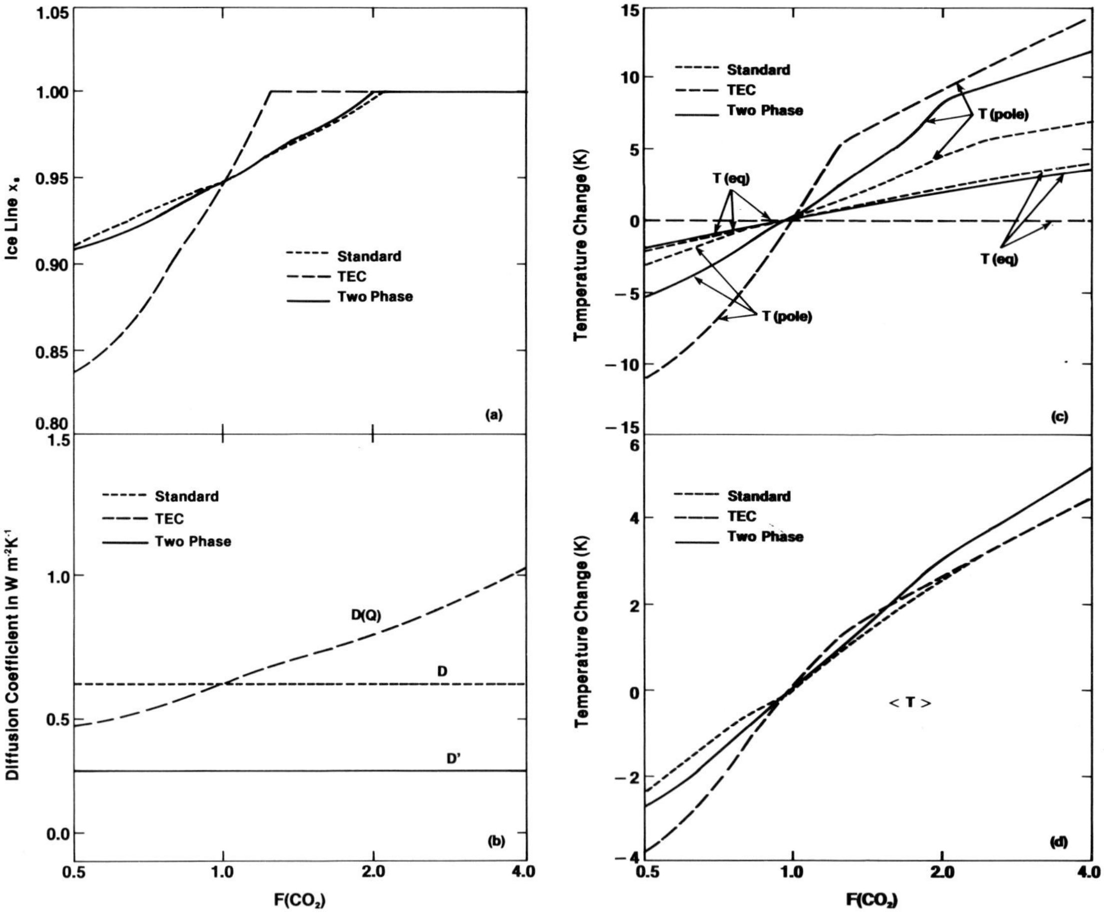

Fig. 2. Comparison of changes induced by variations in the concentration of atmospheric CO_2, relative to its current value, derived from three types of EBM: (a) location of the ice-line, (b) diffusion coefficient, (c) change in equatorial and polar temperature, and (d) global mean temperature.

ice-line coincides for any two models, then A + B <T> must agree. In particular, this happens for all models once the ice-line disappears, for $Q/Q_0 > 1.03$. This accounts for the common value of $\Delta\langle T\rangle$ between the TEC and Standard models for large Q, and explains the slightly larger value of $\Delta\langle T\rangle$ in the two phase model. Since B (two phase)/B (Standard) = 1.14, $\Delta\langle T\rangle$ (two phase) = $1.14\Delta\langle T\rangle$ (Standard). Obviously, the TEC model must exhibit polar amplification, since the mean temperature rise must agree but the equatorial response is constrained. Mathematically, that rise occurs because D(Q) grows, so that more energy can be transported poleward in spite of the reduced temperature gradient. In the two phase model a similar change occurs, associated with a rise in the effective diffusion coefficient, $D_{eff}(T)$, as defined in Eq. (13b).

In fact, Figure 1(c) shows that polar amplification in the standard model only occurs because ice-albedo feedback operates. Once ice disappears in the standard model, it is easy to show analytically, using Eqs. (6-8), that $\Delta T(x) \propto T(x)$, i.e. the equator warms more than the poles in the Standard model. This explains the crossover in Figure 1(c), where equatorial warming exceeds polar warming for large Q. The TEC and two phase models continue to exhibit polar amplification even after ice vanishes.

Figure 2 illustrates the sensitivity of the EBMs to variation in the concentration of atmospheric CO_2. Behavior with increasing CO_2 is similar to that found for increasing insolation. Both models that include effects related to latent heat show polar amplification, with the strongest effects in the TEC model, and the two phase model showing intermediate amplification. Note that, for CO_2 doubling, all models produce a rise in global mean temperature of about 2.5 K, as tuned by the choice of C in the parameterization of CO_2 effects, Eq. (4). The slight differences between the models result from slightly different positions for the ice-line with doubled CO_2, and the different value of B in the two phase model.

4. Comparison with Results from GCM Simulations

In their simulations of the effects of atmospheric CO_2 increases on climate, MW calculated a set of models that compared the effects of increasing insolation with increasing CO_2. Their GCM considered idealized, sector geography with swamp oceans, i.e. no heat capacity, and only evaluated annual mean forcing without seasonal effects. Thus, their model is in many respects comparable to the simple EBMs of this study. Some of our results can be directly compared with the GCM simulations; we have intentionally drawn our Figures 3 and 4 to nearly the same scale as corresponding Figures 19 and 3 in MW.

MW found that insolation and CO_2 forcing produced quite similar response, i.e. 2 (and 4) X CO_2 concentration and +2% (and 4%) changes in Q/Q_o produce nearly identical patterns of temperature change with latitude, and they showed that poleward energy transport of latent heat was comparable with transport of sensible heat. A principal result from their study was the pronounced amplification of temperature rise at the poles: $\Delta T(\text{pole}) \approx 4\Delta T(\text{eq})$.

Figure 3 shows the temperature variation with latitude $\Delta T(x)$ found in the three EBM simulations for $F(CO_2) = (2,4)$ and $Q/Q_o = (+2\%, +4\%)$. The standard model produces a relatively flat distribution of temperature rise, with only slight polar amplification, while the TEC model produces strong amplification. Results for $\Delta T(x)$ from the two phase model agree nearly exactly with the GCM result (see Figure 19 of MW).

In Figure 4, we show the net poleward flux of energy for each of the three models for the cases $F(CO_2) = (1,2,4)$. The pattern of response in poleward flux is completely different in the three cases. In the standard model, the flux falls nearly everywhere with rising CO_2 while in the TEC model flux rises at nearly all latitudes. The more complex response in the two phase model agrees very well with the pattern of change in poleward flux with latitude found in the GCM study (see Figure 3 of MW).

However, the agreement of the two phase model with the GCM results cannot be carried through in complete detail, physically. The two phase EBM formulation prescribes a functional form for transport of latent heat in which flux is proportional to a temperature gradient, Eq. 13(a), and the model equations actually impose constraints only on the divergence of flux. In Figure 3 of MW, the poleward flux is separately shown for the sensible and latent heat components. While the net flux in the GCM agrees quite well with the two phase flux, the latent heat flux in tropical regions in the GCM, and in real climate, actually pumps energy against the gradient of temperature and absolute himidity. This occurs by correlations in the three dimensional motion and energy content that can be resolved in the GCM but cannot be accounted for in the simple diffusion model of this paper. Nonetheless, the divergence of latent heat flux in the GCM behaves as in the two phase model, and the variation in total flux with forcing agrees remarkably well between the GCM and two phase models.

Finally, Figure 5 shows the absolute humidity and effective diffusion coefficient, Eq. (13b), of

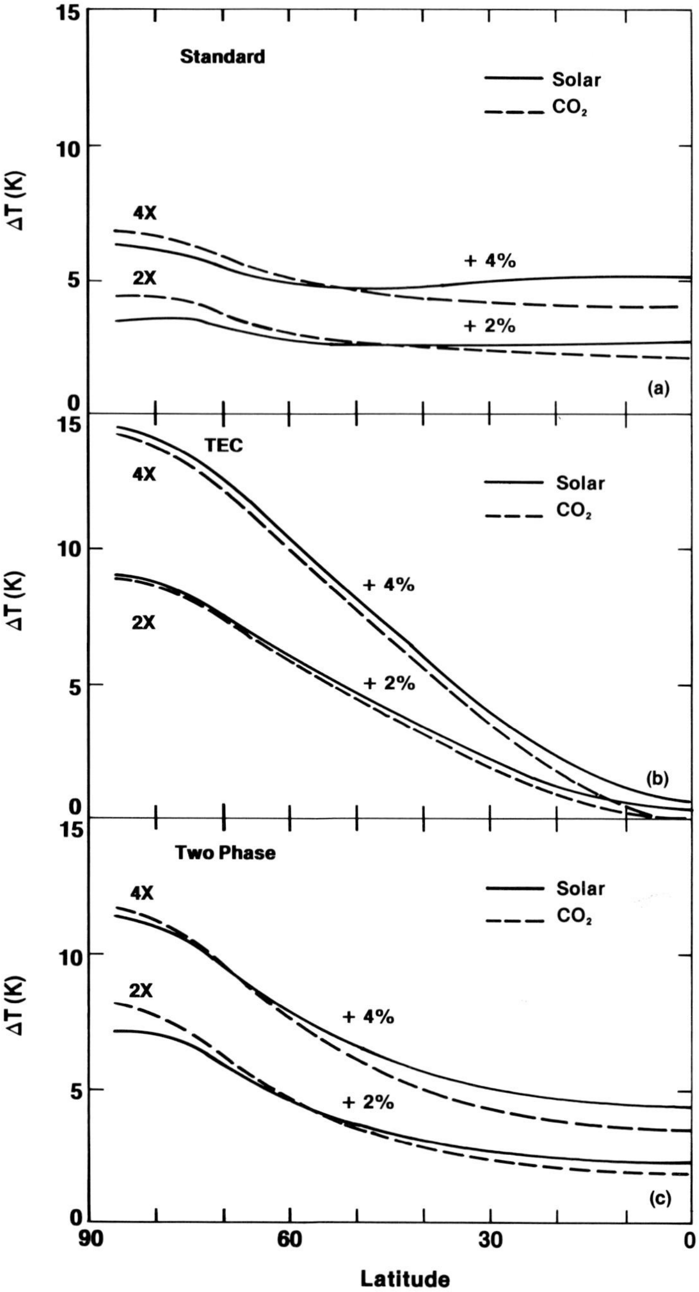

Fig. 3. The latitudinal distribution of temperature change produced by 2% and 4% increases in the solar constant, and increases in atmospheric concentration of CO_2 by factors of 2 and 4, as calculated from the standard (a), TEC (b), and two phase (c), EBM.

the two phase model as a function of latitude for the cases with CO_2 forcing. The distribution of humidity agrees quite well with that found in MW. This produces a variation in $D_{eff}(T)$ such that the mean value roughly agrees with D in the standard model, but the variation from equator to pole is substantial, with $D_{eff}(eq) = 3\ D_{eff}(\text{pole})$. Because of its exponential temperature dependence, $D_{eff}(T)$ rises at all latitudes with

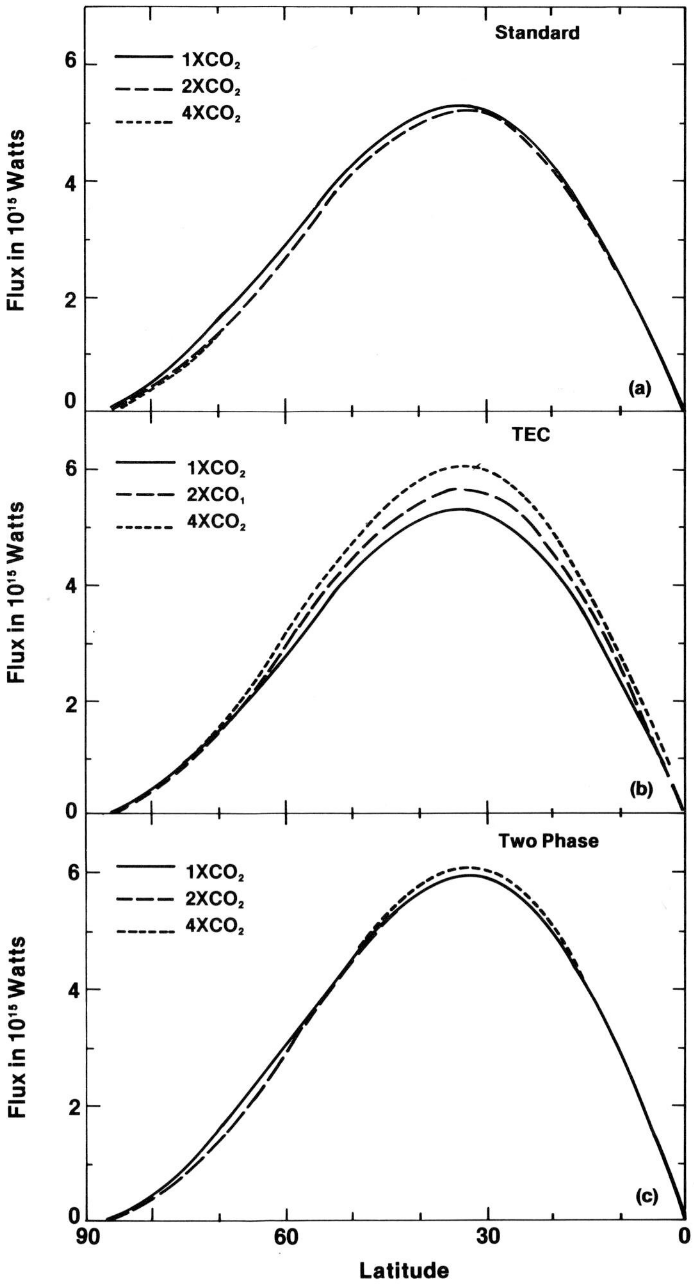

Fig. 4. The latitudinal distribution of poleward flux of energy produced by 2% and 4% increases in the solar constant, and increases in atmospheric concentration of CO_2 by factors of 2 and 4 as calculated from the standard (a), TEC (b), and two phase (c), EBM.

warming, but most rapidly at the equator. Thus, warmer climate becomes more nearly isothermal in the two phase model. With rising temperature, the ability of the system to transport energy rises because the latent heat content rises exponentially.

5. Conclusions

From these models, it is clear that effects of water vapor strongly influence climate sensitivity of EBMs. In particular, they strongly enhance polar amplification of thermal response to both solar and CO_2 forcing beyond that produced solely by ice-albedo feedback and geometrical effects. In this respect, the two phase model agrees far better with the GCM results of MW than does the standard model.

The behavior of the effective diffusion coefficient, $D_{eff}(T)$, characterizing the two phase model explains polar amplification. Since $D_{eff}(T)$ rises exponentially with temperature, the temperature distribution becomes more isothermal with warming. Of course, such an analysis applies to the behavior of the model, but extension to real climate is plausible. In a warmer climate the latent heat content of the atmosphere rises exponentially, so any form of dynamical motion is more effective at transporting energy.

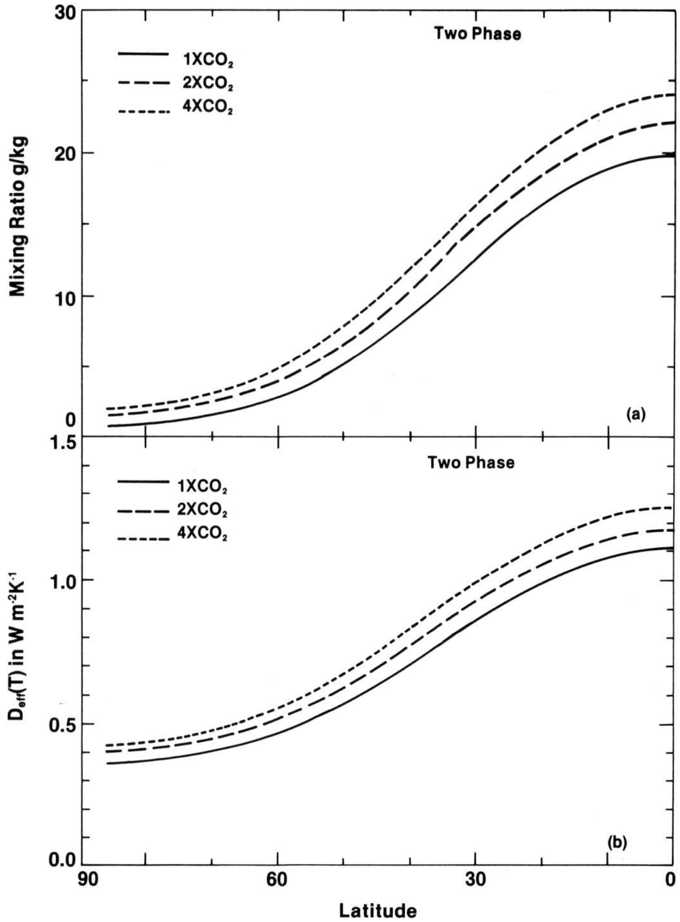

Fig. 5. The latitudinal distribution of (a) water vapor mixing ratio and (b) effective diffusion coefficient, Eq. (13b), as calculated in the two phase EBM for atmospheric concentrations of CO_2 (1, 2 and 4) times its present value.

Of course these simple physical models serve only to highlight specific effects to the extent that they can be analyzed in an EBM; their consequences for true climate require a far more complex analysis in a more fundamental and complete model, such as a GCM. Nonetheless, they have value as understandable examples of analyzable mechanisms. Physically, such behavior produces a tendency to global isothermality with warming.

Results of the TEC model show that limiting equatorial response by evaporative buffering, as proposed by ND, does not limit global response: rather, such buffering strongly amplifies polar response. This result arises from considerations of global energy balance.

6. Comparison with the MLEBM results of Peng, Chou, and Arking

Since presentation of this paper at the Ewing Symposium, our attention has been directed to the recent work of Peng, Chou, and Arking (1982) and Chou, Peng, and Arking (1982) who have developed a Multi-layer EBM which they used to examine climate response to solar and CO_2 forcing. Their model includes highly resolved, vertical radiative transport, and parameterized dynamical energy transfer involving both horizontal and vertical transport. Their results are far more detailed in content, since they have the ability to analyze vertical components of response, and to treat the dependence of CO_2 and water vapor radiative effects as a function of changing vertical distributions with latitude. Their model treats horizontal transport in a far more non-linear fashion than ours, even in the absence of water vapor transport, since they explicitly include parameterizations modeling dependences of baroclinic instability on the solution. Also, they use an albedo prescription that is less sensitive than that used by us.

The most common grounds for comparison between Chou et al. (1982) and this work are their Figure 8, for poleward flux, with our Figure 4, and their Figure 10, for change in surface temperature, with our Figure 3. Qualitatively, the responses are very similar: both models show pronounced polar amplification. However, the magnitude of high latitude temperature change for doubled atmospheric CO_2 found by Chou et al. is about 60% that found here with the two phase model or in the GCM simulation of MW. Although it is difficult to be certain of attribution in such different models, probably the major source of discrepancy arises from the less sensitive albedo response used by them.

Overall, they find that, by comparing models with fixed and variable poleward transport by latent heat effects, only about 20% of the enhanced polar temperature response can be attributed to enhanced transport by water vapor. This is unlike earlier GCM results of MW, or our results, which suggest that enhanced poleward transport by water vapor is of comparable magnitude with ice-albedo feedback. In our models with large variation from current conditions, polar ice disappears but polar amplification persists. Again, part of that discrepancy arises from the smaller overall response found by Chou, et al. than found by us or MW, and which, probably, can be attributed to different specifications for ice-albedo variation. Nonetheless, all of these models highlight the crucial role of water's influence on climate sensitivity through its effect on ice-albedo change, radiative transfer, and meridional energy transport.

References

Chou, M. D., L. Peng and A. Arking, Climate studies with a multi-layer energy balance model. Part II: The role of feedback mechanisms in the CO_2 problem, J. Atmos. Sci., 39, 2657-2666, 1982.

Flannery, B. P., An energy balance model incorporating transport of thermal and latent energy, (preprint, Paper II), 1982.

Frakes, L. A., Climates through Geologic Time, Elsevier Scientific Publishing Company, New York, 310 pp., 1979.

Hartmann, D. L., and D. A. Short, On the role of zonal asymmetries in climate change, J. Atmos. Sci., 36, 519-528, 1979.

Held, I. M. and M. J. Suarez, Simple albedo feedback models of the icecaps, Tellus, 26, 613-629, 1974.

Hoffert, M. I., A. J. Callegari, and C. T. Hseih, The role of deep sea heat storage in the secular response to climate forcing, J. Geophys. Res., 85, 6667-6679, 1980.

Hoffert, M. I., B. P. Flannery, A. J. Callegari, C. T. Hseih, and W. Wiscombe, Evaporation-limited tropical temperatures as a constraint on climate sensitivity, J. Atmos. Sci., in press: (Paper I), 1982.

Kandel, R. S., Surface temperature sensitivity to atmospheric CO_2, Nature, 293 (5834), 634-636, 1981.

Kiehl, J. T. and V. Ramanathan, Radiative heating due to increased CO_2: The role of H_2O continuum absorption, J. Atmos. Sci., 39, 2923-2926, 1982.

Manabe, S. and R. T. Wetherald, The effects of doubling the CO_2 concentration on the climate of a general circulation model, J. Atmos. Sci., 32, 3-15, 1975.

Manabe, S. and R. T. Wetherald, On the distribution of climate change resulting from an increase in the CO_2 content of the atmosphere, J. Atmos. Sci., 37, 99-118, 1980. (MW)

Newell, R. E. and T. G. Dopplick, Questions concerning the possible influence of anthropogenic CO_2 on atmospheric temperature, J. Appl. Meteor., 18, 822-825, 1979.

Newell, R. E. and T. G. Dopplick, Reply to Robert

G. Watts' "Discussion of questions concerning the possible influence of anthropogenic CO_2 on atmospheric temperature", J. Appl. Meteor., 20, 114-116, 1981.

North, G. R., R. F. Cahalan, and J. A. Coakley, Energy Balance Climate Models, Rev. Geophys. Space Phys., 19, 91-121, 1981.

Peng, L., M. D. Chou and A. Arking, Climate studies with a multi-layer energy balance model. Part I: Model description and sensitivity to the solar constant, J. Atmos. Sci., 39, 2639-2656, 1982.

Ramanathan, V., The role of ocean-atmosphere interactions in the CO_2 climate problem, J. Atmos. Sci., 38, 918-930, 1981.

Schopf, T. J. M., Paleoceanography, Harvard University Press, Cambridge, Massachusetts, 341 pp., 1980.

Warren, S. G. and S. H. Schneider, Seasonal simulation as a test for uncertainties in the parameterizations of a Budyko-Sellers zonal climate model, J. Atmos. Sci., 36, 1377-1391, 1979.

TEMPORAL VARIATIONS IN LAKE LEVELS SINCE 30,000 YR BP - AN INDEX OF THE GLOBAL HYDROLOGICAL CYCLE

F. Alayne Street-Perrott and Sandra P. Harrison

Tropical Palaeoenvironments Research Group, School of Geography
Mansfield Road, Oxford, U.K., OX1 3TB

Abstract. Fluctuations in the extent of closed lakes provide a detailed record of global and regional variations in the balance between precipitation and evaporation. The temporal sequence of hydrological fluctuations during the last 30,000 ^{14}C yr BP was reconstructed using published data from 198 lake basins between 41°S and 51°N. These data were coded at 1000-yr intervals in terms of lake status (an index of relative depth) and lake-level trend. The broad pattern of glacial/interglacial variations was established by aggregating the status/trend information from particular regions. There is an important contrast in lake behaviour north and south of the Tropic of Cancer. On the poleward side, phases of high lake levels were a very prominent feature of the last 30,000 yr compared with the intertropical belt and the Southern Hemisphere. Longitudinal asymmetry was also most pronounced in northern extratropical latitudes. During the last 13,000 yr, fluctuations on a shorter time scale (10^2 - 10^3 yr) are exhibited by individual lake-level curves from intertropical Africa, where they account for a significant proportion of the climatic variance.

Introduction

Lake-level fluctuations, particularly those recorded in hydrologically closed basins, have been widely recognized as an indicator of climatic change on time scales ranging from 10 to 10^6 yr (Street and Grove, 1976; Street-Perrott and Harrison, in press). At the local scale, the fractional area of a closed drainage basin occupied by a lake is determined by the balance between precipitation and evaporation (P-E) over the catchment and water surface (Street, 1980). Hence, fluctuations in water depth can also be used as a measure of changes in hydrological inputs and outputs. At the global scale, variations in the proportion of the land area covered by lakes reflect the large-scale distribution of P-E (Street-Perrott and Roberts, 1983). By analyzing lake-level fluctuations over a wide area, therefore, it is possible to reconstruct past changes in the atmospheric circulation.

The Tropical Palaeoenvironments Research Group in Oxford maintains a computer data bank of published information on fluctuations in lake depth in basins that have been closed during all or part of the last 30,000 yr (Street and Grove, 1979). Water-level fluctuations within these basins were inferred from a wide variety of geological, geochemical and palaeoecological evidence (Street-Perrott and Harrison, in press). For comparative purposes, therefore, the original data have been standardized by an arbitrary but objective procedure to yield estimates of lake status (an index of relative depth) and trend (or direction of change in lake status). More sophisticated treatment is not at present justified in view of the uneven distribution, diversity and variable time resolution of the data. Following Street and Grove (1979), quantitative measures of lake behaviour have been derived by aggregating data on status and trend from particular regions and latitude bands.

In previous papers, we emphasized the spatial patterns of lake status and trend at different times during the Late Quaternary (Street and Grove, 1976, 1979), interpreting them in terms of changes in the location and intensity of the major features of the general circulation (Street-Perrott and Roberts, 1983; Street-Perrott et al., in press; Street-Perrott and Harrison, in press). In contrast, in this study, attention is focussed on temporal variations in lake behaviour, in the hope that our data will provide a means of validating climatic time series generated by simulation experiments (cf. Schneider and Thompson, 1979, p. 201). Aggregate data are employed to investigate the broad latitudinal and longitudinal contrasts in the sequence of lake-level fluctuations during the last 30,000 yr. Individual lake-level curves from Africa, with a time resolution of 10^2 yr, are then used to illustrate the fine structure of the palaeoclimatic record from the tropics for the last 13,000 yr.

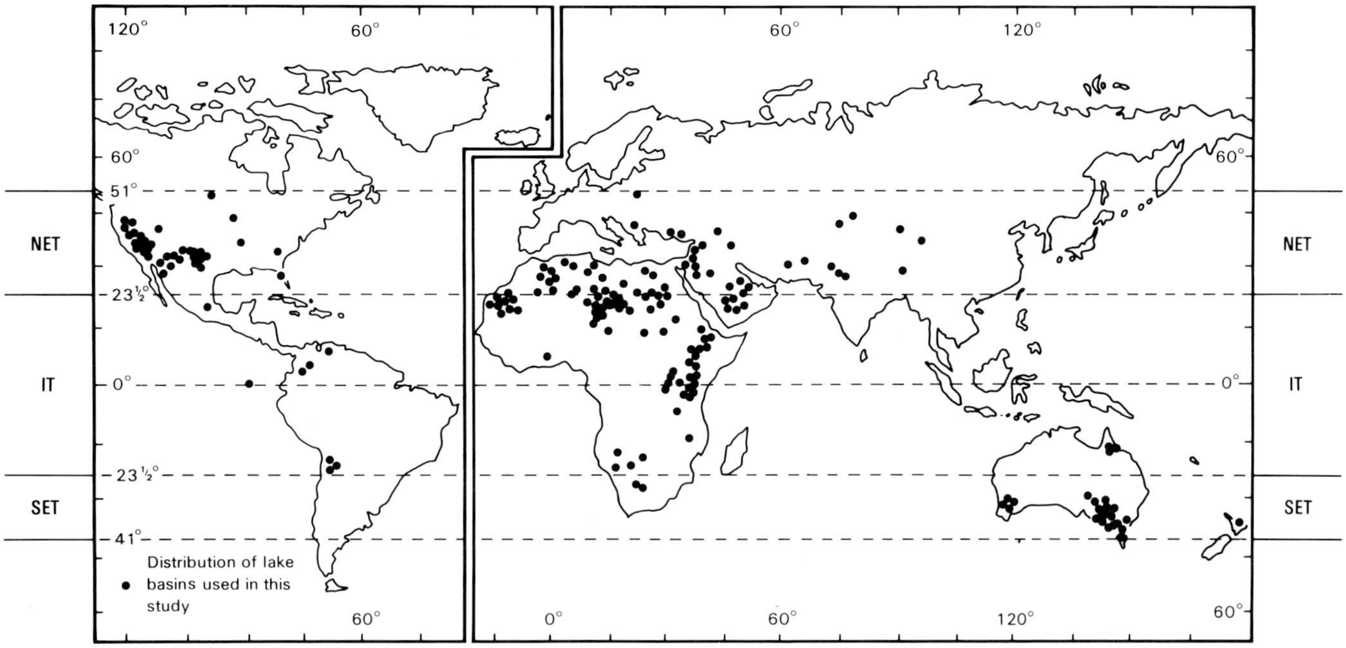

Fig. 1. The distribution of lake basins used in this study.

Methods

Since the initial studies by Street and Grove (1976, 1979), the data bank has been completely recompiled, with a considerable improvement in temporal resolution. The current version comprises 198 lake basins (Figure 1), all of which have chronologies based on ^{14}C, dated tephra layers or historical sources. Water depth, since it is more readily evaluated from stratigraphical, geochemical and palaeoecological data, is used instead of the variable directly controlled by climate, lake area. The water depth within each basin is divided into 3 status classes, each of which has a broadly similar frequency of occurrence in the data set:

low	0 - 15%	of total altitudinal range of fluctuation, including dry lakes
intermediate	15 - 70%	of total altitudinal range of fluctuation
high	70 - 100%	of total altitudinal range of fluctuation, including overflowing lakes.

Lake trend is classified as follows: 1) rising; 2) stable; 3) falling; and 4) unclassifiable/no data. Both status and trend are coded on an ordinal scale at 1000-yr intervals from 30,000 to 0 yr BP (Street-Perrott et al., in press).

The temporal and spatial distribution of published lake-level information is uneven (Figure 1). The basins represented in the data bank cover a latitudinal range from 51°N (Lake Manitoba, Canada) to 41°S (Pulbeena Swamp, Tasmania). However, 56.5% lie between the latitudes 5° and 35°N, and 44% are derived from Africa and the Middle East alone. In part, this pattern reflects the distribution of closed drainage basins (de Martonne and Aufrère, 1928), but it is also a function of the difficulty of obtaining data from inaccessible regions or from certain types of lakes, particularly those in deserts, where the stratigraphic record tends to be both discontinuous and complex and there is a lack of good materials for dating (Street-Perrott and Harrison, in press).

The global data set has been split into three principal subsets. These are referred to below as: NET (northern extratropical: 23 1/2° - 51°N); IT (intertropical: 23 1/2°N - 23 1/2°S); and SET (southern extratropical: 23 1/2° - 41°S). The NET subset has been divided further into two longitudinal blocks, namely Africa-Eurasia and North America. The data-rich belt from 5°S to 45°N has also been subdivided into five 10° latitude bands. At present, a more detailed regional breakdown of the data bank is not justified in view of the number and distribution of basins.

The relationship between status and trend in each subset of the data was assessed using the chi-square test for a 3 x 3 contingency table (Croxton, 1967). It was also expressed diagrammatically by means of a hypothetical 'pulse curve' (Figure 2). Assuming that the sample of paired observations of status and trend in a given subset is not biased by any missing data, the percentage frequency of observations within an individual cell in the contingency table is a measure of the proportion of the last 30,000 yr

STATUS — RAW DATA

TREND	HIGH	INTERMED	LOW
RISING	O = 42 E = 54	O = 84 E = 43	O = 42 E = 70
STABLE	O = 345 E = 319	O = 151 E = 258	O = 496 E = 414
FALLING	O = 56 E = 69	O = 123 E = 56	O = 37 E = 90

STATUS — PERCENT FREQUENCY

TREND	HIGH	INTERMED	LOW
RISING	3.05%	6.10%	3.05%
STABLE	25.07%	10.97%	36.04%
FALLING	4.06%	8.93%	2.68%

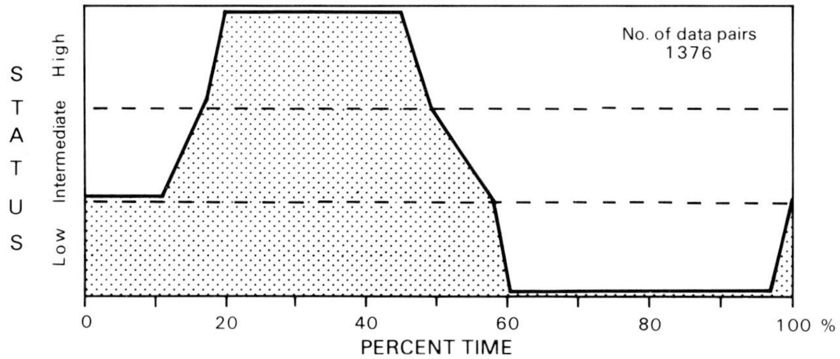

Fig. 2. The construction of a hypothetical pulse curve illustrating lake behaviour from the contingency table of status and trend for the global data set. O = observed values, E = expected values.

spent in the corresponding status-trend class by an average lake in that subset. The percentage frequencies were plotted in an order that represents a complete lake cycle, i.e. intermediate-stable, intermediate-rising, high-rising, high-stable, high-falling, intermediate-falling, low-falling, low-stable, low-rising. Differences in the relative stability of high, intermediate and low lake levels are indicated by the length of the appropriate segment of the graph, while differences in the duration of the rising and falling stages are shown by the slope of the curve (Figure 2).

Results

Aggregate Data

The temporal patterns of lake-level fluctuations in the NET, IT and SET belts are illustrated in Figure 3. The IT histogram shows two distinct phases of moderate to high lake levels, ca. 30,000 - 21,000 and 12,000 - 5000 yr BP. The dating control for the earlier period is poor, but the rapid onset of more arid conditions at 21,000 yr BP is comparatively well dated, as is the return to wetter conditions at ca. 12,000 yr BP. The period after 5000 yr BP was marked by a return to arid conditions, which have persisted to the present time. This pattern is very similar to the African pattern described by Street and Grove (1979), and reflects the predominance of African and Middle Eastern basins in the IT data set.

The SET data set shows a broadly similar sequence of lake-level fluctuations to that experienced within the tropics. The frequency of high lake levels, however, was much lower during both major wet phases. The NET region shows a totally different pattern. Most lakes were high during the period 30,000 - 16,000 yr BP. Figure 3 implies that there was a gradual decline after 16,000 yr BP towards the more arid conditions typical of the present. This generalized pattern, however, conceals marked longitudinal contrasts within the NET belt (see below).

In order to investigate the latitudinal variations in lake levels in more detail, histograms of lake status were constructed for 10° latitude bands from 5°S to 45°N (Figure 4). Prior to 25,000 yr BP, the number of basins in each subset is much too small to place any confidence in the resulting patterns. After that time, however, the quantity of information and the quality of the dating control appear sufficient to justify analysis.

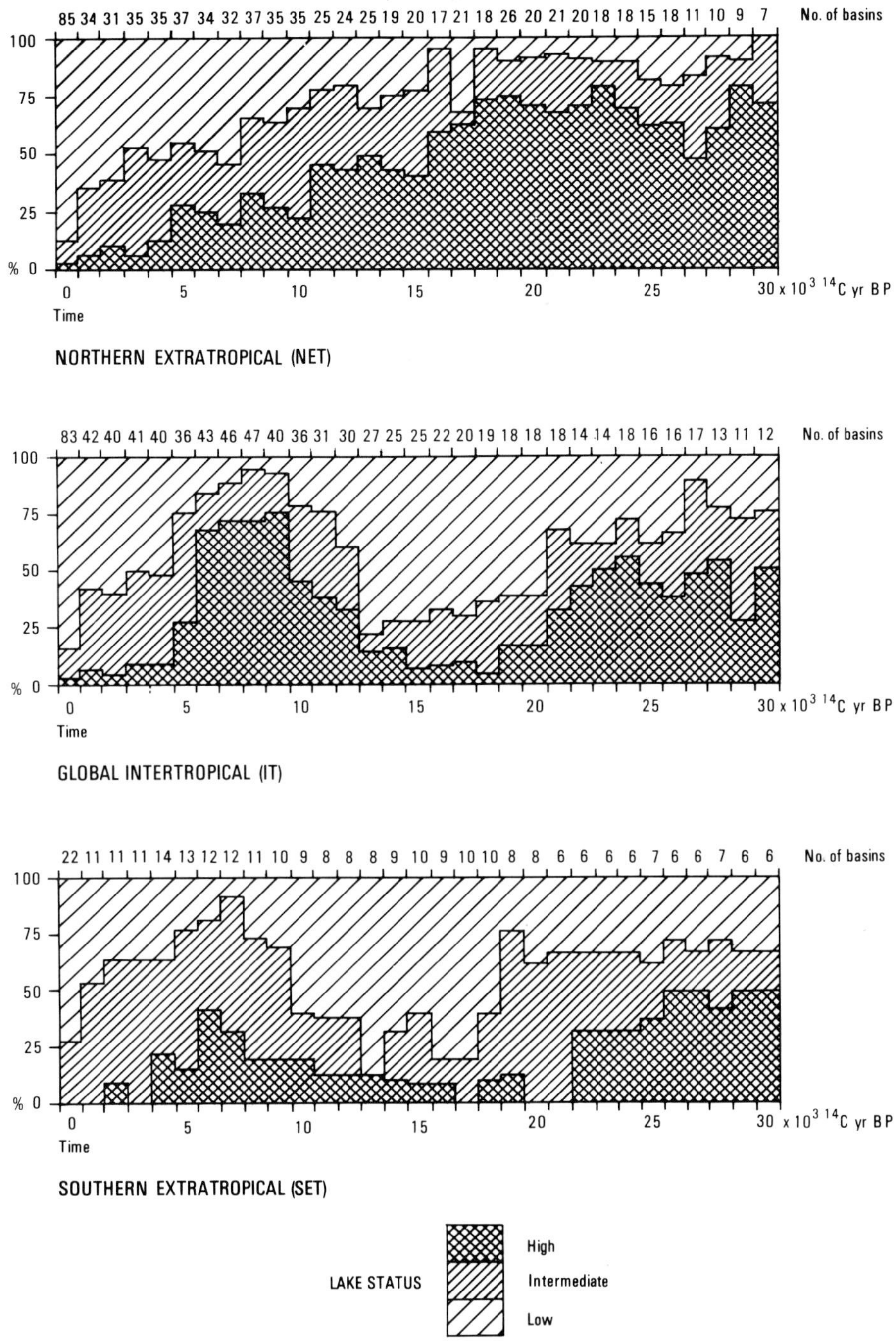

Fig. 3. Histograms showing the temporal variations in lake status in the NET, IT and SET data sets.

The most striking feature of the 10° histograms is the terminal Pleistocene to mid-Holocene pulse of high lake levels in the tropics. This pulse was most pronounced in the zone 5°S - 15°N, becoming progressively more sharply defined towards the axis of the subtropical desert belt. At 15° -25°N, it appears as a distinct 'spike' that dies out between 25° and 35°N. The zone 35° - 45°N was characterized by a markedly different pattern of fluctuations. High lake levels during the glacial gave way to more arid conditions in the Holocene, with only a relatively minor increase in lake levels after 6000 yr BP.

Figure 4 confirms that it is possible to identify distinctive low-latitude and mid-latitude patterns of lake behaviour in the 10° data sets. The former is exemplified by the histogram for 5° - 15°N and the latter by 35° - 45°N. These two patterns appear to be superimposed in the histogram for 25° - 35°N. This belt, therefore, can be regarded as a transitional zone which has been affected by both tropical and mid-latitude influences during the Late Quaternary.

The global data set also reveals longitudinal variations in lake behaviour. These are particularly marked in the NET belt (Figure 5). Although lake levels were consistently high before 16,000 yr BP in both North America and Africa-Eurasia, north of the Tropic of Cancer, there was a marked asymmetry in lake behaviour between the two regions after that time, which may reflect the different retreat histories of the Laurentide and Eurasian ice sheets (Denton and Hughes, 1981). In North America, lake status remained generally high until ca. 11,000 yr BP, after which conditions became more arid. In contrast, there was a dramatic change to drier conditions in Africa-Eurasia around 16,000 yr BP, followed by a return to higher lake levels from 10,000 to 5000 yr BP. Even within the tropics, the northward penetration of the postglacial pulse of wet conditions appears to have been much less in the Americas than in North Africa and South Asia (Street-Perrott and Roberts, 1983; Street-Perrott et al., in press).

The hypothetical pulse curves yield further insights into the temporal and spatial variations in lake behaviour (Figures 2,6,7). Chi-square analysis of the contingency table for the total data set (Figure 2) reveals that several status-trend combinations occur significantly more or less frequently ($p < 0.05$) than would be expected on the assumption that there is no association between status and trend (Table 1). In particular, low-stable conditions occur more frequently than expected, whereas low-falling, low-rising and intermediate-stable are underrepresented. The transitional intermediate states (intermediate-rising and intermediate-falling) are overrepresented. The occurrence of the high states, however, does not differ significantly from expectation. These data suggest that most closed lakes, particularly amplifier lakes (Street, 1980) have alternated between two relatively stable states, high and low, of which low is the more persistent. Water-level maxima that peak at intermediate levels are believed to represent mainly short-lived, rather than less extreme, climatic oscillations. The strong association between intermediate status and rising or falling trend is consistent with this interpretation. The paucity of observations in the categories low-falling and low-rising may result from the poor quality of the lake-level record under conditions of shallow water and high salinity (Street-Perrott and Harrison, in press) and hence have no climatic significance.

The pulse curves for the NET, IT and SET belts are shown in Figure 6. These curves all differ significantly ($p < 0.001$) from the distribution expected if there was no association between status and trend (Table 1). The IT pulse curve is broadly similar to the total data set (Figure

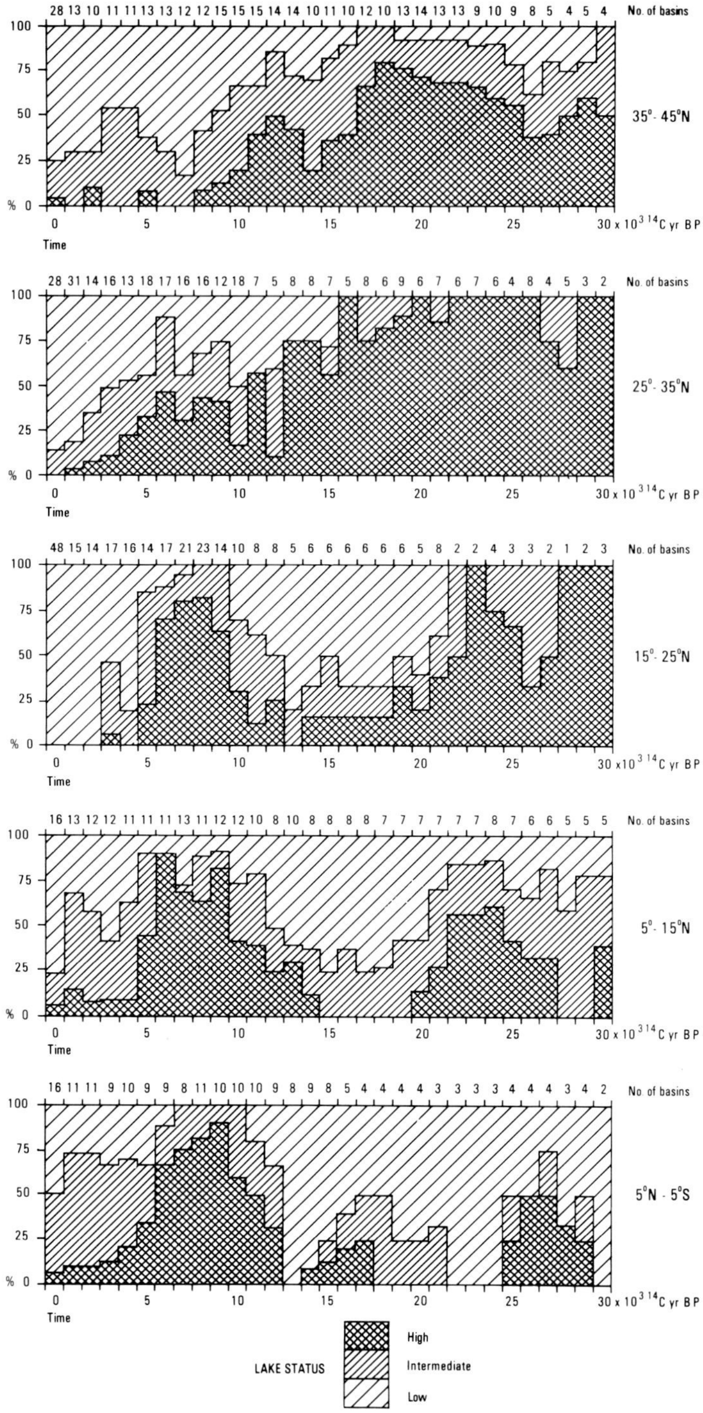

Fig. 4. Histograms showing the temporal variations in lake status for 10° latitude bands between 5°S and 45°N.

2). Low lake levels were persistent in this zone, with phases of high lake levels occurring as more transitory features of the record. Transitions between these two states were generally abrupt.

The form of the NET pulse curve differs significantly from the IT zone (Table 2), in that

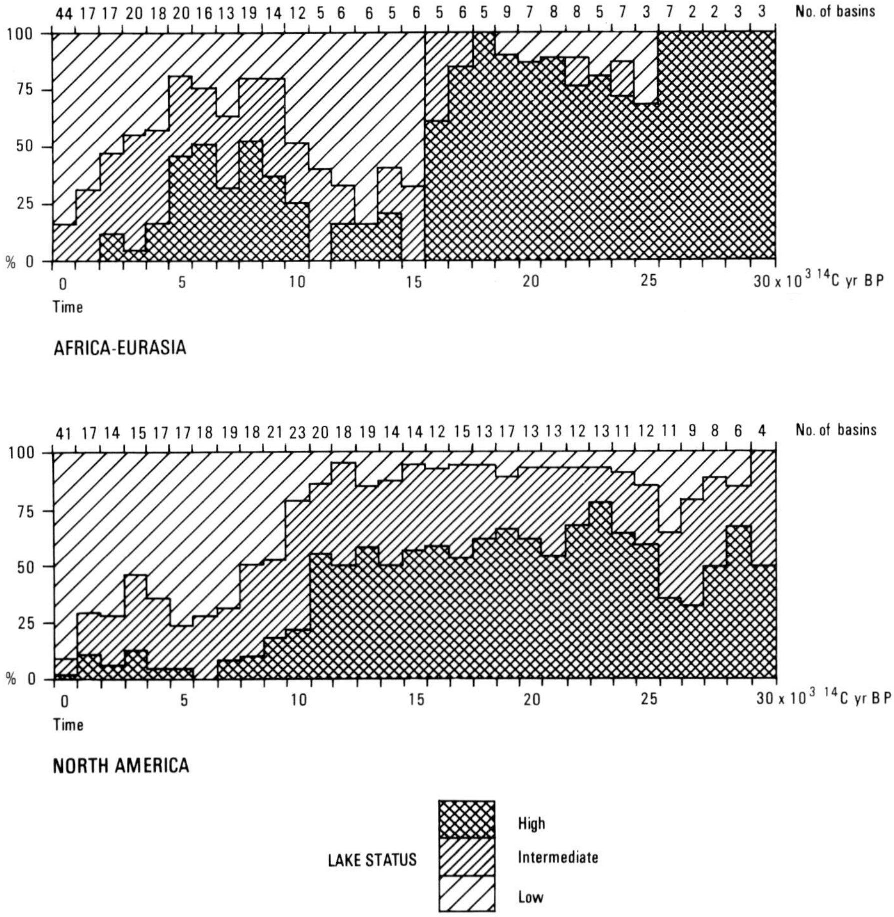

Fig. 5. Histograms showing the temporal variations in lake status in the Africa-Eurasia and North America subsets of the NET belt.

high-stable conditions were much more frequent and low-stable ones much less so ($p < 0.01$). It is interesting to note that in comparison with the other two, the SET pulse curve is more peaked. Given the paucity of observations in this subset, however, further comment does not seem justified at present.

The pulse curves plotted for 10° latitude belts (Figure 7), which also show a significant association between status and trend ($p < 0.001$) (Table 1), exhibit a marked discontinuity in form at 25°N. Equatorward of 25°N, low and intermediate conditions were of long duration and phases of high, stable levels were less persistent. The greatest frequency of wet conditions in this zone was experienced between 5° and 15°N. In contrast, poleward of 25°N, high-stable lake levels were a prominent feature of the lacustrine record. Thus, the pulse curves provide further evidence for the existence of two distinct, low-latitude and mid-latitude, modes of lake behaviour in the northern hemisphere, and confirm the prolonged duration of high levels in the zone 25° - 35°N.

Individual lake-level curves

Curves from individual basins frequently reveal water-level fluctuations on a time scale of 10^2 -10^3 yrs that are obscured in the aggregate data sets. Four lake-level curves from African basins with particularly good stratigraphic resolution and dating control are plotted in Figure 8. These show that large-amplitude fluctuations in lake depth have been experienced in northern intertropical Africa during the last 13,000 yr. Although there are minor differences between individual curves, attributable to the idiosyncracies of basin hydrology or to the errors and uncertainties of ^{14}C dating, the correspondence in the timing of major events is striking. The widespread postglacial wet phase beginning around 13,000 - 12,000 yr BP was interrupted by several dry episodes. These culminated during the intervals 11,000 - 10,000 and 8500 - 6500 yr BP, with a minor oscillation around 6200 - 5800 yr BP. The more arid phase commencing around 5000 yr BP was also broken by brief periods of higher lake levels, but the timing of these episodes

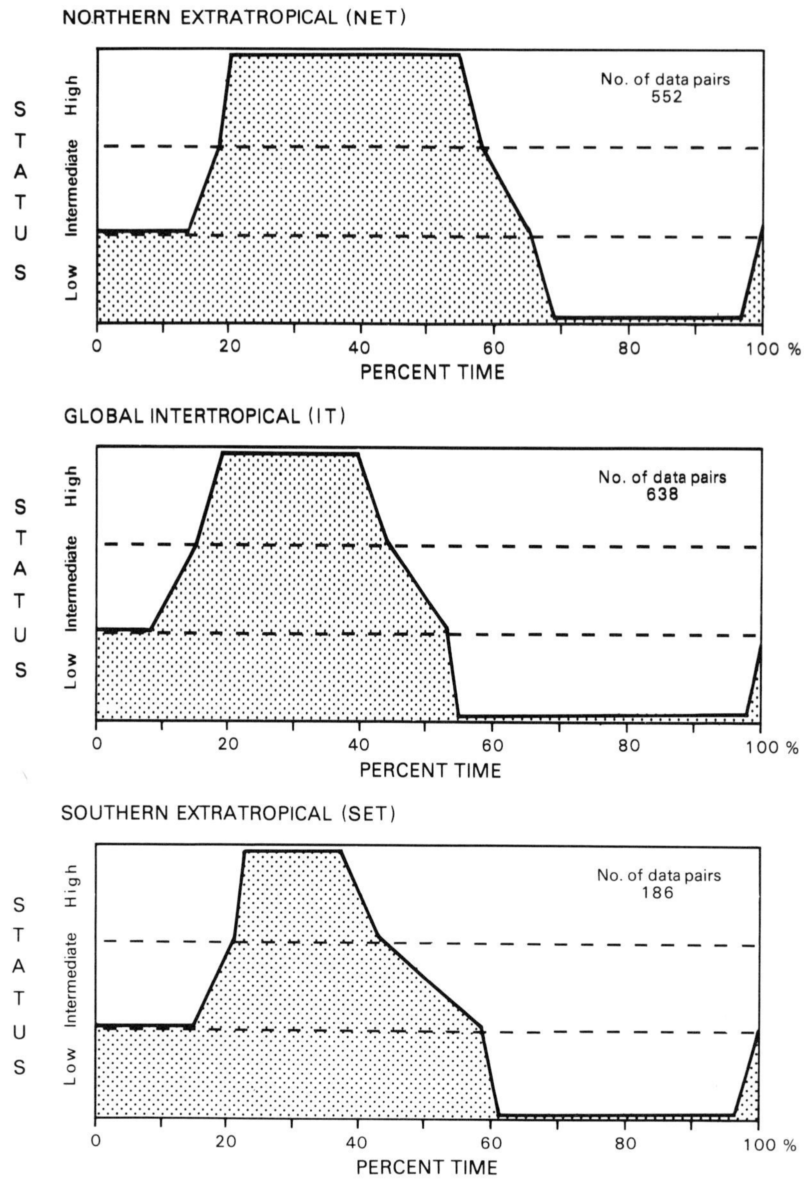

Fig. 6. Hypothetical pulse curves for the NET, IT and SET data sets.

shows no consistency from basin to basin. The curves presented demonstrate that tropical African lakes have responded to climatic mechanisms operating on much shorter time scales than the broad glacial/interglacial cycles. They also illustrate the rapidity with which closed lakes respond to climatic fluctuations, as a result of the strong positive feedbacks operating within the basin hydrological cycle (Szestay, 1974).

Discussion

A number of workers, notably Galloway (1965, 1983) and Brakenridge (1978), have attributed Late Quaternary lake-level fluctuations largely to temperature changes. This 'minevaporal' hypothesis is clearly insufficient to explain the spatial patterns and complex phase relations exhibited by the global data set. For this, it is necessary to invoke large-scale variations in precipitation as well as in evaporation (Street, 1979; Kutzbach, 1983), and hence changes in the location and intensity of the major features of the general circulation.

Atmospheric modelling experiments suggest that the global hydrological cycle is highly sensitive to perturbations in the planetary radiation and energy budget (Manabe and Wetherald, 1975; Wetherald and Manabe, 1975; Charney et al., 1975; Batten, 1975; Manabe and Hahn, 1977; Williams, 1978; Kutzbach, 1981). A variety of different causal mechanisms have been proposed to explain such perturbations (Kutzbach, 1976; Mitchell, 1976; Berger, 1979, 1981). The response of the

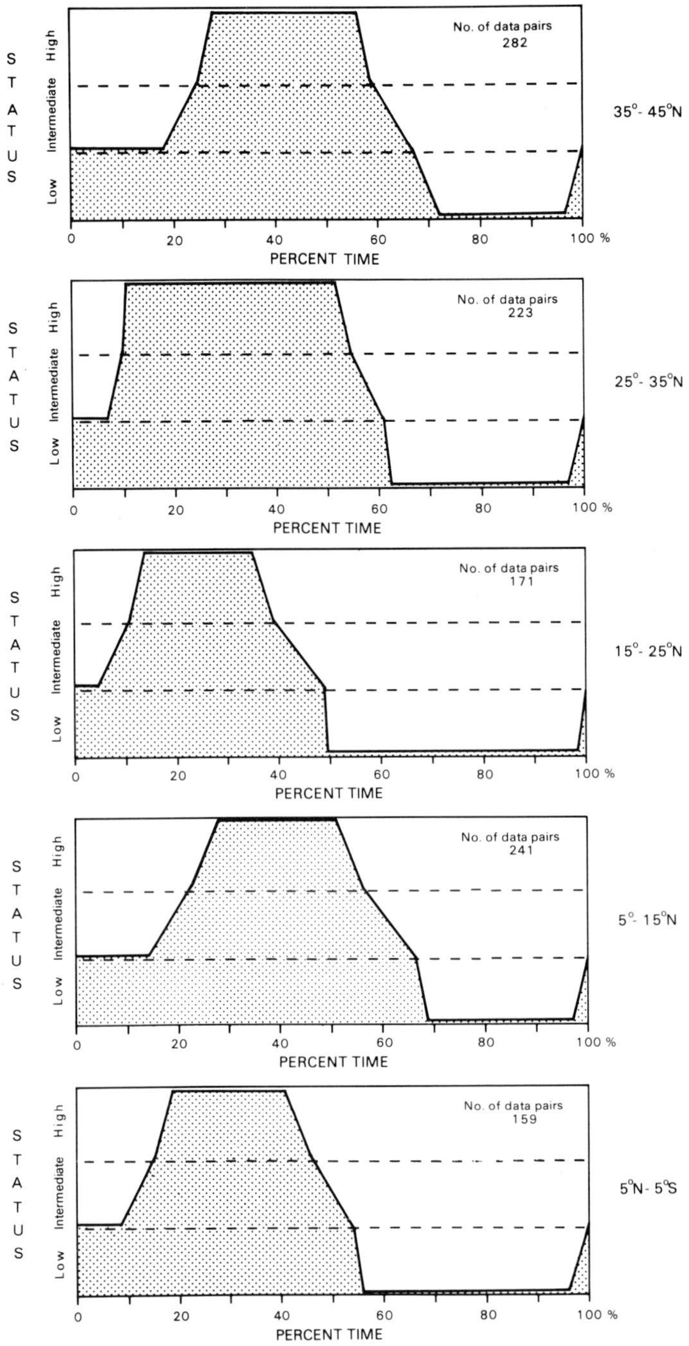

Fig. 7. Hypothetical pulse curves for 10° latitude bands between 5°S and 45°N.

atmosphere to any given forcing function is known to be highly nonlinear, and involves an intricate web of feedbacks and lag effects, many of which are associated with slowly varying components of the climate system such as the cryosphere, oceans and biosphere (Gates, 1981).

The leading contender as an explanation for climatic fluctuations on a glacial/interglacial time scale is undoubtedly the Milankovitch theory (Broecker et al., 1968; Hays et al., 1976; Berger, 1980; Imbrie and Imbrie, 1980). Since the interaction of the periodic astronomical variations in the latitudinal and seasonal distribution of insolation with the planetary geography is extremely complex, the precise identification of

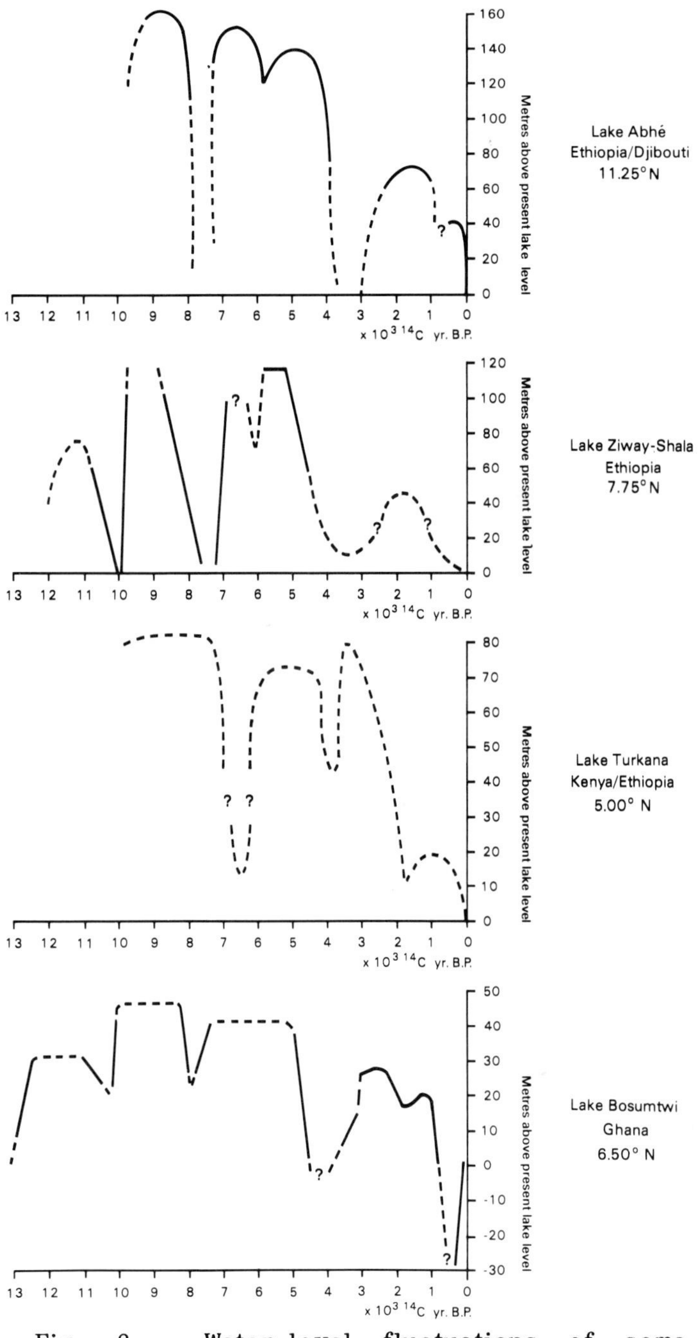

Fig. 8. Water-level fluctuations of some African intertropical lakes, 13,000 - 0 ^{14}C yr BP. (Sources: Lake Abhe: Gasse and Street, 1978; Lake Ziway-Shala: Gillespie et al., in press, Lake Turkana: Owen et al., 1982; Lake Bosumtwi: Talbot and Delibrias, 1980). Dashed lines indicate uncertainty.

Table 1. Contingency table of X^2 values for various subsets of the lake data.

DATA SET		STATUS/TREND CLASSES									
		I/S	I/R	H/R	H/S	H/F	I/F	L/F	L/S	L/R	OVERALL
GLOBAL	O	151.0	84.0	42.0	345.0	56.0	123.0	37.0	496.0	42.0	
	E	258.0	43.0	54.0	319.0	69.0	56.0	90.0	414.0	70.0	N=1376
	X^2	44.8	38.8	2.7	2.0	2.5	79.7	31.4	15.8	11.3	X^2=299.0 S=***
NET	O	75.0	28.0	12.0	189.0	17.0	40.0	20.0	153.0	18.0	
	E	108.0	15.0	23.0	164.7	30.4	20.0	26.6	144.0	20.0	N=552
	X^2	10.1	11.2	5.3	3.5	5.9	20.0	1.6	0.5	0.2	X^2=58.5 S=***
IT	O	49.0	44.0	27.0	129.0	28.0	55.0	11.0	278.0	17.0	
	E	105.8	20.4	25.0	131.5	27.0	21.8	45.1	218.7	42.0	N=638
	X^2	30.6	27.2	0.2	0.1	0.0	50.4	25.8	15.9	14.9	X^2=165.2 S=***
SET	O	27.0	12.0	3.0	27.0	11.0	28.0	6.0	65.0	7.0	
	E	42.9	7.9	4.8	26.2	9.9	16.2	18.9	49.9	9.2	N=186
	X^2	0.0	2.1	grouped	5.9	0.1	8.6	8.8	4.5	grouped	X^2=31.3 S=***
35-45°	O	49.0	20.0	9.0	82.0	8.0	23.0	15.0	69.0	7.0	
	E	65.2	11.7	12.6	70.2	16.1	15.0	14.8	64.5	11.6	N=282
	X^2	4.1	5.9	1.0	1.9	4.1	4.2	0.0	0.0	1.8	X^2=23.4 S=***
25-35°N	O	14.0	7.0	2.0	92.0	7.0	14.0	4.0	76.0	7.0	
	E	28.6	2.5	7.2	82.4	11.3	3.9	9.8	71.0	6.2	N=223
	X^2	7.5	grouped	3.8	1.1	1.6	grouped	3.4	0.3	0.1	X^2=51.1 S=***
15-25°N	O	7	11.0	5.0	36.0	7.0	17.0	1.0	83.0	4.0	
	E	25.8	5.1	5.6	35.4	7.0	5.1	12.9	64.8	10.3	N=171
	X^2	13.6	6.9	0.1	0.0	0.0	28.1	10.9	5.4	3.8	X^2=68.7 S=***
5-15°N	O	33.0	20.0	14.0	55.0	13.0	24.0	7.0	67.0	8.0	
	E	49.5	13.4	14.3	52.7	15.0	14.1	15.0	52.7	14.3	N=241
	X^2	5.5	3.2	0.0	0.1	0.3	6.9	4.3	3.9	2.8	X^2=27.0 S=***
5°S-5°N	O	13.0	11.0	6.0	35.0	8.0	13.0	3.0	63.0	7.0	
	E	25.8	5.6	7.4	34.2	7.4	5.6	11.0	51.0	10.0	N=159
	X^2	6.5	5.1	0.3	0.0	0.0	9.6	5.9	2.7	0.9	X^2=31.0 S=***

Key

H = High
I = Intermediate
L = Low
R = Rising

S = Stable
F = Falling
N = Total number of observations
O = Observed value

E = Expected value
X^2 = Chi-square value
S = Significance with 4 degrees of freedom
** significant at 1% level
*** significant at 0.1% level

Table 2. Contigency table of X^2 values for comparisons of the NET, IT and SET data sets.

REGIONAL SUBSETS		STATUS/TREND CLASSES											
		I/S	I/R	H/R	H/S	H/F	I/F	L/F	L/S	L/R			
NET	O	75.0	28.0	12.0	189.0	17.0	40.0	20.0	153.0	18.0			
	E	60.6	33.7	16.8	138.4	22.5	49.3	14.8	199.0	16.8			
	X^2	3.4	1.0	1.4	18.5	1.3	1.8	1.8	10.6	0.1			
IT	O	49.0	44.0	27.0	129.0	28.0	55.0	11.0	278.0	17.0	N	=	1376
	E	70.0	38.9	19.5	160.0	26.0	57.0	17.2	230.0	19.5	X^2	=	90.1
	X^2	6.3	0.7	2.9	6.0	0.2	0.1	2.2	10.0	0.3	d.f.	=	16
SET	O	27.0	12.0	3.0	27.0	11.0	28.0	6.0	6.0	7.0	S	=	***
	E	20.4	11.4	5.7	46.6	7.6	16.6	5.1	67.0	5.7			
	X^2	2.1	0.1	1.3	8.2	1.5	7.8	0.2	0.1	0.3			
NET	O	75.0	28.0	12.0	189.0	17.0	40.0	20.0	153.0	18.0			
	E	57.5	33.4	18.1	147.5	20.9	44.1	14.4	199.9	16.2	N	=	1190
	X^2	5.3	0.9	2.1	11.7	0.7	0.4	2.2	11.0	0.2	X^2	=	64.2
IT	O	49.0	44.0	27.0	129.0	28.0	55.0	11.0	278.0	17.0	d.f.	=	8
	E	66.5	28.6	20.9	170.5	24.1	50.9	16.6	231.1	18.8	S	=	***
	X^2	4.6	0.8	1.8	10.1	0.6	0.3	1.9	9.5	0.2			
IT	O	49.0	44.0	27.0	129.0	28.0	55.0	11.0	278.0	17.0			
	E	59.8	43.4	23.2	120.8	30.2	64.3	13.2	265.6	18.6	N	=	824
	X^2	1.9	0.0	0.6	0.6	0.2	1.3	0.4	0.6	0.1	X^2	=	24.1
SET	O	27.0	12.0	3.0	27.0	11.0	28.0	6.0	65.0	7.0	d.f.	=	8
	E	17.2	12.6	6.8	35.2	8.8	18.7	3.8	77.4	5.4	S	=	**
	X^2	5.6	0.1	2.1	1.9	0.6	4.7	grouped	2.0	grouped			

Key

H = High
I = Intermediate
L = Low
R = Rising

S = Stable
F = Falling
N = Total number of observations
O = Observed value

E = Expected value
X^2 = Chi-square value
S = Significance with 4 degrees of freedom
** significant at 1% level
*** significant at 0.1% level

orbital controls on continental palaeoclimates is likely to prove a formidable task.

Pioneering modelling experiments by Kutzbach (1981, 1983) and Kutzbach and Otto-Bliesner (1982) suggest that the atmospheric circulation in low latitudes may have responded sensitively to variations in orbital parameters. Using a low resolution general circulation model, they showed that the early Holocene maximum of summer insolation in the Northern Hemisphere could have resulted in stronger heating and increased monsoonal precipitation over North Africa, the Middle East and South Asia (cf. Figure 3). Kutzbach (1981) proposed a similar explanation for the earlier occurrence of high lake levels in the tropics around 30,000 - 25,000 yr BP. For a fuller understanding of the pattern of fluctuations in lake level on the glacial/interglacial time scale, particularly in mid-latitudes, however, it will be necessary to clarify not only the impact of the cryospheric, biospheric and oceanographic feedbacks that modulate the response of the global hydrological cycle to the Milankovitch signal (Street-Perrott et al., 1983), but also the effects of other causal mechanisms.

When one turns to the lake-level record on the 10^3 yr time scale, a different set of causal factors must be invoked (Flohn, 1979). The dramatic fluctuations exhibited by the African lakes shown in Figure 8 are being confirmed by an increasing body of palaeoclimatic data derived not only from Africa (Street-Perrott and Roberts, 1983), but also from other parts of the tropics (Roberts et al., 1981; Singh, 1981; Bradbury et al., 1981). The postglacial lake-level curves from the NET and SET belts do not yet have sufficient temporal resolution to determine whether or not marked fluctuations on this time scale occurred outside the tropics.

At present, paradoxically, there seems to be even less certainty about the causes of wet and dry episodes during the Holocene, for which we have a wealth of empirical evidence, than there is about fluctuations in water balance on the glacial /interglacial time scale. One of the most important ways in which progress could be made in understanding both long-term and short-term palaeohydrological variations, therefore, would be to carry out sensitivity experiments involving the coupling of interactive models of the cryosphere and biosphere to atmospheric models of sufficient sophistication to include a realistic hydrological cycle.

Acknowledgements. The authors are grateful to Andrew McIntyre and an anonymous reviewer for critical comments on this paper. Advice from Michael Bulmer on statistical problems was much appreciated.

References

Batten, E.S., The atmospheric response to a stratospheric dust cloud as simulated by a general circulation model, In Weller, G. and Bowling, S.A. (Eds.), Climate of the Arctic, 24th Alaska Science Conference, University of Alaska, Fairbanks, 150-158, 1975.

Berger, A.L., Spectrum of climatic variations and their causal mechanisms, Geophys. Surv., 3, 351-402, 1979.

Berger, A.L., The Milankovitch astronomical theory of palaeoclimates: a modern review, Vistas in Astron., 24, 103-122, 1980.

Berger, A.L., Spectrum of climatic variations and possible causes, In Berger, A.L. (Ed.), Climatic Variations and Variability: Facts and Theories, D. Reidel, Dordrecht, 411-432, 1981.

Bradbury, J.P., B. Leyden, M. Salgado-Labouriau et al., Late Quaternary environmental history of Lake Valencia, Venezuela, Science, 214, 1299-1305, 1981.

Brakenridge, G.R., Evidence for a cold, dry full-glacial climate in the American Southwest, Quat. Res., 9, 22-40, 1978.

Broecker, W.S., D.L. Thurber, J. Goddard, Ku Teh-Lung, R.K. Matthews, and K.J. Mesolella, Milankovitch hypothesis supported by precise dating of coral reefs and deep-sea sediments, Science, 159, 297-300, 1968.

Charney, J.G., P.H. Stone and W.J. Quirk, Drought in the Sahara: a biogeophysical feedback mechanism, Science, 187, 434-435, 1975.

Croxton, F.E., Applied General Statistics, 3rd Ed., Prentice-Hall, New York, 754 pp., 1967.

Denton, G.H. and T.J. Hughes, The Last Great Ice Sheets, Wiley Interscience, New York, 484 pp., 1981.

Flohn, H., On time scales and causes of abrupt palaeoclimatic events, Quat. Res., 12, 135-149, 1979.

Galloway, R.W., Late Quaternary climates in Australia, Jour. Geol., 73, 603-618, 1965.

Galloway, R.W., Full-glacial southwestern United States: mild and wet or cold and dry?, Quat. Res., 19, 236-248, 1983.

Gasse, F. and Street, F.A., Late Quaternary lake-level fluctuations and environments of the northern Rift Valley and Afar region (Ethiopia and Djibouti), Palaeogeog. Palaeoclim. Palaeoecol., 24, 279-325, 1978.

Gates, W.L., The climate system and its portrayal by climate models: a review of basic principles, In Berger, A.L. (Ed.), Climatic Variations and Variability: Facts and Theories, D. Reidel, Dordrecht, 3-19, 1981.

Gillespie, R., F.A. Street-Perrott, and R. Switsur, Post-glacial arid episodes in Ethiopia - implications for climate prediction, Nature, in press.

Hays, J.D., J. Imbrie and N.J. Shackleton, Variations of the Earth's orbit: pacemaker of the ice ages, Science, 194, 1121-1132, 1976.

Imbrie, J. and J.Z. Imbrie, Modeling the climatic response to orbital variations, Science, 207, 943, 1980.

Kutzbach, J.E., The nature of climate and climatic variations, Quat. Res., 6, 471-480, 1976.

Kutzbach, J.E., Monsoon climate of the early Holocene: climate experiment with the earth's orbital parameters for 9000 years ago, Science, 214, 59-61, 1981.

Kutzbach J.E., Monsoon rains of the Late Pleistocene and Early Holocene: patterns, intensity and possible causes of changes, In Street-Perrott, F.A., M. Beran and R.A.S. Ratcliffe, Variations in the Global Water Budget, D. Reidel, Dordrecht, 371-389, 1983.

Kutzbach, J.E. and B.L. Otto-Bliesner, The sensitivity of the African-Asian monsoonal climate to orbital parameter changes for 9000 years B.P. in a low resolution general circulation model, Jour. Atm. Sci., 39, 1177-1188, 1982.

Manabe, S. and D.G. Hahn, Simulation of the tropical climate of an ice age, Jour. Geophys. Res., 82, 3889-3911, 1977.

Manabe, S. and R.T. Wetherald, The effects of doubling the CO_2 concentrations on the climate of a general circulation model, Jour. Atm. Sci., 32, 3-15, 1975.

de Martonne, E. and L. Aufrere, L'extension des regions privees d'ecoulement vers l'ocean, Ann. de Geogr., 38, 1-24, 1928.

Mitchell, J.M., An overview of climatic variability and its causal mechanisms, Quat. Res., 6, 481-493, 1976.

Owen, R.B., J.W. Barthelme, R.W. Renaut and A. Vincens, Palaeolimnology and archaeology of Holocene deposits north-east of Lake Turkana, Kenya, Nature, 298, 523-529, 1982.

Roberts, N, A., Street-Perrott and A. Perrott, The "late glacial" in the tropics, Quat. New., 35, 1-5, 1981.

Schneider, S. H. and S. L. Thompson, Ice ages and orbital variations: some simple theory and modeling, Quat. Res., 12, 188-203, 1979.

Singh, G., Late Quaternary pollen records and seasonal palaeoclimates of Lake Frome, South Australia, Hydrobiol., 82, 419-430, 1981.

Street, F.A., Late Quaternary precipitation estimates for the Ziway-Shala Basin, Southern Ethiopia, Palaeoecol. Afr., 11, 135-143, 1979.

Street, F.A., The relative importance of climate and hydrogeological factors in influencing lake-level fluctuations, Palaeoecol. Afr., 12, 137-158, 1980.

Street, F.A. and A.T., Environmental and climatic implications of Late Quaternary lake-level fluctuations in Africa, Nature, 261, 385-390, 1976.

Street, F.A. and A.T. Grove, Global maps of lake-level fluctuations since 30,000 B.P., Quat. Res., 12, 83-118, 1979.

Street-Perrott, F.A., M. Beran and R. Ratcliffe, Variations in the Global Water Budget, D. Reidel, Dordrecht, 518 pp. 1983.

Street-Perrott, F. A. and S. P. Harrison, Lake-level fluctuations, In Hecht, A.D. (Ed.), Quantitative Paleoclimate Analysis and Modeling, John Wiley, New York, in press.

Street-Perrott, F.A. and N. Roberts, Fluctuations in closed basin lakes as indicators of past atmospheric circulation patterns, In Street-Perrott, F.A., M. Beran and R.A.S. Ratcliffe (Eds.), Variations in the Global Water Budget, D. Reidel, Dordrecht, 331-345, 1983.

Street-Perrott, F.A., N. Roberts and S. Metcalfe, Geomorphic implications of Late Quaternary climatic changes in the Northern Hemisphere tropics, In Douglas, I. and T. Spencer, (Eds.), Geomorphology and Environmental Change in Tropical Latitudes, George Allen and Unwin, London, in press.

Szestay, K., Water balance and water level fluctuations of lakes, Hydrological Sciences Bulletin, 19, 73-84, 1974.

Talbot, M. R. and G. Delibrias, A new Late Pleistocene-Holocene water-level curve for Lake Bosumtwi, Ghana, Earth Planet. Sci. Letters, 47, 336-344, 1980.

Wetherald, R.T. and S. Manabe, The effects of changing the solar constant on the climate of a general circulation model, Jour. Atm. Sci., 32, 2044-2059, 1975.

Williams, J., A brief comparison of model simulations of glacial period maximum atmospheric circulation, Palaeogeog. Palaeoclim. Palaeoecol., 25, 191-198, 1978.

CLIMATE SENSITIVITY: ANALYSIS OF FEEDBACK MECHANISMS

J. Hansen, A. Lacis, D. Rind, G. Russell

NASA/Goddard Space Flight Center, Institute for Space Studies
2880 Broadway, New York, NY 10025

P. Stone

Center for Meteorology and Physical Oceanography
Massachusetts Institute of Technology, Cambridge, MA 02139

I. Fung

Lamont-Doherty Geological Observatory of Columbia University
Palisades, NY 10964

R. Ruedy, J. Lerner

M/A COM Sigma Data, Inc.
2880 Broadway, New York, NY 10025

Abstract. We study climate sensitivity and feedback processes in three independent ways: (1) by using a three dimensional (3-D) global climate model for experiments in which solar irradiance S_0 is increased 2 percent or CO_2 is doubled, (2) by using the CLIMAP climate boundary conditions to analyze the contributions of different physical processes to the cooling of the last ice age (18K years ago), and (3) by using estimated changes in global temperature and the abundance of atmospheric greenhouse gases to deduce an empirical climate sensitivity for the period 1850-1980.

Our 3-D global climate model yields a warming of ~4°C for either a 2 percent increase of S_0 or doubled CO_2. This indicates a net feedback factor of f = 3-4, because either of these forcings would cause the earth's surface temperature to warm 1.2-1.3°C to restore radiative balance with space, if other factors remained unchanged. Principal positive feedback processes in the model are changes in atmospheric water vapor, clouds and snow/ice cover. Feedback factors calculated for these processes, with atmospheric dynamical feedbacks implicitly incorporated, are respectively $f_{water\ vapor} \sim 1.6$, $f_{clouds} \sim 1.3$ and $f_{snow/ice} \sim 1.1$, with the latter mainly caused by sea ice changes. A number of potential feedbacks, such as land ice cover, vegetation cover and ocean heat transport were held fixed in these experiments.

We calculate land ice, sea ice and vegetation feedbacks for the 18K climate to be $f_{land\ ice} \sim$ 1.2-1.3, $f_{sea\ ice} \sim 1.2$, and $f_{vegetation} \sim$ 1.05-1.1 from their effect on the radiation budget at the top of the atmosphere. This sea ice feedback at 18K is consistent with the smaller $f_{snow/ice} \sim 1.1$ in the S_0 and CO_2 experiments, which applied to a warmer earth with less sea ice. We also obtain an empirical estimate of f = 2-4 for the fast feedback processes (water vapor, clouds, sea ice) operating on 10-100 year time scales by comparing the cooling due to slow or specified changes (land ice, CO_2, vegetation) to the total cooling at 18K.

The temperature increase believed to have occurred in the past 130 years (approximately 0.5°C) is also found to imply a climate sensitivity of 2.5-5°C for doubled CO_2 (f = 2-4), if (1) the temperature increase is due to the added greenhouse gases, (2) the 1850 CO_2 abundance was 270±10 ppm, and (3) the heat perturbation is mixed like a passive tracer in the ocean with vertical mixing coefficient $k \sim 1\ cm^2\ s^{-1}$.

These analyses indicate that f is substantially greater than unity on all time scales. Our best estimate for the current climate due to processes operating on the 10-100 year time scale is f = 2-4, corresponding to a climate sensitivity of 2.5-5°C for doubled CO_2. The physical process contributing the greatest uncertainty to f on this time scale appears to be the cloud feedback.

We show that the ocean's thermal relaxation time depends strongly on f. The e-folding time constant for response of the isolated ocean mixed layer is about 15 years, for the estimated value of f. This time is sufficiently long to allow substantial heat exchange between the mixed layer and deeper layers. For f = 3-4 the response time of the surface temperature to a heating perturbation is of order 100 years, if the perturbation is sufficiently small that it does not alter the rate of heat exchange with the deeper ocean.

The climate sensitivity we have inferred is larger than that stated in the Carbon Dioxide Assessment Committee report (CDAC, 1983). Their result is based on the empirical temperature increase in the past 130 years, but their analysis did not account for the dependence of the ocean response time on climate sensitivity. Their choice of a fixed 15 year response time biased their result to low sensitivities.

We infer that, because of recent increases in atmospheric CO_2 and trace gases, there is a large, rapidly growing gap between current climate and the equilibrium climate for current atmospheric composition. Based on the climate sensitivity we have estimated, the amount of greenhouse gases presently in the atmosphere will cause an eventual global mean warming of about 1°C, making the global temperature at least comparable to that of the Altithermal, the warmest period in the past 100,000 years. Projection of future climate trends on the 10-100 year time scale depends crucially upon improved understanding of ocean dynamics, particularly upon how ocean mixing will respond to climate change at the ocean surface.

Introduction

Over a sufficient length of time, discussed below, thermal radiation from the earth must balance absorbed solar radiation. This energy balance requirement defines the effective radiating temperature of the earth, T_e, from

$$\pi R^2(1 - A)S_o = 4\pi R^2 \sigma T_e^4 \tag{1}$$

or

$$T_e = [S_o(1 - A)/4\sigma]^{1/4} = (s/\sigma)^{1/4} \tag{2}$$

where R is the earth radius, A the earth albedo, S_o the solar irradiance, s the mean flux of absorbed solar radiation per unit area and σ the Stefan-Boltzmann constant. Since $A \sim 0.3$ and $S_o \sim 1367\ W\ m^{-2}$, $s \sim 239\ W\ m^{-2}$ and this requirement of energy balance yields $T_e \sim 255K$. The effective radiating temperature is also the physical temperature at an appropriately defined mean level of emission to space. In the earth's atmosphere this mean level of emission to space is at altitude $H \sim 6$ km. Since the mean tropospheric temperature gradient is ~5.5°C km^{-1}, the surface temperature is $T \sim 288K$, ~33K warmer than T_e.

It is apparent from (2) that for changes of solar irradiance

$$\frac{dT_e}{T_e} = \frac{1}{4}\frac{dS_o}{S_o} = \frac{1}{4}\frac{ds}{s} \quad . \tag{3}$$

Thus if S_o increases by a small percentage δ, T_e increases by $\delta/4$. For example, a 2 percent change in solar irradiance would change T_e by about 0.5 percent, or 1.2-1.3°C. If the atmospheric temperature structure and all other factors remained fixed, the surface temperature would increase by the same amount as T_e. Of course all factors are not fixed, and we therefore define the net feedback factor, f, by

$$\Delta T_{eq} = f\ \Delta T_o \tag{4}$$

where ΔT_{eq} is the equilibrium change of global mean surface air temperature and ΔT_o is the change of surface temperature that would be required to restore radiative equilibrium if no feedbacks occurred.

We use procedures and terminology of feedback studies in electronics (Bode, 1945) to help analyze the contributions of different feedback processes. We define the system gain as the ratio of the net feedback portion of the temperature change to the total temperature change

$$g = \frac{\Delta T_{feedbacks}}{\Delta T_{eq}} \quad . \tag{5}$$

Since

$$\Delta T_{eq} = \Delta T_o + \Delta T_{feedbacks}, \tag{6}$$

it follows that the relation between the feedback factor and gain is

$$f = \frac{1}{1 - g} \quad . \tag{7}$$

In general a number of physical processes contribute to f, and it is common to associate a feedback factor f_i with a given process i, where f_i is the feedback factor which would exist if all other feedbacks were inoperative. If it is assumed that the feedbacks are independent, feedback contributions to the temperature change can be separated into portions identifiable with individual feedbacks,

$$\Delta T_{feedbacks} = \sum_i \Delta T_i, \tag{8}$$

with

$$g = \sum_i \frac{\Delta T_i}{\Delta T_{eq}} = \sum_i g_i \tag{9}$$

and

$$\Delta T_{eq} = \frac{1}{1 - \sum_i g_i} \Delta T_o \quad . \qquad (10)$$

It follows that two feedback gains combine linearly as

$$g = g_1 + g_2 \ , \qquad (11)$$

but the feedback factors combine as

$$f = \frac{f_1 f_2}{f_1 + f_2 - f_1 f_2} \quad . \qquad (12)$$

Thus even when feedback processes are linear and independent the feedback factors are not multiplicative. For example, a feedback process with gain $g_i = 1/3$ operating by itself would cause a 50 percent increase in ΔT_{eq} compared to the no feedback radiative response, i.e., $f_i = 1.5$. If a second feedback process of the same strength is also operating, the net feedback is f = 3 (not 2.25). One implication is that, if strong positive feedback exists, a moderate additional positive feedback may cause a large increase in the net feedback factor and thus in climate sensitivity.

The feedback factor f provides an intuitive quantification of the strength of feedbacks and a convenient way to describe the effect of feedbacks on the transient climate response. The gain g allows clear comparison of the contributions of different mechanisms to total climate change. The above formalism relates f and g and provides a framework for analyzing feedback interactions and climate sensitivity.

A number of physical mechanisms have been identified as causing significant climate feedback (Kellogg and Schneider, 1974). As examples, we mention two of these mechanisms here. Water vapor feedback arises from the ability of the atmosphere to hold more water vapor as temperature increases. The added water vapor increases the infrared opacity of the atmosphere, raising the mean level of infrared emission to space to greater altitude, where it is colder. Because the planetary radiation to space temporarily does not balance absorbed solar energy, the planet must warm to restore energy balance; thus $f_w > 1$ and $g_w > 0$, a condition described as a positive feedback. Ice/snow feedback is also positive; it operates by increasing the amount of solar energy absorbed by the planet as ice melts.

Feedback analyses will be most useful if the feedback factors are independent to first order of the nature of the radiative forcing (at the top of the atmosphere). The similar model responses we obtain in our S_o and CO_2 experiments tend to corroborate this possibility, although there are some significant differences in the feedbacks for solar and CO_2 forcings. We expect the strength of feedbacks to have some dependence on the initial climate state and thus on the magnitude of the climate forcing; for example, the ice/snow albedo feedback is expected to change with climate as the cryospheric region grows or shrinks.

We examine feedback processes quantitatively in the following sections by means of 3-D climate model simulations and analysis of conditions during the last ice age (18K years ago). The 3-D experiments include doubling CO_2 and increasing S_o by 2 percent, forcings of roughly equal magnitude which have also been employed by Manabe and Wetherald (1975) and Wetherald and Manabe (1975). 18K simulations with a 3-D general circulation model have previously been performed by Williams et al. (1974), Gates (1976) and Manabe and Hahn (1977).

Three-Dimensional Climate Model

The global climate model we employ is described and its abilities and limitations for simulating today's climate are documented as model II (Hansen et al., 1983b, hereafter referred to as paper 1). We note here only that the model solves the simultaneous equations for conservation of energy, momentum, mass and water and the equation of state on a coarse grid with horizontal resolution 8° latitude by 10° longitude and with 9 atmospheric layers. The radiation includes the radiatively significant atmospheric gases, aerosols and cloud particles. Cloud cover and height are computed. The diurnal and seasonal cycles are included. The ground hydrology and surface albedo depend upon the local vegetation. Snow depth is computed and snow albedo includes effects of snow age and masking by vegetation.

Ocean temperatures and ice cover are specified climatologically in the documented model II. In the experiments described here, ocean temperatures and ice cover are computed based on energy exchange with the atmosphere, ocean heat transport, and the ocean mixed layer heat capacity. The latter two are specified, but vary seasonally at each gridpoint. Monthly mixed layer depths are climatological, compiled from NODC mechanical bathythermograph data (NOAA, 1974) and from temperature and salinity profiles in the southern ocean (Gordon, 1982). The resulting global-mean seasonal-maximum mixed layer depth is 110m. In our 3-D experiments a 65m maximum is imposed on the mixed layer depth to minimize computer time; this yields a global-mean seasonal-maximum mixed layer depth of 63m. The 65m maximum depth is sufficient to make the mixed layer thermal response time much greater than one year and provide a realistic representation of seasonal temperature variations, so the mixed layer depth limitation should not significantly affect the modeled equilibrium climate.

The ocean heat transport was obtained from the divergence of heat implied by energy conservation at each ocean gridpoint in the documented

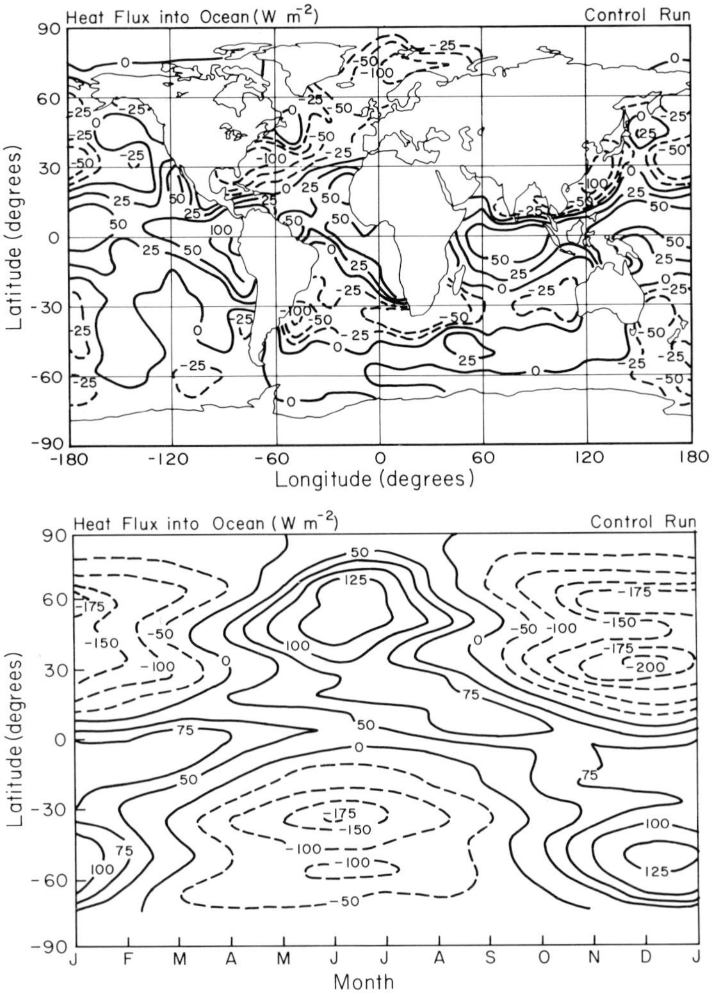

Fig. 1. Specified heat flux into the ocean surface in the 3-D climate model experiments, obtained from the model II run of paper 1 which had specified climatological seasonally-varying ocean surface temperature and ocean ice cover. (a) is the geographical distribution of the annual-mean flux. (b) is the latitude/season distribution of the zonal-mean flux.

model II (paper 1), using the specified mixed layer depths. The geographical distribution of the resulting annual mean heat flux into and out of the ocean surface is shown in Fig. 1a; averaged over the entire hemispheres, it yields 2.4 W m^{-2} into the Southern Hemisphere surface and an equal amount out of the Northern Hemisphere. The gross characteristics of the ocean surface heating and implied ocean heat transport appear to be realistic, with heat input at low latitudes, especially in regions of upwelling cold water, and release at high latitudes, especially in regions of poleward currents. Fig. 15 of paper 1 shows that the longitude-integrated heat transport is consistent with available knowledge of actual transports. A more comprehensive comparison with observations has been made by Miller et al. (1983), who show that the implied annual northward heat flux at the equator is 6.2×10^{14} W. With the ocean heat transport specified in this manner, the control run with computed ocean temperature has a simulated climate nearly the same as the documented model II. It is not identical, as a result of changes in the sea ice coverage which arise when the sea ice is a computed quantity. There is 15 percent less sea ice in the standard control run with computed ocean temperature than in the documented model II, as discussed below. This has local effects, mainly around Antarctica, but otherwise simulated quantities are practically identical to the documented model II climatology.

In our experiments with changed solar irradiance and atmospheric CO_2 we keep the ocean heat transport identical to that in the control run. Thus no ocean transport feedback is permitted in these experiments. Our rationale for this approach as a first step is its simplicity for analysis, and the fact that it permits a realistic atmospheric simulation.

Ocean ice cover is also computed in the experiments described here on the basis of the local heat balance. When the ocean surface loses heat, the mixed layer temperature decreases as far as the freezing point of ocean water, -1.6°C. Further heat loss from the open ocean causes ice to grow horizontally with thickness 1m until the gridbox is covered up to the limit set by the prescription for leads (open water). Still further heat loss causes the ice to thicken. Leads are crudely represented by requiring the fraction of open water in a gridbox to be greater than or equal to $0.1/z_{ice}$, where z_{ice} is the ice thickness in meters (paper 1).

Heat exchange between ocean ice and the mixed layer occurs by conduction in the climate model. A two-slab model is used for ice, with the temperature profile parabolic in each slab. This conduction is inefficient, and, if it were the only mechanism for heat exchange between the mixed layer and the ice, it would at times result in ocean ice coexisting with ocean water far above the freezing point; since this does not occur in nature, other mechanisms (such as lateral heat exchange) must contribute to the heat exchange. Therefore in our standard control run and S_0 and CO_2 experiments we impose the condition that the mixed layer temperature, which represents a mean for an 8° × 10° gridbox, not be allowed to exceed 0°C until all the ice in the gridbox is melted; i.e., if the mixed layer temperature reaches 0°C additional heat input is used to meltice, decreasing its horizontal extent within the gridbox.

The annual mean sea ice cover in out standard control run is shown in Fig. 2b. Evidently there is too little sea ice in the model (15 percent less than the observations of Fig. 2a), especially at longitudes ~100°W and ~50°E in the Southern Hemisphere. Thus we also produced an alternate control run by removing the condition that all heat added to the mixed layer be used to melt ice if the mixed layer temperature reaches 0°C. This alternate control run has about 23 percent greater ocean ice cover (Fig. 2c) than observed, and thus the standard and alternate control runs bracket observations. We use the alternate control run for a second doubled CO_2 experiment, as one means of assessing the role of ocean ice in climate sensitivity.

In the following we first describe our standard S_0 and CO_2 experiments.

S_0 and CO_2 Experiments

S_0 was increased 2 percent and CO_2 was doubled (from 315 ppm to 630 ppm) instantaneously on January 1 of year 1. Both experiments were run for 35 years. In this section we study the equilibrium response of the climate model to the S_0 and CO_2 forcings. The time dependence of the surface air temperature and the heat flux into the planetary surface are briefly noted, but only to verify that equilibrium has been achieved. The time dependence of these experiments is discussed in greater detail in a subsequent section concerned with the transient response of the climate system.

Global Mean Heat Balance and Temperature

Model II (paper 1) has a global annual mean net heat flux into the top of the atmosphere of 7.5 W m^{-2} (~2 percent of the insolation). 2.5 W m^{-2} of this imbalance is due to conversion of potential energy to kinetic energy (which is not reconverted to heat in the model) and computer truncation. The other global 5 W m^{-2} is absorbed by the ocean and ocean ice, at a rate of 7.1 W m^{-2} for the ocean surface area. This portion of the imbalance must be due to inaccuracies such as in the cloud properties, surface albedo, thermal emission calculations, etc. In our control run and experiments with computed ocean temperature we multiply the solar radiation absorbed at the ocean surface by the factor 0.96, which cancels the entire energy imbalance. The radiation correction factor has no appreciable direct effect on model sensitivity since all results are differenced against a control run; however, it does enable physical processes, such as condensation and ice melting, to operate at temperatures as realistic as possible. Together with the specified ocean transports, this allows the control run with computed ocean temperature to have essentially the same ocean temperature and climate as the model II run with fixed climatological ocean temperatures (paper 1).

The global mean heat flux into the planetary surface and surface air temperature are shown in Fig. 3 for the S_0 and CO_2 experiments. The heat flux peaks at ~3 W m^{-2} for both experiments; the radiative imbalance at the top of the atmosphere is essentially the same as this flux into the planetary surface, since the heat capacity of the atmosphere is small. Similar fluxes are expected

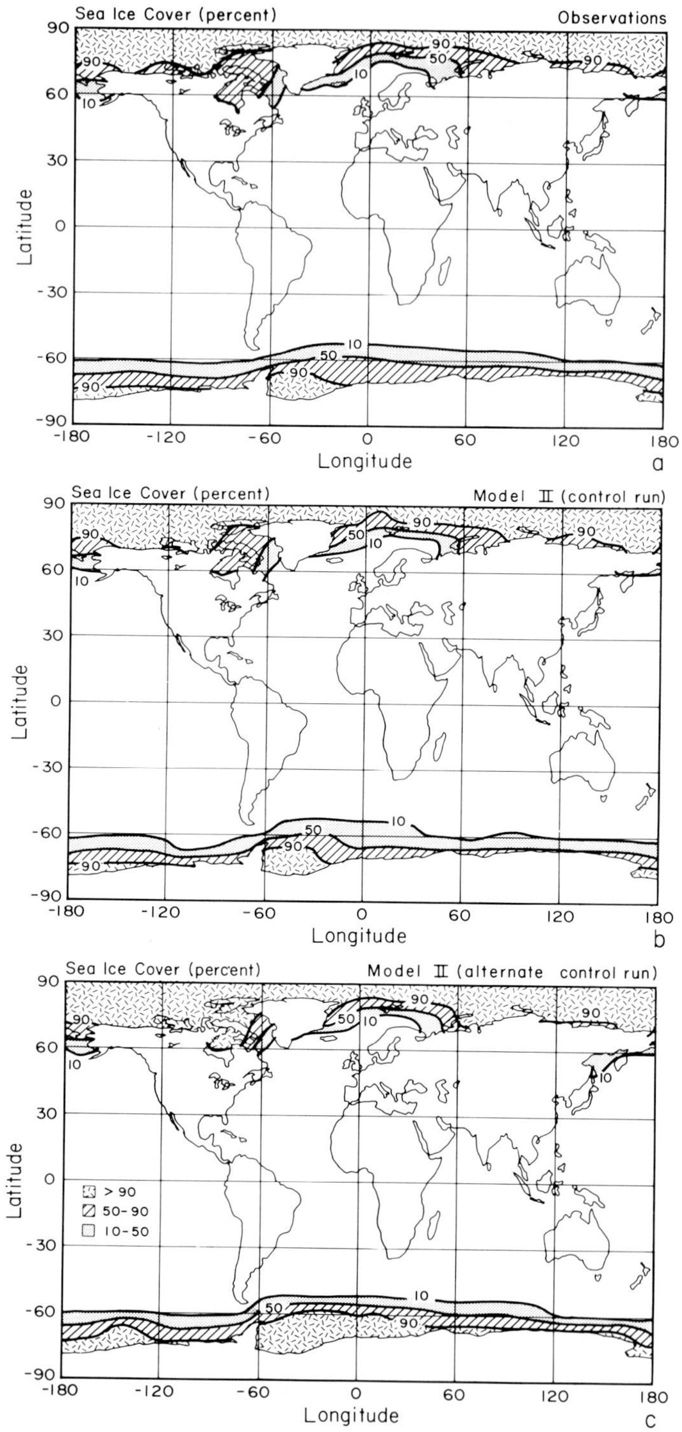

Fig. 2. Annual-mean sea ice cover. (a) observational climatology of Walsh and Johnson (1979) for the northern hemisphere and Alexander and Mobley (1976) for the southern hemisphere. (b) our standard control run of the 3-D climate model. (c) alternate control run, as described in the text.

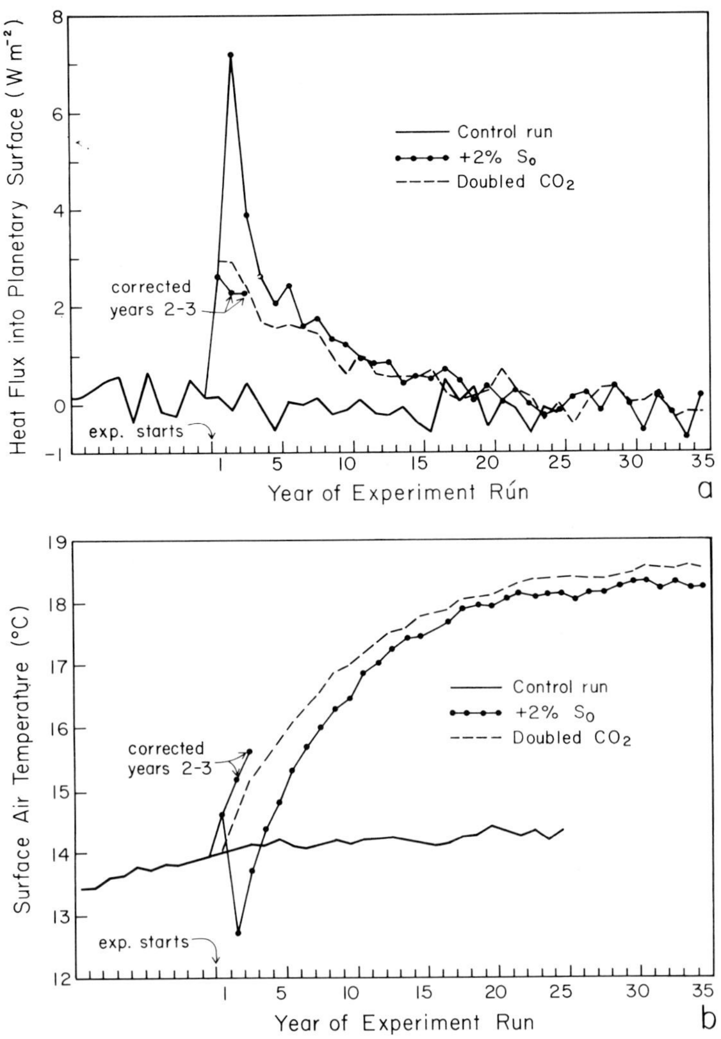

Fig. 3. Global net heat flux into planetary surface (a) and global surface air temperature (b). On April 1 of year 2 in the S_0 experiment the computer was hit by a cosmic ray or some other disturbance which caused improper numbers to be stored in the ground temperature array. This affected the temporal development of that run, but should not influence its equilibrium results. In order to determine the maximum heat flux into the ocean, the S_0 experiment was rerun for years 2 and 3 from March 31 year 2 thus eliminating the computer error for that period.

in the two experiments because of the similar magnitudes of the radiative forcings. The 2 percent S_0 change corresponds to a forcing of 4.8 W m^{-2}. The initial radiative imbalance at the top of the atmosphere due to doubling CO_2 is only ~2.5 W m^{-2}, but after CO_2 cools the stratosphere (within a few months) the global mean radiative forcing is about 4 W m^{-2} (Fig. 4, Hansen et al., 1981). Over the ocean fraction of the globe we find a peak flux into the surface of 4-5 W m^{-2} in both experiments, of order 10 percent greater than the global mean forcing for an all-ocean planet. Thus heating of the air over land with subsequent mixing by the atmosphere increases the net heat flux into the ocean, but not by the ratio of global area to ocean area as assumed by Hansen et al. (1981). Apparently heating over continental areas is balanced substantially by increased cooling to space. A chief implication is that the time constant for the ocean to respond to global heating is longer than obtained from the common practice of averaging the ocean heat capacity over the entire globe (rather than over the ocean area).

The equilibrium global mean warming of the surface air is about 4°C in both the S_0 and CO_2 experiments. This corresponds to a feedback factor f = 3-4, since the no-feedback temperature change required to restore radiative equilibrium with space is ΔT_0 = 1.2-1.3°C. The heat flux and temperature approach their new equilibria with an e-folding time of almost a decade. We show in the section on transient climate response that the e-folding time is proportional to f, and that the value inferred from Fig. 3 is consistent with f = 3-4.

The mechanisms causing the global warmings in these experiments are investigated below, including presentation of the global distribution of key changes. These results are the means for years 26-35 of the control and experiment runs. Fig. 3 indicates that this should provide essentially the equilibrium response, since by that time the heat flux into the ocean is near zero and the temperature trend has flattened out.

Global Temperature Changes

The temperature changes in the S_0 and CO_2 experiments are shown in Fig. 4 for the annual mean surface air temperature as a function of latitude and longitude, the zonal mean surface air temperature as a function of latitude and month, and the annual and zonal mean temperature as a function of altitude and latitude. We discuss the nature and causes of the temperature changes, and then make a more quantitative analysis below using 1-D calculations and the alternate CO_2 experiment with changed sea ice prescription.

The surface air warming is enhanced at high latitudes (Fig. 4, upper panel) partly due to the greater atmospheric stability there which tends to confine the warming to the lower troposphere, as shown by the radiation changes discussed below and the experiment with altered sea ice.

There is a very strong seasonal variation of the surface warming at high latitudes (Fig. 4, middle panel), due to the seasonal change of atmospheric stability and the influence of melting sea ice in the summer which limits the ocean temperature rise. At low latitudes the temperature increase is greatest in the upper troposphere (Fig. 4, lower panel), because the added heating at the surface primarily causes increased evaporation and moist convection, with deposition of

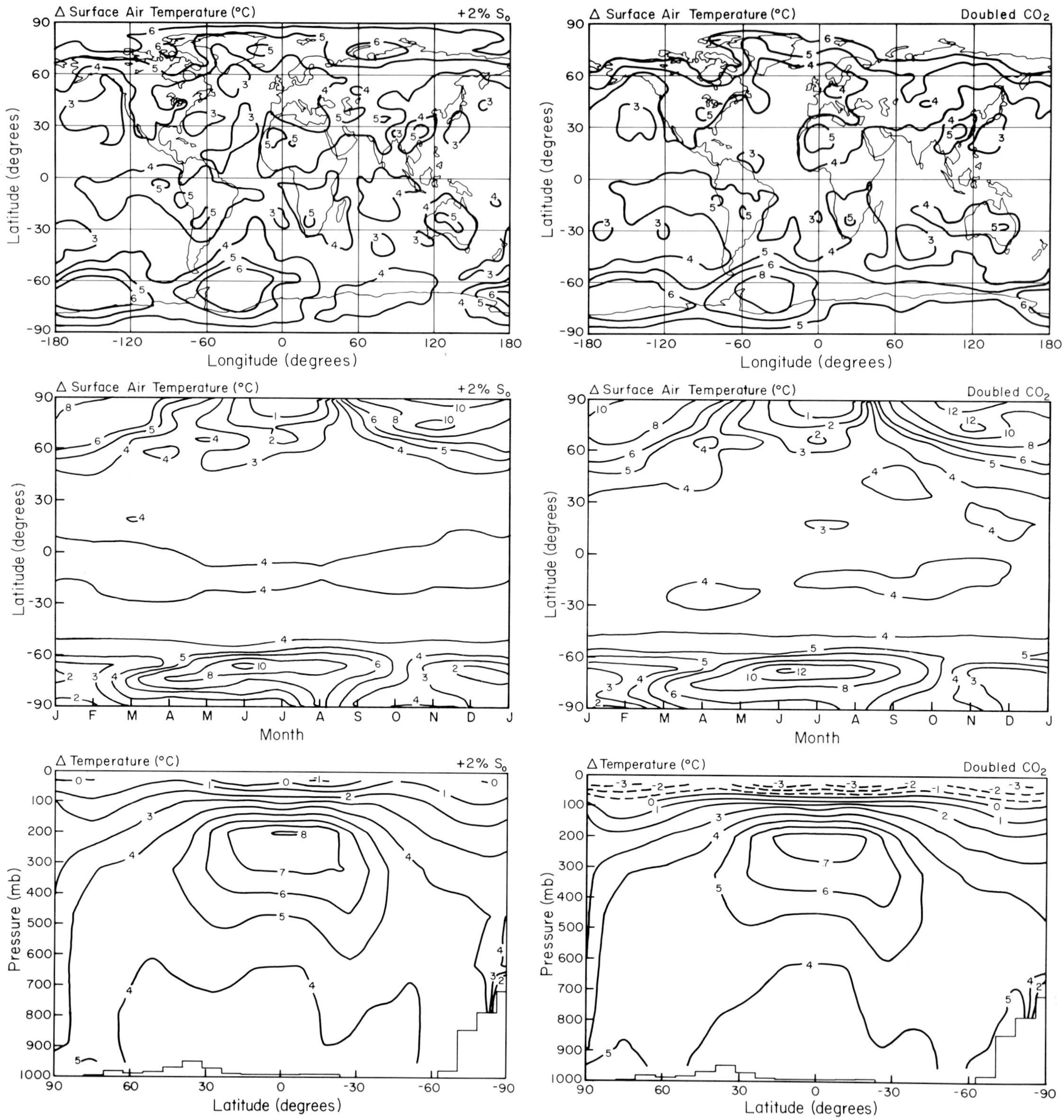

Fig. 4. Air temperature change in the climate model for a two percent increase of solar irradiance (left) and for doubled atmospheric CO_2 (right). The upper graphs show the geographical distribution of annual mean surface air warming, the middle graphs show the seasonal variation of the surface air warming averaged over longitude, and the lower graphs show the altitude distribution of the temperature change averaged over season and longitude.

Fig. 5. Left side: standard deviation of temperature for the last 10 years in the control run. Right side: ratio of temperature change for years 26-35 of the doubled CO_2 experiment to the standard deviation of temperature in the control run.

latent heat and water vapor at high levels.

The statistical significance of these results can be verified from Fig. 5, which shows the standard deviation for the last 10 years of the control run for all the quantities in Fig. 4, and the ratio of the change of the quantity in the doubled CO_2 experiment to the standard deviation. The standard deviation is computed routinely for all of the quantities output from our 3-D model. We only discuss changes in the experiment runs which are far above the level of model fluctuations or 'noise' in the control run.

The patterns of temperature change are remarkably similar in the S_0 and CO_2 experiments, suggesting that the climate response is to first order a function of the magnitude of the radiative forcing. The only major difference is in the temperature change as a function of altitude; increased CO_2 causes substantial stratospheric cooling. This similarity suggests that, to first order, the climate effect due to several forcings including various tropospheric trace gases may be a simple function of the total forcing.

The global mean warming of surface air that we obtain for doubled CO_2 is similar to that obtained by Manabe and Stouffer (1980) for quadrupled CO_2. This large difference in sensitivity of the two models appears to be associated mainly with the feedback mechanisms in the models, as discussed below. The patterns of the temperature changes in the two models show gross similarities, but also significant differences. We defer detailed comparison of the model results until after discussion of the feedback mechanisms.

1-D Analysis of Feedbacks in 3-D Experiments

The processes chiefly responsible for the temperature rise in the 3-D model can be investigated with a 1-D radiative convective (RC) climate model. We use the 1-D model of Lacis et al. (1981) to evaluate the effect of changes in radiative forcing that take place in the 3-D model experiments. As part of the 3-D model diagnostics, we have available global average changes in surface and planetary albedo, and changes in amount and vertical distribution of clouds, water vapor and atmospheric lapse rate. We insert these changes one-by-one, or in combination, into the 1-D model and compute the change in global surface temperature. We employ the usual 'convective adjustment' procedure in our 1-D calculations, but with the global mean temperature profile of the 3-D model as the critical lapse rate in the troposphere. Contrary to usual practice, we allow no feedbacks to operate in the 1-D calculations, making it possible to associate surface temperature changes with individual feedback processes.

There is no a priori guarantee that the net effect of these changes will yield the same warming in the 1-D model as in the 3-D model, because simple global and annual averages of the changes do not account for the nonlinear nature of the physical processes and their 2-D and 3-D interactions. Also, changes in horizontal dynamical transports of heat and moisture are not entered explicitly into the 1-D model; the effects of dynamical feedbacks are included in the radiative factors which they influence, such as the cloud cover and moisture profile, but the dynamical contributions are not identified. Nevertheless, this exercise provides substantial information on climate feedbacks. Determination of how well the 1-D and 3-D results correspond also is a useful test for establishing the value of 1-D global climate models.

The procedure we use to quantify the feedbacks is as follows. The increase of total water vapor in the 3-D model (33 percent in the S_0 experiment) is put in the 1-D model by multiplying the water vapor amount at all levels by the same factor (1.33); the resulting change in the equilibrium surface temperature of the 1-D model defines the second bars in Fig. 6. Next the water vapor at each level in the 1-D model is increased by the amount found in the 3-D experiment; the temperature change obtained in the first (total H_2O amount) test is subtracted from the temperature change obtained in this test to obtain the temperature change credited to the change in water vapor vertical distribution. The change of temperature gradient (lapse rate) between each pair of levels in the 3-D model is inserted in the control 1-D model to estimate the effect of lapse rate change on surface temperature, shown by the fourth bars in Fig. 6. Since the lapse rate changes are due mainly to changes of water vapor, we take the net of these three temperature changes in the 1-D model as our estimate of the water vapor contribution to the total temperature change. The global mean ground albedo change in the 3-D model (defined as the ratio of the global mean upward and downward solar radiation fluxes at the ground) is inserted into the 1-D control run to obtain our estimate of the ice/snow albedo contribution to the temperature change.

Cloud contributions are more difficult to analyze accurately because of the variety of cloud changes that occur in the 3-D model (see below), including changes in cloud overlap, and the fact that the changes do not combine linearly. We first estimate the total cloud impact by changing the cloud amounts at all levels in the 1-D model in proportion to changes obtained in the 3-D model. The total cloud effect on the temperature obtained in this way is subdivided by defining a portion to be due to the cloud cover change (by running the 1-D model with a uniform change of all clouds so as to match the total cloud cover change in the 3-D model) and by assigning the remainder of the total cloud effect to cloud height changes. These assumptions involve some arbitrariness. Nevertheless, the resulting total temperature changes in the 1-D model are found to be within 0.2°C of the global mean temperature changes in the 3-D experiments, providing circumstantial evidence that the procedure takes into account the essential radiative aspects of cloud cover change.

The temperature changes in the 1-D model are shown in Fig. 6 for the standard S_0 and CO_2 experiments, and the CO_2 experiment with alternate sea ice computation. Resulting gains and feedback factors are given in Table 1.

Water vapor feedback. Water vapor provides

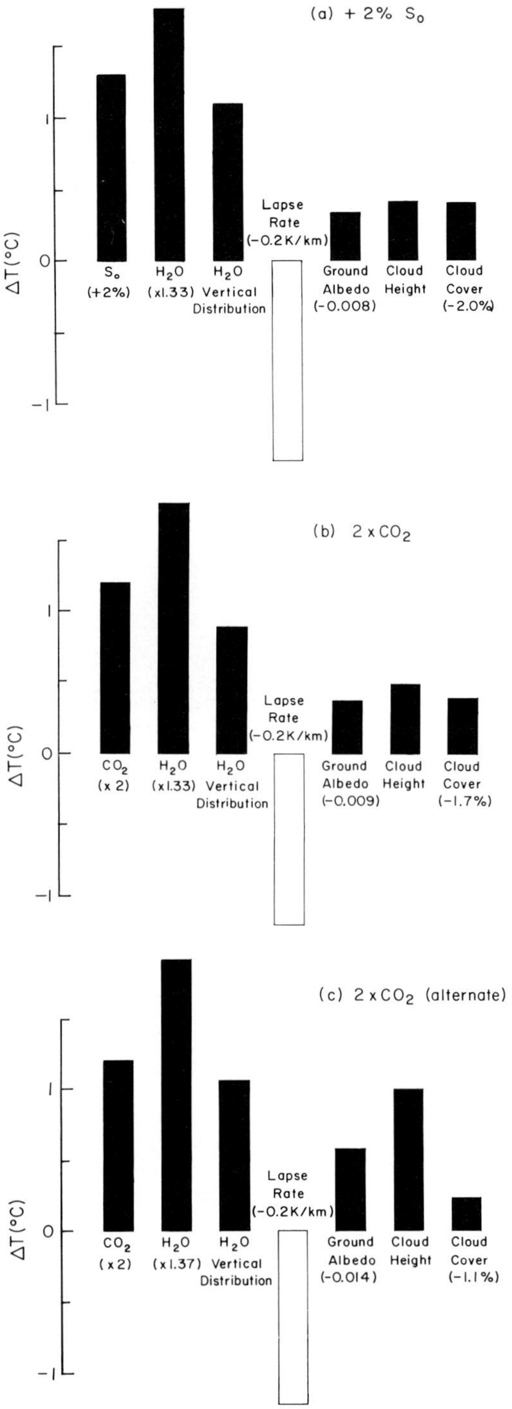

Fig. 6. Contributions to the global mean temperature rise in the S_0 and CO_2 experiments as estimated by inserting changes obtained in the 3-D experiments into 1-D radiative convective model. (a) +2 percent S_0 experiment, (b) doubled CO_2 experiment, and (c) doubled CO_2 experiment for alternate control run with greater sea ice.

the largest feedback, with most of it caused by the increase of water vapor amount. Additional positive feedback results from the water vapor distribution becoming weighted more to higher altitudes, but for the global and hemispheric means this is approximately cancelled by the negative feedback produced by the changes in lapse rate, also due mainly to the added H_2O. The near cancellation of these two effects is not surprising, since the amount of water the atmosphere holds is largely dependent on the mean temperature, and the temperature at which the infrared opacity occurs determines the infrared radiation. This tendency for cancellation suggests that the difficulty in modeling moist convection and the vertical distribution of water vapor may not have a great impact on estimates of global climate sensitivity (excluding the indirect effect on cloud distributions).

The net water vapor gain thus deduced from the 3-D model is $g_W \sim 0.4$, or a feedback factor $f_W \sim 1.6$. The same sensitivity for water vapor is obtained in 1-D models by using fixed relative humidity and fixed critical lapse rate (Manabe and Wetherald, 1967), thus providing some support for that set of assumptions in simple climate models. Relative humidity changed only slightly in our 3-D experiments; for example, in our standard doubled CO_2 experiment the average relative humidity increased 0.015 (1 = 100 percent humidity), with a 0.06 global increase at 200 mb being the largest change at any altitude. This compares with an increase of mean specific humidity of 33 percent. The global mean lapse rate change in the 3-D model ($-0.2°C\ km^{-1}$) is less than the change of the moist adiabatic lapse rate ($-0.5°C\ km^{-1}$), the decrease at low latitudes being partly offset by an increase at high latitudes. And, as explained above, the effect of the lapse rate change on temperature is largely balanced by the effect of the resulting displacement of water vapor to greater altitude.

Snow/ice feedback. Ground albedo decrease also provides a positive feedback. The ground albedo change (upper panel of Fig. 7) is largely due to reduced sea ice. Shielding of the ground by clouds and the atmosphere (middle panel of Fig. 7) makes this feedback several times smaller than it would be in the absence of the atmosphere. However, it is a significant positive feedback and is at least as large in the Southern Hemisphere as in the Northern Hemisphere. The geographic pattern of the temperature increase (Fig. 4) and the coincidence of warming maxima with reduced sea ice confirm that the sea ice effect is a substantial positive feedback.

Further insight into the sea ice feedback is provided by the experiment with alternate prescription for computing sea ice cover. The greater sea ice cover in the control run for this experiment permits a greater surface albedo feedback, as indicated by the analysis with the 1-D model shown in Fig. 6c. These results illustrate the sensitivity of a system which already contains large positive feedbacks, the gain due to increased surface albedo being augmented by in-water vapor and cloud gains.

Based on these experiments, we estimate the sea ice/snow feedback factor as ~1.1. However,

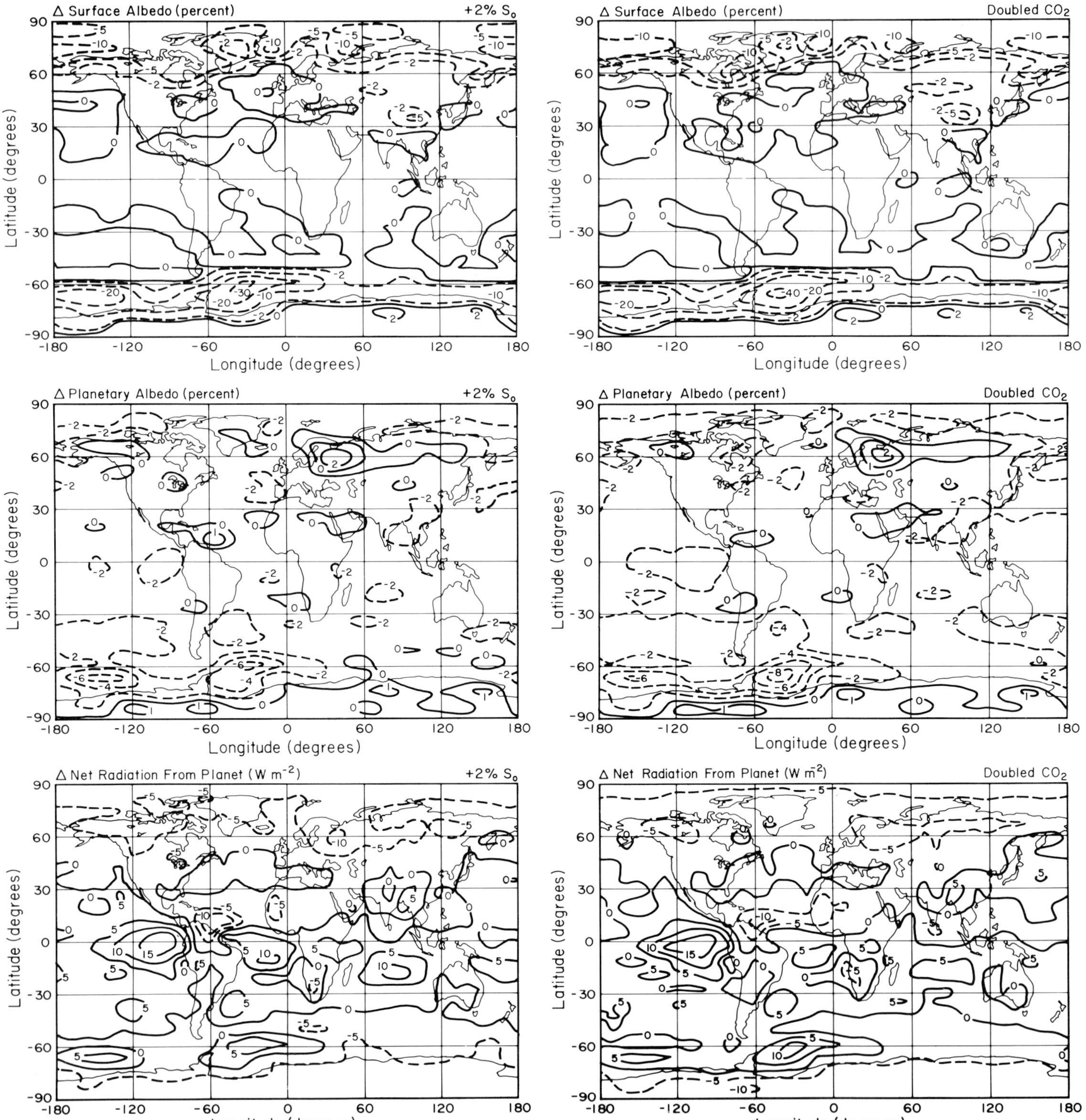

Fig. 7. Annual mean radiation changes in the climate model for two percent increase of S_0 (left) and doubled CO_2 (right). In Figs. 7, 8, 10 and 11 "percent" change refers to the full range of the quantity, e.g., a change from 60 percent albedo to 50 percent albedo is defined to be a 10 percent change.

this value refers to a climate change from today's climate to a climate which is warmer by about 4°C. We expect $f_{snow/ice}$ to decrease as the area of sea ice and snow decreases, so its value is probably somewhat larger in the limit of a small increment about today's climate. Also, the prescription for computing sea ice in our standard experiments (which gives 15 percent too little sea ice for today's climate) probably causes an underestimate of $f_{snow/ice}$, as indicated by the value inferred for f in the experiment with al tered sea ice prescription (which yielded 23 percent too much sea ice for today's climate).

The gain we obtain for ice/snow feedback in our 3-D model (~0.1) agrees well with the value (0.12) obtained by Wang and Stone (1980) from a 1-D radiative convective model. The feedback is much smaller than early estimates such as those of Budyko (1969) and Sellers (1969), who assigned a large albedo increment to ice/snow, but did not account for cloud shielding, vegetation masking of snow, and zenith angle variation of albedo (North, 1975; Lian and Cess, 1977).

Cloud feedback. Cloud changes (Fig. 8) also provide a significant positive feedback in this model, as a result of a small increase in mean cloud height and a small decrease in cloud cover. The gain we obtain for clouds is 0.22 in our standard doubled CO_2 experiment. This happens to be similar to the gain of 0.19 which is obtained in 1-D models if the cloud cover is kept fixed and the cloud height is determined by the assumption of fixed cloud temperature (Cess, 1974). However, a substantial part of the cloud gain in the 3-D model is due to the cloud cover change (Fig. 6). The portion of the cloud gain associated with cloud height change in the S_0 experiment and the standard doubled CO_2 experiment is about midway between the two common assumptions used in simple climate models: fixed clouds altitude (gain = 0) and fixed cloud temperature (gain = 0.2).

The cloud height and cloud cover changes in the 3-D model seem qualitatively plausible. The reduced cloud cover primarily represents reduction of low and middle level clouds, due to increased vertical transport of moisture by convection and the large scale dynamics. The increase of high level cirrus clouds at low latitudes is consistent with the increase of penetrating moist convection at those latitudes. However, the cloud prescription scheme in the model (paper 1) is highly simplified; for example, it does not incorporate a liquid water budget for the cloud droplets or predict changes in cloud optical thickness at a given height. Thus the possibility of an increase in mean cloud optical thickness with the increased water vapor content of the atmosphere is excluded. Indeed, because the cloud optical thickness decreases with increasing altitude (paper 1), the increase of cloud height causes a decrease of optical thickness. This is a positive feedback for low and middle level clouds, but a negative feedback for cirrus clouds, which are a greenhouse material with suboptimal optical thickness. As a crude test of possible effects of changes in cloud optical thickness we let the cloud optical thickness in the 1-D model change in proportion to the absolute humidity: this practically eliminated the positive cloud feedback, i.e., it resulted in $f_{clouds} \sim 1.0$. Clearly, assessment of the cloud contribution to climate sensitivity depends crucially upon development of more realistic representation of cloud formation processes in climate models, as verified by an accurate global cloud climatology.

Summary of model feedbacks. Given the cancellation between the change in lapse rate and change in vertical distribution of water vapor, the processes providing the major radiative feedbacks in this climate model are total atmospheric water vapor, clouds and the surface albedo. Considering the earth from a planetary perspective, it seems likely that these are the principal feedbacks for the earth on a time scale of decades. The albedo of the planet for solar radiation is primarily determined by the clouds and surface, with the main variable component of the latter being the ice/snow cover. The thermal emission of the planet is primarily determined by the atmospheric water vapor and clouds. Thus the processes principally responsible for the earth's radiation balance and temperature are included in the 1-D model, and we have shown that these processes are the source of the primary radiative feedbacks in our 3-D model.

Table 1 summarizes the gains and feedback factors inferred from the changes which occurred in our 3-D model experiments, and the corresponding temperature changes for different combinations of these feedback processes. Note again that effects of dynamical feedbacks are implicitly included in these changes. The temperature changes illustrate the nonlinear way in which feedback processes combine [Eq. (9)]. For example, the ice/snow feedback adds about 1°C to the temperature response, but if the water vapor and cloud feedbacks did not exist the ice/snow feedback would add only a few tenths of a degree to the sensitivity. This nonlinear behavior is a result of the fact that when the ice/snow feedback occurs in the presence of the other (positive) feedbacks, it enhances the water vapor and cloud changes.

Comparison to Manabe and Stouffer. This analysis of the feedbacks in our model provides an indication of the causes of the difference between our climate model sensitivity and that of Manabe and Stouffer (1980). They infer a warming of 2°C for doubled CO_2, based on an experiment with quadrupled CO_2 which yielded 4°C warming. Their model had fixed clouds (altitude and cloud cover), thus $f_{cloud} \equiv 1$. Also their control run had less sea ice than our model, so their $f_{sea\ ice}$ should be between 1 and the value (~1.1) for our model. It is apparent from Table 1 that differences arising from the treatments of these two processes may account for most of the difference in global climate sensitivity.

Another major difference between our model

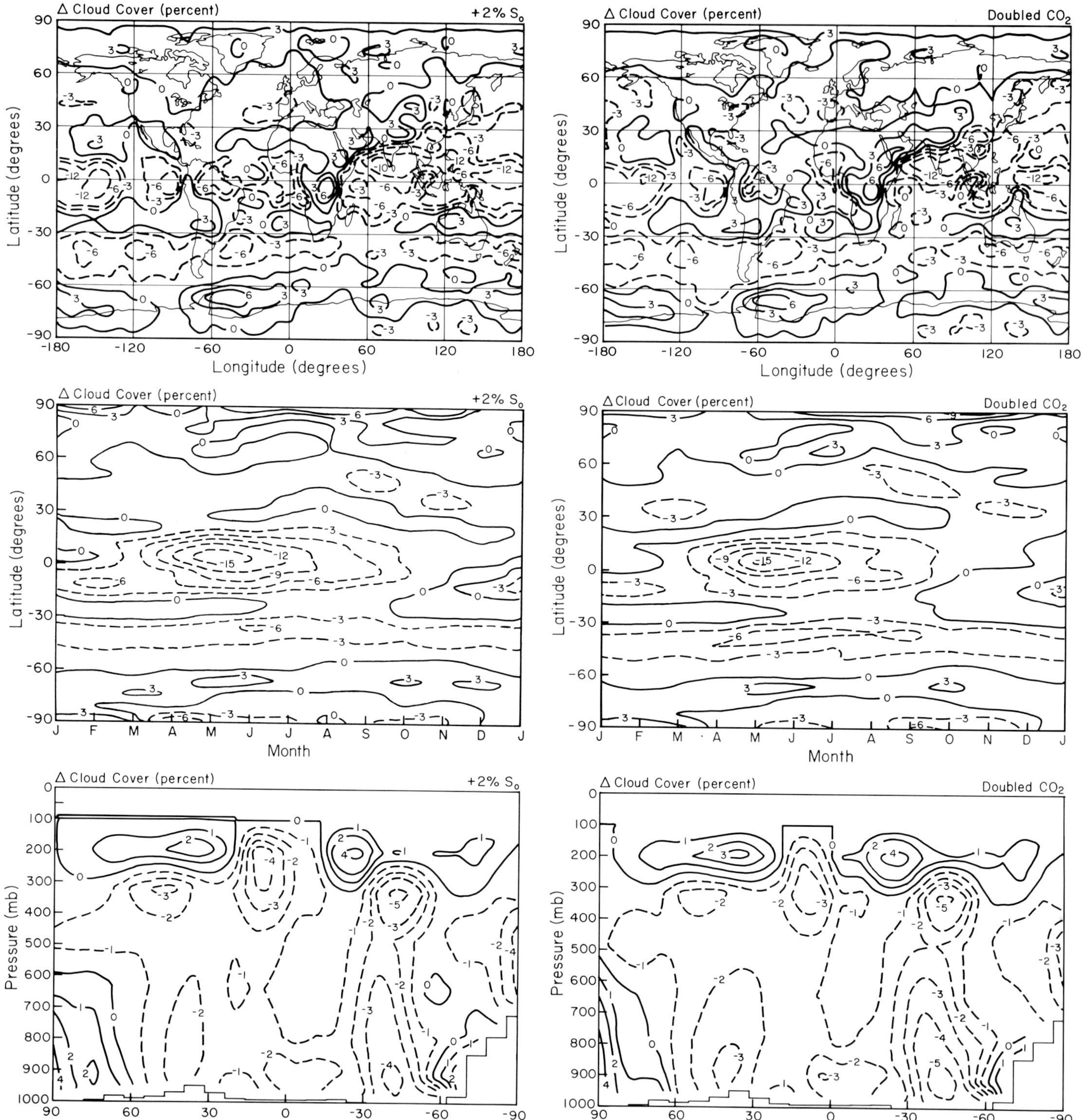

Fig. 8. Cloud changes in the climate model for two percent increase of S_0 (left) and doubled CO_2 (right). The upper graphs show the geographical distribution of annual mean cloud cover change, the middle graphs show the seasonal variation of cloud cover change averaged over longitude, and the lower graphs show the altitude distribution of the cloud cover change averaged over season and longitude.

TABLE 1. Gain (g), feedback factor (f) and equilibrium temperature changes (ΔT) inferred from calculations with 1-D radiative-convective model for global mean changes in the 3-D GCM experiments. The subscripts w, c and s refer to water vapor, clouds and surface albedo. g is obtained as the ratio of the temperature change in the 1-D model (with only the indicated processes included) to the global mean temperature change in the 3-D experiment. f is from $f_i = 1/(1 - g_i)$. ΔT is the equilibrium surface air warming computed with the 1-D model for global mean changes of 3-D model constituents with only the indicated processes included; ΔT_o includes only the indicated radiative forcing, without feedbacks.

	Experiment		
	+2% S_o	Doubled CO_2	Alternate Doubled CO_2
g_w	0.37	0.40	0.37
g_c	0.20	0.22	0.26
g_s	0.09	0.09	0.12
f_w	1.59	1.66	1.58
f_c	1.26	1.29	1.34
f_s	1.09	1.10	1.14
f_{wc}	2.36	2.62	2.67
f_{wcs}	2.96	3.45	3.95
ΔT_o	1.3	1.2	1.2
ΔT_{ow}	2.1	2.0	1.9
ΔT_{oc}	1.7	1.6	1.6
ΔT_{os}	1.5	1.3	1.4
ΔT_{owc}	3.2	3.2	3.2
ΔT_{owcs}	4.0	4.2	4.8

and the model of Manabe and Stouffer is that we include a specified horizontal transport of heat by the ocean. This transport is identical in our control and experiment runs, i.e., the changed climate is not allowed to feed back on the ocean circulation. Of course Manabe and Stouffer do not allow feedback on ocean transport either, since the ocean transport is zero in both experiment and control. However, some other mechanisms must replace oceanic poleward transport of heat in their model, since their high latitude temperatures are at least as warm as in our model (and observations). Enhanced poleward transport of latent and sensible heat by the atmosphere must be the mechanism replacing ocean heat transport in their model. This atmospheric transport is expected to provide a negative feedback (Stone, 1984), and indeed the total atmospheric energy transport did decrease poleward of 50°C latitude in the CO_2 experiments of Manabe and Wetherald (1975, 1980). It is not obvious whether the ocean transport feedback is positive or negative in the real world.

The contribution of ocean heat transport to climate sensitivity, like that of the atmospheric transports, does not appear as an identified component in an energy balance analysis such as in Fig. 6. This is irrelevant for our model, since it has no ocean transport feedback. However, in models which calculate the ocean heat transport or a surrogate energy transport, this feedback is included implicitly as a (positive or negative) portion of identified components of ΔT ($\Delta T_{water\ vapor}$, ΔT_{clouds}, $\Delta T_{ice/snow}$). The portion of these changes due to this feedback process could be identified by running those models with fixed (climatological) ocean heat transport.

Manabe (1983) suggests that our ice/snow feedback is unrealistically large and accounts for most of the difference between our climate model sensitivity and that of Manabe and Stouffer (1980). However, as summarized in Table 2, the amount of sea ice in the control run for our standard CO_2 experiment is actually somewhat less than observed sea ice cover. In our alternate CO_2 experiment, with sea ice cover greater than in observations, the ice/snow feedback increases significantly, suggesting that the ice/snow feedback in our standard experiment may be an underestimate. Also, we show in the next section that the sea ice feedback for the climate change from 18K to today, a warming of about 4°C, is about twice as large as in our doubled CO_2 experiments; this 18K sea ice feedback factor is based on measured changes of sea ice cover. The small ice/snow feedback in Manabe and Stouffer's model may be a result of their model being too warm at high latitudes; indeed, in the Southern Hemisphere (where the sea ice feedback is greatest in our model and in 18K measurements) their control run has almost no ice in the summer. Another likely reason for Manabe and Stouffer's albedo feedback being weaker is the stronger negative feedback in their meridional dynamical flux, as a result of that flux all being carried in the atmosphere. We conclude that our estimate for the sea ice feedback is conservative, i.e., it is more likely to be in error on the low side than on the high side.

We obtain a greater warming at low latitudes (~3-4°C for doubled CO_2) than that found by Manabe and Stouffer (~3°C for quadrupled CO_2). We analyzed the contributions to the warming in our 3-D model as a function of latitude by inserting all zonal-mean radiative changes into the 1-D radiative-convective model. At low latitudes (0-30°) the clouds contribute a positive feedback of about 1-1.5°C; the larger part of this, nearly 1°C, is due to reduction of low level cloud cover with doubled CO_2, with increase of cirrus clouds contributing a smaller positive feedback. At high latitudes (60-90°) the clouds contribute a smaller negative feedback (0-1°C), due to increased low level clouds; this cloud increase (Fig. 8) probably is due to increased evaporation resulting from the reduced sea ice cover. The computed distributions of water vapor may also contribute

TABLE 2. Annual-mean sea ice cover as fraction of global or hemispheric area in several 3-D experiments. In run 1 the sea ice cover is specified to be today's climatology of Alexander and Mobley (1976) for the Southern Hemisphere and Walsh and Johnson (1979) for the Northern Hemisphere. Run 7 specifies the sea ice cover according to CLIMAP data for 18K (CLIMAP, 1981) and run 11 modifies the Southern Hemisphere CLIMAP data as discussed in the section on ice age experiments. In other runs the sea ice cover is computed.

		Sea Ice Cover		
Run	Experiment Description	Globe	Northern Hemisphere	Southern Hemisphere
1	Model II, Run 61 of paper 1; sea ice specified as today's climatology	0.048	0.042	0.054
2	Control run for standard CO_2 and S_0 experiments	0.041	0.039	0.043
3	Standard 2 × CO_2 experiment	0.023	0.028	0.017
4	Standard +2% S_0 experiment	0.025	0.030	0.020
5	Control run for alternate CO_2 experiment	0.060	0.046	0.073
6	Alternate 2 × CO_2 experiment	0.031	0.033	0.029
7	CLIMAP boundary conditions	0.089	0.048	0.131
11	CLIMAP boundary conditions with modified Southern Hemisphere sea ice	0.077	0.048	0.106

to the difference between our result and that of Manabe and Stouffer. For example, in our model low latitude relative humidity at 200 mb increased by 0.085 with doubled CO_2. The cloud and water vapor characteristics depend on the modeling of moist convection and cloud formation; Manabe and Stouffer use the moist adiabatic adjustment of Manabe et al. (1965) and fixed clouds; we use a moist convection formulation which allows more penetrative convection (paper 1) and cloud formation dependent on local saturation. Presently available cloud climatology data has not permitted detailed evaluation of these moist convection and cloud formation schemes.

The high latitude enhancement of the warming is less in our model than in observed temperature trends for the past 100 years (Hansen et al., 1983a). If this observed high latitude enhancement also occurs for large global temperature increases, the smaller high latitude enhancement in our 3-D model suggests the possibility that the 3-D model has either overestimated the low latitude climate sensitivity (probably implicating the low latitude cloud feedback) or underestimated the high latitude sensitivity. If the former case is correct, the global climate sensitivity implied by the 3-D model may be only 2.5-3°C; but if the latter interpretation is correct, the global climate sensitivity may be greater than 4°C. A more precise statement requires the ability to analyze and verify the cloud feedback on a regional basis.

Conclusion. Atmospheric water vapor content provides a large positive feedback, and we find that in our model the effects of changes in lapse rate and water vapor vertical distribution largely cancel (for global or hemispheric means). The existence of the strong positive water vapor feedback implies that moderate additional positive feedback can greatly increase climate sensitivity, because of the nonlinear way in which feedbacks combine. In our model, sufficient ice/snow feedback occurs to increase the global sensitivity to ~2.5°C, and with cloud feedback to ~4°C for doubled CO_2. Although the cloud feedback is very uncertain, our 3-D study suggests that it is in the range from approximately neutral to strongly positive in global mean, and thus that global climate sensitivity is at least 2.5°C for doubled CO_2. The magnitudes of the global ice/snow and cloud feedbacks in our 3-D model are plausible, but confirmation requires improved ability to accurately model the physical processes as well as empirical tests of the climate model on a variety of time scales.

Ice Age Experiments

Records of past climate provide a valuable means to test our understanding of climate feedback mechanisms, even in the absence of a complete understanding of what caused the climate change. In this section we use the comprehensive reconstruction of the last ice age (18,000 years ago) compiled by the CLIMAP project (CLIMAP project members, McIntyre, project leader, 1981; Denton and Hughes, 1981). We first run our climate model with the 18K boundary conditions as specified by CLIMAP; this allows us to estimate the global mean temperature change between 18K and today. We then rerun the model changing feedback processes one-by-one and note their effect on the planetary radiation balance at the top of the atmosphere. This provides a measure of the gain or feedback factor for each process. We also examine the model for radiation balance when all of the known 18K feedbacks are included; this allows some inferences about the model sensitivity and the accuracy of the CLIMAP data. Finally, we compare different contributions to the 18K cooling; by considering the land ice and atmospheric CO_2 changes as slow or specified global climate forcings, we can infer empirically the net feedback factor for processes operative on 10-100 year time scales.

Global maps of the CLIMAP 18K boundary conditions, including the distributions of continental ice, sea ice and sea-surface temperature, are given by CLIMAP (1981) and Denton and Hughes (1981). These boundary conditions, obtained from evidence such as glacial scouring, ocean sediment cores containing detritus rafted by sea ice, and oxygen isotopic abundances in snowfall preserved in Greenland and Antarctic ice sheets, necessarily contain uncertainties. For example, Burckle et al. (1982) suggest that the CLIMAP Southern Hemisphere sea ice cover may be overestimated, and DiLabio and Klassen (1983) argue that the CLIMAP 'maximum extent' ice sheet model may be an overestimate. Questions have also been raised about the accuracy of the ocean surface temperatures, especially at low latitudes (Webster and Streten, 1978). We examine quantitatively the effect of each of these uncertainties on our feedback analyses.

Simulated 18K Climate Patterns

Our 18K simulation was obtained by running climate model II (paper 1) with the CLIMAP (1981) boundary conditions. The boundary conditions included the earth orbital parameters for that time (Berger, 1978). The run was extended for six years, with the results averaged over the last five years to define the 18K simulated climate. The control run was the five year run of model II with today's boundary conditions, which is documented in paper 1.

Temperature. Simulated 18K temperature patterns are shown in Fig. 9. The temperatures in the model, especially of surface air, are constrained strongly by the fixed boundary conditions, and thus their accuracy is dependent mainly on the reliability of the CLIMAP data.

Global surface air temperature in the 18K experiment is 3.6°C cooler than in the control run for today's boundary conditions. Much greater cooling, exceeding 20°C, occurs in southern Canada and northern Europe and cooling of more than 5°C is calculated for most of the Southern Hemisphere sea ice region. Some high latitude regions, including Alaska and parts of Antarctica, are at about the same temperature in the 18K simulation as today; thus there is not universal high latitude enhancement of the climate change.

Temperature changes over the tropical and subtropical oceans are only of the order of 1°C, and include substantial areas that are warmer in the 18K simulation than today. The latter aspect requires verification; diverse areas of the tropics and subtropics experienced mountain glaciation at 18K with snowline descent of about 1km, and pollen data indicate substantial cooling of the order of 5°C at numerous low latitude areas. As indicated by our 3-D model experiment the CLIMAP sea surface temperatures are inconsistent with the observations of tropical cooling, since specification of relatively warm tropical and subtropical ocean temperatures effectively prohibits large cooling over land at these latitudes. We conclude that the low latitude ocean temperatures are probably overestimated in the CLIMAP data. A more quantitative analysis (Rind and Peteet, in preparation) suggests that large areas in the low latitude oceans may be too warm by 2-3°C in the CLIMAP data.

The middle parts of Fig. 9 show that the cooling at 18K occurred especially in the fall and winter. Although the surface air was substantially colder all year at latitude 60°N, this was largely a result of the change in mean surface altitude caused by the presence of ice sheets; the cooling at fixed altitude is considerably less. The zonal mean surface air in the tropics was cooler all year. The lower parts of Fig. 9 show substantial cooling throughout most of the troposphere. At high latitudes the greatest cooling occurs in the lower troposphere.

Radiation. Changes in the planetary radiation budget in the 18K simulation are shown in Fig. 10. The surface albedo increases as much as 0.45 in the regions of ice sheets over northern Europe and southern Canada and about 0.30 in regions of large changes in sea ice coverage. Shielding by the atmosphere and the large zenith angles reduce the impact on planetary albedo to 0.15-0.20 over the ice sheets and 0.05-0.10 over sea ice. The effect of the planetary albedo change on the net radiation from the planet is partially compensated over the ice sheets by reduced thermal emission, but nearly the full effect of the albedo change appears in the net radiation change over sea ice; these conclusions follow from comparison of the middle and lower parts of Fig. 10 and the fact

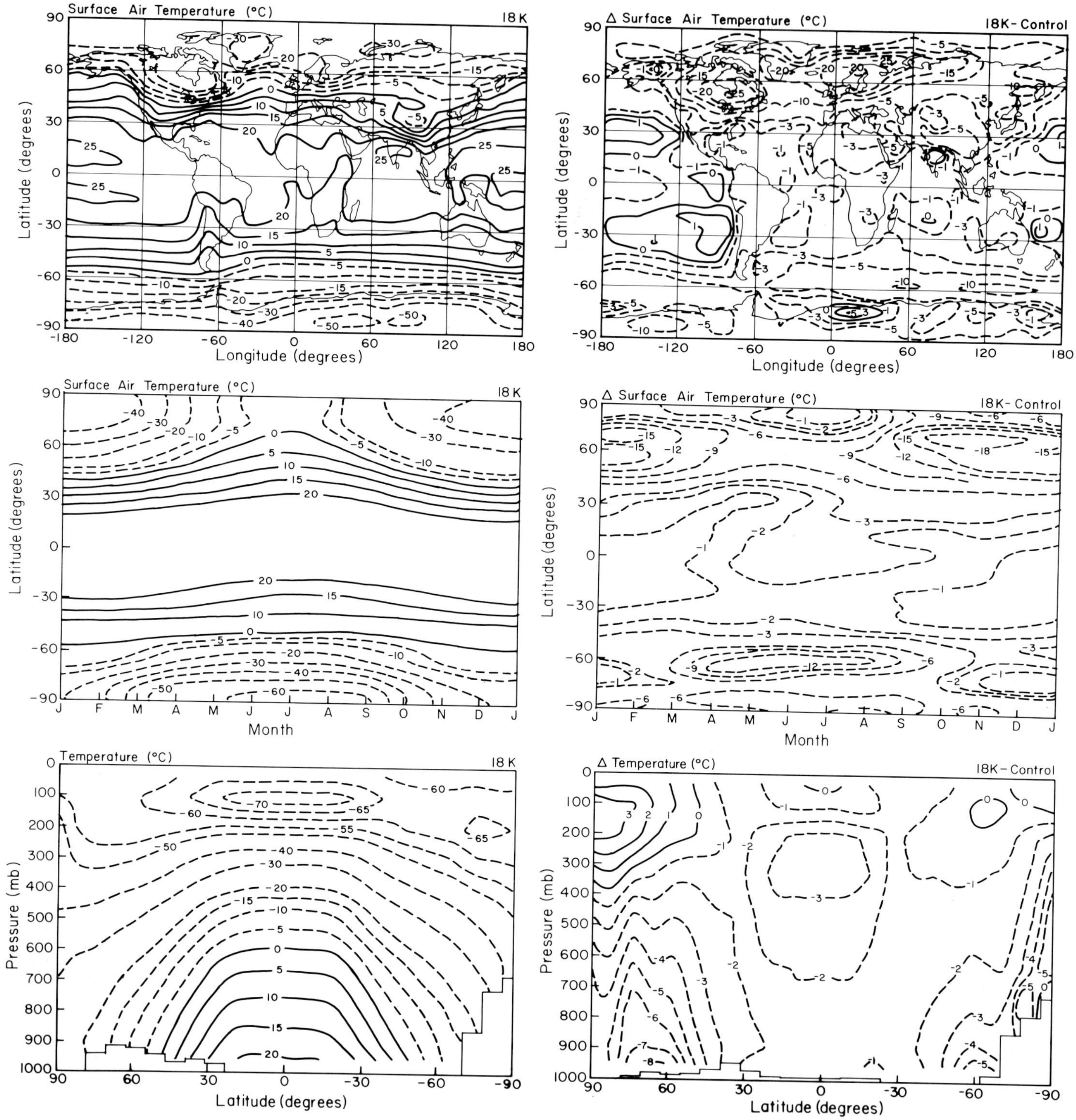

Fig. 9. Air temperature in the climate model experiment with boundary conditions for the ice age 18,000 years ago (left) and the temperature difference between the 18K simulation and the control run for today's climate boundary conditions (right). The control run is described in detail in paper 1.

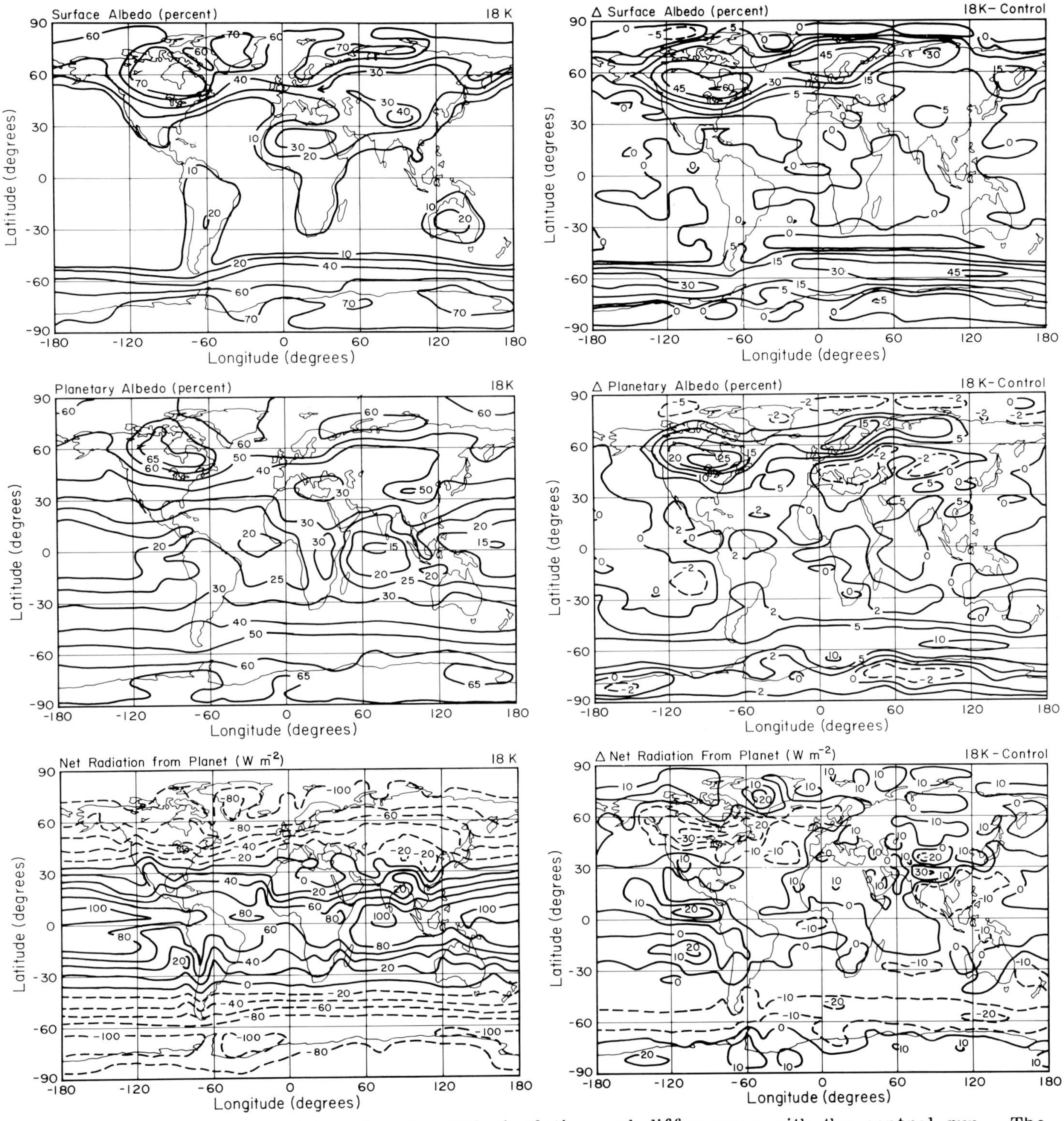

Fig. 10. Radiation quantities for the 18K simulation and differences with the control run. The control run is described in detail in paper 1.

that an albedo change of 0.10 is equivalent to 24 W m^{-2}. Most of the more detailed changes in the geographical patterns of the radiation budget are associated with changes of cloud cover or cloud top altitude.

Clouds. Cloud changes in the 18K simulation are illustrated in Fig. 11. There is a significant reduction of cloud cover in regions with increased sea ice, probably because of the reduced evaporation in those regions. The zonal

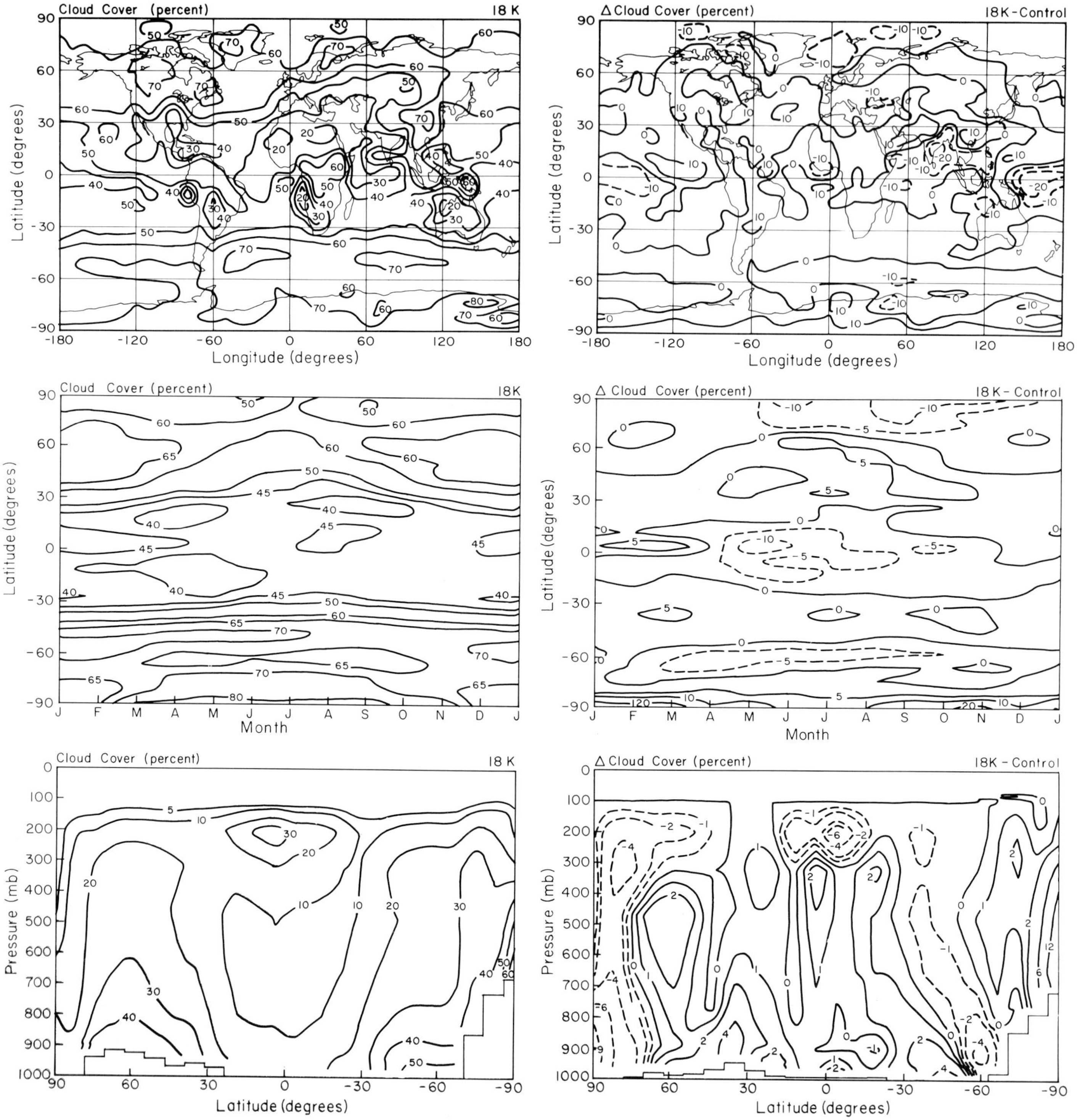

Fig. 11. Cloud cover for the 18K simulation and differences with the control run. The control run is described in detail in paper 1.

mean cloud cover decreases slightly in the tropics during most of the year, increases slightly in the subtropics and increases at Northern Hemisphere midlatitudes in summer. The polar regions exhibit opposite behavior; at the north pole (a region of sea ice) the clouds decrease, while at the south pole (a continental region of high topography) the clouds increase in the 18K experiment. The lowest panel in Fig. 11 shows that the high level (cirrus) clouds are reduced substantially in the 18K simulation, presumably due to the reduction of penetrating moist convection and its associated transport of moisture. Most of these changes are consistent with those in the doubled CO_2 experiment, the cloud changes at 18K being the opposite of those which occur for the warmer doubled CO_2 climate.

Summary. The global mean surface air cooling of the Wisconsin ice age (compared to today) is computed from the CLIMAP boundary conditions to be ~4°C. Thus the mean temperature change between 18K and today is very similar to the projected warming for doubled CO_2. Below we analyze the contributions of different feedback processes to this global climate change.

18K Feedback Factors

We perform two types of experiments to study the feedback processes at 18K. In experiments of the first type, a given factor is modified (say the sea ice cover is changed) and the model is run for several years with the atmosphere free to adjust to the change, but with the ocean temperature and other boundary conditions fixed. Thus the only substantial feedback factors allowed to operate are water vapor and clouds (snow over land and ice can also change, but this represents only a small part of the ice/snow feedback). Experiments of this type enable us to relate surface temperature changes with flux imbalances at the top of the atmosphere under conditions of radiative/dynamic equilibrium in the atmosphere. Results of this type of experiment are summarized in the first part of Table 3 (experiments 8-14) along with the 18K control run (experiment 7). The method for converting the flux imbalance at the top of the atmosphere in these experiments to gain or feedback factors is described below in conjunction with experiment 8.

Experiments of the second type [labeled with a star (*) and tabulated in the lower part of Table 3] provide a faster, but more approximate, method for evaluating feedbacks which can be applied to certain types of radiative forcing. In the starred experiments we determine the radiative forcing by changing a factor in the control run (say sea ice cover) and calculating the instantaneous change in the planetary radiation balance at the top of the atmosphere. The atmospheric temperature and other radiative constituents and boundary conditions are not allowed to change; thus no feedbacks operate in these experiments. The flux change at the top of the atmosphere, ΔF, defines a change of planetary effective temperature

$$\Delta T_e(°C) = (\sigma T_e^3)^{-1} \Delta F(W\ m^{-2}) \sim 0.27 \Delta F(W\ m^{-2}) \quad (13)$$

for T_e = 255K. This relation provides a good estimate of the no-feedback contribution to the equilibrium surface temperature change, if the radiative perturbation does not appreciably alter the vertical temperature structure. This procedure is applicable to solar flux, surface albedo and certain tropospheric gas perturbations (Hansen et al., 1982), but does not work as simply for CO_2 perturbations, because CO_2 cools the stratosphere (Fig. 4 of Hansen et al., 1981).

Although (13) provides a useful estimate of the (no feedback) surface temperature change resulting from a given radiative imbalance at the top of the atmosphere, it is a rough estimate because the radiation to space comes from a broad range of wavelengths and altitudes. In order to account for this spectral dependence, we used the 1-D radiative convective model for the following experiment. A flux of 1 W m^{-2} was arbitrarily added to the ocean surface, and the lapse rate, water vapor and other radiative constituents were kept fixed. The surface temperature increase at equilibrium was 0.29°C, implying

$$\Delta T_s(°C) \sim 0.29\ \Delta F(W\ m^{-2}) . \quad (14)$$

The coefficient in (14) is preferable to that in (13), for radiative perturbations which uniformly affect surface and atmospheric temperatures.

Water vapor and cloud feedbacks. Although we do not have measurements of the water vapor and cloud distribution for 18K, we can use experiment 8 to determine the combined water vapor/cloud feedback factor in our 3-D model for the 18K simulation. In this experiment the ocean surface temperature was arbitrarily decreased by 2°C everywhere. As shown in Table 3, the global mean surface temperature decreased by 2°C and the net radiation flux to space decreased by 2.7 W m^{-2}. Thus, with the sea ice and land ice fixed, the model sensitivity $\Delta T/\Delta F$ for the combined water vapor and cloud feedbacks is 0.76°C $(W/m^{-2})^{-1}$. If no feedbacks were allowed to operate, the sensitivity would be ~0.29°C $(W/m^2)^{-1}$, cf. equation (14). Thus, since the atmospheric feedbacks are the only ones allowed to operate in experiment 8, we infer that the combined water vapor and cloud feedback factor in our model for 18K is f_{wc} ~ 2.6 and g_{wc} ~ 0.6. This is practically the same as the combined water vapor and cloud feedbacks for the doubled CO_2 experiment [Table 1 and equation (12)].

Experiment 8 can be used to convert the flux imbalances at the top of the atmosphere in the other unstarred experiments in Table 3 to equilibrium surface temperature changes. Thus, if the ocean temperature were free to change and water vapor and clouds were the only operative feedbacks, a flux imbalance ΔF at the top of the atmosphere would vanish as the surface temperature changed by an amount $\Delta T = 0.76\ \Delta F$.

Sea ice and land ice feedbacks. Experiments 9 and 10, in which the 18K distributions of sea ice and land ice were replaced with today's distributions, show that both the sea ice and land ice changes made major contributions to the ice age cooling (Table 3). The CLIMAP sea ice and land ice distributions each affect the global ground albedo by ~0.02. Atmospheric shielding and zenith angle effects reduce the impact on plane-

TABLE 3. Changes in planetary radiation balance in climate model experiments. The control run for experiments 7, 11, and 13 is climate model II documented in paper 1. σ is the standard deviation of the annual mean about the 5 year mean for this control run; standard deviations for the other control runs were of similar magnitude. Experiment 7 was run for 6 years, with the results averaged over the final 5 years. The other unstarred experiments were run for 4 years and averaged over the final 3 years. The starred experiments were run for 3 years and averaged over 3 years. T_s is surface air temperature, F net radiation at the top of the atmosphere, A_g ground albedo and A_p planetary albedo. Global values are shown, with the numbers in parenthesis being the results for the Northern Hemisphere and Southern Hemisphere respectively. Experiment 7', the control run for experiment 12, was identical to experiment 7 except that the Koppen vegetation of Fig. 12(a) was substituted for Matthews (1983) 1° x 1° vegetation used in model II.

Experiment	Description	Control Run	ΔT_s(°C)	$100\times\Delta A_g$	$100\times\Delta A_p$	ΔF (W m^{-2})
7	18K boundary conditions	model II (paper 1)	-3.6(-4.6,-2.5)	4.1(5.0,3.1)	1.9(2.2,1.6)	-1.6(-0.1,-3.2)
8	ocean temperature reduced by 2°C	experiment 7	-2.0(-2.0,-2.0)	0.4(0.4,0.4)	0.5(0.5,0.5)	2.7(2.7,2.6)
9	today's sea ice	experiment 7	0.6(0.5,0.7)	-1.8(-0.9,-2.7)	-0.6(-0.3,-0.8)	1.7(0.6,2.8)
10	today's land ice	experiment 7	0.9(1.7,0.1)	-1.9(-3.6,-0.2)	-0.9(-1.3,-0.5)	1.9(2.4,1.4)
11	reduced 18K sea ice	model II (paper 1)	-3.5(-4.6,-2.4)	3.7(5.0,2.4)	1.8(2.2,1.4)	-1.4(0.1,-2.9)
12	18K vegetation	experiment 7'	-0.1(-0.3,0.0)	0.6(1.2,0.0)	0.3(0.5, 0.1)	-0.9(-1.4,-0.4)
13	modified 18K boundary conditions	model II (paper 1)	-3.7(-5.0,-2.4)	3.4(4.9,1.9)	1.7(2.2,1.3)	-2.1(-0.2,-3.9)
14	CO_2 (315→200 ppm)	experiment 7	-0.2(-0.3,-0.1)	0.1(0.1,0.1)	0.2(0.2,0.1)	-2.2(-2.3,-2.1)
9*	today's sea ice	experiment 7	-	-1.6(-0.8,-2.6)	-0.5(-0.4,-0.7)	3.1(2.0,4.3)
10*	today's land ice	experiment 7	-	-1.5(-2.7,-0.4)	-0.6(-1.1,0.0)	3.6(6.4,0.7)
11*	today's sea ice	experiment 11	-	-1.3(-0.9,-1.8)	-0.5(-0.4,-0.6)	2.6(1.8,3.3)
	$\sigma_{control}$		0.04(0.03,0.06)	0.05(0.07,0.03)	0.09(0.09.0.10)	0.3(0.3,0.4)

tary albedo to 0.006 for the sea ice change and 0.009 for the land ice change. The impact on the net radiation balance with space is between 1.5 and 2.0 W m^{-2} in each case, for these experiments in which the atmosphere was allowed to adjust to the changed sea ice and land ice.

The radiation imbalances in these experiments of the first type can be used to estimate the gain factors for these two feedback processes. Based on the conversion factor 0.76°C/(W m^{-2}), the flux imbalances in experiments 9 and 10 yield equilibrium surface temperature changes of $\Delta T_{sea\ ice}$ = 1.9K and $\Delta T_{land\ ice}$ = 2.3K. Since the feedback factor in these experiments is f_{wc} = 2.6, the radiative forcings produced by the sea ice and land ice changes in the absence of feedbacks are $\Delta T_{sea\ ice} = 1.9K/f_{wc} = 0.7K$ and $\Delta T_{land\ ice}$ = 0.9K, respectively. Thus the gain factors for sea ice and land ice changes, for the climate change from 18K to today, can be estimated as

$$g_{sea\ ice} \sim 0.7/\Delta T \sim 0.14\text{–}0.20$$
$$g_{land\ ice} \sim 0.9/\Delta T \sim 0.18\text{–}0.25\ , \tag{15}$$

where ΔT is the change of global mean surface air temperature in °C between 18K and today. Experiment 7 yields ΔT = 3.6°C, but indications that CLIMAP low latitude ocean temperatures are too warm (see above) suggest $\Delta T \sim$ 5°C; the range given for g refers to ΔT = 3.6-5°C.

In experiments 9* and 10* the 18K distributions of sea ice and land ice in experiment 1 were replaced with today's distributions, but only for diagnostic calculation of the planetary radiation balance; all other quantities in the diagnostic calculation were from experiment 7. Based on the radiative forcings computed at the top of the atmosphere and Eq. (13) we estimate the gain factors, $g_i = \Delta T_i/\Delta T$, for the sea ice and land ice changes to be

$$g_{sea\ ice} \sim \frac{0.27 \times 3.1}{\Delta T} \sim 0.17\text{–}0.23$$
$$g_{land\ ice} \sim \frac{0.27 \times 3.6}{\Delta T} \sim 0.19\text{–}0.27\ . \tag{16}$$

These gain factors include the effect of ice on thermal emission and planetary albedo. The fact that the gains estimated from (16) exceed those from (15) indicates that the emission from the added snow and ice surfaces on the average is from a somewhat higher temperature than the effective temperature, 255K.

The accuracy of these feedback gains depends primarily on the accuracy of the CLIMAP boundary conditions. Indeed, it is possible that the CLIMAP sea ice distribution is too extensive. Burckle et al. (1982), on the basis of satellite measurements of sea ice coverage and present sediment distributions, suggest that the sediment boundaries which CLIMAP had assumed to be the summer sea ice limit in the Southern Hemisphere are in fact more representative of the spring sea ice limit. Experiments 11 and 11* test the effect of this reduced sea ice coverage. In experiment 1 the CLIMAP February and August sea ice coverage were used as the extremes and interpolated sinusoidally to other months. For experiments 11 and 11* the winter (August) coverage was left unchanged, but the CLIMAP Southern Hemisphere February coverage was used for the spring (November) with linear extrapolation to February, and linear interpolation between the February and August extremes.

Experiment 11* implies that the sea ice gain estimated in experiments 9 and 9* should be reduced by 15-20 percent, if the arguments of Burckle et al. (1982) are correct. Although there is uncertainty about the true 18K sea ice distribution, it seems likely that the original CLIMAP data is somewhat an overestimate. On the basis of experiments 9, 9* and 11* our best estimate of the sea ice gain for the climate change from 18K to today is $g_{sea\ ice} \sim 0.15$ and thus a feedback factor $f_{sea\ ice} \sim 1.2$. This is larger than the snow/ice feedback obtained in the S_o and CO_2 experiments. However, the area of the sea ice cover change is about twice as large in the 18K experiment (-18.4×10^6 km^2 for the annual mean with our revised CLIMAP sea ice) than in these other experiments (7.8×10^6 km^2, 9.2×10^6 km^2 and 14.8×10^6 km^2 in the S_o, CO_2 and alternate CO_2 experiments, respectively). Thus, the gains obtained from the ice age and the warmer climate experiments are consistent.

It also has been argued (DiLabio and Klassen, 1983) that the CLIMAP land ice cover is an overestimate, because the ice sheet peripheries probably did not all achieve maximal extent simultaneously. This possibility was recognized by the CLIMAP investigators, who therefore also presented a minimal extent ice sheet model for 18K (Denton and Hughes, 1981; CLIMAP, 1981). In this minimal ice model the area by which the ice sheets of 18K exceeded those of today is reduced to five-sixths of the value in the standard CLIMAP model. We conclude that the land ice gain for the climate change from 18K to today is 0.15-0.25. The corresponding feedback factor is 1.2-1.3.

Vegetation feedback. We also investigated the vegetation feedback, which Cess (1978) has estimated to provide a large positive feedback. We used the Koppen (1936) scheme, which relates annual and monthly mean temperature and rainfall to vegetation type, to infer expected global vegetation distributions for the GCM runs representing today's climate (model II in paper 1) and the 18K climate. The resulting vegetation distribution from the run with today's boundary conditions (Fig. 12a) suggests that the model and Koppen scheme can do a fair job of 'predicting' vegetation, in the case of today's climate for which the scheme was derived. Discrepancies with observed vegetation (Matthews, 1983) exist, e.g., there is too much rainforest on the east coast of Africa and too little boreal forest in central Asia, but the overall patterns are realistic.

The vegetation distribution obtained for 18K (Fig. 12b) from the Koppen scheme and our 18K experiment has more desert than today, less rainforest and less boreal forest. These changes are qualitatively consistent with empirical evidence of tropical aridity during the last glacial maximum based on a variety of paleoclimate indicators, such as pollen (Flenley, 1979), fauna (Vuilleumier, 1971), geomorphology (Sarnthein, 1978) and lake levels (Street and Grove, 1979). However, the magnitude of the desert and rainforest changes is smaller than suggested by the paleoclimate evidence. The smaller changes may result from (a) the CLIMAP tropical ocean temperatures being too warm, as discussed above, which would tend to cause an overestimate of rainforest and underestimate of desert area; (b) the lower atmospheric CO_2 abundance of 18K (Shackleton et al., 1983), since CO_2 'fertilization' effects are not included in the Koppen scheme.

In experiment 12 today's vegetation was replaced with the Koppen 18K vegetation (Fig. 12b). The land, land ice and other boundary conditions were identical to those in the control run. In this experiment the modified vegetation directly affects the planetary albedo and also indirectly affects it through the masking depth for snow (paper 1). The 18K Kopppen vegetation of Fig. 12b increased the global ground albedo by 0.006 and the planetary albedo by 0.003 (Table 3) and left a flux imbalance of -0.9 W m^{-2} at the top of the atmosphere. Based on the same analysis as for ice above, the no-feedback temperature change due to vegetation is 0.3°C, yielding $g_{vegetation}$ = 0.06-0.08. Because of the imprecisions in the Koppen 18K vegetation, we broaden the estimated gain to $g_{vegetation}$ = 0.05-0.09, and thus $f_{vegetation}$ = 1.05-1.1. Examination of global maps shows that the greatest impact of the changed vegetation was the replacement of European and Asian evergreen forests with tundra and grassland; the greatly reduced masking depths produced annual ground albedo increases of 0.1 or more over large areas, with the largest changes in spring. For reasons stated above, we also examined an 18K run with ocean temperatures reduced by 2°C; this reduced the number of gridboxes with rainforest from 10 to 5 in South America and from 7 to 2 in Africa, compared to Fig. 12b, in better agreement with paleoclimate evidence cited above. This additional vegetation change did not significantly change the global albedo or flux at the top of the atmosphere.

We conclude that the vegetation feedback factor between 18K and today is $f_{vegetation} \sim$ 1.05-1.1. This is much smaller than suggested by Cess (1978), but consistent with the analysis of Dickinson (1984). We find a somewhat larger feedback than Dickinson obtained, 0.003 change of planetary albedo compared to his 0.002, apparently due to the change of vegetation masking of snow-covered ground.

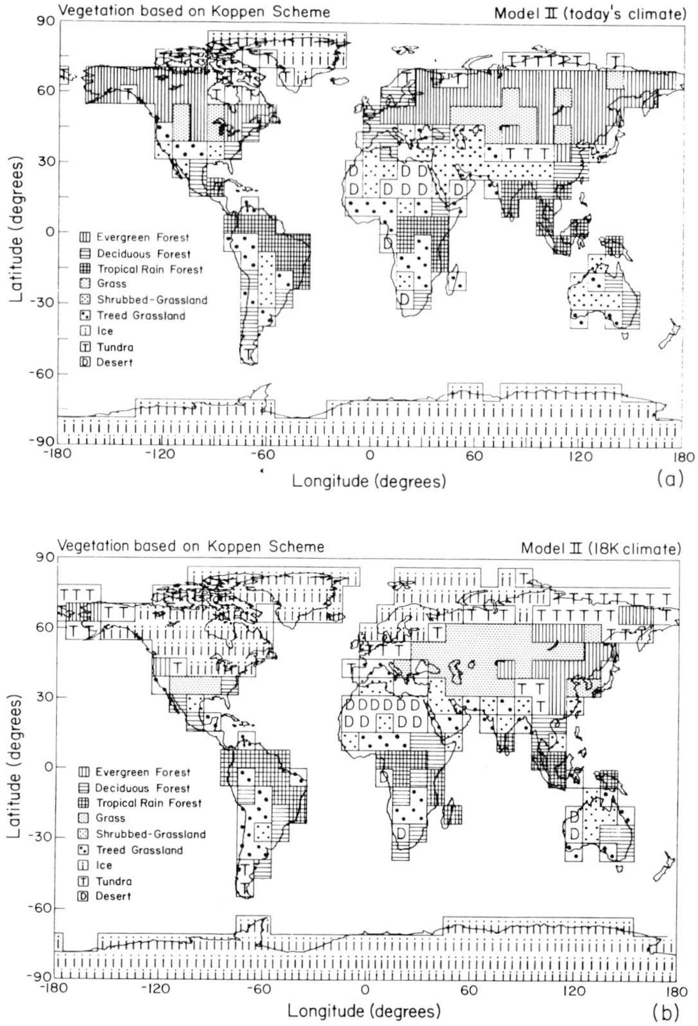

Fig. 12. Vegetation types (for gridboxes with more than 30 percent land) inferred from 3-D model simulations and the Koppen (1936) scheme, which relates annual and monthly mean temperature and precipitation to vegetation type. (a) is the control run for today's climate (paper 1), while (b) is the 18K simulation (experiment 7).

18K Radiation Balance

The simulated 18K climate (experiment 7) is close to radiation balance, the imbalance (Table 3) being 1.6 W m^{-2} at the top of the atmosphere, compared to the control run (model II) for today's climate. This imbalance is small compared to the amount of solar energy absorbed by the planet (~240 W m^{-2}). However, in reality even more precise radiation balance must have existed averaged over sufficient time, because the ice age lasted much longer than the thermal relaxation time of the ocean. (Melting the ice sheets in 10K years would require a global mean imbalance of only ~0.1 W m^{-2}.) Although the model calculations contain imprecisions comparable in magnitude to the radiation imbalance, we expect these to be largely cancelled by the procedure of differencing with the control run. This type of study should become a powerful tool in the future, as the accuracy of the reconstructed ice age boundary conditions improves and as the climate models become more realistic.

Taken at face value, the radiation imbalance in the 18K experiment 7 implies an imprecision in either some of the assumed boundary conditions for 18K or in the climate model sensitivity. The sense of the imbalance is such that the planet would cool further (to -4.8°C, based on the ΔF in Table 3), if the ocean temperature were computed rather than specified. Before studying this imbalance further, we make three modifications to the 18K simulation. First, the Southern Hemisphere sea ice cover is reduced as discussed above; this reduces the radiation imbalance. Second, the vegetation is replaced by the 18K vegetation of Fig. 12b; this slightly reduces the radiation imbalance. Third, the amount of atmospheric CO_2 is reduced in accord with evidence (Neftel et al., 1982) that the 18K CO_2 amount was only ~200 ppm; this significantly aggravates the radiation imbalance.

These three changes are all included in experiment 13, the sea ice and vegetation changes being those tested in experiments 11 and 12. The CO_2 decrease was 75 ppm from the control run value of 315 ppm; this is equivalent to the change from an estimated preindustrial abundance of 270 ppm to an ice age abundance of ~200 ppm. With these changes the radiation imbalance with space becomes 2.1 W m^{-2}. This imbalance would carry the surface temperature to -5.3°C if the constraint on ocean temperature were released.

Two principal candidates we can identify for redressing the 18K radiation imbalance are the CLIMAP sea surface temperatures and the cloud feedback in the climate model. The imbalance is removed if the CLIMAP ocean temperature is 1.5°C too warm (experiment 8, Table 3). The possibility that the CLIMAP sea surface temperatures may be too warm is suggested by the discussion above. The imbalance is also removed if it is assumed that clouds provide no feedback, rather than the positive feedback which they cause in this model; this conclusion is based on the estimate that the clouds cause 30-40 percent of the combined water vapor/cloud feedback (experiment 8), as is the case in the S_0 and CO_2 experiments.

One plausible solution is the combination of a reduction of low latitude ocean temperature by ~1°C and a cloud feedback factor between 1 and 1.3. An alternative is a reduction of low latitude ocean temperature by ~1°C and a greater value for the 18K CO_2 abundance; indeed, recent analyses of Shackleton et al. (1983) suggest a mean 18K CO_2 abundance ~240 ppm. It is also possible that there were other presently unsuspected changes of boundary conditions.

There are presently too many uncertainties in the climate boundary conditions and climate model to permit identification of the cause of the radiation imbalance in the 18K simulation. However, as the boundary conditions and climate models become more accurate, this approach should yield valuable checks on paleoclimate data and climate models. In the meantime, the data permits some general conclusions about the strength of climate feedback processes.

Conclusions from 18K Experiments

The above calculations suggest the following major contributions to the global cooling at 18K, as shown schematically in Fig. 13:

$$\begin{aligned} \Delta T_{\text{water vapor + clouds}} &\equiv \Delta T_{wc} \sim 1.4\text{-}2.2°C \\ \Delta T_{\text{land ice}} &\equiv \Delta T_l \sim 0.7\text{-}0.9°C \\ \Delta T_{\text{sea ice}} &\equiv \Delta T_s \sim 0.6\text{-}0.7°C \qquad (17) \\ \Delta T_{CO_2} &\sim 0.3\text{-}0.6°C \\ \Delta T_{\text{vegetation}} &\equiv \Delta T_v \sim 0.3°C \end{aligned}$$

These estimates are the product of the gain for each process and the total cooling at 18K. But note that the uncertainty in the total ΔT cancels in obtaining ΔT_l, ΔT_s, ΔT_{CO_2} and ΔT_v; thus these ΔT_i are more fundamental and accurate than the corresponding g_i. The ΔT_i represent the isolated radiative forcings, which can be computed accurately, for the assumed changes in these radiative constituents between 18K and today. ΔT_{wc} = 2.2°C is obtained from experiment 8 which yielded $g_{wc} \sim 0.6$. The cloud portion of ΔT_{wc} is uncertain because of the rudimentary state of cloud modeling; however even with no cloud feedback the water vapor contribution (~1.4°C) is a large part of the total ice age cooling. ΔT_l = 0.9°C is based on the CLIMAP maximal ice sheet extent; it is ~0.7°C for the minimal extent model. ΔT_s = 0.7°C is based on CLIMAP sea ice; it is 0.6°C with the reduced sea ice cover in the Southern Hemisphere in experiments 11 and 11*. ΔT_{CO_2} = 0.6°C refers to a CO_2 change from 200 ppm (at 18K) to 300 ppm (at say 1900); this is reduced to 0.3°C if the CO_2 amount was 225 ppm at 18K and 275 ppm at 1900.

The sum of the temperature contributions in Fig. 13 slightly exceeds the computed cooling ΔT = 3.7°C at 18K. This is a restatement of the radiation imbalance which exists in the model when the CLIMAP boundary conditions are used with ΔCO_2 of 50-100 ppm. If the model ocean temperature were allowed to change to achieve radiation balance, it would balance at a global mean 18K cooling of ~5.3°C [model sensitivity = 0.76°C/(W m^{-2})]. We conclude that either the CLIMAP 18K ocean temperatures are too warm by ~1.5°C or we have overestimated one or more of the contributions to the 18K cooling in (17).

It is apparent from Fig. 13 that feedback processes account for most of the 18K cooling. The water vapor, cloud and sea ice contributions represent at least half of the total cooling. On long time scales the land ice portion of the cooling also may be regarded as a feedback, though it operates on a very regional scale and may be a complex function of a variety of factors such as the position of land areas, ocean currents and the meteorological situation. Even the CO_2 portion of the cooling, or at least part of it, may be a feedback, i.e., in response to the change of climate.

Variations in the amount of absorbed solar radiation due to Milankovitch (earth orbital) changes in the seasonal and latitudinal distributions of solar irradiance, which occur on time scales of several thousand years, can provide a global mean forcing of up to a few tenths of a degree. In view of the strength of the climate feedbacks, it is plausible for the Milankovitch and CO_2 forcings to 'drive' glacial to interglacial climate variations. However, discussion of the sequence of causes and mechanisms of glacial to interglacial climate change is beyond the scope of this paper.

We can use the contributions to the 18K cooling summarized in Fig. 13 to obtain an empirical estimate of the climate feedback factor due to the processes operating on 10-100 year time scales, taking the land ice, CO_2 and vegetation changes from 18K to today as slow (or at least specified). The global mean radiative forcing due to the difference in 18K and today's orbital parameters is negligible compared to the other forcing summarized in Fig. 13. The feedback factor for the fast (water vapor, clouds, sea ice) processes is

$$f(\text{fast processes}) = \frac{\Delta T(\text{total})}{\Delta T(\text{slow processes})} \qquad (18)$$

ΔT(total) is ~3.7°C for the CLIMAP boundary conditions, but may be ~5°C, if CLIMAP low latitude ocean temperatures are 1-2°C too warm. Using the nominal CLIMAP boundary conditions and intermediate estimates for $\Delta T_l \sim 0.4$, $\Delta T_{CO_2} \sim 0.45$ and $\Delta T_V \sim 0.3$, yields f(fast processes) ~ 2.4. Using ΔT(total) ~ 5°C and these nominal radiative forcings yields f ~ 3.2. Allowing for the more extreme combinations of the forcings and ΔT(total), we conclude that

$$f(\text{fast processes}) \sim 2\text{-}4 \qquad (19)$$

This feedback factor, f ~ 2-4, corresponds to a climate sensitivity of 2.5-5°C for doubled CO_2. Note that this result is independent of our climate model sensitivity: it depends on the total ΔT at 18K (fixed by CLIMAP) and on the assumption that land ice, CO_2 and vegetation are the only major slowly changing boundary conditions. Of course some vegetation and CO_2 feedbacks may occur in less than 100 years but for projecting future climate it is normal to take these as specified boundary conditions.

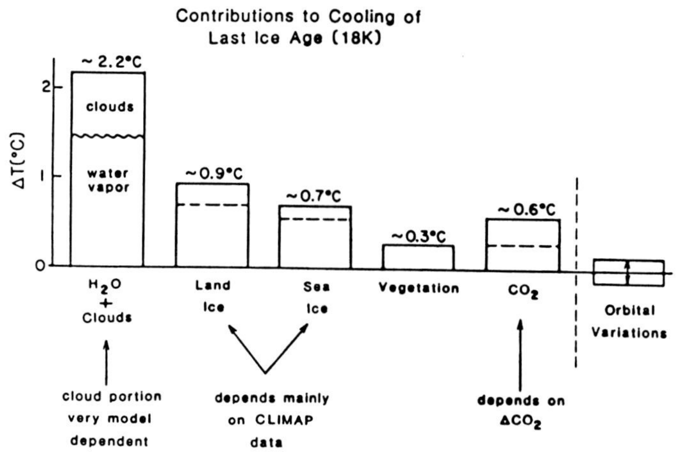

Fig. 13. Contributions to the global mean temperature difference between the Wisconsin ice age and today's climate as evaluated with a 3-D climate model and assumed boundary conditions. The cloud and water vapor portions were not separated, but based on other 3-D experiments the cloud part is estimated as 30-40 percent of their sum. The dashed line for land ice refers to the 'minimal extent' model of CLIMAP, and the dashed line for sea ice refers to the reduced sea ice cover discussed in the text. The solid line for CO_2 refers to ΔCO_2 ~ 100 ppm (300 ppm → 200 ppm) and the dashed line to ΔCO_2 ~ 50 ppm.

Finally, note that a given sensitivity for fast feedback processes, say 4°C for doubled CO_2, does not mean that the climate necessarily would warm by 4°C in 10 or even 100 years. Although water vapor, cloud and sea ice feedbacks respond rapidly to climate change, the speed of the climate response to a changed forcing depends on the rate at which heat is supplied to the ocean and on transport processes in the ocean.

Transient Response

Surface Response Time

The time required for the surface temperature to approach its new equilibrium value in response to a change in climate forcing depends on the feedback factor, f. The following example helps clarify this relationship.

Let the solar flux absorbed by a simple blackbody planet (f ≡ 1) change suddenly from $F_o = \sigma T_o^4$ to $F_o + \Delta F \equiv \sigma T_1^4$, with $\Delta F << F_o$. The rate of change of heat in the climate system is

$$\frac{d(cT)}{dt} = \sigma T_1^4 - \sigma T^4 \sim -4\sigma T_o^3(T - T_1) \qquad (20)$$

where c is the heat capacity per unit area and T is the time varying temperature. Since $T_1 - T_o << T_o$, the solution is

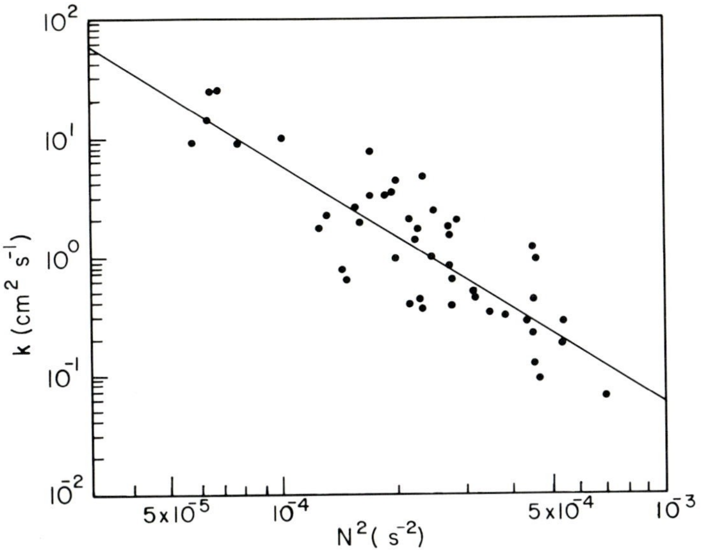

Fig. 14. Relationship between the effective diffusion coefficient (k) and the stability (N^2, where N is the Brunt-Vaisala frequency) at the base of the winter mixed layer for the GEOSECS tritium stations north of 20°S. The regression line fit (Eq. 25) has correlation coefficient 0.85 with the points for individual stations.

$$T - T_o = (T_1 - T_o)[1 - \exp(-t/\tau_b)] \qquad (21)$$

where the blackbody no-feedback e-folding time is

$$\tau_b = c/4\sigma T_o^3 \quad . \qquad (22)$$

For a planet with effective temperature 255 K and heat capacity provided by 63m of water (as in our 3-D experiments), τ_b is approximately 2.2 years. Thus, this planet with f = 1 exponentially approaches its new equilibrium temperature with e-folding time 2.2 years.

Feedbacks modify the response time since they come into play only gradually as the warming occurs, the initial flux of heat into the ocean being independent of feedbacks. It is apparent that the actual e-folding time for a simple mixed layer heat capacity is

$$\tau = f\tau_b \qquad (23)$$

An analytic derivation of (23) is given in Appendix A. The proportionality of the mixed layer response time to f is apparent in Fig. 3; the e-folding time for that model, which has f ~ 3.5 and a 63m mixed layer, is ~8 years.

The 63m mixed layer depth in our 3-D experiments was chosen as the minimum needed to obtain a realistic seasonal cycle of temperature, thus minimizing computer time needed to reach equilibrium. However, the global-mean annual-maximum mixed layer depth from our compilation of observations (see above) is ~110m, and thus the isolated ocean mixed layer has a thermal response time of ~15 years if the climate sensitivity is 4°C for doubled CO_2. Even if the climate sensitivity is 2-3°C for doubled CO_2, the (isolated) mixed layer response time is about 10 years.

In order to determine the effect of deep ocean layers on the surface response time, it is useful to express the heat flux into the ocean as a function of the difference between current surface temperature and the equilibrium temperature for current atmospheric composition. In Appendix A we show that

$$F(W\ m^{-2}) = \frac{F_o}{\Delta T_{eq}(2{*}CO_2)} (\Delta T_{eq} - \Delta T) \qquad (24)$$

where ΔT is the ocean surface temperature departure from an arbitrary reference state and ΔT_{eq} is the equilibrium temperature departure for current atmospheric composition. $\Delta T_{eq}(2{*}CO_2)$ is the equilibrium sensitivity for doubled CO_2; for our 3-D climate model it is 4.2°C. F_o is the flux into the ocean in the model when CO_2 is doubled and the stratospheric temperature has equilibrated; our 3-D model yields F_o = 4.3 W m^{-2}.

The long response time of the isolated mixed layer allows a portion of the thermal inertia of the deeper ocean to come into play in delaying surface temperature equilibrium. Exchange between the mixed layer and deeper ocean occurs by means of convective overturning in the North Atlantic and Antarctic oceans and principally by nearly horizontal motion along isopycnal (constant density) surfaces at lower latitudes. Realistic modeling of heat perturbations is thus rather complex, especially since changes of surface heating (and other climate variables) may alter the ocean mixing. However, we can obtain a crude estimate for the surface response time by assuming that <u>small</u> positive heat perturbations behave as a passive tracer; numerical experiments of Bryan et al. (1984) support this assumption. Measurements of transient tracers in the ocean, such as the tritium sprinkled on the ocean surface by atmospheric atomic testing, provide a quantitative indication of the rate of exchange of water between the mixed layer and the upper thermocline (see, e.g., Ostlund et al., 1976).

We estimate an effective thermocline diffusion coefficient (k) at each GEOSECS measurement station from the criterion that the modeled and observed penetration depths (Broecker et al., 1980) be equal at each station. The resulting diffusion coefficients are well correlated (inversely) with the stability at the base of the winter mixed layer (Fig. 14). In particular, we find a correlation coefficient of 0.85 between k and $1/N^4$, where N is the Brunt-Vaisala frequency at the base of the mixed layer. The global distribution of N^2 was obtained from the ocean data set of Levitus (1982).

The empirical relation between k and stability,

$$k = 5 \times 10^{-8}/N^4, \qquad (25)$$

and the global ocean data set of Levitus (1982) yield the global distribution of k at the base of the mixed layer shown in Fig. 15a. There is a low rate of exchange ($k < 0.2\ cm^2\ s^{-1}$) in the eastern equatorial Pacific where upwelling and the resulting high stability at the base of the mixed layer inhibit vertical mixing, but rapid exchange ($k > 10\ cm^2\ s^{-1}$) in the Greenland - Norwegian Sea area of convective overturning.

The e-folding time for the mixed layer temperature (time to reach 63 percent of the equilibrium response) is shown in Fig. 15b. This is calculated from the geographically varying k and annual-maximum mixed layer depth, assuming a sudden doubling of CO_2 and an equilibrium sensitivity of 4.2°C everywhere. The (63 percent) response time is about 20-50 years at low latitudes, where the shallow mixed layer and small k allow the mixed layer temperatures to come into equilibrium relatively quickly. At high latitudes, where the deep winter mixed layer and large k result in a larger thermal inertia, the response time is about 200-400 years. The time for the global area-weighted mixed layer temperature to reach 63 percent of its equilibrium response is 124 years.

We estimate an equivalent k for use in a global 1-D model by choosing that value of k which fits the global (area-weighted) mean perturbation of the mixed layer temperature as a function of time obtained with the above calculation in which k and mixed layer depth vary geographically. We find that $k \sim 1\ cm^2\ s^{-1}$ provides a reasonable global fit to the area-weighted local calculations for either a step function change of CO_2 or exponentially increasing CO_2 amount. Other analyses of the tracer data yield empirical values of 1-2 $cm^2\ s^{-1}$ for the effective rate of exchange between the mixed layer and deeper ocean (Broecker et al., 1980).

The delay time due to the ocean thermal inertia is graphically displayed in Fig. 16. Equation (24) provides a good approximation of the time dependence of the heat flux into the ocean in our 3-D climate experiment with doubled CO_2, as shown by comparison of Figs. 3b and 16. Note that in our calculation with a mixed layer depth of 110 m, $k = 1\ cm^2\ s^{-1}$, and $\Delta T_{eq} = 4.2°C$. The time required to reach a response of 2.65°C is 102 years. This is in rough agreement with the 124 years obtained above with the 3-D calculation.

The ocean delay time is proportional to f for an isolated mixed layer [eq. (23) and Appendix A], but depends more strongly on f if mixing into the deeper ocean is included. Our 1-D calculation with $k = 1\ cm^2\ s^{-1}$ and mixed layer depth 110 m yields an e-folding time of 55 years for ΔT_{eq} = 3°C and 27 years for ΔT_{eq} = 2°C. Thus our ocean response time is consistent with that of Bryan et al. (1982), who obtained a response time of about 25 years for a climate model with sensitivity ~2°C for doubled CO_2.

Although our calculations were made with a simple diffusion model, the conclusion that the ocean surface temperature response time is highly nonlinear in ΔT_{eq} (or f) is clearly more general. The surface response time increases faster than linearly with f when the deeper ocean is included, because as f increases greater ocean depths come into play. Thus more realistic modeling of ocean transport processes should not modify these conclusions for small climate perturbations.

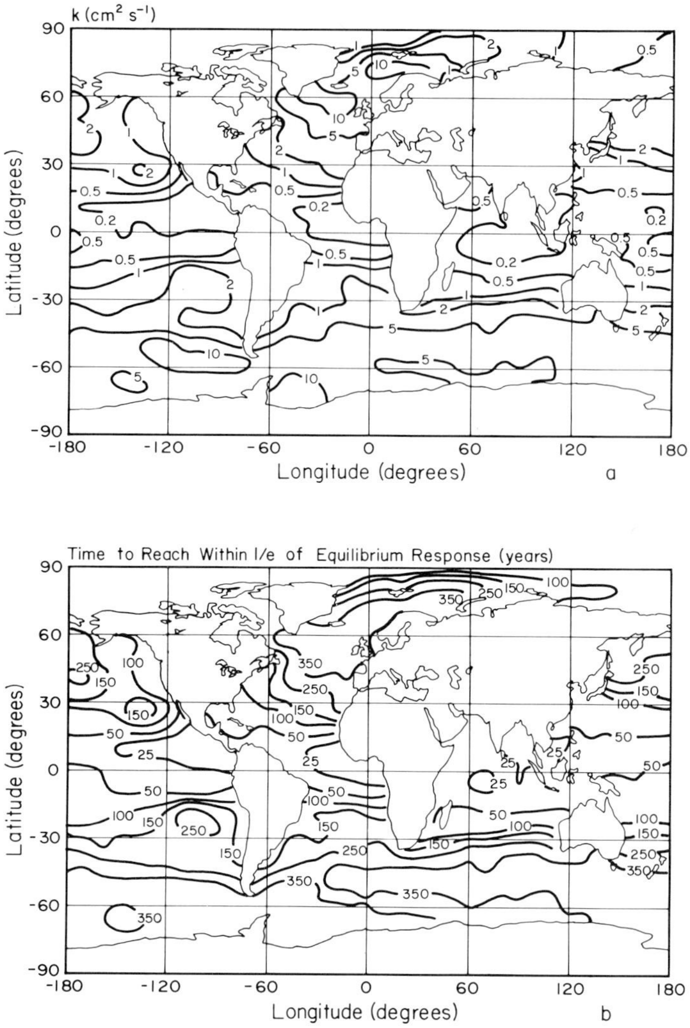

Fig. 15. (a) Geographic distribution of effective thermocline diffusion coefficient, k ($cm^2 s^{-1}$). k is derived from Eq. 25 and the global distribution of N^2, the latter obtained from the ocean data of Levitus (1982). (b) Geographic distribution of the 63 percent response time for surface temperature response to doubled CO_2 in the atmosphere. Only geographic variability of k and mixed layer depth are accounted for. $\Delta T_{eq}(2{*}CO_2)$ is taken as 4.2°C everywhere. The flux into the ocean is from Eq. (24).

Our calculations of ocean response time neglect ocean circulation feedbacks on climate. The relationship between k and stability, equation (25), provides one way to examine the temperature feedback. By using that relation with

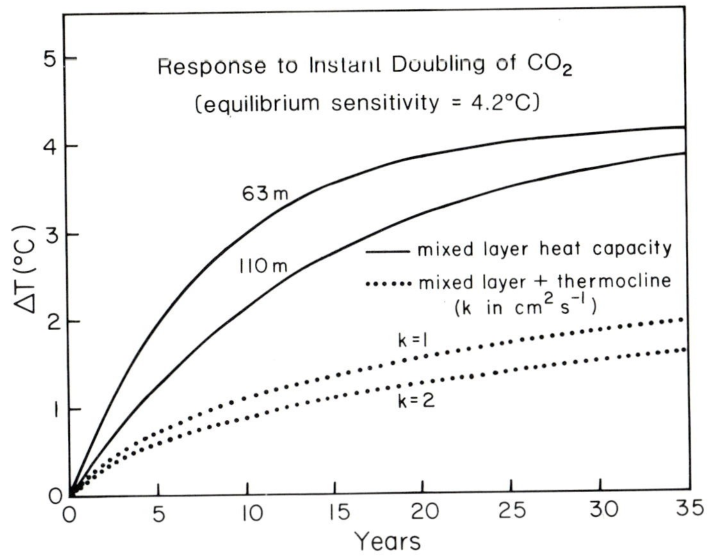

Fig. 16. Transient response to step function doubling of atmospheric CO_2 from 315 ppm to 630 ppm, computed from (24) with three representations of the ocean. The 63 m mixed layer corresponds to the mean mixed layer depth in the 3-D experiments; 110 m is the global-mean annual maximum mixed layer depth obtained from global ocean data. The curves including diffusion beneath the mixed layer are not exponentials (Appendix A).

our 1-D ocean diffusion model, we find that the time required to reach a given transient response is decreased, typically by several percent. Real ocean transports may be more sensitive to surface warming, as well as to related mechanisms such as melting of sea ice and ice sheets and changing winds, precipitation and evaporation. It is easy to construct scenarios in which the ocean feedbacks are much greater, especially in the areas of deep water formation, but not enough information is available for reliable calculation of ocean/climate feedbacks.

Finally, we note that the ocean surface thermal response time reported in the literature generally has been 8-25 years (Hunt and Wells, 1979; Hoffert et al., 1980; Cess and Goldenberg, 1981; Schneider and Thompson, 1981; Bryan et al., 1982). The 3-D ocean model result of Bryan et al. is consistent with our model when we employ a climate sensitivity of 2°C for doubled CO_2, as noted above. The discrepancy between our model response time and that of the other models arises from both the climate sensitivities employed and the choice of ocean model parameters. Key parameters are: mixed layer depth (we use 110m since any depth mixed during the year should be included), rate of exchange with deeper ocean (we use diffusion with $k = 1$ cm^2 s^{-1}, the minimum global value suggested by transient tracers, cf., Broecker et al., 1980) and the atmosphere-ocean heat flux [we use (24) which has initial value 4.3 W m^{-2} over the ocean area for doubled CO_2 and is consistent with other 3-D models]. Obviously the use of a 1-D box-diffusion model is a gross oversimplification of ocean transports. As an intermediate step between this and a 3-D ocean general circulation model, it may be valuable to study the problem with a model which ventilates the thermocline by means of transport along isopycnal surfaces. The agreement between the results from the 3-D ocean model of Bryan et al. and our model with a similar climate sensitivity suggests that our approach yields a response time of the correct order.

Impact on Empirically-Derived Climate Sensitivity

The delay in surface temperature response due to the ocean must be included if one attempts to deduce climate sensitivity from empirical data on time scales of order 10^2 years or less. Furthermore, in such an analysis it must be recognized that the lag caused by the ocean is not a constant, independent of climate sensitivity.

We computed the expected warming due to increase of CO_2 between 1850 and 1980 as a function of the equilibrium climate sensitivity. Results are shown in Fig. 17a for five choices of the 1850 CO_2 abundance (270±20 ppm), with CO_2 increasing linearly to 315 ppm in 1958 and then based on Keeling et al. (1982) measurments to 338 ppm in 1980. For simplicity a one-dimensional ocean was employed with mixed layer depth 110m and $k = 1$ cm^2 s^{-1}. However, we obtained a practically identical graph when we used a simple three-dimensional ocean with the mixed layer depth varying geographically according to the data of Levitus (1982) and k varying as in Fig. 15a.

Use of Fig. 17a is as follows. If we take 270 ppm as the 1850 CO_2 abundance (WMO, 1983) and assume that the estimated global warming of 0.5°C between 1850 and 1980 (CDAC, 1983) was due to the CO_2 growth, the implied climate sensitivity is 4°C for doubled CO_2 (f = 3-4). Results for other choices of the 1850 CO_2 abundance or global warming can be read from the figure.

Undoubtedly some other greenhouse gases also have increased in the past 130 years. Chloroflourocarbons, for example, are of recent anthropogenic origin. CH_4 and N_2O are presently increasing at rates of 1-2 percent yr^{-1} and 0.2-0.3 percent per yr^{-1}, respectively (Ehhalt, et al., 1983; Weiss, 1981; CDAC, 1983). We estimate the influence of these gases on the empirical climate sensitivity by using the trace gas scenarios in Table 4. Although the CH_4 and N_2O histories are uncertain, the chlorofluorocarbons provide most of the non-CO_2 greenhouse effect, at least in the past 10-20 years (Lacis et al., 1981), and their release rates are known. CH_4 may have increased slowly for the past several hundred years (Craig and Chou, 1982), but the reported rate of increase would not affect the results much. O_3 is also a potent greenhouse gas, but information on its past history is not adequate to permit its effect to be included.

The climate sensitivity implied by the assumed global warming since 1850, including the effect of trace gases in addition to CO_2, is shown in Fig. 17b. If the 1850 CO_2 abundance was 270±10 ppm, as concluded by WMO (1983), a warming of 0.5°C requires a climate sensitivity 2.5-5°C for doubled CO_2. The range for the implied climate sensitivity increases if uncertainty in the amount of warming is also included. For example, a warming of 0.4-0.6°C and an 1850 CO_2 abundance of 270±10 ppm yield a climate sensitivity of 2-7°C.

Although other climate forcings, such as volcanic aerosols and solar irradiance, may affect this analysis, we do not have information adequate to establish substantially different magnitudes of these forcings prior to and subsequent to 1850.

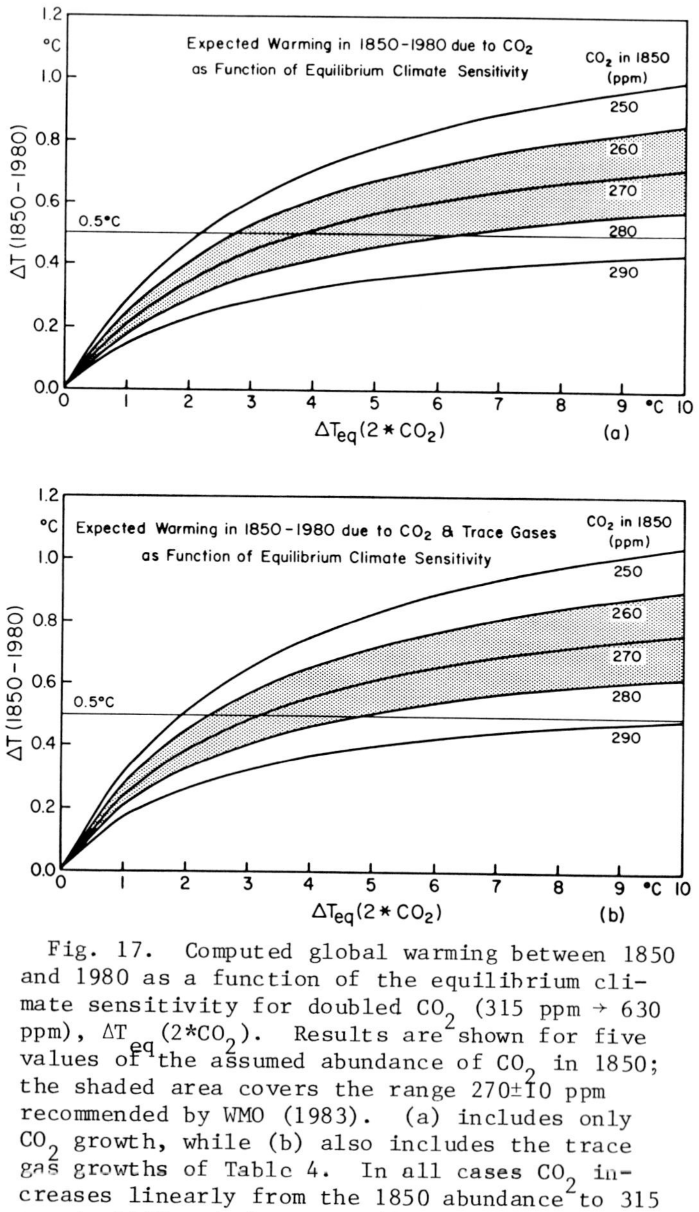

Fig. 17. Computed global warming between 1850 and 1980 as a function of the equilibrium climate sensitivity for doubled CO_2 (315 ppm → 630 ppm), $\Delta T_{eq}(2*CO_2)$. Results are shown for five values of the assumed abundance of CO_2 in 1850; the shaded area covers the range 270±10 ppm recommended by WMO (1983). (a) includes only CO_2 growth, while (b) also includes the trace gas growths of Table 4. In all cases CO_2 increases linearly from the 1850 abundance to 315 ppm in 1958 and then according to measurements of Keeling et al. (1982).

TABLE 4. Trace gas abundances employed in our calculations of the transient climate response for Figs. 17b and 18. CO_2 increases linearly for 1850-1958 and as observed by Keeling et al. (1982) for 1958-1980; ΔCO_2 increases about 2 percent yr^{-1} in the future. The chlorofluorocarbon abundances are based on estimated release rates to date, 150 year and 75 year lifetimes for CCl_2F_2 and CCl_3F, respectively, and constant future emissions at the mean release rate for 1971-1980. The CH_4 increase is about 1 percent yr^{-1} for 1970-1980 and 1.5 percent yr^{-1} after 1980. The N_2O increase is 0.2 percent yr^{-1} for 1970-1980 and 0.3 percent yr^{-1} after 1980.

Date	CO_2 (ppm)	CCl_2F_2 (ppt)	CCl_3F (ppt)	CH_4 (ppb)	N_2O (ppb)
1850	270	0	0	1400	295
1900	291	0	0	1400	295
1950	312	7	1	1400	295
1960	317	33	11	1416	295
1970	326	126	62	1500	295
1980	338	308	178	1650	301
1990	353	479	280	1815	307
2000	372	638	369	1996	313
2010	396	787	447	2196	320

The climate sensitivity we have inferred is larger than obtained by CDAC (1983) from analysis of the same time period (1850-1980) with the same assumed temperature rise. The chief reason is that CDAC did not account for the dependence of the ocean response time on climate sensitivity [equation (23) and Appendix A]. Their choice of a 15 year response time, independent of ΔT_{eq} or f, biased their result to low sensitivities.

We conclude that the commonly assumed empirical temperature increase for the period 1850-1980 (0.5°C) suggests a climate sensitivity of 2.5-5°C (f=2-4) for doubled CO_2. The significance of this conclusion is limited by uncertainties in past atmospheric composition, the true global mean temperature change and its cause, and the rate at which the ocean takes up heat. However, knowledge of these factors may improve in the future, which will make this a powerful technique for investigating climate sensitivity.

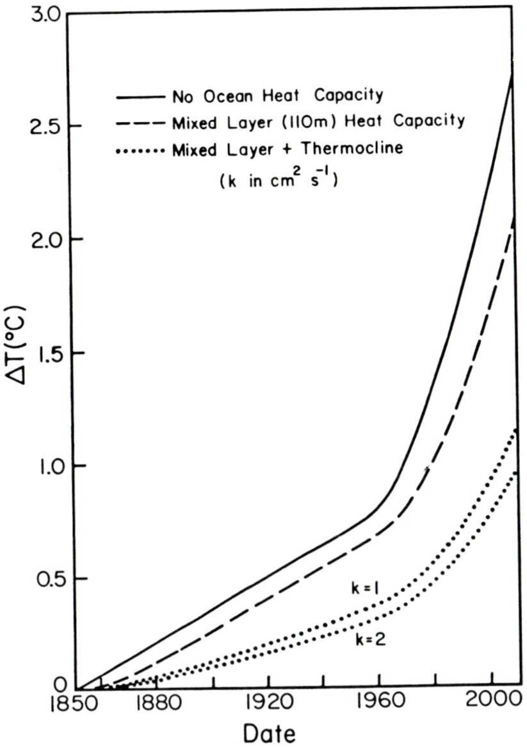

Fig. 18. Global mean warming computed for the CO_2 and trace gas scenarios in Table 4.

Growing Gap Between Current and Equilibrium Climate

One implication of the long surface temperature response time is that our current climate may be substantially out of equilibrium with current atmospheric composition, as a result of the growth of atmospheric CO_2 and trace gases during recent decades. For example, in the last 25 years CO_2 increased from 315 ppm to 340 ppm and the chloroflourocarbons from near zero to their present abundance. Since the growth rates increased during the period, the gas added during the past 25 years has been present on the average about 10 years. 10 years is short compared to the surface temperature response time, even if the climate sensitivity is only 2.5°C for doubled CO_2.

We illustrate the magnitude of this disequilibrium by making some calculations with the 1-D model specified to give the climate sensitivity of our 3-D model, 4.2°C for doubled CO_2, and with the changing atmospheric composition of Table 4. Fig. 18 shows the modeled surface temperature during the past century (1) for instant equilibrium with changing atmospheric composition, (2) with thermal lag due to the mixed layer included, and (3) with the thermocline's heat capacity included via eddy diffusion.

We infer that there is a large and growing gap between current climate and the equilibrium climate for current atmospheric composition. Based on the estimate in Fig. 18, we already have in the pipeline a future additional warming of almost 1°C, even if CO_2 and trace gases cease to increase at this time. A warming of this magnitude will elevate global mean temperature to a level at least comparable to that of the Altithermal (NAS, 1975, adapted in Fig. 1 of Hansen et al., 1984) about 6,000 years ago, the warmest period in the past 100,000 years.

The rate of warming computed after 1970 is much greater than in 1850-1980. This is because (1) ΔCO_2 is ~0.4 ppm yr^{-1} in 1850-1960, but >1 ppm yr^{-1} after 1970, and (2) trace gases, especially chlorofluorocarbons, add substantially to the warming after 1970. The surface warming computed for the period 1970-1990 is ~0.25°C; this is almost twice the standard deviation of the 5-year-smoothed global temperature (Hansen et al., 1981). But note that the equilibrium temperature increases by 0.75°C in the period 1970-1990, if the climate sensitivity is ~4°C for doubled CO_2. Thus our calculations indicate that the gap between current climate and the equilibrium climate for current atmospheric composition may grow rapidly in the immediate future, if greenhouse gases continue to increase at or near present rates.

As this gap grows, is it possible that a point will be reached at which the current climate "jumps" to the equilibrium climate? If exchange between the mixed layer and deeper ocean were reduced greatly, the equilibrium climate could be approached in as little as 10-20 years, the relaxation time of the mixed layer. Indeed, the stability of the upper ocean layers seems likely to increase as the greenhouse warming heats the ocean surface, especially if the warming leads to an increased melting of ice which adds fresh water to the mixed layer. Regions of deep-water formation, such as the North Atlantic Ocean, may be particularly sensitive to changes in surface climate. However, it is difficult to predict the net effect of greenhouse warming on ocean mixing, because changes of precipitation, evaporation and atmospheric winds, in addition to temperature, will affect ocean mixing and transport. If possible, it would be useful to examine paleoclimate records for evidence of sudden climate warmings on 10-20 year time scales, since there may have been cases in the past when the long thermal response time of the ocean allowed gaps between actual and equilibrium climates to build up.

Even if it does not lead to a dramatic jump to a new climate state, the gap between current climate and the equilibrium climate for current atmospheric composition may have important climatic effects as it grows larger. For example, it seems possible that in the summer, when zonal winds are weak, continental regions may tend partly toward their equilibrium climate, thus causing a relatively greater warming in that season. Also, in examining the climate effects of recent and future large volcanoes, such as the

1982 El Chichon eruption, the cooling effect of stratospheric aerosols must be compared to the warming by trace gases which have not yet achieved their equilibrium effect; it is not obvious that a global cooling of several tenths of a degree (Robock, 1983) should actually be expected. These problems should be studied by using a global model in which the atmospheric composition changes with time in accord with measurements, and in which the atmosphere, land and ocean each have realistic response times.

Summary

Climate Sensitivity Inferred from 3-D Models

Our analysis of climate feedbacks in 3-D models points strongly toward a net climate feedback factor of f ~ 2-4 for processes operative on 10-100 year time scales. The water vapor and sea ice feedbacks, which are believed to be reasonably well understood, together produce a feedback f ~ 2. The clouds in our model produce a feedback factor ~1.3, increasing the net feedback to f = 3-4 as a result of the nonlinear way in which feedbacks combine.

Present information on cloud processes is inadequate to permit confirmation of the cloud feedback. However, some aspects of the cloud changes in the model which contribute to the positive feedback appear to be realistic, e.g., the increase in tropical cirrus cloud cover and the increase of mean cloud altitude in conjunction with more penetrating moist convection in a warmer climate. It seems likely that clouds provide at least a small positive feedback. More realistic cloud modeling, as verified by detailed global cloud observations, is crucial for improving estimates of climate sensitivity based on climate models.

Climate Sensitivity Inferred from Paleoclimate Data

Analysis of the processes contributing to the cooling of the last ice age shows that feedbacks provide most of the cooling. The paleoclimate studies serve as proof of the importance of feedback processes and permit quantitative evaluation of the magnitude of certain feedbacks. The CLIMAP data allow us to evaluate individually the magnitudes of the land ice (f ~ 1.2-1.3) and sea ice (f ~ 1.2) feedbacks for the climate change from 18K to today, and to establish that the vegetation feedback was smaller but significant (f ~ 1.05-1.1).

We obtain an empirical estimate of f = 2.5-5 for the fast feedback processes (water vapor, clouds, sea ice) at 18K by assuming that the major radiative feedback processes have been identified (as seems likely from consideration of the radiative factors which determine the planetary energy balance with space) and grouping the slow or specified changes of the ice sheets and CO_2 as the principal climate forcings. This estimate for the fast feedback factor is consistent with the feedback in our 3-D model experiments, providing support that the model sensitivity is of the correct order.

The strength of the feedback processes at 18K implies that only relatively small climate forcings or fluctuations are needed to cause glacial to interglacial climate change. We do not try to identify the sequence of mechanisms of the glacial to interglacial changes, but it seems likely that both the direct effect of solar radiation (Milankovitch) changes on the planetary energy balance and induced changes of atmospheric composition, especially CO_2, are involved.

Climate Sensitivity Inferred from Recent Temperature Trends

The temperature increase believed to have occurred in the past 130 years (~0.5°C) implies a climate sensitivity 2.5-5°C for doubled CO_2 (f = 2-4), if (1) added greenhouse gases are responsible for the temperature increase, (2) the 1850 CO_2 abundance was 270±10 ppm, and (3) the heat perturbation is mixed like a passive tracer in the ocean. This technique inherently yields a broad range for the inferred climate sensitivity, because the response time for the ocean increases with increasing climate sensitivity.

Thus the 3-D climate model, the 18K study and the empirical evidence from recent temperature trends yield generally consistent estimates of climate sensitivity. Our best estimate of the equilibrium climate sensitivity for processes occurring on the 10-100 year time scale is a global mean warming of 2.5-5°C for doubled CO_2.

Transient Climate Response

The rate at which the ocean surface can take up or release heat is limited by the fact that feedbacks come into play in conjunction with climate change, not in conjunction with climate forcing. Thus the (isolated) ocean mixed layer thermal relaxation time, commonly taken as 3-5 years, must be multiplied by the feedback factor f. This, in turn, allows the thermal inertia of deeper parts of the ocean to be effective. If the equilibrium climate sensitivity is ~4°C for doubled CO_2 and if small heat perturbations behave like observed passive tracers, the response time of surface temperature to a change of climate forcing is of order 100 years. If the equilibrium sensitivity is 2.5°C, this response time is about 40 years.

We conclude, based on the long surface temperature response time, that there is a large growing gap between current climate and the equilibrium climate for current atmospheric composition. Our projections indicate that within a few decades the equilibrium global temperature will reach a level well above that which has been experienced by modern man.

Is there a point at which the perturbation of surface climate will be large enough to substantially affect the rate of exchange of heat between the mixed layer and deeper ocean, possibly causing a rapid trend toward the equilibrium climate? This question is similar to one asked by Representative Gore (1982): "Is there a point where we trigger the dynamics of this (greenhouse) process, and if so, when do we reach that stage?". With present understanding of the climate system, particularly physical oceanography, we can not answer these questions.

Appendix A: Influence of Feedbacks on Transient Reponse

Consider a planet for which the absorbed fraction of incident solar radiation (1 minus albedo) is a linear function of the temperature, say $x + yT$. If the planet emits as a blackbody its temperature is determined by the condition of radiative equilibrium

$$s_o a_o = \sigma T_o{}^4 \tag{A1}$$

with s_o the mean solar irradiance and $a_o = x+yT_o$.

Now suppose the solar irradiance changes suddenly by a small amount Δs. At the new equilibrium

$$(s_o + \Delta s)(a_o + \Delta a_{eq}) = \sigma(T_o + \Delta T_{eq})^4 \quad . \tag{A2}$$

Neglecting second order terms (since $\Delta s << s_o$) and using (A1) yields

$$\Delta s a_o + s_o \Delta a_{eq} = 4\sigma T_o{}^3 \Delta T_{eq} \quad . \tag{A3}$$

If there were no feedbacks ($\Delta a_{eq} = 0$), the temperature change at equilibrium would be

$$\Delta T_{eq}(\text{no feedbacks}) \equiv \Delta T_o = \frac{\Delta s a_o}{4\sigma T_o{}^3} \quad . \tag{A4}$$

Thus we can rewrite (A3) as

$$\Delta T_{eq} = \Delta T_o + g\Delta T_{eq} \tag{A5}$$

where

$$g = \frac{s_o \Delta a_{eq}}{4\sigma T_o{}^3 \Delta T_{eq}} = \frac{s_o y}{4\sigma T_o{}^3} \tag{A6}$$

Using the relation between gain g and feedback factor f, $f=1/(1-g)$, equation (A5) becomes

$$\Delta T_{eq} = f\Delta T_o \tag{A7}$$

i.e., the equilibrium temperature change exceeds the no-feedback equilibrium temperature change by the factor f.

The heat flux into the planet as a function of time is

$$\begin{aligned} F &= \Delta s a_o + s_o \Delta a - 4\sigma T_o{}^3 \Delta T \\ &= 4\sigma T_o{}^3 (\Delta T_o + g\Delta T - \Delta T) \\ &= \frac{4\sigma T_o{}^3 \Delta T_o}{\Delta T_{eq}} (\Delta T_{eq} - \Delta T) \\ &= \frac{F_o}{\Delta T_{eq}} (\Delta T_{eq} - \Delta T) \end{aligned} \tag{A8}$$

where

$$F_o = 4\sigma T_o{}^3 \Delta T_o = \Delta s a_o \tag{A9}$$

is the flux into the planet at $t = 0$ (i.e., when $\Delta T = 0$). Thus the initial rate of warming is independent of the feedbacks.

The temperature of the planet as a function of time is determined by the equation

$$\frac{dcT}{dt} = \frac{dc\Delta T}{dt} = F \tag{A10}$$

where c is the heat capacity per unit area. If c is constant (e.g. a mixed layer without diffusion into deeper layers), the solution is

$$\Delta T = \Delta T_{eq}(1 - \exp(-t/\tau)), \tag{A11}$$

$$\tau = f \frac{c}{4\sigma T_o{}^3} = f\tau_b \tag{A12}$$

where τ_b is the no-feedback e-folding time [Equation (22)].

Finally, note that these results are much more general than the specific mechanism we chose for the feedback, which was only used as a concrete example. It is apparent from the above that the only assumption required is the linearization of the feedback as a function of temperature.

Acknowledgements. This work depended on essential scientific contributions from several people, in particular: K. Prentice applied the Koppen classification scheme to the control and 18K simulations, H. Brooks and L. Smith derived the 18K monthly sea surface temperatures and sea ice distributions from the February and August CLIMAP data, S. Lebedeff developed the chlorofluorocarbon scenarios, F. Abramopoulos performed calculations with the 1-D ocean model, D. Peteet provided advice and references on paleoclimate vegetation, and L.C. Tsang compiled mixed layer depths from ocean data tapes. We thank R.

Dickinson, L. Ornstein, W. Ruddiman, S. Schneider and C. Wunsch for critically reviewing the manuscript, J. Mendoza and L. DelValle for drafting the figures, and A. Calarco and C. Plamenco for typing several versions of the manuscript. Our climate model development was supported principally by the NASA Climate Program managed by Dr. Robert Schiffer; the applications to CO_2 studies were supported by a grant in 1982-1983 from EPA, for which we are indebted especially to John Hoffman, John Topping and Joseph Cannon.

References

Alexander, R.C., and R.L. Mobley, Monthly average sea-surface temperatures and ice-pack limits on a 1° global grid, Mon. Wea. Rev., 104, 143-148, 1976.

Berger, A.L., Long-term variations of daily insolation and quaternary climatic changes, J. Atmos. Sci., 35, 2362-2367, 1978.

Bode, H.W., Network Analysis and Feedback Amplifier Design, 551 pp., Van Nostrand, New York, 1945.

Broecker, W.S., T.H. Peng and R. Engh, Modeling the carbon system, Radiocarbon, 22, 565-598, 1980.

Bryan, K., F.G. Komro, S. Manabe and M.J. Spelman, Transient climate response to increasing atmospheric carbon dioxide, Science, 215, 56-58, 1982.

Bryan, K., F.G. Komro and C. Rooth, The ocean's transient response to global surface temperature anomalies, Maurice Ewing Series, 5, 1984.

Budyko, M.I., The effect of solar radiation variations on the climate of the earth, Tellus, 21, 611-619, 1969.

Burckle, L.H., D. Robinson and D. Cooke, Reappraisal of sea-ice distribution in Atlantic and Pacific sectors of the Southern Ocean at 18,000 yr BP, Nature, 299, 435-437, 1982.

CDAC, Changing Climate, Report of the Carbon Dioxide Assessment Committee, 496 pp., National Academy Press, Washington, D.C., 1983.

Cess, R.D., Radiative transfer due to atmospheric water vapor: global considerations of the earth's energy balance, J. Quant. Spectrosc. Radiat. Transfer, 14, 861-871, 1974.

Cess, R.D., Biosphere-albedo feedback and climate modeling, J. Atmos. Sci., 35, 1765-1769, 1978.

Cess, R.D. and S.D. Goldenberg, The effect of ocean heat capacity upon global warming due to increasing atmospheric carbon dioxide, J. Geophys. Res., 86, 498-502, 1981.

CLIMAP Project Members, A. McIntyre, project leader: Seasonal reconstruction of the earth's surface at the last glacial maximum, Geol. Soc. Amer., Map and Chart Series, No. 36, 1981.

Craig, H., and C.C. Chou, Methane: the record in polar ice cores, Geophys. Res. Lett., 9, 1221-1224, 1982.

Denton, G.H. and T.J. Hughes, The Last Great Ice Sheets, 484 pp., John Wiley, New York, 1981.

Dickinson, R.E., Vegetation albedo feedbacks, Maurice Ewing Series, 5, 1984.

DiLabio, R.N.W. and R.A. Klassen, Book review of The Last Great Ice Sheets, Bull. Amer. Meteorol. Soc., 64, 161, 1983.

Ehhalt, D.H., R.J. Zander and R.A. Lamontagne, On the temporal increase of tropospheric CH_4, J. Geophys. Res., 88, 8442-8446, 1983.

Flenley, J., The Equatorial Rain Forest: A Geological History, 162 pp., Butterworths, London, 1979.

Gates, W.L., Modeling the Ice-Age Climate, Science, 191, 1138-1144, 1976.

Gordon, A.L., Southern Ocean Atlas, 34 pp., 248 plates, Columbia Univ. Press, New York, 1982.

Gore, A., Record of Joint Hearing on Carbon Dioxide and Climate: The Greenhouse Effect, March 25, 1982, Science and Technology Committee (available from Goddard Institute for Space Studies, New York, 10025), 1982.

Hansen, J., D. Johnson, A. Lacis, S. Lebedeff, P. Lee, D. Rind and G. Russell, Climate impact of increasing atmospheric carbon dioxide, Science, 213, 957-966, 1981.

Hansen, J., D. Johnson, A. Lacis, S. Lebedeff, P. Lee, D. Rind and G. Russell, Climatic effects of atmospheric carbon dioxide, Science, 220, 873-875, 1983a.

Hansen, J., A. Lacis and S. Lebedeff, in Carbon Dioxide Review 1982, Oxford Univ. Press, New York, 284-289, 1982.

Hansen, J., A. Lacis, D. Rind and G. Russell, Climate sensitivity to increasing greenhouse gases, in Sea Level Rise to the Year 2100, Van Nostrand, New York, 1984.

Hansen, J., G. Russell, D. Rind, P. Stone, A Lacis, S. Lebedeff, R. Ruedy and L. Travis, Efficient three-dimensional global models for climate studies: models I and II, Mon. Wea. Rev., 111, 609-662, (paper 1), 1983b.

Hoffert, M.I., A.J. Callegari and C.T. Hsieh, The role of deep sea heat storage in the secular response to climatic forcing, J. Geophys. Res., 85, 6667-6679, 1980.

Hunt, B.G. and N.C. Wells, An assessment of the possible future climatic impact of carbon dioxide increases based on a coupled one-dimensional atmospheric-oceanic model, J. Geophys. Res., 84, 787-791, 1979.

Keeling, C.D., R.B., Bacastow and T.P. Whorf, in Carbon Dioxide Review 1982, ed. W.C. Clark, Oxford Univ. Press, New York, 377-385, 1982.

Kellogg, W.W. and S.H. Schneider, Climate stabilization: for better of worse?, Science, 1163-1172, 1974.

Koppen, W., Das geographische system der klimate, in Handbuch der Klimatologie 1, part C. ed. W. Koppen and G. Geiger, Boentraeger, Berlin, 1936.

Lacis, A., J. Hansen, P. Lee, T. Mitchell and S. Lebedeff, Greenhouse effect of trace gases, 1970-1980, Geophys. Res. Lett., 8, 1035-1038, 1981.

Levitus, S., Climatological Atlas of the World Ocean, NOAA Prof. Paper, No. 13, U.S. Government Printing Office, Washington, D.C., 1982.

Lian, M.S. and R.D. Cess, Energy-balance climate models: a reappraisal of ice-albedo feedback, J. Atmos. Sci., 34, 1058-1062, 1977.

Manabe, S., Carbon dioxide and climate change, Adv. Geophys., 25, 39-82, 1983.

Manabe, S. and D.G. Hahn, Simulation of the tropical climate of an ice age, J. Geophys. Res., 82, 3889-3911, 1977.

Manabe, S., J. Smagorinsky and R.F. Strickler, Simulated climatology of a general circulation model with a hydrologic cycle, Mon. Wea. Rev., 98, 175-212, 1965.

Manabe, S. and R.J. Stouffer, Sensitivity of a global climate model to an increase of CO_2 con centration in the atmosphere, J. Geophys. Res., 85, 5529-5554, 1980.

Manabe, S. and R.T. Wetherald, Thermal equilibrium of the atmosphere with a given distribu tion of relative humidity, J. Atmos. Sci., 24, 241-259, 1967.

Manabe, S. and R.T. Wetherald, The effects of doubling the CO_2 concentration on the climate of a general circulation model, J. Atmos. Sci., 32, 3-15, 1975.

Manabe, S. and R.T. Wetherald, On the distribution of climate change resulting from an increase of CO_2 content of the atmosphere, J. Atmos. Sci., 37, 99-118, 1980.

Matthews, E., Global vegetation and land use: new high-resolution data bases for climate studies, J. Clim. Appl. Meteorol., 22, 474-487, 1983.

Miller, J.R., G.L. Russell and L.C. Tsang, Annual oceanic heat transports computed from an atmospheric model, Dynam. Atmos. Oceans, 7, 95-109, 1983.

NAS, Understanding Climatic Change, 239 pp., National Academy of Sciences, Washington, D.C., 1975.

Neftel, A., H. Oeschger, J. Swander, B. Stauffer and R. Zumbrunn, Ice core sample measurements give atmospheric CO_2 content during the past 40,000 years, Nature, 295, 220-223, 1982.

NOAA, User's Guide to NODC Data Services, Environmental Data Service, Washington, D.C., 1974.

North, G.R., Theory of energy-balance climate models, J. Atmos. Sci., 32, 2033-2043, 1975.

Ostlund, H.G., H.G. Dorsey and R. Brescher, GEOSECS Atlantic radiocarbon and tritium results, Univ. Miami Tritium Laboratory Data Report No. 5, 1976.

Robock, A., The dust cloud of the century, Nature, 301, 373-374, 1983.

Sarnthein, M., Sand deserts during glacial maximum and climatic optimum, Nature, 272, 43-46, 1978.

Schneider, S.H. and S.L. Thompson, Atmospheric CO_2 and climate: importance of the transient response, J. Geophys. Res., 86, 3135-3147, 1981.

Sellers, W.D., A global climate model based on the energy balance of the earth-atmosphere system, J. Appl. Meteorol., 8, 392-400, 1969.

Shackleton, N.J., M.A. Hall, J. Line and C. Shuxi, Carbon isotope data in core V19-30 confirm reduced carbon dioxide concentration in the ice age atmosphere, Nature, 306, 319-322, 1983.

Street, F. and A. Grove, Global maps of lake-level fluctuations since 30,000 yr BP, Quat. Res., 12, 83-118, 1979.

Stone, P.H., Feedbacks between dynamical heat fluxes and temperature structure in the atmosphere, Maurice Ewing Series, 5, 1984.

Vuilleumier, B., Pleistocene changes in the fauna and flora of South America, Science, 173, 771 780, 1971.

Walsh, J. and C. Johnson, An analysis of Arctic sea ice fluctuations, J. Phys. Ocean., 9, 580-591, 1979.

Wang, W.C. and P.H. Stone, Effect of ice-albedo feedback on global sensitivity in a one-dimensional radiative-convective climate model, J. Atmos. Sci., 37, 545-552, 1980.

Webster, P.J. and N.A. Streten, Late Quaternary ice age climates of tropical Australia: interpretations and reconstructions, Quat. Res., 10, 279-309, 1978.

Weiss, R.F., The temporal and spatial distribution of tropospheric nitrous oxide, J. Geophys. Res., 86, 7185-7195, 1981.

Wetherald, R.T. and S. Manabe, The effects of changing the solar constant on the climate of a general circulation model, J. Atmos. Sci., 32, 2044-2059, 1975.

Williams, J., R.G. Barry and W.M. Washington, Simulation of the atmospheric circulation using the NCAR global circulation model with ice age boundary conditions, J. Appl. Meteorol., 13, 305-317, 1974.

WMO, World Meteorological Organization project on research and monitoring of atmospheric CO_2, Report no. 10, R.J. Bojkov ed., Geneva, 42 pp., 1983.

ON THE TRANSIENT RESPONSE PATTERNS OF CLIMATE TO TIME DEPENDENT CONCENTRATIONS OF ATMOSPHERIC CO_2

G. R. North, J. G. Mengel* and D. A. Short

Goddard Laboratory for Atmospheric Science, NASA/Goddard Space Flight Center
Greenbelt, MD 20771, *Applied Research Corporation

Abstract. A one level two dimensional energy balance model with explicit land-sea geography is tuned to fit the seasonal cycle and then applied to various transient response phenomena. Since the model is linear it can be forced by a sequence of frequencies and the results applied to the construction of the response to any waveform (repeating scenario). The familiar example of step function increase in CO_2 concentration leads to highly nonuniform response in the first few years but a rather homogeneous response after about one decade. It is argued that although this model only includes a mixed layer ocean, the mechanics of the damped diffusion mechanism would be such as to produce a homogeneous response even when a lagging deep ocean is present.

Introduction

Several recent papers have presented studies of the transient response of climate to time dependent forcings such as the CO_2 loading of the atmosphere (Hoffert et al., 1980; Schneider and Thompson, 1981; North et al., 1982a,b). Some argument has arisen over the importance of the transient behavior in view of the extraordinary cost of running coupled atmosphere-ocean general circulation models.

To date the only such model to be run employs a highly simplified (beach ball) geography with seashores running along meridians and with a fifty percent distribution of land at each latitude (Bryan et al., 1982). For a step function increase in CO_2 concentration, the model showed a rapid response (≅ 5 years) to the perturbation with the tropical latitudes leading; but after 10 or so years, the latitude dependence tended to wash away leaving a long term uniform response. The main conclusion was that cheaper equilibrium runs could be made if the time scales of the forcing exceed about 10 years.

Based upon a zonal average energy balance model (Schneider and Thompson, 1981), Thompson and Schneider (1982) have suggested that the simplified land-sea distribution of Bryan et al. may be misleading and that even for longer times (decades) the strong latitudinal asymmetries may persist.

We have recently developed a new energy balance model (North et al., 1982a) which may be useful in shedding light upon the present controversy. Our results cannot be definitive, since our very simple model neglects features present in each of the studies cited. On the other hand, our model does attempt to incorporate the full two dimensional geography of the planet, a complication not included in previous works. The present version of our model only includes a constant depth mixed layer (copper plate) ocean with no deeper component, hence the longest time scale in the system is about 5 years. The system responds freely on this time scale and is not constrained by deeper water as in the other studies. If the deeper water has a time scale of hundreds of years our result may still be valid, but for now our conclusions must be considered preliminary to the studies presently underway by several groups.

In this paper we shall present a brief description of our model and describe how it can be easily solved for transient response studies by a linear analysis method. The results include the geographical pattern of the response to various periodic forcings and to the step function forcing. A theoretical discussion is then presented on the possible effects of deeper ocean lagging upon the surface response distributions in diffusive models. Finally the results are summarized and implications discussed.

Model

The model is a two dimensional one level energy balance model described in detail in North et al. (1982a). It is a two dimensional version of a seasonal model, developed earlier by North and Coakley (1979) and also discussed in the review article by North et al. (1981). The model is

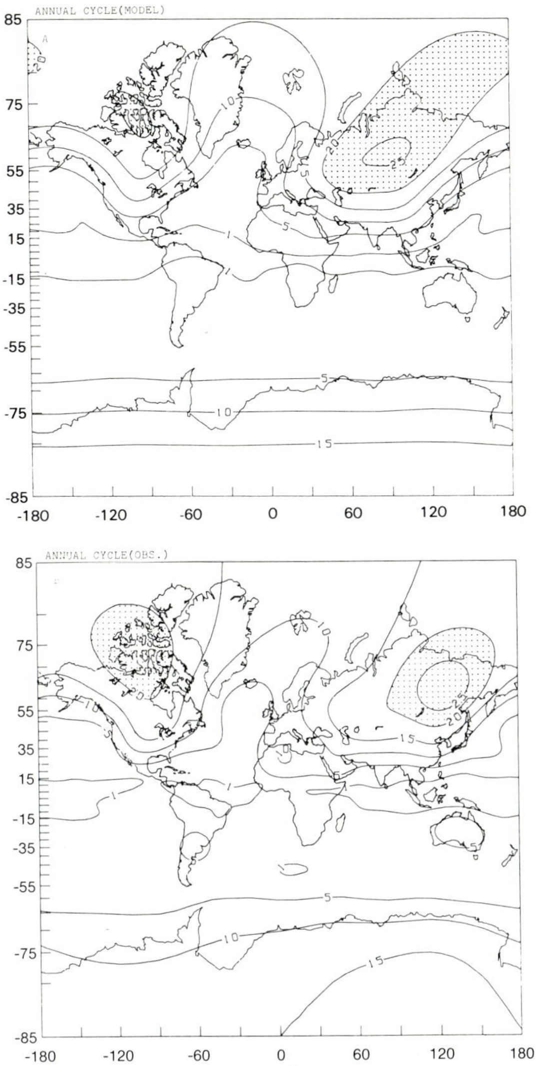

Fig. 1. (a) Amplitude (^{o}C) of the annual harmonic in the model, truncated at spherical harmonic degree 11. (b) Amplitude (^{o}C) of the annual harmonic in the observations. Smoothed by expanding into spherical harmonics and truncating at degree 11, the same resolution as retained in the model. Taken from North et al., 1982a.

defined by the energy-balance equation:

$$C(\hat{r}) \frac{\partial T}{\partial t}(\hat{r},t) - \nabla\cdot(D(x)\nabla T(\hat{r},t)) + A + BT(\hat{r},t) = QS(x,t)a(x) + A_1(t) \qquad (1)$$

where

$\hat{r}$ designates a point on the globe

x = sine of latitude

t = time in years.

$T(\hat{r},t)$ = sea level temperature

Q = solar constant ÷ 4

$C(\hat{r})$ = location dependent effective heat capacity/area (small over land, intermediate over perennial sea ice, large over water in the ratio 1: 9: 60).

$D(x) = 0.81\,(1 - 1.33P_2(x) + 0.67P_4(x))$ a latitude dependent isotropic macro-turbulent thermal diffusion coefficient expanded in Legendre polynomials.

A, B = empirically derived outgoing longwave radiation coefficients (203 W m^{-2}, 2.09 W $m^{-2}{}^{o}C^{-1}$)

S(x,t) = Seasonal distribution of daily averaged sunlight reaching the top of the atmosphere at latitude x.

$a(x) = 0.69 - 0.12P_2(x)$, the coalbedo of the earth-atmosphere system.

$A_1(t)$ = the time dependent external forcing. The numerical values of the parameters in the above definitions are discussed in North et al. (1982a). The physical significance of the various terms in (1) need not be recounted here. The model as given by (1) is a linear system forced by the time dependent terms on the right hand side of the equation.

The model may be solved by expanding $C(\hat{r})$, $T(\hat{r},t)$, and S(x,t) into spherical harmonics and truncating at degree 22, 11, and 11 respectively. The result is a set of 144 coupled linear first order equations in time. If the forcing (RHS) is periodic we can insert each frequency component and solve the resulting set of algebraic equations. As this can be done for any frequency independently, the result for any periodic forcing can be found by the principle of superposition. This is the main technique we shall employ in the present study. A similar use was made with this model in the case of a variable "solar constant" (North et al., 1982b).

As an example consider the seasonal cycle with no other forcing ($A_1(t) = 0$). The factor S(x,t) can be represented with sufficient accuracy (North and Coakley, 1979) by a zero, annual, and semiannual harmonic. The seasonal cycle is thus easily recovered by the procedure just described. It is for this cycle that the values of constants in D(x) and $C(\hat{r})$ have been determined. In this fitting or tuning procedure ~6 parameters are adjusted while 864 amplitudes are computed. We plan to discuss the analysis of errors elsewhere. For present purposes we present the amplitude and phase lag of the annual cycle along with similarly smoothed maps of the observed temperature fields in Figures 1 and 2. The fit with the observational data suggests to us that the model might have some limited role to play in the transient response problem.

Several comments are in order about the

tuning procedure and the uniqueness of parameters. We found that the same level of fit could be obtained for a range of thermal inertia over the mixed layer (C_W) since its main effect is to suppress the seasonal cycle and to force the phase lag over water to be about 1/4 cycle if $C_W/B >> 1$. Our value of $C_W/B = 4.7$ years is somewhat arbitrarily based upon a 75m deep mixed layer. This oceanic radiative relaxation time is not well determined and over some limited range the units of time may be rescaled for easy comparison with other works. Also the value of thermal inertia over land, C_L, ($C_L/B \cong 0.08$ years) can be changed over a range of a factor of two provided a compensating change in D_O is made without significantly affecting the match with observations of the seasonal cycle. In our previous studies (North et at., 1982a) we examined the model's response to external forcing over a similar range of parameter values and found that our (qualitative) conclusions were unaffected. We expect that to remain the case in the present study. Finally, it is necessary to remind the reader that energy-balance models are extremely simplified descriptors of the climate system. Some of the terms in (1) are hardly more than schematic representations of the actual fluxes.

Apparently the seasonal cycle (and perhaps other transient phenomena) is so dominated by the heat balance and the continental distribution that qualitative agreement with observations occurs. By studying such models one hopes to gain insight into the more complex system and perhaps think of better experiments to be performed with more realistic models.

Transient Response

The A_1 (t) term of (1) will be used as a forcing agent for the transient response experiments. This term may be thought of as part of the A + BT terrestrial radiation term and is the term affected by changes in the CO_2 content of a column of air (Augustsson and Ramanathan, 1977). Schneider and Thompson (1981) also allow a temperature dependence of the CO_2 affected term, which in our terminology would add a contribution to B as well as A. We shall choose not to alter B for simplicity. Rather we will consider changes in the parameter A through the term $A_1(t)$. The model's static response to a change in A is

$$\Delta T = \frac{\Delta A_1}{B} \tag{2}$$

as a standard we shall take ΔA_1 (= 2.00 W m^{-1}) to be such that in this case $\Delta T \cong 1°C$. Such a $2\Delta A_1$, corresponds roughly to a doubling of the atmospheric CO_2 content.

Before considering the full two dimensional problem with land-sea geography, consider a homogeneous planet, $C(\hat{r})$ = constant. Also take $A_1(t)$ to be sinusoidal with frequency f (cycles/year).

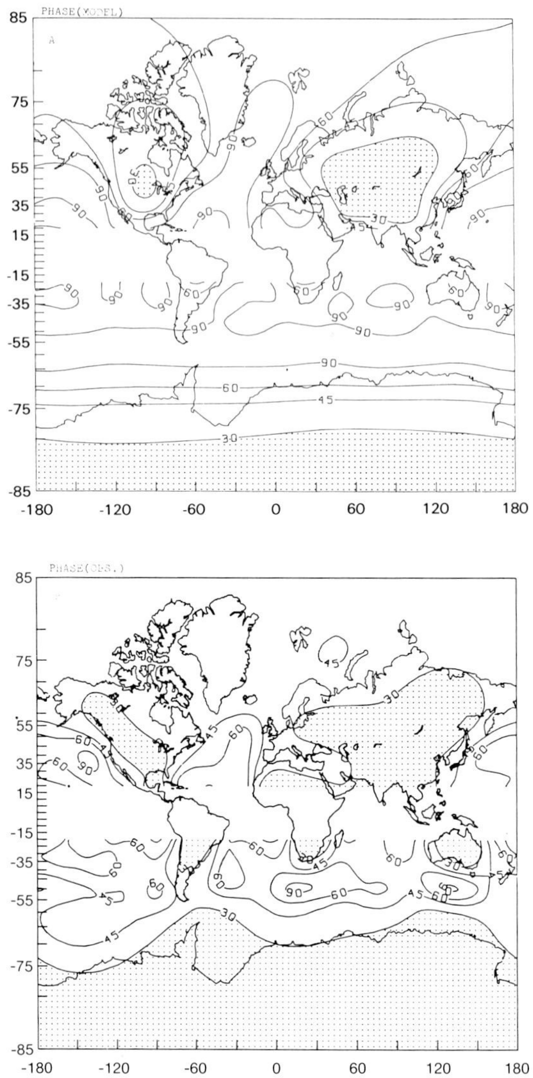

Fig. 2. (a) The phase lag (days) of the thermal response in the first harmonic for the model. Contours are suppressed in the ± 20° latitude band because of the correspondingly small amplitude in that region. (b) The phase lag (days) of the thermal response in the first harmonic for the observations. Contours are suppressed in the ± 20° latitude band because of the correspondingly small amplitude in that region.

We may integrate (1) over the globe to obtain

$$C \frac{dT_o}{dt} + BT_o = A_1 \, e^{2\pi ift} \tag{3}$$

since no other spherical harmonic modes are excited by this forcing, the planet responds equally at all locations sinusoidally with an amplitude

$$\Delta T_f = \frac{A_1/B}{[(2\pi f\tau)^2 + 1]^{1/2}} \tag{4}$$

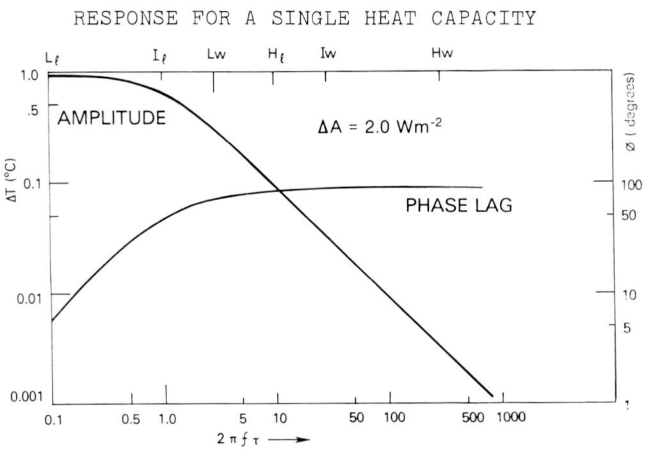

Fig. 3. The amplitude (°C) and phase lag (degrees) for the simple linear global model with time characteristic τ forced at frequency f.

and phase lag

$$\phi = \tan^{-1}(2\pi f\tau) \tag{5}$$

where the characteristic radiative relaxation time is $\tau = C/B$ (4.7 years for mixed layer ocean, 0.08 years for an all land planet). Figure 3 shows plots of (4) and (5) with several relevant frequencies indicated on the abscissa.

Any time dependent forcing function $A_1(t)$ can be expanded

$$A_1(t) = \sum_{n=-\infty}^{\infty} a_n \exp\left(\frac{2\pi i n t}{p}\right) \tag{6}$$

where p is a period large compared to τ. The solution to the generalized version of (3) is

$$T_o(t) = \sum_{n=-\infty}^{\infty} \frac{a_n \exp(2\pi i n t/p)}{2\pi i n/p + B} \quad . \tag{7}$$

In other words, we must solve (3) for each frequency ± 1/p, ± 2/p, ± 3/p, up to some suitable cutoff and then recompose $T_o(t)$. The resulting $T_o(t)$ will also have period p. If $p >> \tau$ we can use the procedure for studying transient behavior of the system in much the same way an engineer studies transient behavior of an electrical circuit with a wave generator (forcing) and an oscilloscope (response).

Now consider the complete system (1) with landsea geography. Because of the r dependence of C(r) many spherical harmonics will be excited, even by the spatially uniform $A_1(t)$. The seasonal cycle is suppressed with no loss of generality. As a preliminary experiment consider forcing the system with a single frequency

$$A_1(t) = \Delta A_1 \, e^{2\pi i f t} \tag{8}$$

where $\Delta A_1 = 4.18$ W m^{-2}, the value which would lead to $\Delta T \rightarrow 2°C$ as $f \rightarrow 0$. Figures 4, 5 show the amplitude of the response for periods of 2 months

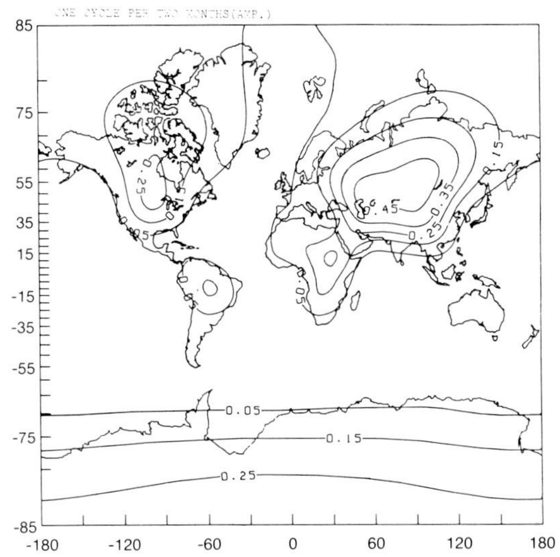

Fig. 4. The linear climate model's response amplitude to a sinusoidal forcing of period 2 months. The amplitude of the forcing (4.18 W m^{-2}) is such that the result would be $\Delta T = 2°C$ uniformly for infinite period.

and one year respectively. Figures 6 and 7 show the amplitude and phase lag for forcing with a frequency of 1/24 cycles/yr. Note that as expected the geographical dependence washes out as $f \rightarrow 0$. Clearly if the forcing is dominated by periods 5 times the oceanic response time the amplitude is ~ 80% of the static response fairly uniformly distributed over the globe. For most of the globe the phase lag is between 20 and 40 degrees (1.33 and 2.66 years). This suggests that for most CO_2 change scenarios envisaged we can be quite satisfied with equilibrium results provided interactions with deep oceanic layers can be neglected (considered in the next section).

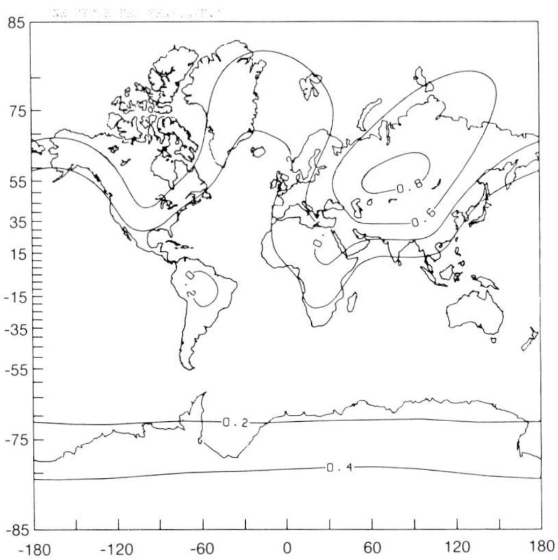

Fig. 5. Same as Figure 4 except $f = 1$ $year^{-1}$.

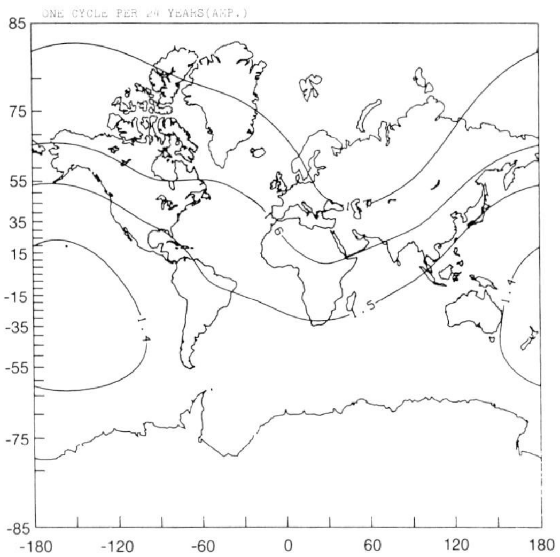

Fig. 6. Same as Figure 4 except f = 24 $year^{-1}$.

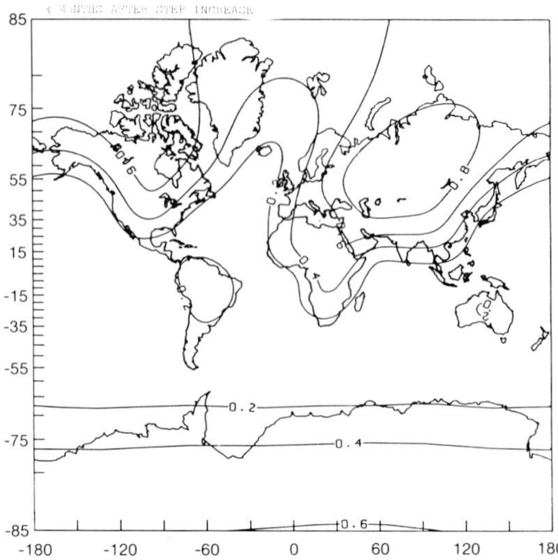

Fig. 8. Temperature field (°C) at Δt = 3 months after onset of the step function increase in CO_2. Equilibrium response is a uniform 2°C.

For comparison with other works we have forced the model with a square wave $A_1(t)$ (jump = 4.18 W m^{-2}) with full period 24 years. This period gives the response 12 years to die out before the next onset phase is applied. Figures 8, 9, 10, 11 show the response after the step warming occurs at intervals of 3 months, 1 year, 5 years and 10 years respectively. Note the similarity of Figures 10 and 6. Apparently 5 to 10 years after onset geographical dependence washes out as expected from the results of forcing at a low frequency. Figure 12 shows how the zonal average temperature responds as a function of latitude for different times after onset, for comparison with Bryan et al. (1982). As Thompson and Schneider (1982) point out, the simplified geography employed by Bryan et al. leads to quite a different latitude dependence of the response at different times after onset than our results. Nevertheless, we agree that such detailed effects tend to wash out after less than a decade. In the next section we present arguments that suggest the spatial uniformization of all low frequency response patterns.

Homogenization of Response Patterns at Low Frequency

Although our model does not contain a deep oceanic layer with a long time constant, we wish to suggest that in models of the same class (i.e., employing similar transport and radiation rules) such low frequency lagging parts do not contribute significantly to nonhomogeneity of the

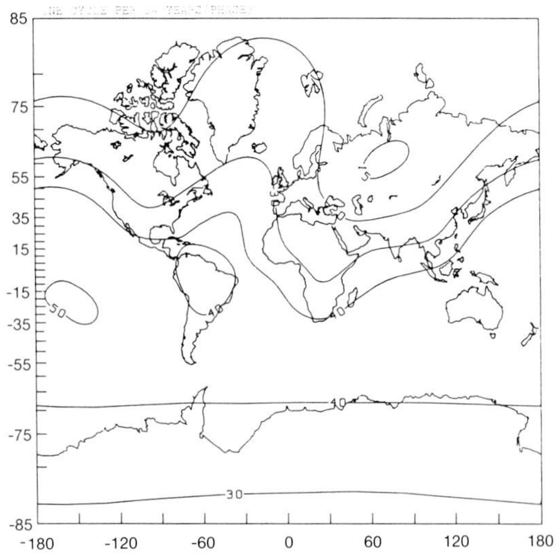

Fig. 7. The phase lag in degrees (24 year = 360°) for a forcing frequency of 24 $year^{-1}$. For reference, 40° = 2.67 years.

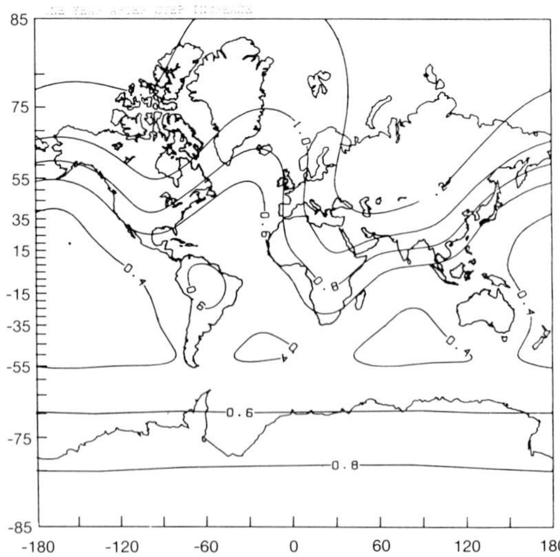

Fig. 9. Same as Figure 8 except Δt = 1 years.

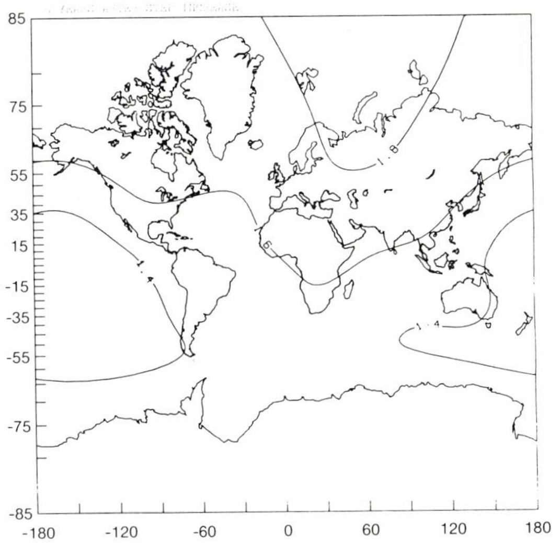

Fig. 10. Same as Figure 8 except Δt = 5 years.

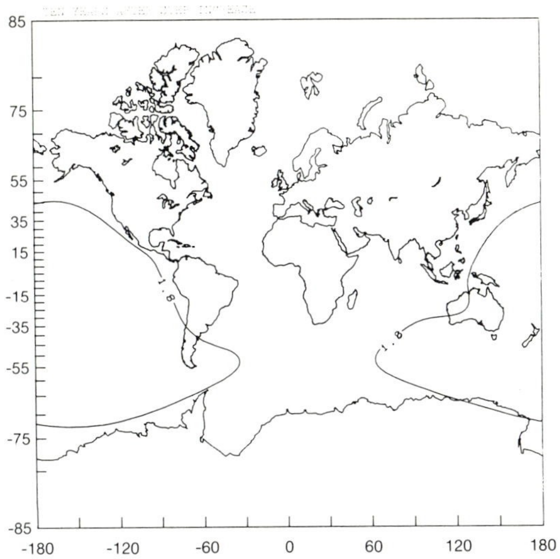

Fig. 11. Same as Figure 8 except Δt = 10 years.

geographical response patterns. As an analog to the system we construct a simple one dimensional diffusive model with earthlike parameters and show that localized low frequency disturbances in the energy balance propagate over a considerable portion of the globe before they are damped by the radiation.

Consider an energy-balance model governed by

$$C \frac{\partial T}{\partial \tau} - D \frac{\partial^2 T}{\partial x^2} + BT = f(x,t) \qquad (9)$$

where B, C, and D are taken as constant and given. The function f(x,t) is localized in space at x = 0 and is periodic in time

$$f(x,t) = f_o\, \delta(x)\, e^{i\omega t} \quad . \qquad (10)$$

Thermal response will propagate diffusively away from the origin both left and right with a damping due to the outgoing radiation. Our goal is to find the characteristic penetration length ℓ as a function of frequency and the other constants in the problem. In particular we wish to know if low frequency imbalances in the deep ocean are felt in the continental interiors. If the penetration length is large compared to a continental dimension we can infer homogenization. The mathematical problem is similar to that of an electromagnetic wave impinging upon a conducting medium.

Consider the trial solution to (9):

$$T(x,t) = \psi(x)\, e^{i\omega t} \qquad (11)$$

whence

$$-D\psi'' + (i\omega c + B)\psi = 0, \; x \neq 0 \qquad (12)$$

and using $\psi \sim e^{-\lambda x}$ we obtain

$$\lambda^2 = \frac{B}{D}(1 + i\omega\tau) \qquad (13)$$

where $\tau = C/B$, is the radiative relaxation time. For $x > 0$ we require the root of (13) with positive real part, which for small $\omega\tau$ is

$$\lambda_+ \cong \frac{B}{D}\left(1 + \frac{i\omega\tau}{2} + \frac{\omega^2\tau^2}{8} + \ldots\right) \quad . \qquad (14)$$

The opposite root is chosen for $x < 0$ and the two solutions can then be matched at the origin and forced to have the appropriate discontinuity in the first derivative so as to satisfy (9) at $x = 0$. In the present case we are simply interested

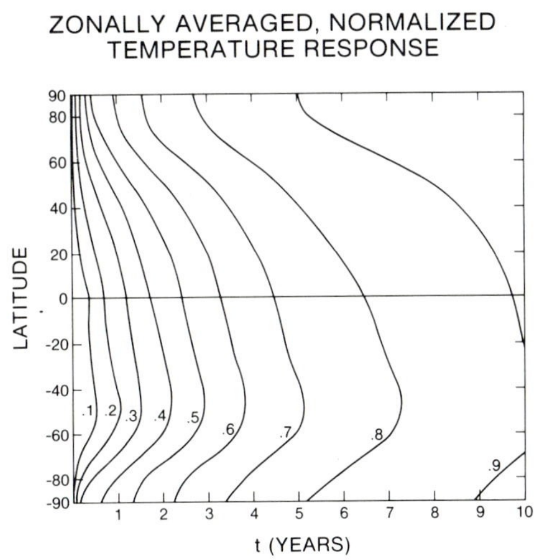

Fig. 12. Response of the model at various latitudes (ordinate) versus time after onset of the step function increase in CO_2. The curves are labeled by the fraction of the equilibrium response. This plot is to be compared with Figure 2b of Bryan et al. (1982), who show a rather different latitudinal dependence apparently due to the differences in geography.

in the decay length,

$$\ell \cong \frac{\sqrt{D/B}}{1+\omega^2\tau^2/8} \tag{15}$$

which is in units of earth radii. The quantity $\sqrt{D/B} \approx 0.70$ and for the characteristic time over land, $\tau \approx 0.10$ years, and if ω, the dominant frequency of the deep oceans is $\sim(2\pi)(.01) \approx .063$ years $^{-1}$ we can neglect the frequency dependence in the denominator of (15) and take ℓ to be 0.70 earth radii.

Note that the oscillitory part of ψ (x) due to the imaginary part of λ_+ is essentially constant in these cases. Based upon this result we might expect that a slowly changing, uniformly cool ocean would transmit its effects into the continental interiors by a distance $\ell \approx 4000$ (± 1000) Km. Thus, the interior of Asia may remain somewhat insulated from such a source, but elsewhere it is likely to be important. As expected, propagation from land into the oceanic interior is much more inhibited at the same frequency because the term

$$\frac{(\omega\tau)^2}{8}$$

in the denominator of (15) is now much larger. It is interesting to look for these effects in the maps already presented. For example, in Figures 3, 4 and 5 as $\omega = 2\pi f$ becomes smaller we can see the homogenization of the response over the globe since the penetration length is increasing to its low frequency limit. Note that even at the lowest frequency (1/24 year^{-1}) in Figure 5 there remains a small bulge over Asia.

The foregoing arguments suggest that the lagging deep layers of the ocean would not contribute much to the inhomogeneity of the global response except possibly in the deep interior of Asia. Naturally, the details of our argument are based upon a diffusive transport mechanism and should not be taken literally. However, the overall success of the model on the annual time scale gives us more than passing confidence in the conclusion.

Conclusions

We have applied a class of very simple energy balance models with diffusive transport to argue that the earth's response to changes in CO_2 concentration will be nearly uniform over the globe. Since our model is linear and employs no ice feedback, we are not concerned here with the well known polar amplification phenomenon but rather with the transient response patterns over land versus over ocean.

By probing the model at several frequencies we can determine its frequency response characteristics. Also by superposition theory we can force the model with any wave form to investigate its transient response. Since the model fits the present seasonal cycle rather well, we argue that it is at least qualitatively useful over a broad range of nearby frequencies. After a step function change in CO_2 we observe rapid response over the continental interiors but homogenization is nearly complete after only one decade as found by Bryan et al. (1982), although our early time response differs from theirs probably because of our geographical detail.

Since the model does not include a deep ocean component we have tried to estimate the effects upon homogeneity due to this low frequency source of imbalance. We have found that for low frequency localized forcing the diffusion mechanism smears the response pattern into continental interiors to a characteristic length of about several thousand kilometers, suggesting that a homogeneous pattern will occur over most of the globe.

These conclusions support the notion that equilibrium computations will be very useful in estimating responses to such phenomena as changes in CO_2 concentration.

References

Augustsson, T., and V. Ramanathan, A radiative-convective model of CO_2 climate problem, J. Atmos. Sci., 34, 448-451, 1977.

Bryan, K., F. G. Komro, S. Manabe, and M. J. Spelman, Transient climate response to increasing atmospheric carbon dioxide, Science, 215, 56-58, 1982.

Hoffert, M. I., A. J. Callegari, and C.-T. Hsieh, The role of deep sea heat storage in the secular response to climatic forcing, J. Geophys. Res., 85, 6667, 1980.

North, G. R., and J. A. Coakley, Differences between seasonal and mean annual energy balance model calculations of climate and climate sensitivity, J. Atmos. Sci., 36, 1189-1204, 1979.

North, G. R., R. F. Cahalan, and J. A. Coakley, Energy balance climate models, Rev. Geophys. and Space Phys., 19, 91-121, 1981.

North, G. R., J. A. Mengel, and D. A. Short, A simple climate model resolving the seasons and the continents: application to the Milankovitch theory of the ice ages, submitted to J. Geophys. Res., 1982a.

North, G. R., J. G. Mengel and D. A. Short, Climatic response to a time varying solar constant in Weather and Climate Responses to Solar Variations, Edited by B. M. McCormac. Colorado Associated University Press, 1983 (in press), 1982b.

Schneider, S. H. and S. L. Thompson, Atmospheric CO_2 and climate: Importance of the transient response, J. Geophys. Res., 86, 3135-3147, 1981.

Thompson, S. L. and S. H. Schneider, Carbon dioxide and climate: The importance of realistic geography in estimating the transient temperature response, Science, 217, 1031-1033, 1982.

SYSTEMATIC ANALYSIS OF CLIMATIC MODEL SENSITIVITY TO PARAMETERS AND PROCESSES

Matthew C. G. Hall and Dan G. Cacuci

Engineering Physics Division, Oak Ridge National Laboratory
Oak Ridge, Tennessee 37830

Abstract. This paper demonstrates the adjoint method of sensitivity analysis on a radiative-convective climate model with a diurnal cycle. A single adjoint calculation, which requires about the same computation time as the original model, suffices to calculate sensitivities of average surface air temperature to all 312 model parameters. The sensitivities accurately predict the effect on average surface air temperature of small variations in the model parameters. Relative sensitivities rank the importance of all the parameters, the most important parameters being those that characterize solar radiation and determine transmission functions. The uncertainties in the model results are expressed formally in terms of all the sensitivities and parameter covariances. For this model, a 10% standard deviation in surface albedo causes the increase in average surface air temperature after doubling CO_2 to have a standard deviation of 0.24 K about the expected value of 1.66 K. For results that cannot readily be compared with observation (for example, the results of a CO_2 doubling experiment), this method of uncertainty analysis is the only systematic way to estimate the reliability of model results.

The radiative-convective model contains complex on-off nonlinear processes of the type found in general circulation models. Therefore, the fact that the adjoint method works successfully and efficiently for the radiative-convective model provides valuable information about subsequent application of the method to general circulation models.

Introduction

An essential part of climatic research consists of interpreting the results of climatic models. Valuable information for this purpose comes from analyzing sensitivities of the results to parameters involved in modeling physical processes. For example, the current concern about the climatic impact of CO_2 stems from the significant sensitivity that climatic models exhibit to the atmospheric concentration of CO_2 (e.g. Manabe and Wetherald, 1980). Also, sensitivities provide valuable insight about the modeling of physical phenomena. For example, the ice-albedo feedback mechanism corresponds to the observed negative sensitivity of surface air temperature to surface albedo. Furthermore, sensitivities quantify the extent that uncertainties in parameters contribute to uncertainties in results of models. For example, subgrid processes need to be parameterized. These parameterizations are highly simplified approximations of complex processes, so the uncertainties in the parameters involved can be large. If the corresponding sensitivities are also large, then the results of the model will have large uncertainties. In fact, it is the need to analyze this type of uncertainty for CO_2 doubling experiments that provides the motivation for this work.

To date, the most common procedure to estimate sensitivities for climatic models has been to change the value of one of the parameters and recalculate the results. For complex models, the large amount of computing time severely restricts the scope of this procedure. Physically meaningful changes in the parameters do not usually shift the climatic mean of the model outside the range of fluctuations of the model's dependent variables, so typically each rerun requires several hours computing time to improve signal-to-noise ratio. In practice this means that the modeler can investigate only a few parameters that he judges a priori to be important, although each rerun does provide sensitivities for all dependent variables everywhere to the one parameter that has been varied.

A way of investigating sensitivities to more parameters is to consider simplified models obtained by reducing the spatial resolution (i.e., number of grid points or wave numbers) of complex three-dimensional models (Held and Suarez, 1975; Aleksandrov and Gates, 1981). Although this makes rerunning less expensive, the parameters must still be selected a priori, and consequently important sensitivities may be missed. Also, reducing the spatial resolution usually requires changes in the way subgrid processes are modeled, so it is not clear whether the respective sensitivities of the simplified and complex models are the same.

Leith (1975) has proposed a method of determining climatic sensitivity based on the fluctuation-dissipation theorem. This theorem relates covariances of the variables in certain statistical systems to the effect that small changes in forcing influences have on the mean value of the variables. But as Leith states, 'the climate ensemble is not a normal equilibrium ensemble as required for the general fluctuation-dissipation theorem to hold exactly'. Recently, Bell (1980) has reported an application of this theorem to a simple model based on the barotropic vorticity equation.

Marchuk (1974) has proposed a method of obtaining sensitivities to boundary conditions for a 'baroclinic model of the atmosphere in the non-adiabatic approximation'. He uses a conjugate system of equations, and expresses perturbations in regional average temperature in terms of the conjugate solution and variations in boundary conditions. His method, though, cannot be applied to models that include nonlinear source terms (e.g. any realistic climatic model), and the method is not designed to calculate sensitivities to parameters other than boundary conditions.

Sensitivity analysis techniques have also been developed in conjunction with applications in other fields. Two major categories of sensitivity analysis techniques are based on direct methods and adjoint methods. These methods yield derivatives of results with respect to parameters.

The direct method (Dickinson and Gelinas, 1975) involves differentiation of the equations describing the model with respect to a parameter. The resulting set of equations is solved for the derivative of all the model variables with respect to that parameter. These derivatives are used to obtain the sensitivity of the result to the parameter. Note that the actual form of the differentiated equations depends on the parameter under consideration. Consequently a different set of equations must be solved to obtain the sensitivity corresponding to each parameter.

Adjoint methods were first developed for sensitivity analysis of specific problems in reactor physics. The mathematical foundations underlying the use of adjoint (and direct) methods have been established by Cacuci (1981a), who presents a general method that can be applied to nonlinear problems. This method starts by expressing a result, customarily called the response, as a functional of the model variables. The sensitivity of this response is defined in terms of a functional derivative. A system of adjoint equations is developed from a differentiated form of the original equations. The sensitivity of the response to all the model parameters can be evaluated in terms of integrals involving the adjoint solution. Note that, in contrast to the direct method, the form of the adjoint equations remains unchanged regardless of the parameter under consideration. Consequently, a single adjoint calculation suffices to obtain sensitivities to all the parameters. The adjoint equations, though, include a source term that depends on the functional defining the response, so that for each response the adjoint equations must be solved anew. In practice, for about the same amount of computation, the adjoint method gives the sensitivity of a functional of the variables to all the parameters, while the direct method gives the sensitivity of all the variables to a single parameter.

This work demonstrates how the adjoint method can be applied to a radiative-convective model (RCM) of the climate. This model includes 312 parameters and contains the nonlinear and on-off phenomena characteristic of radiatively-coupled processes. The next section presents of a brief mathematical description of the RCM. The section after this presents the mathematical definition of sensitivity, the derivation of the adjoint equations satisfied by the adjoint functions, and the expressions of sensitivities in terms of these adjoint functions. Also, uncertainties in the model results are expressed formally in terms of sensitivities and parameter covariances. The section on results presents an analysis of the numerical values obtained for the sensitivities, illustrates the uses of sensitivities, and discusses computational requirements. The summary and conclusions include a discussion of the relevance of this work to future application of the adjoint method to the sensitivity analysis of general circulation models.

The Radiative-Convective Model

The radiative-convective model (RCM) used in this work is the same as described by Hall, Cacuci and Schlesinger (1982). Note, though, that different options are now in effect so that clouds and convective adjustment switch on and off during a diurnal cycle.

This RCM determines the equilibrium vertical temperatures for a single atmospheric column and its underlying surface, subject to prescribed solar insolation, atmospheric composition and surface albedo. The RCM includes models for the transfer of solar and terrestrial radiation, the heat exchange between the earth's surface and atmosphere, the vertical redistribution of heat within the atmosphere, the atmospheric water vapor content, and clouds. The radiative transfer model is that of the Oregon State University two-level atmospheric general circulation model (Schlesinger and Gates, 1980). For the solar radiation, this model includes the effects of water vapor and ozone absorption, molecular scattering and the scattering and absorption by clouds. For the terrestrial radiation, the emission and absorption by water vapor, carbon dioxide and clouds are included. The exchange of heat between the earth's surface and atmosphere is treated as a Newtonian heating with a prescribed heat transfer coefficient multiplied by the temperature difference between the earth's surface and surface air. The vertical redistribution of heat within the atmosphere is modeled following Manabe and Wetherald (1967): convective adjustment prevents the lapse rate from exceeding a prescribed critical value of 6.5 K/km.

Fixed specific humidities in both atmospheric layers make the relative humidity a function of temperature. The lower level relative humidity determines the fractional cloud cover. A diurnal cycle corresponding to a perpetual equinox at 30° latitude makes the solar zenith angle a function of time. This diurnal cycle causes both convective adjustment and clouds to switch on and off periodically. The thermal capacity of the ground is equivalent to a 0.1m depth of water on top of a perfect insulator. This thermal capacity restricts the diurnal variation in ground temperature to about 22°C.

The radiative-convective model proceeds by time integration of the model's set of three nonlinear, coupled differential equations:

$$\frac{du_i}{dt} = f_i(\underset{\sim}{u},\underset{\sim}{\alpha}), \quad i = 1, 2, 3. \tag{1}$$

In (1) the components of the column vector

$$\underset{\sim}{u}(t) = [u_1(t), u_2(t), u_3(t)] \tag{2}$$

are the dependent variables, and the components of the column vector

$$\underset{\sim}{\alpha} = (\alpha_1, \ldots, \alpha_P) \tag{3}$$

are the real scalars representing the parameters in the model. Physically, the variables u_1 and u_2 represent the temperatures of the two layers of the model atmosphere, and u_3 represents the temperature of the earth's surface. The functions f_i, $i = 1, 2, 3$, represent all processes that change these temperatures. In practice, each f_i is a nonlinear function of u_1, u_2, and u_3. The real scalars $\alpha_1, \ldots, \alpha_P$ represent parameters of the model such as specific heats, transmission functions, albedos, and initial conditions. For this study, the total number of model parameters is 312 (i.e., P = 312). Physical and mathematical considerations leading to (1) have been detailed by Hall, Cacuci, and Schlesinger (1982). The initial conditions at time t=a associated with the set of differential equations represented by (1) are

$$u_1(a) = \alpha_i, \quad i = 1, 2, 3. \tag{4}$$

Thus, as indicated by (4), the initial conditions are considered as the first three components of the vector of model parameters $\underset{\sim}{\alpha}$.

The model calculation starts from arbitrary initial temperatures of the ground and two atmospheric layers. The calculation then proceeds for 200 days. Table 1 shows the values of some of the variables during the last day. Convective adjustment occurs from 3:00 PM to 1:00 AM, and clouds exist from 8:00 AM to 1:00 PM. The surface air temperature is a linear extrapolation of the temperatures of the two atmospheric layers (i.e. $-0.5u_1 + 1.5u_2$). The average surface air temperature during the final day is 299.18 K (79°F). The increase in this temperature after doubling atmospheric CO_2 concentration is 1.66 K. The very small value of the fractional cloud over (~ 0.001) is a consequence of tuning the cloud formation parameters to make clouds switch on and off during a diurnal cycle.

Adjoint Method of Sensitivity Analysis

For sensitivity analysis, the results of interest calculated with the model are customarily referred to as responses. The purpose of sensitivity analysis is to assess changes in each response when variations are made in the parameters of the model. The adjoint method of sensitivity analysis presented in this section relies on each response being expressed as a functional of the dependent variables. It would be possible to define the response to be the fields of the dependent variables themselves, but in this case a more sophisticated version of the adjoint method is required (see Cacuci 1981b). Note that if the response R is solely a function of the scalar parameters $\underset{\sim}{\alpha}$ [i.e. $R = R(\underset{\sim}{\alpha})$], then the sensitivity of R to variations $\underset{\sim}{g}$ around the 'base-case' parameter values $\underset{\sim}{\alpha}^\circ$ can be expressed in terms of the partial derivatives $\partial R(\underset{\sim}{\alpha})/\partial\alpha_i$ evaluated at $\underset{\sim}{\alpha}^\circ$. These derivatives can then be used in a first-order Taylor expansion to estimate the change in the value of the response:

$$R(\underset{\sim}{\alpha}^\circ + \underset{\sim}{g}) - R(\underset{\sim}{\alpha}^\circ) = \sum_{i=1}^{P} g_i\{\partial R(\underset{\sim}{\alpha})/\partial\alpha_i\}_{\underset{\sim}{\alpha}^\circ} + 0[g_i^2] \quad . \tag{5}$$

(Here, and in the following, $\{\lambda\}_\mu$ means 'λ evaluated at μ'.)

The response considered in this work is the surface air temperature averaged over the last day of the time-integration. This response can be represented as the functional

$$R(\underset{\sim}{u},\underset{\sim}{\alpha}) = \frac{1}{\tau}\int_a^b \underset{\sim}{d}\cdot\underset{\sim}{u}\; H(t-b+\tau)dt \quad . \tag{6}$$

In (6), b represents the final time value considered in the model, $\underset{\sim}{d} = (-0.5, 1.5, 0)$ is a vector with constant components, H is the customary Heaviside function defined as

$$H(t) = \begin{cases} 1 \text{ if } t > 0 \\ 0 \text{ if } t < 0 \end{cases} , \tag{7}$$

Table 1. Values of Dependent Variables During Last Day of Radiative Convective Model

Day : Hour	Upper Level Atmospheric Temperature (K)	Lower Level Atmospheric Temperature (K)	Surface Air Temperature (K)	Ground Temperature (K)	Convective Adjustment	Fractional Cloud cover
199:00	266.69	275.17	299.41	301.82	yes	0
199:01	226.66	275.14	299.38	300.16	yes	0
199:02	226.64	275.10	299.34	298.65	no	0
199:03	226.61	275.05	299.28	297.26	no	0
199:04	226.58	275.00	299.21	295.98	no	0
199:05	226.56	274.94	299.12	294.80	no	0
199:06	226.53	274.86	299.03	293.71	no	0
199:07	226.52	274.80	298.94	294.18	no	0
199:08	226.51	274.74	298.86	296.23	no	0.00023
199:09	226.51	274.71	298.81	299.59	no	0.00122
199:10	226.51	274.70	298.80	303.84	no	0.00184
199:11	226.52	274.72	298.83	308.47	no	0.00199
199:12	226.52	274.77	298.90	312.96	no	0.00161
199:13	226.53	274.85	299.01	316.80	no	0.00068
199:14	226.53	274.95	299.16	319.54	no	0
199:15	226.57	275.03	299.26	320.87	yes	0
199:16	226.63	275.10	299.33	320.62	yes	0
199:17	226.67	275.15	299.39	318.74	yes	0
199:18	226.70	275.19	299.43	315.59	yes	0
199:19	226.72	275.21	299.46	312.72	yes	0
199:20	226.73	275.22	299.47	310.12	yes	0
199:21	226.73	275.22	299.46	307.75	yes	0
199:22	226.72	275.21	299.46	305.60	yes	0
199:23	226.71	275.20	299.44	303.62	yes	0
200:00	226.69	275.17	299.41	301.82	yes	0

and τ has a constant value of 24 hours.

The response R given by (6) and (7) depends on both $\underset{\sim}{\alpha}$ and $\underset{\sim}{u}(t)$, where $\underset{\sim}{u}(t)$ and $\underset{\sim}{\alpha}$ are implicitly related through the system of nonlinear differential equations (1) and (4). Therefore, the sensitivity can no longer be defined in terms of ordinary partial derivatives, and the more general concepts of functional analysis are used to define the sensitivity.

The most general definition of the sensitivity of a response to variation in the system parameters is the Gateaux-(G-)differential (see e.g., Cacuci, 1981a). The G-differential $VR(\underset{\sim}{u}^\circ,\underset{\sim}{\alpha}^\circ; \underset{\sim}{h},\underset{\sim}{g})$ of $R(\underset{\sim}{u},\underset{\sim}{\alpha})$ at $(\underset{\sim}{u}^\circ,\underset{\sim}{\alpha}^\circ)$ with increment $(\underset{\sim}{h},\underset{\sim}{g})$ is given by

$$VR(\underset{\sim}{u}^\circ,\underset{\sim}{\alpha}^\circ; \underset{\sim}{h},\underset{\sim}{g})$$

$$= \lim \left\{R(\underset{\sim}{u}^\circ + \varepsilon\underset{\sim}{h},\underset{\sim}{\alpha}^\circ + \varepsilon\underset{\sim}{g}) - R(\underset{\sim}{u}^\circ,\underset{\sim}{\alpha}^\circ)\right\}/\varepsilon \quad \varepsilon\to 0 \quad (8)$$

or, equivalently, by

$$VR(\underset{\sim}{u}^\circ,\underset{\sim}{\alpha}^\circ; \underset{\sim}{h},\underset{\sim}{g}) = \left\{(d/d\varepsilon)R(\underset{\sim}{u}^\circ + \varepsilon\underset{\sim}{h},\underset{\sim}{\alpha}^\circ + \varepsilon\underset{\sim}{g})\right\}_{\varepsilon=0}, \quad (9)$$

where $\underset{\sim}{h}(t) = [h_1(t), h_2(t), h_3(t)]$ and $\underset{\sim}{g} = (g_1,\ldots,g_p)$ are column vectors that are at this stage arbitrary, and ε is a real scalar. When $R(\underset{\sim}{u},\underset{\sim}{\alpha})$ is uniformly continuous in u and α at (u°,α°), and when $VR(\underset{\sim}{u}^\circ,\underset{\sim}{\alpha}^\circ; \underset{\sim}{h},\underset{\sim}{g})$ is linear in $\underset{\sim}{h}$ and $\underset{\sim}{g}$, it can be shown (see for example Cacuci 1981a) that the total variation of R is given by

$$R(\underset{\sim}{u}^\circ + \underset{\sim}{h},\underset{\sim}{\alpha}^\circ + \underset{\sim}{g}) - R(\underset{\sim}{u}^\circ,\underset{\sim}{\alpha}^\circ)$$

$$= VR(\underset{\sim}{u}^\circ,\underset{\sim}{\alpha}^\circ; \underset{\sim}{h},\underset{\sim}{g}) + 0[\,\|\underset{\sim}{h}\|^2] + 0[\,\|\underset{\sim}{g}\|^2]\ ,(10)$$

where $\|\ \|$ denotes an appropriate norm, e.g.

$$\|h\|^2 = \sum_{i=1}^{3} \int_a^b [h_i(t)]^2\,dt \ .$$

Note that (10) is a more general form of (5).

The general concepts discussed above are used to determine the sensitivity of the response given by (6). The application to this equation of the definition of the G-differential given in (9) proceeds as follows:

$$VR(\underset{\sim}{u}^\circ,\underset{\sim}{\alpha}^\circ;\ \underset{\sim}{h},\underset{\sim}{g})$$

$$= \left\{(d/d\varepsilon)\frac{1}{\tau}\int_a^b \underset{\sim}{d}\cdot(\underset{\sim}{u}^\circ + \varepsilon\underset{\sim}{h})\ H(t - b + \tau)\ dt\right\}_{\varepsilon=0}$$

$$= \frac{1}{\tau}\int_a^b \underset{\sim}{d}\cdot\underset{\sim}{h}\ H(t - b + \tau)\ dt\ , \qquad (11)$$

where $\underset{\sim}{u}^\circ(t)$ represents the base-case solution of (1) and (4) that is obtained using the base-case parameter values $\underset{\sim}{\alpha}^\circ$.

When performing sensitivity analysis, the vector $\underset{\sim}{g}$ of variations around the base-case parameter values $\underset{\sim}{\alpha}^\circ$ is chosen at the outset. The sensitivity $VR(\underset{\sim}{u}^\circ,\underset{\sim}{\alpha}^\circ;\ \underset{\sim}{h},\underset{\sim}{g})$, though, can be evaluated only after determining the corresponding vector $\underset{\sim}{h}(t)$ of variations around the base-case solution $\underset{\sim}{u}^\circ(t)$. The first-order relationship between $\underset{\sim}{h}$ and $\underset{\sim}{g}$ is obtained by taking G-differentials of (1) and (4). The explicit result of the application of the definition in (9) to (1) and (4) is

$$dh_i/dt - \sum_{j=1}^{3} h_j\{\partial f_i(\underset{\sim}{u},\underset{\sim}{\alpha})/\partial u_j\}\ (u^\circ,\alpha^\circ)$$

$$= \sum_{k=1}^{P} g_k\{\partial f_i(\underset{\sim}{u},\underset{\sim}{\alpha})/\partial\alpha_k\}(u^\circ,\alpha^\circ)\ ,\ i=1,2,3, \quad (12)$$

and

$$h_i(a) = g_i,\quad i = 1,\ 2,\ 3. \qquad (13)$$

In principle, given an arbitrary vector of parameter variations $\underset{\sim}{g}$, (12) and (13) can be solved to obtain $\underset{\sim}{h}(t)$. This value of $\underset{\sim}{h}(t)$ can then be used in (11) to evaluate the sensitivity VR. In practice, though, since there are P linearly independent choices of $\underset{\sim}{g}$, a complete sensitivity analysis using this procedure would involve solving (12) and (13) anew P times. (This procedure would represent a generalization of the direct method mentioned in the introduction.) An alternative procedure to evaluate the sensitivity VR that avoids the need to repeatedly solve (12) and (13) is developed by applying the theory presented by Cacuci (1981a). This theory is based on the use of adjoint operators.

To introduce these adjoint operators, (12) is rewritten in matrix form, i.e.

$$\underset{\sim}{L}\ \underset{\sim}{h} = \underset{\sim}{Q}\ \underset{\sim}{g}\ , \qquad (14)$$

where the components L_{ij} of the 3x3 matrix $\underset{\sim}{L}$ are

$$L_{ij} = \delta_{ij}\ d/dt - \partial f_i/\partial u_j,\ i,j = 1,\ 2,\ 3, \quad (15)$$

and the components Q_{ik} of the 3xP matrix $\underset{\sim}{Q}$ are

$$Q_{ik} = \partial f_i/\partial\alpha_k,\quad i=1,2,3,\quad k=1,\ldots,P. \qquad (16)$$

[The quantity δ_{ij} appearing in (15) is the customary Kronecker delta.] The operator L* that is adjoint to L is introduced through the relationship

$$\int_a^b \underset{\sim}{h}\cdot(\underset{\sim}{L}^*\underset{\sim}{v})\ dt = \int_a^b \underset{\sim}{v}\cdot(\underset{\sim}{L}\ \underset{\sim}{h})\ dt - \left[\underset{\sim}{h}\cdot\underset{\sim}{v}\right]_a^b\ , \quad (17)$$

where $\underset{\sim}{v} = [v_1(t),\ v_2(t),\ v_3(t)]$ is a column vector that at this stage is arbitrary. In view of (17) and (15), the elements L^*_{ij} of the 3x3 matrix $\underset{\sim}{L}^*$ are given by

$$\underset{\sim}{L}^*_{ij} = -\ \delta_{ij}\ d/dt - \partial f_j/\partial u_i,\ i,j = 1,\ 2,\ 3. \quad (18)$$

The vector $\underset{\sim}{v}$ is now chosen by identifying the term on the left-side of (17) with the first term on the right-side of (11). Note that this identification is only possible if the response is defined by a functional. [Cacuci has discussed this point in detail (1981a and b)]. This identification gives the relationship

$$\underset{\sim}{L}^*\underset{\sim}{v} = \frac{1}{\tau}\ H(t - b + \tau)\underset{\sim}{d}\ . \qquad (19)$$

The value of $\underset{\sim}{h}(a)$ is known from (13); the unknown value of $\underset{\sim}{h}(b)$ can be eliminated from (17) by choosing

$$\underset{\sim}{v}(b) = \underset{\sim}{0}\ . \qquad (20)$$

Equations (19) and (20) uniquely determine the adjoint function $\underset{\sim}{v}$. Note that these equations are independent of $\underset{\sim}{h}$ and $\underset{\sim}{g}$, i.e., they are independent of variations in the parameters.

In view of (11), (13), (14), (17), (19), and (20), the sensitivity VR is given by

$$VR = \int_a^b \underset{\sim}{v}\cdot(\underset{\sim}{Q}\ \underset{\sim}{g})\ dt + \sum_{j=1}^{3} g_j v_j(a)\ . \qquad (21)$$

Note that this expression is independent of $\underset{\sim}{h}$. Once the single calculation to determine the adjoint function $\underset{\sim}{v}$ has been performed, (21) can be used to efficiently evaluate the sensitivity VR of R for any vector of variations $\underset{\sim}{g}$ around the base case parameter values $\underset{\sim}{\alpha}^\circ$.

If a variation occurs solely in the nth parameter, then the vector of parameter variations g is denoted by $\underset{\sim}{g}^n$, where

$$\underset{\sim}{g}^n = (0,\ldots,g_n,\ldots,0)\ . \qquad (22)$$

In this case, the corresponding sensitivity VR is denoted by VR^n, where

$$VR^n = VR(\underset{\sim}{\alpha}^\circ, \underset{\sim}{u}^\circ; \underset{\sim}{h}, \underset{\sim}{g}^n) \quad . \qquad (23)$$

The value of VR^n can be used in conjunction with (10) to predict the change in the response when the value of the nth parameter is changed by an amount g_n.

The relative sensitivity of R to $\alpha_n{}^\circ$, denoted by s_n, is defined as the dimensionless quantity

$$s_n = \frac{VR^n}{R} \Big/ \frac{g_n}{\alpha_n{}^\circ} \quad . \qquad (24)$$

In view of (10) and (24), the relative sensitivity, s_n, is a measure of the ratio between the fractional change in R and the corresponding fractional change in $\alpha_n{}^\circ$. In practice, relative sensitivities are used to rank the importance of parameters.

Sensitivities can also be used to calculate the uncertainty in R due to uncertainties in the parameters. If $\underset{\sim}{\alpha}^\circ$ represents the nominal values of the parameters, and $\underset{\sim}{g}$ represents the error in the parameters, then the corresponding error in R is given, to first order in g, by $VR(\underset{\sim}{u}^\circ, \underset{\sim}{\alpha}^\circ; \underset{\sim}{h}, \underset{\sim}{g})$. Consequently, the standard deviation $\sigma(R)$ of R is given by

$$\sigma(R) = (\langle VR^2 \rangle)^{1/2} \quad , \qquad (25)$$

where < > denotes expected value. Since VR is linear in $\underset{\sim}{g}$ [see (21)], VR can be expressed as the sum of VR^n over n, i.e.

$$VR = \sum_{n=1}^{P} VR^n \quad . \qquad (26)$$

Using this relationship, (25) can be recast in terms of relative sensitivities:

$$\sigma(R)/R = [\underset{\sim}{s}^T \underset{\sim}{C} \underset{\sim}{s}]^{1/2} \quad . \qquad (27)$$

In this equation the elements of $\underset{\sim}{s}$ are the relative sensitivities defined in (24), and $\underset{\sim}{C}$ is the relative covariance matrix with elements given by

$$C_{nm} = \langle g_n g_m \rangle / (\alpha_n{}^\circ \alpha_m{}^\circ), \; n,m = 1,\ldots,P. \qquad (28)$$

Note that the square root of the nth diagonal element of $\underset{\sim}{C}$ is the relative standard deviation of the nth parameter.

For a response that is a linear combination of two other responses, i.e.

$$R = \beta R_1 + \gamma R_2 \quad , \qquad (29)$$

application of (9) gives

$$VR = \beta VR_1 + \gamma VR_2 \quad . \qquad (30)$$

This equation can be used to obtain the sensitivities for a CO_2 doubling experiment by taking R_1 as the response with the normal concentration of CO_2, R_2 as the response with double this concentration, and by setting β to -1 and γ to 1.

Results

This section presents the results of using the theory developed in the previous section to evaluate sensitivities of the average surface air temperature to all 312 parameters in the model. Presented in Table 2 are the sensitivities to the solar constant, the surface albedo, the atmospheric CO_2 concentration, and the specific humidity at both atmospheric levels. The results in the column labeled 'predicted change' are the values of VR^n calculated using (21) and (23). The results in the column labeled 'actual change' are the changes in average surface air temperature after actually varying each parameter in turn and rerunning the model.

As shown in Table 2, the relative sensitivity to the solar constant is 0.328. Note that if the Stefan-Boltzmann law were solely responsible for determining the surface air temperature, then the relative sensitivity would be exactly 0.25. The value of 0.328 obtained for this model indicates the existence of a feedback mechanism that enhances temperature changes. Such a feedback mechanism is provided by the prognostically determined cloud cover. Since clouds only occur during the day, a reduction in cloud cover allows more solar radiation to warm the surface, increasing the lower level air temperature. This provides the expected positive feedback. [Note that at night, a reduction in cloud cover would allow more radiation from the ground to escape, reducing the lower level air temperature. Thus nighttime cloud cover causes negative feedback.] With fixed cloud cover, the value of the sensitivity to solar constant is 0.236, verifying the positive feedback of daytime clouds.

For -1% and +0.1% variations in the solar constants, the predicted and actual changes agree to within 10%. This indicates that the surface air temperature varies nearly linearly with solar constant for variations in this range. For a +1% variation, the actual change is 25% less than the predicted change. This is because a 1% increase in solar constant is sufficient to evaporate all the clouds in the model, removing the cloud feedback mechanism that enhances the sensitivity.

Table 2 shows that the sensitivity to surface albedo is negative; this, coupled with the negative sensitivity of ice cover to temperature, causes the ice/albedo feedback mechanism of more complicated models. Table 2 also shows the predicted change resulting from a 100% variation in CO_2 concentration (i.e., a CO_2 doubling experiment). For this variation (100%), the actual change differs from the predicted change by a factor of about one half, although for a 1% variation in CO_2 concentration the agreement be-

Table 2. Sensitivity of Average Surface Air Temperature to Physically Significant Parameters

Average Surface Air Temperature = 299.18 K (Response)					
Parameter Description	**Relative Sensitivity[a]**	**Fractional Variation in Parameter**	**Predicted Change[b] in Response (K)**	**Actual Change[c] in Response (K)**	**Ranking**
Solar Constant	0.328	-1% +0.1% +1%	-0.98 0.098 0.98	-0.91 0.101 0.73	5
Surface Albedo	-0.0295	1% 10%	-0.088 -0.88	-0.091 -0.83	17
Atmospheric CO_2 Concentration	0.0108	1% 100%	0.0323 3.23	0.0327 1.66	35
Upper Level Specific Humidity	-0.0072	1% 10%	-0.0215 -0.215	-0.0203 +0.051	47

[a]Relative Sensitivity = $(VR^n/R)/(g_n/\alpha_n^\circ)$
[b]Predicted Change = VR^n
[c]Actual Change = $R(\underset{\sim}{u}^\circ + \underset{\sim}{h}, \underset{\sim}{\alpha}^\circ + \underset{\sim}{g}) - R(\underset{\sim}{u}, \underset{\sim}{\alpha}^\circ)$

tween predicted and actual change is excellent. At least part of the reason for the factor of one half is that doubling CO_2 eliminates clouds from the model, again removing positive feedback. Also, in this type of model, surface air temperature generally is logarithmically, rather than linearly, dependent on CO_2 the sensitivity would be reduced by a factor in $2 \sim 0.7$.

The sensitivity to the specific humidity at the upper level is positive, confirming the "greenhouse effect" of water vapor. A variation of 10% warms the lower atmosphere enough to evaporate all clouds, once again removing positive feedback, and accounting for the actual change being less than the predicted change. The sensitivity to the specific humidity at the lower level, though, is negative. Although there is also a greenhouse effect of water vapor at the lower level, this effect is masked by cloud cover: an increase in specific humidity increases the relative humidity, increasing cloud cover, and reducing surface air temperature. For a 1% variation in lower level specific humidity, Table 2 shows this reduction in the response, both in the predicted and actual change. The changes arising from a 10% variation though, illustrate the importance of nighttime cloud cover. A 10% increase in specific humidity causes cloud cover to increase to the extent that clouds exist all day and night. Although the increase in cloud cover during the day reduces the surface air temperature, this effect is cancelled by a corresponding increase in cloud cover during the night (when the cloud cover increases the surface air temperature). The actual change is, in fact, positive because of the residual greenhouse effect. This clearly illustrates that the predicted and actual changes only agree for sufficiently small, first order, parameter variations.

In general, the magnitude of the relative sensitivities are a guide to ranking the importance of the parameters. Of the 312 parameters in the model, only 193 have non-zero sensitivities. Of these, only 85 parameters have sensitivities of magnitude greater than 0.001. The eight largest sensitivities all correspond to parameters used to characterize solar radiation and determine trans-

Table 3. Sensitivity of CO_2 Doubling Experiment to Surface Albedo

Increase in Average Surface Air Temperature = 1.66 K (Response)			
Relative Sensitivity	Fractional Variation in Parameter	Predicted Change in Response (K)	Actual Change in Response (K)
1.43	10%	0.24	0.19

mission functions. Almost an order of magnitude smaller are sensitivities used to determine atmospheric composition, surface albedo and stratospheric temperature. The next smallest order of magnitude includes sensitivities of parameters that determine convective adjustment, specific heat of the ground, and the physical properties of water vapor. Sensitivities to initial conditions are at least an order of magnitude smaller again. The last column in Table 2 shows the ranking of the physically significant parameters in relation to all the other parameters.

To use sensitivities for uncertainty analysis requires a knowledge of parameter covariances [see (27) and (28)]. Once covariance information has been obtained, the corresponding sensitivities can be used to estimate the resulting uncertainty in the response. For example, the relative sensitivity of average surface air temperature to the surface albedo is -0.0295. Assuming that the surface albedo (0.1 for this model) has a standard deviation of 10% leads to a standard deviation of 0.295% in average surface air temperature (i.e., 299.18 ± 0.88 K).

Sensitivity and uncertainty analysis can be applied equally well to the response of a CO_2 doubling experiment. Using (32) gives the relative sensitivity of this response (i.e., increase in surface air temperature after doubling CO_2) as 1.43. Table 3 compares the predicted and actual changes in this response for a 10% variation in surface albedo. Assuming again that the surface albedo has a standard deviation of 10% leads to a standard deviation of 14.3% in the response (i.e., 1.66 ± 0.24 K). The large sensitivity for the doubling experiment arises because the 1 x CO_2 calculation has a cloud feedback mechanism that the 2 x CO_2 calculation does not. This emphasizes that a parameter can be relatively unimportant in a 1 x CO_2 experiment, yet be extremely important in a CO_2 doubling experiment.

Uncertainty analysis using a complete set of parameter covariances would enable assessment of the response uncertainty arising from the uncertainties in all the parameters. A knowledge of this response uncertainty is useful in comparative studies between results of the model and observation. If the discrepancy between the model and nature is much larger than the model uncertainty arising from the parameters, then important physical processes must have been modeled inadequately. If, though, all important physical processes and feedback mechanisms have, however crudely, been included, then uncertainty analysis will yield a realistic estimate of the uncertainty in the results of the model. Furthermore, when a comparison between the results of the model and observation is impractical (for example with a CO_2 doubling experiment), uncertainty analysis is the only systematic way to estimate the reliability of the model results.

The total computing time required for this model to reach the final time of 200 days is 7.8s on an IBM 3033. The additional computing time to calculate 312 sensitivities using the adjoint approach is 18.5s, i.e. 5.1s to calculate the adjoint functions and 13.4s to calculate the derivatives $\partial f_i/\partial \alpha_j$, $(i=1,2,3;\ j=1,...,312)$. The additional computing time needed to obtain the same information by rerunning, with each of the parameters changed in turn, would have been about 40 minutes.

Summary and Conclusions

This work demonstrates the adjoint method of sensitivity analysis on a radiative-convective model with a diurnal cycle and prognostically determined clouds. The results for this model show the method to be both efficient and accurate. A single adjoint calculation, involving a similar amount of computation as the original model, suffices to obtain sensitivities of average surface air temperature to all 312 model parameters. These sensitivities accurately predict the effect of small variations in the model parameters. Relative sensitivities rank the importance of all the parameters, the most important parameter being those that characterize solar radiation and determine transmission functions. Sensitivities combined with parameter covariances give the correponding uncertainty in the response (i.e., result of interest) of the model. For this model, a 10% standard deviation in the surface albedo causes the average surface air temperature to have the standard deviation of 0.88 K about the expected value of 299.18 K. For a CO_2 doubling experiment, the same standard deviation in surface albedo causes the increase in average surface air temperature to have a standard deviation of 0.24 about the expected value of 1.66 K. For a CO_2 doubling experiment, uncertainty analysis using sensitivities and covariances for all model parameters is the only systematic way to estimate the reliability of the model results. This estimate, though, will only be realistic if all important physical processes and feedback mechanisms have been represented, even if only approximately, in the model.

Radiative-convective models include many of the complex on-off nonlinear processes that are found in general circulation models (GCM's). This work has shown that the adjoint method can deal with such complex time-dependent processes successfully and efficiently, a finding that provides valuable information for subsequent application of the adjoint method to the sensitivity analysis of a GCM. This subsequent application, though, raises several new issues; GCM's involve many variables, the results can be functions of position, and the solution only reaches a statistical 'quasi-equilibrium.' Further research is underway to investigate these issues.

Acknowledgements. Special thanks are extended to Ann Houston for her expert typing of this manuscript. The work was sponsored by the Carbon Dioxide Research Division, Office of Basic Energy Sciences, U.S. Department of Energy under contract W-7405-eng-26 with Union Carbide Corporation.

References

Aleksandrov, V. V. and W. L. Gates, The performance of a coarse-grid version of the OSU two-level atmospheric GCM, Rep. No. 24, Climatic Research Institute, Oregon State University, 43 pp., 1981.

Bell, T. L., Climate sensitivity from fluctuation dissipation: Some simple model tests, J. Atmos. Sci., 37, 1700-1707, 1980.

Cacuci, D. G., Sensitivity theory for nonlinear systems. I. Nonlinear functional analysis approach, J. Math. Phys., 22, 2794-2802, 1981a.

Cacuci, D. G., Sensitivity theory for nonlinear systems. II. Extensions to additional classes of responses, J. Math. Phys., 22, 2803-2812, 1981b.

Dickinson, R. P. and R. J. Gelinas, Sensitivity analysis of ordinary differential equation systems - a direct method, J. Comp. Phys., 21, 123-143, 1976.

Hall, M.C.G., D.G. Cacuci and M.E. Schlesinger, Sensitivity analysis of a radiative-convective model by the adjoint method, J. Atmos. Sci., 39, 2038-2050, 1982.

Held, I. M. and M. J. Suarez, A two-level primitive equation atmospheric model designed for climatic sensitivity experiments, J. Atmos. Sci., 35, 206-229, 1978.

Leith, C. E., Climatic response and fluctuation dissipation, J. Atmos. Sci., 32, 2022-2026, 1975.

Manabe, S. and R. T. Wetherald, Thermal equilibrium of the atmosphere with a given distribution of relative humidity, J. Atmos. Sci., 24, 241-269, 1967.

Manabe, S. and R. T. Wetherald, On the distribution of climate change resulting from an increase in CO_2 content of the atmosphere, J. Atmos. Sci., 37, 99-118, 1980.

Marchuk, G. I., Onovnye i sopryazhennye uravneniya dinamiki atmosfery i okeana, Meteorol. Gidrol., 2, 9-37, 1974.

Schlesinger, M. E. and W. L. Gates, The January and July performance of the OSU two-level atmospheric general circulation model, J. Atmos. Sci., 37, 1914-1943, 1980.

VEGETATION-ALBEDO FEEDBACKS

Robert E. Dickinson

National Center for Atmospheric Research[1], Boulder, Colorado 80307

Brian Hanson

Department of Geography, University of Minnesota, Minneapolis, Minnesota 55455

Abstract. This paper discusses the role of vegetation in determining the albedo of land surfaces. Changes of vegetation with climate change may, in turn, change planetary albedo. The albedo of vegetated land, strictly speaking, should not be regarded as a fixed parameter at each grid square within a climate model. It varies with the spectral and angular distribution of incident solar radiation. Various physical characteristics of plant canopies contribute to their trapping of light and hence their albedo. Differences in the arrangements of leaves and horizontal variability in the canopy height appear to account for more of the variability in canopy albedos than do the leaf optical properties. As an example of global albedo change due to vegetation change, we have used the CLIMAP (1981) data set to infer the change of albedo between 18000 BP and present. We find that owing to differences in vegetation, the planetary albedo was about 0.2% higher 18,000 years ago than at present. This albedo difference is smaller than those given in previous estimates, but it is still not insignificant.

Introduction

The presence and nature of a plant canopy are of considerable importance in determining the microclimatology of a region. The influence of the plant canopy is felt through its aerodynamic roughness, its influences on evapotranspiration and surface shading, and its albedo. All of these factors modulate the transfer of energy between the surface and the atmosphere and in doing so may influence regional atmospheric climate, but only changes in surface albedo directly influence the net radiation absorbed by the earth-atmosphere system. Solar radiation reflected by the surface largely returns to space, whereas other forms of surface energy transfer simply redistribute energy between the surface and the atmosphere.

The intent of this paper is to discuss the characteristics of plant canopies that determine the albedo of vegetated land and to consider possible climatic feedbacks involving albedo changes consequent to changes in vegetation. As an example, we examine the albedo change implied by the CLIMAP (1981) tabulation of land cover for 18000 BP and present.

The question of the factors determining the albedo of plant canopies has long been studied by micrometeorologists. Reviews of various aspects are available in Monteith (1975, 1976), Norman (1975), Goudriaan (1977), and Ross (1981). However, changes of surface albedo due to land-use change have only recently begun to be recognized as of possible importance for regional and global climate change. Especially influential in promoting this recognition have been papers by Charney (1975), Cess (1978), and Sagan et al. (1979). Charney (1975) and Charney et al. (1977) suggested that hypothetical albedo increases due to desertification of arid lands such as the Sahel of Africa would feed back on the regional hydrological cycle to reduce rainfall and so further increase desertification. Cess (1978) considered the surface albedo change between the last ice age (18000 BP) and present, employing a summary of the surface albedo data used by Gates (1976) for simulating in a general circulation model (GCM) the climate for these two periods. Sagan et al. (1979) estimated the land-use changes over the last millenia and inferred consequent albedo changes. They concluded that humans have made substantial contributions to climate change in the past.

Physical factors determining the albedo of a plant canopy

The earth's land surface includes a wide range of ecosystems and human land-use patterns,

[1]The National Center for Atmospheric Research is sponsored by the National Science Foundation.

whose vegetative covers have differing influences on surface albedo. About one-fourth of the land is desert and semidesert, one-fourth forests, one-fourth grasslands, one-eighth farmlands and cities, and the remainder mostly ice caps and tundra. The processes that determine the albedos of these surfaces are complicated and it is tempting to rely on observational data for specifying surface albedos as prescribed constants in climate models. However, the albedos of vegetated surfaces vary with both the wavelength and the solar zenith angle of the incident solar radiation. Consequently, reflection of solar radiation by land surfaces depends, in part, on the nature of scattering and absorption of the solar beam in the atmosphere. Together with the dependence of surface albedos on solar zenith angle, this dependence of the incident flux on the overlying atmosphere imposes diurnal, seasonal, and latitudinal variations on the effective surface albedo.

The determination of the albedo of vegetation is further complicated by the fact that some of the diffuse radiation incident on the surface was previously reflected from the surface. Such multiple reflections become more important for a relatively cloudy sky and high surface albedo. All of these dependencies can be included in climate models which provide enough detail in their calculated fluxes of solar radiation to have some spectral resolution and which distinguish between direct-beam and diffuse solar fluxes.

Surface albedos are, furthermore, dependent on surface textures, structure, and composition. The albedo of leaf canopies differs remarkably from that of single leaves, with measurements of canopy albedo usually giving values roughly half that of individual leaves, e.g., 0.3 for the leaves versus 0.15 for the canopy. This difference between canopy and leaf albedo results from the typically nearly random distribution of leaf positions and the reflectance of individual leaves being low compared to unity; light passes by outer leaves, and when reflected by more inward leaves is more likely to be absorbed by further encounters than to escape out from the canopy. In other words, canopies act as efficient light traps, especially in the visible spectrum, where canopy albedos can be as low as a few percent. If leaf positions were sufficiently regular that a nearly uniform outer layer of leaves could intercept most of the solar beam, the canopy albedo would be higher, approaching that of individual leaves. Some canopies apparently do have somewhat regular leaf arrangements and consequently higher albedos. Conversely, clustering of leaves as in conifer trees tends to further decrease canopy albedo (Goudriaan, 1977). Orientation of leaves normal to the sun increases canopy albedo, and orientation along the direction of the solar beam decreases canopy albedo.

Other features of canopy structure besides leaf organization contribute to light trapping. In particular, increased canopy roughness lowers albedos. If individual plant elements are sufficiently widely spaced, the canopy is far from being spatially uniform. The heights of the outermost leaves in the canopy then vary widely depending on their relative location within individual plants, and light can penetrate deeply and be trapped. Such trapping by individual plants occurs with the inverted-cone-shaped spruces and firs of northern forests, even for dense forests, and greatly influences the solar zenith dependence of the canopy. Usually taller vegetation presents a rougher surface; the greater light trapping of rougher canopies leads to an inverse correlation between vegetation height and canopy albedo (Stanhill and Fuchs, 1970; Oguntoyinbo, 1970).

An extreme example of a rough vegetation canopy is a field of leafless trees or giant cacti, which for modeling purposes would be treated as an array of poles. Half the light reflected by an isolated pole is downreflected to the underlying surface and, if the albedo of the underlying surface is low, is mostly absorbed. Thus, vertical poles have a much lower effective albedo than they would have if they were flattened and lying on their sides. Poles can have an even more dramatic effect if the underlying surface has a high albedo, by their shading of the underlying surface (as analyzed by Federer, 1971).

The optical properties of individual leaves also enter into determination of canopy albedos. Gates (1980) has summarized the mean integrated reflectances and absorptances of individual leaves for a large number of deciduous plants. For an overhead sun, most reflectances lie between 0.2 and 0.3. Conifer needles have a reflectance of about 0.2. Variations in the optical properties of individual leaves appear to be less important for determining variations in canopy albedo than do variations in the canopy structure. (For sparse canopies, it is also necessary to include the effect of the albedo of the underlying surface.)

Theoretical analyses of the albedos of a plant canopy treat the problem in terms of radiative transfer through a distribution of flat plates with given optical properties. Most attention has been given to horizontally uniform distributions where a one-dimensional treatment is appropriate. If the leaves are, furthermore, randomly distributed in space, the radiative transfer through the canopy can be approximated by a "two stream equation", from which may be obtained canopy albedos. If the canopy is semi-infinite and the fractions of light upscattered and downscattered by an average leaf are assumed equal, the canopy albedo a_c is simply

$$a_c = \frac{\omega k}{(\alpha + k)(1 + \alpha)} \quad (1)$$

where ω is the single scattering albedo of leaves (i.e., the fraction of light incident on a leaf that is not absorbed); $\alpha = (1 - \omega)^{1/2}$; and k is the

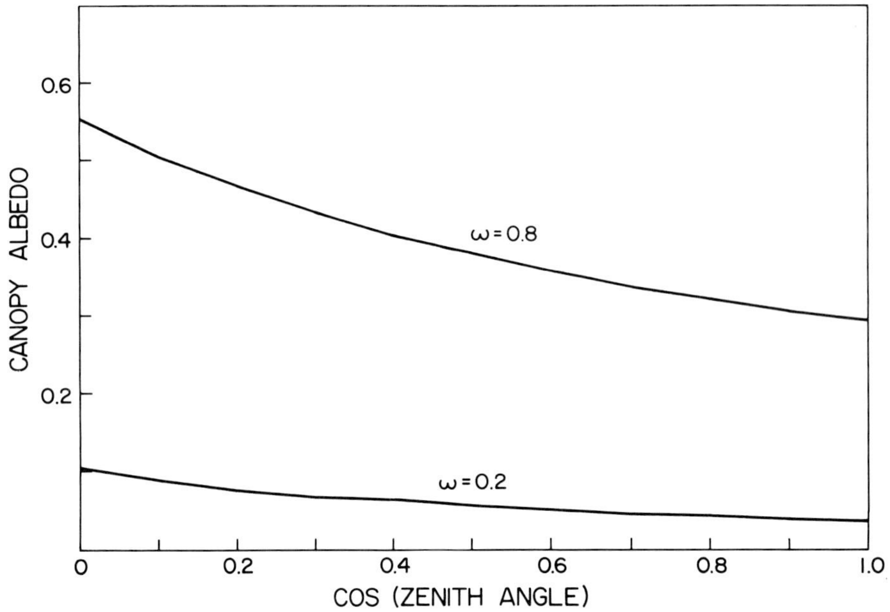

Fig. 1. Theoretical variation of canopy albedo with solar zenith angle for flat-plate single scattering values of 0.2 and 0.8, typical values for leaves with visible and near-infrared solar fluxes. The leaves (flat plates) are semi-infinite in extent and randomly distributed with any spatial position or angular orientation of equal probability.

solar-zenith-angle-dependent light interception factor (i.e., the average leaf surface normal to the solar beam per unit leaf area divided by cos θ, the cosine of the solar zenith).

For example, if the leaves are horizontal, $k = 1$, whereas if they have all orientations of equal probability, $k = 0.5/\cos\theta$. Using this latter more realistic assumption, Fig. 1 shows the solar zenith dependence of albedo inferred from Eq. (1). The two values $\omega = 0.2$ and $\omega = 0.8$ are appropriate, respectively, for the visible and near-infrared fractions of the solar beam. Further details as to the mathematical theory of plant canopy albedos and the derivation of Eq. (1) may be found in Ross (1981) or Dickinson (1982), which treats the more general case of differences between the direct and diffusive beam upward- versus downward-scattered fractions. Variations with solar zenith angle similar to that shown in Fig. 1 have been observed for some crop plants (Idso et al., 1978). However, forests generally show a weaker dependence of albedo on solar zenith angle, with northern spruce-fir forests having an especially weak solar zenith dependence. The trapping of light by surface roughness and irregularity appears to be enhanced at low sun angles.

In summary, the determination of surface albedos for use in three-dimensional climate models requires not only observational data but also considerable insight into the nature of the radiative transfer processes within realistic vegetative canopies. Studies of surface albedo for requirements of remote sensing can also contribute to improving the formulation of surface albedo in climate models. Recent theoretical studies of vegetated land albedo in the context of remote sensing include those of Cooper et al. (1982), and Kimes and Kirchner (1982).

Albedo change from 18000 BP to present as an example of vegetation-albedo feedback

Recently, George Kukla and his colleagues have prepared a chart of land surface features, including vegetation and soils, for 18000 BP and the present (CLIMAP, 1981). "The 18000 BP reconstruction is based on a relatively small number of sites and is therefore nothing more than an educated guess" (Kukla, personal communication). The CLIMAP report also gives preliminary estimates of surface albedos associated with different surface types; separate values are included for moist and dry soil conditions. We have used the 2° by 2° grid cell data on which the charts are based to estimate the albedo change due to vegetation changes in the area between 50°S and 50°N. This latitude range was chosen by Cess (1978) to exclude the northerly latitudes, where the albedo change was due primarily to the 18000 BP ice sheet. Additional albedo change due to vegetation change does occur north of 50°N mostly in the Asian sector (CLIMAP, 1981) which we shall estimate later. The fraction of dry versus moist soil for a given land type, latitude, and season was estimated for our study from geographical and climatological considerations. Errors in the estimate of global albedo change from errors in the moisture change are not expected to be larger than 0.1 of the global albedo change.

Our analysis follows that of Cess (1978), who

Table 1. Summary of land cover changes (in % of globe) from 18000 BP to present (50°N-50°S). The vegetation categories refer to the aggregates of CLIMAP types indicated in Table 2

Water	3.3
Ice	-1.1
Bare Soil	-3.6
All Vegetated Land	1.5
Tropical Forest	1.5
Temperate Forest	1.5
Savanna and Grassland	-3.1
Agriculture	1.5

first raised the question we address here. His analysis, which considered 50°S-50°N surface and planetary albedo changes which he believed to be due to vegetation changes between 18000 BP and present, was based on the GCM simulations of Gates (1976). Gates's simulations, as well as those of Manabe and Hahn (1977), used albedos similar to those prescribed by CLIMAP for 18000 BP, but for current conditions used the surface albedos recommended by Posey and Clapp (1964). The Posey and Clapp albedos for tropical forests (7%) are erroneously low, for reasons given in Dickinson (1980) and Efimova (1980). Hence, it is important to redo Cess's analysis with more consistent and correct albedos for present conditions.

The CLIMAP (1981) data set distinguishes nine dominant vegetation components, each of which has four or five subcomponents, and each subcomponent is given separate dry and wet albedos. There are, likewise, five soil classes, each divided into four subclasses. Each 2° x 2° grid square has been assigned seven basic land-type variables: (1) vegetation type, (2) soil type, (3) terrain roughness and relief, (4) areal proportion of water bodies, (5) percent exposed soil or rock, (6) proportion of exposed permanent ice, and (7) areal proportion of vegetation.

Table 1 shows the gross changes that occurred in the 50°S-50°N area between 18000 BP and present, according to CLIMAP (1981). The lower water cover during 18000 BP was primarily a result of ice cap formation removing water from the oceans. Most of the excess ice shown in the table for 18000 BP represents the southern edge of the Laurentide ice sheet, but some also represents increased mountain glaciers. A decrease in forests accounts for the smaller percentage of vegetated land in 18000 BP.

Table 2 summarizes the range of albedos for different land types, using the CLIMAP algorithms with our estimates of soil moisture. Various types of nontropical forests have been aggregated as have shrubs and grasslands. Most of the ex-

Table 2. Average albedos and ranges obtained from aggregating CLIMAP land types

LAND TYPES	ALBEDO (%)
VEGETATION	
1. Evergreen forest (mostly tropical)	12.5 ± 3.0
2. Mixed forest (evergreen dominant) 3. Mixed forest (deciduous dominant) 4. Deciduous	14.7 ± 3.0
5. Shrub and low tree 6. Tall grass and vegetation 7. Short vegetation, grass, pasture	17.5 ± 3.0
8. Crop, agriculture	14.5 ± 4.0
SOIL	
2. Medium dark	16.5 ± 3.0
3. Medium light	24.0 ± 5.0

posed soil occurs in the two categories shown.

The changes in surface albedo due to changes in ice, water, and vegetation cover are shown in Fig. 2. Our calculations did not make detailed point-by-point comparisons. Rather we determined total albedo change and land cover change for each latitude belt, then assumed an average land albedo of 0.16 to estimate the albedo change due to ice and water change. CLIMAP assigns albedo values of 0.08 and 0.70 to water and snow, respectively. The surface albedos of individual latitudes were then averaged, weighted according to area and incident solar radiation, and converted to albedo changes at the top of the atmosphere, following Cess (1978). The global average surface albedo changes obtained from the data shown in Fig. 2 for a July sun are shown in Table 3. The CLIMAP land data do not include any seasonally varying factors. The albedo change connected with vegetated land change corresponds to a 5% decrease in albedo over 0.09 of the globe (about three-tenths of the land area), e.g., as would occur for a change of 4% of the globe from savanna and grassland (average albedo 18%) to forest and agriculture (average albedo 13%) plus a change of 3% of the globe from a medium light soil (23% albedo) to savanna and 1% of the globe from medium light soil to forest and agriculture. These conversions are, roughly speaking, what is given in the CLIMAP data set.

The albedo changes due to vegetation change poleward of 50°N and 50°S were estimated from the albedo maps for 18000 BP and present time given by CLIMAP (1981). We find an additional 0.23% change in global surface albedo from these regions. Adding this factor to the 0.45% in Table 3 gives a total global surface albedo change of 0.68%.

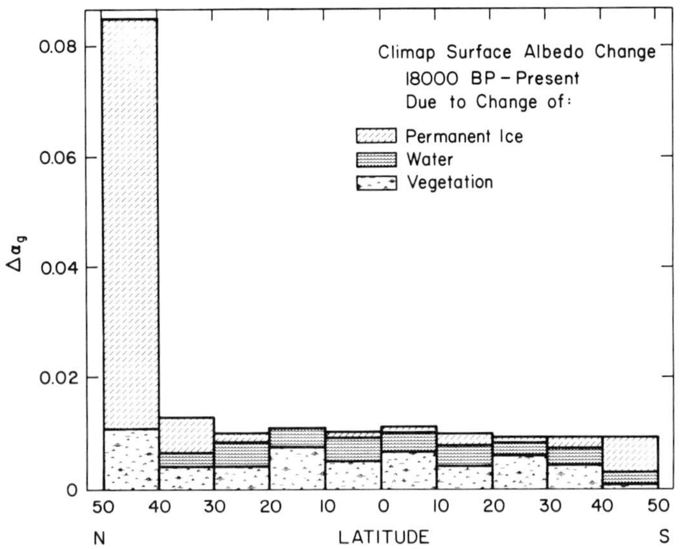

Fig. 2. Surface albedo changes from 18000 BP to present for 10° latitude belts as inferred from the CLIMAP (1981) data set, and distinguishing between permanent ice, water, and vegetation changes. The effect of temporary snow cover has not been included.

Table 3. Decrease of global average July surface albedo from 18000 BP to present for land changes between 50^oN and 50^oS

Changed surface	$\Delta\alpha_g$ (%)
Water	0.29
Vegetation	0.45
Permanent ice	0.73
July total	1.46

The surface albedo change of 0.68% due to vegetation is converted to planetary albedo by multiplying it by the factor 0.43 (Cess, 1978). This factor corrects for the masking of the surface by atmospheric reflection and absorption, including that due to a 50% cloud cover. The planetary albedo change $\Delta\alpha_p^{VEG}$ thus obtained is

$$\Delta\alpha_p^{VEG} = 0.29\% \quad . \qquad (2)$$

Weighting with a January sun gives a planetary albedo change of only about half as much, due to the maximum solar incidence where there are lower land fractions, i.e., in the Southern Hemisphere. However, the January value is somewhat more uncertain since we have not attempted to evaluate contributions due to the interactions between vegetation and snow cover in middle latitudes of the Northern Hemisphere. Other months presumably had albedo change between that of January and July.

A calculation similar to ours has recently been made by Efimova (1980) who found, for an annual average, a planetary albedo change of 0.34%. The difference between her result and ours is perhaps within the uncertainty in prescription of land cover changes and albedos. The value obtained by Cess of 1% planetary albedo change is largely due to the differences between the Posey-Clapp and CLIMAP albedos for similar land covers, as already pointed out by Efimova.

Discussion

There are many intriguing possibilities for changes in vegetative cover due to climate change, which would provide climatic feedback by changing global albedo. Possibilities for increasing albedo include retreat of the Arctic tree line with conifers giving way to tundra, conversion of grasslands to desert, and conversion of tropical forests to savanna. Deforestation may have been strongly promoted over the last millenia by human activities (Sagan et al., 1979). However, the likely albedo changes resulting from widespread deforestation do not imply large global climatic impacts (Dickinson, 1980). Changes from grasslands to more desert-like conditions may also lead to only modest albedo changes, if not accompanied by drastic changes in soil, such as to a quartz sand.

Thus, it is not surprising that our analysis of the CLIMAP data set finds only a 0.22% summer-winter average decrease in global planetary albedo due to change between 18000 BP and present, or equivalently about 0.75 W/m^2 decrease. Following Cess (1978) in assuming a 5°C global temperature increase between 18000 BP and present, we infer an increase of reflected solar radiation of about 0.15 W/m^2 per °C decrease of global temperature. Using Cess's albedo change estimates for surfaces poleward of 50°S-50°N, we infer a total planetary albedo decrease of about 2% from 18000 BP to present. The increase of albedo with increase snow and ice during the last ice age is generally considered as a major feedback term with regard to the global radiation balance. The apparent vegetation-albedo feedback is about 0.1 as large as the ice-albedo feedback, and is still significant. Hence, the conclusion of Cess that vegetation-albedo feedback is an important climatic mechanism is still qualitatively confirmed.

It is interesting to speculate as to the possible effects on GCM climate simulations of the much greater albedo change obtained by switching between the Posey-Clapp and the CLIMAP albedo prescriptions. Further improvements in the specification of surface albedos and their change since 18000 BP would be valuable for future GCM studies. Theoretical analyses of the physical factors determining the albedo of vegetated land, as discussed earlier in this paper, should also contribute to further improvements in the determination of land surface albedos within the framework of climate models.

The discussion up to now has focused on global albedo changes resulting directly from land cover change. However, some indirect mechanisms may turn out to be of at least comparable importance. For example, changes in vegetation may change sensible and latent fluxes to the atmosphere which, in turn, could modify moist convection and cloud cover (Dickinson, 1980). Forest fire frequencies and, consequently, atmospheric content of carbon aerosol may be changed. Conversely, changes in the radiative properties of the atmosphere will change the spectral and angular distribution of incident solar radiation and so to some extent the surface albedo.

Acknowledgments. We thank R. Cess for helpful discussions and E. Box, E. Matthews and G. Kukla for helpful comments on drafts of the manuscript.

References

Cess, R.D., Biosphere-albedo feedback and climate modeling, J. Atmos. Sci., 35, 1765-1768, 1978.

Charney, J.C., Dynamics of deserts and drought in the Sahel, Quart. J. Roy. Met. Soc., 101, 193-202, 1975.

Charney, J.C., W.J. Quirk, S.-H. Chow and J. Kornfield, A comparative study of the effects of albedo change on drought in semiarid regions, J. Atmos. Sci., 34, 1366-1385, 1977.

CLIMAP Project Members, Seasonal reconstructions of the earth's surface at the last glacial maximum, A. McIntyre, Leader, LGM Project, Lamont-Doherty Geological Observatory, Columbia Univ., Palisades, N.Y., 1981.

Cooper, K., J.A. Smith and D. Pitts, Reflectance of a vegetation canopy using the adding method, J. Appl. Opt., 21, 4112-4118, 1982.

Dickinson, R.E., Effects of tropical deforestation on climate, in Blowing in the Wind: Deforestation and Long-Range Implications, pp. 411-441, Pub. 14, Studies in Third World Societies, Dept. of Anthropology, College of William and Mary, Williamsburg, VA., 1980.

Dickinson, R.E., Land surface processes and climate - surface albedos and energy balance, Advances Geophy., 25, 305-353, 1983.

Efimova, N.A., Influence of changes in surface albedo on the thermal regime of the earth, Meteor. i Gidro., 7, 50-56, 1980.

Federer, C.A., Solar radiation absorbed by leafless hardwood forests. Agric. Meteorol., 9, 3-20, 1971.

Gates, D.M. Biosphysical Ecology, 611 pp., Springer-Verlag, New York, 1980.

Gates, W.L., A numerical simulation of ice-age climate with a global general circulation model, J Atmos. Sci., 33, 1844-1873, 1976.

Goudriaan, J., Crop micrometeorology: A simulation study, 249 pp., Pudoc, Wageningen, 1977.

Idso, S.B., J.L. Hatfield, R.J. Reginato, and R. D. Jackson, Wheat yield estimation by albedo measurement, Remote Sensing of Environ., 7, 273-276, 1978.

Jarvis, P.G., G.B. James and J.J. Landsberg, Vegetation and the atmosphere, vol. 2, Case studies, J.L. Monteith, ed., 439 pp., 1976.

Kimes, D.S. and J.A. Kirchner, Radiative transfer model for heterogeneous 3-D scenes, J. Appl. Opt., 21, 4119-4129, 1982.

Manabe, S. and D.G. Hahn, Simulation of the tropical climate of an ice age, J. Geophys. Res., 82, 3889-3911, 1977.

Monteith, J.L., Vegetation and the Atmosphere, Vol. 1, Principles, 278 pp., Academic Press, 1975.

Monteith, J.L., Vegetation and the Atmosphere, Vol. 2, Case Studies, 439 pp., 1976.

Norman, J.M., Radiative transfer in vegetation, in Heat and Mass Transfer in the Biosphere, Part 1: Transfer Processes in the Plant Environment, D.A. DeVries, Ed., 187-205, 1975.

Oguntoyinbo, J.S., Reflection coefficient of natural vegetation, crops and urban surfaces in Nigeria, Quart. J. R. Meteor. Soc., 96, 430-441, 1970.

Posey, J.W. and P.F. Clapp, Global distribution of

normal surface albedo, Geophys. Int., 4, 53-58, 1964.
Ross, J., The radiation regime and architecture of plant stands, Dr. W. Junk Publishers, 391 pp., The Hague, 1981.
Sagan, C., O.B. Toon and J.B. Pollack, Anthropogenic albedo changes and the earth's climate, Science, 206, 1363-1368, 1979.
Stanhill, G. and M. Fuchs, Some results of helicopter measurements of the albedo of different land surfaces, Solar Energy, 13, 54-66, 1970.

ON THE EMPIRICAL VERIFICATION OF MODEL-PREDICTED CO_2-INDUCED CLIMATIC EFFECTS

Stephen H. Schneider

National Center for Atmospheric Research*
P. O. Box 3000, Boulder, CO 80307

Abstract. Historical analogs to a CO_2 warming have been proposed and are critically reviewed. It is concluded that these are crude analogies to any regional CO_2-induced effects. Equilibrium general circulation models have been run and have produced maps of altered soil moisture from fixed levels of increased CO_2. Unfortunately, these equilibrium calculations ignore the possibility of major transient effects. Global energy balance models used to estimate the temperature trends over the twentieth century have shown some promise. If the current crop of climate models is roughly right, the signal-to-noise ratio of global temperature increase from greenhouse gases should become much larger by the end of the century (e.g., Lacis et al., 1981; Thompson and Schneider, 1982b). And if the climatic noise level can be reduced by the technique of subtracting out climatic effects from other known -- or believed known -- natural forcing factors like volcanic dust or solar variations, then detection of a believable CO_2 (plus other trace gases) signal to climatic noise ratio should come even sooner. Many uncertainties remain, which should render predictions of regional, transient climatic anomaly patterns from CO_2 increase highly speculative for some time to come. But the advent of a reasonably clear global temperature signal this decade is a real possibility.

Introduction

Modeling of carbon dioxide effects on climate has taken two general forms: (a) Snapshots of equilibrium climate changes (e.g., maps of surface air temperature T_s or other variables such as soil moisture) as a function of position for a fixed increase in carbon dioxide -- typically a doubling or quadrupling (e.g., Manabe, Wetherald and Stouffer, 1981); or (b) Time series simulations of surface temperature change based on scenarios of carbon dioxide or other time evolving external forcing mechanisms, such as volcanic dust veils or solar constant variation (e.g., Schneider and Mass, 1975; Hansen et al., 1981; Gilliland, 1982).

Verification of "snapshot" modeling (or attempts to estimate empirically regional surface climatic anomaly patterns from CO_2 forcing) have typically been by four different approaches:

1. Simulation of seasonal and regional average climates by a model. This approach is based on the philosophy that a good seasonal model forced by time-varying solar insolation may also be a good CO_2-forced model (e.g., see Schneider and Thompson, 1980 or Manabe and Stouffer, 1980).
2. Warm year analogs. These are based on the supposition that perhaps there is something characteristic about the regional climatic patterns in "warm years", and that the distribution of regional climatic anomalies in the warmest few years of the past century might be repeated if long-term warming were to be created in the future by CO_2 increases (e.g., Williams, 1979; Wigley et al., 1979, and Pittock and Salinger, 1982).
3. Since some paleobotanic and other evidence suggest that the "Altithermal" climatic period between about 5000-9000 years ago was a degree or two Celsius warmer than present, perhaps the regional distribution of climatic anomalies then would be a good analogy to CO_2-induced warming in the future (e.g., Kellogg, 1977; Butzer, 1980).
4. Ancient paleoclimatic analogies. Ice core gas bubble analyses suggest CO_2 variations of 30 percent or so from global cooling maximum to interglacial optimum (e.g., Delmas et al., 1980; Neftel et al., 1982). Could this CO_2 variation help to explain glacial inter-glacial climatic changes? Since the Mesozoic Era (some 65 to 200 million years ago) was perhaps 5-10°C warmer on average than mean global conditions today (Frakes, 1979), and since some evidence suggests that CO_2 levels may have been several times higher then (e.g., Budyko and Ronov, 1979), it is conceivable that this Mesozoic warmth is a verification of CO_2-induced

*The National Center for Atmospheric Research is sponsored by the National Science Foundation.

warming effects. So too, it may be that CO_2 levels were as much as tens to hundreds of times greater in the Archaen Age [about 3 to 4 billion years ago (e.g., Walker, Hays and Kasting, 1981)] than those levels are today. At these early periods liquid water appears to have existed on an earth that was heated by a sun whose luminosity was perhaps twenty to thirty percent less than it is today (e.g., see Sagan and Mullen, 1972). A "super-greenhouse effect" from carbon dioxide (or other trace gases) is one possible explanation of this so-called "faint early sun paradox" (Owen, Cess and Ramanathan, 1979).

Model produced time series of global or zonal temperature trends over the past hundred years or so have been "verified" by comparison to time series of surface temperature observations over the same period. Unfortunately, these models, which tend to be thermodynamic energy balanced models (EBMs) with parameterized dynamics -- which are efficient enough to be run over a hundred year time sequence within reasonable computational budgets -- do not generally distinguish between surface air temperature over land, surface air temperature over water and ocean water temperatures. However, the heat capacities which govern their transient responses to external forcings like CO_2 increase or volcanic eruptions are typically based upon the depth of the mixed layer of the models' oceans, rather than on detailed dynamical mixing processes in atmospheres and oceans which couple these reservoirs of vastly differing heat capacities (e.g., see Schneider and Thompson, 1982). Thus, there is some question as to whether the "match" between model-generated times series and observed time series is appropriate, inasmuch as models are typically generating surface temperatures based on globally averaged values of oceanic heat capacity whereas most surface temperature data are air temperatures over the midnorthern Hemisphere land masses (for example, see Gilliland and Schneider, 1984 or Harvey and Schneider, 1984).

A further complication which relates both to snapshot and time series modeling is this issue of the transient response. For example, can an equilibrium "snapshot" response for a CO_2 doubling experiment with a high resolution dynamical model be applied to a 25 percent increase in CO_2 by merely taking 25 percent of the doubling response? More generally, can a time series of CO_2 increase be mimicked realistically by a time series of successive but discrete equilibria produced by a limited number of snapshot simulations? The answer can be yes only to the extent that transient effects can either be ignored or a global correction applied. Thompson and Schneider (1982) have argued that such effects may not necessarily be ignored or ameliorated by a simple global correction. Thus, it may be necessary to compute the time evolution of the response of detailed regional climatic models to a CO_2 increase scenario in order to estimate the regional climatic changes over time. A sequence of equilibrium snapshots may simply be a poor method to use, even though the latter is considerably less expensive in terms of computer resources than times series simulation with a fully synchronous coupled atmosphere/ocean model.

The issues just outlined will be addressed in expanded discussions in the following sections, not for the purpose of providing an exhaustive survey of work in these subjects, but rather to illustrate by specific examples that empirical verification of a "CO_2 signal" is by no means a straightforward problem. Moreover, it will be concluded that considerably more numerical experimentation and careful and consistent comparison between different models across the climate modeling hierarchy (see Schneider and Dickinson, 1974, for example) is needed. Only such comparisons between models of various complexity and appropriate observational data hold out any real hope that an early empirical "verification" of CO_2-produced climatic change can be widely accepted -- at least before the actual CO_2-induced climatic signal becomes unambiguously large and its associated environmental and societal consequences unavoidable.

Estimating Regional Climatic Responses to CO_2 forcing

Snapshot Modeling

Manabe and coworkers at the Geophysical Fluid Dynamics Laboratory (GFDL), have, for a number of years, used several different three-dimensional general circulation models (GCMs) with simplified thermodynamic oceans to estimate the patterns of regional climatic response to fixed -- doubling or quadrupling -- CO_2 increases (e.g., Manabe and Wetherald, 1975, or Manabe, Wetherald and Stouffer, 1981). Although surface air temperatures increased virtually everywhere -- on the order of 2°C for a CO_2 doubling -- some considerable seasonal and regional departures from global average increases were noted. In particular, high latitudes -- where snow/temperature/albedo feedback effects are important, or in regions where seasonal sea ice is prevalent -- had larger responses than elsewhere. Precipitation and evaporation increased on average, but some latitudes and regions in specific seasons showed marked departure in response from global averages. For example, when CO_2 was quadrupled equilibrium soil moisture in mid-latitude continental areas was decreased significantly in summer (see Figure 1 -- from Manabe, Wetherald and Stouffer, 1981). However, as comparing Figure 1a with Figure 1b shows, a purely internal model change (i.e., spatial resolution) gives a qualitatively similar, but quantitatively different, regional pattern of response. Because only a few years of simulated climate are used in constructing these soil moisture response maps, sampling considerations (see, e.g., Chervin and Schneider, 1976) could add further uncertainty to these results.

Given all these uncertainties, is there any reason at all to take such snapshot simulation responses of GCMs seriously? I believe there is, for some empirical verification of the model's response to large external forcing has been attempted -- and with impressive results. Manabe and Stouffer (1980), for example, plotted (Figure 2) the regional pattern of seasonal temperature range from a seasonal solar forcing simulation along side of observed values. This simulation is based on a fixed oceanic heat capacity at non-ice covered locations from a 68m deep passive thermodynamic mixed layer. Although this depth is a "tunable" parameter which governs the average seasonal temperature range, the geographic distribution of response is not tuned and results from internal model processes. In addition, other independent observations suggest that a 68m mixed layer depth is a reasonable value. Figure 2 shows that this verification exercise does lend confidence to the sensitivity performance of this GFDL GCM. However, as Schneider and Thompson (1980) noted, it is necessary that a valid model reproduce seasonal climatic features, but that this may not be sufficient to engender much confidence that the model will also have a valid response to a different forcing, like CO_2 increase. Nevertheless, the GFDL results were encouraging on the "natural experiment" of seasonal forcing and response, having successfully "passed" this necessary conditions test.

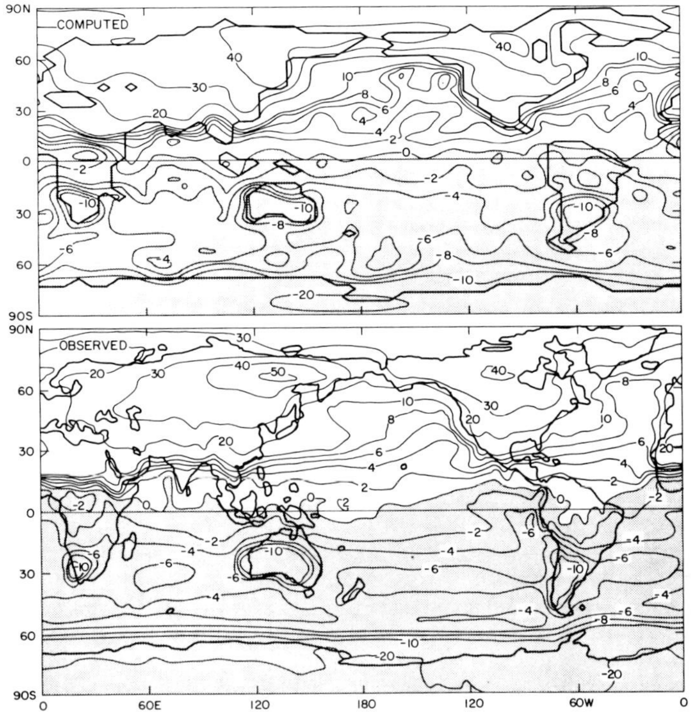

Fig. 2. The difference between the summer maximum and winter minimum surface temperatures computed by the Manabe and Stouffer (1980) simulation and those of observations.

Additional sensitivity tests should be made, particularly intercomparisons of the responses of different models of different resolution and complexity to fixed CO_2 increases. Also, more care is needed in the application of statistical techniques to isolate CO_2 "signal" from model noise in the next generation of simulations.

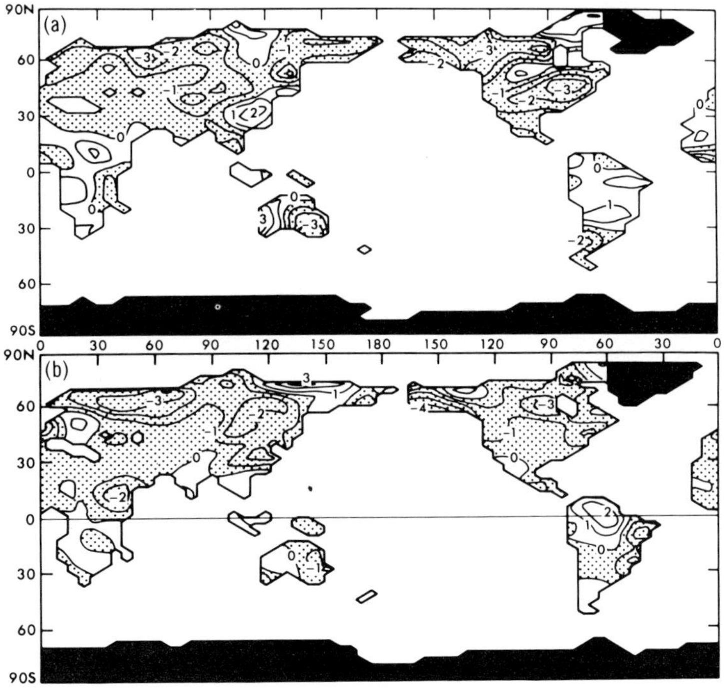

Fig. 1. Geographic distribution of the difference between soil moisture (cm) of a 4-fold CO_2 increase simulation and a control experiment for the June-July-August period. Map (a) is for a 15 wave spectral model and map (b) is for an identical model except for an increase in resolution to 21 waves. (Manabe, Wetherald and Stouffer, 1981).

The Altithermal Analog

Several years ago, William Kellogg (1977) recognized the potential importance of the Altithermal conditions as a CO_2 analog. But this is not a clean analogy for several reasons. First, the Altithermal was not a time of continuous, uniform climatic patterns lasting over several thousands of years. There was considerable shifting of climatic anomalies, with some quite large changes occurring over relatively short times. For example, in a study of July temperature for the past six thousand years in northern Canada it was found that large spatial patterns of warmth and cold could be observed, shifting their centers from time to time every few hundred years or so. While it is fair to say that from about three thousand to six thousand years before present July temperatures averaged all over Canada were warmer than present (e.g., Nichols, 1975) large regional variations occurred (Andrews and Diaz, 1981). Furthermore, this peak in summer warmth in Canada came several thousand years after the mid-Holocene climatic optimum had begun elsewhere on earth, a delay which probably was a result of the cooling effect of the remnant of the Laurentide ice sheet, whose high albedo didn't fully disappear until about 6000 years ago (Bryson et al., 1969).

Kutzbach and Otto-Bliesner (1982) showed that GCM simulations of climatic conditions for 9000 years ago that the Milankovitch orbital cycles created a situation then (high obliquity and planetary perihelion near the summer solstice) where there was several percent more incoming solar radiation in the Northern Hemisphere in summer and a comparable amount of reduced heating in winter. It was this enhanced seasonal forcing, not an overall annual and global increase in heating (like that from extra CO_2), which caused the University of Wisconsin team to simulate enhanced monsoonal activity in Asia and Africa, marked by more summer warmth in continental interiors and greater summer rainfall. The latter results for summer warmth and wetness in the subtropics borne out by a number of paleoclimatic proxies of past activity (e.g., see references in Kutzbach and Otto-Bliesner, 1982). That North America seems to have experienced the Holocene climatic optimum to a lesser extent (and several thousand years later) than had Africa or Asia is also captured by the Kutzbach and Otto-Bliesner simulation, but only when the Laurentide ice sheet remnant of 9000 years ago is also included in their computer simulation. Complexities such as these make the Kellogg or Butzer reconstructions of Altithermal climatic "analogs" to CO_2 highly a speculative exercise.

Even if we were uncertain of the extent to which the Kellogg or Butzer maps are representative of long-term, altithermal-averaged conditions, wouldn't areas of agreement be more likely to be valid? The answer is "not necessarily", of course, since agreement in a few regions of the two maps could still be fortuitous. For example, suppose the worst: each point on each map were of totally random validity. If there are two categories, wetter or drier than now, there would then be a fifty percent probability that any one place on each map were in either the wetter or drier category. In this case of random skill there would be a 25 percent chance that any one place would appear as wetter or drier on both maps. In other words, some 25 percent of the world should be in the "both agree" category by chance sampling of individual points (see, e.g., Chervin and Schneider, 1976 for a discussion of statistical significance of independent grid point values). Moreover, if anomalies at each point are spatially correlated to anomalies elsewhere, estimation of the statistical significance of anomaly fields is even less certain than for point anomalies (Livzey and Chen, 1983).

Are the central North American dry zone or the wetter tropical zones of agreement seen on Figure 4 statistically significant? We cannot know from just two maps. What we need, of course, are more maps constructed from independent data -- or some theoretical reasons why, for example, dryness and warmth should occur together in mid-North America on a warmer earth.

We need to add one extra caveat about warm-earth analogs. Kellogg (1977) himself had pointed out (and Kutzbach and Otto-Bliesner later demonstrated) that unless we can be sure that the Altithermal warm period was caused by an increase in CO_2 (or some equivalent annual, global scale external forcing) we cannot argue convincingly that a global warming induced by burning fossil fuels would result in similar regional climatic shifts -- assuming, of course, we could know reliably what Altithermal seasonal and regional climatic patterns actually were. (And it now appears likely that the extra warmth of the Altithermal was primarily a summer phenomena, not over the whole year.)

With all these caveats, why bother at all looking for a warm-earth analog? The Altithermal analogy does suggest that large, coherent regional shifts in dryness or wetness patterns may well be associated with large-scale temperature changes of only a few degrees Celsius. Even though the important regional details of any CO_2-induced climatic shifts cannot be reliably inferred by analogy alone, analogs can help to provide scenarios of plausible climatic trends from which potential societal impacts might be inferred (Schneider, 1982). This is why they are worth pursuing, provided the user understands the serious limitations inherent in the specific applications of such maps.

Warm Year Analogs in the Twentieth Century

The Altithermal is not the only empirical analogy for a warm earth. Several investigators have looked at a number of warm years in the past half-century, checking to see if any consistent or unusual patterns existed. For example, Williams (1979) examined temperature, rainfall, and pressure patterns in the ten warmest Arctic winters and summers in the past seventy years. Figure 5 (top left) shows the difference in North American summer precipitation between the mean of the ten warmest Arctic-region summers and the remaining summers in the 1900-1969 period. The figure suggests that the North American heartland was largely drier while the South and West became wetter. These results broadly "agree" with Altithermal inferences. Williams also plotted the same variable for winter (Figure 5, top right). Wigley, Jones and Kelly

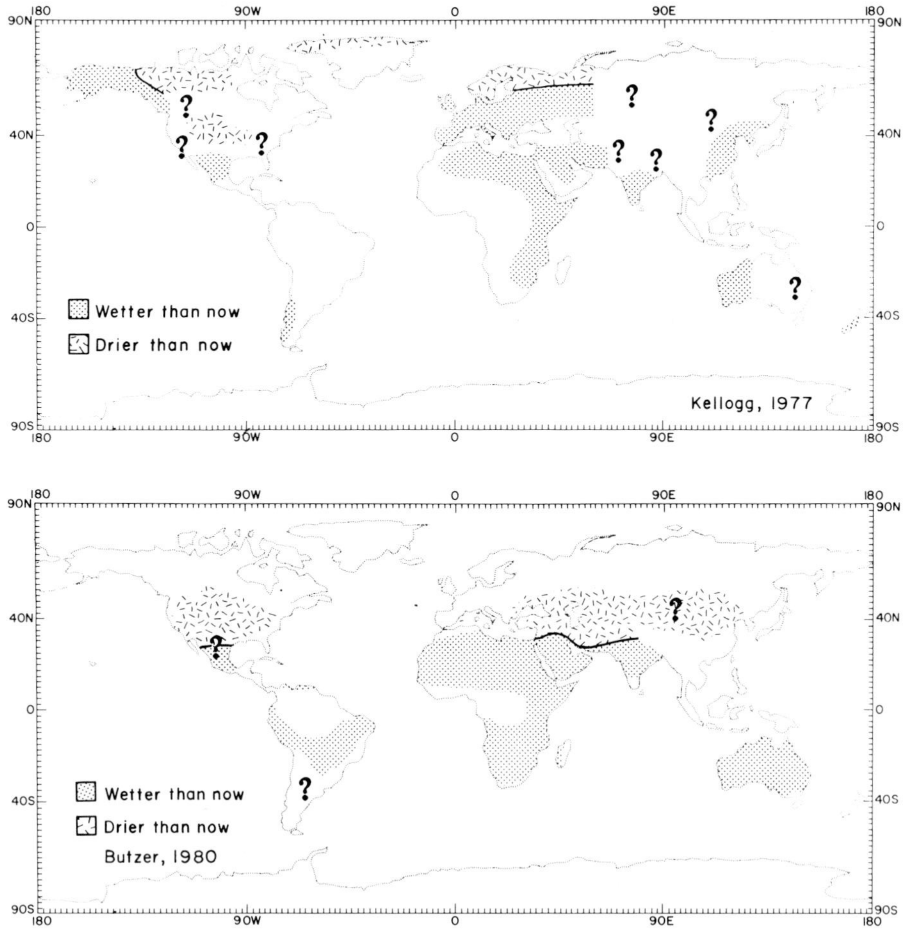

Fig. 3. Maps compiled by Kellogg (1977) (a) and Butzer (1980) (b) representing their estimates of surface moisture condition during the Altithermal period some 5000 to 8000 years ago.

(1979) also used recent data from meteorological networks to search for coherent anomaly patterns in warm years. Figure 5 (bottom) shows their results for precipitation changes from normal for the warmest five years out of the past fifty years. Others have performed similar exercises (e.g., Pittock and Salinger, 1982; Jager and Kellogg, 1983).

Although there is considerable disagreement among the Wigley et al., Kellogg, Butzer, and Williams maps, the mid-North American summer dryness seems a principal common feature, as noted in a review paper by Kellogg (1982). However, three difficulties must be kept in mind. First, as Williams (1979) pointed out, no globally uniform patterns emerge. "Even when the arctic area is warm, there are circulation changes such that large coherent anomalies occur elsewhere, with some regions warming and some cooling." Second, the causes of these sporadic warm years of our century are not CO_2-induced, since they are interleaved between cooler and normal years. Thus, a CO_2-induced warming will not necessarily result in similar seasonal or regional patterns of warmer, cooler, wetter, or drier climates, since CO_2 increase from human activities is an external forcing and individual warm years of the twentieth century could well be internally caused rather than externally forced. Analogously, Lorenz (1979) showed that inferences of cause and effect for longer time "forced" climatic variations can be very different from those described by looking at shorter period "free" climatic variations.

Finally, as mentioned earlier, just because several lines of evidence agree is no guarantee of validity. If all four maps were of random accuracy at every point, then the probability of any point on earth having the same climatic anomaly (drier in mid-North America, for example) on all four charts is still one-sixteenth (that is, one divided by two to the fourth power). In actuality, climatic anomalies do not vary randomly from point to point, but rather are coherent over regions of space. This, as noted earlier, could alter the probability of chance occurrence of the same anomaly in a region on all four maps.

"Reliable predictions of the potential impact of human activities on climate," Williams (1979, p.

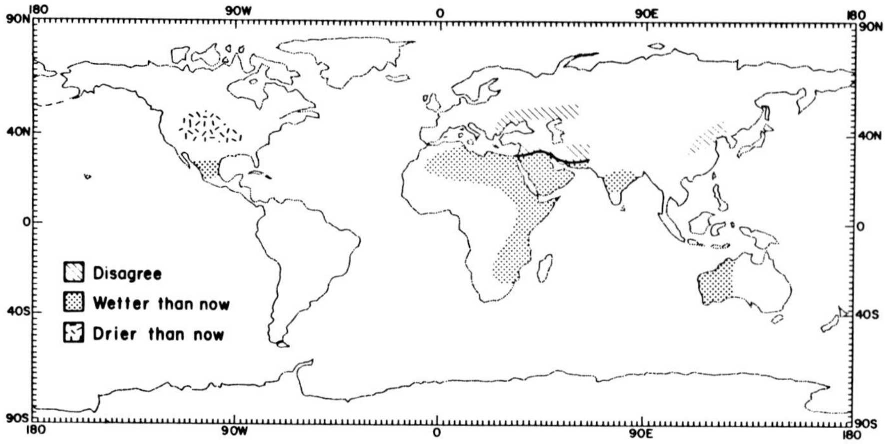

Fig. 4. Kellogg (1982) attempted to compile regions of agreement and disagreement between the two maps seen on Figure 3.

265) writes, "will only be made when a better understanding of the mechanisms of climatic change is achieved." Physical reasonableness is as important as statistical significance tests in engendering confidence that a CO_2-induced climate signal has been detected. This is particularly true when we recognize that statistical significance values depend on physical assumptions as to the number of independent degrees of freedom in both data and model variables (see, e.g., Gilliland and Schneider, 1984).

Direct Verification

CO_2 Forcing and Paleo-climatic Responses

The seasonal cycle provides a powerful, but indirect or surrogate, verification test for climatic models being used to predict the sensitivity of the climate system to CO_2 increases. Successful performance in this indirect test may be necessary before we have confidence in a model, but it is not sufficient, we noted, since CO_2 increase is a different forcing mechanism than seasonal solar forcing. We also need to subject our models to direct verification tests, such as the response of the climatic system in the past to known or suspected variations in CO_2.

CO_2 Variations from Ice Age to Climatic Optimum. In the past the few years two groups--one in Grenoble, France, (Delmas et al., 1980) and another in Bern, Switzerland (Neftel et al., 1982) -- published similar papers on what may yet prove to be among the most interesting recent discoveries dealing with CO_2: CO_2 amounts were tens of percent lower during the last ice age (some 200 ppm) than for the so-called pre-industrial CO_2 level -- (some 260 ppm) (see

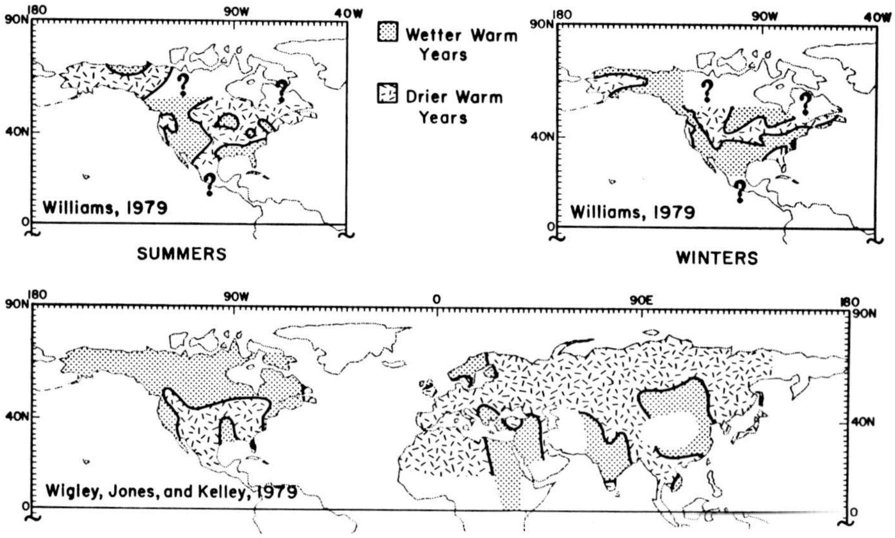

Fig. 5. Warm year analogs of the authors given on the maps, based on twentieth century data sets.

Schneider, 1983). These results were obtained by analysing the CO_2 content of gas bubbles trapped in the ice cores.

Past CO_2 variations from these analyses are at least consistent with the direction of proxy temperature changes: some 3-5°C colder in the late Pleistocene glacial times than today (CLIMAP, 1981; Peterson et al., 1979). Although this consistency could be taken as a direct test of the CO_2 greenhouse theory, it is not conclusive either way for two reasons. First, there is still considerable uncertainty in this ice bubble analysis technique, which is the primary available way to estimate what CO_2 concentrations may have existed over the past few tens of thousands of years. Second, we are certain that other mechanisms, both within and external to climatic system, were also operating (e.g., Hays, Imbrie and Shackleton, 1976; Berger et al., 1983). It is not yet possible to separate unambiguously these simultaneously operating set of forces from the CO_2 effect. Hence, a definitive direct test of the CO_2 - greenhouse warming estimates from climatic models cannot yet be made over these (10^4 year) time scales.

Mesozoic warmth. The mid-Cretaceaus (about 100 million years ago), for example, appears to be one of the warmest -- perhaps 10°C above the present -- periods of the earth's paleoclimatic history (Frakes, 1979). Budyko and Ronov (1979) have estimated that CO_2 levels then were some 8 times greater than at present, and Budyko (1974) has implicated this alleged CO_2 enhancement in the Cretaceaus warmth. However, as Barron, Thompson and Schneider (1981) showed, many other factors -- such as altered continental paleogeography, oceanic heat transport or atmospheric cloudiness -- could well have been major contributors to the Mesozoic warmth. Furthermore, Budyko and Ronov measured the carbonate content of sediments, not Paleo CO_2 levels. They made what they called the "natural approximation" that atmospheric CO_2 and carbonate content were proportional, but such an approximation entirely neglects the problems of carbonate preservation and dissolution which render any such approximations as highly speculative. Thus, once again it could be argued that Mesozoic warmth and suspected high levels of CO_2 are consistent with greenhouse effect theories. However, too many unresolved uncertainties in both paleoclimatic and paleoatmospheric data make verification of model CO_2-climatic theories impossible at present by a Mesozoic analog. But this is a promising area of research.

Faint early sun paradox. The existence of both sedimentary rocks and fossil life in the Archean Age (about 3 to 4 x 10^9 years ago) suggests that the earth's surface temperature was probably stable to within 0-40°C (e.g., see Walker, 1982). At the same time, astrophysicists (e.g., see White, 1977) believe that solar luminosity was some 20-30 percent less in the early Archean than today. Most present climatic models would predict an ice covered earth for such low solar energy output (e.g., see Warren and Schneider, 1979). Sagan and Mullen (1972) suggested that a super-greenhouse effect from gases such as ammonia or methane could explain the faint early sun paradox. Others, such as Owen, Cess and Ramanathan (1979), argued that enchanced CO_2 levels are a more likely explanation. Still others (e.g., Henderson-Sellers and Meadows, 1977, or Rossow et al., 1982) point to hydrospheric variables (i.e., clouds) as their favored explanation. Unfortunately, very little quantitative reliability can be applied to reconstructions of the Archean in the areas of: (1) atmospheric composition; (2) land mass configuration; (3) oceanic depth; (4) planetary rotation rate; (5) geographic extent of liquid water or life; and (6) solar luminosity. Therefore, this ancient paleoclimatic analog is, by itself, of little use to those searching for direct verification of present-day CO_2-greenhouse effect models. Because more data can be obtained for verification, simulation modeling for Mars or Venus today may hold more promise in this regard (e.g., see Pollack, 1979) than does the faint-early sun paradox on earth. On the other hand, it is a fascinating scientific question.

CO_2 and Climatic Changes Since the Industrial Revolution

What of the past hundred years, a period for which we have considerably more knowledge of how climate varied than for the Pleistocene or other ancient paleoclimatic eras? This century is a time in which the very slowly changing orbital forcing, at least, can have played no significant part in observed climatic trends. Unfortunately, before 1958 we have very little reliable evidence of what CO_2 concentration has been. This is based in part on uncertainties over fossil fuel versus biospheric injected CO_2 estimates (e.g., see Broecker et al., 1979 and Moore et al., 1981). Various estimates place the so-called pre-industrial CO_2 concentration between about 260 (Neftel et al., 1982) to 295 (e.g., Oeschger et al., 1980) parts per million, with many recent results suggesting it was closer to the smaller figure (Wigley, 1983; Stanhill, 1982). (In 1982 CO_2 in the air averaged about 340 ppm.) This range of preindustrial levels implies an increase of some 15 to 30 percent in the concentration of atmospheric CO_2 over the past century. If the consensus of most climatic models are roughly correct (see CO_2 Climate Review Panel, 1982), this translates into a global warming of about 0.3 to 0.8°C from 1880 to 1980. This is broadly consistent with the observed global surface air warming seen on Figure 6. Does this finally constitute proof of a direct verification test of climatic models' predictions for CO_2-induced temperature rise? Although some believe this is strong proof, I feel it is simply too soon to

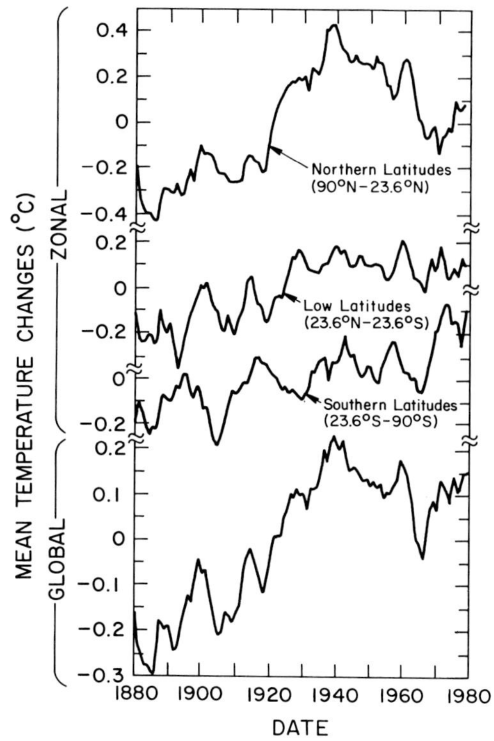

Fig. 6. Surface air temperature trends for the latitudes shown (Hansen et al., 1981).

loudly proclaim "detection" (e.g., see Thompson and Schneider, 1982b). This is because any global temperature signal much smaller than 0.5°C would be hard to detect with less than a decade or two worth of data, since each year the surface air temperature can vary globally by several tenths of a degree.

A number of climatic modeling studies based on semi-empirical methods, (e.g., Hansen et al., 1981; Gilliand, 1982) have shown that recent climatic trends in surface air temperature are not inconsistent with model predictions of CO_2 warming. But the warming is still too small to provide a satisfactory signal-to-noise ratio for the CO_2 causal component, in part because there are still too many plausible competing explanations --e.g., volcanic dust veils or solar constant variations -- of recent global temperature trends. If there alternative, plausible causitive mechanisms are unknown or quantitatively questionable, then we must consider all climatic variations other than our hypothesized CO_2-induced change as "noise". This then creates a high noise level against which any suspected CO_2-induced climatic trend must be compared. This signal-to-noise approach was used by Madden and Ramanathan (1980) who said that we may well have to wait until 2000 A.D. or beyond to detect an unambiguous CO_2 "signal" above the natural background "noise" of the climate. However, as pointed out by Wigley and Jones (1981) the Madden/Ramanathan results only applied at latitude 65°N, and noise levels are usually much smaller in lower latitudes, particularly in summer. Thus, it would be better, they suggested, to look at temperature trends elsewhere than at 65°N latitude, even though climatic models like those of Manabe and co-workers suggest that the equilibrium climatic response to a fixed CO_2 doubling is higher at 65°N than at lower latitudes.

To test the model result that high latitude CO_2-induced signals may be large, Kukla and Gavin, (1981) looked at sea ice records over the past several decades. Although finding no significant trends in the Northern Hemisphere, they found from satellite pictures a decrease in Antarctic summer pack ice of some 2.5 million square kilometers from 1973 to 1978-- about 50 percent of the average coverage. Navy atlases and whaling ship reports dating back some forty years were then consulted. From these it appeared that the ice pack in the 1930s was heavier than the 1970s. Again, these results were claimed to be consistent with observed warming trends, which in turn are consistent with increasing CO_2. But, as Kukla and Gavin properly noted, whether the changes they observed were natural fluctuations or CO_2-induced is not known, and any possible "cause-and-effect" relation cannot yet be established.

All of these attempts to look for regions on earth where a CO_2-induced signal might be more easily detected than in other regions--a so-called precursor indicator--suffer from a common problem: each is attempting to detect a signal derived from an equilibrium climatic model driven by a fixed, large CO_2 increase. But CO_2 is building up steadily over time, which could give rise to a transient CO_2-induced climatic response of very different character from that inferred from equilibrium climatic models.

Equilibrium versus Transient Models

Most climate modelers, at least until recently, did not use a time-evolving scenario of atmospheric CO_2 increases (like those on Figure 7) to force their models. Rather, they computed the equilibrium response of their models to a fixed CO_2 increase -- typically a doubling of the 1958 value. A state-of-the-art estimate for the equilibrium global-surface temperature rise given a CO_2 doubling is, (e.g., see CO_2 Climate Review Panel, 1982) in the range of about 1.5 to 4.0°C. This is as large or larger than most proxies suggest for average Altithermal warming as compared to today (assuming that the botanic proxies for the Altithermal were measuring more than summer warmth). This magnitude of warming at a global level explains why many climatologists have suggested a CO_2 buildup could lead to a climatic change of potentially unprecedented proportions since the end of the last Ice Age. Moreover, any such changes would be, at least on societal timescales, virtually irreversible, as it would likely take many centuries before most of the CO_2 could be removed by the oceans, even if all fossil fuel burning and deforestation activities were abruptly curtailed at some earlier point.

However, estimates of an equilibrium, global-average temperature increase to a steady state CO_2 doubling are, like U.S. EPA automobile gas-mileage ratings, not very useful except "for comparison". Reality will, of course, not see a sudden doubling of CO_2 snapped on one year and held fixed for all time. CO_2 will build up over time like one of those curves on Figure 7. The time-evolving response of the climate system to such a time-evolving CO_2 increase is, of course, its transient response. Possible differences between equilibrium and transient model responses lend considerable uncertainty to the utility of equilibrium-model predictions such as those in Figure 1. Schneider and Thompson (1981) attempted to weigh the seriousness of deliberately neglecting heat capacity of the deep oceans in most models and of using a fixed-CO_2 increase instead of a time-evolving scenario of CO_2 increase.[1] Because the mixed layer of the oceans is generally deeper in high latitudes than low latitudes, a larger column of well-mixed water is in direct contact with the atmosphere in high latitudes. If a CO_2 heating were applied, a deeper mixed layer would heat up more slowly than a shallow mixed layer. Thus, although albedo/temperature feedback processes in high latitudes eventually (that is, in equilibrium) may cause those regions to be warmed more than lower latitudes for a given CO_2 increase, midlatitudes or even some high latitude regions, would not necessarily warm up as fast as lower latitudes during the transition period towards the new equilibrium. In the transition, the larger thermal inertia of the deep high-latitude mixed layer could actually slow the temperature rise relative to tropical waters. Whether this factor causes a serious error in applying results such as Figure 1, Schneider and Thompson noted, depends on how rapidly the CO_2 concentration actually increases over time. If CO_2 increases rapidly, as Rotty (1979) has estimated, neglecting the transient would be more a serious error than if CO_2 increased more slowly, thereby making the results of an equilibrium simulation a closer approximation to the actual time-evolving change (see also the results of Thompson and Schneider, 1982). [Modelers such as Manabe, Wetherald and Stouffer (1981) are perfectly well aware that they neglected the very large oceanic heat storage capacity of the deep oceans in their simulations on Figure 1. The neglect was deliberate, in order to achieve a model response which is near to equilibrium in only a few tens of model simulated years. If deep oceans were included, thousands of simulated years would have been needed to achieve steady state -- a very expensive proposition since the amount of computer time used in any model run is roughly proportional to how many simulated years are calculated.]

The behavioral assumptions underlying the CO_2 buildup rate in the atmosphere also partly dictate which climatic modeling assumptions are

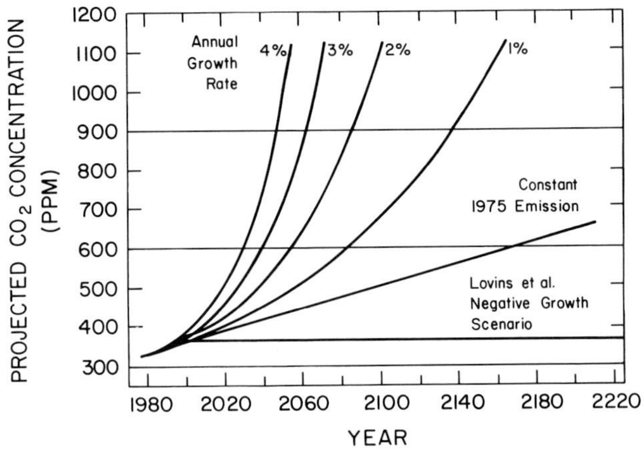

Fig. 7. Various CO_2 scenarios based on specified sets of energy growth assumptions (Lovins et al., 1981).

likely to be most valid. The implication here is that social science concerns must sometimes precede physical scientific investigation, an inversion of traditional research patterns. In this connection, some have suggested (e.g., Ausubel, 1980; Lovins et al., 1981) that a doubling of CO_2 may never occur, since a great increase in coal use may prove economically unfeasible or sufficiently undesirable. But since known reserves of gas and oil could still lead to a further 50 to 70 percent CO_2 increase over the next fifty years, it seems especially important that we de-emphasize the common practice of computing climatic responses to a fixed doubling or quadrupling of CO_2 and focus instead on a much more likely scenario: the transient climatic response to plausible evolving CO_2 concentrations over the next fifty years. In addition, atmospheric water vapor increases with surface temperature increase, accounting for perhaps half of the equilibrium surface temperature response to CO_2 increase. Thus, during the transient warming period oceanic response may well lag that of the atmosphere, thereby temporarily reducing the water vapor increment that would eventually occur after equilibrium was reached.

Still more factors complicate the reliable simulation of time-evolving CO_2-induced climatic changes. Not only are the relative depths of the mixed layers of various parts of the oceans important, but the rates at which waters from below mix with upper layers also affect the warm-up rate of each part of the oceans, given any CO_2-increase scenario (Thompson and Schneider, 1979). In addition, those rates could change with CO_2-induced effects, further altering the long term climatic response to CO_2 (Harvey and Schneider, 1984). Thus, even if the current climatic models were perfectly accurate in predicting the regional equilibrium climatic response to a given fixed CO_2 increase -- such as a doubling -- we do not yet know the effects of neglect of the large thermal inertia of the deep oceans and the fact that annual CO_2 concentrations will increase smoothly over time. Temperature differences from place to place could well evolve very differently in the transient than the temperature differences predicted by equilibrium models or the Altithermal analogy. If so, these analogs or equilibrium calculations could be highly inaccurate predictors of regional patterns of climate change over the next several decades.

Trends in Surface Temperatures: Air Over Land, Air Over Sea and Water Temperatures Are Not the same

Understanding the role of the oceans is rapidly becoming a principal thrust for researchers concerned with the climatic aspects of the CO_2 problem. Indeed, an early attempt to reconstruct trends in ocean surface water temperature (as distinct from surface air temperatures, either over land or sea) by Paltridge and Woodruff (1981) produced results "similar in shape to that derived by (J. Murray) Mitchell from land station data alone, but lags the Mitchell curve by 10-20 years". The delay is comparable to what simple climate models had predicted earlier for the lag between CO_2 forcing and the resulting ocean temperature trends. Unfortunately, Paltridge and Woodruff relied on a historical data set which compiled so-called "bucket temperatures" -- the temperatures of buckets of ocean water hauled out of the seas and recorded by sailors on ships for over a century. Both the geographic coverage and reliability of this record are not of the caliber that climatologists would hope for. (For example, Folland and Kates (1983) showed that a gradual switch from bucket to engine intake temperatures accounts for perhaps half the magnitude of the SST temperature trends computed by Paltridge and Woodruff; the 10-20 year lag, however, was not removed by Folland's correction). But this SST data set is a valuable record, nonetheless, which clearly demonstrates that trends in surface water temperatures, surface air temperatures over water, and surface air temperatures over land can all be distinct and different climatic variables.

It has been the mostly land based surface air temperature records (like Figure 6), which have been used in most direct verification tests between model generated and observed temperature trends (e.g., by Hansen et al., 1981 or Gilliland, 1982). It is important to compare model predictions with observations of all three sets of trends: surface air temperatures over oceans, surface air temperatures at land stations, and surface water temperatures; since these three sets of trends can be different (e.g., see Chen, 1982; Paltridge and Woodruff, 1981; or Gilliland and Schneider, 1984). For all these reasons, one of the greatest uncertainties in climatic model predictions is simulating reliably the time-evolving regional distribution of climatic changes from some CO_2-increase scenario. The most reliable single measure of CO_2-induced climatic change may thus still be the variable that first called attention to the problem: a global surface air temperature increase.

Sea level rise: A direct indicator of CO_2 increase? Another global measure of climate change is sea level variation. Sea levels can change locally because of tectonic activity or isostatic rebound (e.g., Wolcott, 1972) or because of true eustatic (that is, global) ocean water volume changes (e.g., Fairbridge, 1962). Because of the non-eustatic features, local sea level variation trends by themselves have little global meaning. But averages all over the world or at selected places where local effects are small can be indicative of global trends in ocean volume. A warming of the oceans will raise sea levels even if there is no change in global ice volume. This is because warmer water expands, thereby raising sea levels. Measurements from tide gauges in many parts of the world were compiled by Gornitz et al., (1982). They suggest about a 10 cm/100 yrs. trend in sea level rise from 1880 to 1980. After correcting for thermal expansion, Gornitz et al., claim that their results "represent weak indirect evidence for a net melting of the continental ice sheets." A different study of polar ice volume was undertaken by Etkins and Epstein (1982). They concluded that the rotation speed of the earth dropped slightly over the last forty years, mostly because of melting polar ice. Also by calculating the thermal expansion of sea water that would accompany the observed global surface temperature rise of the twentieth century, they too concluded that it alone was not enough to explain all of the increase. "Significant discharges of polar ice must also be occurring", they wrote and their rotation rate arguments were given as added indirect evidence. All this, these various authors presume, is at least consistent with the twentieth century warming trend and the CO_2 buildup. However, care is needed to be certain that the observed temperatures used are sea water, not surface air trends. For this and other reasons it is not yet certain that cause-and-effect should be inferred (Barnett, 1983).

The Combined Greenhouse Effect of Trace Gases

It is also possible that trends in other human activities could work with or against the CO_2-greenhouse effect, further confusing the reliable detection of a CO_2 signal over the next few decades. Table 1 (from Schneider and Londer, 1984) gives some of the possibilities. Already some of these entries have become important in theories of past climatic changes, and some have been used along with CO_2 to explain recent temperature trends, like the 0.15°C or so warming of the globe seen on Figure 6 from 1971 to 1981. For example, it has been known for many years that trace gases like methane or nitrous oxide

Table 1. Summary of Principal Human Activities That Can Influence Climate (Schneider and Londer, 1984)

Activity	Climatic effect	Scale and importance of the effect
Release of carbon dioxide by burning fossil fuels	Increase the atmospheric absorption and emission of terrestrial infrared radiation (greenhouse effect) resulting in warming of lower atmosphere and cooling of the stratosphere	Global: potentially a major influence on climate and biological activity
Release of chlorofluoromethanes, nitrous oxide, carbon tetrachloride, carbon disulfide, etc.	Similar climatic effect as that of carbon dioxide since these, too, are infrared-absorbing and fairly chemically stable trace gases	Global: potentially significant influence on climate
Release of particles (aerosols) from industrial and agricultural practices	These sunlight scattering and absorbing particles probably decrease albedo over land, causing a warming and could increase albedo over water causing a cooling; they also change stability of lower atmosphere; net climatic effects still speculative	Largely regional, since aerosols have an average lifetime of only a few days, but similar regional effects in different parts of the world could have nonnegligible net global effects; stability increase may suppress convective rainfall, but particles could affect cloud properties with more far-reaching effects.
Release of aerosols that acts as condensation and freezing nuclei	Influences growth of cloud droplets and ice crystals; may affect amount of precipitation or albedo of clouds in either direction	Local (at most) regional influences on quantity and quality of precipitation, but unknown and potentially important change to earth's heat balance if cloud albedo is altered.
Release of heat (thermal pollution)	Warms the lower atmosphere directly	Locally important now; could become significant regionally; could modify large-scale circulation
Upward transport of chlorofluoromethanes and nitrous oxide into the stratosphere	Photochemical reaction of their dissociation products probably reduces stratospheric ozone	Global but uncertain influence on climate: less total stratospheric ozone allows more solar radiation to reach the surface but compensates by reducing greenhouse effect as well; however, if ozone concentration increases comparably at lower altitudes, this would lead to potentially very large surface warming; could cause significant biological effects from increased exposure to ultraviolet radiation if total column amount of ozone decrease

Table 1. (continued)

Activity	Climatic effect	Scale and importance of the effect
Release of trace gases (e.g., nitrogen oxides, carbon monoxide or methane) that can increase tropospheric ozone by photochemical reactions	Large atmospheric heating occurs from tropospheric ozone, which enhances both solar and greenhouse heating of lower atmosphere	Local to regional at present, but could become a significant global climatic warming if large-scale fossil fuel use leads to combustion products that significantly increase tropospheric ozone levels; contact with ozone also harms some plants and people
Patterns of land use, e.g., urbanization, agriculture, overgrazing, deforestation, etc.	Changes surface albedo and and evapotranspiration and causes aerosols	Largely regional: net global climatic importance still speculative
Release of radioactive Krypton-85 from nuclear reactors and fuel reprocessing plants	Increases conductivity of lower atmosphere, with possible implications for earth's electric field and precipitation from convective clouds	Global: importance of influence is highly speculative
Large-scale nuclear war.	Could lead to very large injections of soot and cooling lasting several weeks or so, depending on the nature of the exchange and on how many fires were started.	Global, but predominately Northern Hemisphere. Cold and darkness dust causing transient from dust and smoke could wipe out biological productivity, for weeks to months, with severe effects on both natural and agricultural ecosystems of both combatant and non-combatant nations. Reduction of stratospheric ozone and clouds of toxic gases from fires could also be a hazard for several years, as would radioactive fallout and its incorporation into food supplies.

could enhance the greenhouse effect, but their climatic effects have been considered small since their concentrations in the atmosphere are so much less than the two principal IR absorbing gases, water vapor and carbon dioxide. However, Ramanathan (1975) noted that chlorofluorocarbons (CFCs) -- used as spray can propellants or refrigerants -- could do more than interfere with processes responsible for stratospheric ozone. Despite their relatively trace amounts in the atmosphere (measured in units of fractions of parts per billion -- as compared to hundreds of parts per million for CO_2), use of these CFC gases had been growing very rapidly. In addition, they have a very strong greenhouse effect, even in relatively small amounts. Ramanathan calculated that they could be responsible for several tenths of a degree Celsius warming by the end of the century. CFCs are not the only trace gases (other than CO_2) released by human activities with a growing potential to augment any CO_2 greenhouse effect. Nitrous oxide (N_2O) is a by-product of fossil fuel burning and deritrification and may increase significantly as a result of both fossil fuel use and (especially) the growing use of nitrogen fertilizers. Methane (CH_4) is also apparently increasing, probably due to human activities, although its cycling in nature is complex and very much related to biological activity. Moreover, ozone changes themselves can cuase large temperature changes (see Table 1).

Flohn (1980) recognized that these trace greenhouse gases, each associated with human activities causing them to be altered, could, taken together, add a significant increment to a CO_2-induced warming. Flohn calls the combination of these trace greenhouse gases plus CO_2 the virtual CO_2 concentration of the atmosphere.

Earlier we mentioned that Hansen et al. (1981) attempted to match the temperature records of the past century (see Figure 6) to a (one-dimensional column) climatic model-generated temperature based on assumed CO_2, solar and volcanic forcings. Despite the problems with signal-to-noise ratio or surface air (over land or sea) versus surface water temperatures just mentioned, they obtained results which were at least broadly consistent with theoretical expectations. Several months after publishing their first study, the GISS group released another calculation (Lacis et al., 1981); this stressing the combined greenhouse effect of increasing CO_2, CFMs, N_2 and CH_4 over the decade of the 1970s. Interestingly, the computed global equilibrium warming from CH_4, N_2 and CFCs combined was 0.1°C, nearly as large as the 0.14°C warming computed from the CO_2 increase alone during this decade. They estimated, moreover, that the transient effect of oceanic heat capacity would reduce the combined CO_2 plus other trace gases warming for the 1970-1980 period by about half the equilibrium value. Thus, they predicted that there should have been about an 0.1°C warming from all such greenhouse gases during the 1970s. Indeed, there appears to have been such a warming (as seen earlier on Figure 6). But, as Thompson and Schneider (1982b) noted, warming of 0.1°C per decade are not at all uncommon in past temperature records. Thus this coincidence between observed data and model calculation may simply have been chance, not a cause-and-effect relationship. However, the Lacis et al. (1981) estimate for climatic response to increasing greenhouse gases over the decade of the 1980s is larger still, about 0.2-0.3°C -- perhaps twice the decadal average noise level. This prospect prompted the NASA GISS scientists to offer a rather bold prediction: "Global warming due to increased abundance of infrared absorbing gases can be expected to exceed natural variability in the 1980s,..." This statement led Thompson and Schneider (1982b) to comment that if the NASA scientists "are right, then this decade, not the often-cited year 2000, is when a clear warming signal is due. Verification of model predictions by the real atmosphere, of course, takes time; these recent results strongly suggest that the time is closer at hand than previously believed."

Conclusion

If the current crop of climate models is roughly right, the signal-to-noise ratio of global temperature increase from greenhouse gases should become much larger by the end of the century (e.g., Lacis et al., 1981; Thompson and Schneider, 1982b). And if the climatic noise level can be reduced by the technique of subtracting out climatic effects from other known -- or believed known -- natural forcing factors like volcanic dust or solar variations, then detection of a believable CO_2 (plus other trace gases) signal to climatic noise ration should come even sooner. Many uncertainties remain, which should render predictions of regional, transient climatic anomaly patterns from CO_2 increase highly speculative for some time to come. But the advent of a reasonably clear global temperature signal this decade is a real possibility.

References

Andrews, J.T. and H.F. Diaz, Eigenvector analysis of reconstructed Holocene July temperature departures over northern Canada, Quat. Res., 16, 373-389, 1981.

Ausubel, J., Climatic Change and the Carbon Wealth of Nations, International Institute for Applied Systems Analysis, Working Paper WP-80-75, Laxenburg, 1980.

Barnett, T.P., Recent changes in sea level and their possible causes, Climatic Change, 5, 15-78, 1983.

Barron, E.J., S.L. Thompson, and S.H. Schneider, An ice-free Cretaceous? Results from climate model simulations, Science, 212, 501-508, 1981.

Broecker, W.M., T. Takahaski, H.J. Simpson and P.-H. Peng, Fate of fossil fuel carbon dioxide in the global carbon budget, Science, 206, 409-418, 1979.

Bryson, R.A., W.M. Wendland, J.D. Ives and J.T. Andrews, Radiocarbon isochrones on the disintegration of the Laurentide Ice Sheet, Arctic and Alpine Research, 1, 1-14, 1969.

Budyko, M.I., Izmeneniya Klimata, Gidrometeoizdat, Leningrad, 1974.

Budyko, M.I. and A.B. Ronov, Chemical evolution of the atmosphere in the Phanerozoic, Geochem. Internat., 16, 1-9, 1979.

Butzer, K.W., Adaptation to global environmental change, The Professional Geographer, 32, No. 3, 269-278, 1980.

Cess, R.D. and S.D. Goldenberg, The effect of ocean heat capacity upon global warming due to increasing atmospheric carbon dioxide, J. Geophys. Res., 85, 498, 1981.

Chen, R.S., Combined land/sea surface air temperature trends, 1979-1972, Master's Thesis, Department of Meteorology and Physical Oceanography, Massachusetts Institute of Technology, Cambridge, 1982.

Chervin, R.M. and S.H. Schneider, On determining the statistical significance of climate experiments with general circulation models, J. Atmos. Sci., 33, 405-412, 1976.

CLIMAP Project Members, A. McIntyre, Leader, LGM Project, Seasonal Reconstructions of the Earth's Surface at the Last Glacial Maximum, MC-36, 17 maps, 18 pages, 1 microfiche, Geological Society of America, Boulder, 1981.

CO_2/Climate Review Panel, Carbon Dioxide and Climate: A Second Assessment, National Academy Press, Washington, D.C., 1982.

Delmas, R.J., J.-M. Ascencio and M. Legrand, Polar ice evidence that atmospheric CO_2 20,000 yr BP was 50% of present, Nature, 284, 155-157, 1980.

Etkins, R. and E.S. Epstein, The rise of global mean sea level as an indication of climate change, Science, 215, 287-289, 1982.

Fairbridge, R.W., Eustatic changes in sea level, in Physics and Chemistry of the Earth 4, edited by L.H. Ahrens, 99-185, Pergamon, New York and London, 1961.

Flohn, H., Possible Climatic Consequences of a Man-Made Global Warming, International Institute for Applied Systems Analysis, 57, Laxenburg, 1980.

Folland, C.K. and F. Kates, Changes in decadally averaged sea surface temperature over the world, 1861-1980, in Milankovitch and Climate Change, edited by J. Imbrie and A. Berger, Elsevier, Amsterdam (in press).

Frakes, L.A., Climates Throughout Geologic Time, Elseview, Amsterdam, 1979.

Gilliland, R.L., Solar, volcanic and CO_2 forcing of recent climatic changes, Climatic Change, 4, 111-131, 1982.

Gilliland, R.L., and S.H. Schneider, Volcanic CO_2 and solar forcing of Northern and Southern Hemisphere surface air temperatures, submitted to Nature.

Gornitz, V., S. Lebedeff and J. Hansen, Global sea level trend in the past century, Science, 215, 1161-1164, 1982.

Hansen, J., D. Johnson, A. Lacis, S. Lebedeff, P. Lee, D. Rind and G. Russell, Climate impact of increasing atmospheric carbon dioxide, Science, 213, 957-966, 1981.

Harvey, L.D.D. and S.H. Schneider, Sensitivity of transient climate response to ocean model formulation, Part II: Sensitivity experiments with a seasonal, hemispherically averaged coupled atmosphere, land and ocean energy balance model submitted to J. Geophys. Res., 1984.

Hays, J.D., J. Imbrie and N.J. Shackleton, Variations in the earth's orbit: Pacemaker of the ice ages, Science, 194, 1121-1132, 1976.

Henderson-Sellers, A. and A.J. Meadows, Surface temperatures of the early earth, Nature, 270, 589-591, 1977.

Imbrie, J. and A. Berger, Milankovitch and Climate Change, Elsevier, Amsterdam (in press).

Jager, J., and W.W. Kellogg, Anomalies in temperature and rainfall during warm Arctic seasons, Climatic Change, 5, 39-60, 1983.

Kellogg, W.W., Precipitation trends on a warmer earth, in Interpretation of Climate and Photochemical Models, Ozone and Temperature Measurements, edited by R.A. Reck and J.R. Hummel, pp. 35-46, American Institute of Physics, New York, 1982.

Kellogg, W.W., Effects of Human Activities on Global Climate, WMO Technical Note 156, WMO No. 486, World Meteorological Organization, Geneva, 1977.

Kukla, G. and J. Gavin, Summer ice and carbon dioxide, Science, 214, 497-503, 1981.

Kutzbach, J.E. and B.L. Otto-Bliesner, The sensitivity of the African-Asian monsoonal climate to orbital parameter changes for 9000 years BP in a low-resolution general circulation model, J. Atmos. Sci., 39, 1177-1188, 1982.

Lacis, A., J. Hansen, P. Lee, P.T. Mitchell and S. Lebedeff, Greenhouse effect of trace gases, 1970-1980, Geophys. Res. Let., 8, No. 10, 1035-1038, 1981.

Livezey, R.E. and W. Y. Chen, Statistical field significance and its determination by Monte Carlo techniques, Mon. Wea. Rev., 111, 46-59.

Lorenz, E.N., Forced and free variations of weather and climate, J. Atmos. Sci., 36, 1367-1376, 1979.

Lovins, A.B., L.H. Lovins, F. Krause and W. Bach, Least-Cost Energy: Solving the CO_2 Problem, Brick House, Andover, 1981.

Madden, R.A. and V. Ramanathan, Detecting climate change due to increasing carbon dioxide, Science, 209, 763-768, 1980.

Manabe, S., R.T. Wetherald and R.J. Stouffer, Summer dryness due to an increase of atmospheric CO_2 concentration, Climatic Change, 3, 347-386, 1981.

Manabe, S. and R.T. Wetherald, The effects of doubling the CO_2 concentration on the climate of a General Circulation Model, J. Atmos. Sci., 32, 3, 1975.

Manabe, S. and R.J. Stouffer, Sensitivity of a global climate model to an increase of CO_2 concentration in the atmosphere, J. Geophys. Res., 85, 5529-5554, 1980.

Moore B. et al., in Carbon Cycle Modeling, edited by B. Bolin, SCOPE 16, 365-385, Wiley, New York, 1981.

Nichols, H., Palynological and paleoclimatic study of the late Quaternary displacement of the boreal forest-tundra ecotone in Keewatin and Mackenzie, N.W.T., Canada, Occasional Paper No. 15, Institute of Arctic and Alpine Research, University of Colorado, Boulder, Colorado, 67, 1975.

Neftel, A., H. Oeschger, J. Schwander, B. Stauffer and R. Zumbrunn, Ice core sample measurements give atmospheric content during the past 40,000 years, Nature, 295, 220-223, 1982.

Oeschger, H., U. Siegenthaler and M. Heimann, The carbon cycle and its perturbation by man, in Interactions of Energy and Climate, edited by W. Bach, J. Pankrath, and J. Williams, 107-127, Reidel, Dordrecht, 1980.

Owen, T., R.D. Cess and V. Ramanathan, Enhanced CO_2 greenhouse to compensate for reduced solar luminosity on early earth, Nature, 277, 640-641, 1979.

Paltridge, G. and S. Woodruff, Changes in global surface temperature from 1880 to 1977 derived from historical records of sea surface temperature, Mon. Wea. Rev., 109, 2427, 1981.

Peterson, G.M. and T. Webb III, J.E. Kutzbach, T. van der Hammen, T.A. Wijimstra and F.A. Street, The continental record of environmental conditions at 18,000 yr BP: An initial evaluation, Quat. Res., 12, 47-82, 1979.

Pittock, A.B. and M.J. Salinger, Towards regional scenarios for a CO_2-warmed earth, Climatic Change, 4, 23-40, 1982.

Pollack, J.B., Climatic change on the terrestrial planets, Icarus, 37, 479-553, 1979.

Ramanathan, V., Greenhouse effect due to chlorofluorocarbons: Climate implications, Science, 190, 50-52, 1975.

Rossow, W.B., A. Henderson-Sellers and S.K. Weinreich, Cloud feedback: A stabilizing effect for the early earth?, Science, 217, 1245-1247, 1982.

Rotty, R.M., Energy demand and global climate change, in Man's Impact on Climate, edited by W. Bach, J. Pankrath and W.W. Kellogg, 269-283, Elsevier, Amsterdam, 1979.

Sagan, C and G. Mullen, Earth and Mars: Evolution of atmospheres and temperatures, Science, 177, 52-66, 1972.

Schneider, S.H. and C. Mass, Volcanic dust, sunspots, and temperature trends, Science, 190, 741-746, 1975.

Schneider, S.H. and S.L. Thompson, Cosmic conclusions from climatic models: Can they be justified?, Icarus, 41, 456-469, 1980.

Schneider, S. H. and S. L. Thompson, Atmospheric CO_2 and climate: Importance of the transient response, J. Geophys. Res., 86, 3135-3147, 1981.

Schneider, S.H. and R.E. Dickinson, Climate modeling, Rev. Geophys. Space Phys., 12, 447-493, 1974.

Schneider, S.H., Climate impact assessment in the CO_2 context, in Carbon Dioxide Review 1982, edited by W.C. Clark, Oxford University Press, New York, 1982.

Schneider, S.H., The problem of "Pre-Industrial" CO_2 concentration - an editorial, Climatic Change, 5, 311-313, 1983.

Schneider, S.H. and R.S. Londer, The Coevolution of Climate and Life, Sierra Club Books, San Francisco, 576 pp., 1984.

Stanhill, G., The Montsouris series of carbon dioxide concentration measurements, 1877-1910, Climatic Change, 4, 221-237, 1982.

Thompson, S.L. and S.H. Schneider, CO_2 and climate: The importance of realistic geography in estimating the transient response, Science, 217, 1031-1033, 1982a.

Thompson, S.L. and S.H. Schneider, Carbon dioxide and climate: Has a signal been observed yet?, Nature, 295, 645-646, 1982b.

Walcott, R.I., Past sea levels, eustasy and deformation of the earth, Quat. Res., 2, 1-14, 1972.

Walker, J.C.G., Climatic factors on the Archean earth, Paleo., Paleo., Paleo., 40, 1-11, 1982.

Walker, J.C.G., P.B. Hays and J.F. Kasting, A negative feedback mechanism for the long-term stabilization of earth's surface temperature, J. Geophys. Res., 86, 9776-9782, 1981.

Warren, S.G. and S.H. Schneider, Seasonal simulation as a test for uncertainties in the parameterizations of a Budyko-Sellers zonal climate model, J. Atmos. Sci., 36, 1377-1391, 1979.

White, O.R. (ed.), The Solar Output and Its Variation, Colorado Associated University Press, Boulder, 1977.

Wigley, T.M.L. and P.D. Jones, Detecting CO_2-induced climatic change, Nature, 292, 205-208, 1980.

Wigley, T.M.L., The pre-industrial carbon dioxide level, Climatic Change, 5, 315-320, 1983.

Wigley, T.M.L., P. D. Jones, P. M. Kelly, Scenarios for a warm, high CO_2 world, Nature, 283, 17-21, 1979.

SOLAR VARIABILITY AS A SOURCE OF CLIMATE CHANGE

Sabatino Sofia

NASA/Goddard Space Flight Center, Laboratory for Planetary Atmospheres
Greenbelt, MD 20771

Abstract. The current understanding of the physical nature and magnitude of solar variability is reviewed with the purpose of assessing its role as a forcing mechanism for climate. In particular, the discussion is confined to variations of the total solar irradiance, which is directly significant on energetic considerations, as opposed, for example, to variations of the ultraviolet irradiance which, because of the small energy involved, could only affect climate as a trigger mechanism. There is, at the present time, well established evidence of solar variability at the 0.2-0.3 percent level on timescales from tens of days to ~2 years. This variability can be almost fully explained in terms of a modulation of the total irradiance produced as the solar rotation brings into and out of view evolving active regions (i.e., spots and faculae). The most complete current modeling of the active regions modulation indicates that, during the past solar cycle, few months averages of the resulting irradiance did not vary by more than 3 parts in 10^4, and thus it was of little climate consequence. This same modeling, however, successfully produced the general irradiance decrease observed in 1981, which amounted to about 0.1 percent. Competing, simplified models of this modulation for the past cycle, however, lead to larger activity-related modulations, anti-correlated with solar activity. Although the final judgement of which modeling more realistically represents the solar behavior must await further observations currently being made, this simpler model, which predicts an increase of the total irradiance as the solar activity begins to decrease past the maximum early in 1980, does not produce the decline observed in 1981.

There is a second type of modulation of the total irradiance that has characteristic timescales of the order of tens of years, and potentially, larger amplitudes than the active regions modulation. This modulation arises as a consequence of structural variations of the Sun, which have manifested themselves in past changes of the solar radius, and more directly, as a long-term trend in the residuals of the observed values minus the active regions-modulated component under the assumption of a constant non-active photosphere. The physical origin of the structural modulation is the conversion of potential energy into radiative energy and viceversa when the solar outer envelope contracts or expands, probably as a consequence of variations in the dynamo magnetic field. By means of new measurements of the variations of the solar size to be carried out by instruments currently under design, and in conjunction with irradiance measurements to be made with existing instruments, a definitive resolution of the question of solar forcing of climate change is within grasp.

Introduction

In this paper we shall address the issue: "does solar variability, or more precisely, variability of the total solar irradiance, produce a significant forcing of the climate system?" Because the ultimate origin of most of the energetic input in the climate system is solar, the idea of solar forcing of climate has been a tempting one for over a century. However, despite the considerable effort expended working on this question, an answer generally acceptable to a large portion of the scientific community has remained elusive. It is our contention that the principal source for this difficulty is that the techniques almost exclusively used to seek the Sun-climate link have been of a statistical nature. The use of statistics is seductive since it allows, in principle, the possibility of finding correlative links between different phenomena even in the absence of an understanding of the nature of the link. In the Sun-climate case, however, this approach has not been very successful, on two accounts: (a) the record of solar variability is incomplete, inhomogeneous, and affected by large systematic effects, and the climate record, while more complete than the solar one, is also subjected to unknown systematic bias, introduced primarily by very inhomogeneous sampling; and (b) the existence, within the climate system, of large energy reservoirs (land-mass, oceans, atmosphere, etc.) which strongly modulate the climate response to solar variability to such an extent as to make it unrecognizable, at least on timescales short enough so that the integral value of the solar modulation is small when compared to the total energy content of these reservoirs.

The above discussion leads to a natural separation of the problem in three categories, according to the timescales considered. If one considers timescales on the order of a billion years, we encounter the well-known faint-Sun paradox, which has received a great deal of (deserved) recent attention (Crowley, 1983). This problem, while fundamental in regard to a complete understanding of the reactions of a "planetary climate system" to solar change, is beyond the scope of this paper, since its solution is based on the fact that the early climate system of the Earth (atmospheric composition, land-mass distribution, etc.) was very different from the present one. The second regime considers timescales on the order of tens of thousands of years. In this regime, the orbital parameters of the Earth undergo significant changes, and produce the so-called Milankovitch effect. Although the changes of magnitude and yearly averaged geographical distribution of the radiative energy reaching the Earth due to this effect are rather subtle, it appears that the timescales considered are sufficiently long so that the influence of the Earth-based energy reservoirs average out, and the solar signature is quite generally accepted (cf. North et al., 1983). This problem, although fascinating in its own right, involves such small parameter variations, and large timescales, that its relationship to climate change in reasonably well recorded epochs is questionable. Because of this, we will not consider this regime in the present paper.

The subject of this paper is solar variations in the third regime, which spans timescales ranging from a few to hundreds of years. This regime is of a particularly high value since it includes the best (although by no means complete) recorded climate history, and it also contains sufficient climate change (mini ice ages, dust-bowls, etc.) to make hypothesized cause-effects between the Sun and climate meaningful. Moreover, we will concentrate on variations of the <u>total</u> solar irradiance, as opposed to spectral irradiance such as the EUV-UV regime. The reason for this omission is that the total energy content in the solar EUV-UV range is very small, and any plausible climatic effects due to its well-known large variation would be indirect, through modification of the circulation patterns and abundance of absorbing trace chemicals in the Earth's atmosphere. This type of change is of a vastly different nature than the much simpler modulation of total energy deposition on the Earth surface that we envision through the total irradiance variations.

In summary, then, we will discuss the current evidence for the existence of significant variations on the total solar irradiance having timescales in the few to hundreds of years range.

Total Irradiance Modulation Mechanisms

The total solar irradiance, S, (i.e. the amount of energy reaching a unit surface at a distance of one astronomical unit per unit time in the direction of the Earth) can be modified either by a change of the solar luminosity, L, (total energy output of the Sun per unit time), or by redirecting part of the energy output. Moreover, a change of L may occur either as a consequence of a global solar change that leads to relatively long-term changes of the energy output, or through more short-lived, localized modifications of the energy flow that only modify L on a short term. We will discuss hereafter evidence that in fact most of these modulation processes indeed occur, as indicated by a variety of direct and indirect measurements described in the following sections.

Direct Measurements

The solar irradiance (also known as the solar constant) has been the subject of extensive measurements since before the beginning of this century from the ground, and since the early days of the space program from outside the Earth's atmosphere. A summary of these measurements can be found, for example, in the proceedings of a conference on the solar constant (Sofia, 1981). We can summarize these measurements by saying that reliable strings of directly intercomparable data have become available only in the last 3-4 years, as a result of measurements made on Nimbus 7 (Hickey et al., 1981), and especially, on the ACRIM experiment on the Solar Maximum Mission (SMM) (cf. Willson et al., 1981). The quality of these measurements can be appreciated by looking at the figures in these papers. Nearly two years of observation by SMM, in 1980 and 1981, have shown the frequent occurrence of modulation of S at the < 0.2 percent level with duration of days, followed by recovery. During this same period, the average level of S (over, say a few months) had one depression, at the 0.1 percent level, which reached minimum in August 1981, and turned around thereafter (Frazier, 1982). In the first paper reporting a significant stretch of data from the SMM-ACRIM experiment (Willson et al., 1981) it was noted that the two large dips in S around days 53 and 100 of 1980 coincided with central meridian passage of large spot groups. Hudson et al. (1981) further elaborated modeling of these data based entirely on sunspot blockage of radiation. In this picture, the variations of S are due entirely to negative excursions from the undisturbed photospheric level of a spot-free Sun. However, it is well known that the white light faculae that surround spot groups enhance locally the solar energy flow. This component was included in a model of the active regions modulation first proposed by Oster et al. (1982) and described in more detail by Sofia et al. (1982). When we apply this model

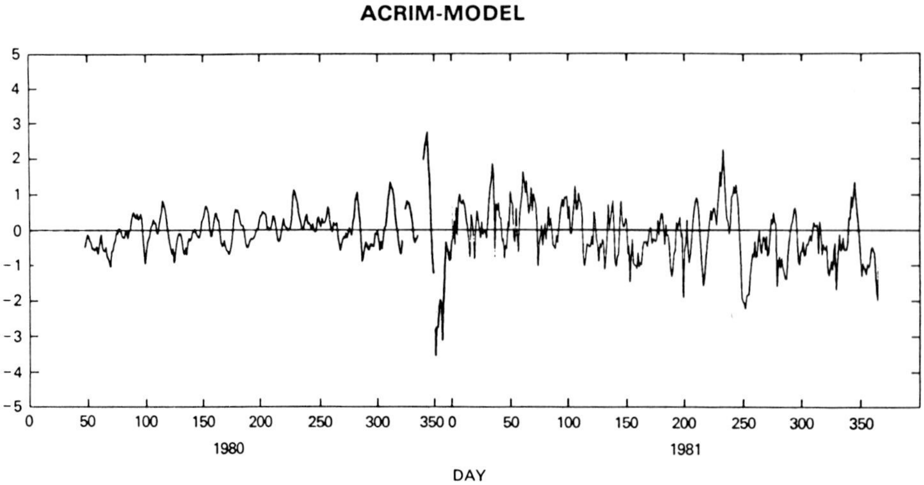

Fig. 1. The residuals (measurements minus model computed by the Oster et al. technique) of the total solar irradiance for the years 1980 and 1981. The amplitude of the residuals is approximately 1 W/m^2, which is ~0.07 percent, and they are primarily due to uncertainties in the active area sizes. The largest departure occurred at the end of 1980 when the attitude control of SMM suffered a catastrophic deterioration. This probably produced the fortuitous oscillation, and altered the nature of the residuals, in the sense that the post-failure values have much higher-frequency components than the pre-failure values. Notice the slight hint of a curvature in the residuals, which are more negative at both ends of the run. The slow decrease of the irradiance which reversed in August 1981 is well modeled, since the residuals do not show such a feature. The ordinate is Watts/m^2.

to the nearly two-years of ACRIM-SMM observations currently available, we find the residuals (observations minus modeling) shown in Figure 1. We see from the residuals that while the fit is not perfect, it does not deteriorate with time, and in particular, it does not increase at the time of the large depression in August 1981. This behavior suggests that the disagreement of the fit is largely brought about (as stated in Sofia et al., 1982) by uncertainties in the measured parameters of the active regions, and is independent of the level of the solar activity, as such activity decreased considerably during the period of observation. An alternative modeling of active regions was subsequently proposed by Hoyt and Eddy (1982). The attractive feature of this model is the prediction (just like in the spot-only model) that the solar irradiance is depressed near solar maximum, and increases to a maximum at solar minimum, i.e., there is an 11-year solar activity-related signature in the total irradiance whose evidence can be sought in the climate history. By contrast, the Oster et al. model produced almost negligible cycle-related variations of the S level during the last solar cycle (Schatten et al., 1982). However, Hoyt and Eddy's model is overly simplified in that it assumes a fixed ratio of facula to spot areas, and only one of two possible contrast values as a function of angular distance from the Sun's center. In fact, since faculae outlive spots by approximately a factor of five, one most often finds faculae not related to current spots, so that neither sizes nor locations are that readily correlated. By contrast, Oster et al. used calcium plages (readily available on a daily basis) as tracers of faculae, and they assumed that facula areas and plage areas are the same. Although this latter assumption may not be strictly true, it is not off by a large percentage, and it allows to include an accurate treatment of the center-to-limb variation of the contrast (Sofia et al., 1982). This leads us to believe that the Schatten et al. (1982) results are more realistic. As stated earlier, though, the final judgement of which model is more realistic must come from further observations.

Active regions modulation, then, appears to account quite satisfactorily for the day-to-day variations of the total irradiance measured by the ACRIM experiment on SMM, up to nearly 2 years. The only departure may be hinted by a slight curvature in the residuals which one may see in Figure 1 if one indulges in some wishful thinking. This curvature may be a hint of longer term modulation of S which can be traced to a modulation of the undisturbed photosphere which, in the present modeling, has been assumed to be constant.

Indirect Methods

The longer timescale modulation of the solar radiant output hinted at by the residuals curvature of Figure 1, is supported by indirect measurements of the solar luminosity. From quite general considerations (cf. Sofia and Endal, 1982)

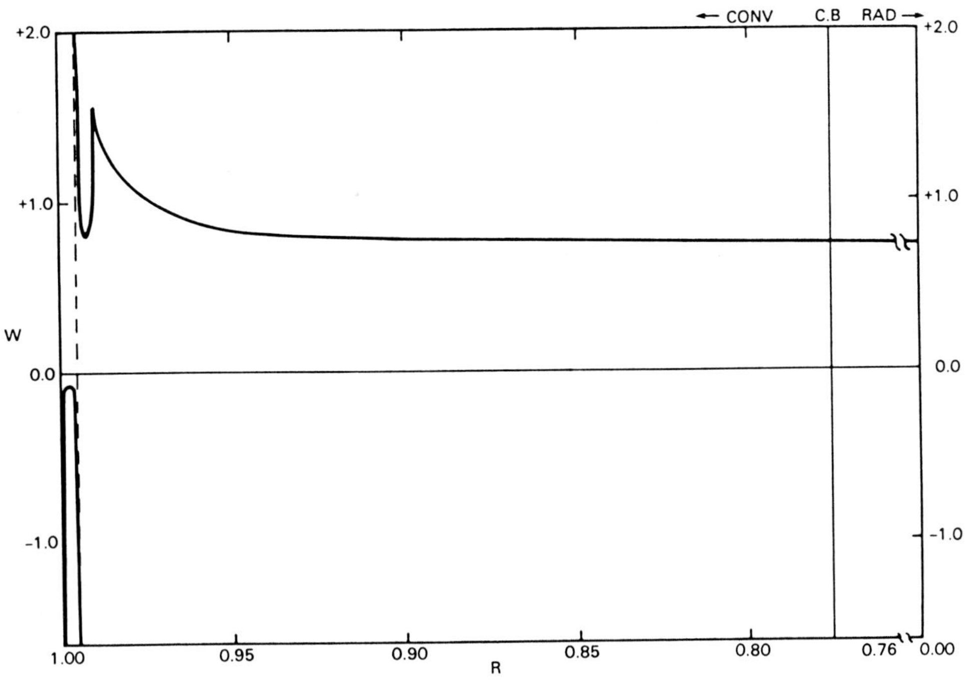

Fig. 2. Parameter $W = \delta \ln R / \delta \ln L$ as a function of radius R in the Sun where the perturbing mechanism operates. Notice that, as long as the perturbation occurs interior to 95 percent of the solar radius, the value is well-behaved, and ~0.7. The base of the solar convection zone, where the dynamo field most likely resides, is at R = 0.75.

it can be shown that a change in L, particularly on timescales below millions of years, must be accompanied by a change in the solar radius R (Sofia et al., 1979). The questions to be answered are: (a) have there been reliably documented variations of the solar radius?, and (b) what is the value of $W = \delta \ln R/\delta \ln L$, which relates these variations? Extensive, and often controversial work has been done relating to both points. To the satisfaction of this author, the current body of evidence supports the occurrence of variations of the solar diameter at the level of ~0.5 arc sec, with timescales of tens to hundreds of years (cf. Sofia et al., 1983; Dunham et al., 1981; Parkinson et al., 1980).

Regarding the value of W, extensive numerical simulations have been carried out including the effects of various perturbations operating at a variety of depths within the solar interior, and at different intervals of time. The results, to date, can be summarized as follows: Let us assume that the perturbing event occurs at time t_0, and that measurements of R and L are made at times t_1 and t_2. If the perturbations are infrequent enough so that for any t_1 and t_2 the inequality

$$t_0 << t_1, t_2$$

holds, then $W = 0.3 \pm 0.1$ for perturbations of all types and amplitudes, occurring anywhere within the Sun. It must be stressed that as the Sun slowly evolves, its R and L change, and the W values obtained for those changes is also the above. If, on the other hand, perturbations occur so frequently that for any t_1 and t_2 we always have $t_1 \leqslant t_0 \leqslant t_2$, then the resulting W, independent of the type and magnitude of the perturbation, depends on the depth within the Sun where the perturbation occurs. This is shown in Figure 2, which summarizes work of Endal and Twigg, 1982; Twigg and Endal, 1981; Sweigart, 1981, and 1983; as well as work in progress at GSFC.

Combining the fact that $\delta R/R \sim 5 \times 10^{-4}$ have occurred, and that W = 0.3 (for infrequent perturbing events) or 0.7 (for frequent perturbing events at depths below 95 percent of the solar radius) we find

$$\frac{\delta L}{L} = \begin{matrix} 0.17 \text{ percent or} \\ 0.07 \text{ percent} \end{matrix}$$

I do not wish to elaborate on the climatic consequences of these changes, as this is the object of several papers in this Symposium. However, from these investigations it appears that global changes as large as 0.3°C for the temperature of the Earth, on timescales of many tens of years, may be produced by structural re-adjustments of the Sun. Since this temperature effect is enhanced at middle to high latitudes, where the edge of the ice-caps is located during ice ages, important climate consequences caused by this type of solar variability are very likely. Moreover, the smallness of this modulation would explain the difficulty of detecting it by statistical means, since a substantially larger modulation could clearly be detected from correlations between solar and climate data.

What do we do next? The first thing to do is to determine the value of W from observations. This requires the simultaneous measurement of ΔR, ΔS and the active regions. From the active regions measurements we can reduce ΔS to ΔL, and this, together with ΔR, gives W. The experimental value of W establishes the nature of the process responsible for the structural perturbations of the Sun. Alternatively, we need to pursue the most complete examination of historical data, particularly the duration of eclipses, to find out the amplitude and timescale (periodicity?) of R changes of the past. Finally, we need to examine the consequences of the resulting temperature changes in climate models and compare them with climate data. It appears that, finally, a definite resolution of the question of solar forcing of climate change is within grasp.

Acknowledgments. I would like to acknowledge the capable help of Ken Schatten and Larry Twigg in running some of the calculations reported in this paper, as well as conversations with them and with A. Sweigart, K. Chan and A. Endal.

References

Crowley, T.J., The geologic record of climatic change, Rev. of Geophys. and Space Phys., in press, 1983.

Dunham, D.W., J.B. Dunham, A.D. Fiala and S. Sofia, Eclipse radius measurements, in Variations of the Solar Constant, edited by S. Sofia (NASA Conference Publication 2191), 117-122, 1981.

Endal, A.S. and L.W. Twigg, The effect of perturbation of convective energy transport on the luminosity and radius of the sun, Astrophys. J., 260, 342-352, 1982.

Frazier, K., Our flickering sun: fluctuating solar flux, Science News, 121, 294, 1982.

Hickey, J.R., B.M. Alton, F.J. Griffin, H. Jacobowitz, P. Pellegrino, E.A. Smith, T.H. Vonder Haar and R.H. Maschhoff, Solar variability indications from Nimbus 7 satellite data, in Variations of the Solar Constant, edited by S. Sofia (NASA Conference Publication 2191), 59-72, 1981.

Hoyt, D.V. and J.A. Eddy, An atlas of variations in the solar constant caused by sunspot blocking and faculae emissions from 1874 to 1981, NCAR Technical Note 194 + STR, 1982.

Hudson, H.S., S. Silva, M. Woodward and R.C. Willson, The effects of sunspots on solar irradiance, Sol. Phys., 76, 211-219, 1982.

North, G.R., J.G. Mengel and D.A. Short, A simple energy balance model resolving the seasons and the continents: application to the astronomical theory of ice ages, J. Geophys. Res., to appear, 1983.

Oster, L.F., K.H. Schatten and S. Sofia, Solar irradiance variations due to active regions, Astrophys. J., 256, 768-773, 1982.

Parkinson, J.H., L.V. Morrison and F.R. Stephenson, The constancy of the solar diameter over the past 250 years, Nature, 288, 548 - 551, 1980.

Schatten, K.H., N. Miller, S. Sofia and L. Oster, Solar irradiance modulation by active regions from 1969 through 1980, Geophys. Res. Lett., 9, 49-51, 1982.

Sofia, S., J. O'Keefe, J.R. Lesh and A.S. Endal, Solar constant: constraints on possible variations derived from solar diameter measurements, Science, 204, 1306-1308, 1979.

Sofia, S., Editor, Variations of the Solar Constant (NASA Conference Publication 2191), 1981.

Sofia, S., L. Oster and K. Schatten, Solar irradiance modulation by active regions during 1980, Solar Phys., 80, 87-98, 1982.

Sofia, S., D.W. Dunham, J.B. Dunham and A.D. Fiala, Solar radius change between 1925 and 1979, Nature, to appear, 1983.

Sweigart, A.V., Effects of changes in convective efficiency on the solar radius and luminosity, in Variations of the Solar Constant, edited by S. Sofia (NASA Conference Publication 2191), 143-164, 1981.

Sweigart, A.V., Effects of core perturbations on the structure of the sun, Astrophys. J., to appear, 1983.

Twigg, L.W. and A.S. Endal, Thermal perturbations of the Sun, in Variations of the Solar Constant, edited by S. Sofia (NASA Conference Publication 2192), 137-142, 1981.

Willson, R.C., S. Goulkis, M. Janssen, H.S. Hudson, and G.A. Chapman, Observations of solar irradiance variability, Science, 211, 700-702, 1981.

SOME ASPECTS OF THE CRYOSPHERE AND ITS ROLE IN CLIMATIC CHANGE

Charles R. Bentley*

Geophysical and Polar Research Center, University of Wisconsin-Madison
Madison, WI 53706

Abstract. The ice cover of the earth plays a direct and important role in climate and climatic change. Sea ice responds rapidly and sensitively, and is important because of effects on albedo, heat exchange between ocean and atmosphere, and oceanic mixing. Glaciers (including ice sheets) interact on a longer time scale, but some mechanisms for significant changes on a human time scale do exist. The possibility of rapid disappearance of the West Antarctic ice sheet with a consequent 5-m rise in sea level has received much attention.

There is some evidence that the extent of the Antarctic sea ice cover had diminished in the last few decades. Evidence related to the ice sheets is ambiguous -- the measured rise in sea level suggests slow melting of the ice sheets, but field glaciological measurements, incomplete though they are, indicate just the opposite. Whatever the effect of climatic warming on the marine West Antarctic ice sheet, it appears to be physically impossible for it to go afloat in as little as 200 years; a more reasonable estimate of the minimum time needed is 500 years.

Introduction

In any consideration of the evolution of present-day polar ice bodies and their possible response to future climatic changes, it is important first to recognize that glacial systems behave in at least four fundamentally different ways. The first is stable steady-state, wherein the system does not change from steady state, i.e., a state in which the dynamical description does not change with time (annually averaged -- we are not concerned with seasonal changes), unless there is a change in some boundary condition. For a small change (e.g., change in surface temperature) the response (e.g., change in marginal extent) is approximately linear and reversible, meaning that doubling the change in the boundary conditions will double the response, and that returning the boundary condition to its original value will result in a return of the glacier to its original configuration. The response does not occur instantaneously, however -- the lag time depends on the size of the glacier and can be many thousands of years for a large ice sheet.

The second, cyclic steady-state, involves inherent cyclic processes whereby the extent of the ice changes in a periodic or pseudo-periodic way due entirely to internal causes, without change in the externally-imposed boundary conditions. The state of the system is constantly changing, although the changes may be quite small most of the time. Both instantaneously and averaged over a length of time long compared with the cyclic period, however, the system may be stable in the sense previously described. The primary example of such behavior is the phenomenon of periodic glacial surging, known to occur in valley glaciers and recognized as at least a possibility in major ice sheets.

The third type of behavior is unstable steady-state, wherein the system can continue without change in dynamical description if boundary conditions do not change, but the response to an external change is irreversible -- the system will move toward a new steady-state position, and will not return to its original configuration.

Finally, transience simply refers to a dynamical description that is currently changing without any current change in boundary conditions, and generally represents a transition toward a steady-state, although in reality there is no guarantee that steady-state behavior will ever be reached.

Of course, real glacial systems are complex and may involve combinations of types of behavior. It is quite possible to have short-term stability superimposed on a longer-term transience, for example. Or, a system may be stable with respect to some boundary conditions and unstable with respect to others. Furthermore, the boundary conditions themselves may be dependent on the system response (feedback); for example, change in surface elevation of an ice sheet can change the surface temperature and mass balance.

It is important to keep the basic distinctions

*Geophysical and Polar Research Center, contribution number 407.

in mind, particularly with respect to cyclic steady-state and unstable steady-state. The word "surge", commonly used to describe the rapid-advance phase of a glacier in cyclic steady state, has recently been used also to include the response of an ice sheet to a push from a position of unstable steady-state. This is unfortunate, particularly in a discussion of interaction with climate, because cyclic steady-state changes proceed, to the extent that the whole cyclic process is stable, without or even despite changes in the (externally-imposed, annually-averaged) environment, whereas displacement from unstable steady-state occurs only because of changes in the environment. The fact that the mechanisms of cycling and unstable response may overlap in some cases only makes it more important to keep the terminology straight. We will use "surge" in connection with cyclic changes only.

The problem of long-term evolution of Arctic and Antarctic ice bodies and their possible responses to future climatic trends can (and should) be attacked in two quite different ways - by examining the empirical evidence of conditions during past climatic epochs, and by theoretical analyses of the interactions between atmosphere, ice, and ocean.

There are several aspects of climatic change to be considered. The most obvious are those directly relating to the atmosphere, such as air temperature, precipitation rates, and cloudiness. Equally or even more important for some mechanisms are the oceanic aspects: water temperature, currents, and sea level. Finally, there are those related to the cryosphere itself, especially the extent and thickness of the sea ice cover, which may interact directly with the margins of the more permanent ice.

We will first look at specific response mechanisms, then examine the evidence for present changes, and finally consider expectations for the future. The discussions of sea-ice cover and ice sheets, which must be separated because of quite different mechanisms and time scales, are carried along in parallel. The sections concerned with sea ice draw heavily on contributions by R.G. Barry to an (as yet) unpublished report of the National Research Council. The author is indebted to Dr. Barry for use of this material.

Response Mechanisms

Sea-Ice Cover

Sea ice is a critically-important element in the world climate system, one that both affects climate and responds sensitively to climatic change. There are three principal processes by which sea ice acts upon the climate. First is the direct effect it has upon the earth's albedo, since snow-covered ice can have an albedo as high at 0.9 compared to 0.1 for water. Secondly, the ice cover acts as an insulating blanket, interfering with free heat exchange between the ocean and the atmosphere. Mid-winter and multi-year ice reduce the heat transfer by two orders of magnitude, so that the areal extent of leads and polynyas characterized by open water and regions of thin ice is of crucial importance. The third process is the change in surface water salinity that occurs whenever the ice either melts, resulting in a freshwater lid that enhances the stability of the water column, or freezes, thus increasing the salinity and the surface density, often to the point of convective over-turning and the production of oceanic deep or bottom water.

Important feedback mechanisms also exist. The most obvious one is regenerative: cooling atmospheric temperatures result from the high albedo of the sea ice cover; the lower temperatures lead to increasing areal extent of the ice and a consequent further lowering of the temperature. Another involves the stability of the water column and is degenerative -- the convective over-turn that results from the freezing of the ice can bring to the surface warmer intermediate water that tends to melt the ice. On the other hand, the existance of the insulating ice cover can have a regenerative effect by reducing vertical mixing, thus decreasing the advection of warmer, deeper water.

The actual mechanisms involved in the interaction between ice, ocean, and atmosphere include not only those already mentioned, which relate to thermal energy transfer, but also the effects of winds and currents, and the impact of freshwater run-off from land on the salinity and temperature structure of the ocean. Different factors may be of different importance depending on the time scale of concern. For example, variations in ice extent in the Greenland Sea have been attributed to wind and atmospheric temperature effects on a short-term basis (weeks or months) (Sanderson, 1975) but to oceanic heat transport for interannual variations (Skov, 1970). A further complication arises from longitudinal differences in the times of maximum and minimum ice extent around the Arctic Ocean and its marginal seas that can be as much as six months (Rogers and van Loon, 1979). Because of the extreme complexity of the interactions, theoretical analyses are still in their early development phase.

Ice Sheets and Glaciers

In considering the response of glaciers and ice sheets to climatic change, one must first define the terrestrial ice bodies (including Arctic glaciers, the Greenland ice sheet, and most of the East Antarctic ice sheet), which rest on a substrate close to or above sea level, and marine ice sheets (principally the major part of the West Antarctic ice sheet), which comprise ice grounded below sea level and the floating ice shelves. The reason for making this distinction is that terrestrial ice bodies interact directly only with the atmosphere, so they probably change their form and extent rather slowly in response to climatic changes, whereas marine ice sheets also interact directly with the ocean as well, and are therefore sensitive to, and capable of rapid response to changes in sea level, ocean temperature, and ocean currents. This must be understood, as it is fundamental to arguments that propose the possibility of catastrophic instability.

In the following sections we will, therefore, first consider terrestrial ice sheets, ignoring smaller ice caps and glaciers that can have only a minor effect on world climate, and then marine ice sheets, comprising the grounded inland ice and the adjoining ice shelves.

Terrestrial ice sheets. The term "ice sheet" implies ice flow from a high central region. In a series of papers, Weertman (1976, and references therein) first showed that an idealized, two-dimensional, ice sheet resting on uniform rock above sea level is in stable steady state, if the position of the snow line, which divides the interior zone of net mass gain from the peripheral zone (if any) in which melting predominates, depends primarily on the distance from the center of the ice sheet. If, however, the position of the snow line depends primarily on elevation above the ice sheet margin, then the ice sheet is unstable and will either grow or shrink indefinitely, depending on the specific climatic conditions. He then developed a model that incorporated both possibilities and examined the effect of changing the surface mass balance, the extent of a basal meltwater layer, and the strength of the solar radiation on the size and rates of growth or decay of terrestrial ice sheets. A common feature of his and similar models is that very large snow accumulation rates would have been needed to produce growth rates of the Pleistocene ice sheets that correspond with the rates of sea level lowering indicated by studies of sea floor sediments, and that unreasonably large melt rates would have been required to explain the rapid retreat of the Laurentide Ice Sheet. This last point, together with glacial geological evidence, led to the conclusion that much more rapidly reacting marine ice sheets, particularly in and around Hudson Bay, probably played an important role in ice sheet growth and retreat (Denton and Hughes, 1981).

However, marine ice sheets are not necessarily required for rapid change. Using a "frictional lubrication model" for basal sliding that includes the important factor of warming and melting of the ice by internal friction, particularly at and near the glacial bed, Budd and Jenssen (1975) found that natural rapid flow in any ice sheet is physically reasonable. Carrying this idea further, Budd and McInnes (1979) found that a numerical model showing surges can be produced by a particular choice in the value of the parameter taken to represent the effect of internal friction. The time between surges calculated from the computer models ranges from 10,000 to 25,000 years, depending on the size of the sector involved. The true values of the controlling parameters are not well enough known to permit the modeling results to be cited as proof of surging; nevertheless, the results are significant in showing that the concept is at least reasonable. Using this model, Budd and Smith (1979) were able to reproduce the general characteristics of the growth and decay history of the Laurentide Ice Sheet in response to variations in the earth's orbital parameters without recourse to marine ice sheets.

CROSS SECTION OF MARINE ICE SHEET

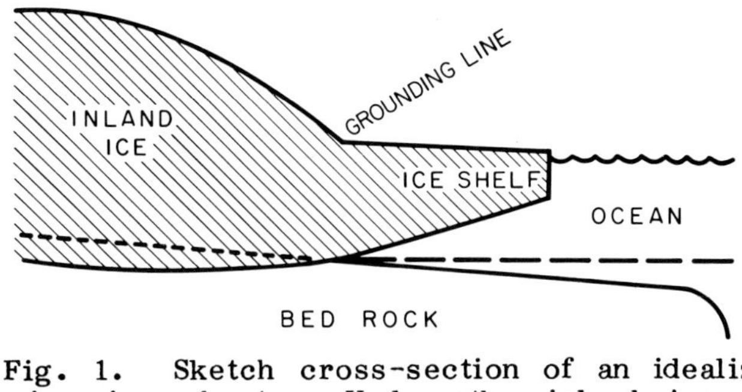

Fig. 1. Sketch cross-section of an idealized marine ice sheet. Under the inland ice the dashed and solid lines depict the undepressed and isostatically depressed bedrock surfaces, respectively.

Marine ice sheets. A marine ice sheet is, in principal, more vulnerable to rapid change than a terrestrial ice sheet simply because its rock bed lies far below sea level. A simplified, two-dimensional, theoretical analysis (Weertman, 1974) suggests that if the sea floor is flat, a marine ice sheet is basically unstable -- it will either extend its grounding lines (the boundaries between the ice shelves and the inland ice) to deep water (i.e., the edge of the continental shelf) or disappear completely, depending on the depth of water at the grounding line (Figure 1). If the water there is deep enough, as it mostly is in West Antarctica, the marginal portions of the inland ice tend to go afloat with a consequent flattening out and spreading over the water. In the process, the elevation of the ice surface diminishes, steepening the surface slope on the adjacent "upstream" part of the inland ice, and thereby accelerating the flow of ice from the marginal zone of the inland ice, into the ice shelf. This causes additional thinning in that marginal zone, since the ice is now flowing out of the region more rapidly than it is being replenished from upstream. As a result, the grounding line retreats farther upstream and the thinning process continues. Conversely, for a shallow sea floor, the tendency is for the floating portion of the ice next to the grounding line to run aground, which slows down its movement, causes it to thicken, and results in a grounding line advance.

The dependence of growth or shrinkage on the depth of the bed below sea level is critical importance, since, if this model is even qualita-

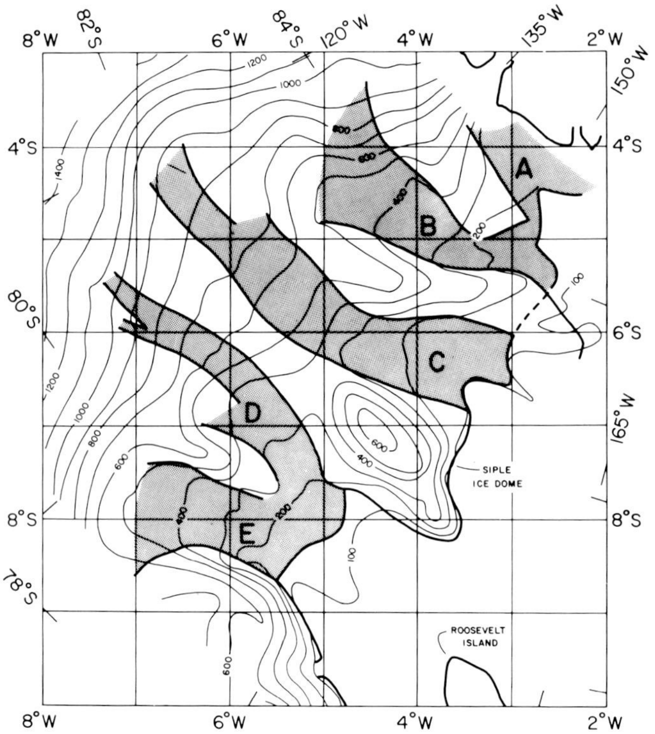

Fig. 2. West Antarctic ice streams (labeled A, B, C, D, E) in the Ross Sea drainage sector. The rectangular "grid coordinates" indicate degrees west of the 0° - 180° meridian and south of the 90°W - 90°E meridian; standard geographic coordinates are indicated by ticks at the edges of the map. Contours are surface elevations in meters above sea level.

tively correct, it provides a mode of direct response of a marine ice sheet to externally imposed sea level changes. This could be of particular importance if there were a relatively shallow sill or shoal near the margin of an extended marine ice sheet. In that case, if sea level should rise sufficiently to produce a bed depth exceeding the critical one on the sill, unstable retreat of the ice sheet would then follow as the grounding line moved back into regions of deeper water behind the sill (Thomas and Bentley, 1978). Such shoals, currently found along the edge of the continental shelf in the Ross Sea, probably marked one margin of the late Pleistocene West Antarctic ice sheet (Stuiver et al., 1981).

The idealized marine ice sheet can be made stable if the bed slopes downward away from the center of the ice sheet, but if the bed slopes the other way, the instability is accentuated. However, the simplified analyses do not take into account the restraints on the movement of the great ice shelves stemming both from drag along their lateral boundaries and from partial blockage at pinning points, spots within the boundaries of the ice shelf where the ice locally is aground. These restraints may be strong enough to stabilize an otherwise unstable marine ice sheet (Thomas et al., 1979). According to these arguments the existence of a marine ice sheet depends upon the existence of the ice shelves.

If the existence of the inland ice depends upon the pinning points, sea level become important again. The beds upon which the pinning points rest commonly lie well below sea level, not much higher than the surrounding sea floor (Jezek and Bentley, 1983). Consequently, a rising sea level could float the ice shelf free of many of them, thus eliminating what could be the critical stabilizing factor. Conversely, a lowering of sea level (or rise in the sea floor) could result in grounding the ice shelf on other shoals, thus blocking ice shelf movement further and contributing to the advancement of the grounding line (Greischar and Bentley, 1980). Since sea level responds immediately to changes in terrestrial ice volume, it provides a direct mechanism linking terrestrial and marine ice sheets.

At present, there is not enough ice in the Northern Hemisphere to threaten the West Antarctic marine ice sheet -- even if it were all to melt, which is highly unlikely, the resulting rise in sea level would be less than 10 m. Only melting of the vast East Antarctic terrestrial ice sheet could change sea level enough, and that is far more unlikely still. Important as sea level change may have been in the past, it does not pose a threat to the West Antarctic ice sheet in the future.

A more serious threat stems from the possibility of a change in ice thickness. In its effect on grounding lines and pinning points, a 100 m thinning of the ice is equivalent to a 90 m rise in sea level. Climatic warming is unlikely to have a significant effect on the upper surfaces of most Antarctic ice shelves, because their mean annual temperature (now less than -20°C; Crary et al., 1962) would remain far below the freezing point and any summer meltwater would simply refreeze as it percolated downward into the underlying cold snow layers (Young, 1981). Any melting on the underside of a shelf, however, would cause immediate and direct thinning. The ice shelves are, therefore, potentially vulnerable to "warm" ocean currents circulating beneath them. We return to this point later.

If, instead of a simple two-dimensinal ice sheet, one considers a three-dimensional model, then ice streams (fast-flowing "glaciers" within ice sheets; see Figure 2) must be included, because their very high activity compared to the relatively sluggish sheet flow elsewhere makes it virtually certain that they will play an important role in any response of the ice (Hughes, 1977). Ice streams occur in terrestrial as well as marine ice sheets, and some progress has been made in modeling an ice stream as an individual unit (Thomas, 1977; Lingle, 1983). However, attempts to model the dynamics of ice sheets that include ice streams are just beginning (e.g., Fastook, this volume). Ultimately, an adequate model must relate movement of the ice sheet grounding line, particularly in the ice streams, to such factors as downdraw of the ice stream drainage basins, the behavior of any fringing ice shelves, rates of growth or shrinkage of other ice sheets, processes of subglacial deposition and erosion (particularly at the grounding line), and the response of the earth's mantle to crustal loads.

The few three-dimensional model studies that have been made of the retreat of marine ice sheets in response to sea level rise at the beginning of the Holocene support the suggestion that a marine ice sheet depends for its existence on the presence of the ice shelves that surround it (e.g., Thomas et al., 1979). The vulnerability of marine ice sheets to climatic warming may then depend upon the vulnerability of its ice shelves. If so, and if rising temperatures, especially of the surrounding ocean water, were to lead to the disappearance of the ice shelves, then rapid shrinkage of the marine ice sheet would follow (Thomas et al., 1979). However, there are widely divergent opinions on the speed with which such ice shelves as those of the West Antarctic would respond to temperature changes (Hughes, 1982; Young, 1981); the presence or absence of sea ice cover may also be important (The Polar Group, 1980).

The possibility of ice sheet surges must also be considered; however, it is not at all clear that an ice sheet characterized by ice stream flow can surge (Weertman and Birchfield, 1982). This is because active ice streams probably are already in the same fast mode of flow that characterizes the surge phase of a surge-type glacier (Weertman, 1962, 1969). Although the speed of movement of an active ice stream is, of course, subject to change, there is no reason to believe that a catastrophic, unstable, additional increase in speed is possible.

Present-Day Natural Status

Sea-Ice

The extent of sea ice in the polar regions is shown in Figures 3 and 4 (CIA, 1978). In Antarctica, the pattern is regular, and the maximum area covered in September (about 2×10^7 km^2) is about eight times as great as the minimum in March. The thickness of Antarctic sea ice is largely unknown, but, since the ice is mostly less than one year old, the mean thickness is surely less than 2 m and probably not highly variable. A particularly interesting feature (not shown in Figure 3) is the giant polynya that, during some winters, exists in the eastern Weddell Sea (Carsey, 1980). A recent analysis by Kukla and Gavin (1981) indicates that the extent of the Antarctic pack ice in summer decreased by 2.5×10^6 km^2 between 1973 and 1980, and perhaps by more since the 1930's. It is quite possible, however, that this change can be attributed entirely to the natural variability of the climate system (Zwally et al., 1983).

In the Northern Hemisphere the seasonal range in sea ice extent is only about 2:1 (Walsh and Johnson, 1979), but the distribution of land causes marked longitudinal asymmetries in the seasonal sea ice zones. For example, at the end of the winter there is ice to the south of 50°N in the Sea of Okhotsk and the Labrador Sea, but open water to the north of 75°N in the Norwegian-Barents Sea (Figure 4). The ice thickness distribution in the Arctic is still not well known; the mean thickness ranges generally from about 2 m on the Siberian side of the basin to 4-6 m off the Canadian Arctic Archipelago, where pressure-ridging is most common and intensive (Hibler, 1981). There is no generally-accepted evidence of long-term change in the areal extent or thickness of the Arctic ice.

Ice Sheets

Greenland. It is still not known whether the total mass of the Greenland ice sheet is increasing, decreasing, or unchanging. Although the mass input in the form of snow accumulation on the surface is fairly well measured, the rate of ice loss is very uncertain. Iceberg discharge is reliably known only for west coast outlet glaciers and even there only for the summer (Carbonnell and Bauer, 1968). LANDSAT imagery of the Jacobshavn Glacier, which accounts for 40% of the discharge from these outlet glaciers, suggests that discharge is considerably reduced during the winter months (T.J. Hughes, personal communication, 1983). Summer meltwater discharged beneath the glaciers is substantial but unmeasured. It is quite possible that the Greenland ice sheet is presently growing in the north and shrinking in the south (Malzer and Seckel, 1976). However, the average rate of growth or shrinkage, if any, is very unlikely to exceed 0.01% of the total ice mass per year; dramatic changes from natural causes do not appear to be occurring.

East Antarctica. Studies of the mass balance of the East Antartic ice sheet suggest a mass increase over most of the interior regions of an amount that corresponds to about a 1 mm per year drop in world sea level change (Hollin, 1970; Allison, 1979), yet actual sea level measurements show a rise of about that amount (Gornitz et al., 1982). The largely unstudied coastal regions have much higher snow input and output rates than the interior, so that, despite a lesser areal extent, they could easily be undergoing a net loss twice as large as the interior gain. Evidence for dramatic rates of surface lowering, as large as 1 m yr^{-1}, has been reported in certain coastal areas (Naruse, 1978), yet in one of these same areas, as in several others, the measured mass balance is strongly positive (Shimizu et al., 1978; Bentley and Jezek, 1982), which should mean surface build-up. These results are suggestive of

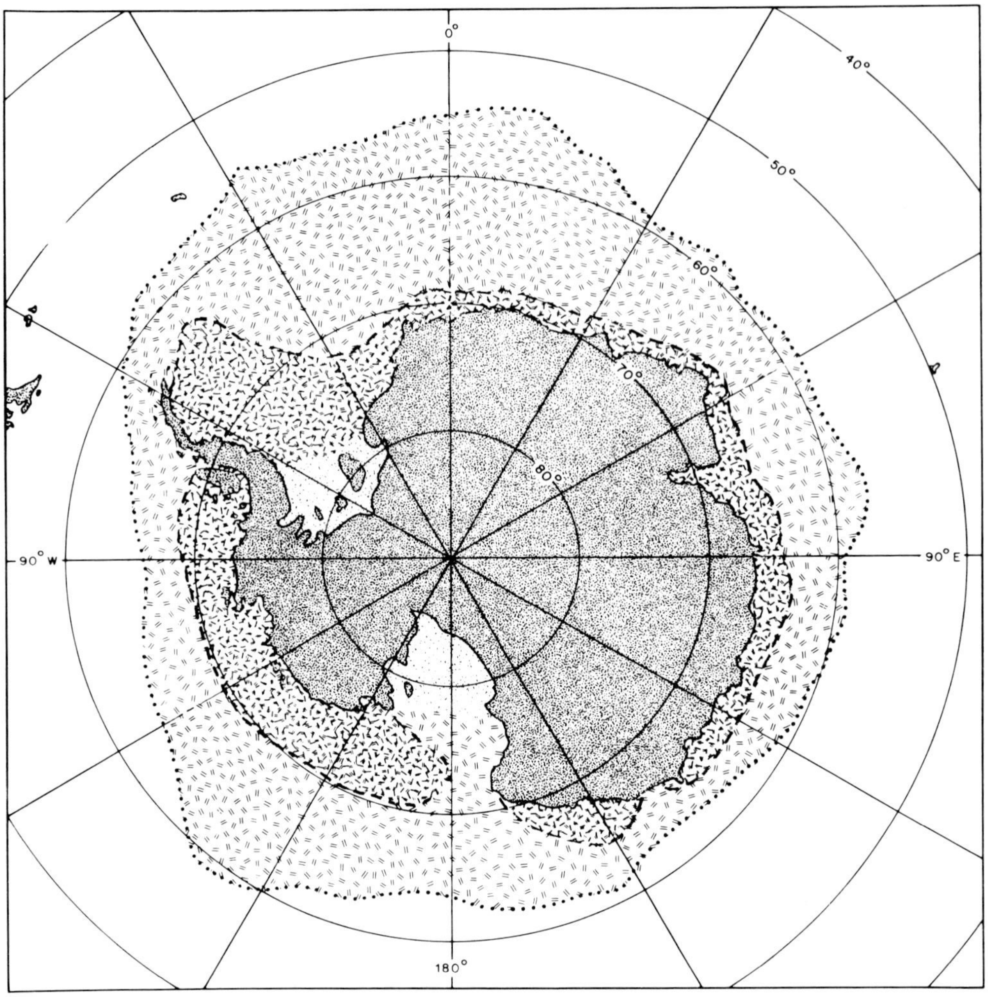

Fig. 3. Map of average winter maximum extent (lightest shading pattern) and average summer minimum extent (darker shading pattern) of sea ice around Antarctic continent (dark stippling). Light stippling shows Ross and Filchner/Ronne ice shelves (other ice shelves are not shown).

interior build-up after a surge although, of course, their correct interpretation is strongly debated. In any case, as in Greenland so a fortiori in East Antarctica -- the percentage change in the total mass is very slow, at most one part in 10^5 per year. A substantial fraction of the East Antarctic ice sheet, however, lies on bedrock far enough below sea level (Steed and Drewry, 1982) that marine ice sheet instability is conceivable, if a sufficiently deep connection to the ocean exists.

West Antarctica. The present mass balance of the West Antarctic ice sheet is uncertain, but it is the subject of widespread interest because of the unstable behavior of which deep-bedded marine ice sheets are probably capable. Most of the West Antarctice ice sheet is marine and deep-bedded -- if the ice were suddenly removed the Weddell, Amundsen, and Ross Seas would all be openly connected by water at least 500 m deep, much of it deeper than 1000 m (Bentley, 1964). Conclusions drawn from analyses of both the current state and that over the last 20,000 years, which encompass the end of the last glacial epoch, have differed in the extreme, from no major changes at all (Drewry, 1979) to drastic shrinkage even at present (Hughes, 1973); however, a growing preponderance of field evidence indicates that major shrinkage has occurred in the past, but is not taking place today (Denton and Hughes, 1983).

Before examining the evidence for present-day changes, it is important to recognize that, in terms of mass balance, the interior of the West Antarctic ice sheet comprises three different major ice-drainage systems, feeding ice into the Ross, Amundsen, and Weddell Seas, respectively (Figure 5). (Minor outflow from mountainous near-coastal areas between the Ross and Amundsen Seas, along the Bellingshausen Sea, and along the Antarctic Peninsula can be ignored for the purposes of this discussion). Ice-drainage systems, like their fluvial equivalents, can have quite different mass balances, and must be con-

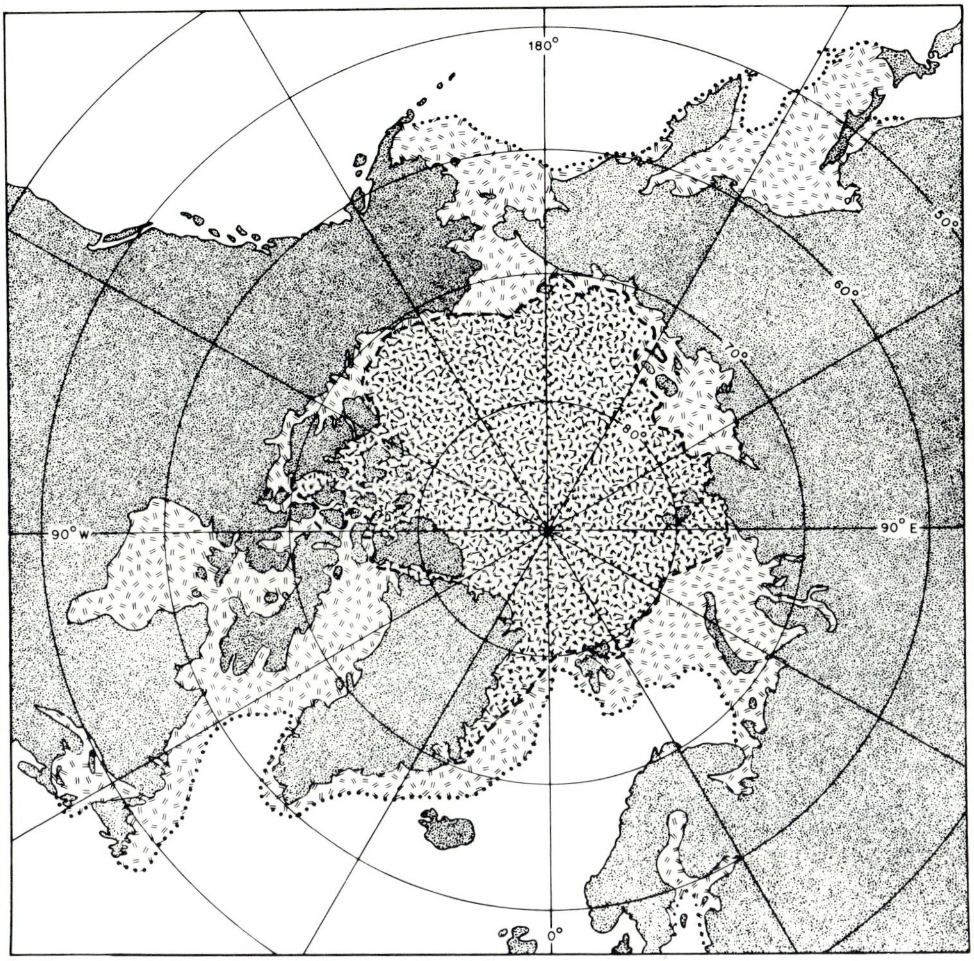

Fig. 4. Map of average winter maximum extent (lightest shading pattern) and average summer minimum extent (darker shading pattern) of Arctic sea ice.

sidered separately. The largest of the three, in terms of both total area and area of inland ice drained, is the Ross Sea system. It also is by far the best studied.

Evidence relating to present-day changes is ambiguous. Field evidence from the Ross Ice Shelf suggests that the ice shelf is nearly in steady state now, but has thinned in the last few hundred years (Bentley and Jezek, 1982). Local mass balance studies near the center of the inland ice sheet show a slow thinning (30 mm of ice per year), part or all of which may be attributable to the slow penetration of surface warming at the end of the last glacial epoch (Whillans, 1976, 1978). On the other hand, calculations using the latest data continue to indicate a strongly positive mass balance for the whole part of the West Antarctic ice sheet that flows into the Ross Ice Shelf. This again suggests post-surge conditions (Bentley and Jezek, 1982). Preliminary numerical studies employing the frictional lubrication model suggest that either an oscillatory regime, with the ice sheet currently in a post-surge state, or a steady-state fast-flow regime, may be appropriate (McInnes et al., in press). Incomplete evidence for the Amundsen Sea drainage system also implies a large excess of mass input over mass output (Crabtree and Doake, 1982).

Evidence from sea level. One way of looking at the mass balance of the earth's ice as a whole is to consider changes in sea level. Since sea level change and its possible attribution to changing ice volume in Antarctica has received some recent attention, it is worth reviewing here.

Emery (1980) calculated a mean global sea level rise of 3 mm yr^{-1} for the past 40 years, an increase over the 1 mm yr^{-1} average for the previous 50 years estimated by Fairbridge (1961), and others. Analysis of this change led Etkins and Epstein (1982) to conclude that the most likely source of the change in sea level is accelerated ice discharge from West Antarctica. However, Gornitz et al. (1982), in a re-analysis of the data set used by Emery (1980), found no significant variations in sea level rise over the last century from the previous average of 1 mm yr^{-1}. They conclude that perhaps half of that can be attributed to ice melt.

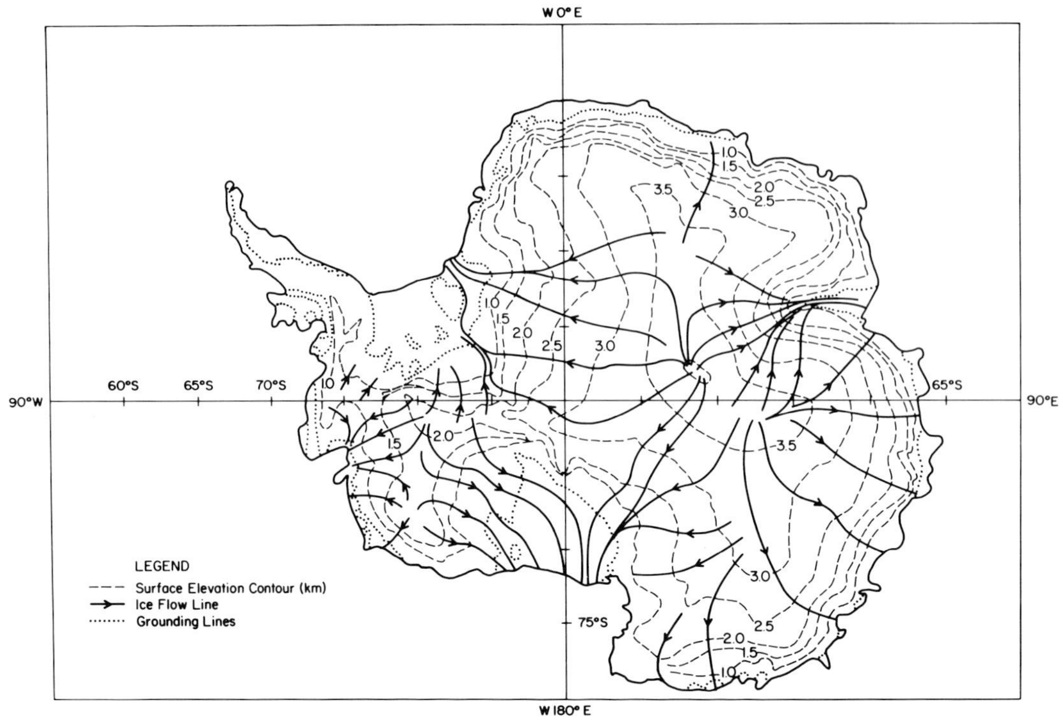

Fig. 5. Map of generalized ice flow lines superimposed on ice surface elevation contours, Antarctica. Elevation contours in kilometers; contour interval: 0.5 km.

As a compromise figure to compare with what we know about Antarctica, we will adopt 1 mm yr^{-1} for sea level rise that might be attributed to ice melt, a rate that would correspond, if unchanged, to disappearance of the West Antarctic ice sheet in 5,000 years (assuming none of the melt is from East Antarctica). This is certainly not a catastrophic rate, yet it has a major glaciological significance if West Antarctica is indeed the source, because a 1 mm yr^{-1} rise in the world ocean is equivalent to a 0.2 m yr^{-1} lowering of the ice sheet. To put this in perspective -- 0.2 m yr^{-1} is approximately the rate of mean annual snow accumulation (measured in water equivalent) on the West Antarctic surface. The implication, then, is that the mass outflow is about twice as great as the mass input, an imbalance that might be directly observable from mass balance studies in Antarctica.

The evidence for positive mass balance from actual West Antarctic measurements outlined above, incomplete though it may be, nevertheless is sufficent to make it very unlikely that the overall mass outflow could be double the input. Only if the output through the Weddell drainage system, where measurements have not yet been made, is many times the input could that be the case.

Could the rise in a sea level come from excess ice loss in East Antarctica? The large Lambert Glacier system exhibits a 3:1 mass build-up (Allison, 1979); the smaller Shirase Glacier basin shows an input:output ratio of about 1.5 (Shimuzu et al., 1978). In both cases the errors are substantial, but taken together, they do not allow a net mass loss. A 2:1 mass excess characterizes the East Antarctic flow through the Ross Ice Shelf (Bentley and Jezek, 1982), and the East Antarctic part of the Weddell Sea drainage system also exhibits a positive balance. Almost everywhere that measurements have been made, the net balances are positive rather than negative. It is true that most of these measurements are for largely interior systems draining through major ice streams or ice shelves; Giovinetto (1970) pointed out long ago that the laterally extensive coastal systems lying between the measured drainage basins could conceivably be supplying the excess water. Nevertheless, as new measurements accumulate since his study, that becomes less and less likely.

There is another possible source of meltwater. Meier (1983) has recently suggested that the combined mass loss from all the smaller glaciers and ice caps of the world since 1880 may have been sufficient to cause the unexplained portion of sea level rise.

Man-Made Changes and Consequences

Modification of the Atmosphere and Its Climatic Consequences

Many workers have pointed out that the known duration of earlier interglacials (10,000 to 15,000 years) suggests that the present one is nearly over; in the natural course of events, temperatures would probably "soon" - sometime in the next few millennia - start to decline towards eventual full-glacial levels. However, it seems increasingly likely that the climatic effects of man's activities, particularly those involving carbon dioxide production, could by the end of the century start to overwhelm any natural cooling trends that may be in progress, causing warming that will last at least several centuries, because of the long residence time of carbon dioxide in the atmosphere.

There is a consensus that the average global warming from a doubling of CO_2 would probably be two or three degrees; modeling also suggests that, because of feedback effects, the warming at high latitudes may be two or three times greater than the average, with a greater increase in the Arctic than in the Antarctic (National Academy of Sciences, 1979, 1982). In view of the limited realism of these models in their inclusion of high-latitude snow, ice, and ocean processes, quantitative prediction for the polar regions must be treated with some caution (The Polar Group, 1980). Nevertheless, some warming undoubtedly will take place, and it probably will be greater at high latitudes than low.

Climatic Warming-Interaction With the Cryosphere

Sea-ice. Supposing that the increasing amount of carbon dioxide in the atmosphere would increase temperatures by several degrees in the polar regions, the next concern is over the resulting effect on the thickness and extent of the sea-ice. Estimates of change in ice thickness due to warming can be based on simple empirical relationships between ice thickness and the number of degree-days below freezing in the Arctic, which lead to a decreasing thickness of about 10% for each 3° rise in winter temperature (Hollin and Barry, 1979; Barry, 1982). This suggests that the extent of the ice is not very sensitive to a winter-time temperature change of the magnitude that could reasonably be expected. On the other hand, other empirical relationships suggests that a rising mean summer-time temperature will move the summer-time southern margin of the Arctic sea ice approximately 1° north for each 1°C in change (Rogers, 1978). If linear extrapolation is valid, it follows that a coverage only half as large as at present would remain at the end of the summer if the temperature increased 3°, and that approximately a 10° temperature change would be needed to remove the summer-time ice cover completely. A temperature increase that large would probably require nearly a quadrupling of atmospheric CO_2 content. However, at least two other theoretical analyses have suggested a much greater sensitivity -- namely that about a 5° increase in temperature would be sufficient to melt the Arctic sea ice completely in mid-summer (Parkinson and Kellogg, 1979; Manabe and Stouffer, 1979), although it would probably reform in winter.

In the Antarctic, because most of the sea-ice is thin and undergoes large seasonal changes in areas, it seems likely that climatic warming would have a relatively greater effect than in the Arctic. One model leads to an estimate of a 2° latitude variation in maximum extent of the ice corresponding to a 1°C change in annual mean temperature (Budd, 1975), and it is likely that interaction with ocean heating would act to augment the simple temperature effect. Assuming that the summer-time extent of the ice would decrease proportionally, it appears that in the Antarctic the doubling of the carbon dioxide content of the atmosphere could quite conceivably diminish the summer-time sea-ice cover, and perhaps the winter-time cover as well, by one half.

Glaciers and terrestrial ice sheets. Mountain glaciers in some arid high polar regions might initially respond to general warming by advancing, as a consequence of increased snowfall; indeed, this may have occurred recently in parts of the Canadian Arctic islands and in Antarctica. In most parts of the world, however, and in the long run, mountain glaciers would shrink.

The Greenland ice sheet would shrink or expand depending on the changes in net mass balance resulting from increased snowfall in the central part and increased melting and ablation at its edges. The East Antarctic ice sheet would probably be little affected until warming had lasted for a long time. Even if snowfall increased while melting was still insignificant, there would be little immediate change because of the very long response time of the ice sheet as a whole. However, as previously mentioned, there are portions of this ice sheet that are grounded below sea level, and they could behave differently from the main body.

The West Antarctic ice sheet. The West Antarctic ice sheet probably would be little affected until or unless its ice shelves, exposed to higher ocean temperatures and less protected by sea ice, were destroyed. As discussed above, it is a matter of controversy what would happen next. The marine ice sheet might then disappear "rapidly" (i.e., in a few centuries), leading to about a 5 m rise in sea level, or there might be little change over the next millenium.

A rough estimate of the temperature rise that might correspond to eventual ice shelf destruction can be made by an analog argument (Mercer, 1978). Along the west coast of the Antarctic Peninsula ice shelves abruptly cease north of the mid-summer surface-water isotherm of -1.5°C, which approximately coincides with the northern limit of sea ice at its minimum extent in late summer and with the 0°-isotherm for a mid-summer air temperature. Mid-summer air temperatures along the Ross and Filchner-Ronne ice barriers and the coast of the Bellingshausen Sea are now about -5°C, so that, by analogy, the critical temperature increase for these bodies is about 5°C. If temperatures rose further and

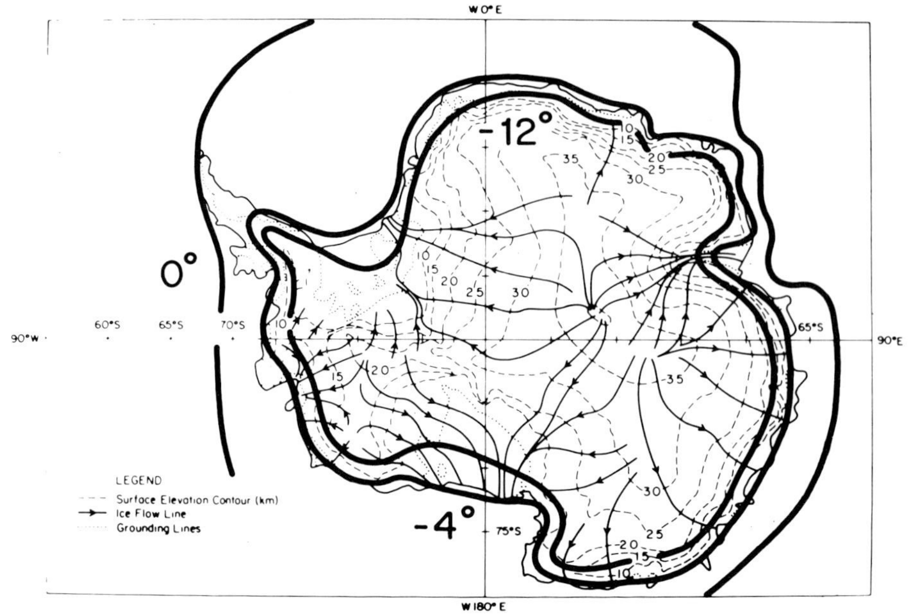

Fig. 6. Reproduction of Figure 5 with addition of mean surface-air isotherms for the month of January (i.e. mid-summer).

were sustained, by this argument, the West Antarctic ice shelves would eventually be destroyed.

The argument is open to question, however, because of the sharp temperature gradient along the ice shelf margins (Figure 6), which implies either that the configuration of the ice shelf primarily determines the position of the isotherms, rather than the reverse, or that warming would only cause a minor retreat of the margins of the ice shelves. The analog argument also is limited in that it provides no rate at which the ice shelves might disappear.

Let us consider specifically the effect on the ice sheet of a rise in temperature. There are three boundary conditions that might change: the temperature of the upper surface, the surface precipitation rate, and the heat and mass exchange at the boundary between the ocean and the ice shelf.

Surface temperature on the ice sheet will change directly as the atmospheric temperature changes. Any direct effect of surface temperature change will be slow and minor. Only if summer temperatures reached 0°C over substantial areas of the ice sheet could melting have a significant impact on the surface mass balance. The lowest lying parts of the ice sheet are the ice shelves, but even on them the mean annual temperature is colder than -20°C, so they will remain cold (i.e., below the melting point) until a temperature change is very large indeed. Even if a temperature change large enough to raise an entire ice shelf to the melting point did occur, it would still take hundreds of years before that temperature change would penetrate deeply into the ice through diffusion and advection processes.

The surface precipitation rate poses more complicated problems. Any atmospheric warming could be accompanied by an increase of precipitation resulting from an increased capacity of the air to carry moisture from the ice-free regions of the Southern Ocean and an increase in the area free of sea ice surrounding the Antarctic continent. On the other hand, changing atmospheric circulation patterns could lead to the opposite effect. These processes have received little attention in recent work on modeling regional features of the general circulation of the atmosphere.

However, a change in precipitation would probably not have a major effect on any rapid changes in the ice sheet. The average accumulation rate on the ice sheet is about 0.2 m yr^{-1}. A rapid shrinkage in the ice sheet would imply thinning rates at least an order of magnitude larger (see discussion below). Therefore, even a doubling or halving of the precipitation rate would have only relatively minor effect, nor could a change be expected to trigger an instability.

Water temperatures and melt/freeze rates beneath the ice shelf. Unfortunately, little is known about how to predict changes in ocean temperature and circulation underneath the ice shelf in response to atmospheric temperature change. The problem is compounded by the fact that little is known about how melting or freezing rates would be modified by changes in ocean temperature and circulation, even if they could be predicted. This complex point is a critical

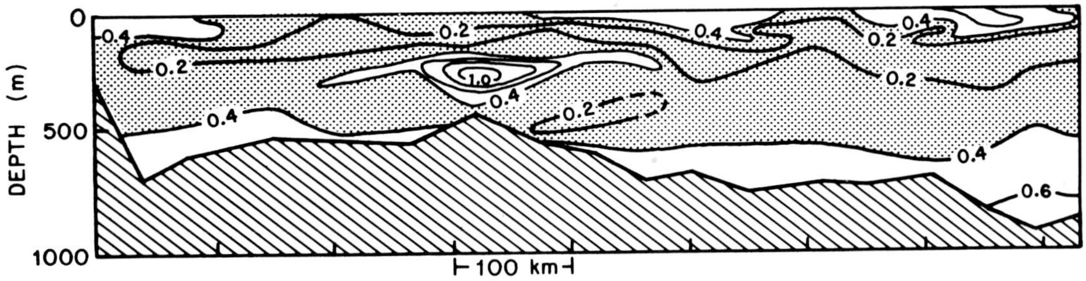

Fig. 7. Temperature above freezing in water along the front of the Ross ice Shelf in December, 1976. Contour interval: 0.2°C. Region of temperatures <0.4°C is stippled. From Jacobs et al., 1979.

weakness in our present ability to predict quantitatively the response of the West Antarctic ice sheet to temperature change.

Nevertheless, we can make some order-of-magnitude estimates. Oceanographic observations extending along about one-third of the front of the Ross Ice Shelf have shown a warm core of water (Figure 7) which has been interpreted by Jacobs et al. (1979) as probably circulating under the ice shelf and causing mean melting from the underside of the grid-western part of the shelf. They estimate that the melt rate may be on the order of the 0.25 m yr^{-1}. If, instead of the 0.5°C temperature difference now available in the warm core, a temperature differential of 2° (or, say, a differential of 1° combined with a doubling of the speed of the current) were to develop, the melt rate could perhaps be increased to 1 m yr^{-1}. In 100 years, thinning of the grid-western part of the shelf by 100 m would cause a large retreat of the grounding line and would unpin most, if not all, of the smaller ice rises. It might thereby trigger an accelerated thinning of the ice shelf and a consequent accelerated retreat of the grounding line. Accelerated bottom melt appears to be by far the most likely mechanism by which climatic warming could affect the ice sheet rapidly.

Maximum rate of removal of West Antarctic ice sheet. A question of crucial practical importance is how fast a "rapid" shrinkage could actually occur. The writer contends that decadal time scales are physically impossible, and that even disappearance in, perhaps, two centuries involves very great problems stemming from the vast volume of ice that has to be removed.

Some order of magnitude calculations based on a 200-yr removal time will clarify the difficulties in getting rid of the West Antarctic ice sheet so rapidly. The area of the inland ice sheet is about 2 x 10^6 km^2, and its mean thickness is roughly 2 km. If we are concerned primarily with sea level change, however, we need consider only the ice that needs to be removed before the remainder is afloat -- about half the total, or 2 x 10^6 km^3. The only rapid exit routes for this ice are through the fast-moving ice streams and ice shelves whose total width is around 1100 km. If we take the average thickness at the exit gates (principally the ice shelves near their fronts) to be 0.3 km, the total cross-sectional area at the exits is 330 km^2, and we can view the ice that needs to be discharged as a horizontal column of ice 6000 km long. To discharge it in 200 years would require an average speed of 30 km yr^{-1}, 30 times faster than the present-day rate.

It is not difficult to show (Bentley, 1983) that such a speed reasonably can be attained only by the mechanism of calving of the ice and its removal as icebergs. Calving models commonly presented involve no ice shelf and a rapidly retreating grounding line, with a more-or-less implicit assumption of removal of the calving ice as fast as it goes afloat. Thus, the rate of ice loss is equal to the cross-sectional area of the exit gate multiplied by the sum of the glacier speed and the grounding line retreat rate. This approach ignores the problem of ice removal, which still exists even for floating ice. If the ice is not removed fast enough, it will have to accumulate in situ, thus blocking the required ice flow and slowing down the retreat of the grounding line.

Again, several approximations can be made. The great bulk of the ice would have to be carried by the ice streams that presently feed into the Ross, Ronne, and Filchner Ice Shelves and Pine Island Bay. The total cross-sectional width of these ice streams at their grounding lines is only about half that of the present exit gate. On the other hand, the ice thickness might be about three times that assumed for the ice shelves. The speed of movement needed thus would be around 20 km/yr. Open water (or pack ice) through which icebergs could move would exist for perhaps one-third of the year; we assume that icebergs would, on the average, be 200 m thick, i.e., one-fifth of the ice thickness at the calving front, and would cover about half the surface area of the sea. The removal speed required, therefore, is (20 x 3 x 5 x 2 = 600) km yr^{-1} = 20 mm s^{-1}. Ocean currents of this speed are common enough, but here that needs to be the speed of the component of the current perpendicular to the shoreline, a situation which is not common except opposite river mouths or in regions of strong coastal upwelling. Probably the only reasonable possibility is the development of gyral circulation such as exists now in the Weddell Sea. Although it is not at all clear that individual gyres would form in each of the bays formed by the retreating ice streams, let us assume that they do. Simple geometrical considerations (Bentley, 1983) lead to the conclusion that maximum current speeds on the order of 0.2 m/sec would be needed, comparable to the main wind-driven currents of the ocean. Although the development of such currents is not impossible, there is no apparent reason why it should occur; surely much stronger currents hardly could be expected.

It can be concluded, then, that discharge of the ice into the Weddell and Ross Seas might barely be possible, although unlikely, in 200 years, but only after removal of the ice shelves, which itself would take at least a century or two, and on the (unsubstantiated) assumptions that calving would keep pace with grounding line retreat and that the required flushing circulation would develop.

So far the outflow into the Amundsen Sea through Pine Island Bay, which Stuiver et al. (1981) showed to have an exceptional potential for instability now, has implicitly been ignored. The problem there is that, due to the relative narrowness of the exit gate, the rate of outflow needed is at least 50 km/yr. Applying the iceberg discharge analysis as above leads to a required current speed of about 0.5 m/sec. This is unreasonably high and implies that ice loss through this exit would at a minimum take several times longer than 200 years.

It is thus the belief of the writer that a 200 year time scale can be rejected on physically sound considerations. In the realm of speculation, we might estimate a minimum time based on the discharge-limited model. Assume that rapid retreat begins in Pine Island Bay. After 200 years the interior ice is, say, one-quarter gone; outflow to the major ice shelves is starting to be diminished seriously. During the next 200 years another quarter of the ice goes out through Pine Island Bay, and the ice shelves disappear. With the whole grounded ice front now exposed, the remaining half of the ice is discharged in another 100 years. Minimum time: 500 years.

References

Allison, I., The mass budget of the Lambert Glacier drainage basin, Antarctica, Journal of Glaciology, 22(87), 223-235, 1979.

Barry, R.G., Snow and ice indicators of possible climatic effects of increasing atmospheric carbon dioxide, IN: Proceedings of the Workshop on First Detection of Carbon Dioxide Effects, DOE/CONF-8106214, U.S. Department of Energy, 207-236, 1982.

Bentley, C.R., The West Antarctic ice sheet: Diagnosis and prognosis, IN: Proceedings: Carbon Dioxide Research Conference: Carbon Dioxide, Science and Consensus Berkeley Springs, W. Va., September 19-23, 1982): IV.3-IV.50, 1983.

Bentley, C.R., The Structure of Antarctica and its ice cover, IN: Research in Geophysics 2 (H. Odishaw, ed.), 335-389 (Solid Earth and Interface Phenomena), Massachusetts Institute of Technology, 1964.

Bentley, C.R. and K. C. Jezek, RISS, RISP AND RIGGS, Post-IGY glaciological investigations of the Ross Ice Shelf in the U.S. program, Journal of the Royal Society of New Zealand 11(4), 355-372, 1981.

Budd, W.F., Antarctic sea-ice variations from satellite remote sensing in relation to climate, J. Glaciology 15(73), 417-27, 1975.

Budd, W.F. and B.J. McInnes, Periodic surging of the Antarctic ice sheet - an assessment by modeling, Hydrol. Sci. Bull 24(1), 95-104, 1979.

Budd, W.F. and I.N. Smith, The growth and retreat of ice sheets in response to orbital radiation changes (abstract), Presented at IUGG XVIII General Assembly, Canberra, Australia, December 1979.

Budd, W.F. and D. Jenssen, Numerical modelling of glacier systems, Symposium on Snow and Ice IAHS-AISH Pub. No. 104, 257-91, 1975.

Carbonnell, M. and A. Bauer, Exploitation des couvertures photographiques aeriennes repetees du front des glaciers velant dans Disko Bugt en Umanak Fjord, Juin-Juillet, 1964, Expedition Glaciologique Internationale au Groenland(EGIG), 1957-1960, 2(3), Meddelelser om Grønland 173(5), 1968.

Carsey, F., Microwave observations of the Southern Ocean, Mon. Weather Rev., 108, 2032-2044, 1980.

CIA, Polar Regions Atlas, National Foreign Assessment Center, CIA, 66 pp, 1978.

Crabtree, R.D. and C.S.M. Doake, Pine Island Glacier and its drainage basin, results from radio echo-sounding, Annals of Glaciology 3, 65-70, 1982.

Crary, A.P., E.S. Robinson, H.F. Bennett and W.W. Boyd, Jr., Glaciological studies of the Ross Ice Shelf, Antarctica, 1957-60, IGY Glaciological Report Series, 6, 193 pp, 1962.

Denton, G.H., and T.J. Hughes, Milankovitch theory of the ice ages; Hypothesis of ice-sheet linkage between regional insolation and global climate, Quaternary Research, 20(2), 125-144, 1983.

Denton, G.H., and T.J. Hughes, The Arctic ice sheet, an outrageous hypothesis, IN: The Last Great Ice Sheets (G.H. Denton and T. Hughes, eds.), Wiley-Interscience, N.Y., 437-467, 1981.

Drewry, D.J., Late Wisconsin reconstruction for the Ross Sea region, Antarctica, Journal of Glaciology, 24(90), 231-244, 1979.

Emery, K.O., Relative sea levels from tide-gauge records, Proceedings of National Academy of Sciences U.S.A., 77(21), 6968-6972, 1980.

Etkins, R., and E.S. Epstein, The rise of global mean sea level as an indication of climate change, Science, 215(4530), 287-289, 1982.

Fairbridge, R.W., Eustatic changes in sea level, Physics and Chemistry of the Earth, 4, 99-185, 1961.

Fastook, J.L., West Antarctica; the sea level controlled marine instability, past and future, Ewing Symposium volume, in press.

Giovinetto, M.B., The Antarctic ice sheet and its probable bi-modal response to climate, International Symposium on Antarctic Glaciological Exploration (ISAGE) (A.J. Gow, C. Keeler, C.C. Langway, and W.F. Weeks, eds.), IASH Publ. No. 86, 347-358, 1970.

Gornitz, V., S. Lebedeff and J. Hansen, Global sea level trend in the past century, Science, 215(4540), 1611-1614, 1982.

Greischar, L.L. and C. R. Bentley, Isostatic equilibrium grounding line between the West Antarctic ice sheet and the Ross Ice Shelf, Nature, 283(5748), 651-654, 1980.

Hibler, W.D., III, Sea ice growth, drift and decay, IN: Dynamics of Snow and Ice Masses (S. Colbeck, ed.), Academic Press, N.Y., 141-209, 1981.

Hollin, J.T., Is the Antarctic ice sheet growing thicker? International Symposium on Antarctic Glaciological Exploration (ISAGE) (A.J. Gow, C. Keeler, C.C. Langway, and W.F. Weeks, eds.), IASH Publ. No. 86, 363-374, 1970.

Hollin, J.T. and R.G. Barry, Empirical and theoretical evidence concerning the response of the earth's ice and snow cover to a global temperature increase, Environment International, 2, Pergamon Press Ltd., 437-444, 1979.

Hughes, T.J., The stability of the West Antarctic ice sheet: What has happened and what will happen, IN: Proceedings: Carbon Dioxide Research Conference: Carbon Dioxide, Science and Consensus (Berkeley Springs, W.Va., September 19-23, 1982), IV.51-IV.73, 1983.

Hughes, T.J., On the disintegration of ice shelves, The role of thinning, Annals of Glaciology, 3, 146-151, 1982.

Hughes, T.J., West Antarctic ice streams, Rev. Geophysics & Space Physics, 15(1), 1-46, 1977.

Hughes, T., Is the West Antarctic ice sheet disintegrating? Journal of Geophysical Research, 78, 7884-7910, 1973.

ICEX, Ice and Climate Experiment, Report of Science and Application Working Group, NASA Goddard Space Flight Center, Greenbelt, MD, 1979.

Jacobs, S.S., A.L. Gordon and J.L. Ardai, Jr., Circulation and melting beneath the Ross Ice Shelf, Science, 203(4379), 439-443, 1979.

Jezek, K.C. and C.R. Bentley, Field studies of bottom crevasses in the Ross Ice Shelf, Antarctica, Journal of Glaciology, 29(101), 118-126, 1983.

Lingle, C.S., A numerical model of interactions between a polar ice stream, the ocean and the solid earth: Application to Ice Stream E, West Antarctica, Ph.D. Thesis, University of Wisconsin-Madison, 165 pp., 1983.

McInnes, B.J., W.F. Budd and U. Radok, Cross-sectional models for West Antarctica and Greenland, Annals of Glaciology, in press.

Malzer, H. and H. Seckel, Das geometrische Nivellement über das Inlandeis-Hohennanderungen zwischen 1959 und 1968 im Ost-West-Profil der EGIG, Zeitschrift fur Gletscherkunde und Glazialgeologie 11(2), 245-252, 1976.

Manabe, S. and R. Stouffer, Study of climatic impact of CO_2 increase with a mathematical model of global climate, Nature, 282, 491-493, 1979.

Meier, M.F., Snow and ice in a changing hydrological world, Hydrological Sciences Bulletin, 28, 3-22, 1983.

Mercer, J.H., West Antarctic ice sheet and CO_2 greenhouse effect: A threat of disaster, Nature, 271, 321-325, 1978.

Naruse, R., Surface flow and strain of the ice sheet measured by a triangulation chain in Mizuho Plateau, Memoirs of the National Institute of Polar Research, Tokyo, Special Issue, 7, 198-226, 1978.

National Academy of Sciences, Carbon Dioxide and Climate: A Scientific Assessment, Report of an Ad Hoc Study Group on Carbon Dioxide and Climate, Climate Research Board, National Academy of Sciences, Washington, D.C., 22 pp, 1979.

National Academy of Sciences, Carbon Dioxide and Climate: A Second Assessment, Report of the CO_2/Climate Review Panel, Climate Board, National Academy of Sciences, Washington, D.C., 72 pp, 1982.

Parkinson, C.L. and W.W. Kellogg, Arctic sea-ice decay simulated for CO_2-induced temperature use, Climatic Change, 2, 149-162, 1979.

Rogers, J.C., Meteorological factors affecting interannual variability of summertime ice extent in the Beaufort Sea, Mon. Weather Rev., 106, 890-897, 1978.

Rogers, J.C. and H. van Loon, The seesaw in winter temperatures between Greenland and northern Europe, II, Some oceanic and atmospheric effects in middle and high latitudes, Mon. Weather Rev., 107, 509-519, 1979.

Sanderson, R.M., Changes in the area of Arctic sea ice, 1966 to 1974, Meteorological Magazine, 104(1240), 313-323, 1975.

Shimizu, H., O. Watanabe, S. Kobayashi, T. Yamada, R. Naruse and Y. Ageta, Glaciological aspects and mass budget of the ice sheet in Mizuho Plateau, Memoirs of the National Institute of Polar Research, Tokyo, Special Issue, 7, 264-274, 1978.

Skov, N.A., The ice cover of the Greenland Sea: An evaluation of oceanographic and meteorological causes for year-to-year variations, Ph.D. thesis, Oregon State University, Corvallis, Oregon, 97 pp, 1970.

Steed, R.H.N. and D.J. Drewry, Radio echo sounding investigations of Wilkes Land, Antarctica, IN: Antarctic Geoscience (C. Craddock, ed.), University of Wisconsin Press, Madison, WI, 969-976, 1982.

Stuiver, M., G.H. Denton, T.J. Hughes and J.L. Fastook, History of the marine ice sheet in West

Antarctica during the last glaciation: A working hypothesis, IN: The Last Great Ice Sheets (G.H. Denton and T. Hughes, eds.) Wiley-Interscience, N.Y., 319-436, 1981.

The Polar Group, Polar atmosphere-ice-ocean processes: A review of polar problems in climate research, Rev. Geophys. & Space Phys., 18, 525-543, 1980.

Thomas, R.H., Calving-bay dynamics and ice-sheet retreat up the St. Lawrence valley system, Geographie Physique Quaternaire, 31, 347-356, 1977.

Thomas, R.H. and C.R. Bentley, A model for Holocene retreat of the West Antarctic ice sheet, Quaternary Research, 10, 150-170, 1978.

Thomas, R.H., T.J.O. Sanderson and K.E. Rose, Effect of climatic warming on the West Antarctic ice sheet, Nature, 277, 355-358, 1979.

Walsh, J.E. and C.M. Johnson, Interannual atmospheric variability and associated fluctuations in Arctic sea ice extent, Journal of Geophysical Research, 84(C11), 6915-6928, 1979.

Weertman, J., Milankovitch solar radiation variations and ice age ice sheets, Nature, 261(5555), 17-20, 1976.

Weertman, J., Stability of the junction of an ice sheet and an ice shelf, Journal of Glaciology, 13(67), 3-11, 1974.

Weertman, J., Water lubrication mechanism of glacier surges, Canadian Journal of Earth Sciences, 6(4), 929-942, 1969.

Weertman, J., Catastrophic glacier advances, Variations of the Regime of Existing Glaciers (W. Ward, ed.), AIHS Publ. No. 58, 31-39, 1962.

Weertman, J. and G.E. Birchfield, Subglacial water flow under ice streams and West Antarctic ice-sheet stability, Annals of Glaciology, 3, 316-320, 1982.

Whillans, I.M., Inland ice sheet thinning due to Holocene warmth, Science, 201, 1014-1016, 1978.

Whillans, I.M., Radio-echo layers and the recent stability of the West Antarctic ice sheet, Nature 264(11), 152-155, 1976.

Young, N.W., Response of ice sheets to environmental changes, IN: Sea Level, Ice and Climatic Change (I. Allison, ed.), IASH Publ. No. 131, 331-360, 1981.

Zwally, H.J., C.L. Parkinson, and J.C. Comiso, Variability of Antarctic sea ice and changes in carbon dioxide, Science, 220(4601), 1005-1012, 1983.

CLIMATE SENSITIVITY AND THE MARGINAL CRYOSPHERE

R. G. Barry

Cooperative Institute for Research in Environmental Sciences
and Department of Geography, University of Colorado, Boulder.

A. Henderson-Sellers and K. P. Shine

Department of Geography, University of Liverpool, England.

Abstract. There is a significant dearth of information on high-latitude cloudiness and the surface properties of the seasonal cryosphere, and on the degree of association between them. Following a survey of available data, some new observational material is presented on cloud conditions in the vicinity of snow cover and sea ice margins, using both 3-D Nephanalysis data and DMSP imagery. 1-D radiative calculations are used to illustrate the urgent need for improved observational data on cloud cover and surface albedo. The presence of cloud cover strongly damps the net effect on planetary albedo of any perturbation in snow/ice extent; an effect which is neglected in simple treatments of the ice-albedo feedback mechanism. The general problem of the appropriate parameterization of high-latitude surface albedo and cloud regimes in GCMs is discussed in the context of the review of observational data. Results from the new GISS GCM illustrate some important problems to be overcome in all climate models.

1. Introduction

The extent and state of the cryosphere are important components of the climate system. In addition to perennial snow and ice, which covers approximately 18 percent of the Earth's surface, the seasonal cryosphere covers an additional 15 percent in January and 9 percent in July. Any comprehensive model of the climate must, ultimately, include prediction of snow cover and sea-ice. The way in which sea-ice is modelled, particularly the seasonal cycle of the marginal sea-ice zone for example, may considerably modify the sensitivity in climate change experiments. Present climate models differ greatly in the method of incorporating and representing sea-ice and clouds in high latitudes and current general circulation models show only limited success in the prediction of the present-day extent of sea-ice in both polar regions. Another area of concern is the question of cloud-cryosphere interaction. Our knowledge of the distribution, type, and thickness of cloud cover in relation to polar surfaces in extremely limited and none of the current GCMs can yet either parameterize or validate high-latitude cloudiness and surface albedo adequately. In view of the emphasis that has been placed upon the ice-albedo feedback effect for climate change, a review of the observational data on cloud-cryosphere parameters at high-latitudes is overdue. A preliminary investigation is made of the performance of the GISS climate model II (Hansen et al., 1983) in high latitudes to illustrate the problems in modelling these features.

2. Observational Data

2.1 Observations of Surface Albedos

Observations of surface energy budget components in the central Arctic in summer (e.g. Vowinckel and Orvig, 1970; Maykut and Untersteiner, 1971) show that solar radiation dominates the regime. The longwave radiation components from the atmosphere and from the surface are, in general, larger than the shortwave flux but the two terms almost balance in net. Figure 1 from Maykut (1983) shows the surface fluxes of sensible and latent heat, radiation and conduction through the sea-ice observed at a site in the Beaufort Sea during the AIDJEX experiment. The net radiation component dominates all other terms. It is interesting to note that the July net radiation exceeds the June levels despite there being lower incoming radiation at the top of the atmosphere. This is a specific striking illustration of the effect of variations in the surface albedo; the July albedo is some 13-14 percent less than that for June (Marshunova, 1966; Chernigovskiy, 1967).

The albedo of ice depends on a number of factors including the depth of any overlying snow,

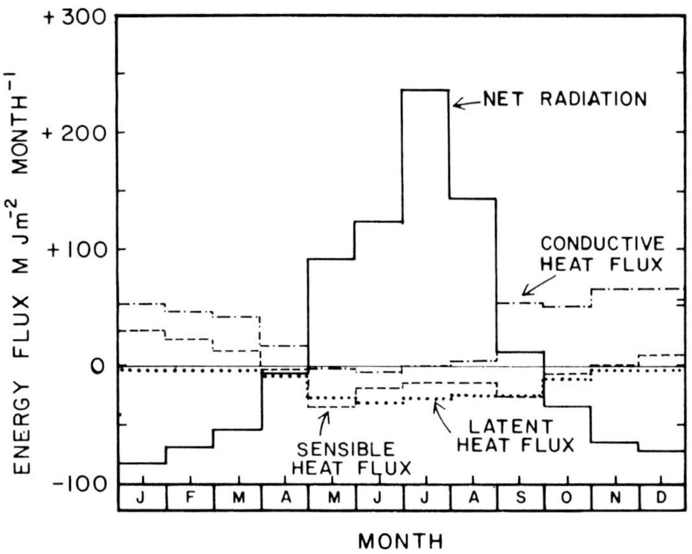

Fig. 1. Area-weighted energy fluxes (MJ^2 month^{-1}) observed over the AIDEX manned array in the Beaufort Sea (after Maykut, 1983).

ice thickness, brine density, melt pond coverage and the extent and latest occurrence of surface drainage. Grenfell and Maykut (1977) review the changes which characterize the Arctic Ocean ice cover: Central Arctic pack ice is predominantly multiyear ice up to 3m thick. About 5-10 percent of the area consists of young sea-ice less than a meter thick, which is created from the freezing of leads. At maximum extent 40-50 percent of the pack ice consists of first year ice (Carsey, 1982). First year and multi-year ice differ in brine content. During formation first year ice traps brine at its upper and lower surfaces which acts to decrease the albedo. Brine drainage, and the flushing effect of salt water during the melt season, causes multi-year ice to have a lower brine content. Internal melting and freezing in multiyear ice creates a greater vapour-bubble density within the ice which acts to increase the albedo since there are a greater number of scattering surfaces. However, the summer melt of surface snow is the single most important albedo change and this is accompanied, in the Arctic, by the formation of melt puddles covering up to 50 percent of the surface. Buzuev et al. (1965) show a roughly linear decrease of ice albedo from 0.75 when it is snow-covered to 0.30 as the ice becomes dark-grey in colour and its surface fully water saturated and extensively covered with thaw holes. According to Buzuev et al., the albedo of an Arctic pack ice surface (α_s) can be estimated from:

$$\alpha_s = (1 - f_m)\ \alpha_i + f_m\ \alpha_m$$

where

α_i = the albedo of the 'dry' ice fraction

α_m = the albedo of melt puddles and thaw holes

f_m = the fractional area occupied by melt puddles and thaw holes.

On this basis, melt puddles should cover at least 50 percent of the ice surface at their peak although some obervations suggest only 20 percent coverage (A.M. Hanson, cited by Carsey, 1982). Langleben (1971) obtained a similar linear decrease over the southern Beaufort Sea with a value of 0.27 for "fully-flooded ice." He finds the relationship $\alpha_s = 0.59 - 0.32\ f_m$. The melt at 78°N in the Soviet sector of the Arctic Basin starts around June 5, on average, and is delayed until about July 5 at the Pole according to Yanes (1962) although the microwave brightness temperature data suggest that in 1974 the melt pattern was determined by synoptic-scale processes rather than exhibiting any latitudinal control (Crane et al., 1982). The period for which Arctic pack ice is snow-free exhibits very large interannual variations. Chernigovskiy (1967) reports that the snow-free period in the central Arctic varies from 0 to 83 days with an average of 36 days. The interannual variations in albedo for the months June, July and August are given as 0.26, 0.28, and 0.28 respectively.

Table 1 shows the broad-band albedos of the ice surface derived from measurements of Grenfell and Maykut (1977), Chernigovskiy, and Hanson (1961). The albedo of the melt ponds depends on both the depth of water and the type of underlying ice (see e.g. Langleben, 1968, 1969 and 1971) and the albedo of the snow depends on the age and depth of the snow. About 10cm of snow is required to make the underlying surface invisible (Wiscombe and Warren, 1980). Surface measurements of albedo are also affected by the prevailing cloudiness (Grenfell and Maykut, 1977).

In polar environments, alterations in the absorbed solar radiation at the surface are fundamental to the rate and extent of the melt. In climate models the importance of the surface albedo of ice may be exaggerated or dampened by the relative magnitude of other model energy fluxes. For example, the thermodynamic ice models of Maykut and Untersteiner (1971) and Semtner (1976) are driven by "standard" values of the surface energy budget taken from a variety of sources. Their albedos are taken from Marshunova (1966) (see observed values given in Table 4). Decreasing these albedos in June, July and August by 0.1 and 0.2 leads to a decrease in the predicted equilibrium ice thickness from 0.288m to 0.105m and zero, respectively. The measurements of summer albedo shown in Table 1 suggest that the albedo values of Marshunova used in these thermodynamic ice models are all too high. Ablation of the ice could be altered by the presence of melt ponds, although Maykut (1983) suggests that the melt puddles have little impact on the mass balance of the ice as absorbed energy goes into deepening the ponds which sub-

TABLE 1. Broadband Albedos for Varying Conditions of the Ice Surface

Surface Description	Source		
	Grenfell and Maykut (1977)[1]	Chernigovskiy in Doronin (1970)	Hanson (1961)
1) Snow on ice	0.63-0.89	0.81-0.85	0.78-0.81
2) Frozen multi-year ice	0.72	---	---
3) Melting multi-year (white surface)	0.69	0.51	0.76-0.78
4) Melt ponds on ice	0.15-0.37	0.33[2]	0.38-0.39
5) First year white ice (melting)	0.47	0.51	---
6) First year blue ice	0.24	---	---

[1]Total albedo from measurements for 0.4-10 μm and calculations for wavelengths > 1.0 μm.
[2]70 percent of surface. For first-year ice, corresponding figure is 0.30.

sequently freeze in winter. However, ablation rate and melt pond coverage are associated since periodically the area of pond coverage decreases as the melt water drains through the ice (Weaver et al., 1976). Maykut and Untersteiner (1971) also showed their model to be sensitive to the ocean heat conduction term. Recent measurements by McPhee and Untersteiner (1982) indicate that this term may be smaller than previously believed, so that the sensitivity of the ice models to albedo changes needs to be re-examined. A smaller oceanic heat flux will allow lower albedo values to give reasonable ice thickness.

The dominant component in the summertime surface energy balance over ice is solar radiation and so it is essential that the large-scale surface albedo be correctly parameterized. In the Arctic region the most important feature is shown here to be the representation of the ice albedo after the snow has melted (i.e., in June and July). The albedo decreases suddenly (30-40 percent) at this time. The albedo value is also a strong function of the presence of melt puddles, leads and polynyi (open water areas). Conditions in the Antarctic are not favorable for the formation of melt puddles which means that the summertime decrease of surface albedo is much less than in the Arctic (Andreas and Ackley, 1982). It is interesting to note that in the coupled ocean-atmosphere-ice models of both Manabe et al. (1979) and Washington et al. (1980) very little circum-Antarctic ice is predicted especially during the summer. It is probable that underestimates of the albedo and the use of prescribed cloud cover contributed to an exaggerated melting of the winter-formed ice by causing sea-surface temperatures to be several degrees too high over most of the Southern Hemisphere (S. Manabe, p.c., 1982).

A more fundamental problem with thermodynamic ice models, however, concerns their treatment of sea ice as corresponding to a single, level flow of multiyear ice. Arctic ice in particular has a thickness distribution that is highly variable in space and time. Hibler and Ackley (1982) show that the inclusion of ice dynamics in a sea ice model greatly increases the formation of summer leads in Antarctic ice and causes a much more rapid and realistic summertime decay of the ice. With these ice dynamics processes included, smaller oceanic heat fluxes are required to obtain a realistic seasonal decay of Antarctic sea ice cover. Further discussion of the simulation of cryospheric processes in GCMs is included in Section 3 below.

It is interesting to examine the degree to which ice albedo forcing by snow cover and by melt pond formation and extent have been incorporated into recent compilations of global surface albedo recommended for use in or calibration of GCMs. The high latitude elements from some of these compilations are listed in Table 2. Kukla and Robinson (1980) assign an albedo of 0.8 to ice except during July and August when it is given as 0.6. North of 80°N it is assumed that there is no open water in the ice. Hummel and Reck (1979) produce seasonal values using different albedos for solid ice and ice-water mixtures taking fractional extents of the different ice types and water from Vowinckel and Orvig (1970). This produces summertime Arctic ice albedos of 0.592 (i.e. close to Kukla and Robinson, 1980) and wintertime values of 0.698. Therefore, by neglecting the formation of melt puddles on pack ice and snow cover on ice, Hummel and Reck (1979) considerably underestimate the seasonal variation of the ice albedo. Robock (1980), in contrast, obtains a seasonal albedo variation of

TABLE 2. Zonally Averaged Surface Albedos at 85°N from Three Difference Surface Albedo Compilations

Reference	J	F	M	A	M	J	J	A	S	O	N	D
Robock, 1980	0.95	0.95	0.82	0.77	0.78	0.49	0.41	0.42	0.73	0.89	0.95	0.95
Hummel Reck, 1979	0.71	0.71	0.71	0.68	0.68	0.68	0.61	0.61	0.61	0.71	0.71	0.71
Kukla and Robinson, 1980	0.8	0.8	0.8	0.8	0.8	0.8	0.6	0.6	0.8	0.8	0.8	0.8

0.54 for his Arctic albedos. He emphasizes the effect of melt water formation on the ice, using a melt-water albedo value of 0.4 whenever the surface temperature is $\geq$273K. His values are similar to the aircraft measurements of Hanson (1961) over the Beaufort Sea. The latter reported an average albedo of 0.40 over 8 tenths ice with 4 tenths puddle cover on the ice.

Data on coverage of melt puddles are sparse. Averages at Soviet "North Pole" drifting stations indicate peak values of 25 percent cover in mid-July (Nazintsev, 1964). Additionally observations of albedo are made from different platforms and in different years e.g. Hanson (1961) and Herman (1977) from aircraft; Vonder Haar (1974) using data from the Nimbus 3 satellite and various surface-based observations. The range of observational data is, however, less wide than the parameterizations currently employed in, and recommended for calibration of GCMs.

2.2 Arctic Cloud Data

Information on Arctic cloud cover (Crane and Barry, 1983) is more limited and spotty than that on surface albedo. Cloud statistics are available from surface observations - at coastal or island stations and from drifting stations (Mohn, 1905; Vowinckel, 1962); from aircraft programs such as the "Ptarmigan" weather reconnaissance flights (Henderson, 1967; Huschke, 1969), the Soviet "flying observatory" programs (Baranov, 1962) and recent Alaskan stratus investigations (Herman, 1977; Jayaweera, 1982; Herman, 1983), and also from satellite-derived products. Satellite observations, however, are hampered by the problem of snow-cloud discrimination in the 1.6 μm band and the 3.7 μm band are now being evaluated.

Total cloud amounts over the Arctic Ocean range from 40-60 percent in winter to 70-90 percent in summer (Huschke, 1969; Vowinckel and Orvig, 1970). In May there is typically a marked increase in low cloud amounts, and since surface melt does not begin until early June (on average) at 75°N (Yanes, 1962) this would appear to be attributable to large-scale moisture advection rather than boundary layer stratus formation. The mean frequency of low cloud in the Arctic Basin in January and July is illustrated in Figure 2 from Huschke (1969). Broadly similar results are shown in summer by Voskresenskiy and Chukanin (1959) for the eastern hemisphere sector and by Vowinckel and Orvig (1970). However, these two studies show strikingly different amounts of winter low cloud, spotlighting the present uncertainty even for mean Arctic cloudiness. The Russian map is in much closer agreement with that of Huschke (Figure 2).

There is little information on the optical properties of Arctic clouds. Aircraft investigations over the Kara Sea and the Laptev Sea indicate that the mean thickness of stratocumulus and stratus in late summer is typically 200-400m (Dergach et al., 1960; Baranov, 1962), whereas Jayaweera and Ohtake (1972) report finding two or three separate stratus layers, each of which may be 300-500m thick, over the Beaufort Sea. Layering is attributed to the trapping of solar radiation within the cloud by Herman and Goody (1976) although Jayaweera (1982) proposes that the lower stratus layers may be formed by the lifting of fog over the sea ice while the upper layers are caused by the large-scale southerly advection of warm moist air. Average liquid water contents are in the range 0.1-0.2 g m^{-3} for summertime stratus with a median drop size of 12 μm (Weller et al., 1975), similar to values in middle latitudes. Herman (1977) cites an extinction optical depth of 7 for the solar spectrum in a 500m thick cloud.

Aircraft data on cloud radiative properties are provided by Koptev and Voskresenskiy (1966) and Herman (1977). Over the Laptev Sea in August-September 1959, stratiform cloud layers up to 500 m thick had albedos of 0.32-0.35 over water and 0.6-0.7 over ice (Koptev and Voskresenskiy, 1966). In one case where a single cloud layer in the Laptev Sea extended across the sea ice margin, the cloud albedo was 0.29-0.32 over the ocean, but 0.72 over 10 tenths ice. In the absence of cloud the sea ice albedo was only 0.60-0.66. Herman (1977) reports reflectances of 0.60-0.75 over the spectral interval (0.29-0.8 μm) for stratus cover over sea ice.

The importance of information on cloud amount and optical properties on the solar radiation budget can be illustrated using a shortwave radiative transfer scheme. The 24 spectral band model of Slingo and Schrecker (1982) is used. The radiative transfer equation is solved using the delta-Eddington two stream approximation and cloud properties are parameterized from Mie

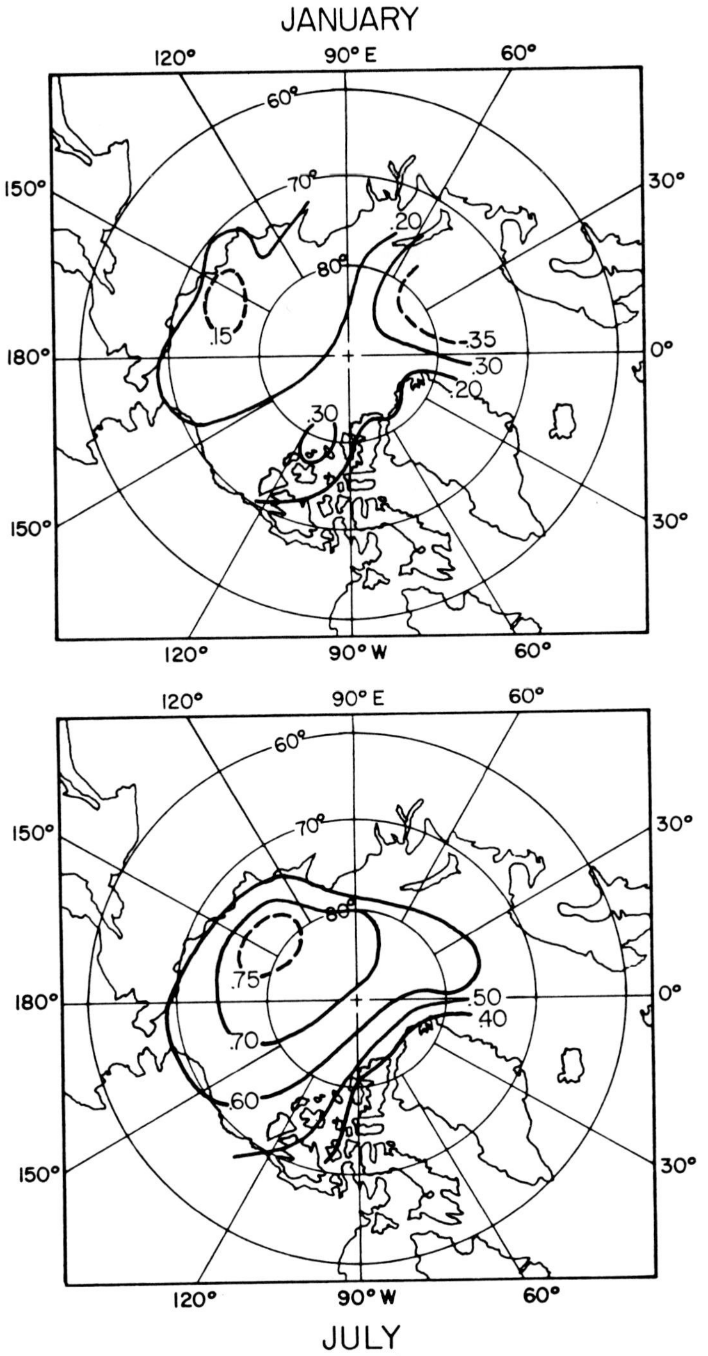

Fig. 2. The mean low cloud fraction in the Arctic Basin in January and July (after Huschke, 1969).

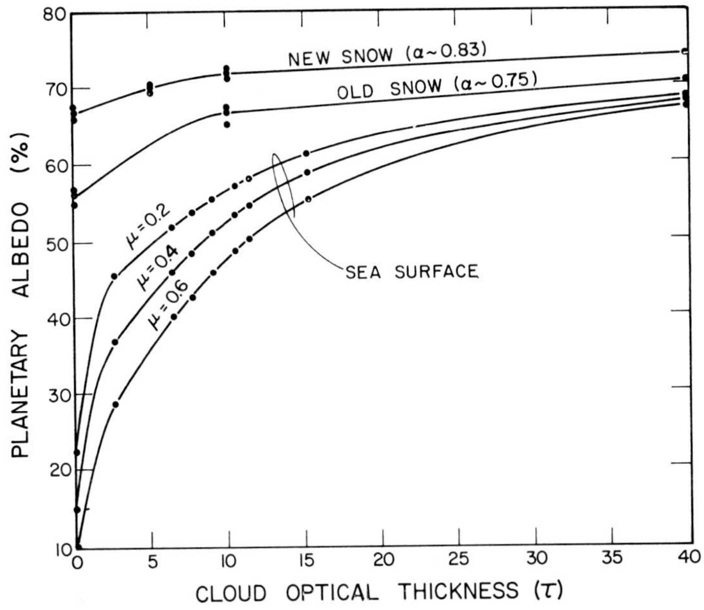

Fig. 3. The influence of cloud optical thickness (τ) on planetary albedo at different zenith angles (μ = cos zenith angle) over a sea surface (albedo = 0.1) and new snow (albedo = 0.83) and old snow surfaces (albedo = 0.75).

$$\tau = \frac{3}{2}\frac{LWP}{r_e}$$

where LWP = liquid water path (g cm^{-2}) and r_e is the equivalent drop size.

scattering calculations in terms of an effective droplet size and the liquid water path. Figure 3 illustrates the variation of planetary albedo with cloud optical thickness (τ) for a sea surface and two snow surfaces. The differences between the planetary albedo over sea and snow surfaces decrease rapidly with optical thickness. Arctic clouds will have τ in the range 5 to 10, whereas lower latitude clouds have optical thicknesses much greater than 10. Climate models generally employ cloud radiative properties independent of latitude. Figure 3 indicates that overestimates of cloud thickness will lead to misrepresentations of the impact of changing surface conditions on the planetary radiation budget.

Although it is not treated here, it should be pointed out that the cloud cover effect on surface net radiation in the central Arctic is predominantly via its modification of the long-wave component. Herman (1980) shows that this is the case in all months except July, based on a comparison of average radiation fluxes at the surface for no clouds and for actual cloudiness. In July, the cloud albedo effect on solar radiation is the most important term.

2.3 Cloud Cryosphere Associations

Until now, the question of cloudiness in relation to underlying surface type has largely been ignored other than in the context of multiple reflection over snow (Schneider and Dickinson, 1976). Yet such relationships are of significance for determinations of planetary albedo. Model calculations show that in the presence of optically thick cloud, for example, there is only a small difference in planetary albedo whether the cloud is over ocean or ice for zenith angles >60° (Henderson-Sellers et al., 1982). During much of the year, surface water and open water areas in high latitudes lead to the formation of boundary layer stratus and it is essential therefore that cloud climatologies be developed that incorporate information on the concurrent surface conditions.

Analyses of variations in cloudiness across snow cover or sea ice boundaries have recently been carried out for both synoptic-scale cloud systems (Carleton, 1981) and for total cloudiness (Henderson-Sellers et al., 1982). The distribution of total cloudiness in Septembers 1977, 1979, and 1981 across the North American-North Atlantic sector of the Arctic and sub-Arctic is shown in Figure 4. These maps suggest a gradient of total cloud amount across the snow and ice limits, with less cloud to the north. Our results also indicate considerably less average cloud cover in the Arctic in this month than the results of Huschke (1969). The cloud data were obtained from the 3-D Nephanalysis product of the U.S. Air Force, but despite the cloud-snow discrimination problem in high latitudes, we consider that the indicated differences are not an artefact of the cloud retrieval scheme used to prepare these data. Jayaweera (1977) reported less cloud over the Beaufort Sea pack ice than at Point Barrow for May-September 1975, although there was anomalous northerly flow during the second half of this season. Hughes (1982) also notes contrasts in the location of the 80 percent isopleth of cloudiness in the North Pacific in January and April 1975-1977. She finds less cloud in higher latitudes in 1976 and 1977 which were years of more extensive Bering Sea ice.

There are several physical reasons for contrasts in cloud regime across snow and ice boundaries. Cold air outbreaks leading to convective cloud growth over open ocean surfaces or the large-scale positioning of baroclinic zones by thermal contrasts and the steering of cyclone tracks (Carleton, 1981; Overland and Pease, 1982) both will tend to increase cloudiness south of the cryospheric margins. Conversely, air moving poleward in summer readily forms stratiform clouds over the ice marginal zone in the Arctic Ocean by cooling to space and diffusive cooling to the surface. Jayaweera (1982) stresses the role of large-scale moisture advection in southerly 700mb flow across the Alaskan coastline in forming thick stratus over the Beaufort Sea. The cloud persists due to the weakness of the solar radiation, the cold moist surface and sluggish synoptic activity (Herman and Goody, 1976). Transects of cloud amount and type across snow and ice margins have been prepared from visual analysis of DMSP visible and infrared imagery, which illustrate some of these regimes (Scharfen, 1982). For the period 24 June-5 July 1979 in the Barents Sea, for example, average cloud cover increased near and north of the sea ice margin, primarily due to low-level cloud formation during a northerly airflow regime in the first part of the time interval. Total cloud also increased some 1000-1500km south of the ice edge due to increased middle and high cloud related to cyclonic activity in the second half of the time period (Figure 5). In several other situations Scharfen found evidence of a northward decrease in cloud cover across the ice or snow line. As yet, however, a climatological average picture based on satellite data has not been established.

The potential importance of cloud-cryosphere interactions can be illustrated using the radiative transfer scheme described above. The extreme condition of cloudiness changes at the ice edge would be that the increase in cloud amount or thickness exactly compensated for the decrease in surface albedo to yield the same planetary albedo across the ice edge. Figure 6 illustrates the necessary conditions. The abscissa and ordinate give the cloud conditions over the ocean, while the contours give the cloud conditions over the ice. Two different surface albedo conditions are shown. In the top diagram, the change in surface albedo across the ice edge is 0.3, and in the lower it is 0.5. For example, in the top diagram a cloud cover of 0.5 over the ice is exactly compensated by 0.8 cloud cover of the same optical thickness over the ocean; Figure 4 indicates that such gradients are plausible, although melt-puddled ice would be necessary to give a low enough surface albedo.

The need for consistent cloud and surface albedo data is illustrated in Figure 7. The two surface albedo sets of Robock (1980) and Kukla and Robinson (1980) (see sub-section 2.1) are used with two different cloud climatologies at 85°N. The seasonally varying data from Vowinckel (1962) and the mean annual data from London (1957). The seasonal variation of cloud amount is clearly important in damping the annual variations of planetary albedo for the Robock (1980) case with lower summertime surface albedos. For the Kukla and Robinson (1980) albedo data, the effect of inserting cloud over an already high albedo surface is small. The sensitivity of the planetary albedo to cloudiness changes is seen to be dependent to a large extent, on surface conditions and their modelling. Shine et al. (1983) discuss these radiative transfer calculations in the context of cloud-cryosphere feedback in more detail.

3. Simulations of Cryospheric Processes in Current Climate Models

3.1 Problems in Parameterization

A review here of cryospheric simulation in current general circulation climate models (GCMs) would be premature for two reasons. Firstly,

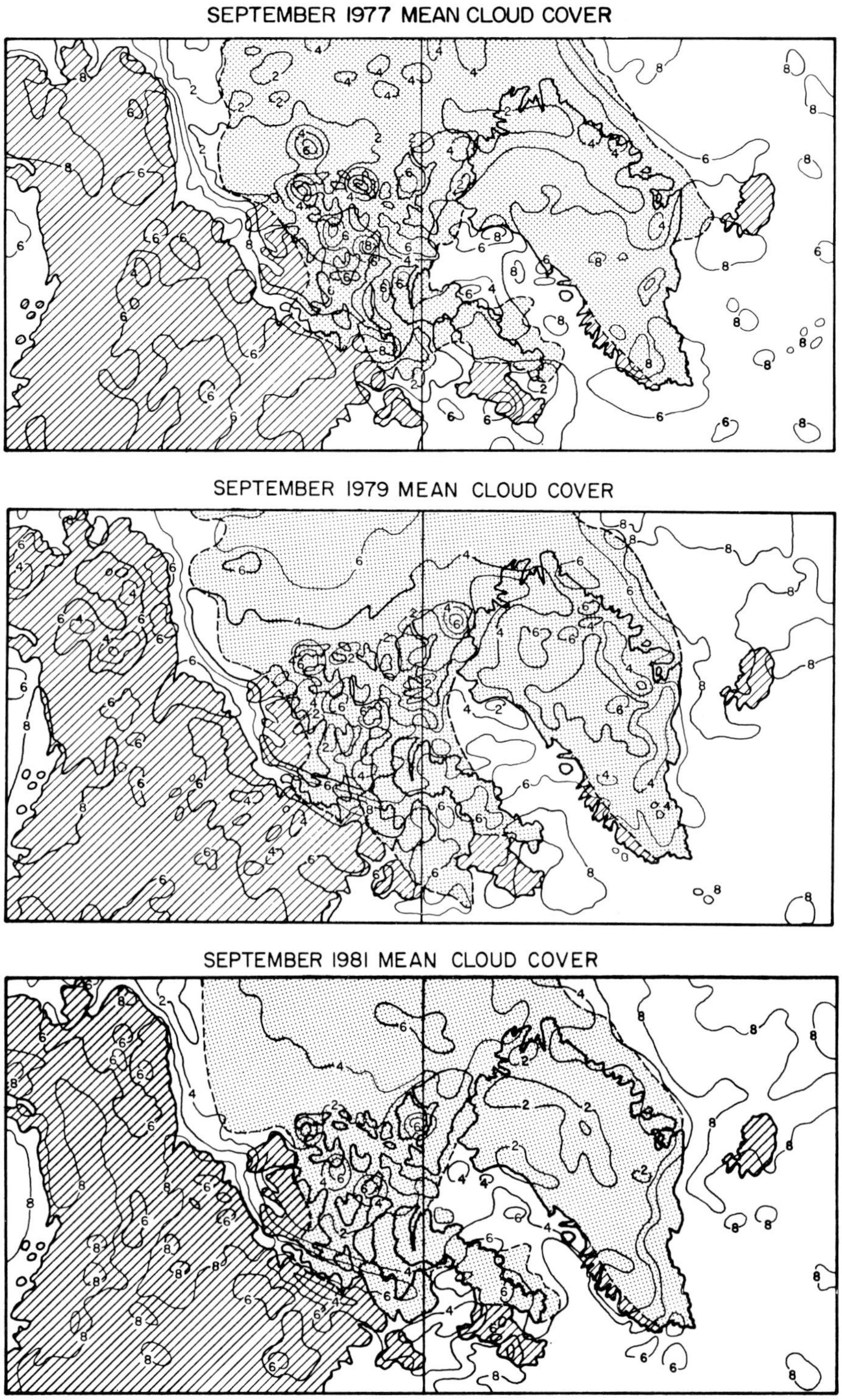

Fig. 4. Mean cloud cover (tenths) determined from the 3-D Neph data and the extent of snow/ice cover (stippled) over North America-Greenland in September a) 1977, b) 1979, and c) 1981. Bare land is shaded obliquely.

inclusion of parameterized cryospheric processes is a very recent addition to most models and secondly, as illustrated in Section 2, there is a dearth of observational data for high latitudes. In this section some of the problems already encountered by GCM modellers in their attempts to include cryospheric processes are considered.

The parameterizations of ice surface albedo used in various GCMs are listed in Table 3. While most models incorporate, in some way, the change between pre-melt and post-melt ice, there is a large variation in the method by which this

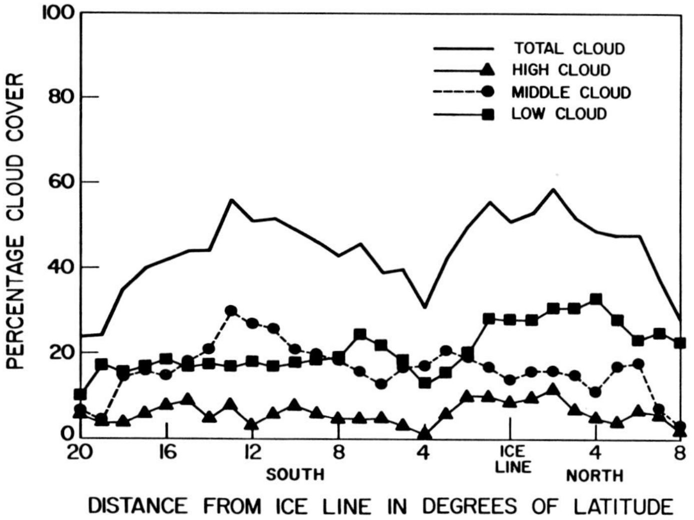

Fig. 5. Transect of cloud amount across the sea ice boundary in the Barents Sea, 24 June - 5 July 1979 from manual interpretation of visible and IR DMSP imagery (after Scharfen, 1982). Cloudiness estimated for 1° x 2° grid boxes along longitudes 40° and 50°E; 1° x 4° boxes north of 70°N.

is done. The U.K. Meteorological Office, GLAS and GFDL models, for example, calculate surface albedos of 0.8, 0.7, and 0.35, respectively, for an identical ice situation. In models using specified ice cover the influence of such albedo differences will be minimal since anomalies in surface energy balance will not be able to feed back on the ice amount or thickness. On the other hand such specifications clearly limit the use of the GCM to near present-day conditions. Differences in albedo parameterization will, in the future, have significant impacts in coupled models.

The choice of the critical temperatures at which the surface albedo changes is difficult to make for GCMs. In the U.K. Meteorological Office model this threshold temperature is related to the freezing point of sea water and set at 271.2K. The top surface of sea ice will, however, have a low salinity and so will only begin to melt at 273K. The difference is unlikely to have a large effect. Strokina (1980) finds that the ice albedo begins to fall sharply for temperatures greater than 268K (i.e., well above the value of 263K used in the GFDL GCM) and the fastest albedo change is at 273K. Similarly, experimental data given in Langleben (1966) show that the major change in observed albedo occurs suddenly as the surface air temperature falls below 273K.

Recently substantial progess has been made in the development of a thermodynamic-dynamic model of sea ice (Hibler and Ackley, 1982) and there have been attempts to integrate the cryosphere into atmospheric models. A simple model of sea-ice cover has been incorporated into GCMs of the Geophysical Fluid Dynamics Laboratory (GFDL) (e.g. Manabe and Wetherald, 1975) and, more recently, atmospheric parameters generated by the Goddard Laboratory of Atmospheric Sciences (GLAS) GCM have been used to drive an ice model (Parkinson and Herman, 1980). Coupled ocean-atmosphere-ice model integrations have been described for the GFDL GCM (Manabe et al., 1979) and the National Center for Atmospheric Research's (NCAR) GCM (Washington et al., 1980). The formation of sea-ice is dependent upon the ocean temperature and the salinity. Globally averaged ocean water has a salinity of approximately 35 °/oo and freezes at a temperature of -1.9°C (~271.1K). Near-surface salinities vary with location around both poles and also as a function of season in response to vertical movement and pycnocline strength. The temperature threshold set for formation of sea-ice varies

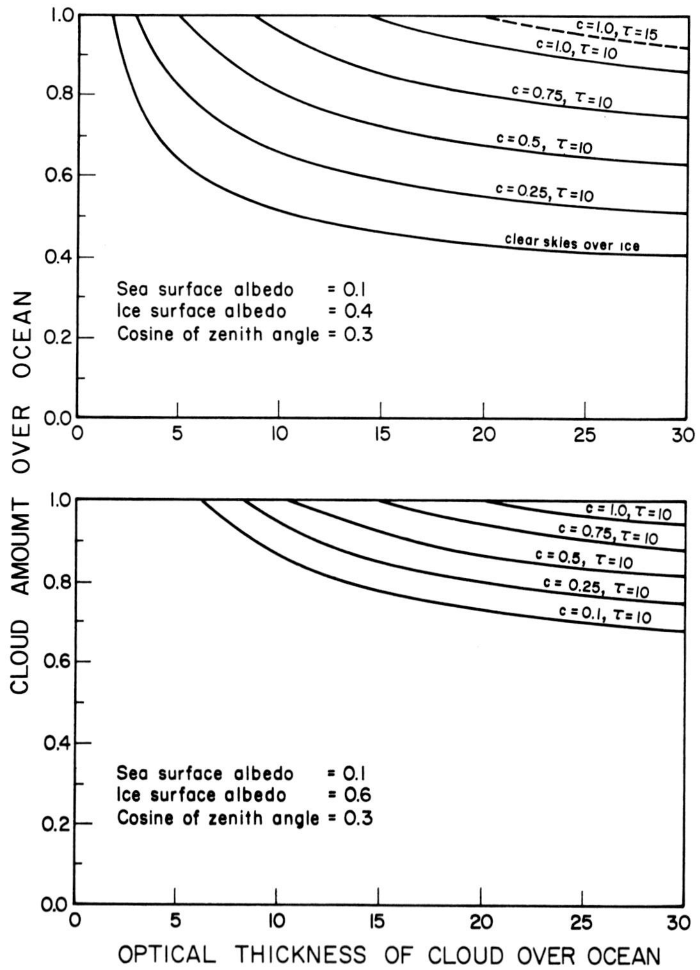

Fig. 6. Cloud fraction (c) and optical thickness (τ) over an ice surface with albedo of 0.4 (above) and 0.6 (below) necessary to give identical planetary albedo of cloud over ocean with albedo of 0.1. (Cosine of solar zenith angle = 0.3).

among GCMs: NCAR: ~271.2K (-1.8°C) (Dickinson et al., 1981, unpubl.); GFDL: 271.0K (-2.0°C) (Manabe and Stouffer, 1980); GISS: 271.4K (-1.6°C) (Hansen et al., 1982, unpubl. p.c.). This range is threshold values needs to be borne in mind in model intercomparisons, although it is smaller than the geographical and seasonal variability.

This discussion of the current state of cryospheric parameterization in GCMs illustrates the difficulty encountered by modelers both in the selection of appropriate parametric values and in any attempt to validate results. These problems are the result of inadequate observational data for the marginal cryosphere. The level of information required by climate modelers can be illustrated by a more detailed examination of the performance of one specific GCM.

3.2 Elements of the High Latitude Climatic Regime as Simulated in the GISS GCM

The basic Goddard Institute for Space Studies (GISS) GCM is described in Hansen et al. (1983); it is demonstrated that the major features of global climate are realistically simulated, despite the coarse resolution of approximately 1000km. The GISS model II version, used for our analysis, has sea surface temperatures computed from specified horizontal heat fluxes in the oceans. Cloud cover and its vertical distribution are also computed, the ground temperature calculation includes diurnal variation and seasonal heat storage, snow depth is calculated and snow albedo includes the effects of snow aging. Sea ice extent, concentration and thickness are computed via the local energy balance.

The persistence of the ice, once formed, depends upon the way energy fluxes are modelled. In current GCMs the prediction of sea ice extent is reasonable in the Arctic but much less successful in the Southern Oceans. When ocean transport is omitted the lack of success becomes greater as in the GFDL simulation described by Manabe and Stouffer (1980). It must be noted that there is considerable interannual variability in the extent of the marginal sea-ice zone (Barry, 1980) and observational data are still being debated, especially for Antarctica. Figure 8, for example, compares the seasonal variation of percentage sea-ice cover in the southern hemisphere. Two simulated curves from the GISS GCM are compared with two sets of observational data. The seasonal cycle is well simulated and the absolute extent of sea-ice predicted by two GISS model runs encompasses observational data. Figure 9 shows the extent and amount of predicted sea-ice in the months of January and July from the GISS climate model. Areas containing any sea ice are hatched and shading categories show the ice concentration. These maps compare favorably with observational data (Barry, 1980). In particular the coverage in the southern hemisphere is quite satisfactory. The Canadian archipelago becomes partially frozen in winter; the effect of penetration of warm water to the north of the British Isles is well simulated, and the east coast of Greenland shows a realistic seasonal fluctuation in pack ice.

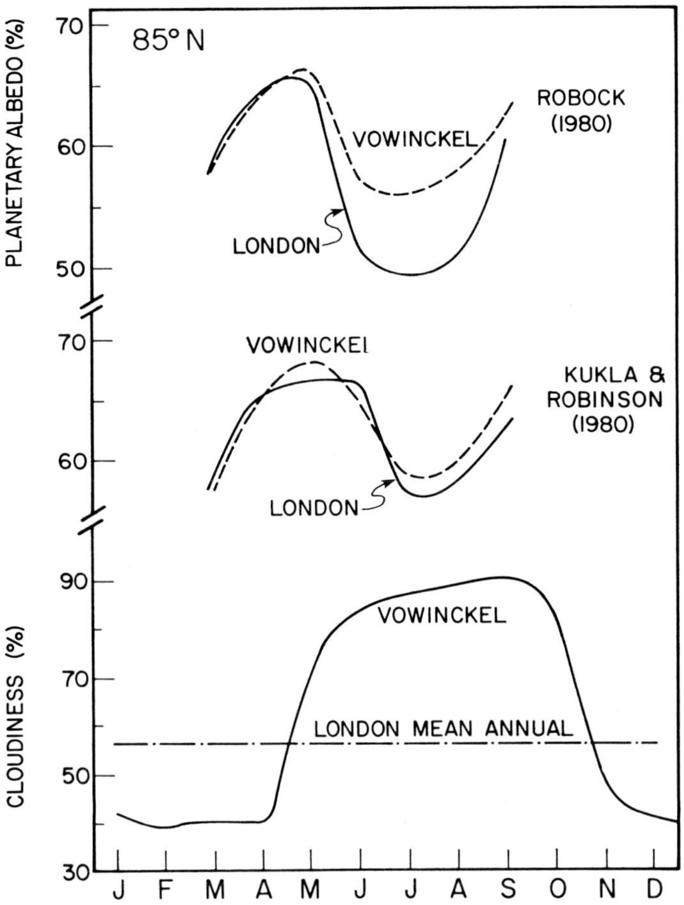

Fig. 7. The effect of different cloud cover data on planetary albedo for two alternative sets of surface albedos. The lower figure shows the mean annual cloudiness at 85°N according to London (1957) and the monthly data of Vowinckel (1962). In the upper figure the planetary albedo is determined using the surface albedos of Robock (1980) and Kukla and Robinson (1980).

The predicted January and July sea-ice areas (Figures 9) are 5-year averaged values taken from years 21-25 of a control simulation. The predicted areal extent of sea-ice is consistent with observational data, the seasonal variation between winter and summer conditions is well simulated in both hemispheres. In seventeen grid elements (8° x 10°) in the southern hemisphere the sea-ice coverage increases to greater than 75 percent between January and July, while in the reverse process from July and January in the northern hemisphere ice cover in two grid elements increases to greater than 75 percent and an additional four elements become totally ice-covered.

TABLE 3. Parameterization of Ice Surface Albedos in General Circulation Models

Model	Example Reference	Albedo (T = surface temp.; ϕ = latitude)
GLAS	Herman & Johnson, 1980	Always 0.7
ANMR (Australia)	McAvaney et al., 1978	Always 0.7
GFDL (i)	Manabe & Wetherald, 1975	0.7 T $\leqslant$ 263 K 0.35 T > 263 K
GFDL (ii)	Manabe & Stouffer, 1980	0.5 ϕ < 55° 0.7 ϕ > 66.5° 0.45 if top surface melting
UK Met Office	Corby et al., 1977	0.8 T < 271.2 K 0.5 T $\geqslant$ 271.2 K
NCAR	Washington et al., 1980	Not given, but addition of snow probably rises bare ice value (Dickinson et al., unpub.)
GISS	Hansen et al., 1983	0.45 bare ocean-ice, higher if snow covered-max for deep total cover is 0.85
CCSAS (USSR)	Marchuk et al., 1979	Always 0.6
UCLA	Arakawa, 1972	Bare ice 0.4; snowcovered 0.7

The two current GISS model parameterizations for prediction of ocean-ice (curves 882 and 824 in Figure 8) encompass the observational data. It should be noted that the observations (from Alexander and Mobley, 1974, and Zwally, et al., 1982) themselves differ in the summer (Dec./Jan./Feb.) though to a smaller extent. Model run 824 tends to overestimate total ice extent. The reason for this is that the ocean mixed layer temperature was permitted to rise above 273K even when sea-ice existed above it and no horizontal energy exchange took place between ice and ocean water within mixed grid boxes. In model run 882 these sub-grid scale horizontal and vertical energy fluxes are incorporated. Figures 9a and 9b are drawn from results from simulation run 882. The reasonableness of the ice-zone simulation suggests that an analysis of the complete surface energy budget may be worthwhile.

The surface energy budget analysis has been undertaken for the north polar region. Table 4 compares eight components of the surface energy budget calculated in the GISS GCM with observations from the central Arctic taken from Maykut and Untersteiner (1971; Tables 1 and 3). The model values listed are zonally averaged for latitude zones of widths 4° and 8° centered at 88°N and 82°N respectively. The predicted values are taken from 5-year averaged statistics from years 21-25 as described before. There is unlikely, therefore, to be any trend in the model values. The observational and predicted values have been rounded to the same number of significant figures. It can be seen that the seasonal cycle seems to be fairly well simulated in both latitude zones. The modelled incoming shortwave radiation and albedo are both slightly lower than the observed values. However this is almost certainly because the observed values are not representative of typical Arctic conditions. As discussed above, the observed surface albedo data in Table 4 follow Marshunova (1966) and are higher than many other albedo measurements. The incoming longwave radiation is well simulated in all but the months of August, September and October when they fall below the observed values. Similarly the outgoing infrared flux is well simulated, especially in the more southern model zone. The net radiation budget is very well simulated during the summer but values are higher than observed in the winter season. Finally, the predicted evaporative fluxes seem to be much closer to the observations than do the

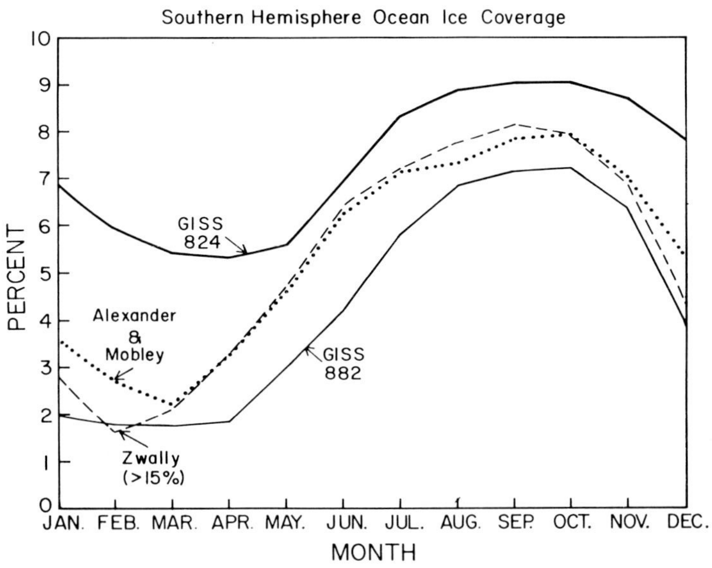

Fig. 8. Annual cycle of southern hemisphere sea-ice cover as a percentage of the global surface. Observational data are from Alexander and Mobley (1974) and from microwave data for concentrations >15% from Zwally et al. (1982). The predicted curves are from two simulations with the GISS climate model: 824 (no vertical or horizontal energy transfers within grid boxes), 5-year average for years 16-20; 882 (energy transfers incorporated), 5-year average for years 21-25.

fluxes of sensible heat. [It must be noted that the observations in Table 4 are not as representative of the real world as the predicted values are of the model's climatology since the former are from measurements restricted in time and space; see Maykut and Untersteiner (1971)]. However, both these predicted energy fluxes resemble the measurements more closely in the highest latitude belt, 88°N, than at 82°N. In particular, the sensible heat flux at 82°N does not even exhibit the seasonal cycle in the observational data whereas this is clearly seen in the polar zone simulation.

The most important conclusion that can be drawn from Table 4 is that the components of the surface energy budget in the model simulations are in good agreement with the observational data set used by Maykut and Untersteiner (1971). Reasonable agreement is seen in both radiant and non-radiant fluxes.

Cloud amount and type is a predicted parameter in the GISS model. The annual average total cloud computed in GISS experiment 882, years 21-25 is ca. 60 percent in the Arctic Ocean and 65-70 percent in the Southern Ocean. In both cases, there is very little high cloud; in the Arctic middle level cloud is ca. 20 percent compared with 35-40 percent over the Southern Ocean. These values seem generally realistic, but the calculated seasonal ranges are too small in comparison with observations.

It is interesting to examine the model cloudiness and sea ice values in the Arctic. For both total cloudiness and low cloud, there is a northward decrease in amounts across the sea ice margins, except in the mid-summer months when the gradient reverses (Figure 10). The former case resembles closely the observational data derived from the 3-D Neph archive (Figure 4). October is chosen from the GISS results although the September output shows like patterns but weaker gradients. A synoptic case matching the computed July pattern shown in Figure 10b was also discussed earlier (Figure 5).

The GISS computations of α_s, α_p and total cloudiness are compared at 65°N in the Norwegian Sea and near the North Pole in April, July and September in Table 5. The latitudinal differences in surface albedo are comparable to those in the lower diagram of Figure 6. Table 5 shows the large damping effect of cloud cover on the latitudinal gradient of planetary albedo (α_p).

4. Conclusions

Better information on surface conditions and cloudiness in polar regions is essential for improving climate modeling parameterizations in high latitudes. Ice surface albedos used in current GCMs differ sufficiently to cause large differences in sea-ice extent if they were incorporated in a predictive model. Such discrepancies are important in view of the predominant role of solar radiation in the seasonal energy budget regimes of snow and ice. Analysis of simulations made with the GISS GCM indicate that satisfactory sea ice regimes can be obtained when realistic surface albedo parameterizations and conductive heat fluxes are used but, ultimately, fully-coupled models will also need to incorporate ice dynamics.

The limited information on high-latitude cloudiness and especially the relationship with concurrent surface conditions has similarly hindered the fuller understanding of ice-albedo feedback processes. In general, it appears that cloudiness considerably damps latitudinal contrasts in planetary albedo across the ice edge. However, it remains to pursue further the question of actual cloud gradients across the cryospheric margins on climatic time-and space-scales and also their total effect on all-wavelength radiative fluxes. The simulations of cloudiness and sea ice made by the GISS GCM show agreement with most of the limited observational data on cloud gradients across the marginal ice zone and their seasonal change; mid-summer cloud amounts over the ice remain problematic, however.

It is now possible to analyze the components of surface energy budgets at high latitudes in climate models. Such an analysis for the GISS GCM gives encouraging results, but it would seem

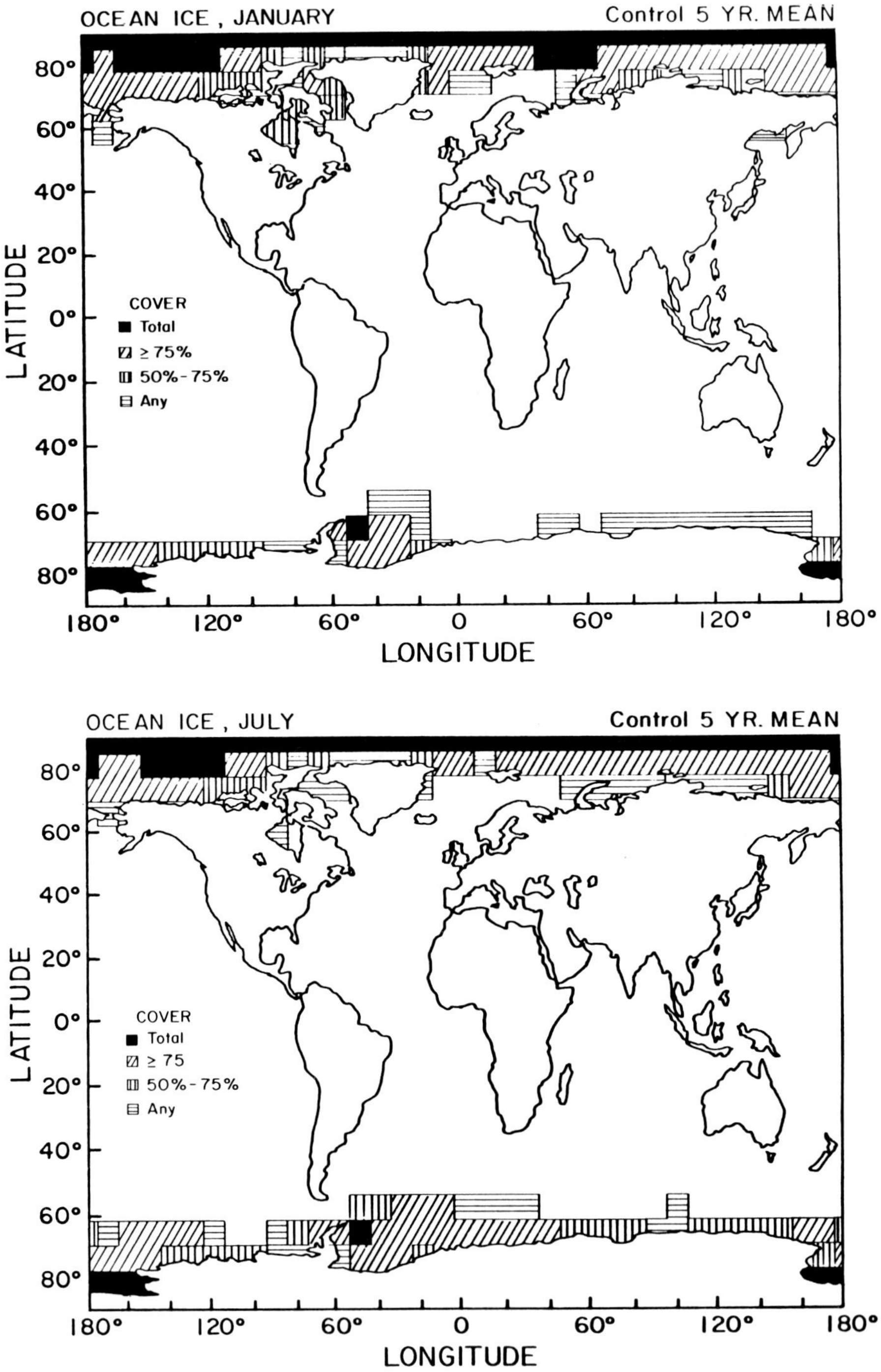

Fig. 9. Predicted sea-ice extent and concentration from the GISS model simulation 882 for a) January, b) July. Areas containing any sea ice are hatched horisontally, vertical hatching = 50-75% ice, oblique hatching $\geq$ 75%, black = total ice cover.

TABLE 4. Surface Energy Budget Components: Observed Values Calculated from Tables 1 and 3 of Maykut and Untersteiner (1971), predictions for latitude bands centered at 88°N and 82°N from GISS GCM (5 year average from years 21-25 of run 882)

		J	F	M	A	M	J	J	A	S	O	N	D
Incoming shortwave $W\ m^{-2}$	88°	0	0	7	122	246	301	196	119	22	0	0	0
	Obs	0	0	30	160	277	310	213	141	60	6	0	0
	82°	0	0	33	131	256	287	210	126	42	1	0	0
Albedo (%)	88°	ND	ND	57	63	72	67	51	47	66	ND	ND	ND
	Obs	ND	ND	83	81	82	78	64	69	84	85	ND	ND
	82°	ND	ND	65	67	76	68	48	48	60	65	ND	ND
Incoming longwave $W\ m^{-2}$	88°	167	161	163	195	238	265	285	272	229	188	171	166
	Obs	163	179	161	187	236	291	299	292	266	217	181	170
	82°	168	165	170	195	232	265	279	267	227	197	178	169
Outgoing longwave $W\ m^{-2}$	88°	181	175	178	237	284	308	317	309	250	205	188	182
	Obs	190	185	192	214	264	306	315	308	277	240	210	196
	82°	184	181	190	236	283	307	315	304	258	218	196	169
Net longwave $W\ m^{-2}$	88°	-14	-14	-15	-42	-46	-43	-32	-37	-21	-17	-17	-16
	Obs	-27	-7	-31	-26	-28	-15	-17	-16	-10	-23	-30	-26
	82°	-16	-16	-20	-41	-51	-42	-36	-37	-31	-21	-18	-17
Net radiation $W\ m^{-2}$	88°	-14	-14	-12	3	23	31	63	26	-13	-17	-17	-16
	Obs	-27	-7	-26	4	21	53	60	28	-1	-22	-30	-25
	82°	-16	-16	-12	3	25	52	74	29	-14	-21	-18	-17
Sensible energy flux $W\ m^{-2}$	88°	-4	-4	-4	-3	-9	-3	-2	-2	-1	-3	-3	-4
	Obs	18	13	11	5	-7	-6	-5	-6	-3	1	9	12
	82°	-11	-11	-9	-8	-11	-9	-9	-9	-9	-13	-12	-11
Evaporative energy flux $W\ m^{-2}$	88°	-2	-2	-2	-2	-9	-11	-5	-7	-1	-1	-1	-2
	Obs	0	0	0	-1	-7	-11	-10	-10	-6	-3	0	0
	82°	-3	-3	-3	-4	-11	-14	-12	-11	-6	-6	-4	-4

(ND = not defined)

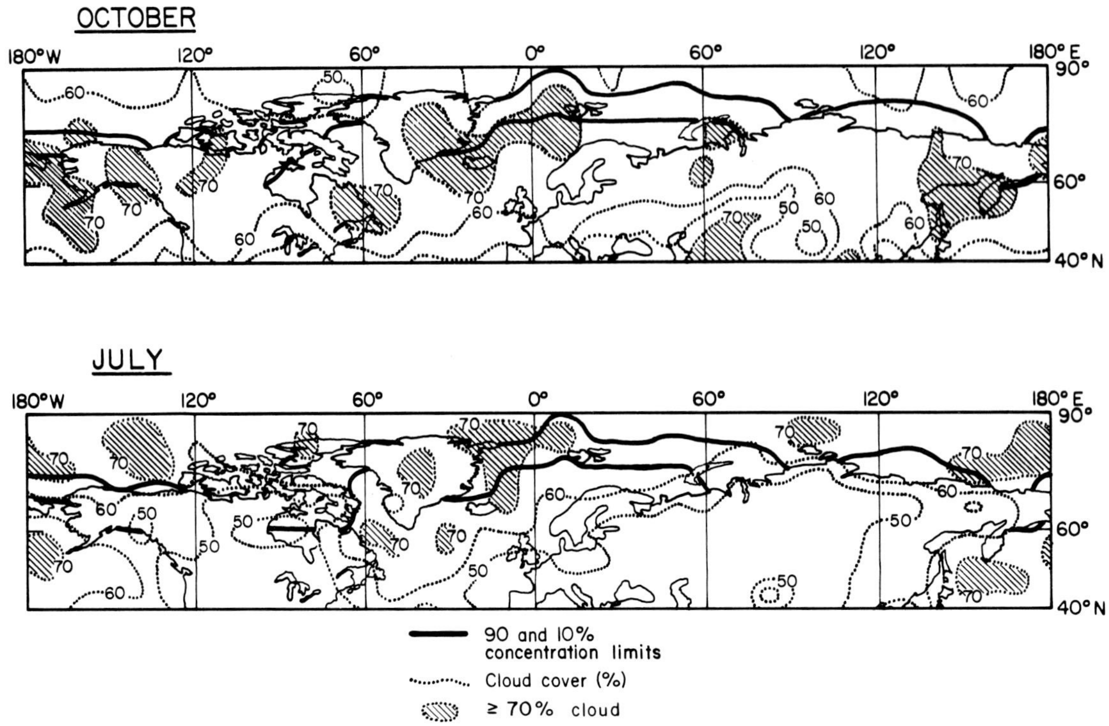

Fig. 10. Predicted total cloud amount (---) and limits of 90% and 10% sea ice concentration (——) from the GISS model simulation 882 for a) October, and b) July. Areas with $\geq$ 70% cloud cover are shaded.

worthwhile to undertake some evaluation of the sensitivity of the marginal cryosphere to cloudiness perturbations in view of the results presented here. We may infer, for example, that changes in polar sea ice extent associated with some external or internal forcing could substantially modify the cloud regime. At present, it would appear that less ice would tend to cause increased cloudiness in both seasons giving an offsetting effect in terms of at least planetary albedo, unless the cloud layers were particularly thin. However, there may also be other complications due to shifts in cyclonic activity. It is important to assess the magnitude and direction of the effects such changes may bring in polar climate. It is also likely that natural fluctuations in polar ice and snow in the past have given rise to similar responses and hence the modeling results may be a useful guide to a better understanding of past changes in polar climate.

TABLE 5. Computed average values of total cloud (c), surface (α_s) and planetary albedo (α_p) from simulation 882 with the GISS GCM. $\Delta\alpha_p$ is the albedo difference between the two regions.

	65°N, Norwegian Sea			85°-90°N			$\Delta\alpha_p$
	α_s	C	α_p	α_s	C	α_p	
April	.10	.65	.45	.65	.50	.58	.13
July	.10	.65	.45	.50	.65	.55	.10
September	.10	.65	.48	.65	.65	.65	.17

Acknowledgments. This study has been supported by NASA grant NAG-5-142 and NSF-ATM-80-18898 to the University of Colorado. Ann Henderson-Sellers conducted part of the study as a National Research Council Research Associate at the Goddard Institute of Space Studies, New York. We are indebted to Jim Hansen for making available simulation results from the GISS GCM for use in this study and to the reviewers for their suggestions.

References

Alexander, R. C. and R. L. Mobley, Monthly average sea-surface temperature and ice-pack limits on a 1° global grid, Report 4-1310-ARPA, Rand Corporation, Santa Monica, California, 30 pp., 1974.

Andreas, E. L. and S. F. Ackley, On the differences in ablation seasons of Arctic and Antarctic sea ice, J Atmos. Sci., 39, 440-447, 1982.

Arakawa, A., Design of the UCLA GCM, Numerical simulation of weather and climate, Tech. Rep. No. 7, Dept. of Met., UCLA, Los Angeles, 1972.

Baranov, A. M., Vysota i vertikal'naya protyazhennost' oblakov nizhnego i srednogo yarusov nad morem Laptevyx, Trudy Arkt. Antarkt. Inst., 239, 85-94, 1962.

Barry, R. G., Meteorology and climatology of the seasonal sea ice zone, Cold Reg. Sci. Technol., 2, 133-150, 1980.

Buzuev, A. Ya., N. P. Shesterikov and A. A. Timerev, Albedo l'da v Arkticheskix Moryax po dannym nablyudeniy s samoleta, Probl. Arkt. Antarkt., No. 20, 49-54, 1965.

Carleton, A. M., Cryosphere-climate interactions based on satellite data analysis of cloud vortex systems in the Northen Hemisphere, Ph. D. thesis, Univ. of Colorado, Boulder, 289, 1981.

Carsey, F. D., Arctic sea ice distribution at end of summer 1973-1976 from satellite microwave data, J. Geophys. Res., 87, 5809-5835, 1982.

Chernigovskiy, N. T., Radiation regime of the central Arctic basin, in Polar Meteorology, WMO Technical Note No. 87, Geneva, 107-115, 1967.

Corby, G. A., A. Gilchrist and R. L. Newson, The United Kingdom Meteorological Office five level general circulation model, Methods in Computational Physics, 17, 67-110, 1977.

Crane, R. G. and R. G. Barry, The influence of clouds on climate with a focus on high latitude interactions, J. Climatol., 3, in press, 1983.

Crane, R. G., R. G. Barry, and H. T. Zwally, Analysis of atmosphere-sea ice interactions in the Arctic Basin using ESMR microwave data, Int. J. Remote Sensing, 3, 259-276, 1982.

Dergach, A. L., G. M. Zabrodskiy and V. G. Morachevskiy, Opyt komplexsnogo issledovaniya oblakov tipa St-Sc i tumanov v Arktike, Akad. Nauk SSSR, Izvestiya, Ser. Geofiz. no. 1, 107-114, 1960.

Dickinson, R. E., J. Jager, W. M. Washington, and R. Wolski, Boundary subroutine for the NCAR global climate model, NCAR Tech. Note 0301/78/01, Boulder, Colorado, (unpublished) 1981.

Doronin, Yu. P., Thermal interaction of the atmosphere and the hydrosphere in the Arctic, Israel Program for Scientific Translations, Jerusalem, 1970.

Grenfell, T.C. and Maykut, G. A., The optical properties of ice and snow in the Arctic Basin, J. Glaciol., 18(80), 445-464, 1977.

Hansen, J. E., G. Russell, D. Rind, P. Stone, A. A. Lacis, S. Lebedeff, R. Ruedy and L. Travis, 1983. Efficient three dimensional global models for climate studies Models I and II, Mon. Wea. Rev., 111, 609-662, 1983.

Hansen, J. E., G. Russell, D. Rind, P. Stone, A. A. Lacis, S. Lebedeff, R. Ruedy and L. Travis, 1982. Unpublished description of 824/882 simulations.

Hanson, K. J., The albedo of sea ice and ice islands in the Arctic Ocean basin, Arctic, 14, 188-196, 1961.

Henderson, P. M., Cloud conditions over the Beaufort Sea, Publ. in Meteorol. no. 86, Dept. of Meteorology, McGill Univ., Montreal, 48, 1967.

Henderson-Sellers, A., R. G. Barry, R. G. Crane, N. A. Hughes, and K. Shine, Cloud-cryosphere interactions near the coastal sea ice boundary: Data retrieval and climate sensitivity, Preprints, First International Conference on Meteorology and Air/Sea Interactions in the Coastal Zone, Amer. Met. Soc., Boston, 378-380, 1982.

Herman, G. F., Solar radiation in summertime Arctic stratus clouds, J. Atmos. Sci., 34, 1423-1432, 1977.

Herman, G. F., Thermal radiation in Arctic stratus clouds, Quart. J. Roy. Met. Soc., 106, 771-780, 1980.

Herman, G. F., Arctic stratus clouds, in Untersteiner et al., eds., The Geophysics of Sea Ice, Plenum Press, New York, 1983, in press.

Herman, G. F. and R. Goody, Formation and persistence of summertime Arctic stratus clouds, J. Atmos. Sci., 33, 1537-1553, 1976.

Herman, G. F. and W. J. Johnson, Arctic and Antarctic climatology of a GLAS GCM, Mon. Wea. Rev., 108, 1974-1991, 1980.

Hibler, W. D., III and S. F. Ackley, On modeling the Weddell Sea pack ice, Annals Glaciol., 3, 125-130, 1982.

Hughes, N. A., Global cloud climatologies: analysis of cloud amount data and investigation of processing and archiving strategies, Ph.D. thesis, Univ. of Liverpool, 1982.

Hummel, J. R. and R. A. Reck, A global surface albedo model, J. Appl. Meteorol., 18, 239-253, 1979.

Huschke, R. E., Arctic cloud statistics from 'air-calibrated' surface weather observations, Rand Corporation, Memo. RM-6173, Santa Monica, California, 79 pp., 1969.

Jayaweera, K. O. L. F. and T. Ohtake, Composition of Arctic clouds, in Weller, G. et al., eds., Studies of the Solar and Terrestrial Radiation

Fluxes over Arctic Pack Ice, Geophysical Inst., Univ. of Alaska, 14-16, 1972.

Jayaweera, K. O. L. F., Characteristics of Arctic stratus clouds over the Beaufort Sea during AIDJEX, AIDJEX Bull., 37, 135-151, 1977.

Jayaweera, K. O. L. F., Meteorological conditions conducive to the formation of stratus cloud over the Beaufort Sea, National Weather Digest, 7, 25-32, 1982.

Koptev, A. and A. I. Voskresenskiy, On the radiation properties of clouds, in Fletcher, J. O., B. Keller and S. M. Olenicoff, eds., Soviet Data on the Arctic Heat Budget and Its Climatic Influence, Rand Corporation, Santa Monica, California, Memo RM-5003-PR, 175-191, 1966.

Kukla, G. and D. Robinson, Annual cycle of surface albedo, Mon. Wea. Rev., 108, 56-58, 1980.

Langleben, M. P., On the factors affecting the rate of ablation of sea ice, Can. J. Earth Sci., 3, 431-439. 1966.

Langleben, M. P., Albedo measurements of an Arctic ice cover from high towers, J. Glaciol., 7, 289-297, 1968.

Langleben, M. P., Albedo and degree of puddling of a melting cover of sea ice, J. Glaciol., 8, 407-412, 1969.

Langleben, M. P., Albedo of melting ice in the southern Beaufort Sea, J. Glaciol., 10, 101-104, 1971.

London, J., A study of the atmospheric heat balance, AFCRE-TR-57-287, College of Engineering, New York University, 99 pp., 1957.

Manabe, S. and R. J. Stouffer, Sensitivity of a global climate model to an increase of CO_2 concentration in the atmosphere, J. Geophys Res., 85, 5529-5584, 1980.

Manabe, S. and R. T. Wetherald, The effects of doubling the CO_2 concentration on the climate of a general circulation model, J. Atmos. Sci., 32, 3-15, 1975.

Manabe, S., K. Bryan and M. J. Spelman, A global ocean-atmosphere model with seasonal variation for future studies of climate sensitivity, Dyn. Atmos. Oceans, 3, 393-426, 1979.

Marchuk, G. I., V. P. Dymnikov, V. N. Lykosov, V. Y. Gabin, I. M. Bobyleva and V. L. Petrov, Numerical simulation of global circulation of the atmosphere, Report of the JOC Study Conference on Climate Models, GARP Pub. 22, Vol. I, 318-370, WMO, Geneva, 1979.

Marshunova, M. S., Principal characteristics of the radiation balance of the underlying surface and the atmosphere in the Arctic, in Fletcher, J. O., B. Keller and S. M. Olenicoff, eds., Soviet Data on the Arctic Heat Budget and its Climatic Influence, Rand Corporation, Santa Monica, California, Memo RM-5003-PR, 51-131, 1966.

Maykut, G. A., Surface heat and mass balance, in Untersteiner, N. et al., eds., "Sea Ice Geophysics," Plenum, (in press), 1983.

Maykut, G. A. and N. Untersteiner, Some results from a time dependent, thermodynamic model of sea ice, J. Geophys. Res., 76, 1550-1575, 1971.

McAvaney, B. J., Bowke, Wa Puri, K., A global spectral model for simulation of the general circulation, J. Atmos. Sci., 35, 1557-1583, 1978.

McPhee, M. A. and N. Untersteiner, Using sea ice to measure vertical heat flux in the ocean, J. Geophys. Res., 87, 2071-2074, 1982.

Mohn, H., Meteorology, in Nansen, F., ed., The Norwegian North Polar Expedition 1893-1896. Scientific Results, Vol VI, 511-522, Longmans, Green and Co., London, 1905 (reprinted Greenwood Press, New York, 1969).

Nazintsev, Yu. L., Teplovoi balans poverkhnosti # nogoletnego ledyanago pokrova v tsentral'noi Arktike, Arkt. Antarkt. nauch-issled. Inst., Trudy, 267, 110-126, 1964.

Parkinson, C. L. and G. F. Herman, Sea ice simulations based on fields generated by the GLAS GCM, Mon. Wea. Rev., 108, 267-285, 1980.

Overland, J. E. and C. H. Pease, Cyclone climatology of the Bering Sea and its relation to sea ice extent, Mon. Wea. Rev., 110, 5-13, 1982.

Robock, A., The seasonal cycle of snow cover, sea ice and surface albedo, Mon. Wea. Rev., 108, 267-285, 1980.

Scharfen, G. R., Snow cover, sea ice and cloudiness characteristics from satellite data, M.A. thesis, University of Colorado, 1982.

Schneider, S. H. and R. E. Dickinson, Parameterization of fractional cloud amounts in climatic models: The important of modeling multiple reflections, J. Appl. Met., 15, 1050-1056, 1976.

Semtner, A. J., A model for the thermodynamic growth of sea ice in numerical investigations of climate, J. Phys. Oceanogr., 6, 379-389, 1976.

Shine, K. P., A. Henderson-Sellers, and R. G. Barry, Albedo-climate feedback: The importance of cloud and cryosphere variability, in Berger, A. and Nicolis, C., ed., New Perspectives in Climate Modeling, Elseview, in press.

Slingo, A. And Schrecker, H. M., On the shortwave radiation properties of stratiform waterclouds, Quart. J. Roy. Met. Soc., 108, 407-426, 1982.

Strokina, A. L., Polar ice cap albedo as a function of air temperature, Soviet Meteorology and Hydrology, 7, 43-46, 1980.

Vonder Haar, T. H., Measurement of albedo on polar snow and ice fields using NIMBUS 3 satellite data, in "Advanced concepts and techniques in the study of snow and ice resources," Monterey, California, December 2-6, 1973, 1974.

Voskresenskiy, A. N. and Chukanin, K. I., Meteorologicheskie usloviya obledeniya v oblakax tipa St i Sc, Trudy Arkt. Antarkt. Inst., 228, 124-134, 1959.

Vowinckel, E., Cloud amount and type over the Arctic, Publ. in Met. 51, McGill University, Montreal, 27, 1962.

Vowinckel, E. and S. Orvig, The climate of the north polar basin, in Orvig, S., ed., "Climates of the Polar regions," World Survey of Climatology, 14, Elsevier, 129-252, 1970.

Washington, W. M., A. J. Semtner, G. A. Meehl, R.

J. Knight and T. A. Meyer, A general circulation experiment with a coupled atmosphere, ocean and sea ice model, J. Phys. Oceanogr., 10, 1887-1908, 1980.

Weaver, R. L., J. D. Jacobs, and R. G. Barry, Fast ice studies in eastern Baffin Island, in Proceedings, Third International Conference on Ports and Ocean Engineering under Arctic Conditions, Vol. 1, 455-466, University of Alaska, Fairbanks, 1976.

Weller, G., S. A. Bowling, B. Holmgren, K. O. L. F., Jayaweera, T. Ohtake, and G. Shaw, The radiation matrix in the Arctic, in Weller, G. and S. A. Bowling, eds., Climate of the Arctic, Univ. of Alaska, Fairbanks, 238-244, 1975.

Wiscombe, W. J. and S. G. Warren, A model for the spectral albedo of snow, I. Pure Snow, J. Atmos. Sci., 37, 2712-2733, 1980.

Yakovlev, G. N., Snezhniy pokrov na dreyfuyushchnix l'dax tsentral'noy Arktiki. Pobl. Arkt. Antarkt., No. 3, 65-76, 1960.

Yanes, A. V., Tayanie snega i l'da v tsentral'nogo Arktike. Probl. Arkt. Antarkt., No. 11, 59-64, 1962.

Zwally, H. J., J. C. Comiso, C. L. Parkinson, W. J. Campbell, F. D. Carsey, and P. Gloerson, Antarctic sea-ice cover 1973-1976 from satellite passive microwave observations, NASA SP-00, (in press), 1982.

THE ROLE OF SEA ICE DYNAMICS IN MODELING CO_2 INCREASES

W.D. Hibler III

USA Cold Regions Research and Engineering Laboratory, Hanover, NH 03755

Abstract. Sensitivity simulations of a hierarchy of Antarctic sea ice models to atmospheric warming are carried out and analyzed. The study includes models with only a thermodynamic ice cover, models with in-situ leads but no ice transport, and a fully coupled dynamic/thermodynamic model that includes transport, leads and strength-thickness coupling. All models employ a 60-m-thick oceanic mixed layer, together with a spatially and temporally varying heat flux into the mixed layer from the deep ocean. The heat flux was generated interactively by using a fixed fraction of the ice growth and cooling rates from the full dynamic/thermodynamic model. The same spatially and temporally varying heat flux fields (with an average of about 13 W m^{-2}) were used in all sensitivity simulations.

Models including full ice dynamics effects are found to be less sensitive to atmospheric warming than thermodynamics-only models, while models with specified lead fractions are more sensitive than thermodynamics-only models. The decreased sensitivity in a fully coupled model is found to depend on the ice velocity field also being modified by the atmospheric warming. In all cases the fully coupled model has a greater seasonal swing of ice extent than the thermodynamics-only model and fixed lead fraction model. All models show an enhanced sensitivity in the summer ice extent as compared to winter.

With regard to air-sea heat exchanges, models including leads and full dynamics show a reduced sea-to-air heat exchange over ice-covered regions under warming conditions, while the thermodynamics-only model shows an increase. In the control simulations the sea-to-air exchanges through the ice cover are about five times larger in the full model than in the thermodynamics-only model.

Sensitivity of the fully coupled model to oceanic forcing and variable thickness parameterization was also investigated. In the standard control run temporally and spatially varying heat fluxes were found to reduce the seasonal swing of ice extent as compared to constant heat flux values. Similarly, in comparison to a 30-m mixed layer, the 60-m mixed layer, reduced the seasonal ice extent variation.

Introduction

An important aspect of model-based CO_2 climate sensitivity studies is the treatment of sea ice in the polar regions. In the recent study by Manabe and Stouffer (1980) the atmospheric warming in the Northern Hemisphere was shown to be enhanced by the reduction of sea ice thickness in the high-CO_2 sensitivity simulations. More recently, simulations by Hansen et al. (1984) have shown significant circulation differences from the Manabe/Stouffer results. Moreover, Hansen et al. (1984) have suggested that some of the differences may be due to a more realistic extent of sea ice in the control runs.

Because of such considerations it has become relevant to ascertain the sensitivity of sea ice models to changes in atmospheric CO_2 increases. An important question in this regard is to what degree inclusion of dynamical processes modifies the sensitivity of sea ice models to changes in the atmospheric forcing. The basic point here is that, in reality, the polar sea ice cover is not static (as is assumed in most model-based CO_2 studies) but is very dynamic. This dynamic character substantially modifies the air-sea heat and momentum exchanges in the polar regions. Considerable progress has been made over the last several years (see e.g. Hibler, 1980a, for a review) in understanding the essential physics of these exchanges and their dependence on the dynamic nature of sea ice. However, to date there has been relatively little work on examining how the dynamic character of sea ice might modify both the sensitivity of the seasonal cycle of sea ice and the air/sea heat exchanges in response to an assumed climatic change.

To address this issue, this paper presents some results from a series of sensitivity simulations of a hierarchy of Antarctic sea ice models to atmospheric warming. The models range from a fully coupled dynamic/thermodynamic sea ice model employing a nonlinear plastic ice interaction to a static thermodynamics-only model. Also, included as a model intermediate between these two is a static model with a seasonally varying prescribed fraction of leads. These model simulations are carried out employing observed atmospheric temperature and wind fields, and for atmospheric forcing modified by an assumed 4°C warming.

The results of these simulations show substantial differences in sensitivity between a full dynamic/thermodynamic model including thickness-strength coupling, and more static models both with and without leads. Analysis of these differences demonstrates several mechanisms by which dynamical processes may modify simulated sea ice extents, thicknesses, and air-sea heat exchanges in model-based CO_2 climate studies. Moreover, different dynamical processes are found to play different roles in the sensitivity to atmospheric warming. Overall, these results emphasize the need for a more complete treatment of dynamical processes in climate sensitivity studies.

Approach

The basic approach in this study is to carry out seasonal equilibrium sensitivity simulations of a modified version of an existing dynamic/thermodynamic sea ice model (Hibler and Ackley, 1983) of the Weddell Sea ice pack. For comparison, simulations are also carried out using simpler versions of this model obtained by removing various aspects of the full physics, but employing the same atmospheric and oceanic forcings. To gain perspective on the motivation for the hierarchy of models employed, it is useful to first qualitatively examine the role different characteristics of a dynamic ice cover might play in climate sensitivity. Once this is completed the fully coupled dynamic/thermodynamic model is described, the treatment of the forcing fields examined, and the detailed hierarchy of models discussed.

Characteristics of a Dynamic Sea Ice Cover Relevant to Climatic Change

The main features of a dynamic sea ice cover (as actually exists in the polar regions) that distinguish it from a static ice cover may be roughly grouped into three main categories. The first is the "variable thickness" feature. This term refers to the fact that over any region of sea ice there will be a variety of thickness categories ranging from leads (areas of open water) to thick ridged ice, with a variety of thicknesses in between. This variable thickness feature is caused by deformation of the ice cover (see e.g. Thorndike et al., 1975), which tends to create leads by divergence and form pressure ridges by convergence. This variable thickness feature causes the actual growth rates of an area of sea ice to be very dependent on the thin ice fraction and hence on the ice dynamics.

A second feature is "ice transport." This term describes the fact that over the long term ice will move from one region to another. This is particularly important for mass balance considerations since it means, that in contrast to a static ice cover, the net growth at a given location will differ substantially from zero over a seasonal cycle (see e.g. Hibler, 1979; Hibler and Ackley, 1983). This in turn means that in regions of net ice growth the atmosphere will likely receive more heat from the ocean than it loses, whereas in regions of heavy ice melt the main effect should be heat loss by the atmosphere. The transport can also play a direct role in ice advance by moving the ice outward, and an indirect role in the retreat by transporting ice to the edge to be melted, thus maintaining a stationary ice edge position even though considerable amounts of decay are occurring there.

The third feature, which plays a pivotal role in any sort of coupled dynamic/thermodynamic model, is ice interaction and a coupling between this interaction and the ice thickness characteristics (Hibler, 1979). This feature distinguishes the ice cover from a passive tracer field which simply follows the surface current of the ocean and allows changes in thickness characteristics to change the ice velocity field. This interaction has two dominant effects. One is a rectifying effect on ice deformation. Basically, sea ice has little resistance to divergence but great resistance to convergence. Consequently, under symmetric but fluctuating forcing fields a sea ice cover will tend to expand, creating more leads and possibly forcing any free ice edge outward. The second effect of ice interaction is to supply a mechanical limit to ice build-up that can override thermodynamic considerations. The idea here is that as the ice becomes thicker it resists compression, so that maximum thicknesses can in principle be controlled by mechanical rather than thermodynamic considerations.

How these dynamic characteristics might change climate sensitivity in the polar regions can be outlined by comparison to simpler models. In climate studies using a passive thermodynamic sea ice cover (Manabe and Stouffer, 1980), CO_2 warming was enhanced by the sea ice in the polar regions. This enhancement was affected both by thinner ice, which has a lower conductivity, and less ice extent, which exposed more open ocean. Consequently, dynamical effects which change either the local air-sea heat exchange or the ice extent may have an effect on climate sensitivity. Leads clearly can play a role. In particular, since the lead has no conductivity there will be a reduction in heat loss for warmer conditions. This is in contrast to the feedback which can occur with a thermodynamic ice slab, where thinner ice gives a lower conductivity, and hence a relative increase in sea-to-air heat loss. With respect to ice extent, dynamical effects may change the ice extent in different ways under warming conditions. A more direct role of ice interaction is to cause the ice thickness to be mechanically determined in certain regions, so that changing the thermodynamic forcing should have minimal effects. It is also possible for warming to change the ice interaction, and hence change the deformation patterns and resultant lead fractions. Finally, the tuning of a purely thermodynamic sea ice model necessary to obtain a realistic seasonal cycle may be quite different from that needed for a model including dynamical effects. Hence, the ice characteristics in a "control" simulation may be unrealistic to begin with.

Description of Dynamic/Thermodynamic Sea Ice Model

The basic model used in this study is a slightly modified version of the Weddell Sea pack ice model employed by Hibler and Ackley (1983). The main features may be briefly outlined as follows. The model calculates a two-dimensional ice velocity field at daily intervals using a momentum balance employing a nonlinear plastic rheology (see e.g. Hibler, 1979). This ice velocity field is then used to drive evolution equations for a two-thickness-level sea ice distribution. This simplified thickness distribution divides the ice into an open water (or lead) category and a thick ice category. The evolution equations also provide for ice build-up, lead formation and ice advection arising from variations in the ice velocity field. In addition, the ice grows and/or decays in response to thermodynamic calculations employing a thermodynamic sea ice model in conjunction with a surface heat budget (see e.g. Hibler, 1980b). The thermodynamic terms include both the ice growth over leads as well as over thick ice. Under melting conditions heat absorbed by leads is used to melt ice until the mixed layer returns to the freezing temperature.

To couple the thickness distribution to the momentum balance, the ice strength is taken to depend on the ice thickness and compactness. This in turn causes compressive deformation to effectively cease after the ice becomes sufficiently thick, although drift and advection may continue. Also, due to the nonlinear nature of the plastic ice interaction, dilation or opening of the pack ice (with attendant lead formation) can occur easily no matter how thick the ice is.

For this study two modifications are made to the model to obtain a more realistic thickness distribution. In the standard two-level model, thermodynamic computations are performed at two levels: over open water (h=0) and at an effective thickness, h_{eff}, representing the mean thickness of ice in the ice-covered portion of the grid cell. In order to incorporate the strong thickness dependence of ice growth rates (e.g. Maykut, 1978) h_{eff} is replaced by a seven-level distribution of thicknesses equally spaced between 0 and 2 h_{eff}. While a more realistic thickness distribution can be obtained from a multi-level model (e.g. Hibler, 1980b) using the ice thickness distribution theory developed by Thorndike et al. (1975), the treatment used here is a useful first-order procedure to include a range of thicknesses in the model thermodynamics.

Growth rates at each of these levels were calculated by using a time-independent thermodynamic sea ice model (see e.g. Semtner, 1976). Following Manabe and Stouffer (1980), the effects of snow cover are approximated by allowing the ice surface albedo to be an estimated snow value (0.75) when the calculated surface temperature of the ice is below freezing, and that of snow-free ice (0.66) when the surface temperature is at the melting point. The albedo of open water is taken to be 0.1. These values were also employed in previous simulations using this dynamic-thermodynamic sea ice model (Hibler and Walsh, 1982; Hibler and Ackley, 1983).

The second modification is to provide for the creation of an additional amount of open water due to shearing deformation. The concept here is that under high compactness conditions ridging and open water creation can occur simultaneously. A consistent way to treat this is to create open water in a manner (see e.g. Rothrock, 1975) energetically commensurate with the assumed yield curve, which in this case is elliptical (Hibler, 1979). Referring to Appendix A of Hibler (1980b), in this two-level case (or open water plus seven thickness levels) the consistency argument results in adding an additional open water creation rate of

$$(1-P)\ 0.5\ \left\{\left[\left(\varepsilon_{xx} + \varepsilon_{yy}\right) 1.25 + \varepsilon_{xy}^2 + 1.5\ \varepsilon_{xx}\ \varepsilon_{yy}\right]^{1/2} - |\varepsilon_{xx} + \varepsilon_{yy}|\ \right\}$$

where

$$P = \frac{(1-A_1)\ [0.5\ (1+A_1) - 0.85]}{0.15\ [0.5(1.85) - 0.85]}$$

and

$$A_1 = \max(0.85,\ A)$$

with A being the compactness of the ice, and ε_{xx}, ε_{yy}, ε_{xy} representing the components of the two-dimensional strain rate tensor. Since these are modifications to the version of the model used by Hibler and Ackley (1983) the sensitivity of the control runs to these modifications is examined in the Appendix.

Forcing Fields

The main atmospheric driving fields for the model consist of daily geostrophic wind fields and surface air temperatures and humidities. These data are also used to calculate empirical incoming long- and short-wave radiation fields (outgoing fields are calculated by the model).

Atmospheric forcing fields to drive the model were largely based on the 1979 Australian analysis which employed FGGE data. The daily geostrophic winds were computed directly from these data.

However, comparison with ice buoy data and sensitivity simulations (see e.g. Hibler and Ackley, 1983) showed the temperature in the analysis fields to be excessively high over the western Weddell Sea. Therefore, for the control runs for both the full and thermodynamic models, these data were reanalyzed by first replacing the temperature field inside the dashed region in Figure 1 with average surface temperatures obtained from two drifting buoys (whose initial positions are shown), and then performing a doubled smoothing operation using a two-dimensional hamming filter. (This filter is a nine-point smoothing convolution operator with the corner points equal to one-fourth of the central point and the middle points equal to one-half of the center point.) This procedure was also applied to the mixing ratios, which were constrained to be the lesser of 0.25 g/kg and the analysis values in this region. These modified temperatures and humidities were also used in the calculation of the long-wave radiation fields.

As a crude approximation to a "high CO_2" climate, all temperatures were increased by 4°C. The incoming long-wave radiation fields were also recalculated using these modified atmospheric surface temperatures. While this simple change in atmospheric temperature does not include any feedback effects between the ice and atmosphere, it is useful for a sensitivity analysis of the ice models.

The momentum effects of the ocean on the model are parameterized by using fixed geostrophic ocean currents taken from Gordon et al. (1978). With respect to ocean thermodynamic effects, a fixed-depth motionless mixed layer together with a heat flux from below are employed. To obtain a somewhat more realistic treatment of the ocean these features were treated differently than done by Hibler and Ackley (1983). In particular, a deeper mixed layer depth of 60 m was assumed rather than the 30 m employed by Hibler and Ackley (1983). A second feature was to use a larger heat flux. However, to get some estimation of the spatial and temporal variations of the heat flux in the model, the ocean heat flux was calculated in an interactive mode by assuming the ocean heat flux to be equal to 25% of any cooling of the mixed layer and/or of the ice growth. This parameterization is based on the fact that cooling and salt expulsion will cause convection and hence will bring up oceanic heat (see e.g. Lemke and Manley, 1984). Both these oceanic modifications reflect a first order attempt to include observed and postulated increased oceanic convective effects in the Antarctic (see e.g. Gordon, 1981, 1984; Martinson et al., 1981).

To prevent heat flux variations from complicating the sensitivity results, the heat flux results from the third year of the standard model control run were used in all other sensitivity and hierarchy simulations. Also, for comparison, sensitivity tests done with constant heat fluxes, a shallower mixed layer depth, and control temperatures are reported in the Appendix.

Description of Hierarchy of Models

The basic approach in constructing a hierarchy of models was to remove progressively greater amounts of physics from the full model. By comparing simulations of these less complete models with one another and with the full model it is then possible to deduce the contribution of different aspects of the dynamics both on the simulated seasonal cycle of sea ice under present conditions and on the sensitivity of the simulated sea ice to atmospheric warming. In addition, the contribution of different dynamical processes to air-sea heat exchanges can be examined.

Keeping in mind the general remarks on dynamics in Section IIa, the following hierarchy of models was selected. The first case consisted of the full dynamic/thermodynamic model, including full thickness-strength coupling, ice deformation, advection, and transport. This model also included the simplified seven-level thickness distribution, as discussed in Section IIb. Secondly, a leads-only model was chosen. In this model a static ice cover was employed with a seasonally varying fraction of leads taken from the third year of the full model. This lead fraction was taken to apply uniformly in space. As in the full model a simplified seven-level ice thickness distribution was used. Basically, this leads-only model includes deformation and variable thickness effects, but not thickness-strength coupling, advection or ice transport. A third model in the hierarchy is a thermodynamics-only model with no leads. The standard case of this model consisted of only one thickness of ice. However, for comparison a thermodynamic model employing a seven-level thickness distribution and a one-level thermodynamic model with a reduced summer albedo were also employed. The seven-level model offers a more nearly parallel comparison with the full model as it omits leads, advection and transport, but does crudely retain the thickness distribution concept for the growth of the non-open-water portion of the ice cover. The one-level case, on the other hand, omits all dynamical effects and is equivalent (with some changes in the albedo) to the sea ice model employed by Manabe and Stouffer (1980).

In the case of the low summer albedo thermo-only model, an albedo of 0.40 for melting sea ice was assumed rather than the 0.66 used in all other cases. The idea here was to parameterize the radiation absorption effects of leads in the summer while omitting the increased growth capacity of leads in the winter.

To examine the seasonal characteristics of these models and to determine their sensitivity to atmospheric warming, two sets of seasonal simulations to quasi-equilibrium were carried out: one with the normal or control temperature, and one with higher temperatures obtained by uni-

TABLE 1. Description of Model Hierarchy for Atmospheric Warming Simulations

Descriptive name	Degree of ice dynamics	Temperature forcing fields	Comments
Full model	Ice interaction, leads advection, transport, seven thickness levels	"Control" and "warm"	
Specified ice velocity field	Leads, advection, transport, seven thickness levels	"Warm"	Ice velocities taken from the control run of full model
Fixed leads	Leads, seven thickness levels	"Control" and "warm"	Zero ice velocities. Leads specified to be uniform spatially with values taken from spatially averaged daily results from full model
Seven-level thermo-only	Seven thickness levels	"Control" and "warm"	
Thermo-only	None	"Control" and "warm"	Only one thickness at each grid cell.
Low summer albedo thermo-only	Summer albedo effect of leads	"Control" and "warm"	Albedo for melting ice set at 0.4 rather than 0.66.

formly adding 4°C to the temperature fields and modifying the incoming long-wave radiation. Initial conditions for the full model with control temperatures consisted of 3 m of ice at all grid cells on 1 January. This model was then integrated for three years with the oceanic heat flux being determined interactively as discussed above. For all other simulations the thickness fields at the end of the first year of the full model control simulation were used as initial conditions and two-year integrations were employed. In addition, in all other simulations the spatially and temporally varying heat flux values from the third year of the full control run were used. Finally, in the case of the fixed-leads simulation, the daily varying spatially averaged lead fraction (for the ice-covered grid cells) from the third year of the full model was applied uniformly at all locations with an ice cover of non-zero thickness.

In addition to this basic hierarchy of three models an "off-line" velocity version of the full model was run using the "warm" temperature fields. This model utilized the ice velocity field from the third-year full-model control run. While this case includes leads, advection and transport, it does not include ice interaction effects, which were modified by the "warm" temperatures in the full model.

To determine the contribution of the thickness distribution and oceanic parameterizations to the seasonal cycle, several two-year sensitivity runs were also made using off-line velocities and the control forcing. Specifically, simulations were run with different constant heat flux values, a 30-m-deep mixed layer, a two-level model only, and a seven-layer model without the extra open water creation. These cases are discussed in the Appendix.

For convenience, the various models are summarized in Table I, which lists the models relevant to climate sensitivity, and Table AI (see Appendix), which lists the model sensitivity studies done only for the control temperatures.

Simulation Results

To carry out numerical simulations employing the hierarchy of models discussed above, a rectangular grid (Fig. 1) employing a nominal 222-km resolution was used. This grid was based on an azimuthal equidistant projection and covers the

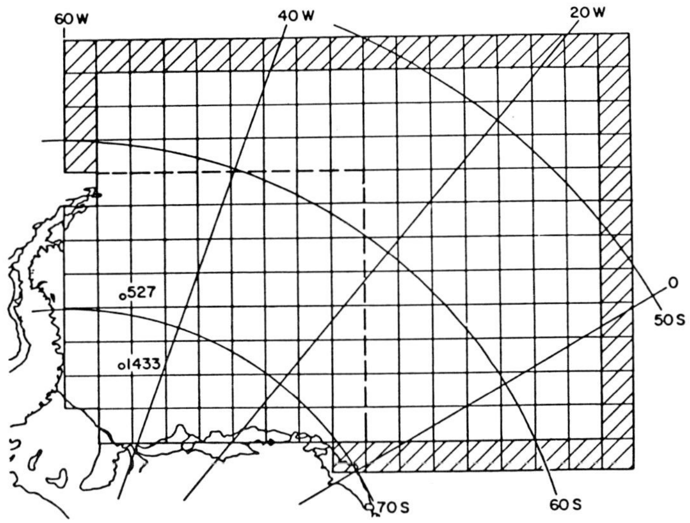

Fig. 1. Grid used for numerical simulations. Dashed lines enclose area where temperature and humidity fields were reanalyzed. Hatched grid cells represent open boundaries. Open circles denote initial positions of two drifting buoys (on 1 Jan 1979) which supplied surface temperature data for the reanalysis.

major part of the Weddell Sea. In the case of the full dynamic/thermodynamic model, inflow and outflow from the grid are allowed at open-boundary grid cells denoted by diagonal shading in Figure 1. Many of the main features of the sensitivity results in this section will be expressed in time series of total thickness over the whole grid and total "ice area" (sum of areas of all grid cells in which there is any ice of non-zero thickness).

Sensitivity of Sea Ice Models to Atmospheric Warming

To examine the effect of different dynamical processes on the sensitivity of sea ice to atmospheric warming, a series of two-year simulations were carried out for the hierarchy of models described in Table I. Sensitivity results from these simulations show the time series of the total ice area and thickness for the control temperatures (Fig. 2) and the "high" temperature forcing (Fig. 3). The models included in these figures are the full model, the fixed-leads model, the thermo-only model, and the seven-level thermo-only model. The observed time series of ice extent is based on fleet weather ice charts prepared by the Navy-NOAA Joint Ice Center, Suitland, Maryland. In addition, for a more direct measure of the relative sensitivity of each model, the ratios of the "warm" and "control" areas and thicknesses are plotted in Figure 4. This figure gives a more direct measure of the relative sensitivity of the different models to atmospheric warming.

As can be seen from Figure 2a, a dominant difference between the full model and the less complete models is a greatly increased seasonal swing of ice extent, especially in summer. The main reason for the difference between the thermo-only model and the full model is the increased heat absorption due to leads in the full

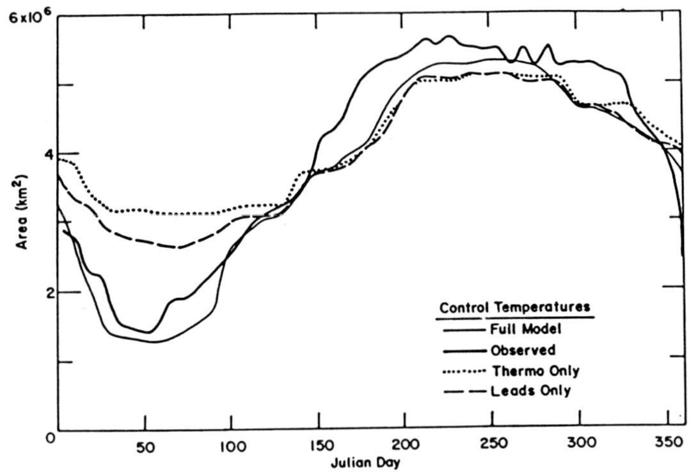

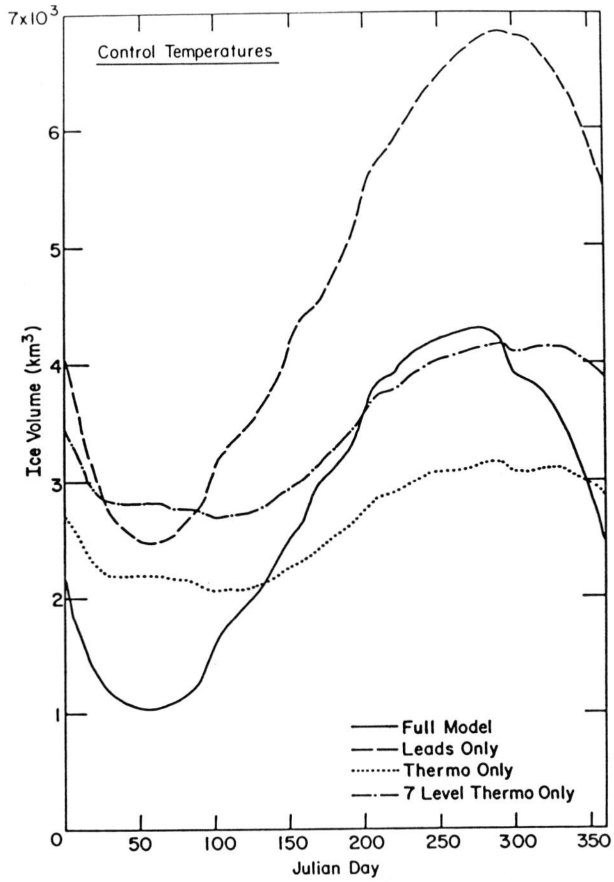

Fig. 2. Time series of spatially averaged ice area (a) and ice volume (b) for a hierarchy of model simulations employing the control temperatures.

model. However, note that this increased summer ice extent reduction is not present in the leads-only case, even though the average lead fraction is identical to that in the full model. There are basically two reasons for this difference. For one, the analysis of the full model shows that the reduced extent is partially due to winds near the ice edge that tend to shift the whole ice edge into open water and thus reduce the extent. Thus transport and advection also play a major role in the ice decay. A second reason (demonstrated in Fig. 2b) is the increased thickness occurring in the leads-only case, which in turn increases the summer extent by making the ice too thick to decay. Much of this increased thickness effect is caused by specifying leads everywhere in conjunction with a static ice cover. In the full model the leads tend

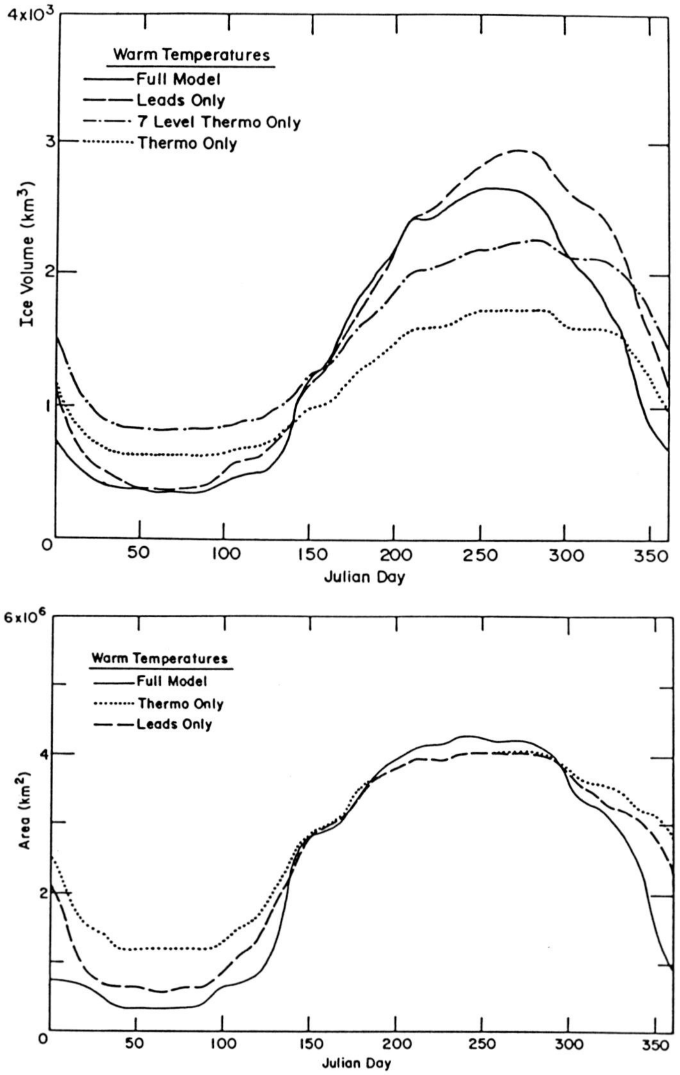

Fig. 3. Time series of spatially averaged ice area (a) and ice volume (b) for a hierarchy of model simulations employing the warm temperatures.

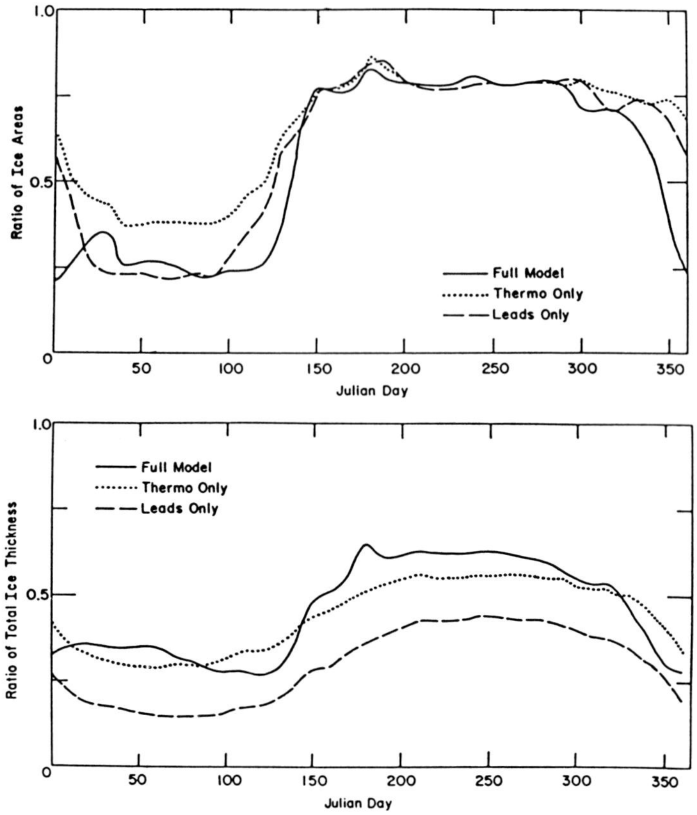

Fig. 4. Relative changes of ice area (a) and ice volume (b) for a hierarchy of models. The ratios are obtained by dividing the results from the warm temperature runs by the control temperature results (from the corresponding model).

to be larger near the ice edge and in regions of offshore ice advection near the coast. Also, with full dynamics, ice production due to leads is compensated for by ice divergence, which tends to thin the ice.

Another smaller, but rather significant difference between the full model and the more static cases (occurring in both the "control" and "high" temperatures) is a slightly enhanced ice advance in winter. Again, this feature is due to advection and transport in the full model, which causes ice near the edge to be advected into warmer water and melt. This in turn causes more rapid cooling of the oceanic boundary layer. The feature is minimized here because the atmospheric temperatures are fixed, but could perhaps be more pronounced in a coupled atmospheric ice model.

It is also notable that the basic character of the thermodynamic model is unchanged by the addition of seven thickness levels. This addition just shifts the thermo-only thickness curve up, but does not change its basic character or its relative sensitivity.

The sensitivity of the models to warming is best seen by examining Figure 4 in conjunction with Figures 2 and 3. With respect to ice edge

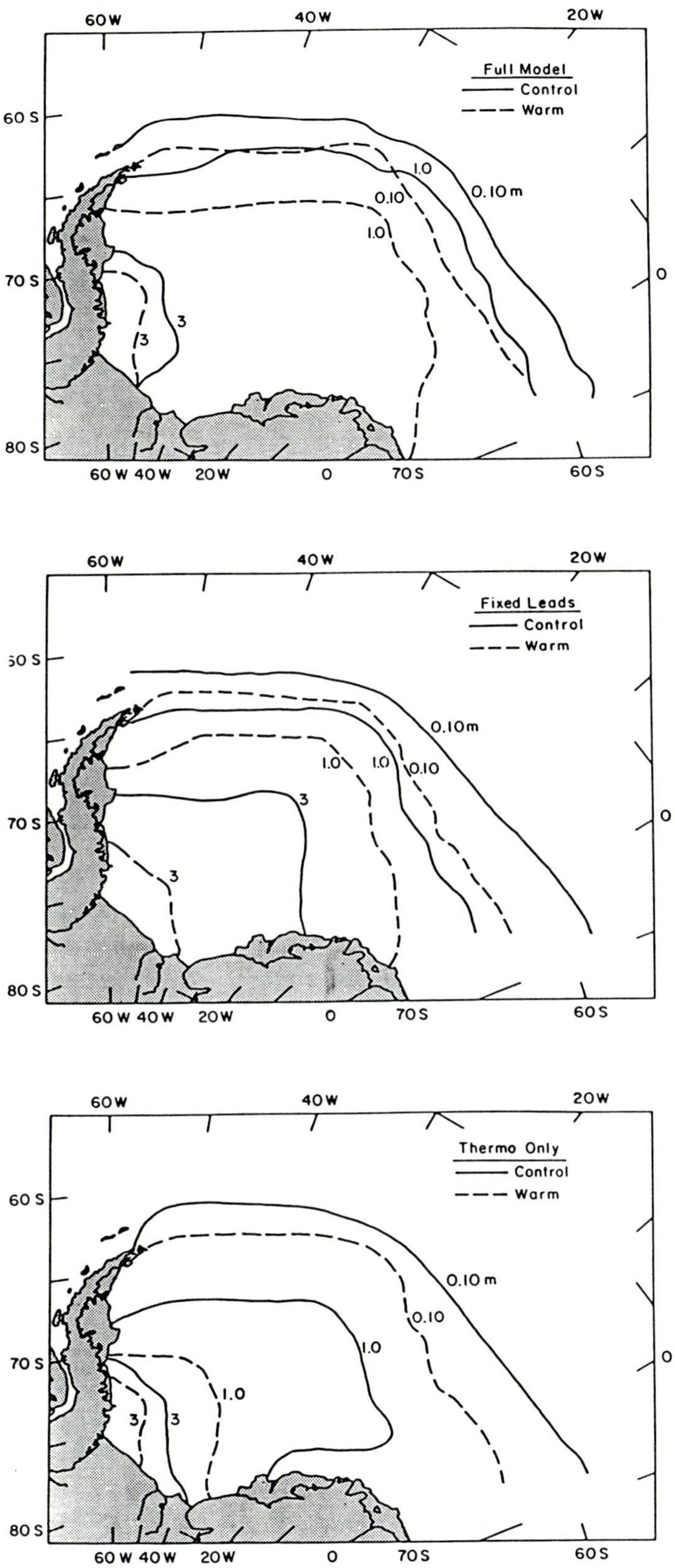

Fig. 5. Contour plots of ice thickness on day 228 for both warm and control temperatures for (a) the coupled model, (b) the leads-only simulation and (c) the thermodynamics-only simulation.

variations, the relative relationship between the three models is similar, i.e., the summer ice extent is more sensitive to atmospheric warming than is the winter extent. However, the ratios show the winter ice extent sensitivity to be about the same for the models, while the summer ice extent decrease is somewhat greater for the models containing leads. This greater sensitivity in the summer appears mainly due to the fact that the higher temperatures simultaneously reduce the growth season while increasing the melt season. As a consequence, to first order it can be argued that the amount of decay is relatively constant, even though the net ice masses are reduced by the higher air temperature. With these assumptions, suppose h and $h - \Delta_W$ are the winter thicknesses for the "control" and "high" temperature cases, respectively. Then $h - \Delta_S$ and $h - \Delta_W - \Delta_S$ will be the summer thicknesses. For $\Delta_W + \Delta_S$ less than h, we see $(h - \Delta_W) / h_W > (h - \Delta_W - \Delta_S)/(h - \Delta_S)$. Note that the reduced summer sensitivity for the thermo-only model would follow because it has a smaller seasonal swing to begin with.

With respect to ice thickness, a particularly prominent difference in sensitivity of the ice thickness ratios occurs between the thermo-only and leads-only models. Specifically, the leads-only case is more sensitive than the thermo-only case to atmospheric warming. This is basically due to the conductivity of the ice being dominated by the leads in the leads-only simulation. Hence, under warming the ice production decreases linearly without feedback. In the thermo-only case, on the other hand, there is a negative feedback, with the thinner ice having a lower conductivity and hence higher relative growth rates. This feature is especially pronounced in regions of thick ice, as is illustrated in Figure 5. This figure compares thickness contours on day 228 for the thermo-only model, the leads-only model and the full model. Note that in regions of very thick ice the dominance of leads in the ice growth for the leads-only model causes the warming to substantially reduce the ice thickness.

Effect of Ice Interaction on Model Sensitivity

While the leads-only case has an increased sensitivity to atmospheric warming, the full model, interestingly enough, has less sensitivity than the thermo-only model. This difference arises because, in addition to the thickness feedback, the full model has feedbacks related to the fact that the lead fractions and ice build-up can also change.

With regard to leads, Figure 6 shows the average compactness time series for the full model. Note that in the "warm" temperature case there is a larger fraction of leads in winter (which in turn increases ice growth), and a smaller fraction in summer (which causes less decay). This increased lead fraction in winter is likely due to

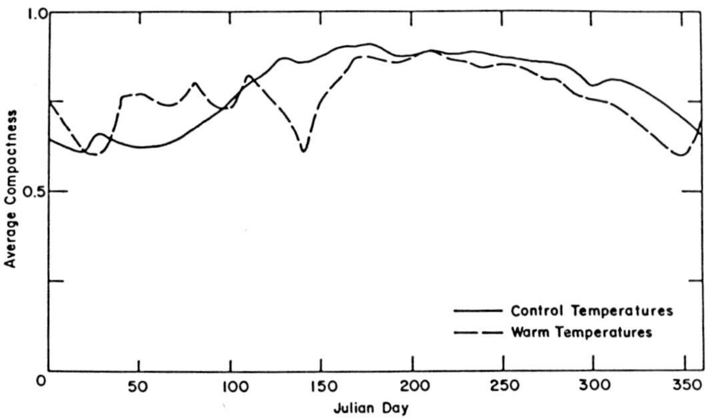

Fig. 6. Spatially averaged compactness time series from the fully coupled model for the control and warm temperatures.

an increased ratio of ice edge perimeter to area. However, the decrease in summer lead fraction appears to be due to ice interaction effects which fill up regions of thin ice. This effect is illustrated in Figure 7 which compares lead fractions from simulations with the full coupled "warm" model and the "warm" model with velocities specified. Basically, with full ice interaction included, the model has a lower fraction of leads in summer and a higher fraction in winter as compared to the control simulation.

However, in addition to leads, direct thickness build-up also plays a role in the full model's sensitivity. The essential concept here is that in regions of ice convergence, mechanical effects supply the main limitation on ice thickness rather than thermodynamics. Consequently, under warming conditions, the ice interaction can be

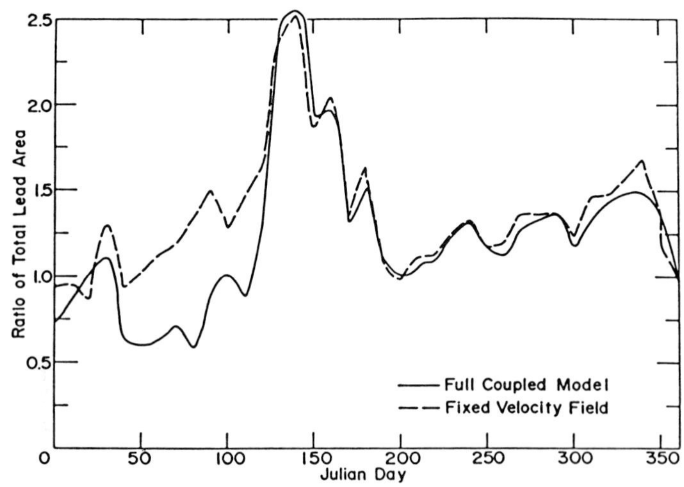

Fig. 7. Effect of dynamically modified ice velocity fields on leads. The two curves show lead fraction ratios for warm temperature simulation results (with and without full dynamic coupling) to the full model results obtained with control temperatures.

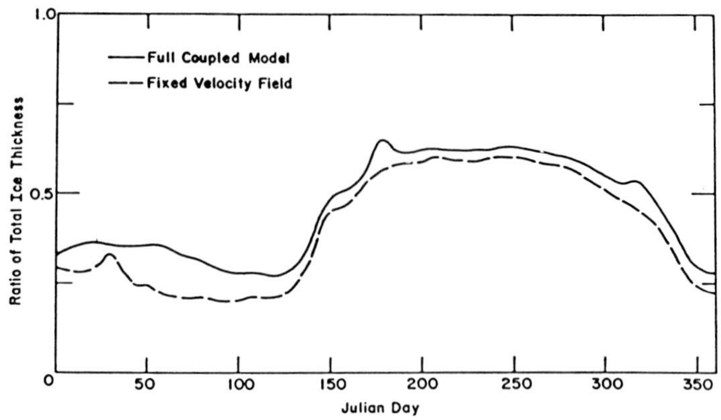

Fig. 8. Effect of ice interaction on ice thickness changes due to atmospheric warming. The two curves show the relative changes with and without dynamically modified ice velocity fields under warming.

modified in such regions, thus allowing greater thicknesses, and hence yielding similar thicknesses to those obtained in the control runs. The importance of this effect is illustrated in Figures 8 and 9. Figure 8 compares ice thickness ratio time series ("warm" to "control" thicknesses) for the full coupled model and the off-line velocity model which does not contain ice interaction feedback effects. Figure 9 makes a similar comparison for the thickness contours on day 228, a period close to the time of maximum thickness.

As can be seen from Figure 8, using a fixed velocity field rather than allowing the ice velocity field to be modified by ice interaction causes the ice thickness to decrease more under a warming (especially in summer). The role of ice interaction in the process is, however, perhaps

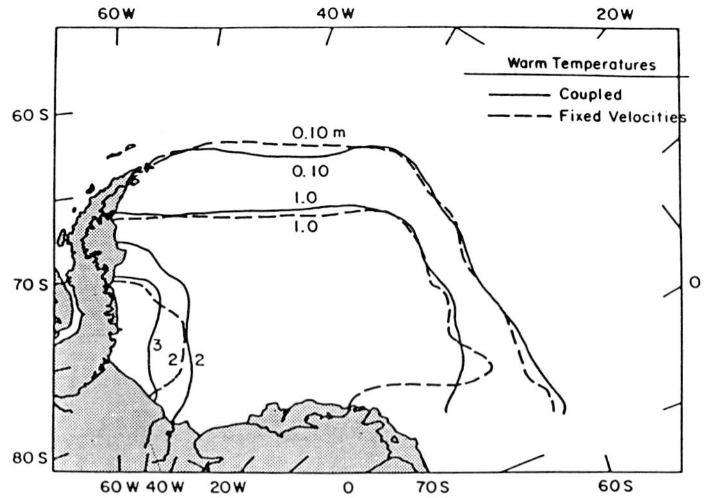

Fig. 9. Contour plots of ice thickness on day 228 obtained from the "warm" fully coupled model and the specified ice velocity field simulation (using warm temperatures).

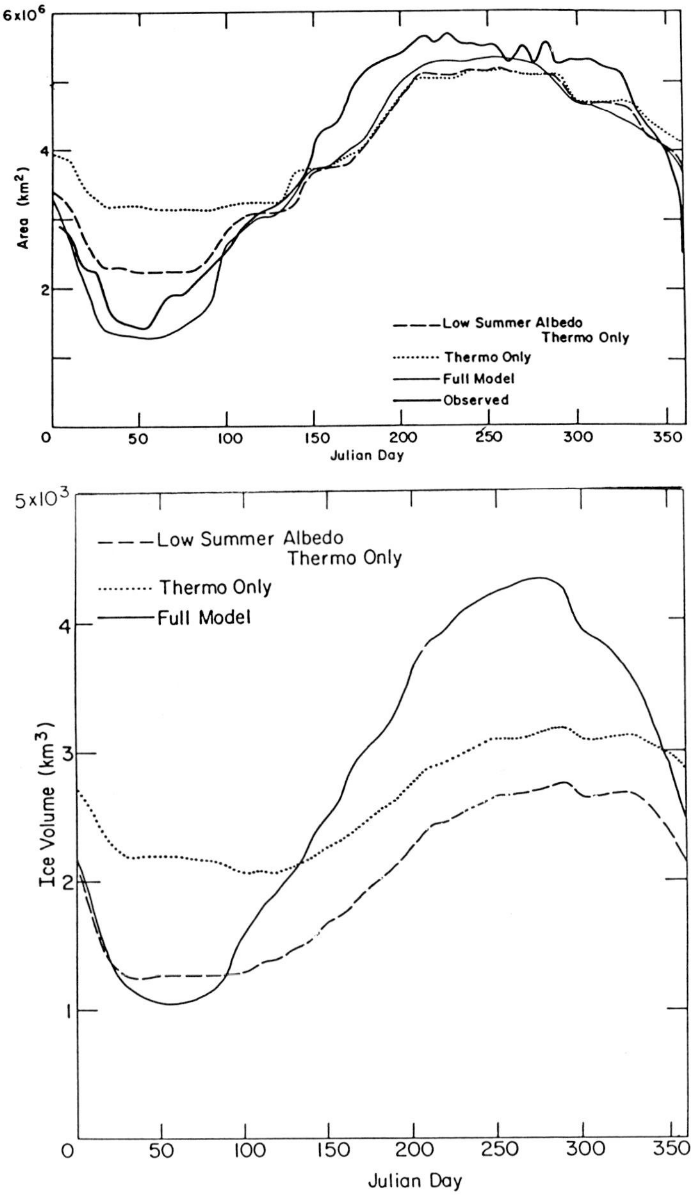

Fig. 10. Effect of a reduced summer albedo on simulated ice area (a) and ice volume (b) for the thermodynamics-only model. The full model results are also shown for comparison.

best shown in Figure 9. In the case of a specified velocity field there is a much greater decrease in thickness near the coast under rising temperatures. With the velocity field modified by the ice interaction, on the other hand, there is less sensitivity because with less growth the ice becomes weaker and thus greater convergence occurs. Overall, this thicker ice build-up in the full model causes a greater summer thickness, and hence causes the model thickness to be less affected by atmospheric warming. In short, the ice thickness in these build-up situations is dictated more by mechanical than by thermodynamic considerations.

It should be noted that such mechanical effects would be expected to be especially pronounced in the Arctic, where substantial spatial variations occur. Such build-up effects have not been included in existing CO_2 Arctic sea ice sensitivity studies (Parkinson and Kellogg, 1979). Overall these results emphasize the need for including full dynamic-thermodynamic coupling in sea ice models used in CO_2 sensitivity studies.

Effects of Albedo Tuning on Thermo-Only Models

A final model in the hierarchy of some interest is a thermodynamics-only model with a reduced summer albedo. Such a procedure is often used in climatic studies to improve the simulated ice extent, and different tunings used in different models (Washington et al., 1980; Manabe and Stouffer, 1980) appear to have substantial effects on the sea ice sensitivity. A reduced summer albedo also has some physical basis in that it crudely models the effect of a reduced albedo due to leads. To examine such a tuning procedure a two-year simulation with an albedo for melting sea ice of 0.4 was carried out. Figure 10 shows the thickness and area results for the control tem-

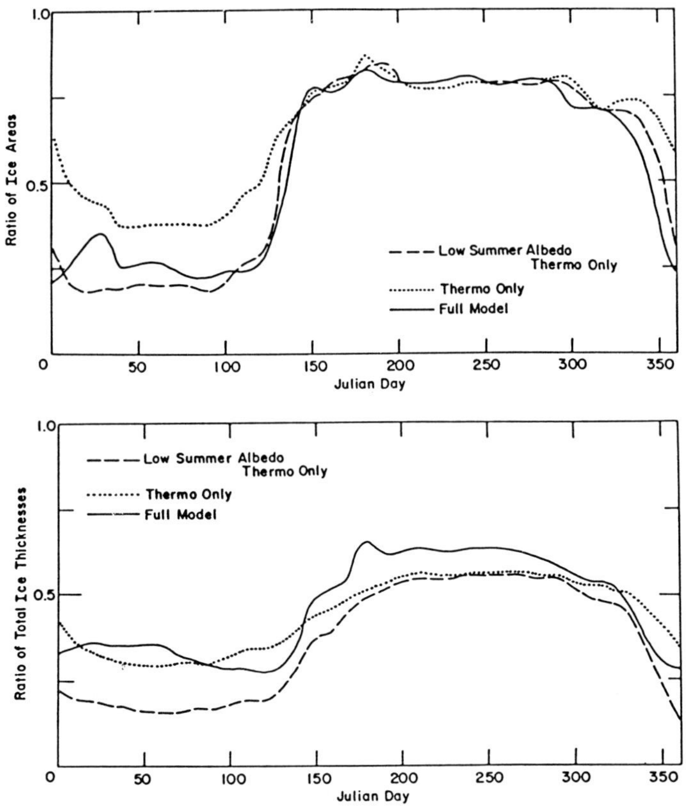

Fig. 11. Effect of a reduced summer albedo on relative change of ice areas (a) and volumes (b) under an atmospheric warming. The full model results are also shown for comparison.

TABLE 2. Net Heat Loss to Atmosphere, Day 60 to Day 240 Over Whole Grid*

	Coupled model	Fixed leads	Thermo-only
Control temperatures	722 m of ice	718 m	588 m
Warm temperatures	833 m	846 m	752 m
% change	122%	118%	128%

* An average thickness change of 1000 m of ice over this time period is equivalent to an average sea to air heat flux of 114 W m^{-2} over the same 180-day period.

peratures while Figure 11 shows the sensitivity of the model by illustrating the ice thickness and area ratios of the "control" to "high" temperature results.

As can be seen from Figure 10, the main feature of the reduced summer albedo is to enhance the seasonal swing of both the ice extent and thickness. Note, however, that the reduced summer extent is still greater than the observed value. The fact that the reduced albedo gives a lower summer extent than the leads-only model is due to a much reduced ice thickness in the thermo-only model since no increased growth effects due to leads are included. The presence of a lower summer albedo tends to reduce the ice thickness in both summer and winter.

With regard to sensitivity the reduced summer albedo tends to give a model with a greater reduction in ice thickness in summer, while the relative reduction in winter tends to be about the same. This greater reduction tends to substantially increase the summer thickness sensitivity compared to a model including full dynamical effects. Also, the summer ice area sensitivity is reduced somewhat, bringing it more into coincidence with the area sensitivity of the leads-only and full models.

Sensitivity of Air-Sea Heat Exchange

The above analysis has dealt with the sensitivity of the sea ice extent and thickness to atmospheric warming. However, a factor of perhaps more relevance to model-based CO_2 studies is the modification to the air-sea heat exchange. This feature was an important facet of the Manabe and Stouffer (1980) simulations in that reduced sea ice extents and thicknesses gave greater sea-to-air heat transfers even though the atmosphere was warmer.

Figure 12 and Tables II and III present the main air-sea heat exchange results from the basic three models in the hierarchy: the full coupled model, the fixed-leads case and the thermo-only model. The heat exchange results are presented in two ways: in terms of the net heat loss to the atmosphere averaged over the whole grid (Table II) and in terms of the net heat loss to the atmosphere over the ice-covered portion of the grid only (Table III and Figure 12). The total loss combines the effect of ice thickness and ice area changes, while the loss over the ice-covered area illustrates thickness and lead feedback effects.

In terms of relative sensitivity over the whole grid, all models show an increase in ocean heat loss under atmospheric warming, with the thermo-only model having the greatest increase. However, if only the ice-covered areas are considered, the results are quite different, a fact which indicates that the whole grid results are dominated by ice extent. In particular, only in the ice-covered region, does the thermo-only model have an increased heat loss under atmospheric warming, while the full and fixed-leads models both have a decreased heat loss.

TABLE 3. Net Ice Growth, Day 60 to Day 280*

	Coupled model	Fixed leads	Thermo-only
Control temperatures	133 m of ice	169 m	27 m
Warm temperatures	87 m	100 m	42 m
% change	65%	59%	156%

* An average thickness change of 1 m of ice per year is equivalent to 49.3 km^3 of ice per year or equivalently a yearly average exchange rate of $0.472 \cdot 10^{12}$ watts.

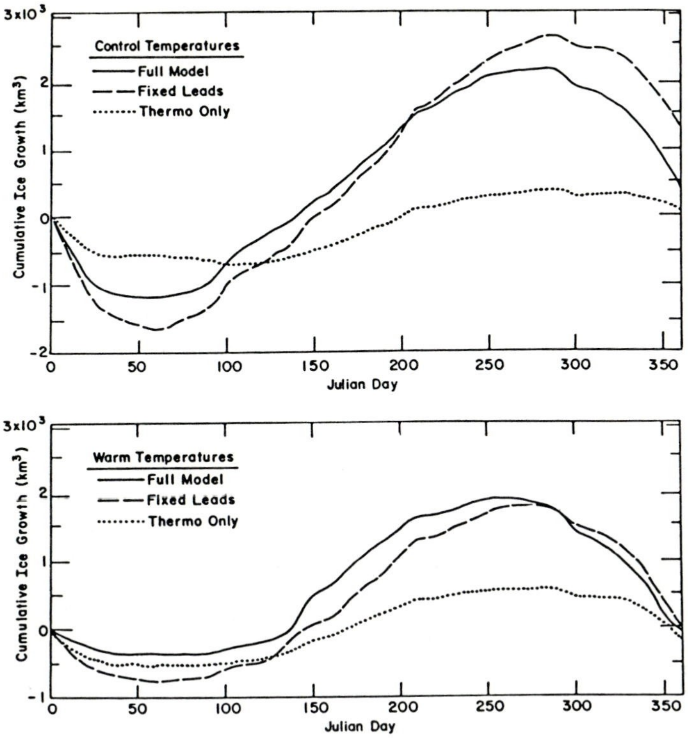

Fig. 12. Heat loss over ice-covered regions for (a) control temperatures and (b) warm temperatures, for a hierarchy of sea ice models.

Thus not only do the dynamics cause a quantitative change in the air-sea heat exchanges, they cause a qualitative modification in that the sign of the heat loss change is different. This decreased heat loss in the two models including dynamical effects is due to the dominance of ice growth in leads. Growth rates over open water tend to decrease as the atmosphere gets warmer and hence there is less heat loss.

In addition to qualitative sign differences there is a very substantial quantitative difference between the heat loss magnitudes of the various models. In the control case, for example, the enhanced seasonal swing of the full model yields a heat loss over ice five times larger than the thermodynamics-only model. The main point here is that a dynamical ice cover will have far greater heat loss to the atmosphere than thermodynamics-only models.

Finally, it should be noted that there are some other subtle, but illustrative differences, between the full model and the fixed-leads case. One difference is that the full model has less ice growth reduction under warming conditions (65% growth reduction). This is likely due to dynamical feedback effects, which create more leads under warming. A second point is that the heat loss over the whole grid is slightly larger for the full model, despite the fact that the heat loss over the ice alone is substantially greater for the leads-only model. This difference is mainly due to the much greater seasonal swing of ice extent of the full model. Thus at the beginning of the growth season the full model has more open water area to fill in with ice.

Concluding Remarks

The simulations of the hierarchy of sea ice models reported here have identified a number of ways in which dynamical processes can modify simulated sea ice extents, thicknesses and air-sea heat exchanges in model-based CO_2 climate sensitivity studies. While the simulations have been of the Weddell Sea pack ice, many of the mechanisms identified by this study should have applicability to the Arctic ice cover as well.

An overall effect resulting from including full dynamics is the reduction of the sensitivity of the average ice thickness to atmospheric warming as compared to thermodynamics-only models. This full model result is in contrast to the inclusion of the lead formation effects of ice dynamics, alone which tend to increase the sensitivity of the modeled ice thickness to atmospheric warming. The two main mechanisms for this reduced sensitivity in the full model are coupling between the ice strength and ice velocity, which causes equilibrium ice thickness near coasts to be dictated by mechanical rather than thermal considerations, and an increase in the fractional lead coverage under atmospheric warming. Both these effects hinge upon the ice velocity field also being modified by atmospheric warming.

The increased sensitivity of a simulation with a fixed fraction of leads to atmospheric warming demonstrates one of the fundamental differences between a variable thickness sea ice cover and static models often used in climatic studies. In a variable thickness model, growth rates are primarily dictated by thin ice and open water. Under warmer conditions there is minimal feedback in that the growth rates are reduced over leads. In a static model, on the other hand, the growth rates are strongly tied to the ice thickness. Consequently, under warming there is a negative feedback in that the ice gets thinner, tends to grow faster, and hence its thickness is less changed. In a full dynamic model including advection, however, there is also a negative feedback effect, due not to ice thickness but to variations in the ice velocity field introduced by modifications of the mechanical behavior of the pack ice.

An interesting common feature of all the ice models under warming is a larger reduction in the summer ice extent as compared to the winter extent. This feature follows from the fact that warming both reduces the growth period and increases the time over which melting occurs. This compensation in turn allows the seasonal variation of thickness and area to remain about the same, even though the maximum thickness and area are smaller. Since the summer extent is less to begin with, such a constant seasonal variation results in a greater percentage decrease in summer.

A particularly pronounced difference between thermo-only models and models including dynamics is the sensitivity of the air-sea heat exchange under warming. Thermo-only models yield thinner ice and hence greater heat loss to the atmosphere. In models with leads, on the other hand, the atmospheric heat loss is reduced by warming, due to the leads dominating the heat exchange characteristics. It should also be noted that in the control runs the amplitude of the air-to-sea heat loss over the ice-covered regions is about five times larger in models including leads than in the thermo-only model.

In the control runs, probably the greatest difference between a full dynamic model and models without advection effects is the enhanced seasonal cycle. The advection enhances the seasonal cycle both in winter and summer. In winter, ice advection toward the edge cools off the ocean faster, allowing a greater advance. In summer, the advection moves the ice into warm water, causing an enhanced melt. It is particularly notable that these advection effects cause a much more enhanced seasonal cycle than can be obtained from including only leads in a static ice cover.

Overall, the simulation results reported here have demonstrated the considerable complexity of different dynamical effects. This complexity makes it difficult to guess the correct ad hoc dynamical modification to add to climate models. Instead the results emphasize the importance of including dynamical processes more completely in climate sensitivity studies via fully coupled dynamic-thermodynamic sea ice models such as that used here (Hibler, 1979, 1980b). Such models are well documented (Hibler, 1980c) and, while more demanding of computer time than ad hoc corrections, are much faster computationally than present atmospheric GCM's. Thus there seems to be no real computational reason for not including such models in GCM-based CO_2 studies.

Appendix: Sensitivity of Full Model to Dynamic and Oceanographic Parameterizations.

To examine the sensitivity of the full model to oceanographic and thickness distribution parameterizations, the series of two-year simulations described in Table AI were carried out. These simulations examine the effect on the full model of different treatments of the oceanic heat flux, the thickness of the oceanic boundary layer, the effect of seven thickness levels, and the sensitivity to extra open water creation. To make

TABLE A1. Simulations for Examination of Oceanic and Thickness Parameterizations*

Description name	Ice thickness treatment	Comments
No extra open water	Same as full model but without extra open water creation.	
Two-level	No extra open water creation, only one level of ice thickness in addition to leads.	
30-m-thick mixed layer	Same as full model (except velocities are specified).	Mixed layer was 60 m deep in full model.
Constant heat flux, 5.818 W m^{-2}	Same as full model.	Heat flux spatially and temporally constant. This heat flux value equal to annual average value for ice-covered region in full model.
Constant heat flux, 13.06 W m^{-2}	Same as full model.	Heat flux spatially and temporally constant. Heat flux value equal to annual average over whole grid for 3rd year of full model.

* All simulations carried out for two years using specified velocities from third year of full model and initialized by thickness and compactness at end of first year of full model simulation. Control temperatures used in all cases.

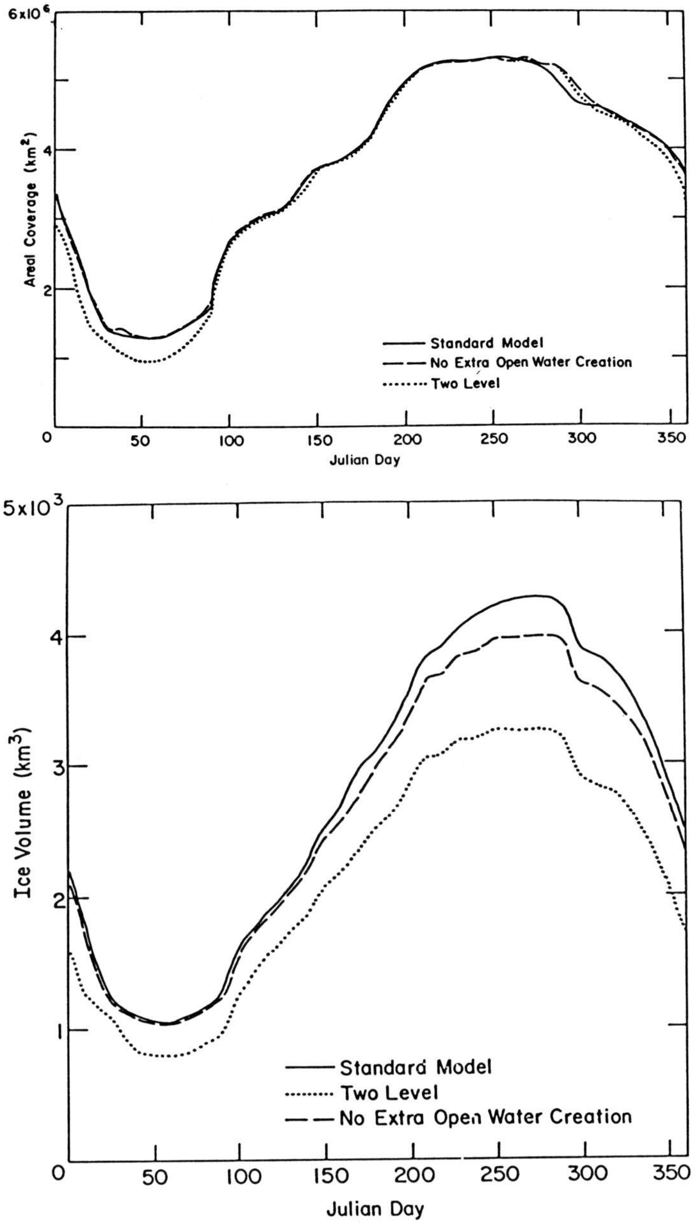

Fig. A1. Sensitivity of simulated seasonal cycle of sea ice to ice thickness distribution parameterizations. Control temperatures were used. Part (a) shows time series of the total area covered by ice within the simulation region while (b) shows the total ice volume.

these sensitivity tests as comparable as possible to the full model, the ice velocity fields were taken from the third and final year of the full-model simulation.

The main results of these comparisons for the thickness distribution are shown in Figure A1. Part of this figure compares ice extent (area of all grid cells covered by ice) while part b shows total ice thickness summed over the whole grid (shown in Figure 1).

While the extra open water creation makes some difference in the simulated winter ice thickness, the main difference is between the two-level and seven-level parameterizations of thickness. The inclusion of a distribution of thickness levels allows a larger growth rate, while the melt rate is relatively little changed. As a consequence the average ice thickness increases over the whole seasonal cycle (although the increase is slightly larger in winter). This also causes some increase in the summer ice extent, since the ice is thicker and cannot be melted away as easily. Note that the effect of extra open water creation mainly increases the winter ice thickness, as it increases the fraction of leads only when the compactness is greater than 0.85. Overall, these sensitivity results reinforce the utility of the two-level model analysis carried out by Hibler and Ackley (1983), but suggest that such two-level models may somewhat underestimate the ice thickness and summer ice extent. It is also possible that inclusion of an explicit snow cover might modify these results.

To interpret the oceanographic sensitivity studies it is useful to first examine the heat flux characteristics generated from the fully coupled standard model. These characteristics are illustrated in Figure A2, which shows contours of the average annual heat flux, and in Figure A3, which shows time series of the average heat flux over the whole grid, and over only the instantaneously ice-covered region. In the case of the ice-covered region, the ice area changes daily, so this averaged flux is based on a spatial region with varying dimensions. Integrating these heat flux time series yields an annual average of 13.06 W m^{-2} over the whole grid and 5.818 W m^{-2} for the ice-covered region. This total grid value is in good agreement with heat flux estimates by Gordon (1984), based on chemical water composition estimates of the winter mixed layer of the Weddell Sea.

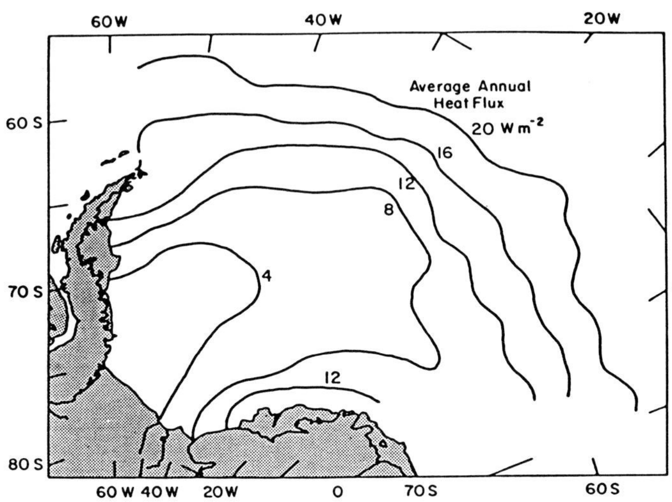

Fig. A2. Contours of average annual oceanic heat flux obtained from the third year of the fully coupled control run.

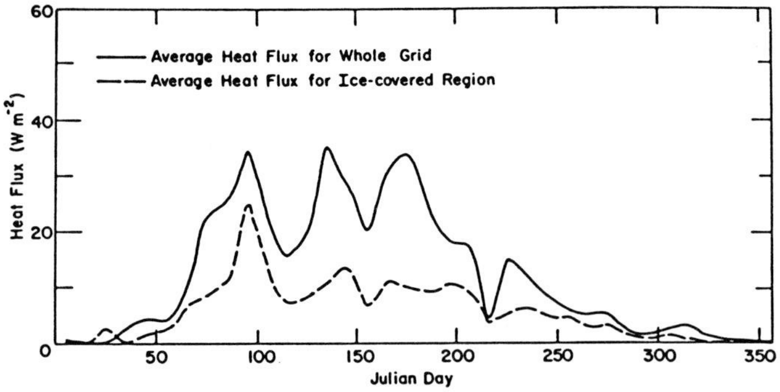

Fig. A3. Time series of spatially averaged heat flux from the third year of the fully coupled control run. The total flux was computed by averaging over the whole grid, while the heat flux for the ice-covered area was computed for the ice-covered region occurring each separate day.

However, an important feature of this estimated heat flux is its spatial and temporal variability induced by tying it to the boundary layer cooling and ice growth rate. Temporally, it has a very strong seasonal cycle with a larger flux during the growth season. Spatially, the heat flux is larger further north, as the open water resides there longer and hence there is more net cooling of the mixed layer. It is notable that the heat fluxes are especially large along the zero meridian, roughly the region where the Weddell Sea Polynya has historically been observed.

These spatial and temporal variations have important ramifications in connection with the simulated ice extent. This fact is illustrated in Figure A4, which compares area and thickness for the oceanographic sensitivity studies. Basically, sensitivity studies were done for two constant heat flux values (equal to the total grid average value and the average value for the ice-covered area) and for a thicker mixed layer (but with the standard spatially and temporally varying heat flux).

The heat flux variations have different effects in winter and summer. Both constant heat flux simulations tend to increase the winter ice extent while decreasing the summer extent. The decrease near the edge is due to the average variable heat flux being larger than the constant case, thus keeping the mixed layer above freezing. The decrease in summer extent, on the other hand (at least in the 5.818 W m^{-2} case), is partially due to the heat flux varying seasonally. Having the heat flux peaking when the ice growth rate is largest tends to minimize heat flux effects by a negative feedback. Specifically, the ice grows slower, which as a consequence causes thinner ice and more leads, which in turn enhances growth. This is especially notable in the ice volume plots, where the larger constant heat flux causes a substantial reduction in ice thickness.

Overall, these heat flux sensitivity studies emphasize the danger of invoking larger constant heat fluxes in thermodynamic calculations in order to get a larger seasonal cycle of sea ice (e.g. Washington et al., 1980; Parkinson and Washington, 1979; Gordon, 1981).

While the variable heat flux tends to reduce the seasonal swing of ice extent, a thinner boundary layer tends to increase the seasonal swing. As seen from the 30-m mixed layer results, the thinner mixed layer primarily affects the extent at the outer edge by causing less heat absorption in summer, and hence reduces the cooling in winter needed to bring the mixed layer to

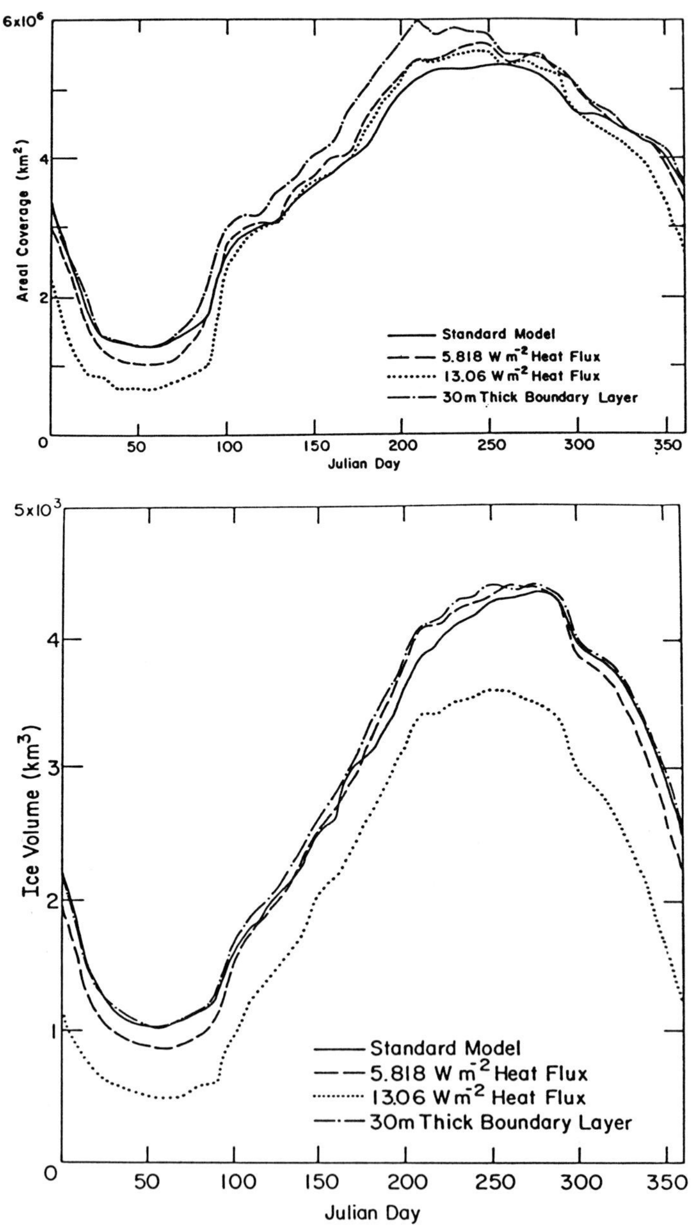

Fig. A4. Sensitivity of simulated seasonal cycle of sea ice area (a) and ice volume (b) to oceanic heat flux and oceanic boundary layer thickness.

freezing. However, while substantially affecting the winter ice edge, the thicker mixed layer has very little effect on the ice thickness, probably due to the large number of leads in the full model near the edge.

In general, these sensitivity results indicate that deeper mixed layers and variable heat flux effects tend to reduce the seasonal swing of ice extent, while less critically affecting the thickness. The variable thickness of sea ice, on the other hand, changes mainly the thickness characteristics. Since, as was discussed in the main text of the paper, the dynamics greatly enhances the seasonal swing of ice extent compared to the thermodynamics-only or fixed leads simulations, these oceanographic sensitivity results tend to support the identification of ice dynamics as an important source of the large seasonal cycle of Antarctic sea ice.

Acknowledgements. This study was made possible by aid from Sandy Smith in file preparation, data transfers, plotting, and numerous other computer oriented tasks. Her help is gratefully acknowledged. Financial support for this study was provided by the Ocean Processes Branch of the National Aeronautics and Space Administration through Agreement Number S-79086-B.

References

Gordon, A.L., The seasonality of Southern Ocean sea ice, J. Geophys. Res., 4193-4197, 1981.

Gordon, A.L., Winter mixed layer entrainment of Weddell deep water, J. Geophys. Res., in press, 1984.

Gordon, A., E. Molinelli, and T. Baker, Large-scale relative dynamic topography of the Southern Ocean, J. Geophys. Res., 83 (C6), 3023-3032, 1978.

Hansen, J., A. Lacis, D. Rind, G. Russell, P. Stone, I. Fung, R. Ruedy and J. Lerner, Climate sensitivity: Analysis of feedback mechanisms, Maurice Ewing Series, 5, 1984.

Hibler, W.D. III, A dynamic thermodynamic sea ice model, J. Phys. Oceanogr., 9, 815-846, 1979.

Hibler, W.D. III, Sea ice growth, drift and decay, Dynamics of Snow and Ice Masses, S. Colbeck, Ed., Academic Press, 141-209, 1980a.

Hibler, W.D. III, Modeling a variable thickness ice cover, Mon. Weather Rev., 108(12), 1943-1973, 1980b.

Hibler, W.D. III Documentation for a two-level dynamic-thermodynamic sea ice model, Spec. Rep. 80-8, U.S. Army Cold Reg. Res. and Eng. Lab., Hanover, NH, 1980c.

Hibler, W.D. III, and J. Walsh, On modeling seasonal and interannual fluctuations of Arctic sea ice, J. Phys. Oceanogr., 12, 1514-1523, 1982.

Hibler, W.D. III, and S.F. Ackley, Numerical simulation of the Weddell Sea pack ice, J. Geophys. Res., 88, 29873-2887, 1983.

Lemke, P. and T.O. Manley, The seasonal variation of the mixed layer under polar sea ice, J. Geophys. Res., in press, 1984.

Manabe, S., and R.J. Stouffer, Sensitivity of a global climate model to an increase of CO_2 concentration in the atmosphere, J. Geophys. Res., 85, 5529-554, 1980.

Martinson, D.G., P.D. Killworth, and A.L. Gordon, A convective model for the Weddell polynya, J. Phys. Oceanogr., 11, 466-488, 1981.

Maykut, G.A., Energy exchange over young sea ice in the central Arctic, J. Geophys. Res., 83, 3646-3658, 1978.

Parkinson, C.L. and W.W. Kellogg, Arctic sea ice decay simulated for a CO_2-induced temperature rise, Climatic Change, 2, 149-162, 1979.

Parkinson, C.L., and W.M. Washington, A large-scale numerical model of sea ice, J. Geophys. Res., 84, 311-337, 1979.

Rothrock, D.A., The energetics of the plastic deformation of pack ice by ridging, J. Geophys. Res., 80, 4514-4519, 1975.

Semtner, A.J., A model for the thermodynamic growth of sea ice in numerical investigations of climate, J. Phys. Oceanogr., 6, 379-389, 1976.

Thorndike, A.S., D.A. Rothrock, G.A. Maykut and R. Colony, The thickness distribution of sea ice, J. Geophys. Res., 80, 4501-4513, 1975.

Washington, W.M., A.J. Semtner, C. Parkinson, and L. Morrison, On the development of a seasonal change sea ice model, J. Phys. Oceanogr., 6, 679-685, 1976.

Washington, W.M., A.J. Semtner, Jr., G.A. Meehl, D.J. Knight and T.A. Mayer, A general circulation experiment with a coupled atmosphere, ocean and sea ice model, J. Phys. Oceanogr., 10, 1887-1908, 1980.

RESPONSE OF ANTARCTIC SEA ICE TO UNIFORM ATMOSPHERIC TEMPERATURE INCREASES

Claire L. Parkinson and Robert A. Bindschadler

Goddard Laboratory for Atmospheric Sciences
NASA/Goddard Space Flight Center, Greenbelt, MD 20771

Abstract. The effect on Antarctic sea ice of hypothetical atmospheric temperature increases is simulated using a thermodynamic/dynamic sea-ice model. The various responses of the modeled ice cover to temperature changes are calculated by running the model with mean-monthly climatological air temperatures as a standard case and then with all temperatures uniformly increased by -1 K, +1 K, +3 K, and +5 K in four subsequent model runs. The results show noticeable regional differences in the response of the ice cover, with a hemispherically-averaged winter ice-edge retreat rate of 1.4 degrees of latitude for each 1 K increase in atmospheric temperatures. These retreat rates are non-linear with respect to temperature change, the sensitivity of the position of the ice edge decreasing as temperatures are further increased. This non-linearity in the response of the ice edge occurs in the response of other ice variables as well, including the total ice area and total ice volume at maximum ice extent. These areas and volumes decrease by roughly half with a 5 K temperature increase. By comparison, for the peak summer period, a 3 K temperature increase suffices to eliminate all ice except in the Amundsen and western Weddell Seas. Another result of warming the atmosphere is an increase in the temporal asymmetry in the simulated annual cycle of the ice cover, showing longer periods of ice growth and shorter periods of ice decay.

Introduction

Increases in the CO_2 content of the atmosphere have been well documented (Broecker, 1975; Keeling et al., 1976a, 1976b; Lacis et al., 1981), and several numerical models have been used to estimate the impact on the atmosphere of a CO_2 increase (e.g., Manabe and Wetherald, 1967, 1975, 1980; Sellers, 1974; Augustsson and Ramanathan, 1977; Hunt and Wells, 1979; Manabe and Stouffer, 1979, 1980; Ramanathan et al., 1979; Hansen et al., 1981). Modeling results consistently predict an increase in atmospheric temperatures, although the magnitude of the increase varies markedly among the various models, with the predicted global average warming generally lying between 1 and 4 K for a CO_2 doubling. Most of the models show the largest increase occurring in the polar or subpolar regions, the polar warming sometimes being several times the global average.

There are several reasons to expect a greater temperature response in the polar regions than elsewhere. Among these are specific positive feedbacks associated with the variable sea ice cover. As air temperatures increase, the lessened ice formation and increased ice melt result in a decreased ice cover and hence a lower overall surface shortwave albedo. This allows greater absorption of solar radiation at the surface, with a consequent further warming. The lessened ice extent also signifies lessened insulation between ocean and atmosphere, so that significantly greater sensible and latent heat exchanges occur from ocean to atmosphere under warmer conditions. This is true especially in winter, when the air is considerably colder than the ocean. The heat exchange factor thereby provides a short-term positive feedback for further increasing surface atmospheric temperatures, although indeed it could provide a long-term negative feedback since more heat becomes lost to space.

Another reason for the expected exaggerated response in the polar regions to a CO_2 warming is the typically stable stratification of the air column in the polar troposphere. By suppressing vertical mixing and hence limiting the amount of turbulence transferring surface heat upward, the stable stratification contributes toward greater warming at the surface (Manabe and Wetherald, 1975).

The question of what would happen to the Arctic ice pack in the event of an atmospheric warming, whether CO_2-induced or otherwise, has been examined by Budyko (1966, 1974) with a heat balance model and by Parkinson and Kellogg (1979) with a thermodynamic/dynamic sea ice model. Budyko concluded that a 4 K rise in summer temperatures would cause the ice pack to disappear and conjectured that under such conditions the ice would not reform in winter. The Parkinson

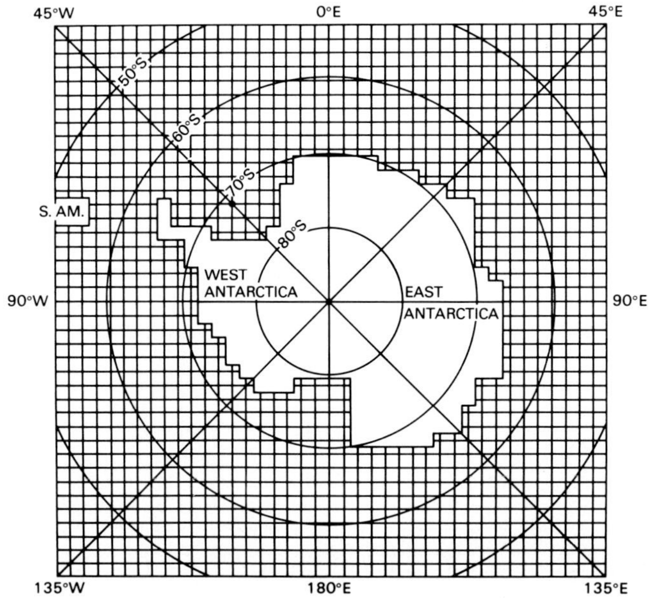

Fig. 1. Grid structure of the model. Time sequences of various sea ice variables are presented later both for the entire grid and individually for grid point (14,28), indicated here by a solid circle.

and Kellogg experiments resulted in the disappearance of sea ice in August and September in the event of a 5 K temperature increase, although ice reappeared in the central Arctic in mid-fall.

We here examine the impact of atmospheric temperature changes on a simulated Antarctic sea ice cover, using the same basic thermodynamic / dynamic model as in the Parkinson and Kellogg study. Comparisons are made among the model's "standard case", which uses mean-monthly climatological inputs, and four perturbed cases with atmospheric temperatures and dew points altered uniformly in space and time by adding -1, +1, +3, and +5 K, respectively. Although somewhat unrealistic, the uniformity in the temperature changes has several clear advantages, such as removing ambiguities in the interpretation of results and allowing calculation of the rate of latitudinal retreat of the ice edge per unit temperature increase.

In an empirical analysis of data from Laurie Island (60°44'S, 45°W), Budd (1975) derived a relationship between the latitudinal extent of sea ice and the variations in annual mean air temperature. With data from only one station, Budd first calculated a value for Laurie Island of the decrease in annual sea ice duration corresponding to each 1 K increase in temperature. He then assumed that this value of 70 days/1 K is representative of the Antarctic, and used it to convert a relationship of the zonally-averaged sea ice duration as a function of latitude to the desired relationship of latitude as a function of temperature variation. Budd found a maximum sensitivity of approximately 2.5° latitude/1 K, occurring at roughly 64°S. This sensitivity decreased away from 64°S such that at 60°S the sensitivity is only 1° latitude/1 K.

Analysis by Kusunoki (1981) of data from near Syowa station (69°S, 40°E) not only does not confirm an ice edge response as large as 2.5° latitude/1 K, but does not confirm any systematic variation of the ice edge with temperature changes. In spite of the latter fact, which presumably resulted from variations in other influencing factors such as oceanic conditions and wind patterns, it is felt that the type of parameter Budd was attempting to obtain is relevant to the issue of sea ice and climate change. Using a numerical model of the temporal behavior of sea ice in the southern hemisphere, we have been able to calculate more directly than Budd the magnitude of the ice edge response to an increase in atmospheric temperatures, and to obtain this sensitivity as a function of both longitude and time of year.

Model Description

The numerical model used for this study is a revised version of the sea ice model described in Parkinson and Washington (1979). The model simulates the seasonal cycle of sea ice thicknesses and concentrations over rectangular grids with a horizontal resolution of roughly 200 km in both the north and south polar regions, the calculations in this study being restricted to the 41 x 41 south polar grid (Figure 1). The model uses an 8-hour timestep and 30-day months.

The model calculations incorporate both dynamic and thermodynamic aspects of the sea ice cover. Ice dynamics are calculated through a momentum equation balancing air stress, water stress, dynamic topography, and Coriolis force, with an adjustment for internal ice resistance. Ice thermodynamics are calculated through energy balances at the interfaces between ice and air, water and ice, and water and air. The energy-balance fluxes include incident and reflected solar radiation, incoming and outgoing longwave radiation, sensible and latent heats, conduction through the ice and snow layers, and a flux of energy from the water to the undersurface of the ice. The latter, termed the ocean heat flux, is set at a constant 25 W m^{-2}, whereas the other fluxes are calculated every timestep.

The 25 W m^{-2} value for the ocean heat flux follows a recommendation of A. Gordon (personal communication, 1976). In agreement with the limited observational data (e.g., Gordon, 1981, and Allison, 1981), this value significantly exceeds the 2 W m^{-2} ocean heat flux often used in Arctic simulations (e.g., by Maykut and Untersteiner, 1971, and by Parkinson and Washington, 1979) and used by Hibler and Ackley (1982) in the Antarctic. Reasons for the higher heat flux in the Antarctic versus the Arctic include the lower fresh water input and the consequent lower stability of the pycnocline (Gordon, 1981).

Atmospheric data, required for the incoming radiation, sensible heat, and latent heat terms in the thermodynamic calculations as well as for the air stress term in the dynamic calculations, are mean-monthly climatological air temperatures, dew points, and u and v components of the geostrophic winds. These values, obtained from Taljaard et al. (1969), are accepted in the standard case as valid for the midpoint of each month and are then interpolated to the 8-hour timestep of the model. Further model details can be found in Parkinson and Washington (1979).

The major revisions made from the 1979 version of the Parkinson and Washington model are the following:

(1) Initial ice concentrations and ocean temperatures, provided for January 1 of year 1 of the simulation, have been made much more realistic than the earlier spatially-uniform 90% for ice concentration and 271.2 K for ocean temperature. Ice concentrations are now initialized from values derived from the satellite imagery of the Nimbus 5 Electrically Scanning Microwave Radiometer (ESMR), described in Zwally et al. (1983a); and ocean temperatures are now initialized from the summertime 20 m depth fields of Gordon and Goldberg (1970) (Figure 2). All initial ice thicknesses remain at 1.5 m, as in Parkinson and Washington (1979).

(2) Snowfall has been eliminated in order to avoid the complicating multiyear cycle which occurs in snow-covered grid squares containing perennial sea ice. In simulation runs including snow, the insulating snow layer slowed the growth of ice to an extent which outweighed the tendency of snow, through its high albedo, to reduce the springtime decay. Over a period of several years, while the snow layer thickened, the maximum and minimum ice thicknesses decreased, with the summer ice eventually disappearing. The multiyear cycle then began again upon the formation of ice in the fall. While this multiyear cycle is of interest in itself, its inclusion in these particular simulations was felt more likely to confuse the analysis than to contribute to the major thrust of understanding the sensitivity of the sea ice cover to changes in atmospheric temperatures. Finally, removal of snowfall has allowed the model results to reach a near equilibrium seasonal cycle within two years, saving considerable computational expense.

(3) The thickness of ice upon initial freezing in a previously ice-free grid square has been increased from 1 to 10 cm following a suggestion of a reviewer of the Parkinson and Washington paper that a 1 cm thickness for initial ice formation is too small (further confirmed by F. Carsey, personal communication, 1979). This change has no direct impact on ice volume, although it does decrease ice concentrations in grid squares in which ice is just beginning to form. Since ice concentrations rapidly approach

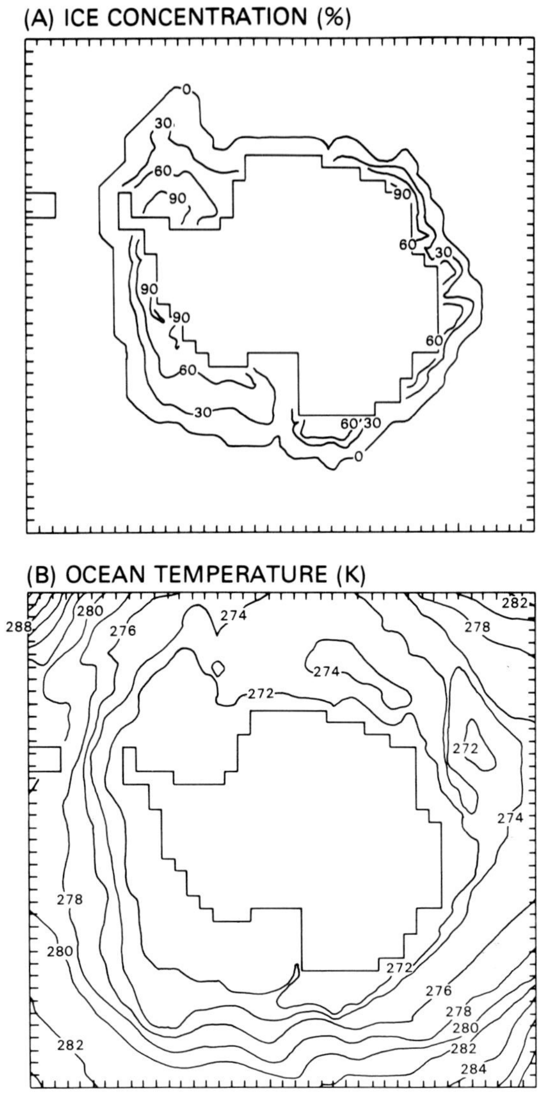

Fig. 2. Fields of initial conditions for (a) ice concentrations and (b) ocean temperatures. Ice concentrations (in percent) for January 1, 1974 were obtained from satellite imagery of the Nimbus 5 Electrically Scanning Microwave Radiometer (Zwally et al., 1983a), and summertime ocean temperatures were obtained from 20 m depth fields of Gordon and Goldberg (1970).

the maximum-allowable 98% in most ice-laden grid squares, the effect of the change from a 1 cm initial ice thickness to a 10 cm initial ice thickness does not extend much beyond the first few timesteps after ice formation.

Results

The model was run for five 38-month simulations: a standard case and four temperature-perturbed cases in which the atmospheric temperatures and dew points were uniformly altered from the standard case by

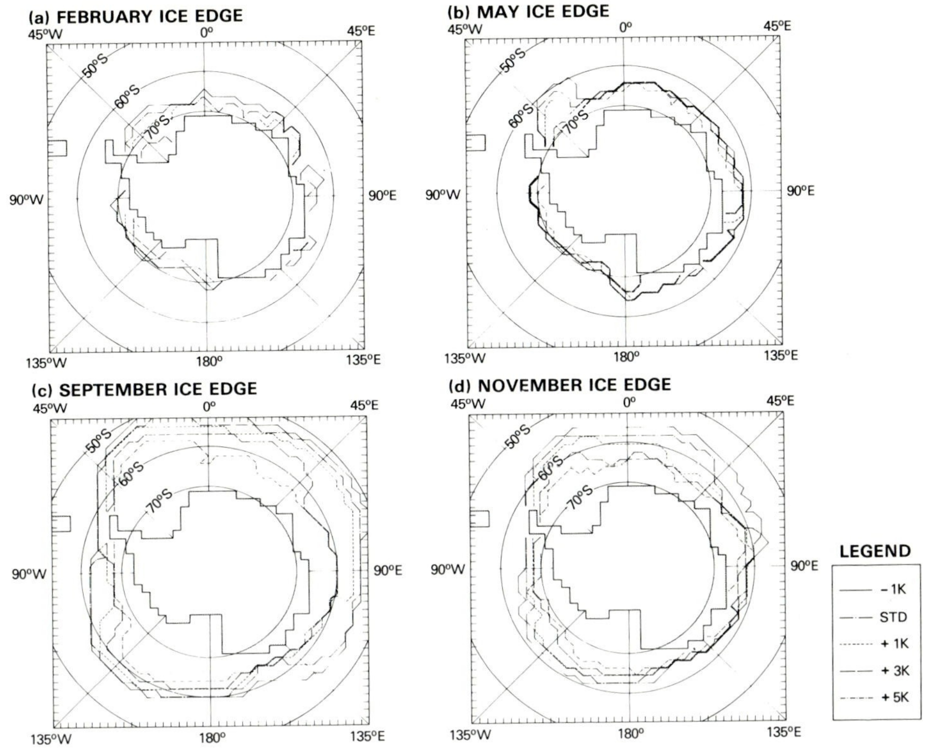

Fig. 3. Location of the simulated sea ice extent in the standard case and in each of four temperature-perturbed model runs. Results are for the midpoints of the following months of year 3 of the simulation: (a) February, (b) May, (c) September, and (d) November.

adding -1, +1, +3, and +5 K. Results are presented in the form of spatial maps of the ice edge and time sequences of various variables at an individual point, summed over the southern ocean, and at selected longitudes. Because satellite observations allow estimates of actual ice concentrations (although not ice thicknesses), some comparisons will be made between the calculated ice extents and the satellite-derived values. These comparisons will be general in nature, since (1) the simulations employ climatological data whereas the satellite observations are for specific years and (2) recent analyses, for instance by Zwally et al. (1983b), confirm the existence of considerable interannual variability in the ice cover.

Spatial Maps

Figure 3 presents the calculated ice edges in mid-February, mid-May, mid-September, and mid-November of year 3 of the simulation for the standard case and for each of the four temperature-perturbed cases. At mid-February, near the peak of summer, the response of the ice cover to the temperature changes is nearly uniform in the western Weddell Sea (30°W - 60°W), where the largest expanse of ice lies. A 3 K temperature increase suffices to eliminate the mid-summer ice around all of East Antarctica, with ice remaining only in the Amundsen (100°W - 140°W) and western Weddell Seas. A 5 K temperature increase suffices to eliminate the ice cover almost entirely, a small amount of ice remaining only off the Thwaites Glacier region in the Amundsen Sea.

As the ice grows in early fall, the warmest cases have the greatest rates of outward expansion of the ice cover so that by mid-May, when most of Antarctica is encircled by ice, the five cases do not differ much in latitudinal extent in any region other than the Weddell Sea from 55°W to 25°W and off George V coast from 135°E to 175°E. In fact in the region of 125°E, where there had been no ice in February in the +3 K and +5 K cases, there is now an equal extent of ice in all five cases (Figure 3b). In the western Weddell Sea, the growth of a tongue of ice is evident in all but the warmest case (+ 5 K), and is increasingly pronounced as the temperature decreases. The far-greater effect of the temperature changes on the ice in the Weddell Sea than around most of the rest of the ice-edge perimeter would have led Budd (1975), by basing his computations on data only from Laurie Island, at 45°W, to overestimate the sensitivity of the total ice edge to temperature changes.

The mid-September results, near the time of greatest ice extent, further demonstrate the pronounced regional nature of the ice edge response to the various temperature increases. At 112°E the September ice edge retreats 6° latitude from the standard to the +1 K case, but is unaffected by the shift from the -1 K to the standard case or by the shifts from the +1 K to the +3 and +5 K cases. At 70°E there is a large impact produced by raising the temperatures from the +1 to the +3 K case but no impact from the additional increase to the +5 K case. By contrast, immediately to the west, over the span 0°E - 45°E, by far the strongest response is between the +3 and +5 K cases. The one location where the modeled ice edge is unaffected by the 5 K temperature increase from the standard case to the +5 K case is off the George V Coast (150°E), where there is a strong response from the -1 K to the standard case. Edge retreat is more uniform in parts of the Weddell, Ross (165°E - 150°W), and Amundsen Seas, although at 100°W, in the eastern Amundsen Sea, the response is similar to that at 112°E, i.e., exhibiting no ice edge response from the -1 K to the standard case, then a strong response to the +1 K temperature increase, and no further response to the +3 and +5 K increases.

Off the East Antarctic shore, the ice edges in the September map suggest an eastward advance of the May-time ice tongue in the Weddell Sea. In September this tongue extends to 5°W for the warmest case (+ 5 K), to 50°E for the +3 K case, to 100°E for the +1 K case, to 140°E for the standard case, and to 155°E for the -1 K case. This progression from case to case at one moment in time (mid-September) parallels, to some extent, the time progression from May to September for the -1 K case. It is tempting to perceive this as an eastward advection of the ice tongue, especially in view of the region's strong westerly winds, but the modeled ice is not actually advected over monthly time scales as fast as the speed of the advancing ice tongue. Instead, the advance of the modeled tongue is probably due mostly to thermodynamic considerations, as is the formation of the modeled Weddell polynya, described by Parkinson (1983).

In mid-November, by which time the decay phase is well underway, the ice edge response to the temperature changes is smallest off Wilkes Land (120°E - 150°E), most uniform in the central Weddell Sea (~ 35°W), and least uniform off Enderby Land (~ 55°E). The strongest retreat from the September ice-edge position is off Wilkes Land for the colder temperature cases, where the latitudinal retreat is 7°S for the -1 K case. Around the remainder of the continent, more typical September-November retreats range from 2° to 4°S.

Obvious candidates for the cause of the regional differences in the ice edge response are the atmospheric temperature fields. All other conditions being equal, a weaker temperature gradient could be expected to yield a stronger ice edge response to temperature increases, since the latitude of the freezing line would shift more dramatically from one model run to another. However, we found no simple relationship between the ice edge responses of Figure 3 and the prescribed air temperature fields. The ice edge position is a complicated function of the model thermodynamics and dynamics, and it shows far less symmetry with respect to the pole than do the atmospheric temperatures (cf. Taljaard et al., 1969).

The marked regional and temporal variability of the response of the ice edge to atmospheric temperature changes shown in Figure 3 highlights the danger of extrapolating the ice response in one area to the response of the ice cover as a whole. Consequently, although time sequences for ice thicknesses and concentrations at one point are presented next for illustrative purposes, time sequences are later presented for variables which have been summed or averaged over the entire southern ocean.

Results at an Individual Point in the Weddell Sea

To illustrate results at an individual location, we present in Figure 4 38-month time sequences of ice thicknesses and ice concentrations for grid square (14,28), which is located in the Weddell Sea at 71.4°S and 45°W (Figure 1). In the standard case, the thickness at this grid square fluctuates from 0.7 m at the summer minimum to 1.4 m at the winter maximum, while the concentration fluctuates from 12 to 98%. The response of both these variables to uniform temperature increases is quite systematic, although not uniform throughout the year.

The response of ice thickness at 71.4°S, 45°W to increased temperatures is greater in summer than in winter, with the interesting consequence that a greater thickness of ice is grown (or advected) into the grid square each fall/winter season in the warmer rather than colder cases (Figure 4a). Greater growth can result because the thinner ice has a larger vertical temperature gradient and hence a stronger conductive heat flux upward through the ice, thereby forcing greater ice accretion in order to maintain an energy balance at the ice/water interface. Also, the thickness response is stronger than linear, with a 5 K temperature increase resulting in a greater thickness decrease than 5 times the decrease for a 1 K temperature increase. Although the timing of the winter maximum thickness is not noticeably affected by the altered temperatures, the timing of summer minimum thickness is affected, with the minima occurring earlier the higher the temperature. In the +3 K case there are 3 months of ice free conditions (mid-January through mid-April), and in the +5 K case there are 4 1/2 months of ice-free conditions (mid-December through the start of May) at this Weddell Sea point.

Initialized from 1 January 1974 ESMR data at 88%, the ice concentration at 71.4°S, 45°W quickly decreases during the summer months, dipping

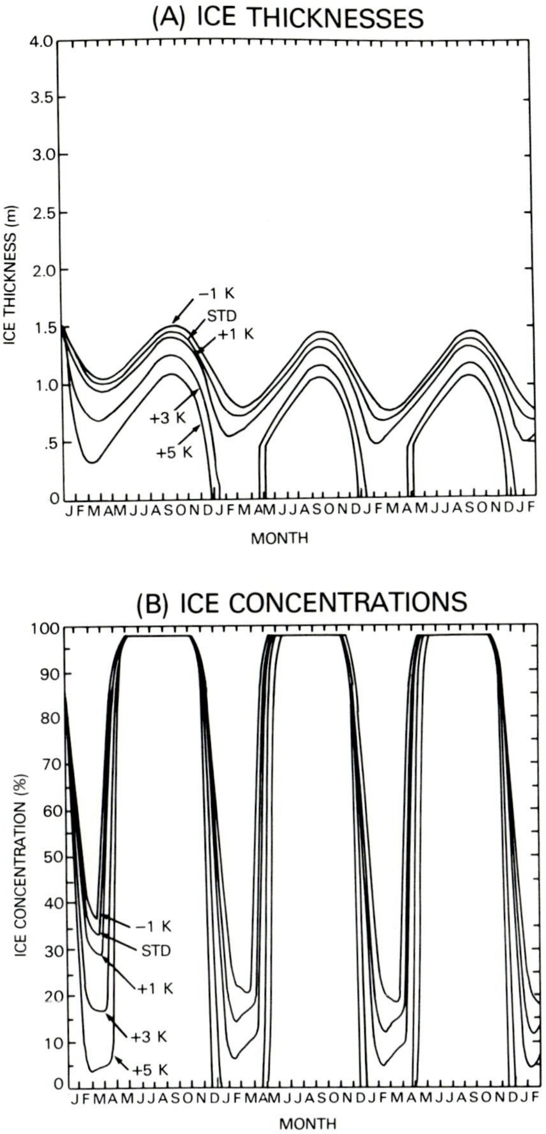

Fig. 4. Time sequences of ice thicknesses and concentrations at grid square (14,28), located in the Weddell Sea at 71.4°S, 45°W. Results are shown for the standard case and four temperature-perturbed cases.

below 40% even in the event of temperatures 1 K lower than the climatological values (Figure 4b). This is a somewhat greater seasonal decrease in ice concentrations than has been calculated from the 1974 ESMR data, which indicate a summer minimum concentration of about 57%. Nonetheless, in spite of the low summertime ice concentrations, the fall increase in ice cover is sufficient in each of the five simulations for the concentration to be at 98% (the maximum allowable) during most of the winter. This being true even with a 5 K temperature increase, the calculated mid-winter ice concentrations at 71.4°S, 45°W are found to be unaffected by the range of temperature adjustments examined here (-1 K to +5 K). Summer concentrations, by contrast, are noticeably affected, with the response being slightly stronger than linear. Interestingly, for the first summer the sea ice minimum comes at the end of the summer period for all cases except the warmest, whereas for the remaining summers it comes at the end only for the -1 K case (Figure 4). Hence, for the equilibrium cycle, the summer minimum is reached in early summer for the case of a warming and in late summer for the case of a cooling.

Results Summed over the Hemisphere

Four variables have been selected to illustrate modeled sea ice responses over the southern ocean as a whole: total ice area, total ice volume, area of ice-laden waters, and average ice thickness (Figure 5). Total ice area and total ice volume, initialized at 4.5×10^6 km^2 and 6750 km^3 respectively, both exhibit a very smooth annual cycle, decreasing until early March, then increasing until early-to-mid-September (Figures 5a and 5b). Thus the growth period from summer minimum to winter maximum is a few weeks longer than the decay period from maximum to minimum, a temporal asymmetry found also in the Nimbus 5 ESMR data.

The values of ice area and ice volume are reduced by more than 50% from the -1 K case to the +5 K case. The highest percentage difference between cases occurs in summer, while the highest absolute difference occurs in winter. Thus, as expected, warmer temperatures result in reduced fall/winter ice growth for the region as a whole. The temperature increases do not definitively impact the timing of summer minima, except for the first summer, when the higher temperatures result in later minimum values for both ice area and ice volume. Similarly, neither the timing of maximum ice area nor the timing of maximum ice volume is greatly affected by the increased temperatures, both maxima remaining near mid-September over the five case studies. Nonetheless, a very slight delay is apparent in attaining the maximum ice area as the temperatures are increased. For both variables, the response to the temperature increases is weaker than linear, the change from the standard case to the +5 K case being less than five times the change from the standard case to the +1 K case.

It should be noted regarding Figures 5a and 5b that the apparent brief disappearance of the ice cover in the final two summers of the +3 and +5 K cases is deceptive, a visual result of the plotting resolution. In actuality, a slight ice cover remains in both cases (cf. Figure 3a), minimum ice area being 38×10^3 km^2 in the +3 K case and 200 km^2 in the +5 K case, with corresponding minimum ice volumes of 18 km^3 and 1 km^3 respectively.

In spite of its importance, especially for climate studies, actual sea ice area was not determined for the southern ocean prior to the advent

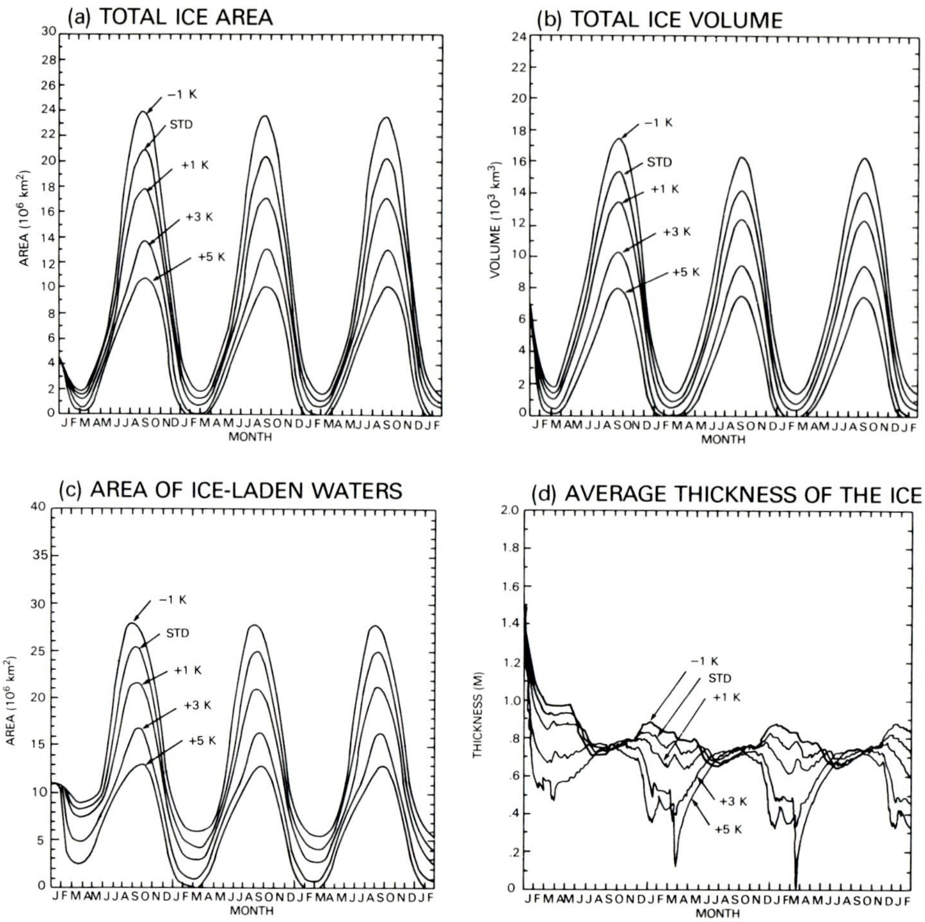

Fig. 5. Time sequences for the total southern ocean, presented for the standard case and four temperature-perturbed cases. The variables plotted are: (a) total ice area; (b) total ice volume; (c) total area of ice-laden waters; and (d) average thickness of the ice cover.

of satellite imagery, and is still not determined on a regular basis. Hence we include also in Figure 5 plots of the area of ice-laden waters, which, being far easier to estimate from incomplete observations, is the more-commonly calculated variable. Sometimes referred to as "ice extent" (e.g., in Zwally et al., 1983a), this area is calculated as the sum of the areas of all grid squares containing sea ice. Hence, in contrast to the ice area and ice volume, the calculated area of ice-laden waters depends directly on the 200 x 200 km grid size. A smaller grid size would tend to produce a smaller area of ice-laden waters, as there would tend to be less excess area included in the grid squares containing the ice edge. Any calculated value is thus an upper limit.

The area of ice-laden waters is initialized for each of the five simulations at 11 x 10^6 km^2. Like the total ice area and total ice volume, this area has monthly values which decrease slightly from year 1 to year 2 and then are very close to attaining an equilibrium yearly cycle by the end of the second year. The response of the area of ice-laden waters to the specified temperature increases is, as with the earlier variables, higher in winter on the basis of absolute magnitude but higher in summer on the basis of percentage change. In contrast to the ice volume, the area of ice-laden waters is more affected by a +1 K change than a -1 K change, although, in agreement with the responses of the total area and total volume, the response of the ice-laden waters to positive temperature increases is weaker than linear. Summer minimum arrives earlier for higher temperatures, whereas winter maximum arrives later, so that the asymmetry, with longer growth periods and shorter decay periods, is stronger with higher temperatures.

Average ice thickness behaves far less systematically than total ice volume, ice area, or area of ice-laden waters. Defined as the area-weighted thickness of all the existent ice at any given moment, the average thickness is not necessarily greater in winter than in summer, even though there is considerably more ice in winter. (E.g., a wintertime situation with 10 x 10^6 km^2 of 1.5 m ice and 10 x 10^6 km^6 of 0.5 m

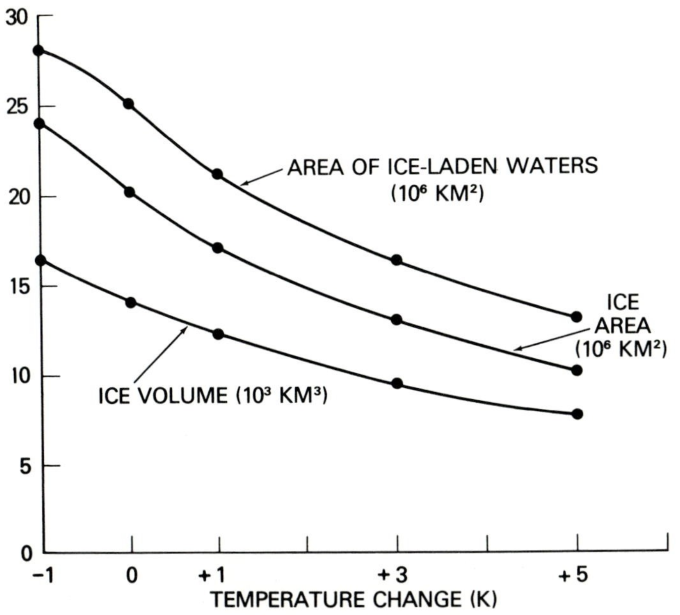

Fig. 6. Modeled response to atmospheric temperature changes of the total sea ice area, sea ice volume, and area of ice-laden waters at the time of maximum ice amount. The data points indicated are obtained from the year 3 values of Figures 5a, 5b, and 5c, respectively.

ice would have the same 1.0 m average ice thickness as a summertime situation with 2 x 10^6 km^2 of 1.0 m ice.) All ice thicknesses are initialized at 1.5 m. The average value quickly drops below 1 m within the first two months and does not again exceed 1 m for any of the five simulations. Wintertime average thicknesses are nearly constant over the five cases, being roughly 0.73 m from July through October, but for the rest of the year the thicknesses for the five cases differ considerably. In the two coldest cases, the average ice thickness is actually somewhat higher in summer than in winter. The springtime peak in average thickness in those two cases can be explained by the preferential springtime melting of the thin ice at the ice edge.

To examine the linearity of the hemispheric response of the sea ice to temperature change, the winter maxima of ice volume, ice area, and area of ice-laden waters (cf. Figure 5) have been plotted versus temperature change (Figure 6). The trends are consistent for the three variables: as temperature increases, the sensitivity of each variable to further warming decreases. This is explained, in part, by the smaller perimeter of the winter ice pack for the warmer temperatures: at higher latitudes, the same latitudinal retreat results in a smaller change in ice amount. This perimeter-effect would suggest even greater sensitivity for larger coolings, but explicit predictions of sea ice responses to large temperature changes should await more complete consideration of environmental feedback mechanisms currently omitted from this study. We note that Budd (1975), by contrast, illustrates in his Figure 3 that the sensitivity of the latitudinal extent of the sea ice cover to temperature change decreases for large coolings as well as for large warmings.

Ice Edge Response

Figure 7 presents time sequences of the latitudinal extent of the ice edge for each of four longitudes: 45°E, 135°E, 135°W and 45°W. The amount of ice edge retreat corresponding to a given temperature increase clearly depends on

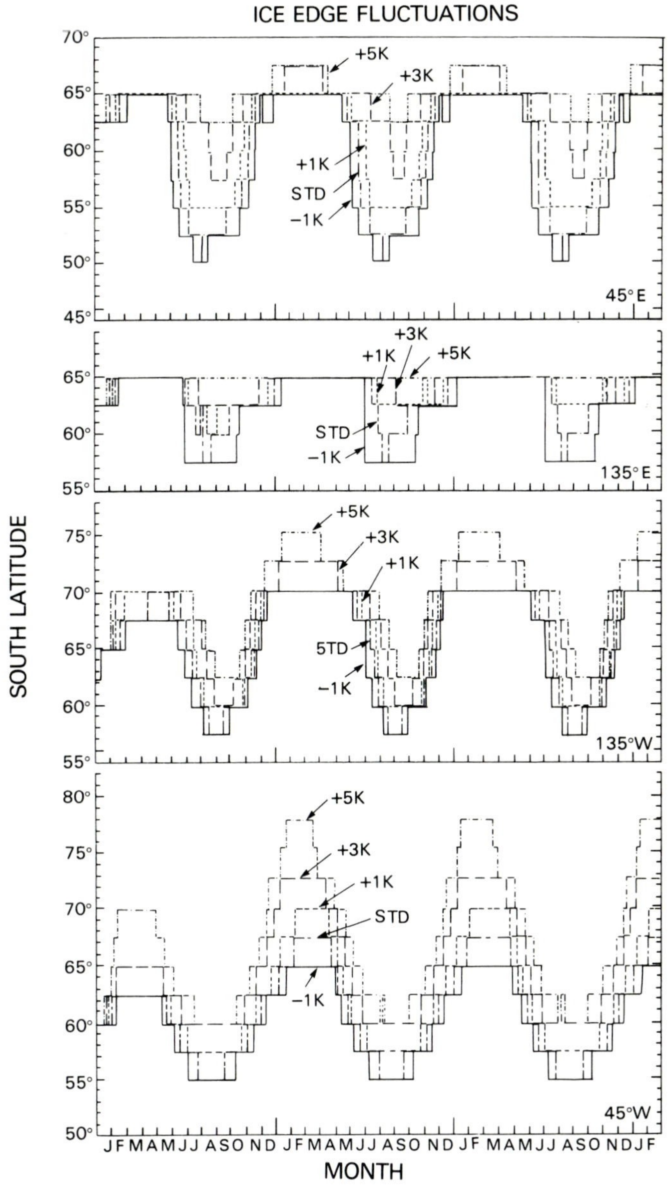

Fig. 7. Time sequences of the latitudinal position of the ice edge at each of four longitudes. Results are shown for the standard case and four temperature-perturbed cases.

both longitude and time, a fact pointed out earlier in the discussion of Figure 3. At 45°W, the ice edge shows the greatest response to temperature changes in summertime, responding, on average, at a rate of 2.1°S/1 K in summer but only 0.8°S/1 K in winter. The timing of minimum ice cover at 45°W shifts to earlier weeks as the temperature is increased, while the timing of maximum cover is shifted to later periods. Thus, although the growth period is longer than the decay period for the warmer temperatures, the decay period is longer than the growth period for the colder temperatures. This same shift in the asymmetry is apparent also at 135°W.

At 135°E, the ice retreats to the continental boundary in summertime for each of the five cases. In wintertime the ice extends outward one, two, or three grid squares from the boundary depending on the temperature conditions. At 45°E, the winter maximum ice extent averages a 2.1°S retreat per 1 K temperature increase. In summer the average rate is 0.4°S/1 K, but the continental boundary reduces the meaningfulness of this value.

To highlight the effect of the temperature changes on the latitude of the ice edge at maximum ice extent, these latitudes are plotted in Figure 8 for the longitudes 45°E, 135°E, 135°W, and 45°W, along with a linear fit through the 20 plotted points. The slope of the linear fit line is 1.25°S/1 K. Figure 9 presents a summary plot of the sensitivity of the ice edge to temperature change. Values are again plotted for the longitudes 45°E, 135°E, 135°W, and 45°W, and are additionally plotted for a hemispheric average.

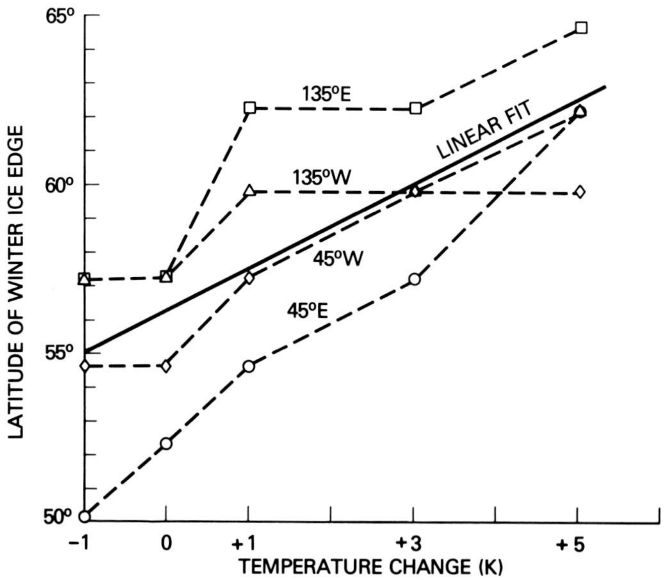

Fig. 8. Latitude of the winter maximum ice edge position at 45°E, 135°E, 135°W, and 45°W, plotted as a function of the atmospheric temperature deviation from the mean-monthly climatological data of the standard case.

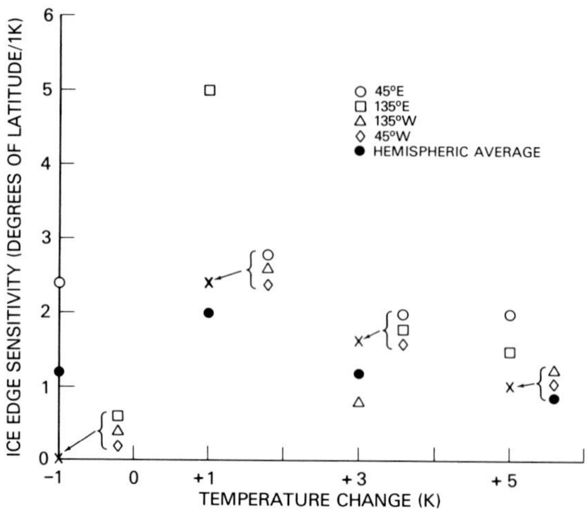

Fig. 9. Sensitivity of the latitudinal position of the maximum ice edge extent to atmospheric temperature changes. At 45°E, 135°E, 135°W, and 45°W these values were computed from the maximum year 3 ice edge positions plotted in Figure 7. The hemispheric average was computed as described in the text.

The latter is calculated from an ice-edge "average latitude", defined as the latitude of the ice edge assuming the ice cover in Figure 3c to be symmetrically distributed about the pole. The ice-edge sensitivity to the temperature changes ranges from zero to 5°S/1 K, with most values lying between 0.8 and 2.4°S/1 K. The large scatter for the +1 K and -1 K cases is an expected consequence of the failure of a 200 km grid to resolve slight variations; the values for the +3 and +5 K cases all fall between 0.8 and 2.0°S/1 K. Although no single number from Figure 9 will have universal applicability, as a general guide, the mean of the four values plotted in Figure 9 for the hemispheric average is 1.4°S/1 K.

Comparison of Modeled Results with Observed Sea Ice Variability

Unfortunately, sea ice data for the 1931-1960 time period of the climatological atmospheric inputs used in the model are too sparse to permit a meaningful comparison with the model results. To allow at least a general comparison with observed sea ice, we have thus examined parameters of the sea ice cover derived from satellite observations of the 1970s and early 1980s. For this we have used the detailed monthly-averaged maps and plots of sea ice conditions for 1973-1976 presented in Zwally et al. (1983a) and the summary plots of sea ice conditions for 1973-1981 presented in Zwally et al. (1983b).

During the period 1973-1981, the Antarctic sea ice cover typically expanded from a minimum extent of about 3 x 10^6 km^2 in late February to a maximum extent of about 19 x 10^6 km^2 in September. The model simulations reproduce well the timing of maximum and minimum extent, but the magnitude of the maximum is better reproduced in the +1 K case than in the standard case (cf. Figure 5). This latter fact can be explained partly by the generally colder temperatures in the climatological data used in the model than in the temperature data of the 1973-1981 period of the ice observations, the 1973-1981 temperatures being available from compilations of the Australian Bureau of Meteorology. From data presented by Jacka (1982) of the mean annual air temperatures at stations distributed around the Antarctic continent, a warming of about 0.7 K is indicated for the period 1958-1978.

Other major characteristics of the observed ice distribution are reproduced in each of the model simulations. These include the prevalence of summer ice in the western hemisphere, the asymmetric pattern of winter ice, and the autumnal development of a tongue of ice in the Weddell Sea. The wintertime asymmetry is such that the ice edge extends farthest equatorward at about the Greenwich meridian.

Another useful comparison is to relate the modeled results to the observed natural variabilities of sea ice and atmospheric temperatures. The 4 - 25 x 10^6 km^2 annual range in area of ice-laden waters for the standard case compares with a range of 4 - 20 x 10^6 km^2 calculated by Zwally et al. (1983b) for 1974, the year used for initializing the ice concentration fields in the model. The 1973-1981 range in maximum area of ice-laden waters was 17 - 20.4 x 10^6 km^2. This 3.4 x 10^6 km^2 range is seen to correspond on Figure 6 to a range in atmospheric temperatures of about 1 K. It thus suggests that mean-annual atmospheric temperature increases would have to exceed 1 K before inducing larger changes in the overall Antarctic sea ice cover than those observed over the past decade. Mean annual atmospheric temperatures presented by Jacka (1982) are given as variations from the 1955-1980 mean for 18 Antarctic coastal stations. Averages of these variations were calculated for each of the 23 years for which data were presented for more than five stations. The maximum magnitude of these averaged temperature variations is 0.8 K. This further demonstrates consistency between the modeled sea ice response to atmospheric temperatures and available sea ice and temperature data.

Final Discussion

Using a numerical model of sea ice, we have calculated the simulated response of the southern ocean sea ice cover to selected uniform increases in surface air temperatures, obtaining an average ice-edge retreat of 1.4°S for each 1 K temperature increase. Numerical experiments such as this serve a useful purpose for considering hypothetical scenarios of future climatic conditions, although the response of the ice in the real world would very likely differ in crucial respects from the response simulated by any particular model. Obvious deficiencies in the present model are the absence of oceanic and atmospheric couplings, since the model accepts values for various important atmospheric and oceanic variables from predetermined data sets and does not adjust those values according to simulated sea ice conditions. Thus this paper deals only with first-order effects of an atmospheric temperature increase on the sea ice and not with resultant effects of sea ice changes on the atmosphere and ocean. Performing the experiments with a fully-coupled ice / atmosphere / ocean model would allow some of the important interactions to be incorporated, although even then there would probably be omissions of interactive processes which in the real world might have a crucial impact on the sea-ice state. Some of the major known climatic feedbacks in the polar regions are discussed by Kellogg (1975).

With the above caveats understood, we can still state conclusions of relevance from the modeled results. The simulation outputs, illustrated in Figure 3, demonstrate clearly the complex spatial and temporal response of Antarctic sea ice to uniform temperature changes, showing that the behavior of the sea ice at one locality or within one region can differ significantly from the behavior of the ice cover as a whole. This suggests both that extrapolations from data at one location or from one region must be viewed with caution, and that any monitoring program or analysis of the southern ocean sea ice must be comprehensive in scope to allow definitive meaningful conclusions to be reached about the ice pack as a whole.

We find a strong response of the southern ocean ice cover to temperature changes, with the wintertime ice amount decreasing by roughly half with a 5 K temperature increase and the summertime ice essentially disappearing with a 5 K temperature increase. This degree of sea ice decrease could be expected to have pronounced ramifications on both the atmosphere and oceans. Among the expected impacts would be ocean warming caused by increased absorption of solar radiation, greater local sensible and latent heat transmission from the ocean to the atmosphere, perhaps inducing increased atmospheric cyclogenesis, and reduced salt rejection from ice formation and ice aging, which could decrease local oceanic overturning and bottom water formation.

Acknowledgments. We thank Drs. William F. Budd, H. Jay Zwally, Robert Cahalan, and Tom Bell for valuable comments on the manuscript, and NASA's Oceanic Processes Branch in the Environmental Observations Division for funding and otherwise supporting the work.

References

Allison, I., Antarctic sea ice growth and oceanic heat flux. In: Sea Level, Ice, and Climatic

Change (ed. I. Allison), International Association of Hydrological Sciences, Guildford, Great Britain, 161-170, 1981.

Augustsson, T. and V. Ramanathan, A radiative-convective model study of the CO_2 climate problem, J. Atmos. Sci., 34, 448-451, 1977.

Broecker, W. S., Climatic Change: Are we on the brink of a pronounced global warming?, Science, 188, 460-463, 1975.

Budd, W. F., Antarctic sea-ice variations from satellite sensing in relation to climate, J. Glaciol., 15, 417-427, 1975.

Budyko, M. I., Polar ice and climate. In: Proc. Symp. on the Arctic Heat Budget and Atmospheric Circulation (ed. J. O. Fletcher), Rand Corporation Memorandum RM-5233-NSF, Santa Monica, California, 3-22, 1966.

Budyko, M. I., Climate and Life (English edition edited by D. H. Miller), International Geophysical Series, 18, Academic Press, New York, 508 pp, 1974.

Gordon, A. L., Seasonality of southern ocean sea ice, J. Geophys. Res., 86, 4193-4197, 1981.

Gordon, A. L. and R. D. Goldberg, Circumpolar characteristics of Antarctic waters, Antarctic Map Folio Series (ed. V. G. Bushnell) Folio 13, American Geographical Society, New York, 1-25, 1970.

Hansen, J., D. Johnson, A. Lacis, S. Lebedeff, P. Lee, D. Rind and G. Russell, Climate impact of increasing atmospheric carbon dioxide, Science, 213, 957-966, 1981.

Hibler, W. D. III and S. F. Ackley, On modeling the Weddell Sea pack ice, Ann. Glaciol., 3, 125-130, 1982.

Hunt, G. B. and N. C. Wells, An assessment of the possible future climatic impact from carbon dioxide increases based on a coupled one-dimensional atmospheric-oceanic model, J. Geophys. Res., 84, 787-791, 1979.

Jacka, T. H., Antarctic temperature and sea ice extent studies, Antarctica: Weather and Climate (preprint volume compiled by N. W. Young), Royal Meteorological Society Australian Branch, Parkville, Australia, 89-98, 1982.

Keeling, C. D., J. A. Adams, C. A. Ekdahl and P. R. Guenther, Atmospheric carbon dioxide variations at the South Pole, Tellus, 28, 552, 1976a.

Keeling, C. D., R. B. Bacastow, A. E. Bainbridge, C. A. Ekdahl, P. R. Guenther, L. S. Waterman, and J. S. Chin, Carbon Dioxide variations at Mauna Loa Observatory, Hawaii, Tellus, 28, 538, 1976b.

Kellogg, W. W., Climatic feedback mechanisms involving the polar regions. In: Climate of the Arctic (eds. G. Weller and S. A. Bowling), Geophysical Institute, University of Alaska, 111-116, 1975.

Kusunoki, K., Variations of sea ice conditions in Lutzow-Holm Bay area, in Antarctica, in the last 20 years. In: Sea Level, Ice, and Climatic Change (ed. I. Allison), International Association of Hydrological Sciences, Guildford, Great Britain, 171-176, 1981.

Lacis, A., J. Hansen, P. Lee, T. Mitchell and S. Lebedeff, Greenhouse effect of trace gases, 1970-1980, Geophysical Research Letters, 8, 1035-1038, 1981.

Manabe, S. and R. J. Stouffer, A CO_2-climate sensitivity study with a mathematical model of the global climate, Nature, 282, 491-493, 1979.

Manabe, S. and R. J. Stouffer, Sensitivity of a global climate model to an increase of CO_2 concentration in the atmosphere, J. Geophys. Res., 85, 5529-5554, 1980.

Manabe, S. and R. T. Wetherald, Thermal equilibrium of the atmosphere with a given distribution of relative humidity, J. Atmos. Sci., 24, 241-259, 1967.

Manabe, S. and R. T. Wetherald, The effects of doubling the CO_2 concentration on the climate of a general circulation model, J. Atmos. Sci., 32, 3-15, 1975.

Manabe, S. and R. T. Wetherald, On the distribution of climate change resulting from an increase in CO_2 content of the atmosphere, J. Atmos. Sci., 37, 99-118, 1980.

Maykut, G. A. and N. Untersteiner, Some results from a time-dependent thermodynamic model of sea ice, J. Geophys. Res., 76, 1550-1575, 1971.

Parkinson, C. L., On the development and cause of the Weddell polynya in a sea ice simulation, J. Phys. Oceanogr., 13, 501-511, 1983.

Parkinson, C. L. and W. W. Kellogg, Arctic sea ice decay simulated for a CO_2-induced temperature rise, Climatic Change, 2, 149-162, 1979.

Parkinson, C. L. and W. M. Washington, A large-scale numerical model of sea ice, J. Geophys. Res., 84, 311-337, 1979.

Ramanathan, V., M. S. Lian and R. D. Cess, Increased atmospheric CO_2: zonal and seasonal estimates of the effect on the radiation energy balance and surface temperature, J. Geophys. Res., 84, 4949-4958, 1979.

Sellers, W. D., A reassessment of the effect of CO_2 variation on a simple global climatic model, J. Appl. Meteorol., 13, 831-833, 1974.

Taljaard, J. J., H. Van Loon, H. L. Crutcher and R. L. Jenne, Climate of the Upper Air, I, Southern Hemisphere, Vol. 1, Temperatures, Dew Points, and Heights at Selected Pressure Levels, NAVAIR 50-IC-55; U.S. Naval Weather Service, Washington, D.C., 135 pp, 1969.

Zwally, H. J., J. C. Comiso, C. L. Parkinson, W. J. Campbell, F. D. Carsey and P. Gloersen, Antarctic Sea Ice, 1973-1976: Satellite Passive - Microwave Observations, NASA SP459, National Aeronautics and Space Administration, Washington, D.C., in press, 1983a.

Zwally, H. J., C. L. Parkinson and J. C. Comiso, Variability of Antarctic sea ice and CO_2 change, Science, 220, 1005-1012, 1983b.

ICE SHEET MARGINS AND ICE SHELVES

Robert H. Thomas*

Jet Propulsion Laboratory, 4800 Oak Grove Drive
Pasadena, CA 91109

Abstract. A possible consequence of climatic warming is an increase in ice sheet melting rates and, ultimately, a perceptible rise in sea level. This would occur very slowly unless thinning at the ice sheet margins triggered a major increase in ice drainage rates. The ice sheet in West Antarctica is the one most susceptible to this type of collapse because it rests on bedrock well below sea level. The most probable outlet for rapid ice discharge would be through the area covered today by the Ross Ice Shelf, which would have to be severely weakened first.

Currently, the ice shelf appears to be either in equilibrium or actually growing thicker. Most of the wastage from the ice shelf is by occasional calving of large tabular icebergs. Weakening could be achieved most rapidly if summer temperatures were to rise to several degrees above the melting point for two or three months each year. Then, widespread surface meltwater would fill crevasses and force them to overdeepen. This would cause accelerated iceberg calving and explosive fragmentation of the resulting icebergs-conditions similar to those found in the Arctic today. Such accelerated iceberg calving could also threaten portions of the ice sheet in East Antarctica. Here, although bedrock is below sea level, the ice sheet appears to be protected by coastal bedrock sills which are too shallow for a tabular iceberg to pass over them, but would not prevent the discharge of small iceberg fragments.

The most vulnerable portion of the ice sheet is probably that drained by Pine Island and Thwaites glaciers in West Antarctica. These glaciers are not protected by large ice shelves and, although there is no evidence for present day retreat, they may be the first glaciers to respond to climatic warming. However, even an order of magnitude increase in the activity of one of these glaciers would cause a sea level increase of less than 1 mm yr^{-1}.

The measurements necessary to monitor ice sheet extent and elevation, summer melt zones and sea ice cover can all be obtained from a satellite in near-polar orbit using well-proven techniques. The establishment of such a program should take highest priority in future research.

Introduction

An ice sheet responds to climatic change by readjusting its shape in an attempt to achieve a state of equilibrium with the changed conditions. By necessity, ice sheet growth takes place rather slowly because it can be achieved only by net retention of some of the accumulated snowfall. Consequently, associated effects which influence mankind -- in particular, changes in sea level -- are also very slow. In contrast, although ice retreat by increased melting is also slow, ice sheet collapse can be as fast as ice can be discharged into the ocean. Currently, the faster glaciers and ice streams in Antarctica move at speeds up to 1 or 2 km yr^{-1}. However, the Jacobshavn Glacier in Greenland has a speed of approximately 8 km yr^{-1}, and it calves near the point where it becomes afloat to produce small icebergs. These icebergs are flushed out to the open sea along a fjord that is 40 km long, 5-7 km wide and shoals at its seaward end to a depth of 50-100m (Weidick, 1969). Through this narrow exit, approximately 16 km^3 of ice per year are discharged to the ocean. The very much wider ice streams of Antarctica have values of basal friction ($\sim 10^{-1}$ bar) that, in some cases, are significantly less than those estimated for Jacobshavn Glacier (~ 1 bar). Consequently, speeds considerably greater than 10 km yr^{-1} could probably be sustained by these ice streams if their movement was not hampered by the ice shelves into which they flow. Under these circumstances, sea-level rise could be appreciable over periods of decades or less.

Current estimates indicate that sea level has risen by approximately 12 cm since 1880, and only part of this rise can be attributed to thermal expansion of the oceans (Gornitz et al., 1982). This suggests a net loss of terrestrial ice, but there is no evidence for systematic thinning of the polar ice sheets during this period. Available data can be used to estimate the mass

---

*Currently assigned to NASA Headquarter, Code EE-8, Washington, D.C. 20546

balance of individual drainage basins within the ice sheets, and these estimates indicate localized thickening or thinning of ice with a bias towards thickening. However, there are vast areas of ice sheet for which no estimates of thickening / thinning rate can be made, and currently we simply do not know whether the large polar ice sheets are growing or shrinking. In this paper, I shall consider the response of ice sheet margins to a warming climate with emphasis on possibly rapid response (tens to hundreds of years). I shall attempt to classify ice sheet regions according to their vulnerability, with some indication of the time sequence of response. I shall conclude by suggesting measurements that should be made to provide an ability, first, to monitor the state of ice sheet equilibrium, and second, to predict ice sheet behavior under prescribed conditions.

There are three ways in which a glacier can quickly change from a state of growth, or near-equilibrium to one of rapid collapse: by surging or by irreversible retreat of the glacier terminus due to instabilities associated with either marine - based glaciers, or fjord glaciers. I shall examine the potential of each of these mechanisms for significantly affecting ice sheet drainage rates.

Ice Sheet Surges

Periodic rapid increase in the activity of individual mountain glaciers is well documented (Meier and Post, 1969), and these glaciers are known as surging glaciers. A surging glacier never reaches equilibrium. During its quiescent phase, it moves too slowly to discharge its total snow accumulation, and it grows in size. At some stage, its mode of flow changes from slow to fast, and it discharges more ice than it accumulates. The resulting "surge" occurs far more rapidly than the earlier growth, and it continues until there is not enough ice within the catchment area to feed the rapid discharge. Then the whole process is repeated. The surging mechanism is not well understood, but it is probably associated with an increase in lubrication by meltwater or mud at the glacier bed. Each surging glacier appears to possess a unique surging period which is probably determined by glacier size, snowfall rate, bed configuration and composition, and meltwater availability at the bed.

There is no evidence for surging of the large Greenland and Antarctic glaciers, and if they do surge, the period between surges will be different for each glacier, and will be measured in thousands of years (Budd and McInnes, 1978). Changing climate may alter the period between surges, but it would be fortuitous if it triggered an individual surge. However, it is appropriate to discuss components of the surging mechanism that may be involved in climate induced ice sheet retreat. Some polar glaciers may permanently be in a fast mode of flow. The polar ice sheets are large enough for such a glacier to acquire a very low surface slope so that slope induced weight stresses parallel to the glacier bed are balanced by the low basal friction associated with "fast mode" glaciers. Then, the speed of the glacier would not be unusually large, and ice discharge could be balanced by total accumulation within the glacier's vast catchment basin. Examples of such glaciers are the ice streams in West Antarctica that flow into the Ross Ice Shelf. These have surface slopes of order 10^{-3}, and speeds of several hundred m yr^{-1}; neighboring portions of the ice sheet which move at only tens of m yr^{-1} have surface slopes that are almost 10^{-2}. It is highly probable that the speeds of the fast ice streams are very susceptible both to changes in surface slope and to changes in the Ross Ice Shelf, which is being pushed seaward by the ice streams. The ice streams that would be most affected by weakening or removal of the ice shelves are those that flow over bedrock well below sea level: marine-based ice streams. These probably can retreat very rapidly and, if the ice sheet upstream is also marine-based, then much of it will collapse to be discharged into the sea as icebergs. Geological data suggest that a large, marine-based portion of the Laurentide Ice Sheet collapsed in less than 200 years about 8,000 years ago with an associated sea level rise of several meters (Andrews et al., 1972).

Marine Ice Sheets and Ice Shelves

Marine ice sheets are ice sheets resting on bedrock that is, for the most part, well below sea level. Most of the ice drainage flows into peripheral ice shelves. It has long been recognized that marine ice sheets might be susceptible to rapid collapse if creep rates within the ice become sufficiently large to cause net thinning of the marginal ice, which would then float free from the seabed allowing the grounding line between ice sheet and floating ice shelf to retreat towards the ice sheet center (Mercer, 1968; Weertman, 1974; Thomas, 1979). The rate of grounding line retreat, and the ultimate fate of the ice sheet would depend on the balance between thinning of the marginal ice by creep, and thickening by snow accumulation and advection of thicker ice from upstream. Creep rates at the grounding line increase with increasing ice thickness, and decrease if movement of the ice shelf seaward is restricted by shear at its margins and compressive stresses upstream from grounded ice rises within the ice shelf (Figure 1). Where the ice calves at, or near to, the grounding line, the rate at which the marginal ice thins by creep is proportional, approximately, to the fourth power of the ice thickness (Weertman, 1957). Under these conditions, ice at the grounding line has a maximum thickness of approximately 400m. If the ice thickness was much greater than this, the high creep rates would result in net thinning of the ice, and retreat of the grounding line to a region with shallower bedrock.

Although portions of the ice sheet in East Antarctic are marine based, the West Antarctic ice sheet is the major existing example of a marine ice sheet. Ice drainage flows mainly into the large, embayed Ross and Ronne ice shelves, and into small peripheral ice shelves which lie between the coast and near coastal shoals on which the ice shelf runs aground to form ice rises. The ice rises and the margins of the embayments restrict seaward movement of the ice shelves and significantly reduce the creep rate compared with that of an unrestricted ice shelf of the same thickness. Consequently, confined ice shelves can be very thick near the grounding line and yet have very small values of creep thinning rate. For instance, interior portions of the Ross and Ronne ice shelves have thicknesses greater than 800m and 1300m respectively, with creep rates of approximately 10^{-3} yr^{-1}, which are typical of unrestricted ice shelves less than 300m thick. In these areas, the grounding line position is sensitively controlled by the characteristics of the ice shelf to seaward. If these characteristics change for some reason, then the creep thinning rates at the grounding line will change, and the grounding line will migrate to a new position. This is illustrated in Figure 1, where ice shelf profiles are shown for three ice shelf configurations: a bounded ice shelf with, and without, an ice rise; and an unrestricted ice shelf. The equilibrium grounding line position for each of these cases is strongly dependent on bedrock topography. For a marine ice sheet, bedrock may slope downward towards the center of the ice sheet due to isostatic depression under the ice sheet load; grounding line retreat may then be irreversible unless halted by a bedrock sill.

Theoretical analyses of the stability (Weertman, 1974) and of the growth and collapse of marine ice sheets (Thomas, 1977; Thomas and Bentley, 1978a; Stuiver et al., 1981) indicate that the rate of retreat (or advance) of the ice sheet grounding line is determined principally by:

(i) Bedrock topography. Retreat is most rapid over regions where the bedrock is deepest below sea level; conversely, rapid growth occurs over shallow seabed. Changes in world sea level, of course, have an immediate effect on grounding line stability.

(ii) Basal conditions of the grounded ice sheet. Ice streams with beds that are well lubricated by either water or mud have very little basal friction, and therefore very low surface slopes. These are susceptible to high rates of grounding line retreat.

(iii) Ice shelf configuration downstream of the grounding line. Retreat rates decrease as forward motion of the ice shelf becomes more restricted by marginal shear and by the presence of grounded ice rises. Thus, a retreating grounding line could be halted by the formation of a sufficiently large and well pinned ice shelf in the region once occupied by the ice sheet margins. The present day Ross and Filchner/Ronne ice shelves are examples; during glacial periods, the embayments they now occupy were probably filled with grounded extensions of the West Antarctic ice sheet. Climatic warming could result in progressive weakening of the ice shelves due to increased melting from both the upper and lower surfaces.

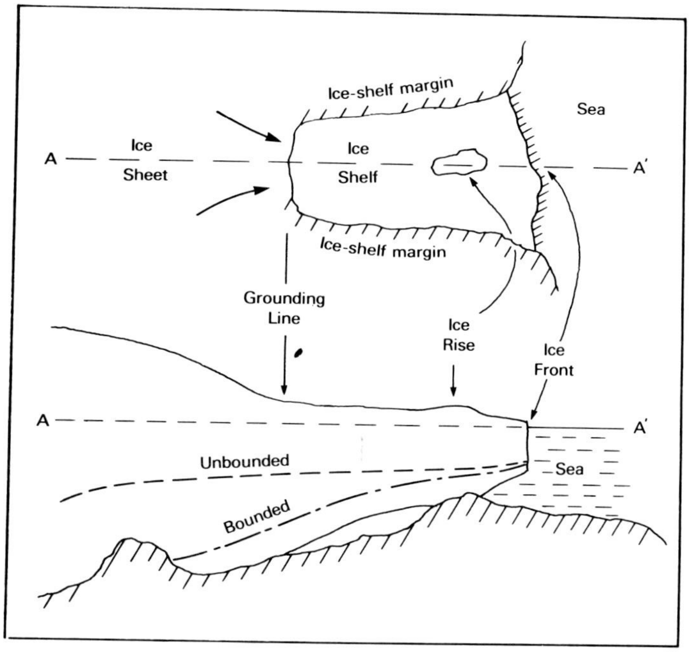

Fig. 1. An embayed ice shelf. Ice flowing from the grounded ice sheet must push the ice shelf past its margins and around the ice rise in order to flow seaward. Removal of the ice rise or the ice shelf margins would permit greatly increased drainage rates and accelerated ice shelf thinning. The grounding line would retreat to a new equilibrium position, as shown by the profiles marked "bounded" (removal of ice rise) and "unbounded" (removal of ice rise and margins).

Clearly, the actual state of equilibrium of a marine ice sheet is determined by all these factors acting together. Other factors, including snowfall rates and ice temperature also have an influence; but, during either rapid retreat or growth, this influence is swamped by that of the factors listed above.

Fjord Glaciers

Glaciers that occupy fjords generally terminate on or near a shallow sill, which may be only a few tens of meters below sea level. Further upstream the fjord bed is considerably deeper and, if the glacier terminus were to retreat into this deeper region, the glacier would probably become unstable with the terminus rapidly retreating toward the upstream end of the fjord. There is historic evidence for such catastrophic retreat of the Muir Glacier in Alaska and the Jacobshavn Glacier in Greenland (Weidick, 1969), and the USGS has been intensively investigating Columbia Glacier, Alaska to determine whether this glacier might undergo a similar retreat (Meter et al., 1980).

A suggested cause for the retreat is enchanced calving of icebergs when the terminus is in deeper water (Meier et al., in press). However, this mechanism can operate only if the icebergs are small enough to pass over the seabed sill marking the equilibrium position of the terminus. Currently, calving from Arctic glaciers is predominantly in the form of comparatively small blocks, whereas in Antarctica calving of large tabular icebergs is most common. The reason for this difference probably lies in the contrast between the climates of both regions. Along most of the Antarctic coastline, the snowline is virtually at sea level, and there is little summer melting. In the Arctic, summer temperatures rise above 0°C for long periods and there is extensive summer melting, so that large volumes of melt water are available to flow into the many crevasses near the termini of active glaciers. This allows the crevasses to penetrate to far greater depths than the few tens of meters associated with dry crevasses (Weertman, 1973). Consequently, the glacier becomes saturated with planes of weakness associated with these deep crevasses, and it readily fractures into small blocks as calving occurs. If this interpretation is correct, then Antarctic drainage basins are protected from this type of rapid retreat until summer temperatures rise sufficiently to cause extensive surface melting in the coastal regions.

Ice Sheet Vulnerability to Climatic Warming

Greenland

Although bedrock beneath central parts of the Greenland ice sheet is depressed below sea level, there is an elevated coastal rim of mountain ranges punctured here and there by major outlet glaciers. There is no indication that any of these glaciers could provide exits for marine ice sheet collapse, and the ice sheet would probably respond only slowly to a warming climate. Clearly, however, surface melt rates would increase. Currently, there is appreciable summer melt over large areas of the ice sheet, and these areas would increase in size and migrate inland as the climate became warmer. The area of summer melt can be mapped from satellite observations of microwave emissions from the ice surface (Zwally and Gloersen, 1977; Stiles and Ulaby, 1980). The necessary measurements have been made since 1973, and will be continued for at least the next decade. These data should be routinely analysed to distinguish wet snow from dry snow over both the Greenland and Antarctic ice sheets. The resulting long time series will give the seasonal and iterannual variability in wet-snow area and, potentially, any long-term trend indicative of a change in climate.

Antarctica

Nearly all of the snow that falls on the Antarctic ice sheet stays there. Some is lost by evaporation, but most of the snow that melts refreezes either in situ or slightly beneath the snow surface. Most of the ice drains seaward along a few major glaciers and ice streams that flow into floating ice shelves. These ice shelves occupy the larger embayments in the coastline and maintain an approximately constant ice front position by the calving of icebergs. Wastage from the ice sheet is principally by iceberg calving and, to a lesser extent, by melting from the underside of the ice shelves.

East Antarctica. In Antarctica, the major potential for a rapid response to a warming climate is by collapse of marine based portions of the ice sheet. Figure 2 shows where bedrock beneath the ice sheet is more than 500m below sea level. These areas are all potentially susceptible to rapid collapse by groundling line retreat unless they are protected by either a bedrock sill or a well pinned ice shelf. Large portions of the East Antarctic ice sheet are marine based, but available evidence indicates that there probably are shallow bedrock sills along the coastal margins of these areas. More data are required to determine whether or not these sills exist. Moreover, if summer temperatures were to become sufficiently high to induce widespread melting, the rate of iceberg calving may accelerate, with explosive disintegration of the icebergs into small fragments which could float over the sills. Under these conditions, rapid retreat by "fjord-glacier instability" might occur.

West Antarctica - (The Ross and Ronne sectors). It can be seen from Figure 2 that most of the bedrock in West Antarctic is well below sea level with connections to the sea via the Ross and Ronne ice shelves, and along the northern coast at D and E in Figure 2. Most of the ice drainage flows into ice shelves, and the ice sheet would be particularly susceptible to any climate change that resulted in removal of these ice shelves. Thickness profiles for two of the ice streams flowing into the Ross Ice Shelf are shown in Figure 3. Also shown are estimates from a simple theoretical model of grounding line retreat under the influence of ice shelf thinning caused by prescribed rates (M) of melting (Thomas et al., 1979). At present, we cannot predict values of M corresponding to a given climatic warming. However, it is probable that a

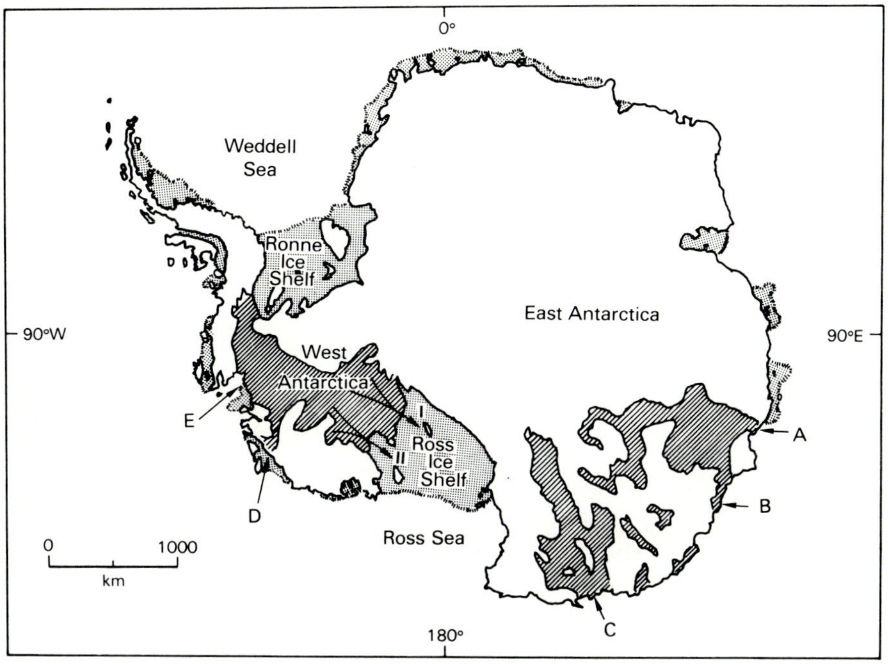

Fig. 2. Antarctica. The floating ice shelves are shaded. The hachured regions are "marine-based" portions of the ice sheet resting on bedrock more than 500 m below sea level. A-E are potential coastal outlets for ice from these regions. I and II are major ice streams flowing from West Antarctica into the Ross Ice Shelf.

melt rate of even 1 m yr^{-1} would require either a major increase in summer air temperatures or a significant alteration in ocean circulation beneath the ice shelf. An upper limit to grounding line retreat rates is provided by the extreme case of "no ice shelf" in Figure 3. This would correspond to instantaneous removal of the Ross Ice Shelf. Such an event would be followed by very rapid collapse of most of the West Antarctic ice sheet and a rise in sea level of approximately 5m. In reality, removal of the ice shelf would take a long time, and some unknown period of sustained climatic warming would be necessary to force M to increase to the values shown in Figure 3. Thereafter, retreat would continue for some hundreds of years, depending on how large the melt rates became. Moreover, initial retreat would be quite slow, even for high melt rates, with a very rapid final retreat. This pattern of response accentuates the need to measure the present behavior of the ice sheets. This would put us in a better position to detect the subtle changes in behavior that will provide early warning of a shift in the state of ice sheet equilibrium.

The time sequence that I have described here is for removal of the Ross Ice Shelf by progressively increased melting. It is important to note that, if Antarctic summers were to become sufficiently warm, the iceberg calving process could change from the present "tabular" mode to one of rapid fragmentation due to weakening by the overdeepening of water filled crevasses. This would accelerate ice shelf demise.

Little is known about the Ronne Ice Shelf and its catchment area. However, it appears to be less vulnerable than the Ross sector for two reasons. Less of the nearby inland ice rests on very deep bedrock (Figure 2), and the Ronne Ice Shelf contains many large ice rises which tend to restrict drainage rates from inland and help to protect the ice shelf from excessive calving. Excessive climatic warming would undoubtedly cause the Ronne Ice Shelf to shrink in size, but the inland catchment area would probably survive until excessive ice drainage through the Ross Ice Shelf and the northern outlets (D and E in Figure 2) resulted in migration of the drainage divide and shrinkage of the Ronne catchment area.

At present there is no evidence for widespread thinning of the Ross sector. The summit region may be thinning a few cm per year (Whillans, 1977) but the entire catchment area probably receives significantly more snow accumulation than it losses by ice drainage (Thomas et al., in press (a)). Moreover, upstream portions of the Ross Ice Shelf may be thickening quite rapidly (Thomas and Bentley, 1978b). The effects of such thickening could be to slow down the big ice streams that flow into the ice shelf, and thus consolidate the inland ice sheet. This trend would have to be reversed before ice sheet collapse could be initiated by climatic warming.

West Antarctica - (The Northern Sector). The northern outlets from West Antarctica (D and E in

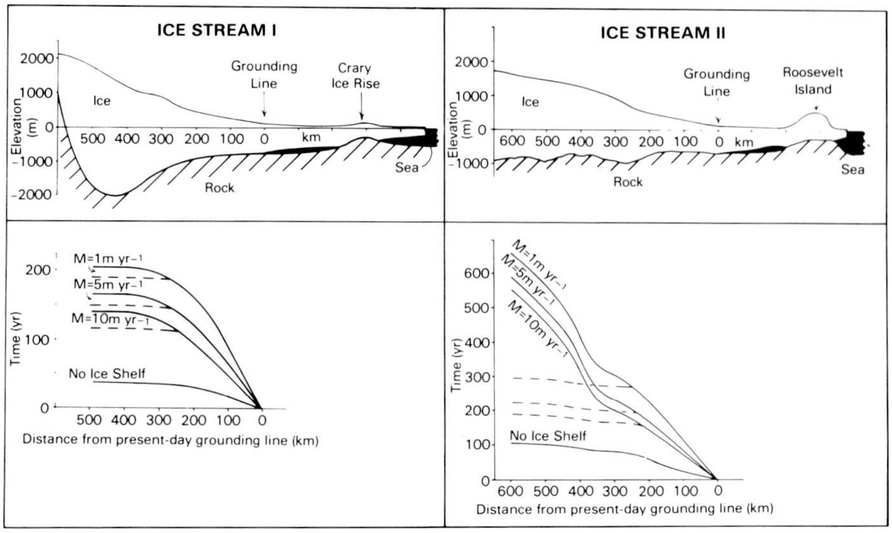

Fig. 3. Thickness profiles and estimated groundling line retreat sequences for the two major ice streams flowing from West Antarctica into the Ross Ice Shelf (Fig. 2). The lower curves are plots of grounding line position against time for different values of M, the ice shelf melting rate. The curve for "no ice shelf" refers to instantaneous removal of the Ross Ice Shelf. The solid lines were calculated assuming an ice shelf continues to exist throughout the retreat sequence; the broken lines are for complete breakup of the ice shelf after it has floated free from the seabed shoals beneath Crary Ice Rise and Roosevelt Island. These retreat sequences take no account of the time taken for M to increase to the high values assumed.

Figure 2) appear to be considerably more vulnerable. Few measurements have been made in these areas, and their present state of mass balance can only be surmised. No data are available for drainage route D in Figure 2, but its deep bedrock channel appears to be comparatively narrow, and the ice flows into an ice shelf containing many ice rises which restrict drainage rates. Nevertheless, this is an area from which more data should be obtained.

Drainage outlet E in Figure 2 includes two major glaciers: Pine Island and Thwaites glaciers. Stuiver et al., (1981, p.430) note that these glaciers are not buttressed by ice shelves and that, unless sufficiently high bedrock sills exist in the glacier channels, this sector of the ice sheet may already be undergoing fast mode recession. Thomas (1979) calculated the critical depth below sea level (d) of bedrock sills necessary to prevent retreat. For Thwaites Glacier, if there is no buttressing ice shelf, d~ 400-550m, depending on how easily the glacier is able to slide over its bed. Measurements of actual bedrock topography have not been made. For Pine Island Glacier, which appears to flow into a small embayed ice shelf, d~ 520-610m. Subsequently, measurements of ice thickness along Pine Island Glacier were made by Crabtree and Doake (1982), and the resulting thickness profile and inferred bedrock topography are shown in Figure 4. For the ice shelf portion of the profile, there is a simple relationship between surface elevation and ice thickness to permit flotation. Crabtree and Doake (1982) noted that there were significant departures from this relationship inland of A in Figure 4, which they consequently designated as the grounding line of Pine Island Glacier. They were then able to simulate the thickness profile of the ice shelf portion as a "steady-state" feature, but only by assuming the ice shelf between A and B in Figure 4 experiences extremely high back pressure, which they contended was caused by the converging ice shelf margins. In fact, the effects of convergence can be explicitly incorporated within Crabtree and Doake's analysis using equations given by Thomas (1973). For Pine Island Glacier, convergence contributes very little effective back pressure, and the observed thickness profile between A and B in Figure 4 cannot be reconciled with steady state unless the high back pressures invoked by Crabtree and Doake can be explained by some other mechanism. If they cannot, it will be difficult to avoid concluding that Pine Island Glacier is currently undergoing fast mode recession.

Rapid retreat is a possibility that cannot be rejected without additional information, but the ice velocity of approximately 2 km^{-1} yr estimated from comparison of aerial and satellite imagery is compatible with steady state, or even thickening within the glacier catchment basin (Crabtree and Doake, 1982). Moreover, a simple explanation for the required back pressure between A and B in Figure 4 is provided if we assume that this portion of the glacier is either partially or totally

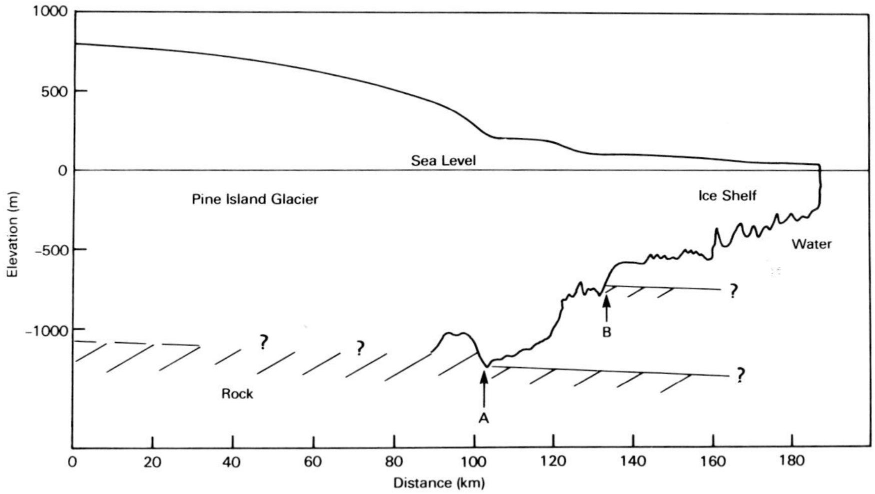

Fig. 4. Thickness profile of Pine Island Glacier (Crabtree and Doake, 1982). The ice shelf lies in an embayment approximately 30km wide. Crabtree and Doake (1982) estimated that the groundling line is at A; I suggest that it is near B, and that the glacier between A and B is sliding over a well-lubricated bed. If this is correct, the glacier is probably in a state of near-equilibrium, as suggested by the mass-balance estimates of Crabtree and Doake (1982).

grounded. It would then represent a lower portion of Pine Island Glacier and the glacier grounding line would be near B in Figure 4, where the ice shelf base is 600-700m below sea level, close to the upper limit of the 520-610m estimated by Thomas (1979). If this is found to be so, the glacier could be in equilibrium, neither receding nor advancing. Nevertheless, the glacier is clearly susceptible to small changes both in climate and in glacier dynamics.

Climatic warming could initiate retreat by weakening the ice shelf, and increased basal lubrication upstream of A would favor an increase in ice velocity and slumping of the glacier surface. Indeed, the region immediately upstream of A is of considerable interest. For a short distance, the glacier surface is steeply sloping, and associated basal shear stresses are very high (> 2 bars). This suggests that most of the glacier movement here is achieved by shear deformation of the ice, rather than by sliding over bedrock. This steeply sloping region probably corresponds to the zone of rapid convergence where ice from the glacier catchment basin is funnelled into the glacier channel. Upstream from this region, ice velocities would then be significantly lower, as suggested by the lower surface slopes. Downstream, the ice between A and B in Figure 4 (if it is grounded, as suggested earlier) is sliding over a bed lubricated by meltwater produced as a byproduct of work done within the zone of convergence. In this region, basal shear stresses are of order 10^{-1} bar, which is similar to those estimated for ice streams flowing from West Antarctica into the Ross Ice Shelf.

If this scenario is correct, the steeply sloping portion of Pine Island Glacier upstream of A is an equilibrium feature associated with ice convergence due to bedrock topography in the catchment area. However, equilibrium depends on the continued survival of the ice shelf, which maintains the grounding line at B. Without the ice shelf, the grounding line would retreat rapidly to A, and then more slowly further inland. Probably, this could only happen if climate were sufficiently warm to prevent reformation of an ice shelf, which would tend to run a ground near B and allow the grounding line to regain its present configuration. I stress that this is only one of many possible scenarios. Crabtree and Doake (1982) may have been correct in assigning the grounding line position to A in Figure 4; if they were, the glacier is probably retreating at the present time. Clearly, this is an area from which more data are urgently needed.

Summary and Recommendations

Early responses of the polar regions to climate warming would be an increase in area of summer melt on the ice sheets and ice shelves, and a decrease in sea-ice cover particularly during winter. For sufficiently large warming, later responses would include breakup of the ice shelves by increased melting and iceberg calving, and an increase in ice drainage rates, particularly from the marine based ice sheets. In attempting to make quantitative assessments of ice sheet response to climate warming it is important to consider seasonal changes in climate. In particular, an increase in summer temperatures sufficient to change a net accumulation zone into a net melting zone probably represents a critical turning point for ice shelf viability. Consequently, it is important that seasonal air temperatures are realistically simulated in climate models.

Actual responses of the polar ice sheets to a given warming cannot at present by reliably predicted, partly because we do not know whether they are currently growing or shrinking. However, we can identify areas of vulnerability. The main response of the Greenland ice sheet would be an increase in surface melting with probably only minor changes in ice drainage rates. Parts of the ice sheet in East Antarctica are marine based, resting on bedrock well below sea level, and these could collapse rapidly if warming became sufficiently pronounced to cause widespread surface melting, flooding of surface crevasses and "explosive" iceberg calving at the coast. Currently, mean annual air temperatures along this part of the East Antarctic coast are approximately -10°C with summer monthly means consistently below 0°C. It is not clear just what climatic conditions are required to promote explosive iceberg fragmentation, but an analogy may be provided by conditions at Jacobshavn Glacier in Greenland. Here, monthly mean temperatures are above 0°C for 5 months with a maximum value of 8°C; the mean annual temperature is -4°C. Thus, a significant summer warming, of about 10°C would be required along the East Antarctic coastline to initiate rapid explosive iceberg calving with the potential for subsequent massive ice retreat.

Most of the West Antarctic ice sheet is marine based and appears to be prevented from collapsing at present by the protective influence of the Ross and Ronne ice shelves. Collapse could be initiated without complete removal of these ice shelves. Much of their protective influence is due to the presence of grounded ice rises within the ice shelves; thinning by increased melting could unground these ice rises and permit major increases in drainage rates from inland. There is no evidence for present day thinning of these ice shelves. Most of the marine based exit channels from West Antarctica are linked to the Ross Sea via the Ross Ice Shelf; outlets to the Ronne ice Shelf are comparatively narrow, and the Ronne Ice Shelf is well pinned by many large ice rises. Thus, the Ross catchment area would probably be the first affected by climatic warming, and it has the greatest potential for rapid collapse along the many broad, deep drainage channels that link West Antarctica to the Ross Sea.

The two major glaciers draining the northern sector of West Antarctica (Pine Island and Thwaites glaciers) are not protected by large ice shelves. They may currently be retreating, but the few available data show no evidence for this. Estimated ice discharge from Pine Island Glacier is of the same order as total upstream snow accumulation. This apparent state of near equilibrium could be due to protection provided by the small, embayed ice shelf into which the glacier flows and by a bedrock sill 600-700m below sea level over which the glacier must flow to become afloat. More data are needed to confirm the existence of this sill. Removal of the small ice shelf would in any event cause the groundling line to retreat a considerable distance into the interior of West Antarctica. Continued ice shelf survival probably depends on summer air temperatures remaining sufficiently low to prevent extensive surface melting, crevasse flooding, and associated iceberg fragmentation. No data are available on the present climate in this area; installation of an automatic weather station in Pine Island Bay is strongly recommended.

It is important to note that, although Pine Island and Thwaites glaciers may be the most vulnerable of the West Antarctic drainage channels, they do not represent the greatest threat to world sea level. They are both about 30 km wide and appear to flow in well defined bedrock channels. Pine Island Glacier is currently moving at ~2 km yr^{-1} near its seaward ice front, where the ice is ~300 m thick. If it were to increase its activity to that of Jacobshavn Glacier, which moves at ~8 km yr^{-1} with ice thickness of ~600 m, then its ice front velocity would be ~16 km yr^{-1}. Even then, it would take more than 10^4 years for the West Antarctic ice sheet to collapse, corresponding to an average sea level rise of ~0.5mm yr^{-1}. Such an increase in ice front velocity should be easily detectable from satellite observations, and it would provide a warning of change. However, rapid sea level rise would probably require a similar increase in the activity of the large ice streams that flow into Ross Ice Shelf and this, in turn, would require removal of most of the ice shelf.

Mean annual temperatures near the seaward margins of the Ross Ice Shelf are currently ~-20°C, with summer mean monthly temperatures below -5°. Significant weakening of the ice shelf will probably require an increase in summer temperatures of at least 5°C, and probably 10°C to initiate widespread surface melting. However, I stress that we know little about processes occuring beneath the ice shelf, where comparatively minor changes in ocean heat content or current velocity may result in major increases in basal melting (MacAyeal, in press).

Finally, I shall briefly describe a program of measurements necessary to:

- identify ice sheet regions vulnerable to climate warming;
- assess the present state of equilibrium of these regions, and monitor trends;
- predict ice sheet behavior under prescribed conditions.

(1) Ice sheet elevations. From these, glacier catchment basins can be delineated and an assessment can be made of basal shear stresses, which are indicative of glacier activity. Repeated

accurate measurements will give an indication of ice thickening/thinning rates. Measurements should be made over most of the polar ice sheet surface. These data are best acquired from space, preferably from a free flying satellite, by radar or laser altimetry (Zwally et al., 1981, and 1983a).

(2) Ice-margin positions. Repeated measurements give an indication of marginal ice velocity and they help locate regions of anomalous activity. Again, the necessary data are best acquired from space using Landsat-type imagery or (preferably) imaging radar. Ice margin positions can also be mapped from satellite altimetry data (Thomas et al., in press (b)).

(3) Sea ice cover. Information on seasonal changes in sea ice cover is important both for climate modelling and for detecting climate change. Routine, long-term measurements are essential to determine natural variability in sea ice extent (Zwally et al., 1983b) and systematic trends indicative of changing climate (Parkinson and Bindschadler, This volume). These data are also obtained from satellites by measuring the microwave radiation from the ocean/ice surface. Observations have been made since 1973 (Zwally et al., in press) and they will continue at least into the next decade.

(4) Summer melt zones on the ice sheets. The size of summer melt zones on the ice sheets, and particularly on the ice shelves, is sensitively controlled by summer air temperatures. Long time series of these data will provide independent warning of climate change and of possible changes in ice shelf calving mechanisms. Necessary data are provided by the passive-microwave measurements discussed above.

(5) Ice thicknesses, velocities and strain rates. These measurements, together with measurements of snow-accumulation rates and other "conventional" glaciological studies, should be made in areas identified as vulnerable to climate change. There is a strong need for greater understanding of iceberg calving mechanisms, particularly the factors that determine whether icebergs are large and tabular or small and fragmented. Other major problem areas include mass and heat exchange beneath ice shelves, and the rapid sliding of glaciers and ice streams.

(6) Modelling studies. This is an area where results are strongly dependent on success of the work described above. For instance, it is difficult to model an ice stream if we do not understand how it slides over its bed. However, the physics that we do understand can and should be incorporated within ice sheet models that are efficient numerically, and sufficiently flexible to permit easy modification as our understanding of the physics improves. Moreover, useful results from the modelling can be obtained at an early stage by empirically incorporating processes that are poorly understood and conducting sensitivity tests to investigate model responses to specific perturbations. Models are also useful aids in designing field experiments to yield better focused research.

References

Andrews, J. T., A. Mears, G. H. Miller and D. R. Pheasant, Holocene late-glacial maximum and marine transgression in the eastern Canadian Arctic, Nature, Physical Science, 239, No. 96, 147-149, 1972.

Budd, W. F. and B. J. McInnes, Modelling surging glaciers and periodic surging of the Antarctic ice sheet, In: Climatic change and variability. A southern perspective, edited by Pittock, A. B., L. A. Frakes, D. Jenssen, J. A. Petersen and J. W. Zillman, Cambridge University Press, 228-233, 1978.

Crabtree, R. D. and C. S. M. Doake, Pine Island Glacier and its drainage basin: results from radio echo-sounding, Annals of Glaciology, 3, 65-70, 1982.

Gornitz, V., S. Lebedeff and J. Hansen, Global sea level trend in the past century, Science, 215, 1611-1614, 1982.

MacAyeal, D. R., Tides, tidally-driven baratropic circulation and the formation of tidal fronts below the Ross Ice Shelf, Antarctica, Symposium on ice and climate modelling, Evanston, Illinois, International Glaciological Society, 1983.

Meier, M. F. and A. Post, What are glacier surges? Canadian Journal of Earth Sciences, 6, No. 4, 807-817, 1969.

Meier, M. F., L. A. Rasmussen, A. Post, C. S. Brown, W. G. Sikonia, R. A. Bindschadler, L. R. Mayo and D. C. Trabant, Predicted timing and disintegration of the lower reach of Columbia glaciers, Annals of Glaciology, in press.

Mercer, J. H., Antarctic ice and Sangamon sea level, International Association of Scientific Hydrology. Publication No. 79, 217-225, 1968.

Parkinson, C. L. and R. A. Bindschadler, Response of Antarctic sea ice to uniform atmospheric temperature increases, This volume.

Stiles, W. H. and F. T. Ulaby, The active and passive microwave response to snow parameters, 1-wetness, Journal of Geophysical Research, 85, 1037-1044, 1980.

Stuiver, M., G. H. Denton, T. J. Hughes and J. L. Fastook, History of the marine ice sheet in West Antarctica during the last glaciation: a working hypothesis, In: The last great ice sheets, edited by G. H. Denton and T. Hughes, Wiley-Interscience, 319-436, 1981.

Thomas, R. H., The creep of ice shelves: theory, Journal of Glaciology, 12, 64, 45-53, 1973.

Thomas, R. H., Calving-bay dynamics and ice sheet retreat up the St. Lawrence Valley system, Geographie Physique Quaternaire, 31, No. 3-4, 347-356, 1977.

Thomas, R. H., Dynamics of marine ice sheets, Journal of Glaciology, 24, No. 90, 167-177, 1979.

Thomas, R. H. and C. R. Bentley, A model for Holocene retreat of the West Antarctic ice sheet, Quaternary Research, 10, No. 2, 150-170, 1978a.

Thomas, R. H. and C. R. Bentley, The equilibrium state of the eastern half of the Ross Ice Shelf, Journal of Glaciology, 20, No. 84, 509-518, 1978b.

Thomas, R. H., T. J. O. Sanderson and K. E. Rose, Effect of climatic warming on the West Antarctic ice sheet, Nature, 277, No. 5695, 355-358, 1979.

Thomas, R. H., D. R. MacAyeal, D. H. Eilers and D. R. Gaylord, Glaciological studies on the Ross Ice Shelf, Antarctica, 1973-1978, In: The Ross Ice Shelf: Glaciology and Geophysics, edited by C. R. Bentley and D. E. Hayes, Antarctic Research Series, 40, in press (a).

Thomas, R. H., T. V. Martin and H. J. Zwally, Mapping ice sheet margins from satellite altimetry data, Annals of Glaciology, in press (b).

Weertman, J., Deformation of floating ice shelves, Journal of Glaciology, 3, No. 21, 38-42, 1957.

Weertman, J., Can a water-filled crevasse reach the bottom surface of a glacier? International Association of Scientific Hydrology, Publication No. 95, 139-145, 1973.

Weertman, J., Stability of the junction of an ice sheet and an ice shelf, Journal of Glaciology, 13, No. 3, 3-11, 1974.

Weidick, A., Investigations of the Holocene deposits around Jakobshavns Isbrae, West Greenland. In: The periglacial environment, past and present, edited by T. L. Pewe, International Association for Quaternary Research, 7th Congress, Alaska, 1965; Montreal, Quebec, McGill-Queen's University Press, 249-262, 1969.

Whillans, I. M., The equation of continuity and its application to the ice sheet near "Byrd" Station, Antarctica, Journal of Glaciology, 18, No. 80, 359-371, 1977.

Zwally, H. J. and P. Gloersen, Passive microwave images of the polar regions and research applications, Polar Record, 18, No. 116, 431-450, 1977.

Zwally, H. J., R. H. Thomas and R. A. Bindschadler, Ice sheet dynamics by satellite laser altimetry, NASA Technical Memorandum, No. 82128, 1981.

Zwally, H. J., R. A. Bindschadler, A. C. Brenner, T. V. Martin, and R. H. Thomas, Surface elevation contours of Greenland and Antarctic ice sheets, Journal of Geophysical Research, 88, No. C3, 1589-1596, 1983a.

Zwally, H. J., C. L. Parkinson and J. Comiso, Variability of Antarctic sea ice and changes in carbon dioxide, Science, 220, No. 4601, 1005-1012, 1983b.

Zwally, H. J., J. C. Comiso, C. L. Parkinson, F. D. Carsey, W. J. Campbell and P. Gloersen, Antarctic sea ice cover 1973-1976 from satellite passive microwave observations, NASA Special Publication, In press.

WEST ANTARCTICA, THE SEA-LEVEL CONTROLLED MARINE INSTABILITY: PAST AND FUTURE

J. L. Fastook

Institute for Quaternary Studies, University of Maine, Orono, ME, 04469

Abstract. A world-wide coupled hydrosphere model, triggered by ablation-controlled terrestrial retreat along the landward margin of the northern hemisphere ice sheets, and amplified by the sea-level controlled marine instability, produces a sea-level curve which compares well with an oxygen isotope sea-level curve. Collapse of Antarctica from its maximum extent 18000 years ago to a configuration similar to its present state is modelled. Implications for further retreat of West Antarctica following the removal of buttressing ice shelf in the Amundsen Sea are also shown.

Modelling results suggest that the Amundsen Sea sector including Pine Island Bay can only be stabilized at its present position in the presence of a buttressing ice shelf. Since no ice shelf presently occupies Pine Island Bay, CO_2-induced climatic warming can create an ablation zone on the surfaces of Thwaites and Pine Island Glaciers, causing surface melting which triggers grounding-line retreat for these two ice streams. Surges accompanying retreat of the grounding lines downdraw the ice drainage basins of the two ice streams. In turn, downdraw causes the Marie Byrd Land ice divide to move into the catchment area of the ice streams draining into the Ross Ice Shelf. The reduced area of the ice drainage basins for the Ross Ice Shelf causes the forward ice velocity to fall behind the iceberg calving rate. Calving bays then carve away the Ross Ice Shelf, freeing them from their bedrock pinning points and allowing their grounding lines to migrate into Bentley Subglacial Trench, along with the grounding lines of Thwaites and Pine Island Glaciers.

Introduction

Concern about possible CO_2-induced warming (Kellogg, 1977) and its effect on the stability of marine West Antarctica (Mercer, 1978), an area which is postulated to have disintegrated more extensively during previous particularly warm interglacials (Mercer, 1968) and may be receding at the present (Hughes, 1972, 1973), has lead to an effort to model and understand the processes controlling the disintegration of marine-based ice sheets (Stuiver et al., 1981; Thomas and Bentley, 1978).

A marine-based ice sheet is one which, upon removal of the ice and consequent isostatic rebound, will leave a bed predominantly below sea level. A marine ice sheet's behavior is controlled largely by four major components: ice streams, ice shelves, calving bays, and ice divides.

The importance of ice streams to the dynamic behavior of a marine ice sheet is evident from an examination of West Antarctica, the one remaining marine ice sheet. Examination of surface contours shows that fully 90% of the catchment area of West Antarctica drains through only 19 major ice streams (Hughes, 1977). The importance of ice shelves is indicated by the fact that of these major ice streams 15 feed into either the Ross or the Filchner-Ronne ice shelves. Only 2, Pine Island glacier and Thwaites glacier on the Amundsen Sea, flow into open sea with no major buttressing ice shelf. West Antarctica has only very small ablation zones (mainly where katabatic winds are warmed and dried as they flow down concave ice stream surfaces) so that the predominant mechanism for the removal of ice is calving from ice shelves. It will be shown how the relative movement of the calving front and the grounding line interact to provide the major negative feedback required to contain the otherwise unstable marine ice sheet system. Whereas several external factors will be found to trigger the instabilities of marine ice sheets, an internal link between catchment basins on opposite sides of an ice divide will be discussed. This link involves movement of ice divides as one ice stream retreats or surges, thereby lowering the ice surface along its flowline and moving the ice divide into the catchment area of the opposite flowline.

The Marine Instability

Instability in a marine ice sheet is focused along the grounding line that separates the floating ice shelf margins from the grounded interior ice domes (Weertman, 1974). Figure 1 shows a schematic drawing of the region near the grounding line where ice comes afloat as it flows

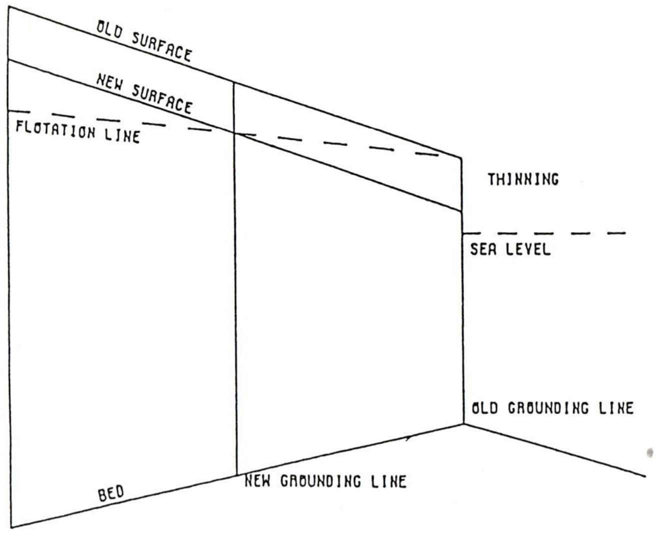

Fig. 1. Schematic diagram of the grounding line region of an ice stream showing the effect of thinning on the position of the grounding line.

into an ice shelf or into open sea. The grounding line is defined by the flotation criteria, which is that for a column of ice, the hydrostatic pressure at the bed due to the ice load be equal to the hydrostatic pressure of the column of water. With ρ_I and ρ_W the densities of ice and water respectively, g the acceleration of gravity, and h_I and h_W the thicknesses of the ice and the water column respectively

$$\rho_I \; g \; h_I = \rho_W \; g \; h_W \tag{1}$$

or

$$h_I = \frac{\rho_W}{\rho_I} h_W \tag{1a}$$

at the grounding line. In Figure 1, the surface defined by this flotation criteria is shown as a dashed line mirroring the bed as the depth increases inland. This condition of the bedrock sloping downward is often encountered due to a combination of differential erosion along ice streams as the bed coupling progressively decreases from the head of the ice stream to the grounding line, and the increased isostatic depression due to a greater ice load as one moves upstream to greater ice thicknesses. Should some mechanism exist whereby thinning of the ice at the grounding line occurs, it produces the solid line labeled "new surface". One can easily see that at the old grounding line the flotation criteria no longer holds. Indeed the ice, now being thinner, lifts off the bed until the new surface and the flotation surface intesect at the point marked "new grounding line". One can also see from Figure 1 how an increase in sea level affects the position of the grounding line. Increasing sea level raises the flotation surface and the grounding line retreats until the old surface and the new flotation surface intersect. Expressing the ice surface in terms of the slope at the grounding line, α, and the flotation surface in terms of the bedrock slope, β, and solving for their intersection one obtains the following expression

$$v_X = \frac{\frac{\rho_W}{\rho_I} v_Z + \dot{h}_I}{\alpha - \beta(1-\rho_W/\rho_I)} \tag{2}$$

where $\dot{h}_I$ is the rate of thinning at the grounding line, v_Z the rate of sea level increase, and v_X the rate of grounding line retreat. The coordinate system is taken with x along the flowline with x=0 at the grounding line. The z direction is positive in the vertical direction with z=0 at present day sea level. The y direction is perpendicular to the flow direction. The rate of thinning at the grounding line, $\dot{h}_I$, can be expressed as

$$\dot{h}_I = u_X \; (\alpha - \beta) - h_I \dot{\varepsilon}_{ZZ} - \dot{a} \tag{3}$$

where u_X is the ice velocity, $\dot{\varepsilon}_{ZZ}$ is the strain rate, and $\dot{a}$ the accumulation rate, all at the grounding line. The first term represents the advection of ice moving as a block across the grounding line. The second term, representing creep thinning, can be expressed in terms of a flow law (Glen, 1955) such that

$$\dot{\varepsilon}_{ZZ} = -\psi \; A^{-n} \left(1/2\rho_I \; g \; h_I[1 - \rho_I/\rho_W] - \frac{2 \; s\tau_S}{w}\right)^n \tag{4}$$

where A and n are flow law parameters, the first term in the brackets represents a thinning stress due to the hydrostatic head of the ice above sea level (Weertman, 1957), and the second term represents the compressive back stress produced by a shear stress τ_S along the two sides of an ice tongue of length s and width w as it pushes into the buttressing ice shelf (Thomas and Bentley, 1978). Here ψ represents the strain-rate field such that

$$\psi = \left(1 + [\frac{\dot{\varepsilon}_{yy}}{\dot{\varepsilon}_{xx}}] + [\frac{\dot{\varepsilon}_{xy}}{\dot{\varepsilon}_{xx}}]^2 + [\frac{\dot{\varepsilon}_{xy}}{\dot{\varepsilon}_{xx}}]^2\right)^{\frac{n-1}{2}}$$

$$\left(1 + [\frac{\varepsilon_{yy}}{\varepsilon_{xx}}]\right) \; / \; \left(2 + [\frac{\varepsilon_{yy}}{\varepsilon_{xx}}]\right)^n \quad . \tag{5}$$

The value of ψ depends primarily on the degree of divergence with extreme values of

$$\frac{2}{3^{(\frac{n+1}{2})}}$$

for no ice shelf ($\tau_s = \dot{\varepsilon}_{xy} = 0$) and maximum divergence ($\dot{\varepsilon}_{xx} = \dot{\varepsilon}_{yy}$), and 2^{-n} for parallel flow ($\tau_s = 0$, $\dot{\varepsilon}_{xy} = 0$ along the centerline, and $\dot{\varepsilon}_{yy} = 0$).

It should be noted that u_x, the ice velocity across the grounding line, is taken in this model to be a function of the thinning rate $\dot{h}_I$, as this velocity is assumed to reflect the mass balance changes along the entire flowband. Integrating the continuity equation from margin to dome and approximating by a summation one obtains.

$$u_x = \frac{1}{(h_I \; w)_{i=0}} \sum_{i=0}^{i=\frac{L}{\Delta x}} (\dot{a} - \dot{h}_I)_i \; w_i \; \Delta x \quad (6)$$

where any bracket term with a subscript i implies that all terms within the bracket are evaluated at step i. This convention will be followed throughout this paper. The thinning rates at each step i are obtained by comparing successive profiles from an ice sheet reconstruction scheme which differ only in their grounding line thicknesses (note that the method for obtaining profiles is secondary here as only differences between profiles are necessary for the mass balance calculations). This reconstruction scheme is based on the CLIMAP reconstruction scheme (Hughes, 1981) and will be discussed at length in the following section.

Equation (3) is solved iteratively by putting all terms on the right-hand side and systematically searching for zeros of the equation by varying the thinning rate $\dot{h}_I$. Having obtained a thinning rate, one can then use equation (2) to obtain a retreat rate, v_x. From this a time to retreat one Δx step is obtained, sea level (if it is rising) is adjusted, length of the shear zone, s, is changed to reflect assumed calving rates, and the process repeats.

The Reconstruction Scheme

The reconstruction scheme used in this model involves separating the effects of isostacy and bed topography from the effects of bed coupling along the flowline. Essentially a finite difference recursion scheme beginning at the margin, each successive ice elevation inland h_{i+1} is given in terms of h_i by

$$h_{i+1} = h_i + \left(\frac{\tau_o}{h - h_R}\right)_i \frac{\Delta x}{\rho_I \; g} \quad (7)$$

where h is the ice surface elevation, h_R is the bedrock elevation, Δx is the finite difference step, ρ_I is the density of ice, and g is the acceleration of gravity. τ_o is the basal shear stress to be discussed later. While this expression neglects the effects of the longitudinal stress gradients and transverse stresses it is a valid approximation for distances greater than about 4 ice thicknesses (Budd, 1971). In this exercise the finite difference step is 100 km. Isostatic depression of the bed and modification of the ice surface proceed as in Weertman (1964) yielding a similar expression in terms of depressed surface, h* and the depressed bed, h_R*

$$h^*_{i+1} = h^*_i + \left(\frac{\tau_o}{h^* - h^*_R}\right)_i \frac{\Delta x}{\rho_I \; g} \quad (8)$$

where h* and h_R* are related to the undepressed ice surface and bed elevation by

$$h^* = h_o + \frac{(h - h_o)}{(1+r)^{1/2}} \quad (9)$$

and

$$h^*_R = h_o + (1+r)^{1/2}(h_R - h_o) - r(h^* - h_o) \quad (10)$$

where h_o is the ice elevation at the grounding line referenced to present day sea level, and r is a measure of the degree of isostatic depression. For complete isostatic equilibrium

$$r = \frac{\rho_I}{\rho_R - \rho_I} = \frac{1}{3} \quad (11)$$

where ρ_R is the density of mantle rock. Incomplete isostatic adjustment will require values for r less than 1/3. In calculating values of r for incomplete isostatic adjustment an exponential rebound scheme with a time constant of 3000 years is used.

Having removed the influence of bed topography h_R and isostacy r in equation (8), variations in τ_o can only reflect variations in ice-bed coupling. The variations are related to variations in ice velocity, which increases for surface accumulation and downslope convergence, and decreases for surface ablation and downslope divergence. By removing the contributions of topography and isostacy τ_o reflects ice-bed coupling on a flat horizontal bed.

Derivations of basal shear stress variations along surface flowlines of an ice sheet begin with the continuity equation. For three-dimensional flow on a flat horizontal bed and the continuity equation is:

$$\int \dot{a} w dx - \int \frac{\partial h}{\partial t} w dx + \int d(uwh) = 0 \quad (12)$$

where $\dot{a}$ is the ice accumulation/ablation rate over a flowband of width w and height h above the ice sheet margin, $(\partial h/\partial t)$ is the change of h in time t when the ice sheet is not in equilibrium, and u is the mean horizontal ice velocity in the flowband, all measured a distance x from the ice margin. Neglecting the $(\partial h/\partial t)$ term (justifiable as long as $(\partial h/\partial t) \ll \dot{a}$) the continuity equation reduces to:

$$\int_L^x \dot{a} w dx \simeq - \int_o^{uwh} d(uwh) \quad (13)$$

where L is the flowline length.

Flow velocity is the sum of the internal creep velocity u_c and the basal sliding velocity u_s. Creep velocity can be computed approximately from the flow law for ice for simple shear (Paterson, 1969, p 89; Glen, 1955):

$$\dot{\epsilon}_{xz} = \left(\frac{\sigma_{xz}}{A}\right)^n \qquad (14)$$

where $\dot{\epsilon}_{xz} = (1/2)\,(du_c/dz)$ is the internal shear strain rate in ice of thickness h, $\sigma_{xz} = \rho_I g(h-z)\alpha$ is the internal shear stress for an ice surface slope α, A is an ice hardness coefficient, and n is an ice viscoplastic exponent.

Sliding velocity u_s can be approximated from the sliding law of ice on a rough bed (Weertman, 1957):

$$u_s = \left(\frac{\tau_o}{B}\right)^m = \left(\frac{\tau_o}{B_o C_o^2}\right)^m \qquad (15)$$

where $\tau_o = \rho_I g h \alpha$ is the basal shear stress, m is an ice viscoplastic exponent, $B=B_o C_o^2$ is an effective sliding constant, B_o is a sliding law constant which depends on the flow law constant of ice at the melting point, the heat of melting for ice, and the coefficient of thermal conductivity for bedrock. $C_o = \Lambda/\Lambda'$ is a parameterization of the bed roughness, where Λ is the controlling obstacle size and Λ' is their spacing. Equation (13) can be integrated using equations (14) and (15):

$$uwh = \left[\frac{2}{n+2}\left(\frac{\rho_I g\alpha}{A}\right)^n wh^{n+2} + \left(\frac{\rho_I g\alpha}{B}\right)^m wh^{m+1}\right] \qquad (16)$$

Ice presumably cannot slide over a frozen bed unless shear rupture occurs when the bed becomes thawed because a basal shear stress cannot be transmitted across a water layer, so $u_s>0$. Furthermore $u_s \gg u_c$ when the bed is thawed because the lowered bed traction reduces the amount of internal shear in the ice. A frozen bed is dry, and equating equations (13) and (16) for continuity when $u_s = 0$ and defining $\Gamma(x)$ in terms of the left hand side of equation (13)

$$\Gamma(x) = \int_x^L \dot{a}(x)w(x)dx \qquad (17)$$

then

$$\Gamma(x) = \frac{2}{n+2}\left(\frac{\rho_I g\alpha}{A}\right)^n wh^{n+2} \qquad (18)$$

The surface slope at a distance x from the margin is:

$$\alpha = \frac{dh}{dx} = \frac{A}{\rho_I g}\left[\frac{(n+2)\,\Gamma(x)}{2w(x)\,h^{n+2}}\right]^{1/n} \qquad (19)$$

where w=w(x) is specified at distance x. Integrating equation (19) for a flat horizontal bed yields:

$$h = \left[h_I^{\frac{2n+2}{2}} + \left(\frac{2n+2}{n}\right) C_D\, \Pi_D(x)\right]^{\frac{n}{2n+2}} \qquad (20)$$

where h_I is the ice thickness at the margin (h_I = 0 for terrestrial margins, h_I = the grounding line ice thickness for marine margins) and C_D is a frozen bed constant defined by:

$$C_D = \left(\frac{n+2}{2}\right)^{1/n} \frac{A}{\rho_I g} \qquad (21)$$

and $\Pi_D(x)$ is a frozen bed function defined by:

$$\Pi_D(x) = \int_o^x \left[\frac{\Gamma(x)}{w(x)}\right]^{1/n} dx \qquad (22)$$

The basal shear stress distribution for a frozen bed is obtained by combining equations (19) and (20):

$$\tau_w=\tau_o=\rho_I\, g\, h\, \alpha = \frac{\rho_I\, g\, C_D \left[\frac{\Gamma(x)}{w(x)}\right]^{1/n}}{\left[h_I^{\frac{2n+2}{n}} + \left(\frac{2n+2}{n}\right) C_D\, \Pi_D(x)\right]^{\frac{1}{n+1}}} \qquad (23)$$

For a thawed bed $u_s \gg u_c$ so that equating equations (13) and (16) yields:

$$\Gamma(x) = \left(\frac{\rho_I g\alpha}{B}\right)^m wh^{m+1} \qquad (24)$$

The surface slope at distance x from the margin is:

$$\alpha = \frac{dh}{dx} = \frac{B}{\rho_I g}\left[\frac{\Gamma(x)}{w(x)h^{m+1}}\right]^{1/m} \qquad (25)$$

Integrating equation (25) for a flat horizontal bed yields:

$$h = \left[h_I^{\frac{2m+1}{m}} + \left(\frac{2m+1}{m}\right) C_w\, \Pi_w(x)\right]^{\frac{m}{2m+1}} \qquad (26)$$

where C_w is a thawed bed constant defined by:

$$C_w = \frac{B}{\rho_I g} \qquad (27)$$

and $\Pi_w(x)$ is a wet bed function defined by:

$$\Pi_w(x) = \int_o^x \left[\frac{\Gamma(x)}{w(x)}\right]^{1/m} dx \qquad (28)$$

The basal shear stress distribution for a thawed

bed is obtained by combining equations (25) and (26):

$$\tau_w = \tau_o = \rho_I g h \alpha$$

$$= \frac{\rho_I g C_w [\frac{\Gamma(x)}{w(x)}]^{1/m}}{[h_I^{\frac{2m+1}{m}} + (\frac{2m+1}{m}) C_w \Pi_w(x)]^{1/2m+1}} \tag{29}$$

In an ice stream the basal shear stress approaches zero as the ice approaches the grounding line. This is presumed to reflect an increasing basal water layer that progressively drowns the controlling obstacles defined in Weertman's theory of basal sliding (Weertman, 1957). Substitution of equation (29) into equation (15) and evaluation of u_s at x=0 (where $\tau_w = 0$ due to the drowning of the controlling obstacles) yields:

$$u_s = \frac{\Gamma(o)}{w(o)\ h_I} \tag{30}$$

Recalling that $\Gamma(x)$ is the integrated mass flux at x, this is just the mass balance velocity.

It should be noted that with the use of a constant accumulation and a constant width that equations (23) and (29) reduce to Hughes' equations (41) and (46) (Hughes, 1981). The use of equations (23) and (29) allows a more realistic treatment by allowing both variable widths and accumulations to be included by a simple numerical integration of equations (18), (22), and (28).

The actual basal shear stress distribution is a combination of the frozen and thawed bed shear stress distributions, weighted to reflect the differing basal conditions:

$$\tau_o = F\ \tau_D + (1-F)\ \tau_w \tag{31}$$

where F is the fraction of the bed which is frozen. This linearization neglects the possibility of more complex interactions through longitudinal and transverse components of the stress tensor. This is a reasonable assumption for distances greater than about 4 ice thicknesses. In principle F can be obtained for paleo-ice sheets from the erosion-deposition characteristics along a flowline. In this modelling exercise a linear variation from a completely frozen bed at the ice-sheet dome to a completely thawed bed at the head of the ice stream is assumed throughout.

A criticism of this approach that might be raised is that the use of equilibrium profiles to obtain the thinning rate is an oversimplification. In justifying this approach there are two major factors. First, the model is calibrated by fitting calculated profiles to existing profiles in Greenland and Antarctica. In this way existing departures from equilibrium (admittedly not as drastic as presumably occurred during grounding line retreat) are built into the model through the flow and sliding law parameters. In principle an iterative process could be used by including the $\partial h/\partial t$ term in equation (17) so that the non-equilibrium state would be expressed through $\Gamma(x)$. Since this would have the effect of modifying the accumulation curve, which in itself is a major unknown for the time period of interest, this was considered to be an unnecessary complication. The second factor involving the use of equilibrium profiles is that their use will produce an upper bound on the magnitude of u, the ice velocity at the grounding line. Assumptions concerning the time response of the Antarctic will tend to reduce the magnitude of this velocity. One can note from equation (3) that the effect of lowering the grounding line velocity is to reduce the thickening advection term, and hence increase the thinning rate at the grounding line, thereby also increasing the retreat rate of the grounding line. Hence a model of this sort will be slightly more stable than one which minimizes the response of the ice sheet far from the grounding line.

Running the Model

Initially the model is taken to be in equilibrium conditions (that is the thinning rate of equation (3) is equal to zero). In the computer solution τ_s and s are specified, h_I is obtained from the grounding line bedrock depth adjusted for appropriate sea-level conditions, and equation (3) with $\dot{h}_I = 0$ is solved for A, the ice hardness parameter at the grounding line. While this is somewhat arbitrary, a wide range of A is known to exist since A is an exponentially activated function of temperature, as well as a function of ice fabric and strain history.

Having balanced equation (3), a perturbation to the metastable initial condition is applied, a thinning rate obtained, and consequent retreat mapped out as a function of time. Many processes can contribute to the unbalancing of equation (3), but we will confine ourselves here to rising sea level, to CO_2-induced effects, and to the interaction of flowlines through their changing catchment areas.

Rising sea level's primary effect is to induce retreat of the grounding line by equation (2) even in the absence of thinning at the grounding line. If the bed depth increases with distance inland, as is frequently the case, this will produce greater ice thicknesses at the grounding line and a consequent increase in the creep thinning term of equation (3) due to the increase hydrostatic head in the first term of equation (4). This, in the absence of restraining effects to be discussed later, leads to accelerating retreat as long as the bed slope is downward inland.

Earlier models of this sort (Thomas and Bentley, 1978, Hughes, 1981) obtained the rise in sea

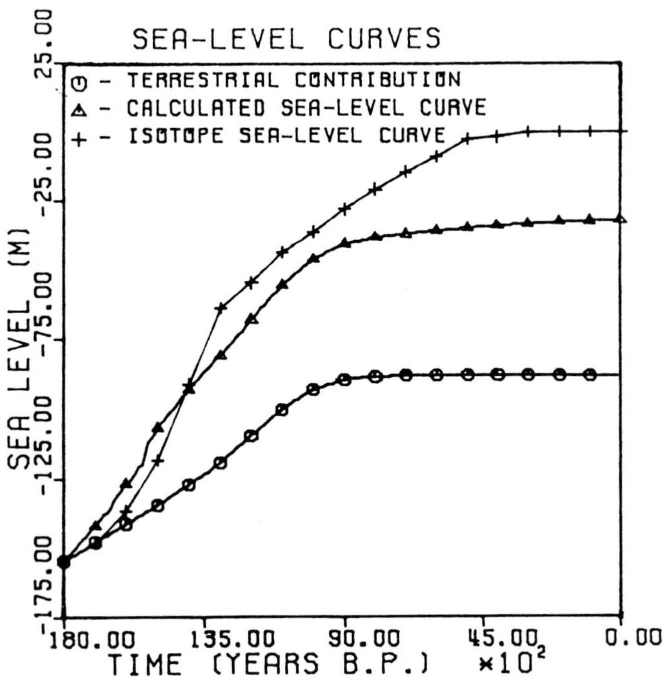

Fig. 2. Model generated sea-level curves compared with an oxygen isotope sea-level curve. Circles show the contribution of northern hemisphere terrestrial retreat, triangles show the combination of terrestrial and marine retreat, and crosses show the isotope sea-level curve.

level from independent sea-level curves obtained from isotopic analysis of ocean sediment cores. This model differs in that the sea-level curve is internally generated. This was done in an attempt to delineate a possible mechanism linking the derived solar insolation in the past (Milankovich, 1941; Vernakar, 1972) with observed isotopic sea level (Ninkovich and Shackleton, 1975) and to explain the good degree of correlation between the solar insolation at mid-northern hemisphere latitudes and the total cryosphere volume (Denton and Hughes, 1983).

The mechanism whereby the world cryosphere responds in concert to a localized insolation signal is postulated as follows. Climatically sensitive terrestrial margins (predominantly on northern hemisphere ice sheets) melt and retreat in response to solar insolation variations. With the assumption of no major long-term reservoir effects in local groundwater flow, this melting increases world sea level, thereby carrying the message of margin retreat to the marine margins of the northern sectors of the northern hemisphere ice sheets as well as to the marine-margined Antarctic Ice Sheet. As these marine margins respond, they too contribute to the rising sea level, thus initiating a positive feedback mechanism that greatly amplifies the effect of the local insolation variations in the critical terrestrial margin latitude band.

This was carried out by coupling an ablation-controlled terrestrial retreat model with the sea-level controlled marine retreat model. Output from the terrestrial retreat model operating on the southern margins of the Laurentide and Eurasian Ice Sheets included approximate meltwater volumes which was assumed to contribute directly to rising sea level. This sea-level message triggered the marine instability along the marine margins of the Laurentide, the Eurasian, and the Antarctic Ice Sheets, producing worldwide retreat of marine margins. The changing ice volume necessary to provide this degree of marine margin retreat was also added to the sea-level message providing a positive feedback mechanism whereby marine margin retreat was greatly amplified. In figure 2 open circles represent the sea-level curve produced by the retreat of the terrestrial margins. Shown by triangles is the calculated sea-level curve obtained by adding the terrestrial retreat component to that due to the retreat of the marine margin. For comparison the crosses show a representative sea-level curve obtained from isotopic data and used in the original CLIMAP experiment (Matthews, reported in Stuiver et al., 1981). While the derived sea-level curve falls short of the actual sea-level curve near the present, this is due to the fact that marine-margin retreat terminates as the ice margin comes above sea level. At this point terrestrial margin ablation mechanisms should take over to continue retreat until the ice sheet disappears. Further details of this modelling exercise will be reported elsewhere.

Climatological effects concentrate in two main mechanisms, the mass balance and the buttressing ice shelf. The affect on the mass balance of an increase in ablation (or decrease in accumulation) over a large portion of the catchment area is to lower the outlet velocity, u_x, thereby lowering the advection term of equation (3). In addition, changes in accumulation-ablation rate at the grounding line can perturb the metastable equilibrium through the third term of equation (3) leading directly to increased thinning. It should be noted that the time response to mass balance changes that affect the flowline as a whole will in general be slower than the response to local accumulation-ablation changes at the grouding line. One should also note that the $\dot{a}$ term of equation (3) represents both surface and bottom accumulation-ablation rates at the grounding line, and consequently could be a response to changing sea water temperature or circulation as well as to local precipitation effects.

A secondary mass balance effect involves the migration of ice divides and the encroachment of one flowband's catchment area into another's. As an ice stream retreats it will lower its ice surface all along the flowband. This lowering forces the ice divide into the catchment area of the flowband opposite the ice divide. If a flowband in equilibrium experiences a reduction in its catchment area and consequent reduction of u_x and the positive advection term of equation

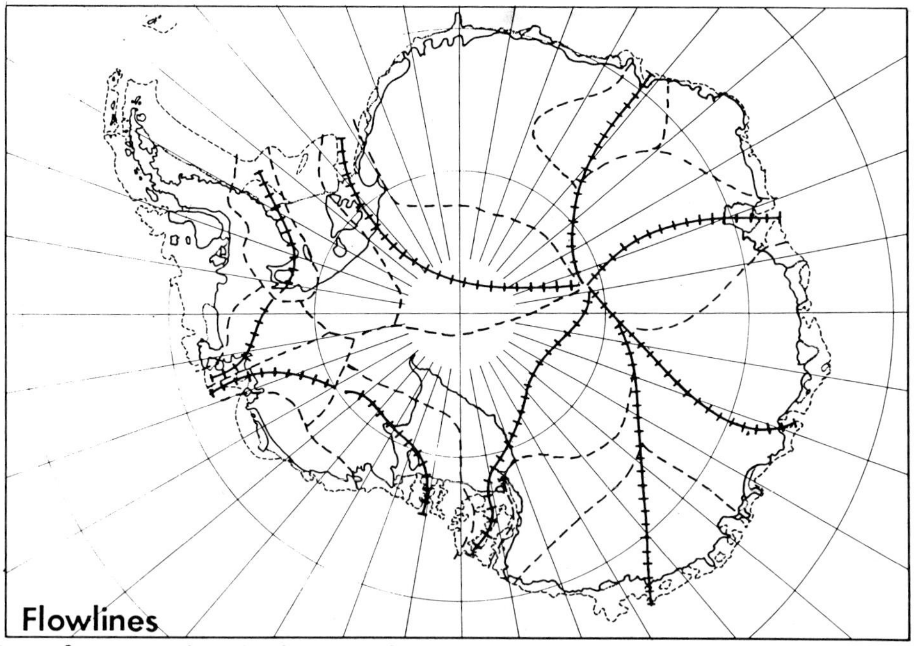

Fig. 3. Map of Antarctica indicating flowlines used in this model of retreat. Flowlines are shown by heavy solid lines with cross hatching at 100 km finite difference steps. Boundaries of the catchment areas for the various flowlines are shown by the heavy dashed lines.

(3), it leads to thinning at and generally retreat of the grounding line. Adjacent flowbands can affect each other in a similar fashion as movement of the grounding line lowers the surface and diverts flow into the retreating flowband.

The buttressing ice shelf, represented by the compressive $2s\tau_s/w$ term of equation (4), has been mentioned as the major negative feedback mechanism controlling retreat. If the calving rate lags behind the retreat rate, as might be the case during vigorous retreat when the ice velocity across the grounding line is high, the effective shear zone length, s, increases, thereby increasing the compressive back stress and reducing the overall creep thinning term. Conversely, an increased calving rate due to climatological factors leads to reduced back stress and increased creep thinning with consequent grounding line retreat. The retreat of the grounding line is controlled by varying the quantity $s\tau_s$. In the following section calving front positions will be shown based on the assumption that τ_s has a value of 1 bar. This is only a crude approximation of the true form of the back stress exerted by the buttressing ice shelf. A more realistic treatment would need to include the effect of ice rises and would result in a smaller variation in the extent of the buttressing ice shelf.

An additional climatological effect is the possible softening of ice either in the shear zone along the buttressing ice shelf (thereby reducing τ_s) or at the grounding line (thereby reducing A). Both possibilities lead to an increased creep thinning term.

Results

Figure 3 shows the flowlines used to reconstruct this retreat history of Antarctic. Flowlines are shown by heavy solid lines with cross hatches at the finite-difference steps. Boundaries of the catchment basins for the various flowlines are indicated by heavy dashed lines. Bed topography was obtained from the General Bathymetric Chart of the Oceans, published by the Canadian Hydrographic Service (Johnson et al., 1980). Accumulations along the flowlines were obtained from an empirical formula (Oerlemans, 1982) which parameterized the accumulation rate in terms of the surface slope, the elevation, and the distance from the ocean.

Figure 4 shows the initial configuration taken to represent the maximum extent of the Antarctic ice sheet at 18000 B.P. The solid outer contour is the maximum grounding line position, taken in general to coincide with the 500 m bathymetry contour. This 500 m contour is shown by the light dashed line. Shown by the heavy hatched line is the assumed position of the calving front. Heavy dashed lines indicate surface elevations at 500 meter contouring intervals.

Figures 5, 6, 7, 8 show surface elevation contours and calving front positions at 16000, 14000, 12000, and 10000 B.P. respectively. These retreat figures have been fitted to geological data (Stuiver et al., 1981) where available and the retreat rate has been controlled by varying the extent of the buttressing ice shelf shown on these figures. Figure 8, representing 10000 B.P., shows no solid outer contour as the

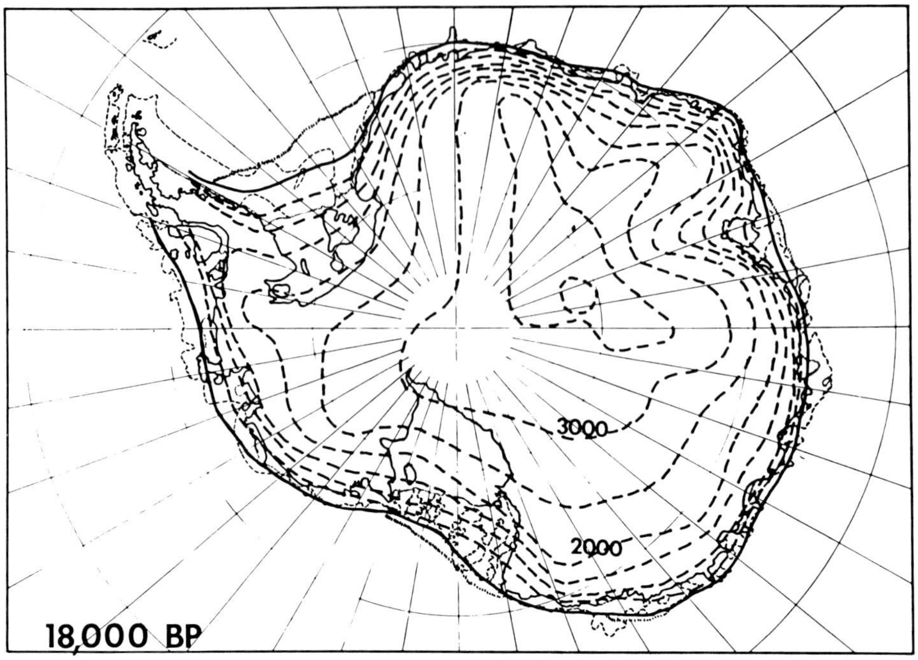

Fig. 4. Map of Antarctica showing reconstructed surface elevation contours at 18000 BP used as the initial configuration for the retreat model. The grounding line, shown by the heavy solid line, is taken to coincide in general with the 500 m bathymetry contour, shown by the light dashed line. The initial ice shelf, primarily in the Ross and Weddell embayments, is shown by a hatched line. Surface elevation contours at 500 m intervals are shown by heavy dashed lines.

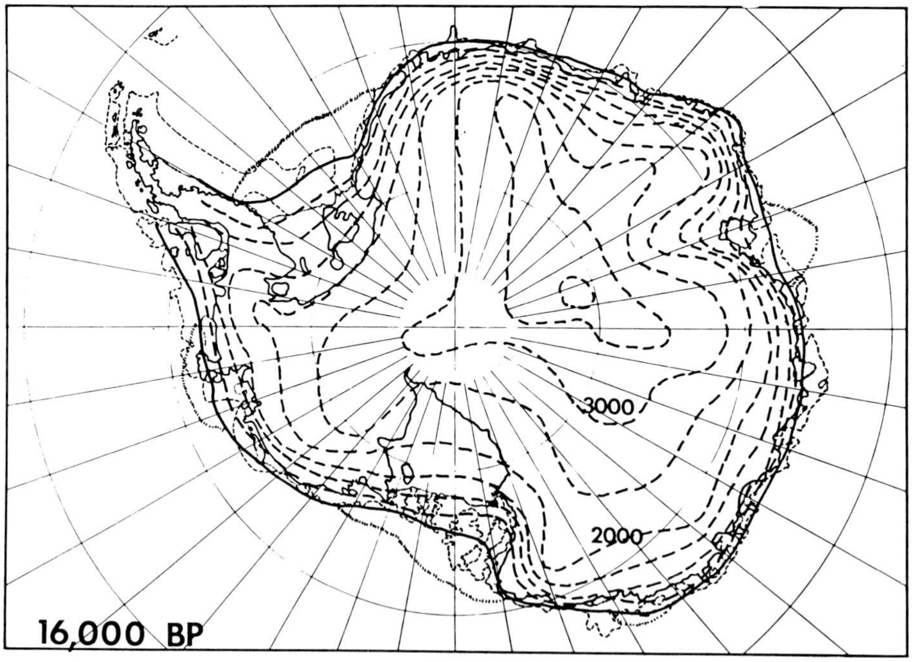

Fig. 5. Surface elevation contours at 16000 BP. Note the increased extent of buttressing ice shelves in the Ross Sea, the Weddell Sea, the Amundsen Sea, and at the outlet of Lambert Glacier.

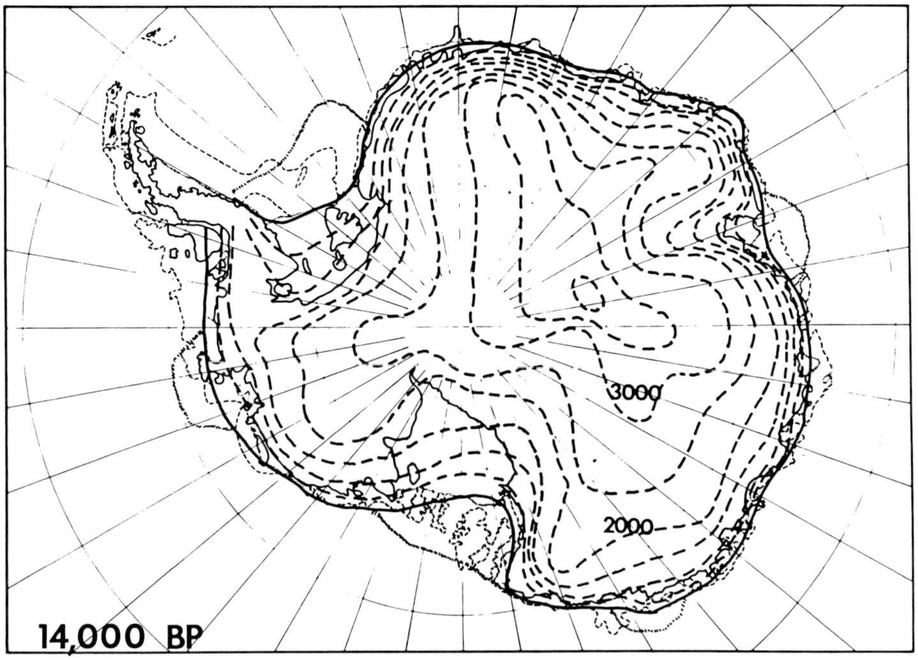

Fig. 6. Surface elevation contours at 14000 BP.

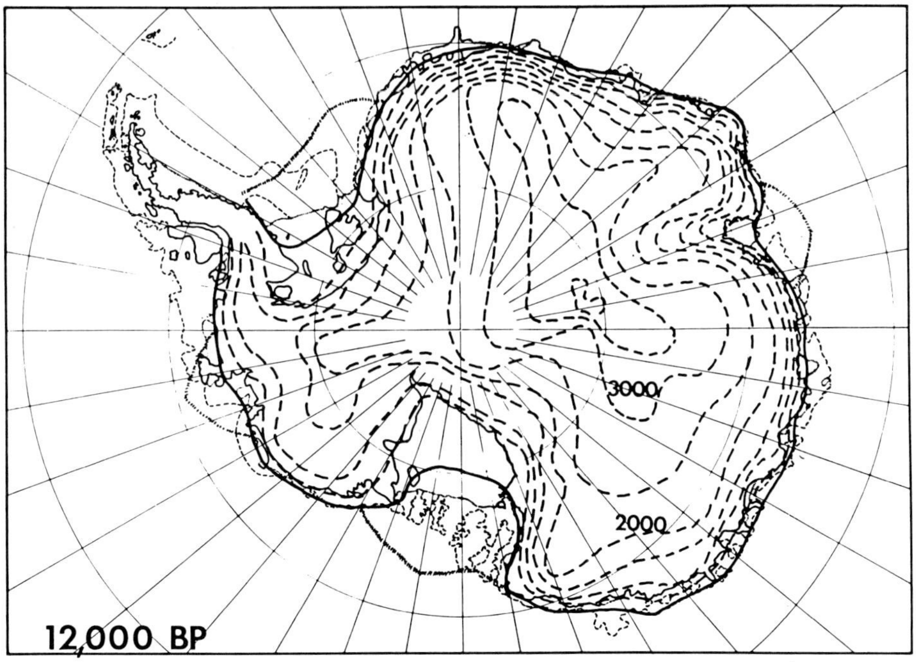

Fig. 7. Surface elevation contours at 12000 BP.

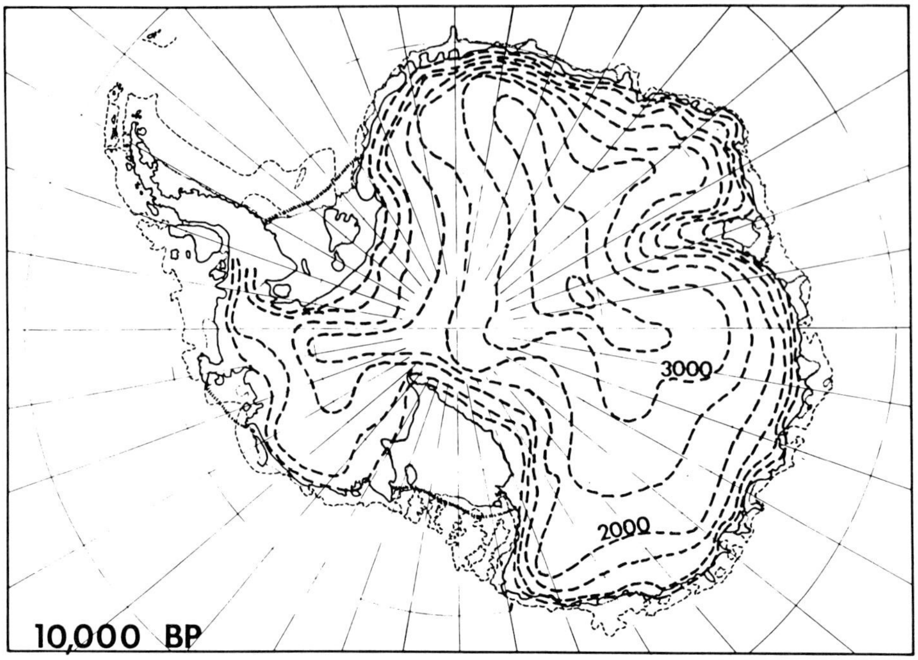

Fig. 8. Surface elevation contours at 10000 BP. The grounding line, now coincident with the present grounding line, is no longer shown. The extent of the buttressing ice shelf in general agrees well with the current extent of the ice shelves, shown by the light hatched line. Note that in the Amundsen Sea a significant ice shelf is necessary to stabilize the grounding line at its current position.

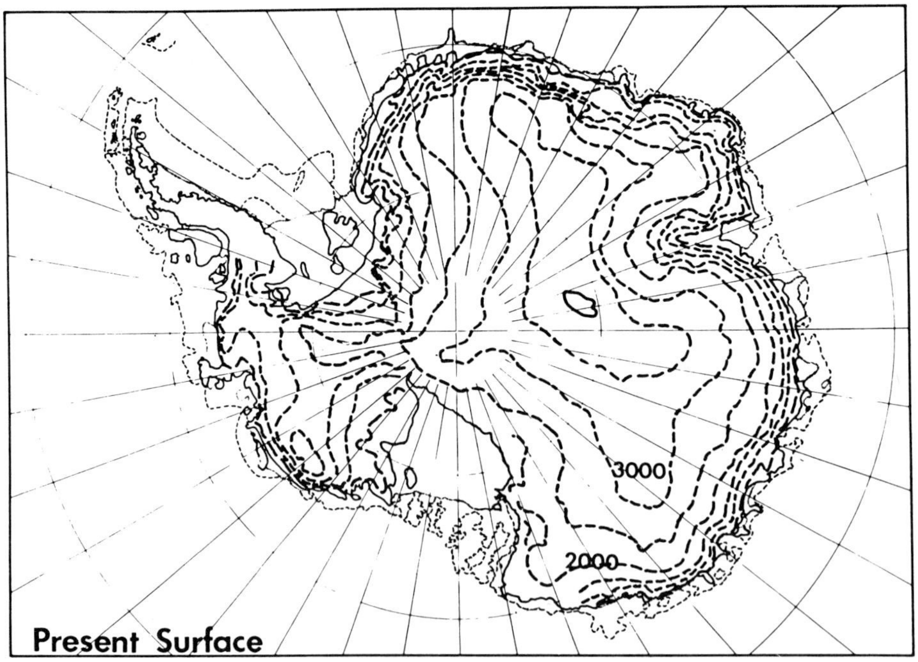

Fig. 9. Surface elevation contours as they exist at the present.

grounding line position coincides with the present configuration. Figure 9 shows the present surface of Antarctica for comparison (Johnson et al., 1980).

Notable in these figures is the relatively stable configuration of East Antarctica. With the exception of the Lambert Glacier in East Antarctica very little movement of the grounding line is observed. This is in general due to high sills surrounding East Antarctica which prevent the marine mechanism from propagating inland.

West Antarctica, on the other hand, shows drastic retreat of the grounding line, particularly in the Ross and Ronne embayments. Shown in Figure 8 are the ice shelf extents necessary to halt retreat at the present configuration. In both the Ross and Ronne embayments generally good agreement between the predicted ice shelf extent (shown by a light hatched line) is observed. A notable exception to this good agreement occurs in the Amundsen Sea sector where an ice shelf of nearly 350 km is needed to stop retreat of Pine Island and Thwaits Glaciers. An ice shelf of this size would extend from Thurston Island on the west to Siple Island on the east, practically filling the Amundsen Sea out to the continental shelf. A possible explanation for this discrepancy is that until recently this ice shelf did exist, and that its disappearance could be the first step in a second stage of West Antarctic retreat that could greatly reduce the extent of West Antarctica. Supporting evidence for this hypothesis can be seen in sediment records from cores taken near the edge of the continental shelf in the Amundsen Sea (Kellogg et al., 1982). The sparce diatom assemblages found in these cores indicates ice shelf coverage of these areas as recently as 1000 years ago, in general agreement with the model prediction.

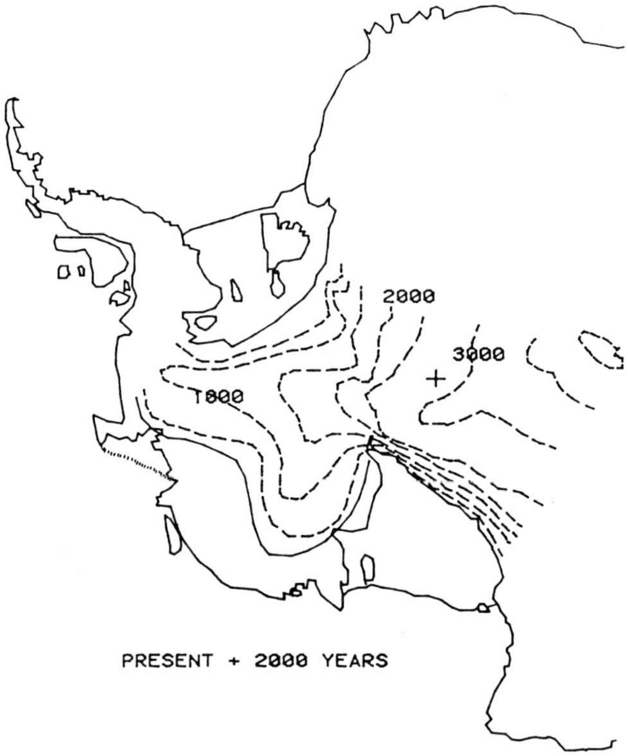

Fig. 11. Surface elevation contours 2000 years after removal of the Amundsen Ice Shelf. The moving ice divide between the Amundsen Sea and the Ross Sea embayments has destabilized those ice streams flowing into the Ross Ice Shelf and they too begin to retreat.

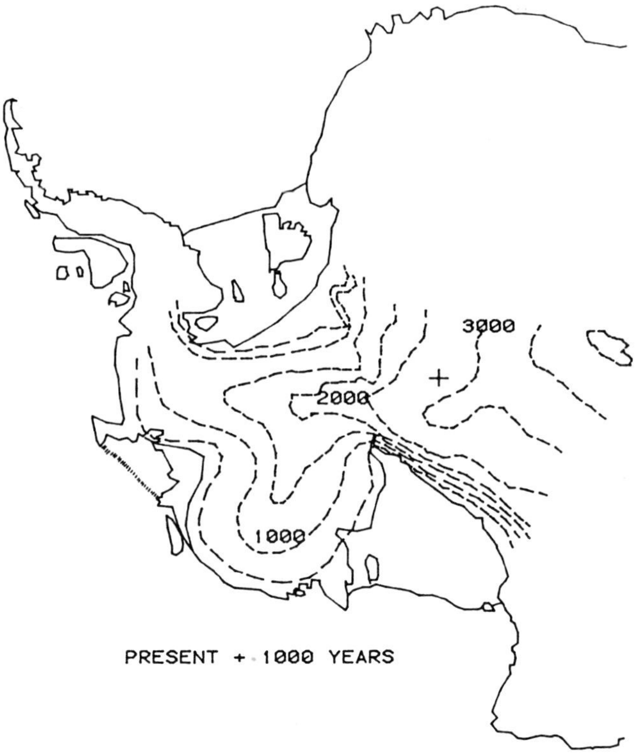

Fig. 10. Map of West Antarctica showing surface elevation contours 1000 years after removing the hypothetical Amundsen Ice Shelf. A new ice shelf forms in essentially the same position but the grounding line's movement into the deep Bentley sub-glacial trench causes retreat to continue.

Figures 10, 11, 12 and 13 show the results of a modelling experiment where the shelf required to hold the grounding line at its present position for Pine Island and Thwaite's glacier is removed. Grounding line positions are shown at 1000, 2000, 4000, and 6000 years after removal of the ice shelf. A conservative estimate of the extent of the buttressing ice shelf produced by the accelerated discharge of ice is shown on these figures by a hatched line. It should be stressed that should this ice shelf not reform, retreat into the interior of West Antarctica would proceed far more rapidly than is indicated by these figures. Modelling exercises with no ice shelf forming retreat far into the Bentley subglacial trench in hundreds of years. An ancillary affect of this

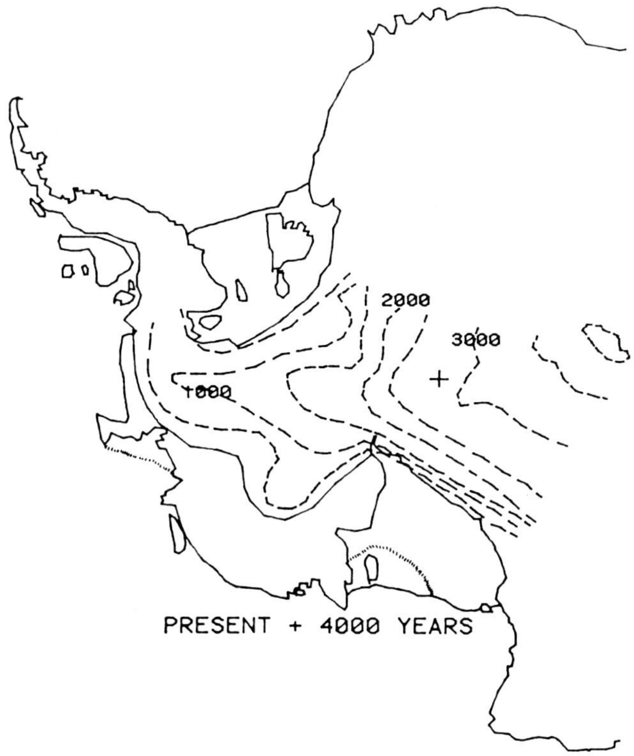

Fig. 12. Surface elevation contours 4000 years after removal of the Amundsen Ice Shelf.

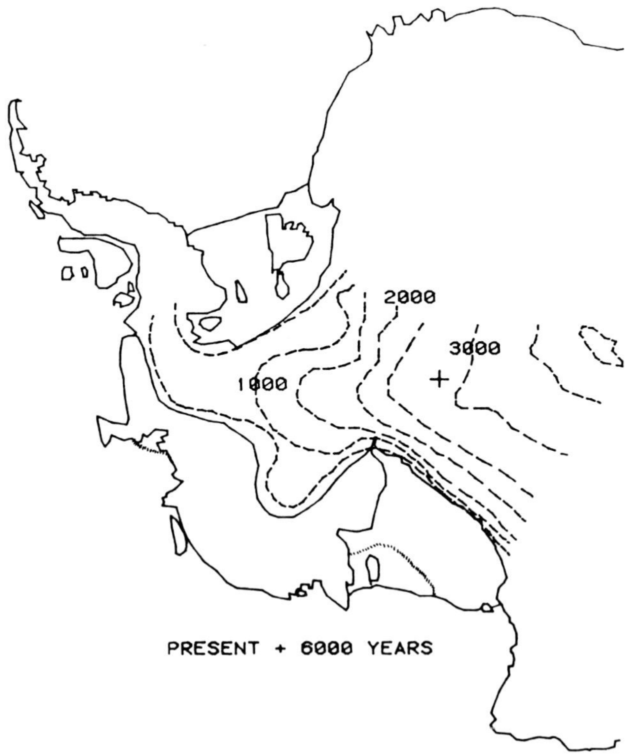

Fig. 13. Surface elevation contours 6000 years after removal of the Amundsen Ice Shelf.

retreat in the Amundsen Sea sector is its affect on the ice streams flowing into the Ross Ice shelf. As has been mentioned migrating ice divides can encroach upon the catchment areas of flowlines on the opposite side of the divide, triggering retreat by reducing outlet velocities. This is shown in Figures 11, 12, and 13 as the migrating ice divide reduces the catchment area of ice streams D and E, causing the calving rate to exceed the ice flux across the grounding line, thereby reducing the back stress and allowing further retreat. The highly fractured surface of ice stream C, coupled with its anomalously low outlet velocity indicates that such a process may already be taking place.

The purpose of this modelling exercise has been to demonstrate that a simple model of the marine instability which generates its own sea-level curve is capable of reproducing the general features of the hypothesized collapse of Antarctica since 18000 B.P. The model predicts a stable configuration in areas which are presently buttressed by major ice shelves. With some reasonable assumptions about the formation of an ice shelf during retreat in the critical Amundsen Sea sector, a scenario for possible future grounding line retreat has been developed. Perhaps the most important result of this work is to emphasize how strongly the stability of any marine ice sheet depends upon the presence of extensive buttressing ice shelves.

Acknowledgements. I would like to acknowledge the National Science Foundation, whose Grants DPP8017072 and DPP8107668 supported this research. I am also indebted to Terry Hughes for valuable discussions and assistance.

References

Budd, W.F., Stress variations with ice flow over undulations, Journal of Glaciology, 10(59), 177-195, 1971.

Glen, J.W., The creep of polycrystalline ice, Proceeding of the Royal Society of London, (Series A), 228, 519-538, 1955.

Denton, G.H. and Hughes, T.J., Milankovitch theory of ice ages: Hypothesis of ice-sheet linkage between regional insolation and global climate, Quat. Res., 20,,125-144, 1983.

Hughes, T.J., ISCAP Bulletin No. 1 - Scientific justification, ice stream cooperative Antarctic project. Ohio State University, Institute of Polar Studies, Columbus, 1972.

Hughes, T.J., Is the West Antarctic Ice Sheet disintegrating? Journal of Geophysical Research, 78, 7884-7910, 1973.

Hughes, T.J., West Antarctic ice streams, Review of Geophysics and Space Physics, 15, 1-46, 1977.

Hughes, T.J., Numerical reconstruction of paleo-ice sheets, In: The Last Great Ice Sheets, (G.H. Denton and T.J. Hughes, Eds.), 221-261, Wiley-Interscience, New York, 1980.

Johnson, G.L., J-R. Vanney, D.J. Drewry, G.de Q. Robin, General Bathymetric Chart of the Oceans (GEBCO), Canadian Hydrographic Service, Ottawa, Canada, 1980.

Kellog, W.W., Effects of human activities on global climate, World Meteorological Organization, 1977.

Kellogg, T.B., D.E. Kellogg and J.B. Andersen, Preliminary results of microfossil analysis of Amundsen Sea sediment cores, Ant. J., in press, 1982.

Matthews, R.K., The unmixed volume signal for the last 220,000 years: Model I, Unpublished CLIMAP manuscript, 1977.

Mercer, J.H., Antarctic ice and Sangamon sea level, International Association of Scientific Hydrology Publication, 79, 217-225, 1968.

Mercer, J.H., West Antarctic Ice Sheet and CO_2 greenhouse effect: A threat of disaster, Nature, 271, 321-325, 1978.

Milankovitch, M.M., Canon of insolation and the ice age problem, Koniglich Serbische Akademie, Beograd, English translation by the Israel Program for Scientific Translations, published for the United States Department of Commerce and the National Science Foundation, Washington, D.C., 1941.

Ninkovich, D. and N.J. Shackleton, Distribution stratigraphic position and age of ash layer "L", in the Panama Basin region, Earth and Planetary Science Letters, 27, 20-34, 1975.

Oerlemans, J., Response of the Antarctic Ice Sheet to a climatic warming: A model study, J. Clim., 2, 1-11, 1982.

Paterson, W.S.B., The Physics of Glaciers, Pergamon Press, England, 250 pp, 1969.

Stuiver, M., G.H. Denton, T. Hughes and J.L. Fastook, The history of the marine ice sheet in West Antarctica: A working hypothesis, In: The Last Great Ice Sheets, (G.H. Denton and T.J. Hughes, Eds.), Wiley-Interscience, New York, 1981.

Thomas, R.H. and C.R. Bentley, Model for Holocene retreat of the West Antarctic Ice Sheet, Quaternary Research, 10(2), 150-170, 1978.

Vernekar, A.D., Long-period global variations of incoming solar radiation, Meteorological Monographs 12, 34, American Meteorological Society, 1972.

Weertman, J., On the sliding of glaciers, Journal of Glaciology 3(62), 205-214, 1957.

Weertman, J., Rate of growth or shrinkage of non-equilibrium ice sheets, Journal of Glaciology, 5, 145-158, 1964.

Weertman, J., Stability of the junction of an ice sheet and an ice shelf, Journal of Glaciology, 13, 3-11, 1974.

NORTH ATLANTIC CLIMATIC OSCILLATIONS REVEALED BY DEEP GREENLAND ICE CORES

W. Dansgaard, S.J. Johnsen, H.B. Clausen, D. Dahl-Jensen,
N. Gundestrup and C.U. Hammer

Geophysical Isotope Laboratory, University of Copenhagen, Haraldsgade 6
DK-22000 Copenhagen, Denmark

H. Oeschger

Physikalisches Institut, Universitat Bern, Sidlerstrasse 5,
CH-3012 Bern, Switzerland

Abstract. Five long-term oxygen isotope (δ) records along ice cores are discussed, in particular two from the Greenland ice sheet that exhibit persistent δ oscillations with a quasi-periodicity of ca. 2550 years. A detailed study of the δ cycles in the Wisconsin glaciation show that they cannot be ascribed to discontinuities in the cores, nor to ice-dynamic instabilities in the ice sheet. In the Holocene, the δ cycles are less pronounced, but they are concurrent with the fluctuating glacier extention elsewhere, which substantiates their climatic significance. An anti-correlation with ^{14}C concentration in atmospheric CO_2, and with ^{10}Be deposition rates on the ice sheets, suggests a connection between climate and solar processes, but a conclusion on this point must await clarification of the terrestrial circulation and mixing processes, and the relationship between the solar outputs of radiation and particulate matter.

Introduction

The American-Danish-Swiss joint effort, Greenland Ice Sheet Program (GISP, 1971-81), emerged from the successful deep ice core drilling by U.S.A. C.R.R.E.L. at Camp Century, NW Greenland, in 1966 (Hansen and Langway, 1966), and from C.C. Langway's (1970) studies on the ice core from the nearby Site 2 that demonstrated the great ice sheets as rich sources of information on past environmental conditions.

The main objectives of GISP (1976) were to extract this paleoenvironmental information by drilling and studying new ice cores and to survey the Greenland ice sheet with a view to ice flow modelling.

Prior to 1978, most of the ice sheet was surveyed by airborne radio-echo sounding equipment (Gudmandsen, 1976). Three ice cores were drilled to 400 m depth by a thermal drill (Ueda and Garfield, 1969), and a number of cores up to 110 m were drilled by other techniques (Rufli et al., 1976; Johnsen et al., 1980) at the locations shown in Fig. 1. The GISP field activities culminated in 1981, when a new deep drill (Gundestrup et al., 1983) penetrated the ice sheet and reached bedrock 2037 m below surface at Dye 3, South Greenland. In parallel with the field activities, extensive stratigraphic, chemical and physical studies were performed on the ice cores, and ice flow models were developed. The initial results were presented at a GISP symposium in May 1982 (Langway et al., 1983).

According to the simple Rayleigh condensation model, the isotopic composition (oxygen-18 or deuterium concentration in the δ scale) of a given polar snow fall depends on several parameters, of which the most important one is cooling of the precipitating air mass, since the last substantial uptake of water vapor from the ocean. As the seasonal or longer term climatic conditions in the source area of the vapor are considerably more stable than those at high latitudes, the isotopic composition of the polar snow is strongly influenced by the temperature at the site and time of snow deposition (Picciotto et al., 1960; Dansgaard, 1964). With some reservations (Johnsen et al., 1972; Johnsen, 1977) $\delta^{18}O$ profiles along polar ice cores therefore exhibit seasonal oscillations (Epstein and Benson, 1959; Hammer et al., 1978) and changes due to general climatic temperature variations (see e.g. Dansgaard et al., 1973; 1975).

However, the δ profiles may also be influenced by varying summer to winter precipitation ratio; by surface elevation changes, which cause surface temperature variations that are not necessarily connected to general climatic changes; and by changing sea-ice cover, i.e. changing distance to the open ocean, which may affect the degree of cooling of precipitating air

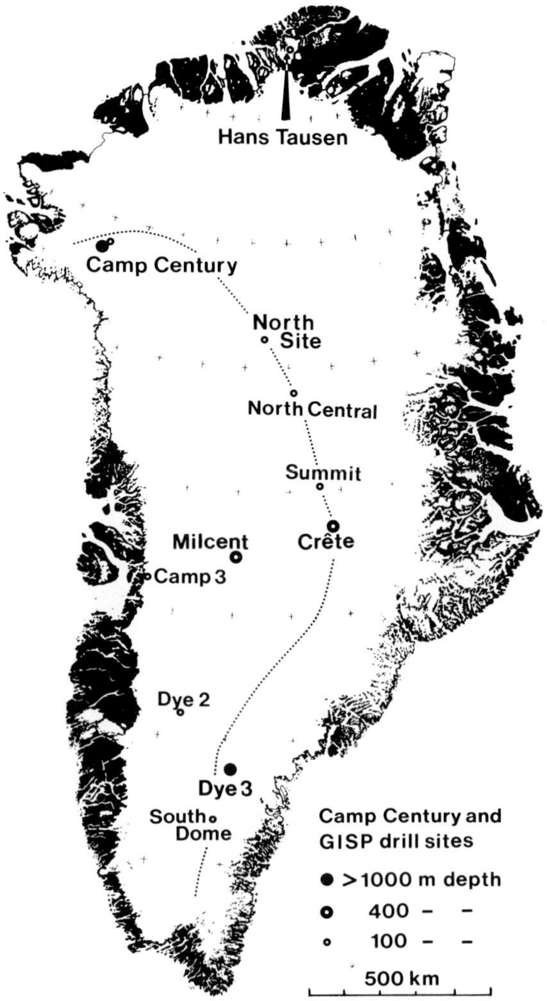

Fig. 1. GISP drill sites. The dotted curve shows the main ice divide that encloses the West greenland discharge area.

masses, but not necessarily the mean air temperature on the ice sheet, at least not to the same degree. In general, it should be kept in mind that the isotopic composition of a given ice core increment may give a hint as to the climatic conditions that prevailed during the relatively short periods of snowfall when the material was deposited, but it tells nothing about the much longer periods of dry weather conditions inbetween.

The concentration of solid particles (mainly continental dust) in Holocene ice in Greenland is of the order of 0.05 mg/kg of ice, which is 6 times higher than in W. Antarctica, but one to two orders of magnitude lower than in Greenland ice from late Wisconsin (Cragin et al., 1977; Thompson, 1977; Petit et al., 1981; Hammer et al., 1983). An important reason for the extremely high Greenland dust contents around the glacial maximum is that the strong meridional temperature gradients (CLIMAP, 1976) and the consequent high storminess at mid-latitudes (Ruddiman and Glober, 1975) created rich sources of aerosols in large, dry areas covered by glacial outwash south of the North American and Eurasian ice sheets (Schultz and Frye, 1968). But, the high content of calcium rich carbonates in late Wisconsin ice suggests that a new source of alkaline aerosol was formed, when the sinking sea level left vast high- and mid-latitude areas of the continental shelves exposed to scouring (Cragin et al., 1977; Hammer et al., 1983).

The latter source was cut off, when the sea level began to rise, but the former areas expanded during the disintegration of the ice sheets and, according to Ruddiman and McIntyre (1981), they stayed dry and cold up to 13 ka B.P. (1 ka (kiloannum) B.P. = 1000 years before present). In the same period (20 to 13 ka B.P.) an intensified supply of icebergs from the disintegrating ice sheets cooled off the North Atlantic surface water (Denton and Hughes, 1981), thus reducing the moisture content of maritime air masses and, hence the precipitation at high latitudes (Ruddiman and McIntyre, 1981). For any given annual fall-out of dust, lower annual accumulation on the Greenland ice sheet is, in itself, a contributory cause for higher dust concentration in the ice.

Even since the Wisconsin glaciation terminated, the deposition of dust in Greenland has varied seasonally with a maximum preferably in spring, probably because the stationary anti-cyclone and the associated high tropospheric winds bring dust from spring-dry areas at mid latitudes. The oscillating deposition of dust provides an alternative means for annual layer identification and, thereby, for absolute dating of ice cores (Hammer et al., 1978).

Also the acidity of the deposited snow varies with the season, which can be used for annual layer identification in periods of no or low volcanic activity (Hammer, 1980). Large and violent volcanic eruptions in the northern hemisphere leave a high acidity signal in the ice due to washout of, particularly, sulfuric acid and hydrochloric acid from the atmosphere (Hammer et al., 1980). However, the seasonal acidity variations, and even the volcanic signals are suppressed in ice deposited in Greenland during most of the Wisconsin glaciation. This ice is alkaline due to the then high alkaline aerosol load that obviously neutralized the atmospheric acids, at least at high northern latitudes (Hammer et al., 1980).

Absolute dating of Holocene ice in Greenland by the stratigraphic methods outlined above is laborious, but very accurate and, in principle, simple. These methods are being applied on the new Dye 3 core, which has so far been absolute dated back to ca. 4 ka B.P., but data on the earlier part of the Holocene are not yet available. As to the Camp Century ice core, annual layer thicknesses measured on essentially all suitable Holocene increments defined the input parameters of a simple steady state ice flow model (Dansgaard and Johnsen, 1969), which was subsequently used to date the younger part of the Camp Century record (Hammer et al., 1978). The dating accuracy was estimated at a few percent back to 8.5 ka B.P., but beyond 10 ka B.P. the steady state assumption behind this time scale is probably unrealistic.

The early attempts to date the Wisconsin part of the Camp Century record by various techniques were discussed by Dansgaard et al. (1982), who outlined a new tentative Camp Century time scale by correlating the major trends in isotope profiles along ice cores and deep sea cores. The new time scale lower glacial than post-glacial accumu-

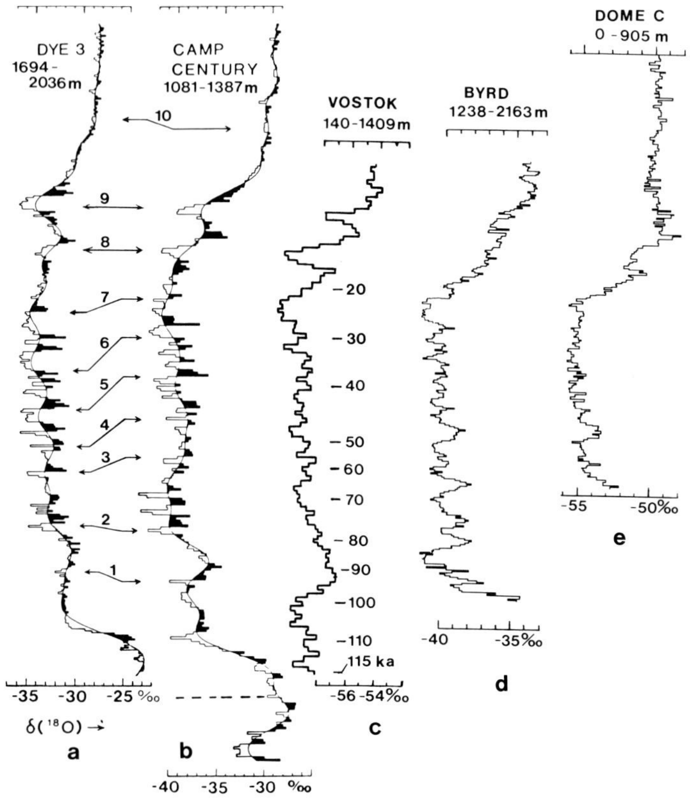

Fig. 2. Continuous $\delta(^{18}O)$ records along five deep ice cores, plotted on linear depth scales and spanning the depth intervals indicated in the upper part of the figure. Each of the samples represent increments of 1 meter in the two Greenland records (a and b), 10 meters in the Vostok record (c); 4 meters in the Byrd Station record (d); and generally 3-4 meters in the Dome C record (e). The arrows between (a) and (b) indicate common high frequency features, used to transfer the new Camp Century time scale to the Dye 3 core, cp. Fig. 3. The Vostok time scale shown to the right of curve c was calculated by Gordienko et al. (1982) on the basis of annual layer measurements by Wilson and Hendy (1981). It does not deviate considerably from the new Camp Century time scale shown in Fig. 3, if the δ maximum around 90 ka B.P. is interpreted as stage 5a.

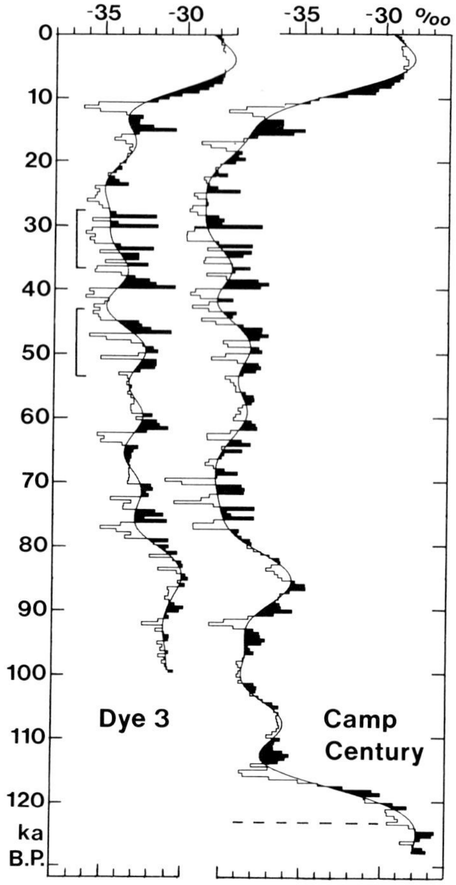

Fig. 3. $\delta(^{18}O)$ profiles along the Dye 3 (0 to 1982 m depth) and the Camp Century (0 to 1370 m depth) ice cores plotted on a common linear time scale as outlined in the text.

lation, and sliding of the ice over the bedrock or at least fast deformation of the ice close to the bedrock. Its accuracy is hardly better than ± 5 ka for ice older than 20 ka, even if the suggested correlation proves to be correct. In turn, the two Greenland δ records (plotted on linear depth scales in Fig. 2a and 2b) were correlated by their common high frequency features indicated by the arrows, and the new Camp Century time scale could then be transferred to the Dye 3 record. In Fig. 3 the datable parts of the two data sets are plotted as functions of time. The smoothed curves are obtained by low pass filtering technique (with sharp cut off of all cycles shorter than 20 steps (10 ka), and less than 10% leakage), combined with a maximum entropy prediction filter (Andersen, 1974) that allows extention of the smoothed curves through the entire range of the raw data.

The new time scale puts all Camp Century δ values higher than those in the Holocene maximum into the 130-120 ka B.P. time interval (Eemian interglacial, or Emiliani (1966) Stage 5e). This is consistent with McIntyre et al.'s (1972) evidence that in this very period subtropical Atlantic

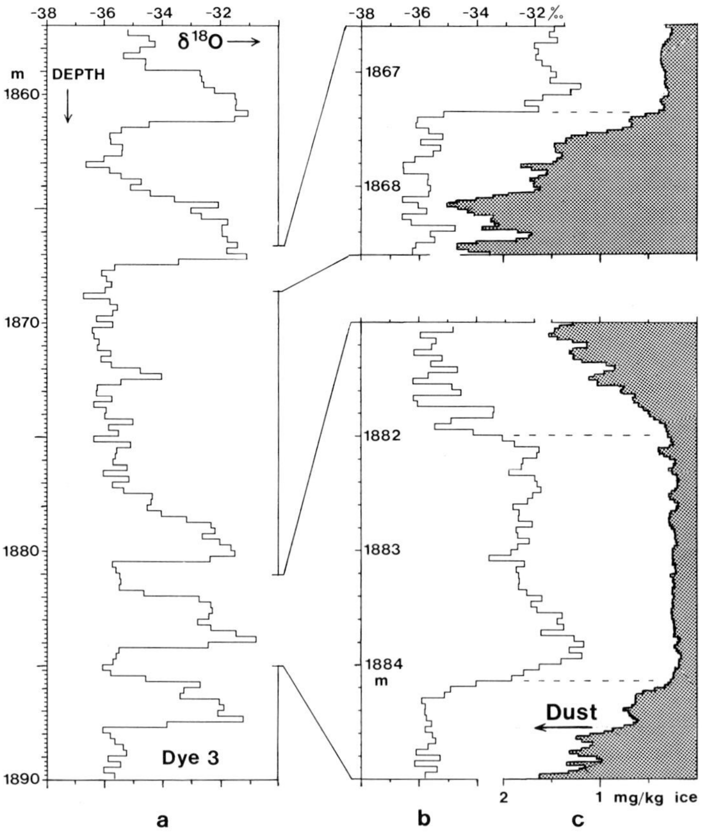

Fig. 4. a: Details of the δ oscillations marked to the left of the Dye 3 record in Fig. 3. b: Further details of some of the δ shifts. c: Concentration of insoluble microparticles. Notice, increasing values towards the left.

water masses advanced to more than 50°N in the Atlantic ocean for the first and only time in the last 225 ka. Accordingly, the deepest parts of the Dye 3 (Fig. 2a) and Devon Island (Paterson et al., 1977) ice cores with more than 2^{0}/oo higher δ's than in the Holocene were probably deposited in the Eemian interglacial.

Obviously, the Pleistocene to Holocene shift in δ is 7^{0}/oo (per mille) at Dye 3, but 11 to 12^{0}/oo at Camp Century. These shifts were discussed by Dansgaard et al. (1983), who suggested that the termination of the Wisconsin glaciation is marked by a δ-increase of generally 6 ± 1^{0}/oo in ice cores from the main parts of the ice sheets in Greenland and Antarctica, cp. Fig. 2c: Vostok, East Antarctica (Gordienko et al., 1983); Fig. 2d: Byrd Station, West Antartica (Johnsen et al., 1972); and Fig. 2e: Dome C, East Antarctica (Lorius et al., 1979). Furthermore, they suggested that the higher δ-shifts found around the Baffin Bay (11-12^{0}/oo at Camp Century; 10^{0}/oo on Agassiz ice cap, Ellesmere Island (Fisher et al., 1983); 15^{0}/oo on Barnes ice cap, Baffin Island (Hooke, 1976) may reflect that the disappearance of the North American ice sheet made the climatic change more drastic in this area than in Southeast Greenland or in Antarctica.

Abrupt Environmental Changes in the Wisconsin

The mid and late Wisconsin from 77 to 10 ka B.P. is characterized by large amplitude oscillations of $\delta^{18}O$ in both of the records in Fig. 3. In the Camp Century record they are present even before 90 ka B.P., yet of lower amplitude. Although the transfer of the new Camp Century time scale to the Dye 3 record was based on the assumption that 9 characteristic features in the two records could be correlated (cp. Fig. 2a and 2b), the synchroneity between the high frequency oscillations is considered real, firstly because the said procedure results in a correlation coefficient of no less than 0.88 (P > 99.98 % for estimated 10 degrees of freedom) between the filter-smoothed curves in Fig. 3; secondly, because the shape and the sequence of the δ peaks and valleys are often so similar in the two records that it cannot be due to a coincidence, cp. for example, the transition from a relatively quiet δ record below arrow No. 2 to the region of vigorous oscillations above it, beginning with a characteristic bipartite δ maximum; and the sequence of a broad peak followed by three sharp ones in the interval between arrows No. 5 and 6, which, in addition, adjoins relatively broad intervals of generally extremely low δ's; cp. also the shape of the δ maxima between arrows No. 8 and 9.

In a large part of the Dye 3 record, the δ's have a tendency to alternate between two levels of approximately -35.5^{0}/oo and -32^{0}/oo. Fig. 4a shows a detailed version of the δ oscillations in the Dye 3 core from 1857 to 1890 m depth, corresponding to times of deposition from approximately 28 to 37 ka B.P., according to the time scale in Fig. 3. Obviously, the peaks are saw tooth shaped with considerably more abrupt δ shifts towards "warmer" values. All of the δ peaks are provided with a secondary peak or, at least, a "shoulder" on the younger side.

The abruptness of the δ shifts may be connected to fast changes of the mean latitude of the polar front in analogy with the event 13 ka B.P. described by Ruddiman and McIntyre (1981). Once started by some kind of external forcing (varying insolation or volcanic activity?), the effect of a displacement of the polar front may be amplified by feed-back mechanisms, like the one suggested by Flohn (1982): "...in the case of a cooling ...meridional temperature gradients will become greater, latitudes of the subtropical anticyclones become lower, intensity of cell winds (trade) increases, equatorial sea surface temperature becomes cold, atmospheric content of CO_2 and H_2O becomes lower, thus resulting in further cooling". Although the importance of the greenhouse effect is not yet clarified, it should be mentioned here that CO_2 measurements on the Dye 3 core show that the CO_2 concentration in the atmosphere varied considerably and in phase with the δ oscillations in the Wisconsin glaciation (Stauffer et al., 1983).

Continuity of the Time Scale

On the face of it, a possible explanation for the abrupt alternations would seem to be that, somehow, two homogeneous ice masses of different isotopic compositions were folded repeatedly into each other, which would invalidate any continuous time scale along the core. However, in view of the completely different ice flow conditions in Southeast and Northwest Greenland, the close correlation between the Dye 3 and Camp Century δ records is strong evidence against the folding hypothesis.

But, this argument does not exclude that the δ shifts indicate "holes" in the time scale. Such "holes" might be caused either by essential cease of Greenland accumulation, or by negative accumulation (due to wind erosion, as observed today in some areas close to Vostok, East Antarctica; cp. Barkov, 1975) through long periods of time till the onset of different climatic conditions, including resumed accumulation of ice of different δ. In both cases, the other environmental characteristics of the ice (concentrations of dust, sea salts etc.) must be expected to shift exactly at the same depths as does the isotopic composition, though not necessarily with the same abruptness, because the various isotopes and impurities have migrated differently since the time of deposition, depending on their diffusion coefficients in the ice. Hence, the continuity problem may be solved by comparing high resolution δ and impurity profiles across the δ shifts:

The sharp δ shift at 1867.35 m depth, and the entire δ oscillation between 1881.0 and 1885.0 m depth are shown in great detail (5 cm samples) in Fig. 4b along with a high resolution dust concentration profile in Fig. 4c (2.5 cm samples). Both of them are typical for the other oscillations studies in detail so far. The dust concentrations (in mg dust per kg ice) were measured in the field by a light scattering technique (Hammer et al., 1983) applied on melted ice soon after the recovery of the ice core increments in question. The technique involved some degree of "cross-talk" between adjoining samples, and it may therefore be more realistic to consider the 5 cm resolution in the δ profile as valid for the dust concentration profile as well.

In general, high dust concentrations correspond to low δ values, and vice versa, in ice from mid and late Wisconsin (Hammer et al., 1983). This is, in part, due to colder climates being dominated by higher storminess at mid latitudes, but changes of the general atmospheric circulation pattern, and of the extent and location of the source areas of the dust may also be important.

However, Fig. 4b and 4c show the following features: (i) by the end of the periods of extremely low δ's the dust concentration begins to decrease (notice the scale at the bottom of Fig. 4c) long before the δ shift; (ii) it reaches the new level before or at the same time as the δ shift is completed; (iii) the dust concentration stays low till after the onset of the less well defined δ shift back to extremely low values, in fact till δ has dropped below $-34^0/00$; and (iv) the total change in dust concentration extends over a longer core increment (more than 0.6 m) than does the corresponding δ shift (less than 0.3 m).

These features indicate that the δ shifts are not due to discontinuities in the time scale, because, as mentioned above, such discontinuities would cause δ and dust concentration shifts at exactly the same depths, and with <u>smaller</u> δ than dust concentration gradients, since the diffusion coefficient of insoluble microparticles in the 0.1-2 μm size range (Hammer et al., 1983) is negligible, and at least much lower than that of $H_2{}^{18}O$.

Periodicity and Persistency of δ Fluctuations

The duration of a given δ oscillation cannot be determined until new techniques are developed to measure annual stratification in Wisconsin ice in Greenland. The new tentative time scale was established by considering the long term δ trends only, and should therefore not be used to estimate the duration of, for example, the δ oscillations shown in Fig. 4, much less the duration of the δ shifts.

However, the Camp Century record contains 18 oscillations from 77 to 30 ka B.P. in the new time scale (Fig. 3), i.e. one per 2600 yrs. This leads back to the assumed 2400 yr periodicity in climate that was previously used (Dansgaard et al., 1971) to correct Dansgaard and Johnsen's (1969) initial Camp Century time scale (the reason why the result differed considerably from the time scale in Fig. 3 is probably that the initial one deviated too much from absolute chronology to serve as a basis for the correction procedure).

Using data from a deep sea core back to 133 ka B.P., Pisias et al. (1973) demonstrated a similar periodicity in ocean surface summer temperatures west of Ireland (53°N, 22°W). Fourier spectral analysis of palaeotemperatures derived from palaeoecological transfer functions (Imbrie and Kipp, 1971) "shows a very significant peak representing 2600 years". In interglacial times, the sea surface temperature oscillations are less pronounced than under glacial conditions, and this feature is also present in the ice cores. It appears from the Camp Century record in Fig. 3 that the δ amplitudes are modulated by the δ level: The highest amplitudes occur at extremely low δ values, whereas the δ cycles are hardly recognizable in the high δ range in the Holocene (the Dye 3 record is not yet completed).

Nevertheless, as shown by Dansgaard et al. (1971) and Fisher (1982) the Holocene Camp

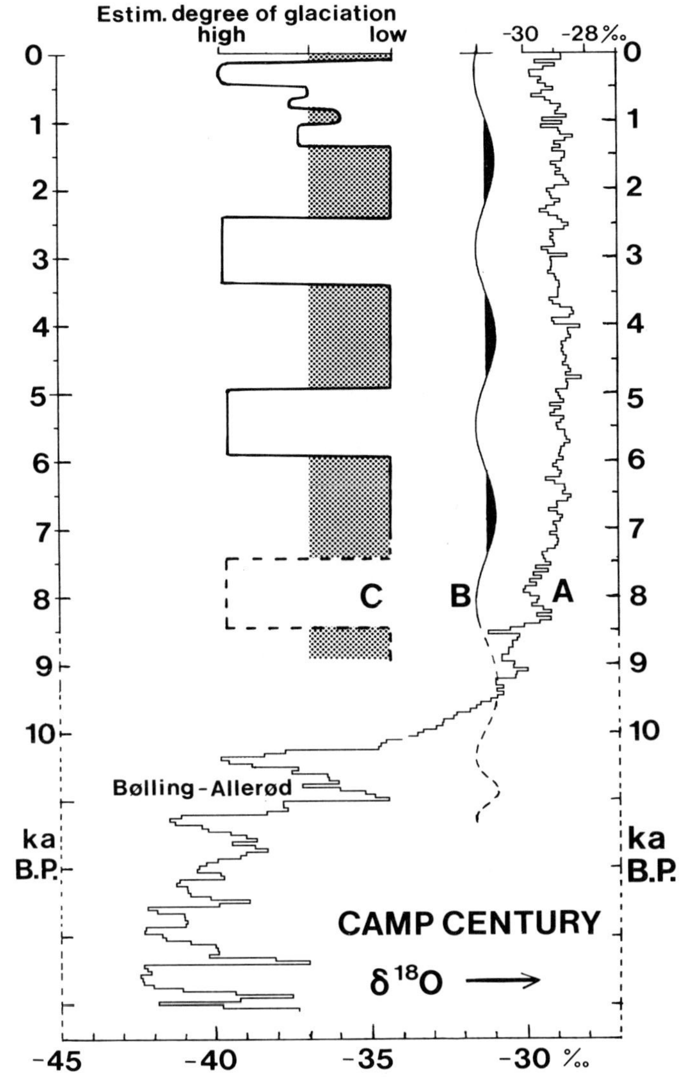

Fig. 5. A: The Holocene and late glacial part of the Camp Century δ record, plotted on a linear time scale that is close to absolute chronology back to 8.5 ka B.P. B: Band pass filtered version calculated for the last 8,500 yrs, and suggested by the dashed curve beyond this range, where the time scale is less accurate due to poorly known strain and accumulation rate. C: Estimated degree of worldwide glaciation (from Denton and Karlen, 1973).

Century δ record does contain an approximately 2500 yr cyclic component. The Holocene part of the Camp Century record is shown in the upper part of Fig. 5A. It has been dated by stratigraphic methods (Hammer et al., 1978) to an accuracy that is probably better than 2% back to 8.5 ka B.P. Fig. 6 shows a logarithmic power density spectrum of this part of the record calculated by the maximum entropy method (Ulrych and Bishop, 1975; Johnsen and Andersen, 1977). The 7200 yrs peak contains 50% of the power, but this is essentially a reflection of the long-term δ trend depicting the post-glacial climatic optimum in the middle of the Holocene. Apart from this trend, the 2550 yr oscillation is dominating with 17% of the power. A band-pass (2200 to 2800 yrs periods) filtered version of the raw data is shown with an amplification factor of 2 in Fig. 5B.

In times of transition between glacial and interglacial conditions the situation is less clear, i.e. because the time scale is most difficult to verify during drastic changes of the glaciological environment. Nevertheless, when comparing the curves in Fig. 5A and 5B, it is tempting to extend the latter one beyond the well-dated time interval as shown by the dashed curve, i.e. to add at least 1.5 more δ oscillations with maxima at approximately "9.3" and "10.9" ka B.P. on the dashed time scale. As shown by Siegenthaler et al. (1983), the latter maximum most likely corresponds to the late glacial Bolling-Allerod interstadial known from Western Europe (Iversen, 1954; Nilsson, 1968; Eicher and Siegenthaler, 1976; Watts, 1980) and the North Atlantic ocean (Ruddiman and McIntyre, 1981).

Assuming a constant δ periodicity of 2550 yrs and following the procedure of Denton and Karlen (1973), addition of two full periods to the oldest well-defined maximum in Fig. 5B (6.8 ka B.P.) puts an age of 11.9 ka B.P. on the middle of Bolling-Allerod. Furthermore, addition of a quarter of a period to the latter maximum gives 13.2 ka B.P. for the abrupt δ shift from extremely low preBolling δ values, which is in agreement with ^{14}C dating of the onset of Bolling elsewhere (Wegmuller and Welten, 1973) and of a radical change of the circulation in the North Atlantic ocean - atmosphere system, including a drastic northward displacement of the polar front 13 ka B.P. (Ruddiman and McIntyre, 1981). This suggests that the dashed time scale is ca. 2 ka off around the "11" ka B.P. mark, which is quite possible in view of the drastic changes of the ice flow conditions (accumulation, ablation, mechani-

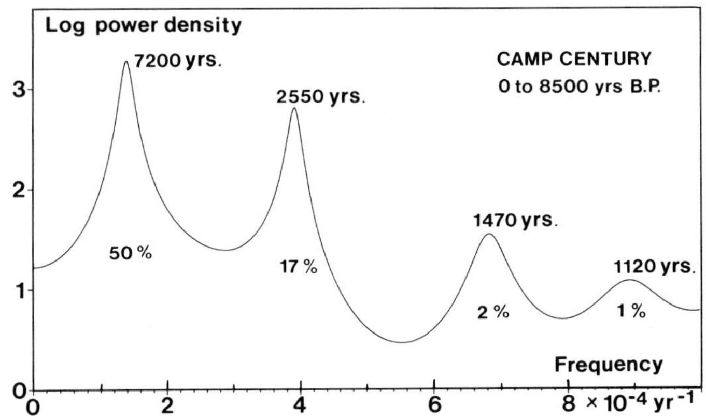

Fig. 6. Logarithmic power density spectrum calculated on Camp Century δ values back to 8,500 yrs by the maximum entropy method.

cal properties of the ice, cp. Hammer et al., 1978; Shoji and Langway, 1983; Gundestrup and Hansen, 1983) during the period of transition.

Hence, there are good reasons to believe that δ of Greenland snow oscillates persistently with quasi-periodicity of approximately 2550 yrs, and that the amplitude of the δ cycles is modulated by the degree of glaciation (and perhaps the latitude), being small under warm conditions and high under cold conditions.

Climatic Significance of the δ Cycles

The amplitude of the δ cycles in Greenland and on the Canadian Islands (Paterson et al., 1977); Fisher, 1982) was an order of magnitude higher in mid and late Wisconsin, than in the Holocene. The corresponding shifts in surface temperatures may also have been much higher, though it is difficult to tell how much, cf. the remarks in the 5'th paragraph of section 1 about δ profiles being influenced by factors that are not directly related to the local mean temperature of the troposphere.

One of these factors is surface elevation change that might happen in case of fast run-off of huge amounts of ice (surge) as a result of internal instability in the ice flow system, in analogy with observed surges of minor glaciers (Post, 1960; 1969; Dolgushin and Osipova, 1975). On the face of it, the abrupt ca. $5^{0}/_{00}$ δ increases in mid and late Wisconsin (Fig. 3) might be due to surges associated with several hundred meter lowerings of the surface elevation and, consequently, several degrees centigrade warmings of the surface.

On the other hand, as the ice flow conditions in Southeast and Northwest Greenland are quite different, it is not likely that the entire ice sheet could surge at one and the same time as a result of an internal instability alone. Therefore, general surges would have to be initiated by some external impact. For example, a substantial rise in sea level, caused by a great surge in Antarctica, would be extremely dangerous to the stability of at least the marginal zones of an expanded Greenland ice sheet, if they were resting on sub-sea level areas and fringed by ice shelves like most of West Antarctica today. The West Antarctica ice sheet is considered most vulnerable to surging, because it is grounded far below sea level (Wilson, 1964; Mercer, 1978). Therefore, it is remarkable that the abrupt mid-Wisconsin δ changes are not considerably higher in the Byrd Station profile (Fig. 2d), than in any other δ profile, including those along the Dome C and the Vostok (Fig. 2c and 2e) ice cores. This does not support the idea of surges as a cause for the fast δ increases.

Furthermore, it is difficult to envisage how the South Greenland ice sheet, which rests on high elevation bedrock, could surge to a degree that lowered the surface elevation close to the ice divide by many hundred meters. Even if the western edge of the ice sheet were located 170 km west of its present position, i.e. at the submerged moraine ridges found 50 km off-shore, the surface elevation at the ice divide would not be significantly higher than at present, according to 3-dimensional ice sheet model calculations (Reeh, 1982; 1983). A possible surge might disintegrate the marginal zone of an ice sheet advanced that far, but the edge would retreat to higher elevations, and the surge would be stopped by the blocking effect of the mountains. Far inland, including the area where the Dye 3 ice was formed, the stabilized ice sheet would be left with essentially unchanged surface elevation.

Finally, if the repeated δ shifts to higher values were to be explained by surges, the ice sheet must have rebuilt itself in between, i.e. the reverse δ trends would have to be ascribed to increasing surface elevation. However, build-up of an ice sheet is a very slow process (Weertman, 1964; Oerlemans, 1981) and although the δ decreases following the five peaks in Fig. 4a are more smooth than the δ increases, the five peaks terminate by abrupt drops in δ that correspond to considerable thickening of the central part of the ice sheet within a time frame of the order of a century. This implies a much higher accumulation rate than today, which is unrealistic under full glacial conditions.

Hence, surges are implausible as an explanation for the high frequency δ oscillations in the Wisconsin, that are rather to be explained by general climatic changes at high northern latitudes. This is the case with their successors at the end of the glaciation (Siegenthaler et al., 1983), and in the Holocene as well, judging from the concurrence of the curve in Fig. 5B with Denton and Karlen's (1973) independent estimate on the degree of glaciation (or rather deglaciation - notice the reversed scale on top of Fig. 5C), which is based on mainly geological evidence of worldwide (before 7 ka B.P. mainly Swedish) Holocene glacier fluctuations. According to Fig. 5C, considerable glacier expansion took place in periods of approximately 1000 years duration around 5500, 3000 and 500 years ago.

For obvious reasons, Denton and Karlen's estimate is most accurate and detailed in the latter period that spanned most of our own millennium, and culminated in the "Little Ice Age". It is interesting that the glacier advance lasted more than 1000 years, and was interrupted by halts and minor retreats, whereas the main retreat began around A.D. 1890 and was nearly completed 50 years later. In other words, the best documented glaciation cycle in Fig. 5C is saw-tooth shaped like the mid Wisconsin δ oscillations in Fig. 4a (the shape of the Holocene δ oscillations is doubtful, because the signal to noise ratio is small).

According to the conclusion in section 4, the abrupt temperature rise in the 1920's may thus be the latest member of a very long series of similar events occurring once every ca. 2550 years to an extent that is modulated by the degree of glaciation and dependent on the latitude: The 1920-30 temperature rise was 4°C in North Greenland, 2°C in South Greenland (Dansgaard et al., 1975), but only half a degree as an average over the northern hemisphere (Mitchell, 1963). Data from the southern hemisphere are sparse, but recent estimates (Hansen et al., 1981) suggest an 0.3°C warming through the last 100 years, interrupted by a minor cooling in the 1920's.

This suggests that short-term climatic changes in the southern hemisphere are not concurrent with, or at least that they are less abrupt than those in the northern hemisphere, probably due to the different land-sea distribution in the two sub-polar belts. There is also evidence that the post glacial climatic optimum occurred already between 11 and 8 ka B.P. in the southern hemisphere (see Lorius et al., 1979, and references therein).

It is still doubtful, if the southern hemisphere climate has a 2550 yr cycle. The Vostok record (Fig. 2c) is not yet detailed enough to show it, but the Byrd and Dome C records (Fig. 2d and 2e) both contain several oscillations that might reflect damped versions of this cycle.

Why?

The coupled atmosphere-ocean-cryosphere system may have a built-in feed-back mechanism that favors a 2550 yr cycle in the high latitude climate, but climate models are not yet developed to a stage that allows any conclusion on this point.

Another possibility is that some periodic external forcing is responsible for the climatic cycle, for example varying volcanic activity that modulates the load of aerosols in the stratosphere and, thereby, the solar radiation flux in the troposphere. However, although 30% of the climatic variability in the last 1400 yrs may be ascribed to varying volcanic activity (Hammer et al., 1980), there is no evidence that this activity is periodic. The only cyclic variations of the solar radiation flux in the atmosphere known at present are the Milankovich effects (cp. e.g. Berger, 1978), but they are too slow to be of interest in this context.

Since the solar radiation is the only important input of energy to the climatic system, it is most obvious to seek an explanation in solar processes (Eddy, 1977). Unfortunately, we know much less about the solar radiation output than about the emission of solar particulate matter in the past. The latter has modulated the production rate of cosmogenic isotopes (reduced production rates in times of high sunspot activity; deVries, 1958; Stuiver, 1961; Stuiver and Quay, 1981). There is evidence that the ^{14}C concentration in atmopsheric CO_2, has varied in a non-random way in the past (see e.g. Neftel et al., 1981), and that its Holocene variations are in antiphase with climatic temperature changes (Suess, 1968; Sonett and Suess, 1983), including the 2550 yr cycle (Fisher, 1982).

Furthermore, the deposition rate of another cosmogenic isotope, ^{10}Be, on the ice sheets also seems to be in antiphase with climatic changes: Raisbeck et al. (1981) found 50% elevated ^{10}Be concentrations in Dome C ice from the 17'th century, i.e. more than can be explained by the estimated 25% lower accumulation rate in this cold period (Lorius et al., 1979). Beer et al. (1983a) have shown a similar feature in 17'th century ice from Milcent (70.4°N, 44.6°W), Greenland, where no significant accumulation changes occurred (Reeh et al., 1978). Since this period coincides with the Maunder sunspot minimum, increased ^{10}Be production rate is at least a contributory reason for the high concentrations. The ^{10}Be data so far obtained from the Dye 3 ice cores show a close anticorrelation with the δ's in late Wisconsin to early Holocene (Oeschger et al., 1983). The Younger Dryas value is 3 times higher than those from Bolling-Allerod and Preborial time. A similar ^{10}Be to δ anticorrelation has most recently (Beer et al., 1983b) been demonstrated to hold for the drastic δ oscillations in Wisconsin ice.

This is additional evidence that low solar activity in terms of particle emission is associated with cold climatic conditions. Hence, "whatever might cause changes in the ^{14}C level of the atmosphere CO_2 might well have an influence upon the global climate" (Suess, 1980).

But, as long as the long-term relationship between the solar outputs of radiation and particles is an open question (as to the short-term, i.e. a few years, relationship, see Eddy et al., 1982) and as long as the influence of shifting atmospheric circulation and mixing patterns is unknown, correlation of past cosmogenic isotope deposition rates with climatic changes contains several elements of speculation.

Acknowledgement. We are grateful for the financial, logistic and other support provided by U.S. National Science Foundation, Division of Polar Programs; Swiss National Science Foundation; Commission for Scientific Research in Greenland; Danish Natural Science Research Council; Carlsberg Foundation; and Nordic Cultural Foundation. Furthermore, we thank Polar Ice Coring Office, University of Nebraska, for logistic management, and the numerous individuals, who helped us in the field and in the laboratories and workshops involved.

References

Andersen, N.O., On the calculation of filter coefficients for maximum entropy spectral analysis, Geophysics, 39, 69-72, 1974.

Barkov, N.I., Snow accumulation characteristics in the Vostok Station area, 1970 through 1973, Antarc. Journ., U.S., 10, No. 2, 55-56, 1975.

Beer, J., H. Oeschger, M. Andree, G. Bonani, M. Suter, W. Wolfli and C.C. Langway, Temporal variations in the ^{10}Be concentration levels found in the Dye 3 ice core, Greenland, Proc. Symp. Ice and Climate Modeling, Evanston, Ill. U.S.A., June 1983, to be published in Ann. Glaciol., 1983a.

Beer, J., H. Oeschger, M. Andree, G. Bonani, M. Sutter, W. Wolfli and C.C. Langway, Temporal variations in the ^{10}Be concentration levels found in the Dye 3 ice core, Greenland, Proc. Symp. Ice and Climate Modeling, Evanston, Ill. U.S.A., June 1983, to be published in Ann. Glaciol., 1983b.

Berger, A.L., Long-term variations of caloric insolation in Quaternary climatic changes Quat. Res., 9, 139-167, 1978.

CLIMAP Project Members, The surface of the ice-age earth, Science, 191, 1131-1137, 1976.

Cragin, J.H., M.M. Herron, C.C. Langway, Jr. and G. Klouda, Composition of atmospheric precipitation during the late Cenozoic era, in Polar Oceans (ed. M.J. Dunbar), Proc. Polar Oceans Conf., McGill Univ., Montreal, May 1974, Calgary, Alberta, Arctic Inst. North Am., 617-631, 1977.

Dansgaard, W. Stable isotopes in precipitation, Tellus, 16 436-468, 1964.

Dansgaard, W. and S.J. Johnsen, A flow model and a time scale for the ice core from Camp Century, Greenland, J. Glaciol., 8, 215-223, 1969.

Dansgaard, W. S.J. Johnsen, H.B. Clausen and C.C. Langway, Jr., Climatic record revealed by the Camp Century ice core, in The late Cenezoic glacial ages (ed. K.K. Turekian), Yale Univ. Press, New Haven, Conn., 37-56, 1971.

Dansgaard, W., S.J. Johnsen, H.B. Clausen and N. Gundestrup, Stable isotope glaciology, Meddel. om Grønland, 197, No. 2, 53 pp, 1973.

Dansgaard, W., S.J. Johnsen, N. Reeh, N. Gundestrup, H.B. Clausen and C.U. Hammer, Climatic changes, norsemen and modern man, Nature, 255, 24-28, 1975.

Dansgaard, W., H.B. Clausen, N. Gundestrup, C.U. Hammer, S.J. Johnsen, P.M. Kristinsdottir and N. Reeh, A new Greenland deep ice core, Science, 218, 1273-1277, 1982.

Dansgaard, W., H.B. Clausen, N. Gundestrup, S. J. Johnsen and C. Rygner, Dating and climatic interpretation of two deep Greenland ice cores, Am. Geophys. Union Meeting, Philadelphia, May-June, 1982, to be published, A.G.U. Special vol., 1983.

Denton, G.H. and T. Hughes, The last great ice sheets, Wiley Interscience Press, New York, 1981.

Denton, G.H. and W. Karlén, Holocene climatic changes, their pattern and possible cause, Quat. Res., 3, 155-205, 1973.

Dolgushin, L.D. and G.B. Osipova, Glacier surges and the problem of their forecasting, Int. Ass. Hydrol. Sci., 104, 292-304, 1975.

Eddy, J.A., Climate and the changing Sun, Climatic Change, 1, 173-190, 1977.

Eddy, J.A., R.L. Gilliland and D.V. Hoyt, Changes in the solar constant and climatic effects, Nature, 300, 689-693, 1983.

Emiliani, C., Paleotemperature analysis of the Caribbean cores P6304-8 and P6304-9, and a generalized temperature curve for the last 425,000 years, J. Geol., 74, 109-126, 1966.

Epstein, S. and C. Benson, Oxygen isotope studies, Trans. Am. Geophys. Union, 40, 81-84, 1959.

Eicher, U. and U. Siegenthaler, Palynological and oxygen isotope investigations on late-glacial sediment cores from Swiss lakes, Boreas, 5, 109-117, 1976.

Fisher, D.A., Carbon-14 production compared to oxygen isotope records from Camp Century, Greenland and Devon Island, Canada, Climatic Change, 4, 419-426, 1982.

Fisher, D.A., R.M. Koerner, W.S.B. Paterson, W. Dansgaard, N. Gundestrup, and N. Reeh, Effect of wind scouring on climatic records from ice-core oxygen-isotope profiles, Nature, 301, 205-209, 1983.

Flohn, H., Oceanic upwelling as a key for abrupt climatic change, J. Meteor. Soc. Japan, 60, 268-273, 1982.

G.I.S.P., Greenland Ice Sheet Program, Science Plan, 2'nd Ed., Univ. Nebraska, Lincoln, Nebraska, 1976.

Gordienko, F.G., V.M. Kotlyakov, E.S. Korotkevich, N.I. Barkov and S.D. Nikolejev, New results of oxygen-isotope studies of the ice core from Vostok Station down to the depth of 1412 m, Data of glaciological studies, Acad. of Sci. of U.S.S.R., Publ. No. 46, 168-171, 1983.

Gudmandsen, P., Studies of ice by means of radio echo sounding, Report 162, Electromagnetic Inst., Tech. Univ. of Denmark, Lyngby, 1976.

Gundestrup, N. and L.B. Hansen, Logging of the Dye-3 deep hole, in preparation, 1983.

Gundestrup, N., S.J. Johnsen and N. Reeh, ISTUK, a new deep core drill system, Proc. 2'nd Intern. Symp. Drilling Technol., to be published by U.S. Army Cold Regions and Engineering Laboratory, Hanover, N.H., 1983.

Hammer, C.U., Acidity of polar ice cores in relation to absolute dating, past volcanism, and radio-echoes, J. Glaciol., 25, 359-372, 1980.

Hammer, C.U., H.B. Clausen, W. Dansgaard, N. Gundestrup, S.J. Johnsen and N. Reeh, Dating of Greenland ice cores by flow models, isotopes, volcanic debris, and continental dust, J. Galciol., 20, 3-26, 1978.

Hammer, C.U., H.B. Clausen and W. Dansgaard, Greenland ice sheet evidence of post-glacial volcanism and its climatic impact, Nature, 288, 230-235, 1980.

Hammer, C.U., H.B. Clausen, W. Dansgaard, A. Neftel, P. Kristinsdottir and E. Johnsen, Continuous impurity analysis along the Dye-3 deep core, Am. Geophys. Union meeting, Philadelphia

May-June, 1982, to be published, A.G.U. Special vol., 1983.

Hansen B.L. and C.C. Langway, Jr., Deep core drilling in ice and core analyses at Camp Century, Greenland, 1961-66, U.S. Antarctic Journ., 1, 207-208, 1966.

Hansen, J., D. Johnson, A. Lacis, S. Lebedeff, P. Lee, D. Rind and G. Russell, Climatic impact of increasing atmospheric carbon dioxide, Science, 213, 957-966, 1981.

Hooke, R. le B., Pleistocene ice at the base of the Barnes Ice Cap, Baffin Island, N.W.T., Canada J. Glaciol., 17, 49-60, 1976.

Imbrie, J. and N.G. Kipp, A micropaleontological method for quantitative paleoclimatology: Application to a late Pleistocene Caribbean core, In The Late Cenozoic Glacial Ages (ed. K.K. Turekian), Yale Univ. Press, 71-181, 1971.

Iversen, J., The late-glacial flora of Denmark and its relation to climate and soil, Geol. Surv. of Denmark, Ser, II, No. 80, 87 pp, 1954.

Johnsen, S.J., Stable isotope homogenization of polar firn and ice, Internat. Union og Geodesy and Geophysics, Internat. Ass. Sci. Hydrol., Commission of Snow and Ice. Symp. on Isotopes and impurities in snow and ice, Grenoble Aug./Sept., 1975. IAHS-AISH publication No. 118, 210-219, 1977.

Johnsen, S.J., W. Dansgaard, H.B. Clausen and C.C. Langway, Jr., Oxygen isotope profiles through the Antarctic and Greenland ice sheets, Nature, 235, 429-434, 1972.

Johnsen, S.J. and N.O. Andersen, On power estimation in maximum entropy spectral analysis, Geophysics, 43, No. 4, 681-690, 1977.

Johnsen, S.J., W. Dansgaard, N. Gundestrup, S.B. Hansen, J.O. Nielsen and N. Reeh, A fast lightweight core drill, J. Glaciol., 25, 169-174, 1980.

Langway, C.C., Jr., Stratigraphic analysis of a deep core from Greenland, Geol. Soc Am., Special Paper, 125, 186 pp. 1970.

Langway, C.C., Jr., W. Dansgaard and H. Oeschger (eds.), Proc. Symp. GISP 1971-1981, Am. Geophys. Union Meeting, Philadelphia May-June, 1982, To be published, A.G.U. Special vol., 1983.

Lorius, C., L. Merlivat, J. Jouzel and M. Pourchet, A 30,000 yr isotope climatic record from Antarctic ice, Nature, 280, 644-648, 1979.

McIntyre, A., W.F. Ruddiman, and R. Jantzen, Southward penetration of the North Atlanatic polar front: Faunal and floral evidence of large-scale surface water mass movements over the last 225,000 years, Deep Sea Res., 19, 61-77, 1972.

Mercer, J.H., West Antarctic ice sheet and CO_2 greenhouse effect: a threat of disaster, Nature, 271, 321-325, 1978.

Mitchell, J.M., Jr., On the world-wide pattern of secular temperature change, in Changes of Climate, Arid zone research XX, UNESCO, Paris, 161-181, 1963.

Neftel, A., H. Oeschger and H.E. Suess, Secular non-random variations of cosmogenic carbon-14 in the terrestrial atmosphere, Earth and Planet. Sci. Let., 56, 127-147, 1981.

Nilsson, E., Södra Sveriges senkvartara historie (English summary), Kungl. Svenska Vetenskaps Skad. Handlingar, Ser. 4, vol 12, No. 1, 177 pp, 1968.

Oerlemans, J., Some basic experiments with a vertically-integrated ice sheet model, Tellus, 33, 1-11, 1981.

Oeschger, H., J. Beer, U. Siegenthaler, B. Stauffer, W. Dansgaard and C.C. Langway, Jr., Late-glacial climate history from ice cores, This volume, - , 1983.

Paterson, W.S.B., R.M. Koerner, D.A. Fisher, S.J. Johnsen, H.B. Clausen, W. Dansgaard, P. Bucher and H. Oeschger, An oxygen-isotope climatic record from the Devon Island ice cap, Arctic Canada, Nature, 266, 508-511, 1977.

Petit, J.-R., M. Briat and A. Roger, Ice age aerosol content from East-Antarctic ice core samples and past wind strength, Nature, 293, 391-394, 1981.

Picciotto, E., X. de Maere and I. Friedman, Isotopic composition and temperature of formation of Antarctic snows, Nature, 187, 857-858, 1960.

Pisias, N.G., J.P. Dauphin and C. Sancetta, Spectral analysis of late Pleistocene-Holocene sediments, Quat. Res., 3, 3-9, 1973.

Post, A.S., The exceptional advances of the Muldrow, Black Rapids and Susitna glaciers, J. Geophys. Res., 65, 3703-3712, 1960.

Post, A.S., Distribution of surging glaciers in western North America, J. Glaciol., 8, 229-240, 1969.

Raisbeck, G.M., F. Yiou, F. Fruneau, J.M. Loiseaux, M. Lieuvin, J.C. Ravel and C. Lorius, Cosmogenic ^{10}Be concentrations in Antarctic ice during the past 30,000 years, Nature, 292, 825-826, 1981.

Reeh, N., A plasticity theory approach to the steady state shape of a three-dimensional ice sheet, J. Glaciol., 28, 431-455, 1982.

Reeh, N., Reconstruction of the Greenland and Innuitian ice sheets under glacial conditions by 3-dimensional perfectly plastic ice sheet modeling, Symp. on Ice and Climate Modeling, Evanston, Ill., U.S.A., June-July 1983. To be published in Ann. of Glaciol., 1983.

Reeh, N., H. B. Clausen, W. Dansgaard, N. Gundestrup, C.U. Hammer and S.J. Johnsen, Secular trends of accumulation rates at three Greenland stations, J. Glaciol., 20, 27-30, 1978.

Ruddiman, W.F. and L.K. Glover, Subpolar North Atlantic Circulation at 9,300 yr B.P.: Faunal Evidence, Quat. Res., 5, 361-389, 1975.

Ruddiman, W.F. and A. McIntyre, The North Atlantic ocean during the last deglaciation, Palaeogeogr., Palaeoclimatol., Palaeoecol., 35, 145-214, 1981.

Rufli, H., B. Stauffer and H. Oeschger, Lightweight 50 m core drill for firn and ice, Proc. Symp. Ice-core drilling (J.F. Splettstoesser ed.), Univ. of Nebraska Press, Lincoln, Nebraska, 139-153, 1976.

Schultz, C.B. and J.C. Frye, eds., Loess and related eolian deposits of the world: University of Nebaska Press, 355 pp, 1968.

Shoji, H. and C.C. Langway, Jr., Mechanical properties of fresh ice core from Dye-3, Greenland, Am. Geophys. Union Meeting, Philadelphia May-June, 1982, To be published, A.G.U. Special vol., 1983.

Siegenthaler, U., U. Eicher, H. Oeschger and W. Dansgaard, Lake sediments as continental ^{18}O records from the transition glacial-postgalcial, Proc. Symp. ice and Climate Modeling, Evanston, Ill. U.S.A., June 1983, To be published in Ann. of Glaciol., 1983.

Sonett, C.P. and H.E. Suess, Correlation of Brislecone pine ring widths with atmospheric carbon-14 variations: a climate-sun relation. Preprint to be published, 1983.

Stauffer, B., H. Hofer, H. Oeschger, J. Schwander and U. Siegenthaler, Atmospheric CO_2 concentration during the last glaciation, Proc. Symp. Ice and Climate Modeling, Evanston, Ill., U.S.A., June 1983, To be published in Ann. Glaciol., 1983.

Stuiver, M., Variations in radiocarbon concentration and sunspot activity, J. Geophys. Res., 66, 273-276, 1961.

Stuiver, M. and P.D. Quay, Changes in Atmospheric carbon-14 attributed to a variable Sun, Science, 207, 11-19, 1980.

Suess, H.E., Climatic change, solar activity, and the cosmic-ray production rate of the natural radiocarbon, Meteorol. Monographs, 8, 146-150, 1968.

Suess, H.E., The radiocarbon record in tree rings of the last 8000 years, Radiocarbon, 22, 200-209, 1980.

Thompson, L.G., Variations in microparticle concentration, size distribution elemental composition found in Camp Century, Greenland, and Byrd Station, Antarctica, deep cores, Proc. Symp. Isotopes and Impurities in Snow and ice, Intern. Ass. Hydrol. Sci., Publ. No. 118, 351-364, 1977.

Ueda, H.T. and D.E. Garfield, The USA CRREL drill for thermal coring in ice, J. Glaciol., 8, 311-314, 1969.

Ulrych, T.J. and T.N. Bishop, Maximum entropy spectral analysis and autoregresive decomposition, Rev. Geophys. Space Physics, 13, 183- , 1975.

deVries, H., Variation in concentration of radiocarbon with time and location on earth, Koninkl. Nederl. Akad. van Wetenschappen, Amsterdam, Proc. Ser. B, 61, 94-102, 1958.

Watts, W.A., Regional variation in the response of vegetation to late glacial climatic events in Europe, In Studies in the lateglacial of North-west Europe (ed. J. J. Lowe, J. M. Gray and J. E. Robinson), Pergamon Press, Oxford, 1-22, 1980.

Weertman, J., Rate of growth and shrinkage of non-equilibrium ice sheets, J. Glaciol., 5, 145-158, 1964.

Wegmuller, S. and M. Welten, Spatgaziale Bimstufflagen des Laacher Vulkanismus in Gebiet der westliched Schweiz und der Dauphine (F)., Eclogae Geol. Helv., 66, 533-541, 1973.

Wilson, A.T., Origin of ice ages: an ice shelf theory for Pleistocene glaciation, Nature, 201, 147-149, 1964.

Wilson, A.T. and C.H. Hendy, The chemical stratigraphy of polar ice sheets - a method of dating ice cores, J. Glaciol., 27, 3-9, 1981.

LATE GLACIAL CLIMATE HISTORY FROM ICE CORES

H. Oeschger, J. Beer, U. Siegenthaler and B. Stauffer

Physics Institute, University of Bern, Switzerland

W. Dansgaard

Geophysical Isotope Laboratory, University of Copenhagen, Denmark

C. C. Langway

Dept. of Geological Sciences, State University of New York at Buffalo, Amherst, N. Y., USA

Abstract. Ice cores contain information on climatic variations and their causes. Recent results obtained on the new deep ice core drilled in 1981 at Dye 3, South Greenland, in the frame of the US-Danish-Swiss Greenland Ice Sheet Program, are:

- Comparison of the $\delta^{18}O$ variations in the Greenland ice cores with those in European lake carbonate exhibits strong similarities and provides time marks (13,000, 11,000, 10,000 B.P.) for the Late Glacial section of the ice cores;
- CO_2 concentration measurements in the occluded air indicate low (180-200 ppm) CO_2 concentration 30,000 to 15,000 B.P. and an increase to ca. 300 ppm around 13,000 B.P. The CO_2 increase might reflect a change in the ocean circulation at the end of the last glaciation and could have contributed to the establishment of the Holocene environmental conditions;
- ^{10}Be concentration measurements on samples covering the last 50,000 years show a correlation with $\delta^{18}O$, low $\delta^{18}O$ values corresponding to high ^{10}Be concentrations (atoms per g of ice). It is likely that this correlation primarily reflects changes in the rate of precipitation in the northern hemisphere.

Based on the ice core information, climatic events during the Glacial-Postglacial transition are discussed.

Introduction

In the last 5 to 10 years the potential of ice core studies as a means for reconstructing the history of the environmental system and for better understanding its mechanisms has been recognized by a broad spectrum of disciplines involved in earth and planetary sciences. Reasons for this development are:

- the successful ice core drilling in polar areas, e.g. the drilling through the Greenland ice sheet in 1981 at Dye 3, South Greenland, in the frame of the US-Danish-Swiss Greenland Ice Sheet Program (GISP) [AGU, 1983];
- recent progress in the development of highly sensitive field and laboratory analytical techniques. Examples are spectroscopic CO_2 analysis, accelerator mass spectrometry, electrical conductivity measurements and ion chromatography, enabling the study of new types of environmental parameters;
- strenghtened attempts to understand environmental system processes in view of possible natural and/or anthropogenic climatic changes.

In this paper we discuss information obtained from ice core analyses covering about the last 60,000 years of the Dye 3 ice core. First, the $\delta^{18}O$ records of the Greenland ice core and of ^{14}C dated lake sediments from Central Europe are compared. Close similarities between the late glacial climatic events in the two regions are revealed, and permit the definition of time horizons for the Greenland $\delta^{18}O$ profiles. Then CO_2 concentrations of the air trapped in the ice are presented and the implications regarding carbon cycle and climate are discussed. Furthermore, ^{10}Be data are given together with possible explanations for changes observed. Based on the information on these three parameters, conclusions and more speculative ideas regarding the climatic evolution during the Glacial-Postglacial transition, are drawn.

Comparison of the $\delta^{18}O$ Records from Dye 3 and from Central European Lake Sediments

A continuous $\delta^{18}O$ profile has been measured along the newly recovered deep ice core from

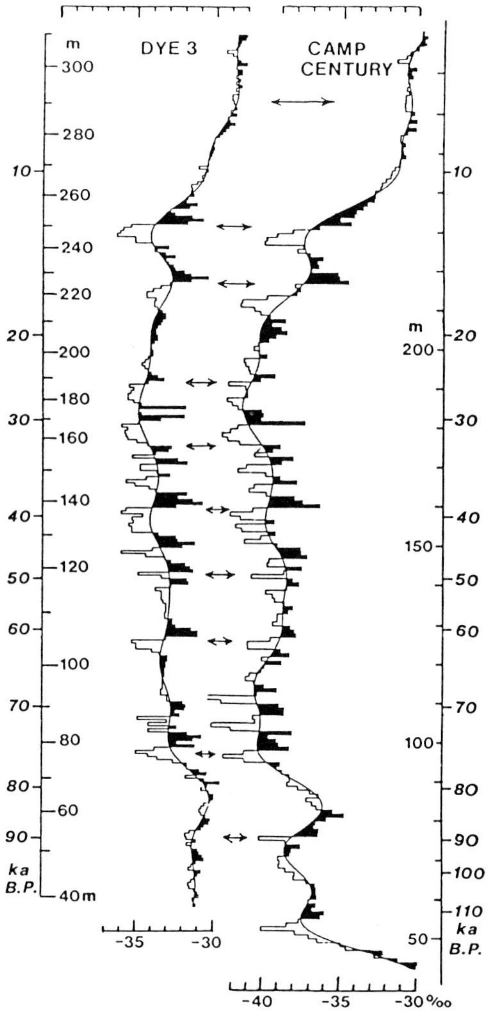

Fig. 1. Comparison of the deepest few hundred meters of two $\delta^{18}O$ profiles through the ice sheet at Dye 3, SE Greenland and at Camp Century, NW Greenland.

Southern Greenland (Fig. 1). Dansgaard et al. (1982) compared the features of this $\delta^{18}O$ profile with those observed for the Camp Century ice core drilled in the 1960's. It appears that for the late Wisconsin essentially all the oscillations in the Dye 3 record can be correlated with similar features in the Camp Century record. The authors concluded that, in view of the distance of more than 1400 km between the drill sites and of the different ice flow conditions, the extreme oscillations in $\delta^{18}O$ are to be attributed to climatic changes in the Arctic.

$\delta^{18}O$ variations of precipitation are recorded not only in polar ice, but also in authigenic carbonate in lake sediments. Late Glacial $\delta^{18}O$ profiles of carbonate from European lake sediments show significant variations which can be correlated with climatic changes as inferred from pollen analysis. Fig. 2 shows, as an example, the $\delta^{18}O$ record of a sediment profile from Lake Gerzensee, Switzerland (Eicher, 1980). At several other sites in Central Europe, $\delta^{18}O$ profiles have been obtained which exhibit the same major features (Eicher and Siegenthaler, 1976; Eicher et al., 1981). A first major warming around 13,000 B.P. (^{14}C age) accompanied by a rapid increase in carbonate $\delta^{18}O$, corresponding to the Bolling-Allerod warm phase (pollen zones Ib-II); a transition to the Younger Dryas cold phase (pollen zone III) at about 11,000 B.P. Each of the three climate changes is marked by an abrupt $\delta^{18}O$ shift which indicates that the climate in Central and NW Europe changed drastically at those times.

That part of the Dye 3 $\delta^{18}O$ profile which represents the Glacial-Postglacial transition period is compared in Fig. 3 with the Gerzensee $\delta^{18}O$ profile. The similarity of the $\delta^{18}O$ fluctuations between 1785 m and 1815 m depth with those observed in the lake sediments is striking: correlation of the major fluctuations is strongly suggested. Also, some of the minor fluctuations seem to be correlated, however, a more detailed study, including the $\delta^{18}O$ record from Camp Century and those of additional $\delta^{18}O$ lake marl records, is necessary to confirm this. This strongly suggests that the two $\delta^{18}O$ profiles reflect the same evolution of climatic events. We therefore assume, for the Greenland $\delta^{18}O$ record, the approximate time marks on the ^{14}C time scale of 13,000 B.P., 11,000 B.P. and 10,000

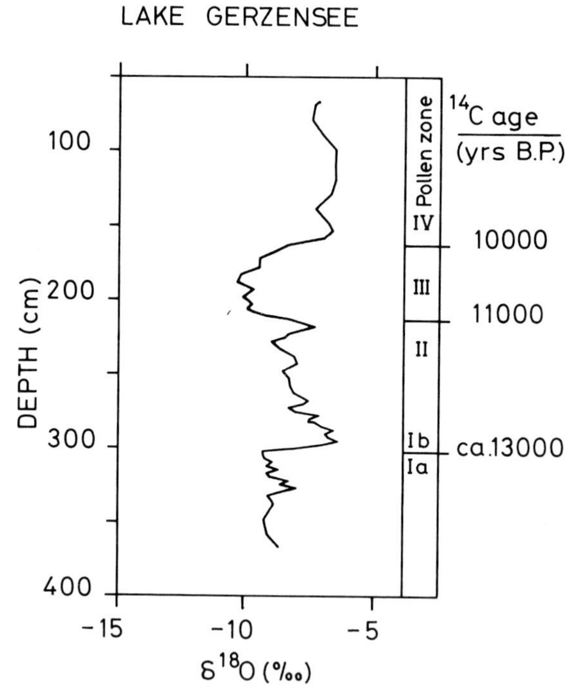

Fig. 2. $\delta^{18}O$ vs. depth in marl sediment of Lake Gerzensee, Switzerland, (from Eicher, 1980). The profile section covers the Late Glacial and Early Postglacial period, see approximate time scale at the right. The three pollen zone boundaries at ca. 13,000, 11,000 and 10,000 B.P. are marked by large $\delta^{18}O$ shifts.

B.P., as indicated in Fig. 2. This time scale should, of course, be verified independently, which will be possible using accelerator mass spectrometry for ^{14}C dating of the CO_2 in air bubbles occluded in the ice. The comparison of our time scale shown in Fig. 3, with ice flow calculations assuming constant precipitation rate, suggests that between 10,000 and 13,000 B.P. precipitation rates were about a factor of two to three lower than in the Holocene.

The Milankovitch hypothesis, that changing seasonal solar irradiation due to variations of the earth's orbital parameters is the primary cause for the glaciation cycles, has obtained considerable support in the past few years e.g. by spectral analyses of palaeoclimatic records (Imbrie and Imbrie, 1980). The corresponding insolation variations are, however, relatively slow compared to the climate fluctuations revealed in the Greenland ice and in European lake sediments. These events, however, can be linked to oceanic changes, namely advances and retreats of cold polar water in the North Atlantic Ocean, as documented by faunal and stratigraphic sediment studies (Ruddiman and McIntyre, 1981; Duplessy et al., 1981). The periods of southward advance of polar water coincide with the cold phases, those of polar water retreat with the warm phases on the European continent. This strongly suggests that the continental climatic shifts were an effect of changes of the surface conditions of the North Atlantic Ocean.

CO_2 Concentrations Along the Dye 3 Ice core

The second parameter to be discussed is the CO_2 content of the air occluded in the Dye 3

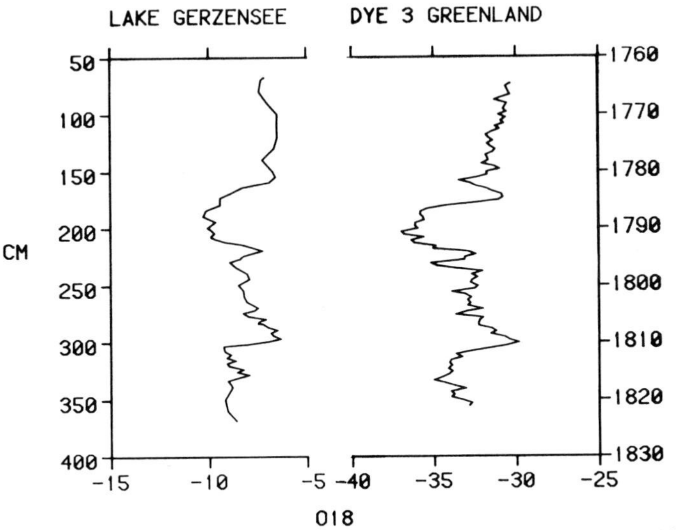

Fig. 3. Comparison of a section of the $\delta^{18}O$ profile from the Dye 3 ice core (right) with the $\delta^{18}O$ record in lake carbonate from Gerzensee (left). The strong similarities suggest that both records represent the same sequence of climatic events and thus the same time period.

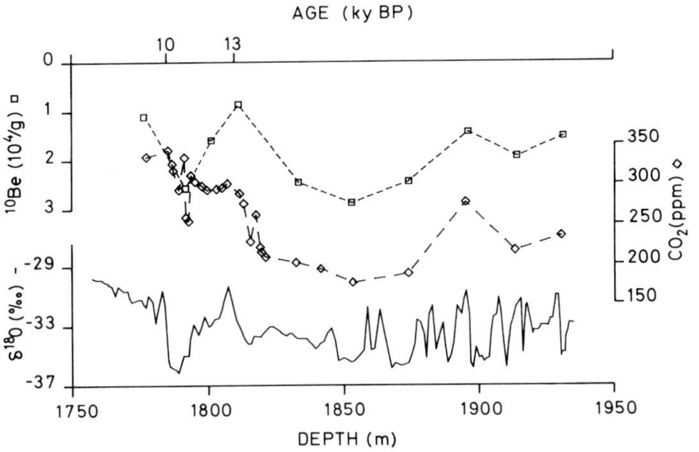

Fig. 4. ^{10}Be concentrations (10^4 atoms per g of ice), CO_2 concentrations (ppm) and $\delta^{18}O$ as measured in the Dye 3 ice core. Top: Tentative time marks, as suggested by the comparison with European lake sediments (Fig. 3).

ice. During the transition from firn to ice, at Dye 3 presently occurring in a depth interval from 65 to 80 m, air parcels are occluded as bubbles in the freshly formed ice. Thus ice continuously collects air samples. As has been shown by different groups (Raynaud and Delmas, 1977; Scholander et al., 1961; Stauffer, 1981) the concentration of the major constituents of the occluded air (N_2, O_2, Ar) closely reflects the atmospheric composition, if summer melting can be excluded. We therefore assume that the CO_2 concentration of this air also corresponds to the original atmospheric value as long as no melt layers are present. If melting occurs, higher CO_2 concentrations are observed because CO_2 is strongly soluble in liquid water and during refreezing gets trapped preferentially compared to the main air gases. The air in the bubbles represents the composition of the atmosphere at the time of air occlusion; it is therefore younger than the surrounding ice.

In Fig. 4 the CO_2 concentrations observed in the Dye 3 core between 1780 and 1940 m depth, corresponding approximately to the period 9,000 to 60,000 B.P., are represented, together with $\delta^{18}O$ and ^{10}Be data. CO_2 concentrations were determined by crushing ice samples of ca. 1 cm^3 and measuring the CO_2 content of the escaping air by means of an infrared laser spectrometer. The precision of the analyses is about ±6 ppm. It can be assumed that during the glacial period no summer melting occurred so that the CO_2 composition of the ice age atmosphere is reflected without alterations. This does not hold for the Holocene ice which is affect ed by melting. CO_2 concentrations have been measured in only a small part of the Holocene ice. The data obtained so far indicate that all the Holocene ice is affected by melting (Stauffer et al., 1983).

The following main results are observed (Fig.

4). For the late Wisconsin (1829 m to 1870 m, corresponding to about 14,000 B.P. to 30,000 B.P.) low CO_2 values in the range of 180 to 200 ppm are obtained, in agreement with earlier measurements on the ice cores from Camp Century, Greenland, and Byrd Station, Antarctica (Neftel et al., 1982), as well as on other Antarctic ice cores (Delmas et al., 1980). The CO_2 increase as observed in the Dye 3 core occurs at almost the same depth as the $\delta^{18}O$ increase indicating the first warming at the end of the last glaciation, ca. 13,000 B.P. The depth at which the CO_2 shift takes place cannot, based on the present data, be exactly defined; it is a few m below the $\delta^{18}O$ shift (at Dye 3). The age differences between the trapped air and the surrounding ice is at present ca. 100 years and was probably somewhat larger 13,000 B.P. because of smaller accumulation rate and lower temperatures. The annual layer thickness can be estimated from Fig. 3 to 1 cm or slightly less, so that the depth differences between the shifts of $\delta^{18}O$ and CO_2 would be between 1 and 3 m if both changes actually occurred at the same time. We therefore conclude that the two parameters changed almost simultaneously with a possible time difference of a few centuries at maximum.

The question of the time relationship between CO_2 and $\delta^{18}O$ increase is fundamental for the understanding of the climatic mechanisms which lead to the end of last glaciation, as well as for the general question of the sensitivity of global climate to CO_2 concentration changes. It deserves a more detailed treatment based on more experimental work and requires an improved understanding of the gas enclosure mechanism in polar ice.

For the period 13,000 to 11,000 B.P. relatively constant CO_2 values in a band between 300 and 320 ppm are observed. These values may be slightly affected by melt layer ice which is enriched in CO_2. At about 11,000 B.P., the beginning of the Younger Dryas cold period, two low CO_2 concentration values of about 250 ppm are observed; followed again by a value slightly above 300 ppm and a series of increasing values from 280 to 330 ppm at the Younger Dryas-Preboreal transition. It is of interest that a Younger Dryas effect on the CO_2 concentration, though relatively small, seems to exist. Additional measurements are needed to clearly establish the detailed CO_2 concentration during the Younger Dryas and the adjacent periods.

Upon considering the CO_2 data from Camp Century, Greenland, and Byrd Station, Antarctica, we conclude that at the end of the last glaciation, the CO_2 concentration of air occluded in polar ice from Greenland, as well as from Antarctica, increased from 180-200 ppm to 280-300 ppm. The transitions are parallel to the $\delta^{18}O$ shifts. Since the changes of the atmospheric CO_2 concentration have to be global, they can be used as time marks in ice core studies. Further detailed CO_2 measurements on ice cores will show whether the change in the CO_2 concentration is synchronous with the $\delta^{18}O$ change in Antarctica as well. This would suggest that the Glacial-Postglacial transition in the two hemispheres occurred synchronously.

The detailed Dye 3 measurements indicate that the atmospheric CO_2 increase and the first warming occurred nearly at the same time (around 13,000 B.P.), with a maximum time difference of a few centuries. During the climatic deterioriation of the Younger Dryas period, which almost reestablished glacial conditions, lower CO_2 concentrations have also been observed. To confirm these observations, more detailed measurements are needed.

Possible Causes for the Atmospheric CO_2 Variations

The question arises as to the mechanisms responsible for the atmospheric CO_2 increase about 13,000 B.P. On the time scale of interest ($\sim 10^4$y), the main carbon reservoirs with which the atmosphere exchanges actively are the ocean and the biosphere. In view of the large carbon content of the ocean, about 60 times that of the atmosphere, the atmospheric CO_2 concentration is mainly regulated by the chemical conditions in the ocean surface waters. Siegenthaler et al. (1980) have estimated that at 18,000 B.P., the atmospheric CO_2 concentration would have been about 15 ppm lower than in the Holocene due to the changes of ocean temperature and volume. This is much smaller than the observed change of nearly 100 ppm. Atmospheric pCO_2 depends strongly on total CO_2 and alkalinity of surface waters, which in turn are influenced by biological processes in the sea. Changes in pCO_2 are probably linked to the production of organic matter. Based on ^{13}C measurements on deep-sea sediment and on observations on changes of lysocline depth, Broecker (1982) estimated that the physico-chemical conditions in the glacial ocean were such that pCO_2 was 80 ppm lower during glacial than during postglacial time. The main mechanisms responsible for these changes would be deposition of organic matter on the continental shelves, newly water-covered due to the sea level rise during the Glacial-Postglacial transition and of deposition of carbonate in the deep sea. While Broecker's hypothesis explains the observed changes in a plausible and consistent way in the long run, his time scales are several thousand years, and thus too long to account for the rapid variations suggested by the ice core observations.

Superimposed on a variation of the type proposed by Broecker, we suggest a different mechanism for the rapid CO_2 changes. This is based on the observation that the phosphate concentration in surface water, while being nearly zero in large areas of the world ocean, is not so depleted in the Equatorial Pacific and in the

Antarctic Ocean; which are approximately 20% and 70%, respectively, of their deep water values. This suggests that marine bioproductivity is not limited by the availability of phosphorus. If the upwelling of deep water was slower, the organisms would then have time enough to consume all the phosporus available; the consequence being that both the total CO_2 and pCO_2 concentrations would be reduced. A rough estimate shows that a complete depletion of phosphate in all surface waters would involve a drop of pCO_2 from 300 ppm to approximately 200 ppm. A possible explanation for the relatively rapid CO_2 increase around 13,000 B.P. is as follows: a transition of the ocean from a relatively stagnant state, with low productivity due to complete nutrient consumption (therefore, low pCO_2 in the surface water), to a state with more rapid vertical mixing, thus, preventing the consumption of all phosphate, therefore implying higher total CO_2 and pCO_2 in part of the surface water. Indeed, Worthington (1968) postulated a rapid influx of meltwater in the period of deglaciation, resulting in a stabilized low salinity layer at the ocean surface-near waters leading to limited vertical mixing. As discussed in detail by Ruddiman and McIntyre (1981), the effects of this meltwater input process was most significant in the North Atlantic Ocean and led to a drastically reduced biological production. Around 13,000 B.P. the polar front in the North Atlantic Ocean was affected by glacier meltwater and iceberg production, and retreated rapidly initiating vertical exchange of oceanic water layers. This brought nutrient-rich deep water to the surface. According to our hypothesis, this could have induced the observed rapid CO_2 increase at the Oldest Dryas-Bolling transition.

The meltwater input was a phenomenon associated with deglaciation which lasted several millenia, and our results suggest low CO_2 during that period. At present, there are not enough data to know how large the CO_2 concentration was in earlier phases of glacial time. However, the higher values observed in the lowest part of the profile (see Fig. 4) suggest that the CO_2 concentration was not uniformly low during the Ice Age, but that fluctuations occurred. This has indeed been confirmed by very recent measurements (Stauffer et al., 1983).

Nevertheless, there are several open questions regarding the suggested mechanisms. The evidence for ocean circulation changes comes mainly from the North Atlantic, whereas the key region for incomplete nutrient consumption is the Antarctic Ocean. How was this region affected by the meltwater input in the North or by other oceanic changes? Is there any evidence of a change in an oceanic nutrient concentration around Antarctica? What was the effect of greater sea-ice coverage and, thus, reduced open sea surface during glacial conditions?

In spite of these questions, it is clear that (1) atmospheric CO_2 variations <u>can</u> be caused by changes of ocean circulation in regions where phosphorus concentration is significantly different from zero, and (2) events in the ocean <u>did occur</u> rapidly in late glacial time and that these events may have involved large-scale circulation changes.

Climatic Implications of the Atmospheric CO_2 Increase During the Glacial-Postglacial Transition

Estimates of global temperature change based on different types of climate models indicate an almost logarithmic dependence of temperature on the atmospheric CO_2 concentration. The mean global temperature change corresponding to a CO_2 increase by a factor of 1.5 from 200 ppm to 300 ppm, as calculated by climate models, is about $1.5 \pm 0.5°C$. Climate models also indicate a strongly increased warming over ice-covered areas. The warming which probably occurred as a result of the global CO_2 increase, therefore, was probably amplified in regions of sea ice and on the continental shelves, leading to additional melting of continental ice and the disappearance of sea ice.

A CO_2 induced warming at the end of the last glaciation would have provided an interhemispheric coupling. Such a strong interhemispheric coupling is, however, a necessity if the climatic change in both hemispheres occurred simultaneously. The rising CO_2 level in the end phase of the last glaciation influenced the radiation balance of the earth and amplified the general warming tendency. In addition, one might speculate that it contributed to the regrowth of vegetation due to CO_2 fertilization, which involved an albedo decrease over the continents, thus, providing an indirect CO_2-climate effect.

Taken together, these effects, resulting from a global CO_2 increase, may have significantly contributed to the establishment of the present interglacial climatic conditions.

The ^{10}Be Record in the Dye 3 Ice Core

The development of accelerator mass spectrometry rendered feasible the measurements of the number of ^{10}Be atoms in one kg of precipitation. ^{10}Be ($t_{1/2} = 1.5 \times 10^6 y$) is mainly produced by spallation reactions on atmospheric nitrogen and oxygen. The freshly produced ^{10}Be becomes attached to aerosols within a short time, and after a mean residence time in the atmosphere of ca. 1 to 2 years, is deposited on the earth's surface by precipitation. ^{10}Be removed from the atmosphere is stratigraphically preserved in polar ice cores. Its concentration shows variations due to production rate changes and variations in atmospheric mixing and deposition. Recent ^{10}Be measurements on ice samples from Dye 3, covering the last seven sunspot cycles, seem to reflect the known variation of cosmic ray interaction with the atmosphere over the 11 year sunspot cycle (Beer et al., 1983).

For the Maunder minimum period (1640 to 1710) of the quiet sun Raisbeck et al. (1981) found ^{10}Be concentrations in Antarctic ice higher by a factor of 1.5 than observed for other periods.

More detailed measurements on ice samples from Milcent which we performed using the Zurich accelerator mass spectrometer, have confirmed this result (Beer et al., 1983). Besides production rate, other factors may influence ^{10}Be concentration. In periods of major climatic change, changes in the ^{10}Be concentration in precipitation may also have been caused by changes in the atmospheric mixing and circulation mode, or by variations in precipitation rates.

The results from the Dye 3 core, covering the last 30,000 years, show a correlation both with $\delta^{18}O$ and CO_2 (Fig. 4): low ^{10}Be concentrations are found in samples with high $\delta^{18}O$ and high CO_2 concentrations. Several explanations are possible. At present, we favor the explanation that the variations of ^{10}Be concentration, expressed as atoms per g of water, mainly reflect changes of the precipitation rate. For the end of the Glaciation, ice age theories point at the possibility of reduced precipitation leading to a starvation of the ice sheets. The high ^{10}Be concentrations between 1820 and 1870 m depth might well reflect this phenomenon. Interestingly enough, ^{10}Be and CO_2 correlate with $\delta^{18}O$, even for relatively short term fluctuations, e.g. a low ^{10}Be concentration at 1885 m depth corresponds to high $\delta^{18}O$ values, and a high ^{10}Be concentration at 1785 m depth to low $\delta^{18}O$ (Younger Dryas) values. This again supports the explanation that part of the observed ^{10}Be fluctuations between 10,000 and 30,000 B.P. are caused by changes of the precipitation rate, although we do not exclude the possibility of contributions due to variations of the production rate associated with changing solar and terrestrial magnetism; note that solar-induced ^{10}Be variations, by a factor of 1.5, have indeed been found for the Maunder minimum.

Conclusions

Measurements of $\delta^{18}O$, CO_2 and ^{10}Be on ice cores provide new insight into the processes at the glacial-postglacial transition. Changes of the position of the oceanic polar front in the North Atlantic (Ruddiman and McIntyre, 1981), in the period 16,000 to 10,000 B.P., probably influenced the climatic events in Europe and Greenland as reflected in the climatic parameters recorded in ice cores, lake sediments and peat bogs. Around 13,000 B.P. a first warming took place, expressed by the retreat of the deglacial polar front. A significant increase of $\delta^{18}O$ values in Greenland ice and in Central European lake sediments, combined with a shift of the vegetation from tundra to forest in Europe was noticed. This first warming was accompanied by a CO_2 increase which caused a significant change in the global radiation balance, equivalent to a global warming of ca. 1.5°C. This phenomenon could have lead to a retreat of the sea ice cover in both hemispheres, providing a positive feedback to the developing climatic transition. In addition, the CO_2 increase may have supported the vegetation regrowth which caused a reduction of the global albedo, creating another positive feedback. At 11,000 B.P., a reestablishment of glacial conditions with a strong advance of the deglacial polar front occurred, possibly caused by a rapid disintegration of continental ice masses (Ruddiman and MacIntyre, 1981). This is reflected in the $\delta^{18}O$ profiles, both in Greenland ice and in European lake sediments. Deep sea cores show a decrease of foraminiferal productivity and a decrease in upwelling of nutrient-rich deep water. Some ice core samples of this period also show low CO_2 concentrations. At 10,000 B.P. the second (final) warming took place. This is reflected by high $\delta^{18}O$ values, high CO_2 concentrations and relatively low ^{10}Be concentrations in the ice core, and in high foraminiferal activity in the North Atlantic Ocean, indicating a strong retreat of the deglacial polar front.

By means of more core drillings and further detailed CO_2 and $\delta^{18}O$ analyses it should become possible, using the CO_2 increase as a time mark, to check if the transition Glacial-Postglacial was synchronous in both hemispheres.

The atmospheric CO_2 concentration is regulated by changes in the ocean's total CO_2 concentration, alkalinity, temperature and the content of nutrients (PO_4 and NO_3); the latter controls the biospheric activity in the surface ocean and, thus, the reduction of the CO_2 partial pressure relative to that of the average ocean water. Broecker (1982) has emphasized the role of phosphorous storage in the sediments as an important mechanism controlling the atmospheric CO_2 content in order to explain the fast CO_2 rise, within a few centuries, at 13,000 B.P. However, this process is too slow to explain the observed rapid CO_2 change. We propose that CO_2 changes were a consequence of a change in the ocean's turnover rates; a slow turnover (ice age) leaving enough time for a complete nutrient consumption involving low pCO_2, and a faster turnover (Holocene) leading to incomplete nutrient consumption with a correspondingly higher pCO_2. It will be interesting to determine, by means of precise CO_2 measurements on ice cores, whether significant CO_2 variations also occurred during the Glacial and in the Holocene.

The combined detailed analysis of parameters recorded in polar ice reveals valuable information on terrestrial and planetary processes. The main aspects are:

- The ice core information is of very high resolution, enabling the reconstruction of rapid climatic changes superimposed on the low frequency climatic functions which are probably a main result of the changes in the orbital elements of Earth. Responsible for the short-term fluctuations seem to be changes in ocean circulation with physical and chemical implications, rapid desintegration of continental ice and possibly changes of solar properties combined with fluctuations in electromagnetic and plasma emission.

- Natural glacier ice is the only known terrestrial archive which has the ability to continuously collect and store air samples representing the composition of the ancient atmosphere.
- Results of the CO_2 concentration variations measured in ice cores indicate that a signifi cant role may be played by atmospheric CO_2 concentration levels. This record is even more important since at present anthropogenic activities, over periods of decades to a few centuries, lead to considerable changes in the concentrations of atmospheric CO_2 and other trace gases.
- Accelerator mass spectrometry enables detailed high resolution measurements of cosmic ray produced radio isotopes deposited in natural ice. Combined interpretation of their data with data on soluble and particulate matter concentrations should enable us to disentangle the modulating terrestrial and solar effects. In this respect, the discussion of ice core parameters as related to climatic change given in this paper far from covers the whole spectrum of data contained in the ice. Equally important results can be obtained from other measurements, such as soluble and particulate matter, or electrical conductivity studies that are currently in progress.

Acknowledgments. The authors express their gratitude to the many motivated individuals who greatly contributed to the results contained in this paper: in the development and construction of the drilling and analytical equipment, the drilling operation, the analytical work in the field and in the laboratory and in data reduction and interpretation. Our efforts in ice core research have become possible only through many years of support by the U.S. and Swiss National Science Foundations, the Danish Commission for Scientific Research in Greenland, the Danish Natural Science Research Council and the U.S. Department of Energy.

References

AGU, Proceedings of the Spring meeting 1982, Philadelphia, Union session: Greenland Ice Sheet Program, (in press), 1983.

Beer, J., M. Andree, H. Oeschger, B. Stauffer, R. Balzer, G. Bonani, Ch. Stoller, M. Suter, W. Wolfli and R. Finkel, Temporal ^{10}Be variations in ice, Radiocarbon, 25, 269-278, 1983.

Beer, J., U. Siegenthaler, H. Oeschger, M. Andree, G. Bonani, M. Suter, W. Wolfli, R.C. Finkel and C.C. Langway, Temporal ^{10}Be variations, Proceedings of the 18th International Cosmic Ray Conference, Bangalone, Lake volume (in press), 1983.

Berger, W.H., R.F. Johnson, and J.S. Killigley, Unmixing of the deep-sea record and the deglacial meltwater spike, Nature, 269, 661-663, 1977.

Broecker, W.S., Ocean chemistry during glacial time, Geochim. Cosmochim. Acta, 46, 1689-1705, 1982.

Dansgaard W., H.B. Clausen, N. Gundestrup, C.U. Hammer, S.F. Johnsen, P.M. Kristinsdottier, and N. Reeh, A new Greenland deep ice core, Science, 218, 1273-1277, 1982.

Delmas, R.J., J.M. Ascencio, and M. Legrand, Polar ice evidence that atmospheric CO_2 20,000 yr B.P. was 50% of present, Nature, 284, (5752), 155-157, 1980.

Duplessy, J.C., G. Delibrias, J.L. Turon, C. Pujol, and J. Duprat, Deglacial warming of the northeastern North Atlantic Ocean: correlation with the palaeo-climatic evolution of the European continent, Palaeogeogr., Palaeoclimat. Palaeoecol., 35, 121-144, 1981.

Eicher, U., Pollen- und Sauerstoffisotopenanalysen an spatglazialen Profilen vom Gerzensee, Faulenseemoos und vom Regenmoos ob Boltigen, Mitt. Naturforsch. Ges. Bern, 37, 65-80, 1980.

Eicher, U., U. Siegenthaler, Palynological and oxygen isotope investigations on Late-Glacial sediment cores from Swiss Lakes, Boreas, 5, 109-117, 1976.

Eicher, U., U. Siegenthaler, and S. Wegmuller, Pollen and oxygen isotope analyses on Late- and Post-Glacial sediments of the Tourbiere de Chirens (Dauphine, France), Quat. Res., 15, 160-170, 1981.

Imbrie, J., and J.Z. Imbrie, Modeling the climatic response to orbital variations, Science, 207, 943-953, 1980.

Neftel, A., H. Oeschger, J. Schwander, B. Stauffer, and R. Zumbrunn, Ice core sample measurements give atmospheric CO_2 content during the past 40,000 yr, Nature, 295, 220-223, 1982.

Raisbeck, G.M., F. Yiou, M. Fruneau, J.M. Loiseaux, M. Lieuvin, J.C. Ravel and C. Lorius, ^{10}Be concentrations in Antarctic ice during past 30,000 years, Nature, 292, 825-826, 1981.

Raynaud, D., and R. Delmas, Composition des gaz contenus dans la glace polaire, Proceedings of the Int. Symp. on Isotopes and Impurities in Snow and Ice, IAHS-Publication No. 118, 377-381, 1977.

Ruddiman, W.F., and A. McIntyre, The North Atlantic during the last deglaciation, Palaeocol., 35, 145-214, 1981.

Scholander, P.F., E.A. Hemmingsen, L.K. Coachman, and D.C. Nutt, Composition of gas bubbles in Greenland icebergs, Glaciology, 3, 813-822, 1961.

Siegenthaler, U., H. Oeschger, and M. Heimann, ^{14}C variations caused by changes in the global carbon cycle, Radiocarbon, 22, 177-191, 1980.

Stauffer, B., Die Zusammensetzung der Luft in naturlichem Eis, Z.f. Gletscherkunde und Glazialgeologie, 17, 57-78, 1981.

Stauffer, B., H. Hofer, H. Oeschger, J. Schwander and U. Siegenthaler, Atmospheric CO_2 concentration during the last glaciation, Proc. Symposium on Ice and Climate Modelling, Annuals of Glaciology, (in press), 1983.

Stauffer, B., A. Neftel, H. Oeschger, and J. Schwander, CO_2 concentration in air extracted from Greenland ice samples, Proceedings AGU spring Meeting 1982, Philadelphia, (in press), 1983.

Worthington, L.V., Genesis and evolution of water masses, Meteorol. Monographs, 8, 63-67, 1968.

SIMULTANEOUS CLIMATIC CHANGE IN BOTH HEMISPHERES AND SIMILAR BIPOLAR INTERGLACIAL WARMING: EVIDENCE AND IMPLICATIONS*

John H. Mercer

Institute of Polar Studies and Department of Geology and Mineralogy, Ohio State University, Columbus, Ohio 43210

Abstract. Evidence now seems conclusive that orbital changes in some way controlled the succession of glacial and interglacial ages, at least during the second half of the Pleistocene. However, the mechanisms involved are not yet understood. In particular the simultaneity in the Northern and Southern hemispheres of the last interglacial ~125,000 BP, the last glacial maxima ~20,000 and ~14,500 BP, and the following abrupt and rapid warming 14-11,000 BP, so far defies satisfactory explanation. Ability to forecast the relative magnitude of future CO_2-induced climatic warming in the Northern and Southern hemispheres is needed in order to foretell the response of the cryosphere; climatic models give contrasting prognostications. Inconclusive evidence for deglaciation of West Antarctica during the last interglacial suggests that the Antarctic area then warmed by as much as did the Arctic. In many ways the last interglacial was probably an analogue of a future CO_2-induced warm interval; proof or disproof of the suspected exceptional level of last interglacial warmth in Antarctica is needed in order to estimate the sensitivity of the global climate system to CO_2 input.

Introduction

Although there no longer seems much doubt that changing orbital parameters have exerted a controlling influence on glacial-interglacial cycles, the mechanisms that have enabled the resultant small changes in radiation to cause such major changes in climate that have been simultaneous in both hemispheres are still little understood. An essential part of the data base needed for solution of this problem is an accurate chronology of major climatic fluctuations. The interval from about 20,000 to 10,000 BP, which is accessible to reliable dating by the radiocarbon method, was a time of rapid climatic change, comprising the last full-glacial maxima and the glacial-interglacial transition. The magnitude and ^{14}C-dated timing of glacier variations in middle latitudes during this interval give a proxy record of climatic fluctuations in both hemispheres; paleoclimatic reconstructions from marine cores younger than 20,000 years old are less satisfactory because sites with sufficiently high sedimentation rates are scarce, and because ^{14}C age determinations from marine carbonates are less reliable than are those from terrestrial materials, especially wood.

For the future, an important aspect of the sensitivity of climate to rising levels of CO_2 will be the difference, if any, between the Northern and Southern hemispheres, in the resultant warming at high latitudes, as this will determine the major responses of the cryosphere. In attempting to forecast this, a key question is whether the Antarctic warmed as much as did the Arctic during the last interglacial, which is the most recent and, therefore, the most easily studied and the best dated of possible past analogues of a future CO_2-induced warm interval.

Globally Simultaneous Climatic Change: a Fly in the Ointment of the Milankovitch Theory

Imbrie and Imbrie (1979) give a detailed account of the history and current status of the astronomic theory of ice ages which Imbrie (1982) summarizes. They point out that Croll's (1875) version required not only that glaciations had occurred at different times in the Northern and Southern hemispheres, but also that neither hemisphere had experienced a glaciation in the last 80,000 years. At the time opinion was divided as to whether or not this had been so, but the means to answer the questions were not available. By the end of the 19th century geologists realized that the last glaciation had ended nearer 10,000 than 80,000 years ago. For this and other reasons, Croll's theory fell into disrepute, but was later revised and modified by Milankovitch (1941), in which form it was widely accepted. Although during the 1960s it again fell out of favor, especially with those who had accepted at face value finite ^{14}C age determinations beyond the reliable range of the method, it was revived on receiving

*Institute of Polar Studies Contribution No. C-495

strong support, perhaps final vindication, from isotopic studies of paleomagnetically-dated ocean cores (Hays et al., 1976). According to Imbrie and Imbrie (1980), the conclusion is now inescapable that, at least for the past 730,000 years, the climate system has responded to orbital forcing at the frequencies of variation in obliquity and precession. However, the growing body of evidence that the glacial-interglacial cycles in the two hemispheres have been synchronous (within the resolving power of the dating mechanisms) remains what Broecker (1978) calls "a fly in the insolation ointment", leaving many workers with a nagging doubt about the Milankovitch mechanism (Gribbin, 1978).

Some believe that the key to the problem is to take into account the pre-eminent importance for glacial development of the zone between latitudes 50°N and 65°N; the Northern Hemisphere dog is then seen as having wagged the Southern Hemisphere tail, chiefly through the positive feedback of changing Northern Hemisphere albedo as snowfields and ice cover waxed and waned. However, as Broecker (in press) points out, general circulation models suggest that the climatic effect of these albedo changes would be confined almost entirely to the Northern Hemisphere.

Gribbin (1978) and Berger (1980, 1981) accept an explanation that J.D. Hays tentatively put forward during an unpublished discussion at the WMO-IAMAP Symposium on Long-Term Climatic Fluctuations held in 1975 in Norwich, England. Hays' solution was that because of the contrast in the distribution of land and sea in the two hemispheres, for ice-age climates to develop the Northern Hemisphere needs less radiation in summer to enlarge the area of permanent snow, whereas the Southern Hemisphere needs less radiation in winter to enlarge the area of frozen sea; the Milankovitch mechanism tends to produce such conditions simultaneously. However, as Hays soon realized, this hypothesis is inadequate for two reasons: first, because over most of the latitudinal belt in the Southern Ocean where the bulk of sea ice now forms, solar radiation in winter is so small that any modest decrease caused by orbital variations will have little effect; and second, because when winter radiation is minimal, summer radiation is high, and the effect of this on the sea ice will greatly outweigh that of any winter cooling.

Kerr (1983) gives a brief account of the symposium on the Milankovitch theory, held at Lamont-Doherty Geological Observatory in December 1982. According to Kerr, Held suggested that North Atlantic Deep Water linked the hemispheres; this water mass, which originates near Greenland, travels southward the length of the Atlantic Ocean and then rises to near the surface in the Southern Ocean as a relatively warm component of the Antarctic Circumpolar Current. Held here follows Corliss (1982) who points out that production of North Atlantic Deep Water is greatly reduced during glacial intervals, and that the resultant change in composition of the Antarctic Circumpolar Current should have affected the heat budget of the Southern Ocean and, consequently, the position of the Polar Front and the distribution of sea ice. This mechanism is unsatisfactory as an explanation for globally simultaneous changes in temperature, as it would surely have produced changes in the Southern Hemisphere that were detectably later and also less abrupt than the primary changes in the Northern Hemisphere.

Broecker (1982) suggests that the climates of the two hemispheres are linked by changing levels of CO_2 in the atmosphere; as sea level falls and rises in response to the waxing and waning of Northern Hemisphere ice sheets, CO_2 is exchanged between ocean shelf sediments and the atmosphere, resulting in globally simultaneous changes in the greenhouse warming effect. However, Broecker (in press) points to the difficulty of explaining why the faunal and floral changes in the Southern Ocean lead the ^{18}O change (interpreted as the ice volume signal) in the world ocean, and why the ice age terminations in the Southern Hemisphere are so abrupt. He emphasizes the need to learn more about the links between climate, ocean-atmosphere chemistry, and deep-sea ventilation.

Variations of Glaciers in Middle Latitudes of the Northern Hemisphere

North America

For the last glacial maximum and the advance towards it, a disproportionate number of dates from middle latitudes of the Northern Hemisphere, including all those from wood, are from North America. This is because in many sectors of the Middle West the ice advanced into boreal forest, whereas in Europe a belt of tundra bordered the ice. Mayewski et al. (1981) summarize the results obtained by many workers in North America over several decades.

East of the Mississippi the southern margins of the Erie and Michigan lobes of the Laurentide Ice Sheet oscillated near their outermost positions for about 4000 years, according to dated samples of wood from trees overrun by advancing ice. Some parts of the ice margin reached their maxima ~21,000 BP, and others between 19,800 and 17,000 BP. After ~17,000 BP the lobes rapidly receded during the Erie Interstade, and then readvanced several hundred km to reach a lesser maximum 100 km north of their outer limits after 14,780±192 BP and before 14,300±450 BP (Mayewski et al., 1981).

At about the same time as this major readvance of the Erie and Michigan lobes, the Des Moines lobe in Iowa, whose source was in the western part of the Laurentide Ice Sheet, and the Vashon lobe of the Cordilleran Ice Sheet in Washington, advanced to their outermost positions of the last glaciation. The Des Moines lobe reached its maximum before 13,775±300 BP (organic muck on end moraine) and after ~14,150 BP (average of 7 dates, ranging from ~13,820 to ~14,700 BP, from wood beneath till or outwash) (Ruhe, 1969). On the Pacific coast the Vashon lobe is estimated to have reached its maximum 14,500 - 14,000 BP: bracketing ages are 15,000±400 BP and 15,100±300 BP (wood), and 13,650±550 BP (peat), all from about 80 km behind the outer limit of the ice (Mullineaux at., 1965).

Between 14,000 and 9500 BP several readvances interrupted the general recession of the Laurentide Ice Sheet's southern margin, most of them confined to one lobe and thus of doubtful climatic significance. However, both the Erie and Huron lobes readvanced during the Port Huron stade ~13,000 BP (wood 13,100±110 years old in Ontario predates the maximum). Saarnisto (1974) concludes that the rate of retreat of the Laurentide Ice in the Great Lakes area slowed markedly between 11,000 and 10,100 BP, and that this was a climatically caused event coeval with the European Younger Dryas stade; however, his chronology needs to be substantiated (Mayewski et al., 1981). After 10,000 BP, the southern margin of the ice sheet receded rapidly, interrupted only by short-lived local readvances, and the ice sheet had disappeared from the mainland of North America by 6500 BP.

The chronology of mountain glacier variations in middle latitudes of North America is poorly known. In the St. Elias Mountains at lat 60°N on the Alaska-Yukon border, glaciers reached their maxima ~14,000 BP, and then rapidly disintegrated, shrinking to near their present margins by 11,300 BP (Denton, 1974).

Europe

In Europe the timing of the last glacial maximum, in the few places where it has been closely determined, in general agrees with that in North America. No age determinations have been obtained from wood for full-glacial or late-glacial variations; all are from peat, organic sediments, moss or shells. Andersen (1981) summarizes the results of many workers, and lists the important ^{14}C age determinations.

In Germany two samples from a peat band covered by outwash sand from the Brandenburg moraine that marks the southernmost extent of the ice during the last glaciation are 20,270±1000 and 20,475±600 years old (Cepek, 1965). In Great Britain the glacier in Yorkshire reached its greatest extent after 18,500 BP (dates from moss under till (Penny et al., 1969)), and the Irish Sea glacier before then (date from moss on till (Shotton, 1977)). The age of the later major readvance to the Pomeranian moraines in Germany has been estimated at ~15,000 BP, but is poorly dated (Andersen, 1981). The climatic significance of recessional readvances in Scandinavia between 13,000 and 11,800 BP (Mangerud, 1980) is uncertain as they occurred during the apparently uninterrupted deglacial warming of the eastern North Atlantic 13,000 - 11,000 BP, determined by Ruddiman and McIntyre (1981).

About 11,000 BP the last climatically-caused regional standstill or readvance in northwest Europe began: the Younger Dryas stade. There is no doubt that this event resulted from a drop in temperature; not only do changes in the flora (Watts, 1980) and beetles (Coope and Joachim, 1980) imply cooling, but glaciers re-formed on previously deglaciated mountains in Wales and northwest England (Sissons, 1980). Outlet glaciers on the western side of the Scandinavian Ice Sheet advanced down the Norwegian fjords, while the eastern margin stabilized or slowed its retreat in Sweden and Finland (Mangerud, 1980). The definition of the Pleistocene-Holocene boundary -- 10,000 BP on the ^{14}C timescale -- is based on the abrupt end of the Younger Dryas stade about that time. The brief but severe cooling in northwest Europe during the Younger Dryas stade is widely believed to have been a global event, but in fact there is no reliable evidence that glaciers readvanced at that time outside the northern North Atlantic area. The Sumas stade in the Pacific Northwest, formerly the strongest contender in the Northern Hemisphere for a Younger Dryas equivalent (Armstrong et al., 1965), is now known to have preceded it by several centuries (Armstrong, 1975). Mercer (1969) suggests that the Younger Dryas stade was caused, not by external factors, but by the injection into the northern North Atlantic Ocean of large amounts of ice produced by the break-up in the Arctic Ocean of a cold, Antarctic type of ice cover (ice shelves and grounded marine ice sheets) that had formed during the severe conditions of full-glacial time. This hypothesis has recently received support from paleoceanographic studies by Ruddiman and McIntyre (1981).

Variations of Glaciers in Middle Latitudes of the Southern Hemisphere

In middle latitudes of the Southern Hemisphere, age determinations for the last glacial maximum and the following deglaciation have been obtained in Australia, New Zealand and South America.

Australia and New Zealand

In Australia wood in proglacial silt below outwash gravel in Tasmania is 18,800±500 years old, pointing to a maximum stand shortly after (Kiernan, 1983). In New Zealand, the last glaciation on the west side of the mountains of South Island culminated after 18,600±290 BP and probably no later than 18,000 BP (Suggate and Moar, 1970), and on the east side of the mountains shortly after 19,200±550 BP (Soons and Burrows, 1978); in neither case has an end moraine been directly dated. Glaciers then receded an unknown distance before starting their final major readvance which, based on a single age determination of 14,100±220 BP, is thought to have culminated ~14,500 - 14,000 BP; the ice retreated rapidly after 14,000 BP (Suggate and Moar, 1970).

Burrows and Russell (1975) believe that a later readvance on the east side of the mountains culminated between 11,900 and 10,000 BP; but,the evidence is permissive only. Wardle (1978) reports that the Waiho Loop, a prominent end moraine 10 km in front of the Franz Josef Glacier on the west side of the mountains, was formed after 11,450±200 BP (wood beneath till 2 km behind the moraine). At a site close by, wood covered by till has recently been dated at 12,120±400 BP (sample GX-9981 (unpublished)). In Tasmania, MacPhail (1979) interprets the vegetational succession as showing uninterrupted warming between 11,500 and 9500 BP.

South America

In southern South America at lat 41°S in Chile, an end moraine of the piedmont glacier tongue that occupied Lago Rupanco covers peat 19,450±350 years old, suggesting that the moraine is 19,000 -19,500 years old (Mercer, 1970). This is the most reliable date for the last glacial maximum in the Southern Hemisphere. By 18,900±370 BP the adjacent glacier in the Lago Llanquihue basin was receding from its maximum extent (Porter, 1981). Mercer (1976) and Porter (1981) conclude, from the ages of wood and peat covered by lacustrine sediments, that a later, final major readvance culminated about 13,000 BP (about the same time as the Port Huron readvance of the Erie and Huron lobes in North America). Later, however, a site on the east coast of the island of Chiloe near lat 42°S was studied that less equivocally records a glacial readvance. Here, an end moraine, formed by ice flowing from the mainland and composed of outwash sediments beneath till, covers wood dated at 14,355±700, 14,970±210 and 15,600±560 BP (samples GX-8686, I-12,996 and GX-9978) (unpublished), pointing to a probable age of 15,000 to 14,500 years for the moraine. Furthermore, evidence for extensive deglaciation of southern South America by 13,000 BP conflicts with a similar date for culmination of the preceding major readvance. This evidence comes from both sides of the southern Andes. In Chile at lat 40°S Lago Ranco, which was occupied by a piedmont ice lobe in full-glacial time, was deglaciated by 13,000 BP; wood at successively lower levels in "postglacial" lake sediments is 12,200±400 (Mercer, 1976), 12,810±190 (sample I-12,995)(unpublished), and 13,900±560 (sample GX-9979) (unpublished) years old. South of lat 46°S several piedmont lakes on the east side of the cordillera now drain westward to the Pacific Ocean through the mountains. In full- and late-glacial times the trans-Andean valleys were ice-filled and the lakes, or the glaciers occupying them, drained eastward to the Atlantic Ocean. Radiocarbon dating shows that the abandoned spillway at the eastern end of glacial Lago Pueyrredon (lat 47°S) was dry by 12,810±445 BP (sample GX-8682, unpublished), and has remained so ever since.

Glaciers continued to recede after 12,800 BP. Two ^{14}C dates show that by 11,000 BP the Tempano Glacier, an icefield outlet glacier near lat 48°S, had withdrawn behind its AD 1968 borders, where it remained until Neoglacial time (Mercer, 1970). It did not readvance between 11,000 and 10,000 BP, the interval corresponding to the European Younger Dryas stade; this behavior conflicts with climatic reconstructions from pollen stratigraphy in southern Chile (Heusser and Streeter, 1980), which show a pronounced reversion to colder and wetter climate at that time. However, it conforms with the conclusion of Lorius et al. (1979), from the oxygen isotopic stratigraphy of an ice core from Dome C, Antarctica, that the warmest part of the past 11,000 years was the interval 11,000 - 8000 BP, and with the conclusion of Hays et al. (1976) that subantarctic surface water was warmest 9400±600 BP.

Polar Temperatures during the Last Interglacial: Relevance to Forecasts of CO_2-Induced Climatic Warming

To determine the likely effects on the cryosphere of the expected climatic warming caused by rising levels of atmospheric CO_2, particular attention must be paid to the Southern Hemisphere, because the eventual warmth-induced deglaciation of West Antarctica might be the first major adverse result. Therefore an important question is: during the last interglacial, did the Antarctic warm by as much as did the Arctic? If it did, and if the global warming resulted from Milankovitch mechanisms, then probably (but not certainly) the same would apply during a future CO_2-induced warming, because the feedback mechanisms that magnify polar warming would be the same. High atmospheric levels of CO_2 probably contributed to the warmth of the last interglacial: Shackleton et al. (1983) show that values then exceeded those of the Holocene.

Approximately equal warming of the polar regions of both hemispheres as a consequence of rising atmospheric CO_2 levels is shown by the global climatic model results of Hansen et al. (1984), but not by Manabe and Stouffer's (1980) model which shows much greater warming in the Arctic than in the Antarctic after an increase in atmospheric CO_2. However, Manabe (1983) points out that his 1980 model tends to underestimate the warming of the Southern Hemisphere because the simulated surface temperature of the Southern Ocean is too high and the extent of sea ice is too small.

According to Dansgaard and Duplessy (1981), only in high latitudes was the last interglacial warmer than the Holocene; for a time the Southern Ocean and the northern North Atlantic were both a few degrees warmer than they are today. Indirect evidence for exceptional warming of the Antarctic during the last interglacial has come from an inferred major decrease in its ice cover that has not occurred during the present interglacial. As Robin (1958) points out, ice shelves are cold glaciers and are absent from those parts of the Antarctic Peninsula where sea surface temperature rises above - 1.5°C in summer; Robin and Adie (1964) conclude that a climatic warming that raised the water temperature above this critical level would remove all ice shelves and consequently all ice grounded below sea level, resulting in the deglaciation of most of West Antarctica. Mercer (1968) calculates that this deglaciation would add a layer of water ~4-4.5 m thick to the present ocean. Noting that this is close to estimates of 6±3 m for eustatic sea level during the last interglacial, he suggests that the high sea level resulted mainly from deglaciation of West Antarctica after temperatures there had risen above the critical level for ice shelves. He concludes that deglaciation of West Antarctica would begin if the present climate of the north-western Antarctic Peninsula spread south to the fronts of the Ross and Filchner/Ronne ice shelves. This would entail a rise of ~5°C in summer air temperature at the ice shelf fronts. Complete removal of all ice grounded below sea level in West Antarctica would require further climatic warming; how much would depend on the temperature of the water circulating under the ice shelves. In any case, if West Antarctica was in fact largely deglaciated during the last interglacial, the Antarctic must have been warmer than it is now by at least as much as was the Arctic, where no evidence has been found for summers more than a few degrees warmer than today's.

Conclusions

For estimating the sensitivity of the climate system to rising levels of atmospheric CO_2, accurate knowledge of the temperatures that were reached during the last interglacial in the polar regions of both hemispheres is essential. Although warming during the last interglacial is generally thought to have been caused chiefly by orbital factors (the abnormally high concentration of atmospheric CO_2 demonstrated by Shackleton et al. (1983) must have contributed) - the mechanism is not fully understood. Therefore, if the Antarctic warmed as much as or more than did the Arctic during the last interglacial, it would not necessarily do so during a future CO_2-induced climatic warming. Nevertheless such an outcome would seem more likely; it is, therefore, important to try to determine for sure whether or not the West Antarctic ice sheet was temporarily absent during the last interglacial, and if it was, how much warmer high southern latitudes then were than they are today. This would also be important in attempting to evaluate the relative realism of different models of the CO_2-greenhouse effect.

By analogy with the course of the present interglacial, during which temperatures in middle and high southern latitudes apparently reached today's by 11,000 BP, the Southern Hemisphere was probably warmest near the beginning of the last interglacial, when large volumes of ice still survived in the Northern Hemisphere (Mercer, 1981). Evidence for highest temperatures in the South should therefore be sought in features or sediments dating not from the time when sea level was highest, but from the time when it was 50 m or so below present levels and rapidly rising.

If the high stand of the sea during the last interglacial at ~5 m above its present level did result from warmth-induced deglaciation of West Antarctica at that time -- and no other satisfactory explanation has been offered -- the implication is that both the last interglacial ~125,000 years ago and the current interglacial were more or less simultaneous in the Northern and Southern hemispheres. The same appears to be true for the last glacial maximum, the few dates for which in the Southern Hemisphere fall between ~20,000 and 18,000 BP, which is well within the interval between 21,000 and 17,000 BP when many lobes on the southern margin of the Laurentide Ice Sheet oscillated near their outer limits. However, such a prolonged interval of expanded Northern Hemisphere glaciers provides a poor basis for a comparison of the timing of climatic changes in the two hemispheres. The later, final full-glacial readvance, which was a more sharply defined event, is more suitable for hemispheric comparisons of timing. In both hemispheres, this readvance culminated between 15,000 and 14,000 BP, and was followed by rapid glacial recession and other evidence for warming. No explanation has yet been offered for this major cooling, during which full-glacial conditions returned to middle latitudes of both hemispheres after ~4000 years when summer insolation at lat 50°N was higher than it is today, and constantly rising.

Thus the fly remains in the ointment of the Milankovitch theory: the last interglacial and the last glaciation both appear to have affected the Northern and Southern hemispheres simultaneously, and with comparable severity. Although there may no longer be any doubt about the theory's general validity, the mechanisms by which orbital variations affect the global climate remain mysterious.

References

Andersen, B.G., Late Weichselian ice sheets in Eurasia and Greenland, in The Last Great Ice Sheets, edited by G.H. Denton and T.J. Hughes, pp. 1-65, John Wiley and Sons, New York, 1981.

Armstrong, J.E., D.R. Crandell, D.J. Easterbrook and J.B. Noble, Late Pleistocene stratigraphy and chronology in southwestern British Columbia and northwestern Washington, Geol. Soc. Amer. Bull., 76, 321-330, 1965.

Armstrong, J.E., Quaternary geology, stratigraphic studies and reevaluation of terrain inventory maps, Fraser Lowland, British Columbia, Geol. Surv. Canada. Paper 75-1, Part A, 377-380, 1975.

Berger, A.L., The Milankovitch astronomical theory of paleoclimates: a modern review, Vistas Astron., 24, 103-122, 1980.

Berger, A.L., The astronomical theory of paleoclimates, in Climatic Variations and Variability: Facts and Theories, edited by A. Berger, pp. 501-525, D. Reidel, Dordrecht, Boston and London, 1981.

Broecker, W.S., The cause of glacial to interglacial climatic change, in Evolution of Planetary Atmospheres and Climatology of the Earth, pp. 165-190, Centre National d'Etudes Spatiales (France), 1978.

Broecker, W.S., Glacial to interglacial changes in ocean chemistry, Progr. Oceanogr., 11, 151-197, 1982.

Broecker, W.S., Terminations, in the proceedings of the Milankovitch Symposium, held at Lamont-Doherty Geological Observatory, December, 1982, in press.

Burrows, C.J., and J.B. Russell, Moraines of the Upper Rakaia Valley, J. Roy. Soc. N.Z., 5, 463-477, 1975.

Cepek, A.G., Geologische Ergebnisse der ersten Radiokarbondatierungen von Interstadialen im Lausitzer Urstromtal, Geologie, 14, 625-657, 1965.

Coope, G.R. and M.J. Joachim, Lateglacial environmental changes interpreted from fossil Coleoptera from St. Bees, Cumbria, NW England, in Studies in the Lateglacial of North-West Europe, edited by J.J. Lowe, J.M. Gray and J.E. Robinson, pp. 55-68, Pergamon Press, Oxford, 1980.

Corliss, B.H., Linkage of North Atlantic and Southern Ocean deep-water circulation during glacial intervals, Nature, 298, 458-460, 1982.

Croll, J., Climate and Time in the Geological Relations, Appleby and Co., New York, 1875.

Dansgaard, W. and J.C. Duplessy, The Eemian Interglacial and its termination, Boreas, 10, 219-228, 1981.

Denton, G.H., Quaternary glaciations of White River valley, Alaska, with regional synthesis for the northern St. Elias Mountains, Alaska and Yukon Territory, Geol. Soc. Amer. Bull., 85, 871-892, 1974.

Gribbin, J., Astronomical influences, in Climatic Change, edited by J. Gribbin, pp. 131-154, Cambridge Univ. Press, Cambridge, 1978.

Hansen, J.E., A. Lacis, D. Rind, G. Russell, P. Stone, I. Fung, J. Lerner and R. Ruedy, Climate sensitivity experiments with a three-dimensional model: Analysis of feedback mechanisms, Climate Processes and Climate Sensitivities, ed. J.E. Hansen and T. Takahashi, 1984.

Hays, J.D., J. Imbrie, and N.J. Shackleton, Variations in the Earth's orbit: pacemaker of the Ice Ages, Science, 194, 1121-1132, 1976.

Heusser, C.J., and S.S. Streeter, A temperature and precipitation record of the last 16,000 years in southern Chile, Science, 210, 1345-1347, 1980.

Imbrie, J., Astronomical theory of the Pleistocene Ice Ages: a brief historical review, Icarus, 50, 408-422, 1982.

Imbrie, J., and K.P. Imbrie, Ice Ages: Solving the Mystery, Enslow, New York, 1979.

Imbrie, J., and J.Z. Imbrie, Modeling the climatic response to orbital variation, Science, 207, 943-953, 1980.

Kerr, R.A., Orbital variation - Ice Age link strengthened, Science, 219, 272-274, 1983.

Kiernan, K., Weathering evidence for an additional glacial stage in Tasmania, Austral. Geogr. Stud., 21, 197-220, 1983.

Lorius, C., L. Merlivat, J. Jouzel and M. Pourchet, A 30,000-yr isotope climatic record from Antarctic ice, Nature, 280, 644-648, 1979.

MacPhail, M.K., Vegetation and climates in southern Tasmania since the last glaciation, Quat. Res., 11, 306-341, 1979.

Manabe, S., Carbon dioxide and climatic change, Advances Geophys., 25, 39-82, 1983.

Manabe, S. and R.J. Stouffer, Sensitivity of a global climate model to an increase of CO_2 concentration in the atmosphere, J. Geophys. Res., 85, 5529-5554, 1980.

Mangerud, J., Ice-front variations of different parts of the Scandinavian ice sheet, 13,000 - 10,000 years BP, in Studies in the Lateglacial of North-West Europe, edited by J.J. Lowe, J.M. Gray and J.E. Robinson, pp. 23-30, Pergamon Press, Oxford, 1980.

Mayewski, P.A., G.H. Denton and T.J. Hughes, Late Wisconsin ice sheets of North America, In The Last Great Ice Sheets, edited by G.H. Denton & T.J. Hughes, pp. 67-178, John Wiley & Sons, New York, 1981.

Mercer, J.H., Antarctic ice and Sangamon sea level, Int. Ass. Sci. Hydrol. Commission of Snow and Ice, Gen. Assemb. Berne, Publ., 79, 217-225, 1968.

Mercer, J.H., The Allerod oscillation: A European climatic anomaly?, Arctic Alpine Res., 1, 227-234, 1969.

Mercer, J.H., Variations of some Patagonian glaciers since the Late-Glacial: II, Amer. J. Sci., 269, 1-25, 1970.

Mercer, J.H., Glacial history of southernmost South America, Quat. Res., 6, 125-166, 1976.

Mercer, J.H., West Antarctic ice volume: the interplay of sea level and temperature, and a strandline test for absence of the ice sheet during the last interglacial, in Sea Level, Ice and Climatic Change, edited by I. Allison, pp. 323-330, Internat. Assoc. Hydrol. Sci. Publ. 131, 1981.

Milankovitch, M., Kanon der Erdbestrahlung und seine Andwendung auf das Eiszeitenproblem, Royal Serv. Acad. Spec. Publ. 133, 1-633, 1941.

Mullineaux, D.R., H.H. Waldron, and M. Rubin, Stratigraphy and chronology of Late Interglacial and early Vashon Glacial time in the Seattle area, Washington, U.S. Geol. Surv. Bull., 1194-0, 1965.

Penny, L.F., G.R. Coope and J.A. Catt, Age and insect fauna of the Dimlington silts, East Yorkshire, Nature, 224, 65-67, 1969.

Porter, S.C., Pleistocene glaciation in the southern lake district of Chile, Quat. Res., 16, 263-292, 1981.

Robin, G. de Q., Glaciology II. Norwegian-British-Swedish Antarctic Expedition 1949-52, Scientific Results 5, p. 1-134, Norsk Polarinstitutt, Oslo, 1958.

Robin, G. de Q., and R.J. Adie, The Ice Cover, in Antarctic Research, edited by R. Priestley, R.J. Adie, and G. de Q. Robin, pp. 100-117, Butterworths, London, 1964.

Ruddiman, W.F. and A. McIntyre, The North Atlantic Ocean during the last glaciation, Palaeogeogr., Palaeoclimatol., Palaeoecol., 35, 145-214, 1981.

Ruhe, R.V., Quaternary Landscapes in Iowa, Iowa State Univ. Press, Ames, 1969.

Saarnisto, M., The deglaciation history of the Lake Superior region and its climatic implications, Quat. Res., 4, 316-339, 1974.

Shackleton, N.J., M.A. Hall, J. Line, and Cang Shuxi, Carbon isotope data in core V 19-30 confirm reduced carbon dioxide concentration in the ice age atmosphere, Nature, 306, 319-322, 1983.

Shotton, F.W., The Devensian Stage: its development, limits and substages, Philos. Trans. Roy. Soc. London, Ser. B., 280, 107-118, 1977.

Sissons, J.B., Palaeoclimatic inferences from Loch Lomond Advance glaciers, in Studies of the Lateglacial of North-west Europe, edited by J.J. Lowe, J.M. Gray, and J.E. Robinson, pp. 31-43, Pergamon Press, Oxford, 1980.

Soons, J.M. and C.J. Burrows, Dates for Otiran deposits, including plant microfossils and macrofossils, from Rakaia Valley, N.Z.J. Geol. Geophys., 21, 607-615, 1978.

Suggate, R.P. and N.T. Moar, Revision of the chronology of the late Otira Glacial, N.Z.J. Geol. Geophys., 13, 742-746, 1970.

Wardle, P., Further radiocarbon dates from Westland National Park and the Omeroa River mouth, New Zealand, N.Z.J. Bot., 16, 147-152, 1978.

Watts, W.A., Regional variation in the response of vegetation to Lateglacial climatic events in Europe, in Studies in the Lateglacial of North-West Europe, edited by J.J. Lowe, J.M. Gray and J.E. Robinson, pp. 1-21, Pergamon Press, Oxford, 1980.

IS THERE A TIE BETWEEN ATMOSPHERIC CO_2 CONTENT AND OCEAN CIRCULATION?

Wallace S. Broecker and Taro Takahashi

Lamont-Doherty Geological Observatory of Columbia University
Palisades, New York 10964

Abstract. The possibility that the CO_2 partial pressure of the surface waters of the ocean (and hence also of the atmosphere) is dependent on the rates of oceanic mixing is explored. The purpose of this exercise is to ascertain whether the abrupt rise in atmospheric CO_2 content at the end of the last glacial period could have been caused by a reorganization of deep sea ventilation. We also have our eye on the possibility that future warming of the planet induced by anthropogenic CO_2 will lead to a positive feedback (i.e., polar warming, reduced deep sea ventilation rate, increase in the p_{CO_2} of surface ocean water). Our approach is to consider two end member models for the distribution of CO_2 in the sea. One is thermodynamic, the other is biological. The surface water p_{CO_2} difference between these end member models is nearly a factor of two. We show that the situation for the real ocean likely lies between these extremes and explore how the ratio of deep sea ventilation rate to air-sea CO_2 exchange rate might push the system closer to one extreme or the other. It appears that decreases in the ventilation rate will push the ocean closer to the thermodynamic end member and hence raise the atmosphere's p_{CO_2}. Such a decrease in ventilation rate may have occurred at the end of glacial time. A further decrease may be induced during the next hundred or so years by the buildup of CO_2 in the atmosphere.

Introduction

Recent measurements on the gases extracted from polar ice suggest that the atmospheric CO_2 concentration increased by some 80 ppm at the termination of the last major glaciation (Berner et al., 1979; Delmas et al., 1980; Neftel et al., 1982). Several explanations can be posed for this increase:

1) it is an artifact of the ice lithification process,
2) it reflects a warming of the ocean surface at the close of glacial time,
3) it is caused by a change in the amounts of C, N, and P dissolved in the ocean due to a pulse of sedimentation at the close of glacial time,
4) it is the result of a redistribution within the sea of the dissolved C, N and P caused by a change in deep sea ventilation at the close of glacial time.

While the first explanation cannot be ruled out, the case against it is becoming ever stronger. Consistent estimates of the CO_2 to air ratio for glacial ice have now been obtained from two Greenland (Camp Century and Dye 3) and three Antarctic (Dome C, D10 and Byrd) ice borings. The records in all five cores are consistent with an atmospheric CO_2 content during peak glacial time about 80 ppm lower than that for late Holocene time. The Bern group (Oeschger, personal communication) has obtained nearly the same preanthropogenic atmospheric CO_2 content (i.e. 265 ppm) for gas from ice from localities, where mean annual temperatures range from -22°C to -55°C.

The ocean temperature change explanation can be rejected because the faunal based paleotemperature estimates of CLIMAP (≃1.7°C increase between full glacial and Holocene time) and the $^{18}O/^{16}O$ based temperature estimates (<3°C increase between full glacial and Holocene time) are far smaller than the 8° to 12°C temperature rise required to generate the observed CO_2 increase. As most workers in the field consider the faunal and isotope evidence to be quite firm, until some strong contrary arguments are raised this explanation must be put aside.

Two hypotheses involving whole ocean chemistry changes have been published. Broecker (1981, 1982-a, 1982-b) and Broecker and Peng (1984) have suggested that the accumulation of sediment on shelves and in estuaries during the sea level transgression of early postglacial time may have removed as much as 30% of the ocean's NO_3 and PO_4. Were this N and P bound in organic matter with roughly the Redfield C:N:P ratio, then not only would this removal produce the observed CO_2 rise but also the early postglacial $CaCO_3$ preservation event observed in marine sediments (Berger, 1978) and the $^{13}C/^{12}C$ trends observed for benthic and planktic foraminifera (Shackleton, 1977).

Berger (1982) and Berger and Keir (1984) have proposed an alternate whole ocean chemistry scenario. They postulate that growth of coral during the Holocene transgression lowered the alkalinity to ΣCO_2 ratio in sea water and thus generated a rise in CO_2 pressure. While the Berger mechanism could produce a CO_2 rise it would lead to a $CaCO_3$ dissolution (rather than a $CaCO_3$ preservation) event in the deep sea. Also, it would not produce the increase in $^{13}C/^{12}C$ ratio observed for benthic forams at the close of glacial time.

Both of these whole ocean chemistry scenarios are open to the criticism that they require large amounts of shallow water deposition over a very short period of time. While not impossible geologically, both certainly stretch the evidence. Because of this, other explanations for the CO_2 change have been sought.

The remaining category has to do with changes of the distribution of nutrient constituents within the sea. In today's ocean quite large differences in ΣCO_2, NO_3 and PO_4 content are found from place to place and depth to depth. In the case of NO_3 and PO_4 these differences are solely the result of photosynthesis and respiration. For ΣCO_2, $CaCO_3$ formation and dissolution and CO_2 transport through the atmosphere from one region of the surface ocean to another are also important. The question is whether there are ways in which these distributions might have changed such that a different mean surface water (and hence atmospheric) p_{CO_2} could have existed.

Broecker (1982-a and 1982-b) suggested one such scenario. Were the ratio of C/P and N/P in the organic residues falling from the surface to deep sea to have decreased by about 25% at the close of glacial time then this change would have produced the observed CO_2 pressure rise. Unfortunately we know almost nothing about the factors controlling the C/P and C/N ratio in organic matter formed in the sea. We don't even know much about how these ratios vary from place to place in today's sea. Thus, we have no basis to say whether or not they could have changed between glacial and postglacial time. To claim that a Redfield ratio change was responsible for the CO_2 increase at the close of glacial time would be much like claiming that a change in solar output was responsible for warming at the close of glacial time. Both are possible. Neither can be substantiated at this time.

A closely related scenario involves a change in the ratio of $CaCO_3$ produced per unit of NO_3 and PO_4 upwelled to the surface ocean. If this ratio substantially increased at the close of glacial time then the observed CO_2 rise could be generated. Since some groups of marine organisms produce $CaCO_3$ while others do not, the ecologic changes which accompanied deglaciation might have also changed the oceanwide ratio of $CaCO_3$ production to nutrient upwelling. There are several drawbacks to this scenario. The first is that like the coral growth scenario presented above it would be unable to explain the $CaCO_3$ preservation event of early post glacial time. Second, it calls for higher $CaCO_3$ accumulation rates for Holocene than for late glacial time (if anything the opposite is observed). Finally, it would require a substantial global rise in the lysocline at the close of glacial time. Such a rise is not observed. Thus, this alternate to the Redfield ratio change scenario does not show much promise.

There is one other way in which the distribution of nutrient substances in the sea might be changed. If the pattern and rate of deep sea ventilation were different in the past, then the distribution of the constituents of interest might have been different. Evidence that the ventilation of the deep sea differed in late glacial time from that of today is being accumulated (e.g. Boyle and Keigwin, 1982; Curry and Lohmann, 1982). It is not at all clear, however, that ventilation changes would necessarily lead to a change in atmospheric CO_2 content. The concepts connected with the relationship between circulation and chemistry lie at the heart of the field of physical oceanography. The same gaps in knowledge which prevent us from creating a meaningful general circulation model for the ocean also stand in our way when we try to determine whether a change in ventilation would change the atmospheric CO_2 pressure.

Contrast between a Thermodynamic and a Redfield Ocean

In order to understand why the p_{CO_2} over the ocean might be sensitive to the pattern and/or rate of deep ventilation, it is instructive to consider two simple, yet highly contrasting two-box representations of the ocean. We will call one the thermodynamic ocean and the other the Redfield ocean. Both oceans have the same simplified geometry (see Figures 1 and 2). Both have the same bulk chemistry. Both have the same PO_4 and NO_3 distributions (i.e., none in the upper box). To simplify the situation we assume that neither has $CaCO_3$ production or dissolution and that neither has salinity variations. The important point is that the two models differ in their ΣCO_2 distribution. This difference is created by the adoption of two limiting cases.

In the thermodynamic model (Figure 1), the atmospheric CO_2 is assumed to be in thermodynamic equilibrium with both the cold and warm surface ocean waters. Accordingly, transport of CO_2 from the cold reservoir outcrop to the warm surface reservoir is assumed to occur through the atmosphere by maintaining equality between the CO_2 partial pressures for the two reservoirs. Under this circumstance the difference between the ΣCO_2 concentrations of the surface and deep reservoirs is entirely thermodynamic; it depends only on the temperature difference between the reservoirs. No matter how much carbon is removed from the warm reservoir with the organic debris, CO_2 transport through the atmosphere restores the thermodynamic equilibrium ΣCO_2 concentration.

By contrast in the Redfield ocean (Figure 2), CO_2 transport through the atmosphere is assumed to be ineffective. In this case the difference between the ΣCO_2 content of the warm and cold reservoirs is controlled by carbon to nitrogen and carbon to phosphorus ratios in the organic debris falling from the surface to deep reservoir. In the real ocean the N/P ratio in the residues is nearly equal to the NO_3/PO_4 ratio in the bulk ocean. Thus it is not immediately clear whether the critical nutrient is NO_3 or PO_4. Fortunately for our purposes the distinction is not necessary. Because of the removal of ΣCO_2 from the warm reservoir its CO_2 pressure will be lower than the CO_2 pressure for the deep reservoir. Hence, there will be a tendency for CO_2 to flow through the atmosphere from the cold to the warm reservoir. In the Redfield ocean, the assumption is that the gas exchange rate between air and sea is slow compared to the transport of water between the reservoirs.

The equations defining the carbon chemistry of the thermodynamic and Redfield oceans are given in Appendix A.

The Kinetic Ocean

The characteristic of the model ocean which determines whether it will be closer to the thermodynamic case or to the Redfield case is the ratio of the rate of CO_2 gas exchange between surface water and atmosphere and the rate of exchange of water between the cold and warm reservoirs. The higher this ratio the greater the tendency toward thermodynamic equilibrium. The lower this ratio, the greater the tendency toward a Redfield ocean.

To see whether conditions in the ocean should tend toward one of these extremes or the other we will consider a third model which we call the kinetic ocean (Figure 3). From studies of both the distribution of natural and of bomb-testing radiocarbon we know the invasion rate, I, of CO_2 into the sea to be 20/300 or 6.667×10^{-2} moles $CO_2/m^2 \cdot yr\ \mu atm$. Thus, the net CO_2 exchange flux across the air-sea interface, E, is given by:

$$E = I[(p_{CO_2})_a - (p_{CO_2})_{sw}] \text{ moles } CO_2/m^2 yr$$

where p_{CO_2} is in μatm, and the subscripts a and sw denote the atmosphere and the surface ocean water respectively. From studies of the distribution of natural radiocarbon in the sea we know that the flux of water across the main thermocline must be about 5 m^3/m^2 yr. Finally, we know from the distribution of temperature in surface waters that the outcrops of the deep sea constitute about 9% of the surface ocean. The equations defining the carbon chemistry of the kinetic model ocean are given in Appendix A.

With these values we can compute distribution

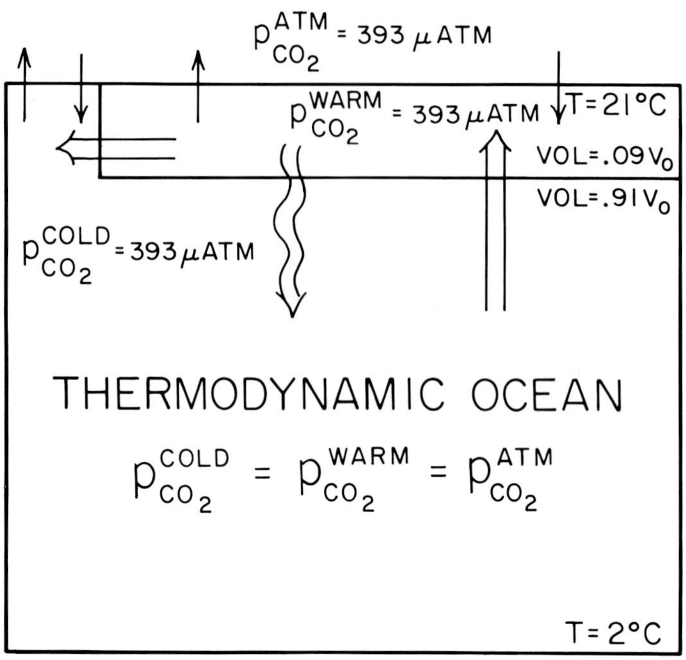

Fig. 1. Two reservoir model of the ocean in which the transport of CO_2 through the atmosphere (from the cold reservoir outcrop to the warm reservoir) is assumed to be rapid compared to the water transports of carbon between the reservoirs. In such a system the distribution of carbon between the warm and cold reservoirs is independent of the photosynthesis-respiration cycle. Rather the carbon distributes such that the partial pressures of CO_2 in the cold reservoir outcrop water and in the warm surface water are both equal to that in the atmosphere, hence the thermodynamic ocean. Assuming that the alkalinity and salinity of the two reservoirs are identical, the total carbon concentration (ΣCO_2) in the cold reservoir (2°C) turns out to be higher due to its low temperature than that in the warm reservoir (21°C). However, as summarized in Table 1, the difference between the ΣCO_2 concentrations in the cold and warm reservoirs in this model is less than half as great as that for the Redfield model (Fig. 2), i.e. 140 versus 286 μm/kg. While the photosynthesis-respiration cycle works to increase this difference to the Redfield value, the transport of CO_2 through the atmosphere from the cold to the warm reservoir prevents this from happening. The fine arrows represent ocean-atmosphere CO_2 transfers, the broad arrows water exchange between the reservoirs and the broad wavy arrow the rain of organic debris from the warm to the cold reservoir.

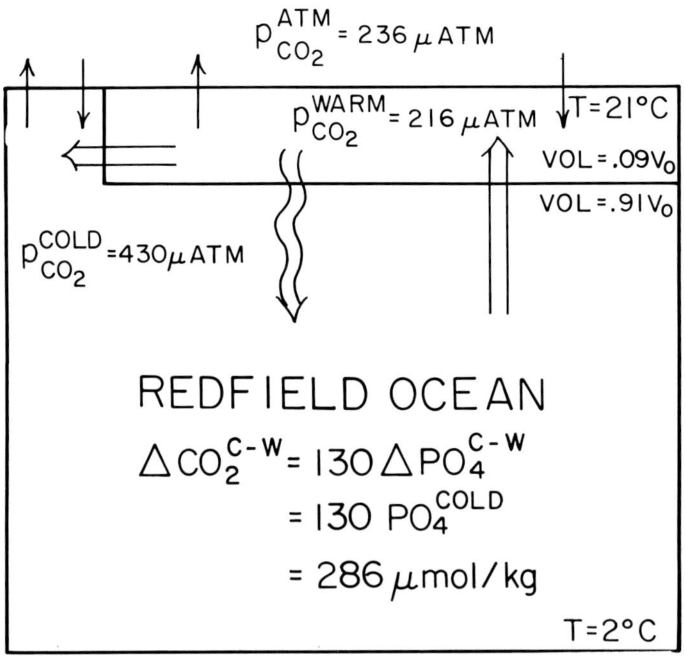

Fig. 2. Two reservoir model of the ocean in which the transport of CO_2 through the atmosphere (from the cold outcrop to the warm reservoir) is assumed to be slow compared to the water transports of carbon between the reservoirs. In such a system the distribution of carbon between the warm and cold reservoirs is controlled by the photosynthesis-respiration cycle, hence the Redfield ocean. Assuming that the NO_3 and PO_4 content of the warm reservoir are close to zero (i.e. the critical nutrients supplied by upwelling are fully utilized by plants and returned to cold reservoir via organic debris), the difference between the ΣCO_2 concentration in the cold and warm reservoir is given by the C/P ratio in organic debris (i.e., ≃130) times the PO_4 content of deep water (2.2 μm/kg) or 286 μm/kg. In our simplified ocean we neglect salinity differences and calcium carbonate production and dissolution. We also neglect the alkalinity difference caused by the uptake (during photosynthesis) and release (during respiration) of HNO_3 from sea water. Thus the two reservoirs in our model ocean have the same alkalinity (2374 μeq/kg). The pCO_2 pressures for the two reservoirs can then be calculated from the salinity (35.0°/oo), alkalinity (2374 μeq/kg), ΣCO_2 contents (2241 μm/kg for the cold reservoir and 1955 μm/kg for the warm reservoir), temperatures (2°C for the cold reservoir and 21°C for the warm reservoir) and the relative volume (0.91 and 0.09 the total ocean volume for the cold and warm reservoirs respectively). The chemical compositions for the model are summarized in Table 1. The atmospheric pCO_2 is calculated assuming that the outcrop area for the deep sea represents 0.09 that of the total sea surface and that the piston velocities for CO_2 invasion are the same for the cold and warm regions of the sea surface.

of ΣCO_2 and p_{CO_2} for the kinetic ocean. As can be seen in Figure 3 and Table 1, the result lies roughly halfway between that for the thermodynamic and the Redfield oceans. The important point is that the two main kinetic factors, the gas exchange rate and the cross thermocline water transfer rate roughly balance one another yielding a kinetic ocean in limbo between the two extremes. Thus, at least in the simple two-reservoir model ocean, changes in circulation rate (or gas exchange rate) between glacial and interglacial time would lead to shifts in the atmospheric CO_2 content. For example, had the ratio of glacial ventilation rate to glacial gas exchange rate been higher than today's, then the glacial ocean would have been more Redfield in character than today's, and hence the atmospheric CO_2 content would have been lower. The effect of changes in the ratio of the water upwelling rate, U, to the gas exchange rate, I, on the p_{CO_2} values in the atmosphere and the warm and cold water reservoirs is shown in Fig. 4. A change of 50% in the U/I ratio yields a change of 20 to 30 μatm in the atmospheric p_{CO_2}.

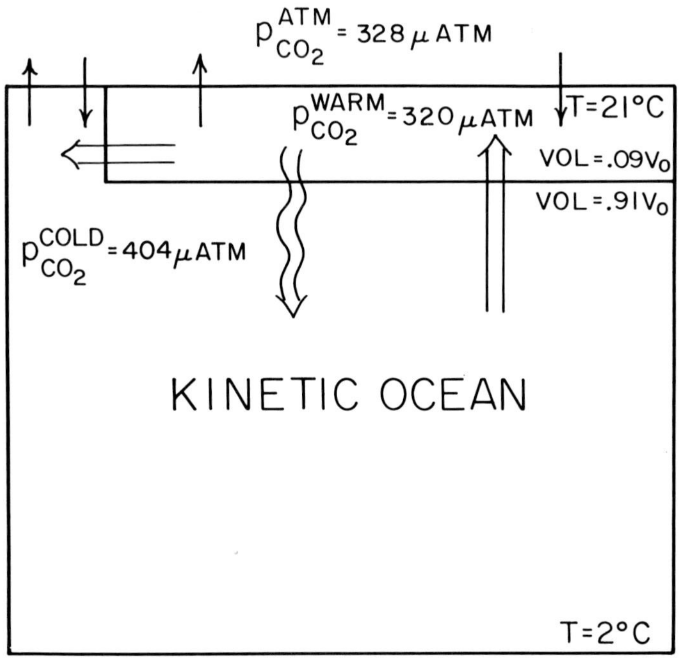

Fig. 3. Two reservoir model in which the rates of CO_2 invasion and water exchange (i.e. the kinetic ocean) are used to estimate where the ocean lies relative to the thermodynamic and the Redfield extremes. The water exchange rate, U, is taken to be 5 m^3/m^2 yr. The area of the interface between the atmosphere and the two water reservoirs is taken to be 0.91 and 0.09 the surface area of the ocean. The CO_2 gas exchange rate constant, I, is taken to be 6.667 x 10^{-2} moles CO_2/m^2 yr μatm. The results as summarized in Table 1 show that the kinetic ocean has characteristics roughly half way between those for the end members.

TABLE 1. Characteristics of the three model oceans

	Thermodynamic Ocean		Kinetic Ocean		Redfield Ocean	
	$p^{atm}_{CO_2}$ = 393 µatm		$p^{atm}_{CO_2}$ = 328 µatm		$p^{atm}_{CO_2}$ = 236 µatm	
	Warm	Cold	Warm	Cold	Warm	Cold
Temperature (°C)	21	2	21	2	21	2
Volume of water (%)	9	91	9	91	9	91
Surface Area (%)	91	9	91	9	91	9
Salinity (°/oo)	35	35	35	35	35	35
Alkalinity*(µeq/kg)	2374.0	2374.0	2374.0	2374.0	2374.0	2374.0
PO_4 (µm/kg)**	0	2.2	0	2.2	0	2.2
ΣCO_2 (µm/kg)+	2087.6	2227.5	2044.1	2231.8	1954.7	2240.7
HCO_3^- (µm/kg)++	1873.1	2097.5	1804.0	2103.6	1659.7	2115.9
$CO_3^=$ (µm/kg)++	202.1	107.1	230.0	104.7	288.2	99.8
CO_2 (µm/kg)++	12.4	22.9	10.1	23.5	6.8	25.0
pCO_2 (µatm)++	393	393	320	404	216	430

* The alkalinity value represents the global mean value for the oceans computed by Takahashi et al. (1981-a).

**The mean phosphate concentration in the world oceans has been estimated to be 2.0µm/kg and that in the world deep oceans 2.2 µm/kg on the basis of the data compiled by Takahashi et al. (1981-b). The global deep ocean average is used in this paper.

\+ The mean total CO_2 concentration in the three model oceans is fixed at 2215.0 µm/kg. This value is somewhat smaller than the observed mean value of 2254 µm/kg for the world oceans (Takahashi et al., 1981-a). The adjustment was made to bring the atmosphere P_{CO_2} value for the kinetic model closer to the observed value.

++These values have been computed using the following thermodynamic constants; the solubility of CO_2 in seawater by Weiss (1974), the first and second dissociation constants for carbonic acid by Mehrbach et al. (1973), the first dissociation constant of boric acid by Lyman (1956), the first and second dissociation constants of phosphoric acid by Kester and Pytkowicz (1967), and the dissociation constant of water by Millero (1979).

Evidence from the Distributions of ΣCO_2, NO_3 and PO_4 in Today's Sea

The next issue to be considered is the extent to which the model ocean represents the real ocean. In this regard let us first consider the surface outcrop of the cold water sphere which plays so important a role in the model ocean. We imply in the model that the cold waters are stirred so rapidly between the interior and the surface outcrop that no chemical gradients exist. While not identical to the average for deep water, the cold surface waters in the Antarctic region have NO_3 and PO_4 contents which are about two thirds those for average deep water (Figure 5). In this region of the ocean plants apparently are not able to keep pace with the supply of these fertilizers; the majority of the nutrients brought to the surface in this region are not fixed into residues which fall to the deeps.

In the real ocean the boundary between the two realms of interest is gradational. Surface waters warmer than 16°C are free of NO_3 with the exceptions of equatorial and coastal upwelling zones. From 16° to 6°C the NO_3 content of Antarctic surface waters rises. This increase is nearly a linear with decreasing temperature (Figure 5). Surface waters colder than 6°C in the Antarctic Ocean have high nitrate contents. In our model we split the transition zone volume placing part in one reservoir and part in the other.

Having established that the inclusion of an outcrop for the cold ocean in the Antarctic has validity let us next seek evidence as to whether the real ocean has a more thermodynamic or more Redfield distribution of ΣCO_2. One test is to compare the CO_2 partial pressure of Antarctic surface waters with those for warm ocean surface waters. In the Redfield ocean a large difference is predicted whereas in the thermodynamic ocean no difference at all is predicted. As can be seen in the map of ocean surface CO_2 pressure (Figure 6), the p_{CO_2} values for the Antarctic ocean surface waters are nearly equal to those for the temperate ocean and atmosphere. In this regard then, the real ocean appears to be thermodynamic! The high nutrient content cold surface waters of the Antarctic have nearly the same mean p_{CO_2} as the low nutrient content warm surface waters.

Another test is through the relationship between ΣCO_2 and NO_3 or PO_4 in today's sea. In a Redfield ocean the warm waters of the sea should lie along the ΣCO_2 to NO_3 or PO_4 trend defined

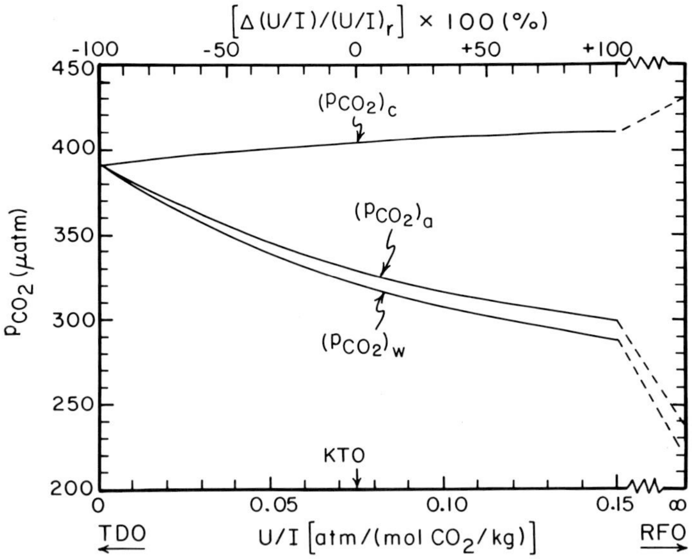

Fig. 4. The effect of change in the ratio of the water upwelling rate, U, to the CO_2 gas exchange rate constant, I, on the atmospheric pCO_2, $(pCO_2)_a$, and the warm and cold reservoir pCO_2, $(pCO_2)_w$ and $(pCO_2)_c$, in the kinetic ocean model. The thermodynamic ocean (marked with TDO) is represented by a U/I value of zero, and the Redfield ocean (marked with RFO) by a U/I value of infinity. The U/I value used for the kinetic ocean in Table 1 is shown by an arrow with KTO. The upper horizontal axis indicates the percentage change in U/I from this reference value, $(U/I)_r$, in which the values for U and I are taken to be 5.1×10^3 kg/m^2 yr and 6.667×10^4 μm CO_2/m^2 yr μatm respectively.

by the cold waters. In a thermodynamic ocean the warm water points should deviate toward higher ΣCO_2 values than predicted by the extension of the cold water trends to zero nutrient content (Figure 7).

For this analysis we choose PO_4 instead of NO_3. The reason is that NO_3 is consumed (i.e. converted to N_2 gas) in the parts of the ocean interior and in sea floor sediments which are O_2 free. There is currently no way to quantify this NO_3 utilization. As no similar fate exists for PO_4, it proves more reliable for our purposes.

Salinity changes and $CaCO_3$ production and dissolution significantly alter the ΣCO_2 content of sea water. Thus, we remove these influences by normalization to a constant salinity and to a constant Ca content of sea salt. This is done as follows:

$$\Sigma CO_2{*} = \Sigma CO_{2N} - \frac{(ALK_N - ALK^{STD})}{2} - \frac{(NO_{3N} - NO_3^{STD})}{2}$$

where

$$\Sigma CO_{2N} = \Sigma CO_2 \frac{35.00}{S}$$

$$ALK_N = ALK \frac{35.00}{S}$$

$$NO_{3N} = NO_3 \frac{35.00}{S}$$

$$ALK^{STD} = 2318 \ \mu eq/kg$$

$$NO_3^{STD} = 16 \ \mu m/kg$$

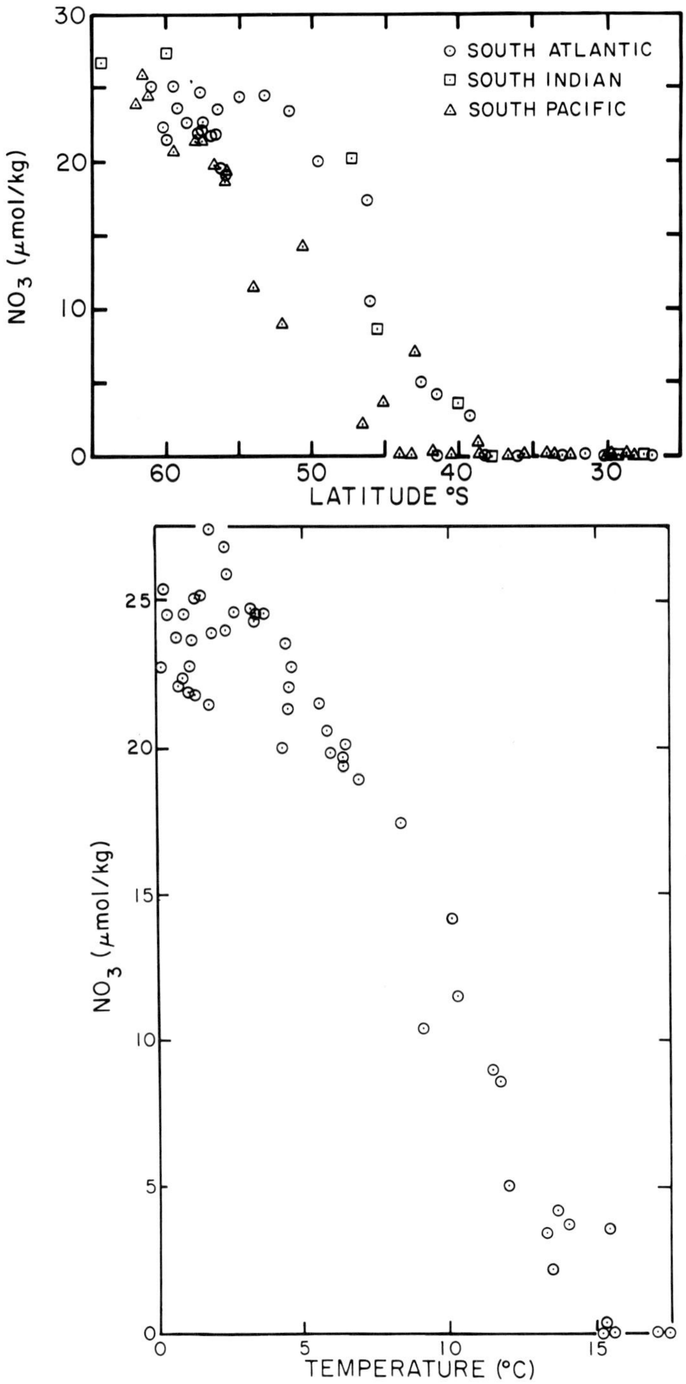

Fig. 5. Plots of nitrate content of southern ocean surface waters as a function of latitude (upper panel) and of temperature (lower panel) observed during the GEOSECS Program, 1972-1978.

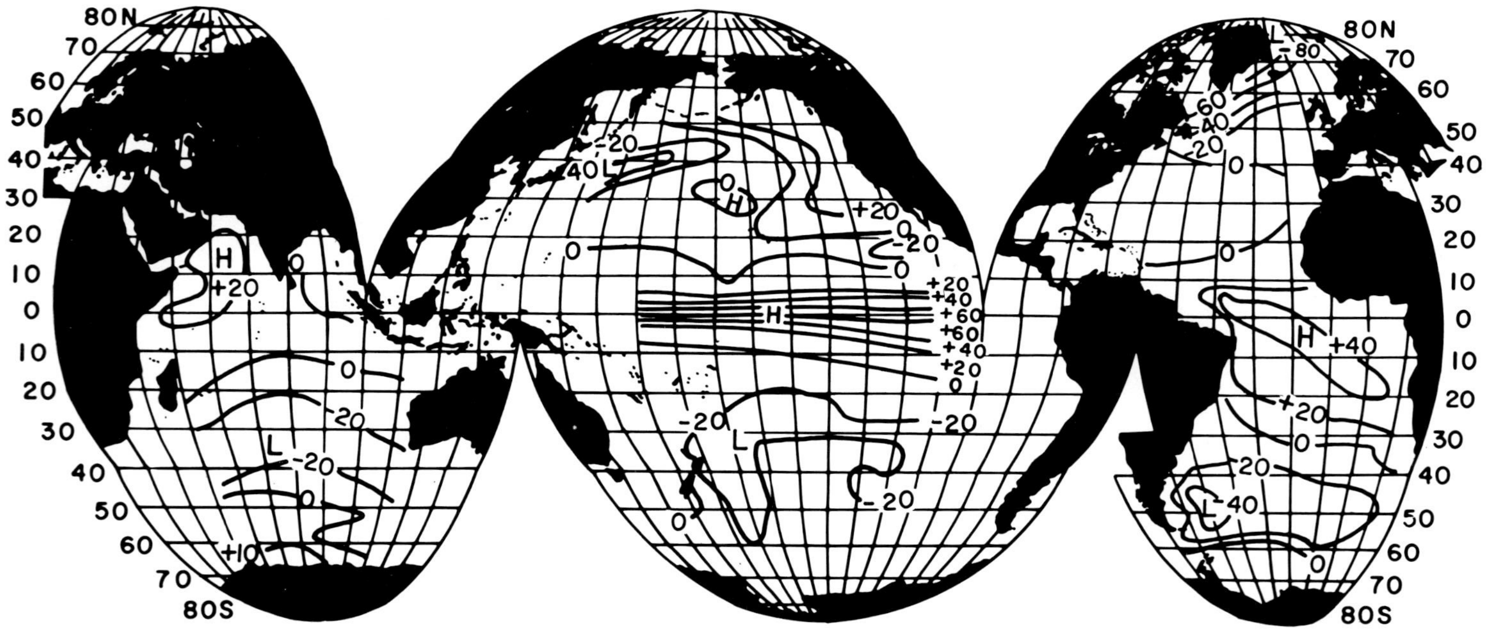

Fig. 6. Map showing the distribution of the difference (in microatmospheres) between the CO_2 partial pressure in the surface ocean water and that in the atmosphere. The positive values indicate that the ocean water is a source for CO_2 and the negative values indicate that it is a sink for CO_2. The atmospheric and surface ocean data were obtained during the GEOSECS Expeditions in the Atlantic, 1972-73, in the Pacific, 1973-74 and in the Indian Ocean, 1977-78. The high latitude values are for summertime conditions.

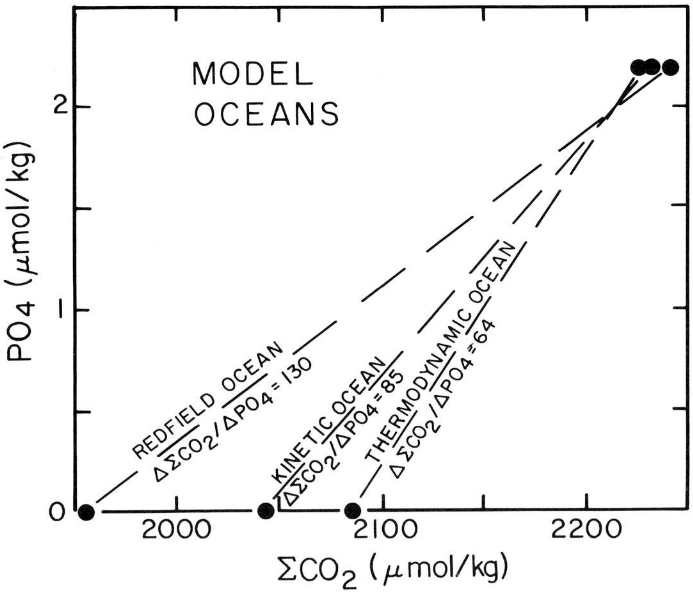

Fig. 7. Plot showing the relationship between the concentrations of phosphate and total dissolved inorganic carbon in the three two-box models presented in this paper. The important point is that, for the thermodynamic and kinetic models, the warm surface waters which are depleted of PO_4 show a large deviation from a line with the Redfield slope passing through the deep water value.

The NO_3 term is required to correct for the effects on sea water alkalinity of the addition (during respiration) and removal (during photosynthesis) of nitric acid. The nitrate corrected alkalinity change is then a measure of the alteration of Ca content by $CaCO_3$ formation and dissolution. The PO_4 values are also normalized to constant salinity. The impact of this correction is however quite small.

Shown in Figure 8 are the PO_4 - ΣCO_2* trends for a number of GEOSECS and TTO stations. There is no indication that the values for nitrate free surface water fall off the trends for deep waters. On these plots we show a reference line with a slope of 130, which represents the mean global trend. Table 2 summarizes the evidence for this choice for the C/P ratio in marine organic residues. The average ΣCO_2* value for nitrate free surface water is taken to be 1975 μm/kg (which as shown in Figure 9 is the average ΣCO_2* value obtained from the GEOSECS and TTO data sets). The average PO_4 value for nitrate free surface water is 0.15 μm/kg. Hence in the ΣCO_2*-PO_4 diagrams we plot warm surface water at the coordinates 1975 μm/kg-0.15 μm/kg.

Thus we have a situation where the p_{CO_2} values for surface ocean water seem to be telling us that the ocean has its dissolved inorganic carbon distributed in accord with the thermodynamic model and where the ΣCO_2*-PO_4 trends seem to be telling us that its dissolved inorganic carbon is distributed in accord with the Redfield model.

Within the framework of our models these two observations are in direct conflict. Thus not only does this tell us that the real ocean is more complicated than the model ocean but also that these complications tend to mask its true character. While we do not claim to understand the reason for this masking, we can think of possible explanations. For example, if part of the nutrients in the water reaching the surface at the high latitude outcrop of the Redfield model were allowed to leave with organic residues, the CO_2 pressure for the outcrop would drop. The trend followed by the p_{CO_2} with decreasing PO_4 content is shown in Figure 10. As can be seen a decrease from 2.2 to 1.3 μm/kg PO_4 would bring the Redfield model's cold reservoir outcrop p_{CO_2} to the same value as that for the model's warm reservoir. Because of the lower temperature at the cold outcrop, further removal of PO_4 would reduce the outcrop p_{CO_2} below that for the warm reservoir. As mentioned above and shown in Figure 10 the surface waters of the Antarctic contain less PO_4 (and NO_3) than mean deep water. Thus we see that by modifying the Redfield model we could create uniform CO_2 pressures along its surface.

One might conclude then, that if the real ocean is akin to this modified Redfield ocean, its surface p_{CO_2} must be insensitive to the rates of ocean mixing. This is true only as long as the fraction of the nutrients removed at the outcrop of the cold reservoir remains constant. In this case a change in mixing rate would not alter the atmospheric p_{CO_2}. However, as the factors controlling the fraction of nutrients removed from the outcropping Antarctic waters must depend on the rates of mixing, changes in ventilation would upset the coincidence which currently allows the ocean to be both thermodynamic (in the sense that its surface p_{CO_2} pressure is nearly uniform) and Redfield (in the sense that ΣCO_2* and PO_4 fall along the 130-1 line).

It should also be pointed out that in assuming our model oceans to be free of salinity differences and differences created by the $CaCO_3$ cycle in the ocean, we introduce some bias. In Table 3 we compare the compositions for surface waters observed during the GEOSECS program at all the stations occupied in the Antarctic with those at selected stations occupied in the temperate zones. These results show what we have mentioned before; the p_{CO_2} values are on the average nearly the same and the Antarctic waters have high nutrient contents. In addition, it can be seen that the salinities for Antarctic waters are much lower than those for temperate waters. Despite this difference in salinity, the alkalinity of Antarctic surface waters is only slightly lower than that for temperate waters.

In Table 4 we show the impact the salinity

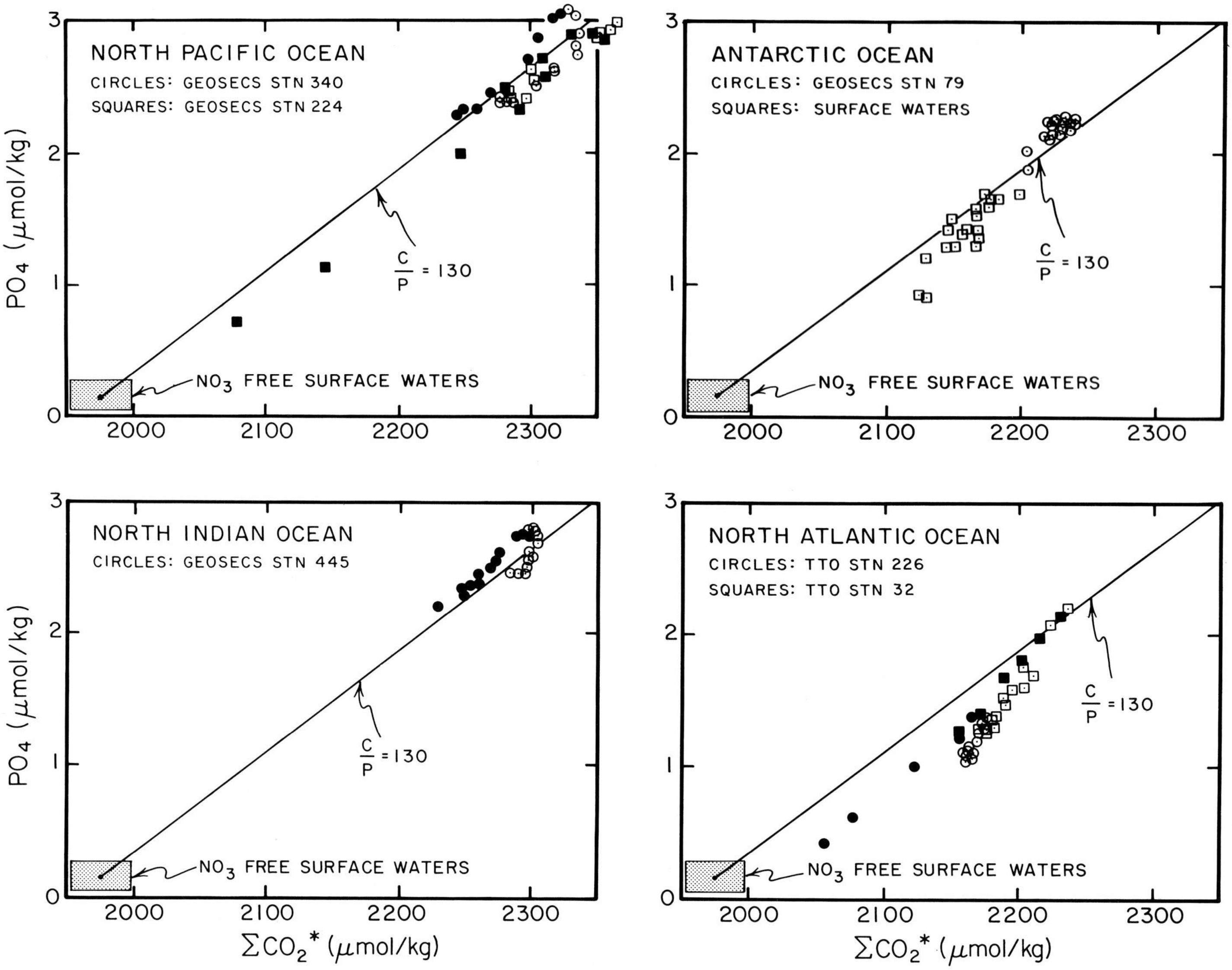

Fig. 8. Plots of ΣCO_2* against PO_4 for four separate regions of the ocean. On each diagram the range for nutrient free surface waters is shown. Also shown is a reference line representing the mean global ocean trend with a C/P slope of 130 passing through the average concentration for nitrate free surface water. Where appropriate the samples beneath the water column phosphate maximum are shown by open symbols and those from above are shown by closed symbols.

and $CaCO_3$ normalization would have on the observed values. For representative stations we perform a salinity normalization (to 35°/oo). For the Antarctic waters which have salinities less than 35°/oo this normalization raises the P_{CO_2} value and for temperate waters, which have salinities above 35°/oo, it lowers the P_{CO_2}. In addition, we also perform a $CaCO_3$ normalization for these same stations. This normalization raises the P_{CO_2} value for Antarctic waters by about 10% and those for temperate waters by about 3%. At the level of our present inquiry these differences are sufficiently small so as not to effect the conclusions. However, it should be kept in mind that

TABLE 2. Redfield ratio estimates based on the analysis of chemical gradients along isopycnal horizons in the sea (Broecker et al., in press)

	P	N	C	$-O_2$
ATLANTIC (thermocline)	1	17.8	130	177
INDIAN (thermocline)	1	14.5	131	170
INDIAN (deep)	1	13.8	125	164
PACIFIC (deep)	1	14.5	131	170

TABLE 3. The chemical composition of surface ocean waters in the Antarctic and temperate oceans observed during the GEOSECS program, 1972-1978. The PCO_2 values have been computed using the data presented in this table.

Stn.	Lat.	Long.	T °C	S o/oo	NO_3 µm/kg	PO_4 µm/kg	SiO_3 µm/kg	ΣCO_2 µm/kg	Alk µeq/kg	PCO_2 µatm
				ANTARCTIC SURFACE WATERS (Pacific Sector)						
280	56°S	170°E	6.5	34.06	19.2	1.30	2.9	2109	2297	330
282	58°S	170°E	5.6	33.88	21.5	1.49	3.4	2098	2303	289
285	61°S	170°E	2.2	33.96	25.9	1.65	15.9	2137	2304	322
286	66°S	174°E	0.3	33.91	24.5	1.48	52.7	2160	2307	344
287	69°S	173°W	-1.1	33.38	25.2	1.68	54.9	2122	2280	296
290	58°S	174°W	4.5	33.96	21.4	1.52	1.4	2115	2301	310
291	56°S	176°W	6.9	34.16	18.9	1.29	2.8	2085	2259	358
				(Atlantic Sector)						
76	58°S	66°W	4.5	33.94	22.1	1.29	2.4	2097	2288	297
78	61°S	63°W	1.3	33.83	25.1	1.69	20.0	2132	2295	316
79	60°S	45°W	1.8	34.06	21.5	1.38	41.4	2133	2320	285
82	56°S	25°W	1.0	33.73	24.5	1.65	45.1	2130	2288	322
85	57°S	17°W	1.1	33.56	22.7	1.42	22.0	2102	2277	284
87	59°S	9°W	0.2	33.56	22.7	1.36	50.2	2124	2290	296
89	60°S	0°	1.0	34.10	22.3	1.31	55.5	2149	2329	292
90	56°S	4°E	1.2	33.84	23.6	1.52	36.9	2130	2300	304
91	50°S	11°E	4.3	33.84	20.0	1.29	0.5	2102	2296	290
92	46°S	15°E	7.4	33.82	17.4	1.20	1.0	2073	2270	216
				(Indian Sector)						
429	48°S	58°E	6.5	33.73	20.2	1.48	5.4	2093	2271	343
430	60°S	61°E	1.8	33.84	27.4	1.87	37.9	2149	2290	374
431	64°S	84°E	2.4	33.79	26.8	1.76	44.6	2152	2294	380
432	59°S	93°E	1.8	33.87	23.9	1.57	29.3	2143	2306	326
433	53°S	103°E	3.3	33.91	24.4	1.62	12.0	2142	2295	364
				TEMPERATE SURFACE WATERS (North Pacific)						
212	30°N	160°W	25.2	35.47	0.0	0.02	1.8	1973	2344	316
213	31°N	168°W	26.8	35.52	0.1	0.00	2.2	1976	2351	334
214	32°N	177°W	25.8	34.75	0.1	0.04	3.8	1951	2296	340
215	37°N	177°W	20.1	34.33	0.3	0.09	5.2	1945	2270	286
223	35°N	152°E	22.5	34.52	0.0	0.00	2.4	1926	2290	268
224	34°N	142°E	23.9	34.50	0.0	0.00	2.2	1919	2285	280
225	33°N	182°E	23.2	34.51	0.0	0.02	3.4	1918	2283	277
226	31°N	171°E	24.7	35.06	0.1	0.02	3.3	1938	2305	300
227	25°N	170°E	26.6	35.21	0.0	0.03	1.7	1929	2313	305
				(South Pacific)						
267	19°S	171°W	28.3	34.93	0.1	0.15	0.4	1915	2312	307
268	21°S	173°E	27.9	35.41	0.1	0.16	1.4	1917	2339	282
269	24°S	175°W	27.3	35.42	0.1	0.05	1.2	1949	2339	315
273	30°S	176°W	23.7	35.53	0.1	0.08	0.0	2001	2368	309
278	37°S	180°	19.8	35.67	0.0	0.13	1.6	2017	2355	294
303	38°S	170°W	18.2	34.93	0.0	0.22	8.2	2020	2309	325
305	36°S	167°W	20.3	35.31	0.0	0.12	5.1	2001	2334	299
306	31°S	164°W	22.2	35.30	0.0	0.06	6.4	1972	2336	282
308	30°S	160°W	23.4	35.36	0.1	0.05	2.4	1989	2339	317
310	27°S	157°W	24.8	35.51	0.0	0.10	2.1	1989	2370	305
314	24°S	154°W	26.0	35.61	0.0	0.11	0.5	1997	2363	341
317	24°S	127°W	25.9	36.56	0.0	0.24	3.6	2036	2427	333

TABLE 3. (Continued)

Stn.	Lat.	Long.	T °C	S °/oo	NO_3 μm/kg	PO_4 μm/kg	SiO_3 μm/kg	ΣCO_2 μm/kg	Alk μeq/kg	P_{CO_2} μatm
					(North Atlantic)					
121	36°N	68°W	19.1	36.43	0.7	0.05	0.9	2059	2400	299
120	33°N	57°W	17.1	36.36	1.0	0.04	0.7	2086	2378	291
119	32°N	51°W	21.1	36.69	0.0	0.01	0.8	2058	2411	314
118	31°N	46°W	20.2	36.78	0.0	0.01	0.8	2064	2418	304
117	31°N	39°W	19.4	36.69	0.0	0.00	0.7	2070	2407	311
116	30°N	30°W	18.3	36.57	0.3	0.01	0.7	2070	2395	309
115	28°N	26°W	19.1	36.97	0.1	0.01	0.8	2098	2432	321
114	21°N	22°W	20.0	36.80	0.3	0.05	0.5	2086	2409	342
31	27°N	54°W	27.4	36.90	0.1	0.06	1.0	2041	2426	366
32	24°N	54°W	26.8	36.91	0.1	0.04	1.4	2040	2427	354
33	21°N	54°W	24.9	37.29	0.1	0.04	0.8	2086	2467	351
					(South Atlantic)					
55	18°S	31°W	25.7	37.26	0.1	0.14	1.0	2074	2456	358
56	21°S	33°W	24.7	37.07	0.1	0.08	1.1	2056	2433	342
57	24°S	35°W	23.6	36.70	0.1	0.04	1.1	2047	2421	323
58	27°S	37°W	21.8	36.41	0.1	0.03	0.6	2044	2400	314
59	30°S	39°W	21.0	36.25	0.0	0.05	0.6	2040	2385	313
60	33°S	43°W	18.5	36.00	0.0	0.09	1.0	2045	2382	288
102	32°S	9°E	22.3	35.75	0.1	0.08	2.1	2016	2351	331
103	24°S	8°E	23.4	35.59	0.1	0.17	1.4	2037	2360	369
105	20°S	2°E	24.7	35.92	0.0	0.20	0.2	2033	2375	367
					(South Indian)					
427	27°S	57°E	25.6	35.24	0.0	0.13	2.2	1959	2308	337
428	38°S	58°E	18.8	35.33	0.0	0.20	0.7	2023	2324	321
436	29°S	110°E	24.5	35.73	0.0	0.06	4.0	1992	2349	330
437	24°S	105°E	24.4	35.73	0.0	0.07	3.4	1994	2347	332
438	20°S	101°E	26.6	35.28	0.0	0.08	1.7	1957	2311	349
452	20°S	80°E	26.8	34.75	0.0	0.15	2.1	1924	2282	331
453	23°S	74°E	24.7	35.53	0.0	0.13	2.2	1974	2331	327
454	27°S	67°E	24.5	35.51	0.0	0.11	2.1	1974	2329	325

TABLE 4. Impact of salinity and $CaCO_3$ normalizations on the apparent P_{CO_2} of sample surface water from Table 3. The subscript N indicates normalization to a salinity of 35.00°/oo. $P_{CO_2})_N$ was computed using ALK_N and ΣCO_{2N}, and $P_{CO_2}{}^*$ was computed using ALK^{STD} and $\Sigma CO_2{}^*$ at salinity 35.00°/oo and temperature indicated in the second column, where $\Sigma CO_2{}^* = \Sigma CO_{2N} - (ALK_N - 2318 + NO_{3N} - 16)$.

	T °C	S °/oo	As Measured ΣCO_2 μm/kg	ALK μeq/kg	P_{CO_2} μatm	Salinity Normalized ΣCO_{2N} μm/kg	ALK_N μeq/kg	$P_{CO_2})_N$ μatm	Salinity and $CaCO_3$ Normalized ΣCO_2^* μm/kg	ALK^{STD} μeq/kg	$P_{CO_2}^*$ μatm
Antarctic Stn 90	1.2	33.84	2130	2300	303	2203	2379	319	2168	2318	356
No. Pac. Stn 212	25.2	35.47	1973	2344	315	1947	2313	309	1958	2318	319
So. Pac. Stn 273	23.7	35.53	2001	2368	308	1971	2333	301	1971	2318	317
So. Ind. Stn 454	24.4	35.51	1974	2329	324	1946	2295	318	1966	2318	321
No. Atl. Stn 119	21.1	36.69	2058	2411	314	1963	2300	291	1980	2318	295
So. Atl.	24.0	36.70	2047	2421	328	1952	2309	304	1964	2318	311

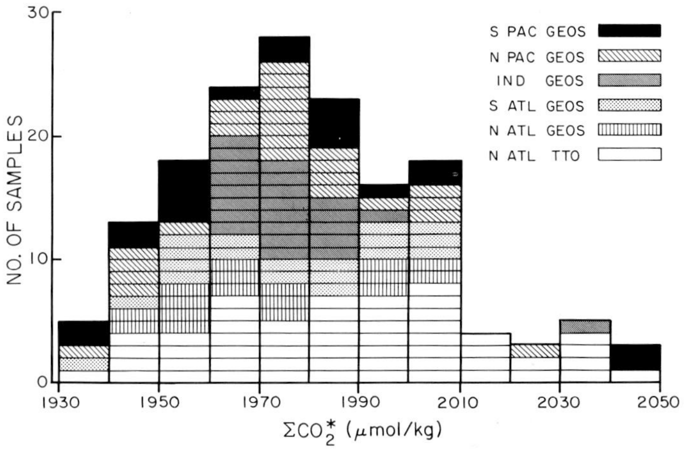

Fig. 9. Histogram showing the ΣCO_2* values obtained for nutrient depleted surface water observed during the GEOSECS and TTO North Atlantic programs.

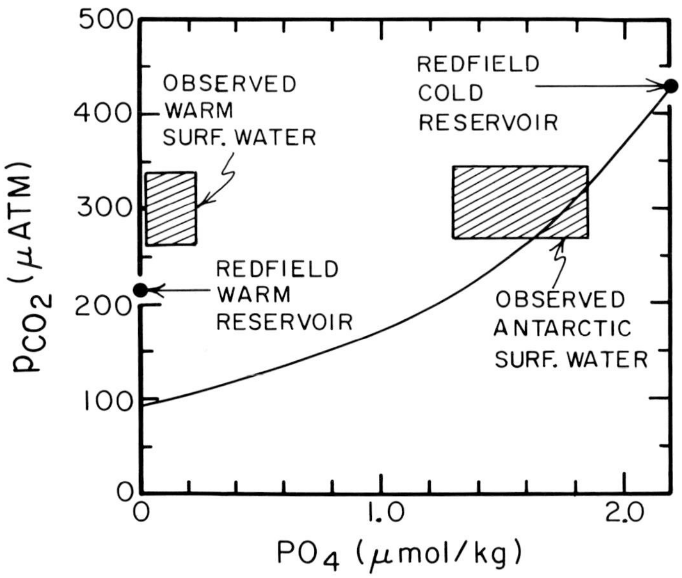

Fig. 10. The relationship between p_{CO_2} and PO_4 concentration in cold water at 2°C. This shows the effect on p_{CO_2} of utilization of PO_4 in the outcropping cold water of the Redfield model. The reduction of p_{CO_2} from the Redfield ocean value of 430 μatm with decreasing PO_4 is due to biological utilization of CO_2 and PO_4. It is to be noted that, at the observed mean PO_4 value of 1.5 μm/kg in the Antarctic surface water, the calculated p_{CO_2} value is 245 μatm, which is similar to the warm water value of 216 μatm at 21°C. The difference between the warm water p_{CO_2} value at 21°C (indicated by an arrow) for zero PO_4 and the zero phosphate intercept of the curve (at 2°C) represents the temperature effect alone. The shaded areas indicate the observed range of values in the Antarctic surface water and warm surface waters in the world oceans.

the distribution of CO_2 partial pressures in the sea is influenced by the H_2O evaporation - condensation and $CaCO_3$ formation - dissolution cycles. They may ultimately play a role in models of changing atmospheric CO_2 content.

Summary. In this preliminary consideration of the connection between ocean mixing and atmospheric CO_2 content we have attempted to demonstrate that because of the interaction between the Redfield and thermodynamic tendencies for ΣCO_2 distribution in the sea, it is possible that the CO_2 content of the atmosphere changes with deep sea ventilation rate. We then go on to show that the situation in today's ocean is sufficiently complicated that conclusions drawn from the two box models used to investigate the sensitivity of the system may be misleading. Far more thought will have to go into this problem before any meaningful answers can be given.

Appendix A

Model Ocean Calculations. For the three ocean models presented, the alkalinity and salinity of the warm and cold waters are assumed to be equal and constant. The mean alkalinity value adopted is that for the world ocean computed by Takahashi et al. (1981-a). The mean concentration for total CO_2 in the ocean is the same in all three models, and the value used is so adjusted to yield an atmospheric p_{CO_2} value for the kinetic ocean model close to today's atmospheric p_{CO_2} value. This mean total CO_2 concentration is similar to but somewhat smaller than the mean world ocean value of 2254 μm/kg computed by Takahashi et al. (1981-a). The mean phosphate content for the deep ocean reservoir is set at 2.2 μm/kg in all three models. The surface ocean is phosphate free in all three models.

The following sets of equations were solved by iteration, and the results are presented in Table 1.

A) Equations defining the thermodynamic ocean:

1) $(pCO_2)_w = (pCO_2)_a$
2) $0.91\ (\Sigma CO_2)_c + 0.09\ (\Sigma CO_2)_w = 2215$ μm/kg
3) $(pCO_2)_w = f[(ALK)_w, (\Sigma CO_2)_w, S, T]$
4) $(pCO_2)_c = f[(ALK)_c, (\Sigma CO_2)_c, (PO_4)_c, S, T]$
5) $(ALK)_w = (ALK)_c = 2374$ μeq/kg

where the subcripts a, w and c indicate respectively the atmosphere, warm and cold waters. f[ ] indicates the thermodynamic relationship between p_{CO_2}, ALK, ΣCO_2, PO_4, salinity and temperature, and includes the dissociation constants of carbonic, boric and phosphoric acids and of water as referenced in Table 1. The concentration of boric acid is 410.6 μm/kg, which represents that for mean global ocean water. ALK and ΣCO_2 are defined by:

$ALK = (HCO_3^-) + 2(CO_3^=) + (H_2BO_3^-) + (H_2PO_4^-) + 2\,(HPO_4^=) + 3\,(PO_4^{3-}) + (OH^-) - (H^+)$

$\Sigma CO_2 = (CO_2) + (HCO_3^-) + (CO_3^=)$.

B) Equations defining the Redfield ocean:

7) $(pCO_2)_a = 0.91(pCO_2)_w + 0.09\,(pCO_2)_c$

8) $(\Sigma CO_2)_c - (\Sigma CO_2)_w = 130\,(PO_4)_c = 286$ μm/kg

9) $0.91(\Sigma CO_2)_c + 0.09(\Sigma CO_2)_w = 2215$ μm/kg

10) $(pCO_2)_w = f[(ALK)_w, (\Sigma CO_2)_w, S, T]$

11) $(pCO_2)_c = f[(ALK)_c, (\Sigma CO_2)_c, (PO_4)_c, S, T]$

C) Equations defining the kinetic ocean:

12) $(pCO_2)_a - (pCO_2)_w = (U/I)((\Sigma CO_2)_w + 130\,(PO_4)_c - (\Sigma CO_2)_c)$

13) $(pCO_2)_a - (pCO_2)_c = (0.91\,U/0.09\,I)((\Sigma CO_2)_c - (\Sigma CO_2)_w - 130(PO_4)_c)$

14) $0.91(\Sigma CO_2)_c + 0.09(\Sigma CO_2)_w = 2215$ μm/kg

15) $(pCO_2)_w = f[(ALK)_w, (\Sigma CO_2)_w, S, T]$

16) $(pCO_2)_c = f[(ALK)_c, (\Sigma CO_2)_c, (PO_4)_c, S, T]$

Where U is the cold water upwelling rate (assumed to be 5 $m^3/m^2 \cdot yr$ or 5.125×10^3 $kg/m^2 \cdot yr$) and I is the gas exchange rate constant across the air-sea interface (assumed to be 20/300 or 6.667×10^{-2} moles CO_2/m^2 yr μatm).

Acknowledgements. The excellent work of the Bainbridge-Williams seagoing analytical group in obtaining the high quality set of oxygen, nitrate, phosphate, total dissolved carbon and alkalinity data for the GEOSECS and TTO Programs made this study possible. Financial support was provided by the Department of Energy Program, Grant Nos. 79 EV-10038, and DE-AC02-81ER60000A, and by the National Science Foundation, Grant No. OCE-81-10646. Lamont-Doherty Geological Observatory Contribution No. 0000.

References

Berger, W.H., Deep-sea carbonate: pteropod distribution and the aragonite compensation depth, Deep Sea Res. 25, 447-452, 1978.

Berger, W.H., Increase of carbon dioxide in the atmosphere during deglaciation: the coral reef hypothesis, Naturwissenschaften, 69, 87-88, 1982.

Berger, W.H. and R.S. Keir, Glacial-Holocene changes in atmospheric CO_2 and the deep-sea record, (In this volume) 1984.

Berner, W., B. Staufer and H. Oeschger, Past atmospheric composition and climate, gas parameters measured on ice cores, Nature, 275, 53-55, 1979.

Boyle, E.A. and L.D. Keigwin, Deep circulation of the North Atlantic over the last 200,000 years: Geochemical evidence, Science, 218, 784-787, 1982.

Broecker, W.S., Glacial to interglacial changes in ocean and atmosphere chemistry, In "Climatic variations and variability: Facts and Theories", A. Berger, editor, D. Reidel Publishing Co., 109-120, 1981.

Broecker, W.S., Glacial to interglacial changes in ocean chemistry. Progress in Oceanography, 11, 151-157,1982-a.

Broecker, W.S., Ocean chemistry during glacial time, Geochim. et Cosmochim. Acta, 46, 1689-1705, 1982-b.

Broecker, W.S. and T.-H. Peng, The Climate-chemistry connection, (In this volume), 1984.

Broecker, W.S., T. Takahashi and S. Langer. Redfield ratio estimates based on chemical data from isopycnal surfaces. Submitted to Jour. Mar. Res.

Curry, W.B. and G.P. Lohmann, Carbon isotopic changes in benthic foraminifera from the western South Atlantic: Reconstruction of glacial abyssal circulation patterns. Quat. Res. 18, 218-235, 1982.

Delmas, R.J., J.-M. Ascencio and M. Legrand, Polar ice evidence that atmospheric CO_2 29,000 yr. BP was 50% of the present, Nature 284, 155-157, 1980.

Kester, D.R., and R.M. Pytkowicz, Determination of apparent dissociation constants of phosphoric sea water, Limnol. & Oceanogr., 12, 243-252, 1967.

Lyman, J. Buffer mechansism of seawater, Ph.D. Thesis, University of Calif., Los Angeles, 196 pp., 1956.

Millero, F.J., The thermodynamics of the carbonate system in sea water, Geochim. et Cosmochim. Acta, 43, 1651-1661, 1979.

Mehrbach, C., C.H. Culberson, J.E. Hawley and R.M. Pytkowicz, Measurement of the apparent dissociation constants of carbonic acid in seawater at atmospheric pressure, Limnol. and Oceanogr., 18, 897-907, 1973.

Neftel, A., H. Oeschger, J. Swander, B. Stauffer and R. Zumbrunn, Ice core sample measurements give atmospheric CO_2 content during the past 40,000 years, Nature, 295, 220-223, 1982.

Shackleton, N.J., Tropical rain forest history and the equatorial Pacific carbonate dissolution cycles, In "The Fate of Fossil Fuel CO_2 in the Oceans", N.R. Anderson and A. Malahoff, editors, Plenum Press, New York, 401-428, 1977.

Takahashi, T., W.S. Broecker, and A.E. Bainbridge, The alkalinity and total carbon dioxide concentration in the world oceans, In "Carbon Cycle Modelling", B. Bolin editor, SCOPE Vol. 16, 271-286, J. Wiley & Sons, N.Y., 1981-a.

Takahashi, T., W.S. Broecker, and A.E. Bainbridge, Supplement to the alkalinity and total carbon dioxide concentration in the world oceans, In "Carbon Cycle Modelling", B. Bolin editor, SCOPE Vol. 16, 159-199, J. Wiley & Sons, N.Y., 1981-b.

Weiss, R.F., Carbon dioxide in water and seawater: the solubility of a non-ideal gas, Marine Chem. 2, 203-215, 1974.

THE CLIMATE-CHEMISTRY CONNECTION

Wallace S. Broecker

Lamont-Doherty Geological Observatory of Columbia University,
Palisades, New York 10964

Tsung-Hung Peng

Environmental Sciences Division, Oak Ridge National Laboratory
Oak Ridge, Tennessee 37830

Abstract. The impacts on ocean chemistry of a variety of scenarios capable of producing the 90×10^{-6} atmosphere glacial to interglacial increase in CO_2 are explored. It is shown that each scenario produces a unique combination of ^{13}C, Cd, O_2 and $CaCO_3$ dissolution responses. Thus, once we have reliable ocean wide records of these parameters, it should be possible to greatly reduce the list of those scenarios capable of explaining this CO_2 change. If our present information regarding these records is correct, then the only acceptable hypotheses are 1) shelf deposition, 2) Redfield ratio change plus forest growth or 3) phosphorite deposition plus forest growth. When global cadmium records become available further cuts will be possible.

Introduction

While there has been much discussion about the influence of the anthropogenic "greenhouse" gases on future climates, relatively little of substance is to be found regarding the interaction between paleoclimate and paleochemistry. This situation has prevailed not because of a lack of awareness of the potential chemical changes to bring about climate change but rather because of a lack of paleochemical data. In the absence of such data the situation is much the same as that with regard to the sun; i.e., while fluctuations in its energy output may have caused climate to change, at present we have no way to say anything regarding the amplitude or timing of these fluctuations.

The discovery by Oeschger and his colleagues in Bern (Berner et al., 1979; Neftel et al., 1982) and by Lorius and his colleagues in Grenoble (Delmas et al., 1980) that a record of atmospheric CO_2 content over the last 100,000 years or more is available from borings in the polar ice caps suddenly altered this long-standing situation. These results suggest that the CO_2 content of the atmosphere during peak glacial time was about two thirds its Holocene value. The global average surface temperature change produced by such a CO_2 content change would be about 1.5°C. [Based on GCM model results which suggest a global warming of 3°C for a doubling of the atmospheric CO_2 content]. If the CLIMAP estimate (CLIMAP McIntyre leader, 1981) of about 2°C for the surface ocean temperature change between glacial and interglacial time is correct then the CO_2 induced temperature change must certainly have been a significant contributor to the major glacial to interglacial climate transition. Thus, it is important to consider the cause of the CO_2 changes. Are they, like albedo, a by-product of cycles driven by some other cause or could an interaction between atmospheric CO_2 content and some other climate linked phenomena produce a self sustained cycle? In view of the current thinking that orbital cycles are the primary cause for the glacial cycles experienced by the earth over the last million years, it is especially important to carefully consider the ramifications of the new wrinkle provided by ice core research.

Before proceeding with this analysis we hasten to mention that when considering atmospheric CO_2 variations on time scales longer than several hundred years we must focus our attention on the ocean. The ocean currently contains about 60 times more carbon (as HCO_3^-, $CO_3^=$ and CO_2) than does the atmosphere. The atmosphere equilibrates its CO_2 content with the ocean's mixed layer on the time scale of a few years. The ocean itself is stirred completely on the time scale of 1000 years. Thus, any atmospheric CO_2 content change which persists for several thousand or more years must be dictated by the ocean.

The CO_2-temperature Connection

We know from laboratory experiments that when sea water is isochemically heated its CO_2 partial pressure rises about 4% per °C. At the Holocene atmospheric CO_2 partial pressure of

LDGO Contribution #3590

about 300 x 10^{-6} atm, this corresponds to 12 x 10^{-6} atm/°C. In the actual ocean-atmosphere system the increase in CO_2 pressure due to a uniform 1°C warming would be 10×10^{-6} atm. rather than 12×10^{-6} atm. The reason is that in order to increase the atmosphere's CO_2 content the ocean would have to give up some of its dissolved inorganic carbon. On the other hand, climate models indicate that to raise the earth's surface temperature by 1°C via the greenhouse effect requires a CO_2 pressure change of about 25% (i.e. in today's atmosphere by about 75 x 10^{-6} atm). Thus the CO_2 feedback created by an ocean warming would be only about 13% (10/12 x 4/25 x 100). If the atmospheric CO_2 content rise at the end of glacial time observed in ice cores was caused by a surface ocean warming, then this warming must have been much larger than that produced by the CO_2 rise itself.

If the CLIMAP estimate of about 2°C for the peak glacial to Holocene surface ocean temperature change is anywhere near correct the observed CO_2 change must have some other origin. As pointed out by Broecker (1981) the 20 x 10^{-6} atm CO_2 rise due to a 2°C warming would be largely compensated by the drop in ocean's mean salinity resulting from melting of the glacial ice sheets. Further, Shackleton (1977a) has pointed out that the removal of CO_2 from ocean-removal of CO_2 from ocean-atmosphere system required for the expansion of forests during early Holocene time would likely have caused a CO_2 pressure decrease even greater than that induced by the small oceanic warming.

Thus we are left with two possibilities. The first is that the CLIMAP estimate of the glacial to interglacial surface ocean temperature change is greatly in error. For example, if the surface ocean warmed about 12°C at the close of glacial time and if the CO_2 requirement for the expansion of the forests in early Holocene time is taken to be as small as is consistent with paleobotanical evidence then the CO_2 change observed in ice cores could be explained. This is an unlikely scenario. Even if the CLIMAP faunal and floral results are disregarded, the similarity of the amplitude of glacial to interglacial $^{18}O/^{16}O$ changes for benthic (i.e., bottom dwelling) and plankton (i.e., surface dwelling) foraminifera appears to preclude the possibility of temperature change any larger than about 3°C (Shackleton 1977b). [This isotopic argument is based on the assumption that most (i.e., >80%) of the $\delta^{18}O$ shift from glacial to interglacial time recorded by benthic foraminifera is the result of storage of ^{18}O depleted water in continental ice sheets. This assumption is based on the fact that the deep sea is currently sufficiently close to its freezing point that it could not in all likelihood have been more than about two degrees colder during peak glacial time. If so, at least 80% of the benthic ^{18}O signal must be due to ice volume rather than temperature change. In recent years evidence from uplifted coral reefs (Broecker 1978) and from the deconvolution of the past glacial sea level record for many points on the globe (Peltier 1982) suggest that the ice volume was about a factor of two too small to produce the required 80% of the observed $\delta^{18}O$ change for benthic forams. One possible solution to this seeming dilemma is to postulate that the deep sea temperature was 5 to 6°C warmer during peak Holocene time than it is today. If this proves to be the case then the isotopic argument for small surface ocean temperature changes falls and we are left with only the CLIMAP factor analysis result. We point this out to alert the reader to the possibility that the temperature explanation for the CO_2 change may not be dead after all]. The second possibility is that the cause of CO_2 change lies in a change in ocean chemistry. It is this avenue we choose to pursue.

The CO_2-nutrient Connection

Broecker (1981, 1982) has pointed out that the CO_2 gas content of surface ocean water (and hence the CO_2 content of the atmosphere) is very sensitive to the ratio of dissolved PO_4 to dissolved ΣCO_2 in mean sea water and to the ratio of carbon to phosphorus in the organic residues falling from the surface to the deep sea. Together these two ratios determine the fraction of the dissolved carbon reaching surface water which is removed as organic residues. This removal leads to a reduction in the CO_2 partial pressure for surface waters.

With this sensitivity in mind Broecker (1981, 1982) proposed two nutrient-based explanations for the CO_2 increase observed in ice cores at the close of glacial time. The first involved the deposition of organic material in shelf sediments during the marine transgression which accompanied deglaciation. The second involved a change in the C/P ratio in falling organic residues caused by the ecologic changes which accompanied the transition from glacial to post glacial time. There are actually a number of chemical scenarios which could lead to a rise in CO_2 pressure at the close of glacial time. They are listed in Table 1. Fortunately there are a number of parameters recorded by deep sea sediments which give us the wherewithal to reduce this list. As can be seen in Table 1 each of the seven scenarios produces a different set of changes in these parameters. Before going on let us discuss in turn each of these recorders and what we know about its glacial-interglacial record.

1) $^{13}C/^{12}C$ ratio in benthic forams. Shackleton (1977a) was the first to obtain $\delta^{13}C$ records for benthic forams. His measurements on Meteor core 12392 from the eastern North Atlantic yielded a rectangular cycle with interglacial values about 1°/oo higher than glacial values. Assuming that these changes were the

TABLE 1. Comparison of the impacts of various scenarios leading to a 90 x 10^{-6} atm glacial to interglacial CO_2 change on ocean nutrient chemistry (left hand block) and on ocean "recordables" (right hand block).

Glacial to Interglacial Change	Whole Ocean $\Delta\Sigma CO_2$ %	Whole Ocean ΔALK %	Whole Ocean ΔNO_3 %	Whole Ocean ΔPO_4 %	Whole Ocean $\Delta\delta^{13}C$ °/oo	Warm Surf $\Delta\delta^{13}C$ °/oo	Deep Ocean $\Delta CO_3^{=}$ μm/kg	Deep Ocean ΔO_2 μm/kg	Whole Ocean ΔCd %	ATM ΔpCO_2 10^{-6} atm
Observed										
Target	-	-	-	-	+0.7	+0.1	+20	?	?	+90
Hypothetical Scenarios										
Forest	+5	0	0	0	-0.7	-0.8	-25	0	0	+90
Shelf deposition	-3	0	-40	-40	+0.7	0.0	+25	+100	-40	+90
Phosphorite deposition	0	0	0	-30	0.0	-0.7	0	+ 70	0	+90
Coral growth	-8	-15	0	0	0.0	0.0	-40	0	0	+90
Decrease in C/P in organic residues*	0	0	0	0	0.0	-0.5	0	+ 80	0	+90
Forest shrinkage & C/P change	+5	0	0	0	+0.5	0.0**	+10	+ 80	0	+90
Increase in $CaCO_3/P$ in raining debris+	0	0	0	0	0.0	0.0	0	0	0	+90

*From 115 in glacial to 85 during interglacial time
+From 16.7 in glacial to 50 during interglacial time
**Depending on the timing of the changes a spike in ^{13}C may occur (see figure 7).

result of changes in the $\delta^{13}C$ of the ΣCO_2 in bottom water rather than to some effect associated with the life cycle of the foraminifera species <u>Planulina</u> <u>wuellerstorfi</u> and <u>Uvigerina</u> SPP, which he measured, Shackleton concluded that ΣCO_2 in the deep waters of the Atlantic had undergone a $\delta^{13}C$ change of similar magnitude to that for the foraminifera. Shackleton measured cores from other parts of the ocean and found similar shifts across the boundary between late glacial and post glacial time. He concluded that there had been a change in the $\delta^{13}C$ for average ocean carbon and attributed it to the expansion of the forest and soil reservoirs during early post glacial time. Broecker (1981) averaged Shackleton's results and concluded that the increase in the $\delta^{13}C$ of ocean ΣCO_2 at the end of glacial time was about 0.7°/oo. Subsequent measurements by Berger and Shackleton on cores from the Pacific cast doubt on the magnitude of this shift. When adequate data are available the increase may prove to be smaller than 0.7°/oo. In any case the magnitude of this change fixes the amount of organic carbon removed from the ocean at the end of glacial time. $CaCO_3$ production and dissolution do not significantly alter the whole ocean $^{13}C/^{12}C$ ratio. No other mechanism for a sudden change in whole ocean $\delta^{13}C$ has been proposed. While telling us how much carbon was removed from the sea to form organic compounds the whole ocean $\delta^{13}C$ change tells us nothing about where the organic carbon went. It could have gone to build the Holocene forest and soil reservoirs or it could have gone into shelf sediments.

2) $^{13}C/^{12}C$ in planktonic forams. Broecker (1981, 1982) summarized the published ^{13}C results for individual surface dwelling foram species and found that while the records were very noisy they suggest that the glacial to interglacial change was less than 0.2°/oo. He points out that this is to be expected if the organic carbon removed from the sea at the close of glacial time went to marine sediments and carried with it N and P in roughly the so-called Redfield proportions (i.e., P:N:C ≈ 1:16: 105). He also points out that a combination of removal of organic carbon to the forest reservoir <u>and</u> a de-

crease in the C/P and C/N ratios in the organic residues falling from the surface ocean could lead to the same constancy (provided the magnitudes of these opposing influences just balanced). Since Broecker's compilation of these ^{13}C results to our knowledge no new records have been published nor has the source of the bothersome noise been found. As in the case for the benthic record more data are needed.

3) Cadmium content of benthic forams. The concentrations of dissolved PO_4 and Cd are well correlated in today's ocean (Boyle et al., 1976; Bruland et al., 1978). Both go to near zero values in warm surface water. Their ratio in all other waters is almost constant. As for NO_3 and PO_4, near surface waters show a fairly large Cd:P ratio variation. Below a few hundred meters these discrepancies vanish. Thus, in today's sea, Cd provides a good stand-in for PO_4. Hester and Boyle (1982) have shown that the Cd concentration of benthic forams found in core tops is linearly related to the Cd content of bottom water. Boyle and Keigwin (1982) have studied cores from the north Atlantic and shown a two fold decrease in the Cd content of benthic forams between glacial and post glacial time. Two interpretations of this change exist. The first is that the Cd content of the ocean decreased at the close of glacial time. If so then one might conclude that organic residues bearing P and Cd (in roughly their oceanic proportions) had been lost from the sea. In this case the Cd decrease would provide an estimate of the P decrease and would lend support to the shelf removal hypothesis. Boyle however feels that the decrease instead reflects primarily a change in the pattern of ocean mixing at the close of glacial time leading to the strong chemical gradients found today between the deep Atlantic and the deep Pacific Oceans. Until cadmium data from Pacific cores is in hand there is no way to say how much of the cadmium change observed for the Atlantic was due to a change in ocean cadmium content and how much was due to a change in the pattern of deep sea ventilation. [Boyle's (personal communication) new Pacific data shows that there probably is a small global C_d inventory shift but that this change is proportionally less than the $\delta^{13}C$ shift. About 60% of the North Atlantic C_d change is likely due to circulation changes]. The important point is that once the glacial to interglacial change in whole ocean cadmium content is established it will tell us something about the degree to which the ocean's phosphorus content changed. Of course as the shelf sediment C:P:Cd ratio may be quite different from the Redfield values questions will remain.

4) The $CaCO_3$ dissolution record. At most places on the sea floor the extent to which calcite or aragonite dissolve has varied with time. The most likely reason for these variations is that the $CO_3^=$ ion content of deep water has changed with time. The most dramatic of these changes is an aragonite preservation event at the transition between glacial and post glacial time. This event has been documented in both the Atlantic and Pacific Oceans and involves a drop in the lysocline for aragonite of over one kilometer (Berger 1977). The duration of this event was several thousands of years. When Shackleton (1977a) discovered what he considered to be a whole ocean $\delta^{13}C$ shift and attributed it to an extraction of CO_2 from the oceans needed for the expansion of the forests reservoirs at the onset of Holocene time, he pointed out that this removal should lead to a rise in oceanic $CO_3^=$ ion content and hence to the excess preservation of $CaCO_3$. This observation became a major point in Broecker's (1981, 1982) theory. The only reasonable explanation for a pronounced preservation event of several thousand year duration is the removal of carbon from the ocean in the form of organic matter.

5) The O_2 record for deep sea water. Were it possible to assess the change in the O_2 content of deep water then another strong diagnostic tool would be available. Any change in the $\Sigma CO_2/PO_4$ ratio in sea water or in the C/P ratio in organic debris falling from the surface ocean would change the residual dissolved O_2 content of the deep sea. As the O_2 content in today's deep sea lies in the range 50-250 μm/kg, the 70 to 100 μm/kg changes shown for several of the scenarios in Table 1 are large. While Streeter and Shackleton (1979) have claimed that the glacial to interglacial change in the species composition of benthic forams in the deep Atlantic points to an increase in the dissolved oxygen content of Atlantic deep water at the close of glacial time, the tie between the abundance of a given benthic species and dissolved oxygen concentration awaits verification. Also no record exists for the Pacific Ocean.

Thus we see that while the potential to differentiate among the variety of possible scenario combinations is there, resolution awaits the establishment of a reliable whole ocean $\delta^{13}C$ record, of a reliable warm surface ocean $\delta^{13}C$ record [An alternate way to obtain this information is through the ^{13}C record for CO_2 from polar ice cores. Oeschger's group in Bern, Switzerland is developing the wherewithal to make such measurements.], of a reliable whole ocean cadmium record, and of a reliable deep ocean O_2 record. It also awaits better documentation of the changes in lysocline depth.

Cyclic Models of Ocean-Atmosphere Chemistry

Further information can be obtained by examining the shape of the time histories for the various parameters. This is particularly true in connection with the relationship between the timing of the $CaCO_3$ dissolution record and the record for ^{13}C and Cd. While the time constants for the adjustment of the whole ocean ^{13}C, and the whole ocean P content are on the order of

100,000 years, that for $CO_3^=$ is on the order of 5,000 years. In order to illustrate this point we have generated a set of cyclic models. In all these models we adopt a 50,000 year rectangular glacial-interglacial cycle. Associated with the abrupt changes from one type of climate to another we invoke one or more of the scenarios in Table 1.

The calculations were carried out using the simple two reservoir ocean shown in Figure 1. In this model carbon and phosphorus are added from rivers and lost as residues of organisms. We neglect NO_3 assuming that its ratio to PO_4 in the ocean is maintained by ecologic feedback. The calculations are kept simple by assuming the ocean's alkalinity is matched by bicarbonate and carbonate ion alone (i.e., the ocean contains no borate). The ventilation time of the deep sea is taken as 1000 years. Rivers are assumed to supply an amount of water equal to 1/50,000th of that in the sea each year. All the phosphorus reaching the surface reservoir is assumed to be converted to organic residues containing 100 carbon atoms per phosphorus atom. These residues fall to the deep sea where 99% are destroyed by oxidation. The C and P contained in these residues are returned to solution. The remaining 1% of the residues are assumed to accumulate intact in the sediment. Although in our first order model we assume the fraction of residues destroyed in the deep sea to remain constant with time, the efficiency of residue oxidation very likely changes with the O_2 content of deep water. A more advanced model should take this dependence into account as it would surely influence the nature of the response to perturbations. For each phosphorus atom reaching the surface reservoir the steady state model assumes that 25 carbon atoms are incorporated into $CaCO_3$ which falls to the deep reservoir. In this case the fraction of the $CaCO_3$ dissolved is assumed to vary with the $CO_3^=$ concentration in the deep reservoir. The relationship between fraction of $CaCO_3$ dissolved and $CO_3^=$ ion concentration is shown in Table 2 along with other parameters characterizing the steady state operation of this model. One of its major deficiencies in this formulation is that it fails to take into account the dissolution of existing sedimentary calcite in response to perturbations which produce "destruction" events.

The first scenario we consider is the one proposed by Shackleton based on his ^{13}C data for benthic forams. It involves removal of CO_2 from the ocean to form forests in early post glacial time. To make the cycles symmetrical we release an equal amount of CO_2 via the demise of forests at the onset of each glacial episode (see figure 2). While this scenario generates the observed pattern of benthic ^{13}C change (i.e. about 1°/oo higher during the interglacial time) and the early postglacial preservation event observed in deep sea sediments, it generates a P_{CO_2} change of the wrong sign and a ^{13}C change

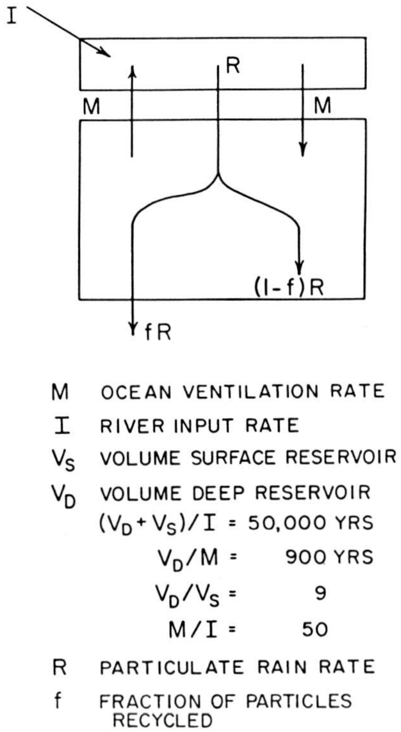

Fig. 1. Two box model of the ocean: The volume associated with the ocean's thermocline is split between the surface and deep sea yielding the nine to one volume split. Each reservoir is assumed to be well mixed. Water is exchanged between the two reservoirs. River input carries carbon and phosphorus to the sea. These constituents are lost to the sediments as undecomposed residues of the surface ocean life cycle. The P_{CO_2} for the atmosphere is assumed to be the same as that for the surface reservoir.

for surface water equal in magnitude to that for deep water. As it produces no PO_4 change it would not produce a Cd change. It would also not produce a change in the dissolved oxygen content of deep water.

The second scenario (see figure 3) involves deposition of organic residues on the continental shelfs and in estuaries during the sea level transgression of early post glacial time. Again, to make the cycle symmetrical an identical amount of carbon and phosphorus are returned to the sea via erosion at the onset of glacial time. As pointed out in previous papers this scenario produces all the desired responses.

The third scenario involves the deposition of phosphorite at the onset of glacials (see Figure 4). [Phosphorites consist mainly of calcium phosphate. Thus, the deposition of phosphorites should cause a reduction in the alkalinity in ocean water. This change is so small that its inclusion would not alter the model result]. While it produces the correct P_{CO_2} response it generates no early post glacial $CaCO_3$ preserva-

TABLE 2. Steady state conditions for two reservoir model (the superscript °'s refer to the steady state values).

$[PO_4]^{\circ}_{rivers}$	$= 1.0\ \mu m/liter$
$[PO_4]^{\circ}_{surface}$	$= 0.0\ \mu m/liter$
$[PO_4]^{\circ}_{deep}$	$= 2.0\ \mu m/liter$
τ_{PO_4} in ocean	= 100,000 yrs
$[\Sigma CO_2]^{\circ}_{rivers}$	$= 940\ \mu m/liter$
$[\Sigma CO_2]^{\circ}_{surface}$	$= 1985\ \mu m/liter$
$[\Sigma CO_2]^{\circ}_{deep}$	$= 2235\ \mu m/liter$
$\tau_{\Sigma CO_2}$ in ocean	= 130,000 yrs
$[ALK]^{\circ}_{rivers}$	$= 1680\ \mu eq/liter$
$[ALK]^{\circ}_{surface}$	$= 2235\ \mu eq/liter$
$[ALK]^{\circ}_{deep}$	$= 2335\ \mu eq/liter$

$$P^{atm}_{CO_2} = \frac{(2[\Sigma CO_2]^{\circ}_{surface} - [ALK]^{\circ}_{surface})^2}{(2[\Sigma CO_2]_{surface} - [ALK]_{surface})^2}\, P^{atm\,\circ}_{CO_2}$$

$$P^{atm}_{CO_2} = \frac{(2 \times 1985 - 2235)^2}{(2[\Sigma CO_2]_{surface} - [ALK]_{surface})^2}\, 300 \times 10^{-6}\ atm$$

$\delta^{13}C^{\circ}_{river}$	$= -0.3\ ^{o}/oo$
$\delta^{13}C^{\circ}_{surf}$	$= 1.9\ ^{o}/oo$
$\delta^{13}C^{\circ}_{deep}$	$= 0.1\ ^{o}/oo$
$\delta^{13}C_{org\ res.}$	$= \delta^{13}C_{surf} - 20^{o}/oo$
$\delta^{13}C_{calcite}$	$= \delta^{13}C_{surf}$
$R_{phosphorus}$	$= [PO_4]_{deep}M$
$R_{org.\ res.}$	$= 100[PO_4]_{deep}M$
$R_{calcite}$	$= 25[PO_4]_{deep}M$
where M	= the exchange rate between the surface and deep reservoirs.
$f_{org.\ res.}$	$= 0.01$
$f_{calcite}$*	$= f_{calcite})^{\circ}\ \dfrac{\Delta_D - 60}{\Delta^{\circ}_D - 60}$
where Δ_D	$= [ALK]_{deep} - [\Sigma CO_2]_{deep}$
Δ°_D	$= 100\ \mu m/liter$
$f_{calcite})^{\circ}$	$= 0.30$

*Fraction of calcite rain preserved in the sediments. This fraction is related to the $CO_3^{=}$ ion content of the deep sea which in turn is related to the difference between the alkalinity and ΣCO_2 content of the sea water.

tion event nor does it give the desired ^{13}C response; surface waters show a ^{13}C change and deep waters do not. It should be pointed out that by properly combining scenarios 1 and 3 responses identical to those for scenario 2 could be obtained.

Another chemical mechanism by which the CO_2 content of the atmosphere could be altered was proposed by Berger (1982). It involves the growth of coral reefs; thus our fourth scenario involves the growth of coral at the onset of interglacial time coupled with the destruction of

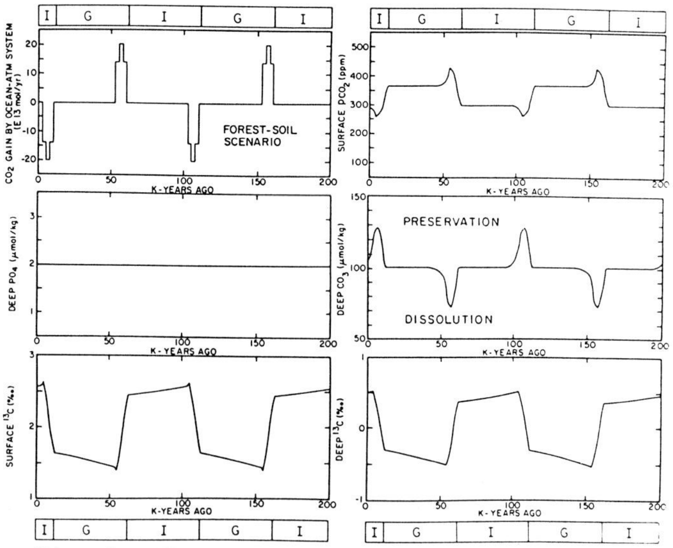

Fig. 2. Forest scenario: The onset of each interglacial (I) period is marked by the withdrawal of CO_2 to form the excess forests characterizing these episodes. This CO_2 is returned to the ocean-atmosphere system at the beginning of each glacial (G) episode. The timing of the cycles is set so as to roughly correspond to the real world climate history where the transition from the last glacial to the present interglacial occurred about 12,000 years ago.

an equivalent amount of $CaCO_3$ at the onset of glacial time. As shown in Figure 5 this scenario produces no ^{13}C changes in surface or deep waters and generates a very large dissolution

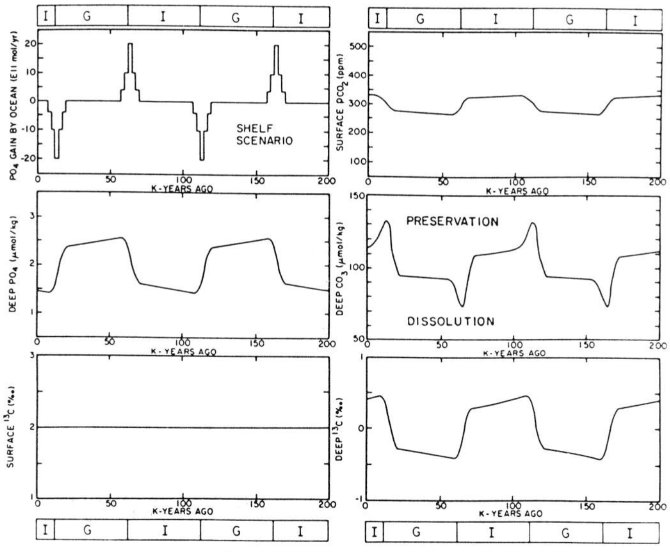

Fig. 3. Shelf deposition scenario: Organic residues with 100 carbon atoms per phosphorus atom are deposited in shelf sediments during the onset of interglacial episodes. Similar amounts are released by erosion at the onset of each glacial episode.

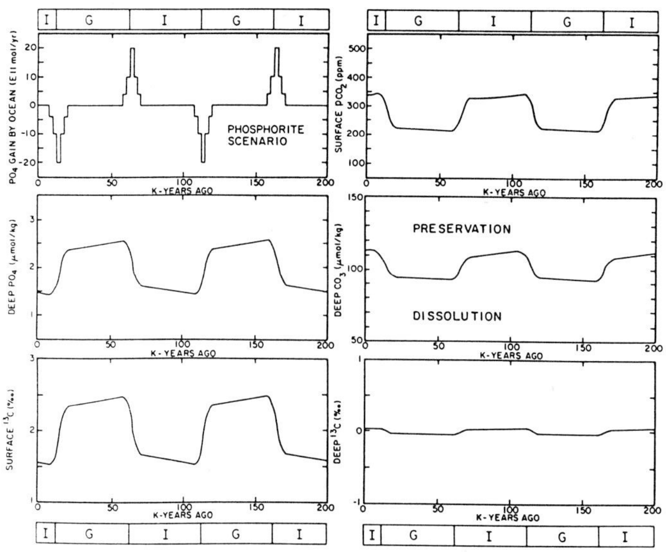

Fig. 4. Phosphorite deposition scenario: Phosphorites are assumed to form in great abundance at the onset of each interglacial episode and to be preferentially destroyed at the onset of each glacial episode.

event in the deep sea corresponding to period of coral growth. Further, the increase in CO_2 content persists only as long as coral growth goes on.

The fifth scenario involves changes in C/P ratio in the residues falling from the surface ocean (see Figure 6). The ratio is assumed to be higher than average during glacials and lower than average during interglacials. While pro-

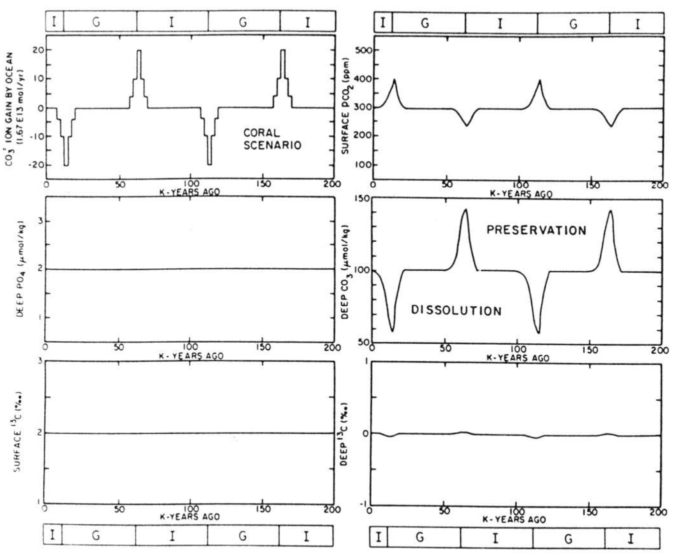

Fig. 5. Coral scenario: Coral is assumed to form in great abundance during the early stages of each interglacial episode and to be preferentially destroyed during the early stages of each interglacial episode.

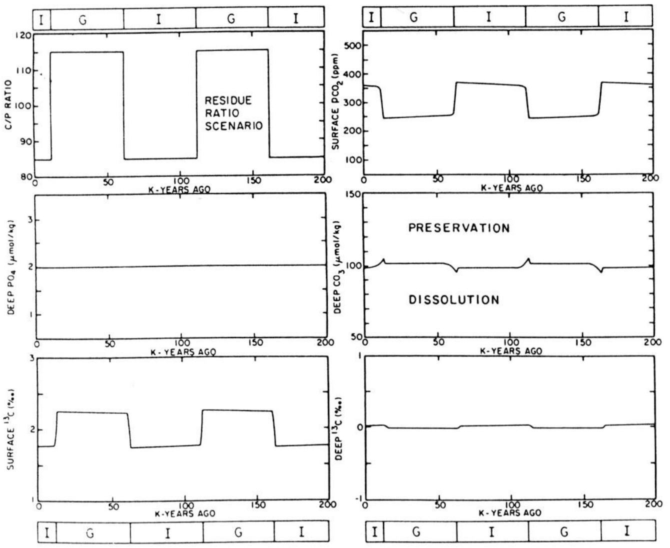

Fig. 6. Residue composition scenario: The C/P ratio of the organic residues falling to the deep sea is assumed to change from 115 during glacial episodes to about 85 during interglacial episodes.

ducing the desired p_{CO_2} response this scenario generates no preservation event and it generates a ^{13}C response in surface water and none in deep water.

The sixth scenario combines the first and the fifth (i.e., forest expansion during early interglacial time and a decrease in the C/P ratio in organic residues at the close of glacial time). As shown in Figure 7 it produces the desired responses for p_{CO_2}, ^{13}C and $CaCO_3$ preservation. It would also, like the shelf scenario, produce an O_2 rise in deep sea water at the close of glacial time. It would differ from the shelf hypotheses however, in that it produces no PO_4 change (and hence no cadmium change) in deep water.

The last scenario involves a change in the ratio of $CaCO_3$ to organic residues falling to the deep sea (see Figure 8). The amount of $CaCO_3$ produced per unit of phosphorus reaching the surface ocean is assumed to be much smaller during glacial time than during interglacial time. While producing a rise in p_{CO_2} at the close of glacial time this scenario produces no ^{13}C changes nor any $CaCO_3$ preservation event in early interglacial time.

One might ask why we do not call on changes in river input of P or C or on changes in the rate at which the deep sea is ventilated to generate the desired CO_2 change. The reason is that such scenarios cannot produce the sudden rise in the atmospheric CO_2 pressure observed in ice cores. Even if abrupt changes in river composition or in ventilation rate are invoked the p_{CO_2} response is very slow (the time constant for this change would be similar to the residence times for carbon and phosphorus in the sea). Figure 9 shows the response to changes in the river input rate of carbon, phosphorus and alkalinity from 30% higher than the mean during glacials to 30% lower than the mean during in-

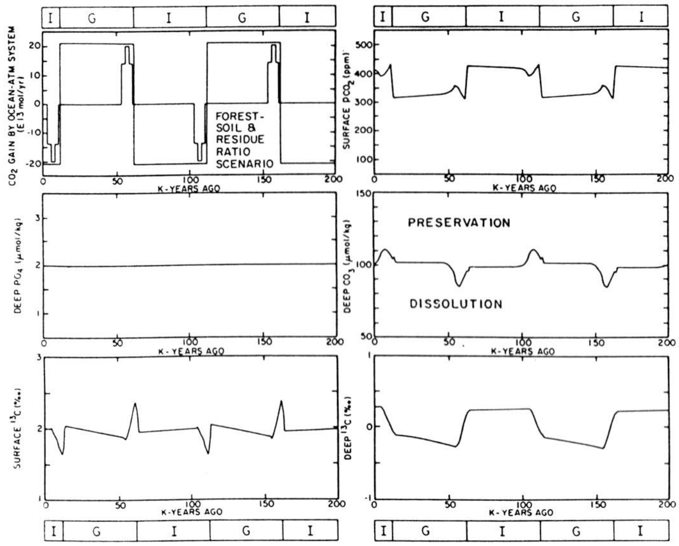

Fig. 7. Combined forest-resudue ratio scenario: The ratio of C to P in the organic residues is assumed to change sharply at the transitions between glacial and interglacial episodes. Storage of carbon in forest is assumed to occur early in interglacial time. The separation of these events in time produces the ^{13}C spikes in surface water ΣCO_2. The scale for the C/P ratio change is the same as that in Figure 6.

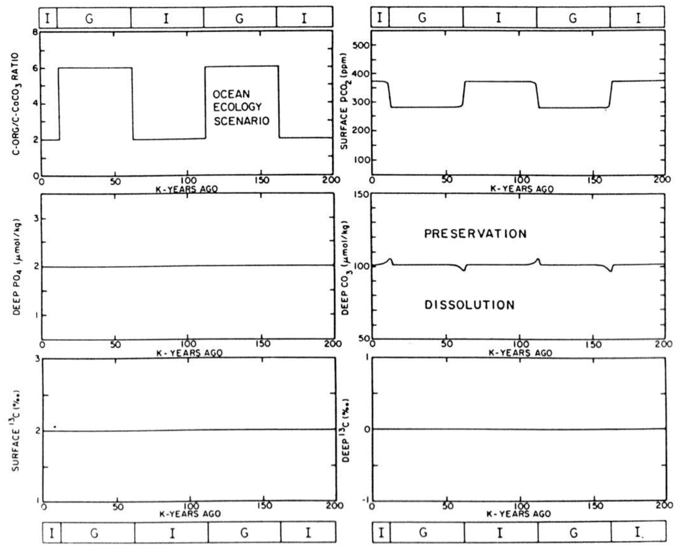

Fig. 8. Ocean ecology scenario: The ratio of carbon in $CaCO_3$ to carbon in organic matter in the residues falling from the surface ocean is assumed to change from 6 during glacial episodes to 2 during interglacial episodes.

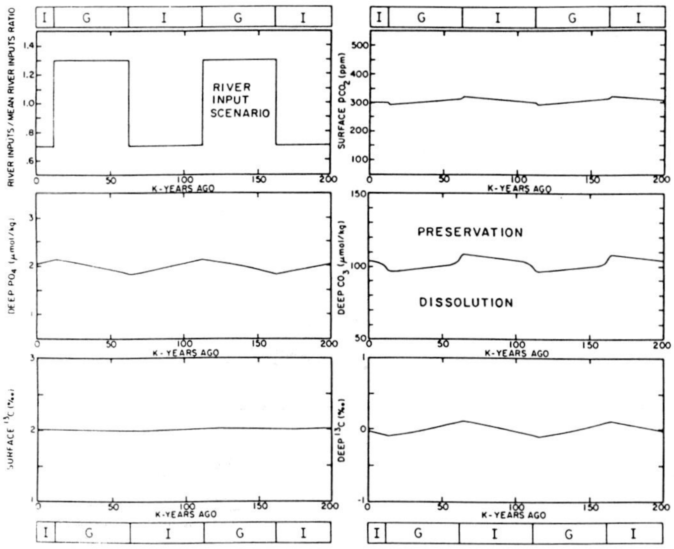

Fig. 9. River input change scenario: The amounts of carbon, phosphorus and alkalinity added by rivers to the sea are changed abruptly at the transitions between glacial and interglacial time.

terglacials. Figure 10 shows the response to changes in ventilation rate from 30% above the mean rate during glacial time to 30% below the mean rate during interglacial time. As can be seen the p_{CO_2} changes are gradual.

Another question which comes to mind has to do with our decision to make the cycles symmetrical. For the shelf deposition, the phosphorite formation and the coral formation scenarios

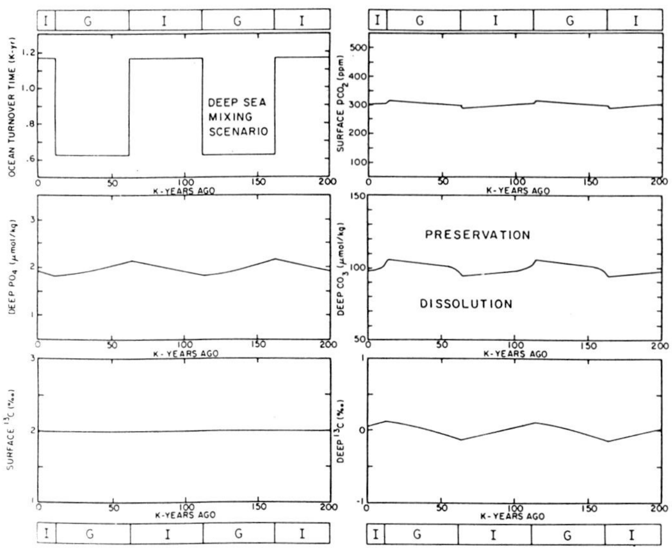

Fig. 10. Deep sea ventilation rate scenario: The ventilation time for the deep sea is changed abruptly at the transitions between glacial and interglacial time.

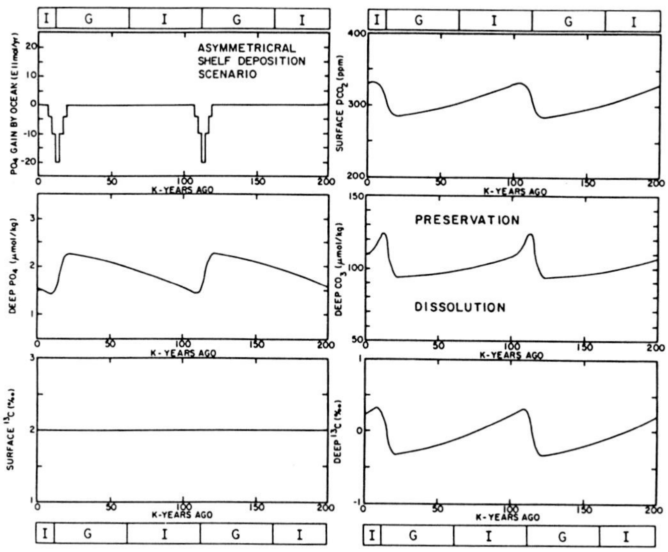

Fig. 11. Assymetrical shelf deposition scenario: Rerun of the model in Figure 3 without the erosion events at the onset of the glacial cycle. The excess C and P loss during the early postglacial shelf deposition events is balanced by a larger river input rate (uniform with time).

we introduced erosional events at the beginning of each glacial epoch to counterbalance the depositional events. Actually we could have achieved the same characteristics for the transition from glacial to interglacial time without including these erosion events. In Figure 11 we repeat the shelf scenario without the matching erosional events. In this case the loss of phosphorus and carbon during the deposition events is balanced by river input. As can be seen the glacial to interglacial transitions have much the same character as in Figure 3. As expected, the sharp changes associated with the onset of glaciation disappear. As the records for the beginning of the last glacial episode (≈70,000 years ago) are even less complete than the records for the beginning of the present interglaciation (≈11,000 years ago) it is not clear whether the symmetrical or assymetrial cycle is the more appropriate.

So far all the scenarios we have discussed involve chemical changes driven by climatic changes. In these cases CO_2 acts as a feedback amplifier rather than a causal agent. Were there no climatic changes there would be no chemical changes. The question arises as to whether a scenario can be devised in which chemical changes are a primary component of the causal scenario for glacial to interglacial climate oscillations. While it is possible to create such scenarios, it is difficult to create any which yield the abrupt change in CO_2 pressure observed for the close of the last glacial episode. We should hasten to mention that this sharp transition from full glacial to interglacial time plagues all attempts to model the glacial record. Existing explanations for these rapid melting events are ad hoc in origin.

Conclusions

The wherewithal exists to distinguish among the various chemical scenarios capable of explaining the rise in atmospheric CO_2 content which occurred at the close of the last glacial period. In addition to the CO_2 record, we need a whole ocean ^{13}C record, a surface ocean (or atmosphere) $\delta^{13}C$ record, a whole ocean cadmium record, and a whole ocean lysocline record. At present none of these records (including that for CO_2) are complete enough or accurate enough to permit firm conclusions to be drawn. The challenge of the next decade is to exploit these indicators and to seek new ones. Of utmost importance in this regard is the development of a paleooxygen indicator.

The scenarios which have been proposed to date all involve climatic forcing of the changes in ocean-atmosphere chemistry. It is possible of course that the chemical changes are active participants rather than passive bystanders in climate cycles. Inventive thoughts as to how orbital cycles, deep sea ventilation patterns and ocean-atmosphere chemistry could interact to produce the observed climatic oscillations are surely needed.

Acknowledgements. Discussion with and help from Taro Takahashi in connection with this work are much appreciated. Financial support was provided by National Science Foundation grant OCE80-19448 to W.S. Broecker at Lamont-Doherty Geological Observatory. T.-H. Peng's work at Oak Ridge National Laboratories was sponsored jointly by National Science Foundation Ecosystem Studies Program under interagency agreement No. DEB-8115316 and the Carbon Dioxide Research Division, Office of Energy Research, U.S. Dept. of Energy, under contract W-7405-eng-26 with Union Carbide Corp. Environmental Sciences Division publication No. 2256.

References

Berger, W.H., Deep-sea carbonate and the deglaciation preservation spike in pteropods spike in pteropods and foraminifera, Nature, 269, 301-303, 1977.

Berger, W.H., Increase of carbon dioxide in the atmosphere during deglaciation: the coral reef hypothesis, Naturwissenschaften, 69, p. 87, 1982.

Berner, W., B. Stauffer and H. Oeschger., Past atmospheric composition and climate, gas parameters measured on ice cores, Nature, 275, 53-55, 1979.

Boyle, E., F. Sclater and J.M. Edmond., On the marine geochemistry of cadmium, Nature, 263, 42-44, 1976.

Boyle, E. and L.D. Keigwin., Deep circulation of the North Atlantic over the last 150,000; years; geochemical evidence, Science, 217, 1982.

Broecker, W.S., The cause of glacial to interglacial change. In: "Evolution of Planetary Atmosphere and Climatology of the Earth". Centre National d'Etudes Spatial (France), p. 165-177, 1978.

Broecker, W.S., Glacial to interglacial changes in ocean and atmosphere chemistry. In: Climatic variations and variability: Facts and Theories. A. Berger, editor, D. Reidel Publishing Co., 109-120, 1981.

Broecker, W.S., Glacial to interglacial changes in ocean chemistry. Progress in Oceanography, VII, 151-157, 1982.

Bruland, K.W., G.A. Knauer and J.H. Martin. Cadmium in northeast Pacific waters. Limnol. and Oceanog., 23, 618-625, 1978.

Delmas, R.J., J-M. Ascencio and M. Legrand, Polar ice evidence that atmospheric CO_2 29,000 yr BP was 50% of the present, NATURE, 284, 155-157, 1980.

CLIMAP Project Members, A. McIntyre, Leader LGM Project., Seasonal reconstruction of the earth's surface at the last glacial maximum, Geol. Soc. Amer. Map and Chart Series, No. 36, 1981.

Hester, K. and E. Boyle., Water chemistry control of the cadmium content of some recent benthic foraminifera, Nature, 298, 260-262, 1982.

Neftel, A., H. Oeschger, J. Swander, B. Stauffer and R. Zumbrunn, Ice core sample measurements give atmospheric CO_2 content during the past 40,000 years. Nature, 295, 220-223, 1982.

Peltier, R., Dynamics of the ice age earth. Advances in Geophysics, 24, 1-146, 1982.

Shackelton, N.J., Tropical rain forest history and the equatorial Pacific carbonate dissolution cycles. In: The Fate of Fossil Fuel CO_2 in the Oceans, eds. N.R. Anderson and A. Malahoff, Plenum Press, N.Y., p. 401-428. 1977a.

Shackleton, N.J., The oxygen isotope record of the last Pleistocene: Phil. Trans. Roy. Soc. Lond. B. 280, 169-182, 1977b.

Streeter, S.S. and N.J. Shackleton, Paleocirculation of the deep North Atlantic: A 150,000 year record of benthic foraminifera and ^{18}O. Science, 203, 168-171, 1979.

GLACIAL-HOLOCENE CHANGES IN ATMOSPHERIC CO_2 AND THE DEEP-SEA RECORD

W. H. Berger and R. S. Keir

Scripps Institution of Oceanography, University of California, San Diego
La Jolla, California 92093

Abstract. An increase in atmospheric pCO_2 during deglaciation, of about 100 ppm, has been ascribed to a change in the chemistry of the ocean (Broecker, 1982, Progr. Ocn. 11, 151). It appears that Broecker's phosphate-extraction model has to be modified by decoupling a fertility decrease of the ocean from organic carbon buildup on the shelves, because of constraints from the amplitude and timing of the foraminiferal $\delta^{13}C$ signal. Denitrification and carbon/carbonate rain ratios are possible mechanisms. Buildup of shelf carbonates, during deglacial transgression, may deliver a considerable amount of CO_2 to the atmosphere, through decreasing the alkalinity of upper ocean waters. This mechanism relies heavily on a decrease in oceanic mixing rate, to retard equilibration with the deep sea. The Holocene increase in the dissolution of deep sea carbonates may yield constraints on shelf carbonate buildup (basin-shelf transfer). Very few cores from the deep-sea floor have been studied in sufficient detail to produce tests for these various hypotheses.

Introduction

The pressure of atmospheric CO_2 above the surface waters of the ocean depends on the alkalinity of the seawater, its total content of carbon dioxide, and its temperature. The saturation of seawater with calcium carbonate (aragonite and calcite) likewise depends on alkalinity, total CO_2, and temperature. Both the patterns of carbonate saturation and those of ocean temperatures have fluctuated through geologic time, with changes being rather sudden during certain periods. The best evidence in this regard has come from the stratigraphy of carbonate preservation on the deep-sea floor (Arrhenius, 1952; Oba, 1969; Berger, 1973; Broecker and Broecker, 1974; Thompson, 1976) and from oxygen isotopes in foraminifera (Emiliani, 1955; Shackleton and Opdyke, 1973). More recently, the carbon isotope record has been added to this evidence (Broecker, 1973; Shackleton, 1977). These signals suggested that the carbon dioxide content of the atmosphere changed during climatic cycles (Berger, 1977a; see Figure 1), although no direct evidence for CO_2 changes was available until recently. Speculations on this topic have been quite common; they go back to the last century.

The relationships between ocean chemistry and the carbonate record on the deep-sea floor are rather complicated (e.g., Broecker, 1974). Essentially, it has not been possible to make meaningful quantitative statements about changes in atmospheric pCO_2, based on the deep-sea record. The new data from polar ice cores (Berner *et al*., 1980; Delmas *et al*., 1980; Neftel *et al*., 1982) have profoundly changed this situation. We now have at hand a target for modeling, representing a best guess of atmospheric pCO_2 for the last 20,000 years or so (Figure 2). The constraint thus provided allows, in conjunction with the signals in the deep-sea record, a realistic assessment of widely different geochemical models.

The Models

The Phosphate-Extraction Model. The first step in relating the ice-core data to the deep-sea record is to realize that the ocean with its large reservoir of CO_2 (~60 atmospheric carbon masses, or ACM) drives the pCO_2 of the atmosphere on the time-scale of interest, and that the deep-sea record (oxygen isotopes, carbon isotopes, and carbonate preservation) provides useful constraints. Broecker (1981, 1982a,b) took this step in constructing a geochemical model which can produce a pCO_2 change across deglaciation of the kind seen in the ice-cores (Figure 2).

His model (Broecker, 1981, 1982a,b) is based on the observation that the ocean as a whole and the atmosphere are at present at a disequilibrium with respect to the partitioning of CO_2. The pCO_2 at the surface of the ocean (and hence the atmospheric CO_2 content) is too low, when compared with the overall alkalinity and total CO_2 content of the ocean. The fundamental reason for this situation is photosynthesis, in the sunlit zone of the oceans. Photosynthesis extracts carbon within this zone and the associated downward flux of

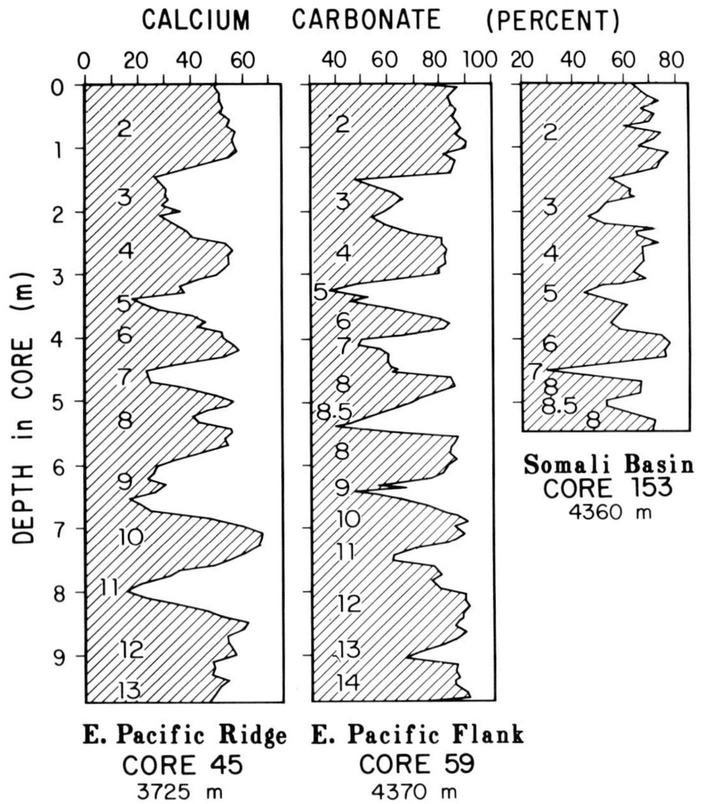

Fig. 1. Geologic evidence for changes in the carbon chemistry of the ocean. Pleistocene carbonate cycles of Arrhenius (1952) (Pacific cores) and of Olausson (1960). From Berger and Vincent (1981). Numbers for peaks and valleys are those of Arrhenius (1952); they identify the carbonate stages.

organic particles removes a portion of this carbon from the sunlit layer. Thus, the total CO_2 is decreased here, and the pCO_2 is at a correspondingly reduced level compared to equilibrium. It is true that the flux of carbonate particles downward from the sunlit zone has a compensating effect, by providing for a net decrease in alkalinity. However, the flux of organic carbon exceeds that of carbonate by a factor of about four on a mole-for-mole basis (Li et al., 1969).

Thus, ocean photosynthesis is responsible for a lower-than-overall-equilibrium pCO_2 in the atmosphere. The corollary is that, were the rate of photosynthesis increased, atmospheric CO_2 would drop even lower, and were it decreased, the pCO_2 would rise. From this basic concept, Broecker develops the model of a high-fertility glacial ocean, and a change to lowered fertility in postglacial time. His tool for changing productivity (that is, rate of carbon fixation) is a varying concentration of phosphate in the sea. In his model, phosphorus is extracted into shelf sediments during sea level rise (deglaciation) and is released to the ocean during sea level fall (glacial buildup). This lowers the amount of upwelling phosphate, which governs the rate of photosynthesis in the sunlit layer:

$$CO_2 + H_2O \rightarrow CH_2O + O_2$$

The organic matter is oxidized in the waters below, producing CO_2 which eventually comes back to the surface through mixing and upwelling. However, during the time that the particles and the deep-produced CO_2 reside below the surface, they cannot be seen by the atmosphere. In fact, they raise the total CO_2 of the deep ocean by several percent over the equilibrium condition.

The Denitrification Model. Phosphate-extraction (and -release) is not the only way to alter the fertility of the sea. Besides phosphorus, a great many micronutrients are depleted in the sunlit layer of the ocean, and any one of these can become limiting under certain conditions. Available nitrogen (mainly nitrate) apparently is extracted preferentially over phosphorus within newly upwelled water in the California Current System (Thomas and Seibert, 1974). Iron is known to be limiting under certain conditions such as in the Sargasso Sea (Menzel and Ryther, 1961), and so may be other trace metals involved in chlorophyll formation and in vitamin synthesis. Thus, one can build fertility-type models relying on changes in nitrate concentration, or in trace element abundance, to produce the pCO_2 changes observed.

The basic form of the Broecker model need not be affected by a switch in the choice of limiting nutrient. The performance of the model would be affected, however. The reason is that the phosphate-extraction model relies on buildup of organic carbon. For each mole of phosphorus sequestered, and hence made unavailable to the ocean, one hundred moles of carbon are also extracted from the ocean-atmosphere system. (This ratio, or more exactly the ratio chain C/N/P = 106/16/1, is known as the "Redfield" ratio; see Redfield et al., 1963). The large-scale extraction of organic carbon during the sea level rise lowers the total CO_2 in the system, which tends to pull the pCO_2 down. However, from the ice cores we know that pCO_2 rises at that time. Thus, the effect from phosphate-extraction, via photosynthesis, must both compensate for the loss of organic carbon and provide the increased atmospheric CO_2. The efficiency of the model is thereby decreased by about 40% (Broecker, 1982a, Keir and Berger, 1983).

If, instead of phosphate, we choose nitrate as the limiting factor, the efficiency of Broecker's model can be increased by a factor of 1.5 or so. In this scenario, nitrate is destroyed in newly generated oxygen-deficient environments on the shelf (estuaries, lagoons, swamps) by denitrifying bacteria. The process is ubiquitous and readily observed (Berner, 1980), and the rate is such that denitrification could destroy all ocean nitrate in a time span of a few thousand years (assuming that fixation rates must balance denitrification). A fractional increase of denitrification of, say, 10

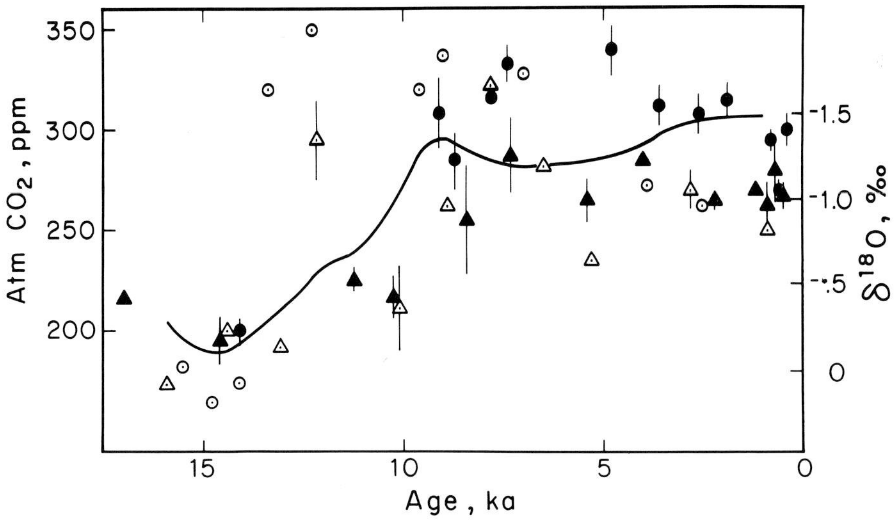

Fig. 2. Ice-core evidence for a change in atmospheric CO_2 during deglaciation. Circles: Greenland cores. Triangles: Antarctic cores. Open: Delmas et al. (1980). Filled: Neftel et al. (1982). Line is $\delta^{18}O$ for box cores from the western equatorial Pacific (Berger, 1982b).

to 30%, without a concomitant increase in fixation, could substantially lower ocean fertility on the time scale considered. In fact, a decrease in the nitrate content of the ocean of 30% would produce the pCO_2 change observed in the ice cores, when substituting N for P in Broecker's model. The two fertility-type models are sketched in Figure 3A and B.

Neither model has much hard data to back it up, such as amount of postglacial organic matter on shelves (and P content), or evidence for changes in the extent of anaerobic conditions. Given the ad hoc nature of the two hypotheses, all we can say is that the second works better than the first.

The C/P Model. There are yet other ways to modify Broecker's fertility model. Broecker (1982a) quotes P. Weyl as suggesting that the C/P ratio of settling organics was higher by 30% in glacial time, compared with the present. It is as though phosphorus was somehow more efficient in drawing down the total CO_2 of sunlit waters in the past. The effect is exactly the same as keeping phosphorus efficiency constant, and changing concentrations, except for doing away with the necessity of shelf-carbon buildup. Thus, the Weyl modification has a very salutary effect on the efficiency of the Broecker model. Broecker (1982a) presents evidence that the C/P ratio of organic matter oxidized within the deep waters of Atlantic and Pacific Oceans differs substantially (97/1 *versus* 135/1). If we assume that the Atlantic situation is analogous to a low-fertility postglacial ocean, and the Pacific to a high-fertility glacial one, then Weyl's hypothesis is supported. The reason for the difference, presumably, is the differential lability of phosphorus-containing debris, under the different chemical conditions on the seafloor (Eh, pH). The idea that changes in foodweb structure are responsible (P. Weyl as cited in Broecker, 1982a) is not attractive: it seems unlikely that the use of phosphate by organisms should be less efficient (Atlantic) where phosphate is less abundant in surface waters to begin with.

The Rain Ratio Model. We have, so far, said little about the precipitation and dissolution of carbonate, which affects both total CO_2 and alkalinity. In the models mentioned (phosphate extraction, denitrification, change in Redfield ratio) these processes are kept constant. However, we know that within the particle flux (or "rain") crossing the thermocline the ratio between carbonate and organic carbon varies substantially as a function of fertility. Within eastern boundary currents, production of dinoflagellates and diatoms greatly dominates over that of the calcareous coccoliths, and the foraminifera. In the low fertility regions, the central subtropical gyres, coccolith production is relatively more important. Results from sediment traps underscore this general observation.

Sediment traps in the Sargasso Sea and other low fertility areas show a ratio of organic to carbonate carbon of 0.6 to be typical of deep water particulate fluxes and a value of about 2 in shallow traps (<1 km depth, Honjo, 1980). In contrast, the ratio of trapped organic to carbonate carbon is generally larger in high fertility areas, being about 2 to 2.5 at all depths in the Panama Basin (Spencer, 1981) and ranging from 6 to 10 in the Southern Ocean (Wefer et al, 1982).

From these relationships it follows that the drawdown of total CO_2 by photosynthesis will be much better compensated by associated carbonate production in a low-fertility ocean than in one with high fertility. Put differently, compensation, from carbonate precipitation, of

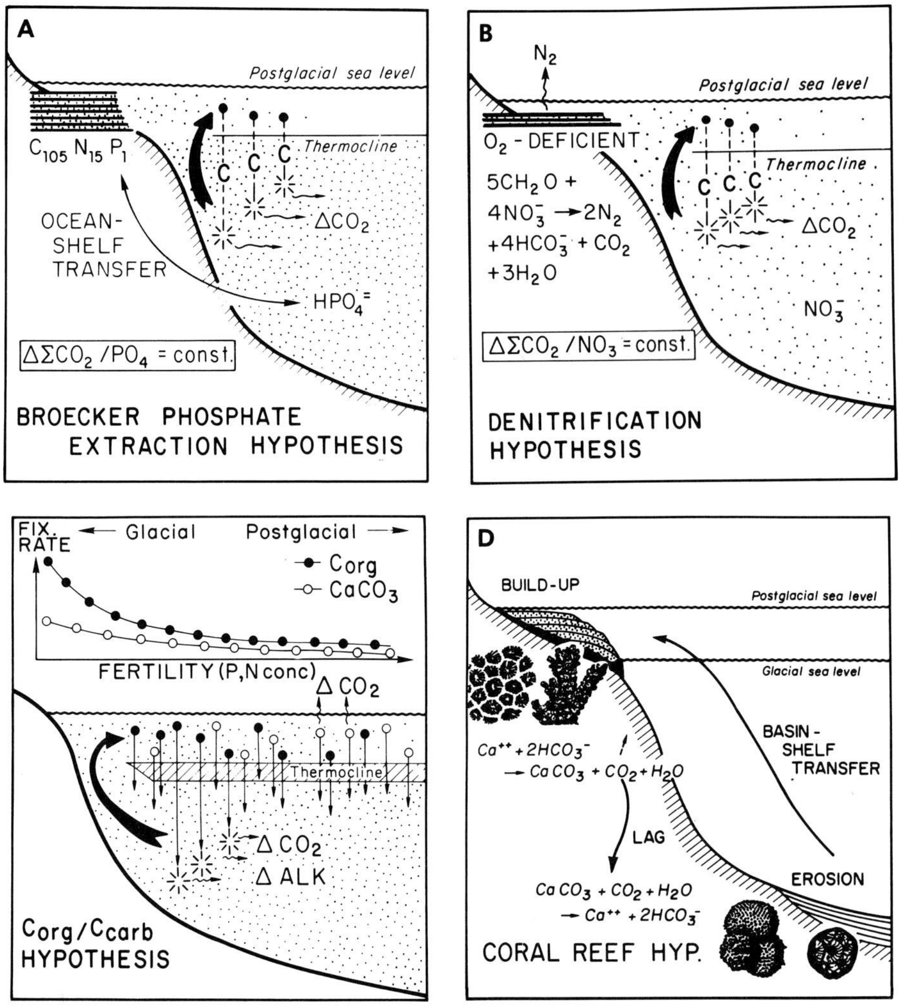

Fig. 3. Four ways to change the carbon chemistry of the ocean. Fig. 3A, extraction of phosphorus by buildup of organic carbon on shelves (Broecker, 1981). Fig. 3B, extraction of nitrate by denitrification on newly flooded shelf. Fig. 3C, change in carbon/carbonate rain ratio. Fig. 3D, buildup of carbonate on shelf (Berger, 1982a).

pCO_2-change produced by photosynthesis is less efficient for high productivity than for low. Thus, for a given drop in fertility, from glacial to postglacial time, the pCO_2 rise is greater than it would be for a constant carbonate/organics rain ratio (Figure 3C).

The Coral-Reef Model. The last model in our line-up (Figure 3D) only considers carbonate precipitation, and only on the shelf. It is the only model which does not invoke fertility changes at all, but only the geometric-geographic effects of sealevel fluctuation. It has but one feature in common with the original Broecker model of Figure 1A: the concept that the atmospheric CO_2 concentration depends on the chemical state of the ocean.

The coral-reef hypothesis (Berger, 1982a,c; Berger and Killingley, 1982) rests on two basic concepts: basin-shelf exchange of carbonate sediments, and deglacial mixing lag. The first item refers to the notion, well-used in marine geology, that carbonate tends to accumulate in the shallowest regions of the seafloor which are available at the time. Thus, whenever sea level rises and the ocean transgresses over the shelves, shelf carbonates (including coral-reefs) start to build up. The basic equation is:

$$Ca^{++} + 2HCO_3^- \rightarrow CaCO_3 + CO_2 + H_2O$$

In turn, the deep sea receives less carbonate than before. A decrease in saturation results from the greater ease with which carbonate precipitates on flooded shelves, compared with precipitation in the deep sea. The net effect is a slight concomitant increase in the pCO_2. This net effect, however, is not the one of chief interest here. On a short time scale, and during times of rapid sealevel rise, it is of interest how quickly the deep sea will respond to shelf-carbonate buildup, by reducing deep carbonate accumulation.

If we postulate a lag in the response of the deep sea to the coral-reef growth of the order of 1000 years or more, then the drop in alkalinity caused by shelf-carbonate buildup can increase the atmospheric pCO_2 considerably. It is entirely reasonable to assume that deep-ocean mixing slowed during deglaciation, when great masses of freshwater were loaded on top of the sea, and when the surface waters warmed over large areas (Worthington, 1968; Berger et al., 1977). If so, shelf-carbonate buildup which exceeds steady-state precipitation of carbonate will produce a fast rise of pCO_2 and an overshoot. Because of the imbalance represented by the overshoot, subsequent deep-ocean overturn will greatly reduce atmospheric CO_2: the overshoot is followed by a rebound (Berger and Killingley, 1982). The pronounced fluctuations in climate during deglaciation, with a wavelength near 2000 years, may have been amplified by such pCO_2 instability.

In sum, there seems to be no essential difficulty, as far as modeling, in producing atmospheric pCO_2 changes of the desired magnitude by manipulating the carbon chemistry of the ocean in various ways. Let us now turn to the constraints supplied by the deep-sea record: they should facilitate a choice between the models or combinations thereof.

The Signals

The Oxygen Isotope Record and the Time-Scale.

The Ariadne Thread of Pleistocene stratigraphy is the oxygen isotope record in deep-sea sediments (Emiliani, 1955, 1966, 1978; Broecker and van Donk, 1970; Shackleton and Opdyke, 1973). The $\delta^{18}O$ record provides time-markers, as well as an estimate for sea level change, which can be used as an input function to the models. The convention is that, barring substantial temperature change in the region studied, an 0.1 ‰ change in the $\delta^{18}O$ of planktonic foraminifera corresponds to 10 m of sealevel change (Shackleton and Opdyke, 1973).

In the present context a time-scale for deglaciation is of special interest. There are competing scales with substantial differences (see Ruddiman and McIntyre, 1981). The best available scale for the Pacific is given in Figure 2. It is based on the analysis of 8 cores, above the lysocline, in the western equatorial Pacific. There are about 80 oxygen isotope determinations, and two dozen radiocarbon dates contained in the curve. Isotopic analysis was done on *Pulleniatina obliquiloculata*, a planktonic foraminifera living in the deeper portions of the mixed layer. Radiocarbon data are based on bulk carbonate.

Assuming that the stable isotope curves from the ice cores (which record changes in climate) may be correlated with those of the deep-sea sediments (which record mainly changes in mass of continental ice), we can now put the pCO_2 record into a time frame. Admittedly, this time frame is not necessarily correct. Nevertheless, it is one that is relevant to the discussion if we are to compare the sequence of events recorded in ice cores with that seen in deep-sea cores.

Although there is a fair amount of scatter in the CO_2 data, there are several remarkable features which emerge in Figure 2. First, the concentration of atmospheric CO_2 clearly appears to be 90±30 ppm greater for the last 7000 years than for a period 15 thousand years ago. The values for the most recent samples (about 1000 years old) range from 260 to 300 ppm, which is similar to various estimates of the "pre-industrial" atmosphere before fossil fuel burning and land deforestation. The variation of atmospheric CO_2 appears to follow to some extent the variation of $\delta^{18}O$ and hence ice volume. There is also an indication of perhaps a 35 ppm decrease between about 8000 years ago and the pre-industrial present. There is some indication of a brief drawdown in pCO_2 near 11,000 years ago, after an early overshoot, a suggestion which may be confirmed by as yet unpublished data (H. Oeschger, pers. comm., Oct. 1982). If overshoot and rebound are real, the change of pCO_2 during deglaciation has attributes of a transient: it is then not just a simple change from one steady state to another.

Accumulation rates. The deep-sea signals important in the carbon cycle are accumulation rates of carbonate and carbon, carbon isotopes, and carbonate preservation stratigraphy. Variations in accumulation rates are not at present resolvable on the time scale here considered. The reasons are mixing (bioturbation), local disturbance (miniturbidites and -hiatuses) and sampling problems (compression or extension during coring). Every core that has been looked at in sufficient detail bears evidence of some or all of such problems, to the best of our knowledge. Explicit or implicit claims to the possession of perfect cores may safely be disregarded. There is only one test: several cores must deliver nearly identical records, preferably from separate areas of sedimentation. The latter two signals, carbon isotopes and preservation stratigraphy, must deliver the main criteria allowing a choice among the models, at present.

Carbon isotopes: Background. The interpretation of $\delta^{13}C$ records is quite difficult, because the $^{13}C/^{12}C$ ratio in calcareous shells is rarely in equilibrium with the ratio in the surrounding water, and because the amount of isotopic disequilibrium differs between species and between members of the same foraminiferal species. This is true both for planktonic and benthic forms (for review see Berger, 1981; Douglas and Woodruff, 1981; Vincent and Berger, 1981). After the problem of varying disequilibrium has been taken care of by appropriate statistics (or ignored) changes in the $\delta^{13}C$ of foraminiferal shells are usually taken to reflect changes in the isotopic composition of the dissolved carbon in the water surrounding the shell. There are three main sources of variation for this composition: verti-

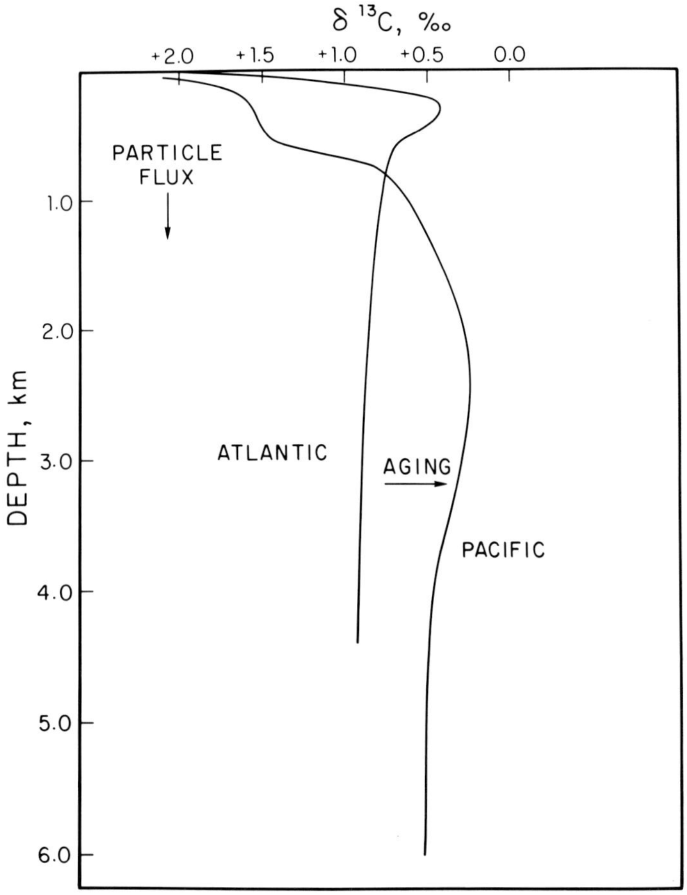

Fig. 4. Factors controlling the $\delta^{13}C$ in the ocean's dissolved inorganic carbon. 1. The organic particle flux maintains a shallow-deep gradient. 2. Aging of deep waters adds ^{12}C-rich CO_2 (several percent of ΣCO_2). 3. Scale-change (not shown) through addition or extraction of organic carbon, from or to external reservoirs.

cal fractionation, basin-basin fractionation, and buildup or collapse of transient carbon reservoirs.

The effects of the first two processes, vertical and horizontal fractionation, are illustrated in the $\delta^{13}C$ profiles shown in Figure 4. Two profiles are given, one from the Atlantic (Duplessy, 1972) and one from the Pacific (Kroopnick et al, 1977). The first impression - quite correctly - is that of an overall similarity of the $\delta^{13}C$ curves with micronutrient profiles. This correspondence arises because, just like in micronutrients, the profiles are governed by the formation and redissolution of organic particles. The particles have $\delta^{13}C$ values of about -20 ‰. Where they form, they remove ^{12}C preferentially from the water: the remaining total CO_2 becomes heavier in ^{13}C (and the water becomes nutrient-poor). Where the particles decay, by oxidation, they release ^{12}C in excess of the overall $^{12}C/^{13}C$ ratio: the total CO_2 becomes richer in ^{12}C (and the water is enriched in nutrients). As this process continues within deep water, through time, $\delta^{13}C$ values are continually lowered. Thus, the deep water of the Pacific, which is older than that of the Atlantic, has the lower $\delta^{13}C$ values.

A fertile ocean, with a strong particle flux, will show strong vertical fractionation. Thus, the difference between $\delta^{13}C$ values in planktonic and benthic foraminifera should be high. If phosphate is ultimately limiting to productivity, then this planktonic-benthic difference is a measure of the availability of phosphate (Broecker, 1973). An ocean with strongly asymmetric bottom water circulation will produce strong basin-basin differences. At present, the Atlantic is ventilated at a much faster rate with new deep and bottom waters than the Pacific: Atlantic $\delta^{13}C$ values are high, those in the Pacific low. Whenever Atlantic deep water sources were diminished, during the course of Pleistocene climatic fluctuations, the difference between Pacific, and Atlantic waters also was diminished, and *vice versa*. In order to interpret a $\delta^{13}C$ signal from benthic foraminifera in terms of global fertility, we must first establish the difference between the two major ocean basins, over the time interval considered.

The overall position of a $\delta^{13}C$-profile can be changed by subtracting organic matter from the ocean or adding organics to it. The buildup or collapse of terrestrial carbon reservoirs, with a $\delta^{13}C$ value of -25 ‰ is slightly more efficient in this process than marine carbon reservoirs (-20 ‰). The ocean has about 60 atmospheric carbon masses, the biosphere (mainly forests) has 1.5 or so. If one-half of the biosphere has to be rebuilt during deglaciation, then the ocean's $\delta^{13}C$ profiles will be shifted by

$$(0.75/60) \times 25 ‰ = 0.31 ‰$$

Such an argument was presented by Shackleton (1977), based on the findings of paleobotanists (Street and Grove, 1976; and others). Shackleton postulated a factor-of-three increase, of total terrestrial biomass and associated humus during deglaciation, and obtained a value of 0.7 ‰ for the shift of $\delta^{13}C$ in the ocean. Broecker (1981, 1982a, b) accepted the evidence presented by Shackleton for the 0.7 ‰ shift in $\delta^{13}C$, in benthic foraminifera, but attributed it to shelf-carbon buildup rather than forest growth. Also, by reducing ocean fertility, Broecker straightens the water-column $\delta^{13}C$ profiles. In the sunlit zone, a positive shift from carbon buildup is thus compensated by a negative shift toward average values. Planktonic foraminifera, then, should show a much smaller change in $\delta^{13}C$ than benthics, according to Broecker. There is some evidence that this is so.

Carbon Isotopes: Data. In attempting to give a representative $\delta^{13}C$ signal for deglaciation, we are faced with the difficulty of basin-to-basin variations, as mentioned. Also, there is again

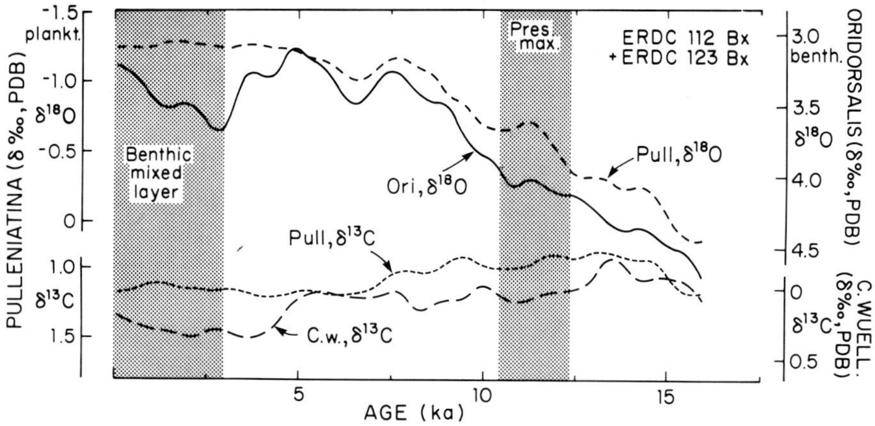

Fig. 5. Oxygen and carbon isotope stratigraphies, west-equatorial Pacific. Pulleniatina scale refers to planktonic foraminifera data, stacked and fixed at the present-day Pulleniatina level. Oridorsalis and C. wuellerstorfi scales refer to benthic foram data, stacked, and fixed at the respective present-day levels. From Keir and Berger (1983), based on data of Vincent et al. (1981a, b).

the problem of resolution. Cores with low sedimentation rates, from pelagic areas, must be selected and stacked to deliver reliable signals on the scale desired. Cores with high sedimentation rates (such as METEOR 12392, used by Shackleton, 1977) are from continental slope areas, where nutrient budgets (and hence $\delta^{13}C$ signals) have a strong regional overprint.

Shackleton (1977) presents $\delta^{13}C$ signals from three cores, and comparisons of recent and glacial values for another three. The METEOR benthic signal (off NW Africa) shows a difference of 0.8 ‰ between glacial and recent. However, the shift in $\delta^{13}C$ comes late during deglaciation, when compared with the $\delta^{18}O$ signal. A slowdown of the Canary Current with decreased supply of organics to the sediment and a corresponding change in the chemistry of interstitial water near the sediment surface may be responsible. Alternatively a turning-on of North Atlantic Deep Water could reasonably be invoked to account for much of the change (see Figure 1 in Boyle and Keigwin, 1982). The signal from V22-174 (Shackleton, 1977), from the Mid-Atlantic Ridge at 10°S, shows a $\delta^{13}C$ shift in both planktonics and benthics. Again, the main shift comes late, in the benthics (about 0.5 ‰) and it could well be due to an increase in NADW production. The corresponding change in the planktonics is gradual and not much less than in the benthics (~0.4 ‰). It may simply reflect a spreading out of subtropical gyre with its nutrient depleted waters. The gyre was more confined during glacial time (see CLIMAP, 1976). The third signal, from Y71-6-12 MG2 (Shackleton, 1977), has very low resolution (10 cm of record for the Holocene). The two Holocene values for the benthic foram, *Uvigerina* (-0.74 ‰ and -0.92 ‰) are distinctly heavier than the preceding glacial values (-1.09 ‰, -1.30 ‰). However, much of the change again is late, that is, the Holocene value of -0.74 ‰ is needed for a large effect.

Our own data from the west-equatorial Pacific also indicate that the shift in $\delta^{13}C$ toward heavier values occurs within the Holocene rather than within the glacial-Holocene transition (Figure 5). Thus, the $\delta^{13}C$ record, as far as known, does not support Broecker's shelf-carbon buildup during deglaciation, as a means to produce late glacial pCO_2 changes. Also, Shackleton's hypothesis of a change of 0.7 ‰ during deglacial forest buildup is not supported by these data. We must await better geographic coverage to settle this issue. Incidentally, on examining Figure 5 it will be noted that the last several thousand years of the Holocene are marked by cooling of deep waters, and an increase in their oxygenation (heavier $\delta^{13}C$), in the west-equatorial Pacific. It looks as though deep circulation may have become more vigorous since the Altithermal.

Preservation stratigraphy. We have seen that the $\delta^{13}C$ signal depends on changes in the organic carbon budget. It is a tracer for changes in total CO_2, as produced by oxidation and reduction of carbon. Another such tracer is the preservation of carbonate on the seafloor. In the present ocean, the depth levels of equal preservation (as measured by the partial dissolution of foraminiferal shells) differ by up to 2000 m between North Atlantic and North Pacific. The reason is that the older deep waters of the Pacific have accumulated excess CO_2 from the oxidation of organic matter, and this CO_2 causes a decrease in the degree of saturation of $CaCO_3$ in ocean water. Since the CO_2 is derived from organic matter with $\delta^{13}C = -20$ ‰, there is a close correspondence between saturation and the $\delta^{13}C$ of the total CO_2. When $\delta^{13}C$ changes to heavier values, we should expect better preservation (that is, a deepening of levels of saturation).

Shackleton (1977) recognized this principle, and proposed that the deglacial preservation maximum seen in Indo-Pacific cores (Oba, 1969; Thompson and Saito, 1974; Luz and Shackleton, 1975;

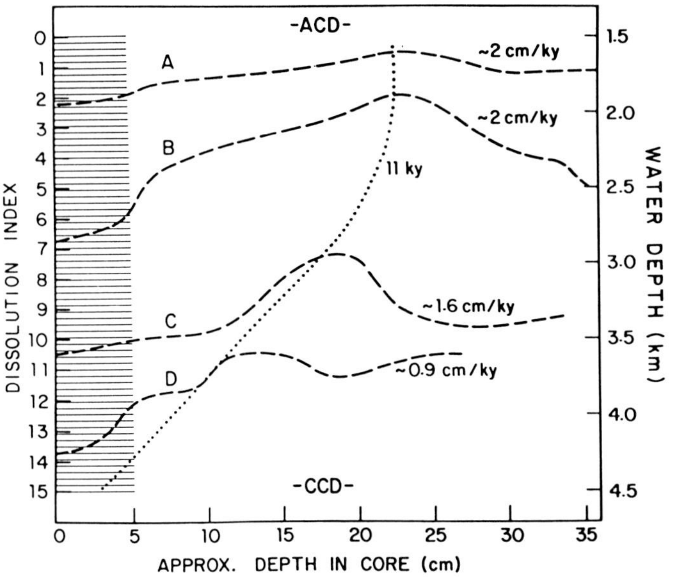

Fig. 6. Sketch of preservation stratigraphies in the west-equatorial Pacific, as a function of water depth. Each graph represents several cores (A, 5; B, 3; C, 6; D, 4 cores). Water depth valid for present indices (zero core depth). ACD, present aragonite compensation depth. CCD, present calcite compensation depth.

Thompson, 1976; Berger and Killingley, 1977) and in Atlantic cores (Chen, 1968; Broecker, 1971; Thiede, 1971, 1973; Kudrass, 1973; Diester-Haass et al., 1973) is closely related to the deglacial $\delta^{13}C$ step he saw in his records. As the rainforest builds up, in his model, CO_2 leaves the ocean, preservation levels deepen, and the $\delta^{13}C$ increases.

A series of averaged records of preservation of foraminifera in the west-equatorial Pacific is shown in Figure 6. It is based on dissolution indices of the type given in Berger and von Rad (1972), which rely on percentages of sensitive species (here: Globigerina rubescens) and on abundance of damaged foraminiferal shells. The position of the preservation maximum is shown by a dotted curve. It straddles deglaciation. The same is true for all former deglaciations of the late Pleistocene observed in west-equatorial Pacific cores (see Figure 6 in Thompson, 1976). The age of the preservation spike is 11 kyr (^{14}C) (Berger, 1982b). The previous estimate of 14 kyr (Berger, 1977b) was based on Atlantic core data whose ^{14}C dates appear to be too old, perhaps from terrigenous contamination and on Core ERDC 92 (Peng et al., 1979) from the west-equatorial Pacific. The ^{14}C stratigraphy of this core does not entirely agree with its oxygen isotope stratigraphy and with that of other nearby cores.

The overall dependence of carbonate preservation on the depth of deposition is seen in the reduction of sedimentation rates with depth (from ~2 to ~1 cm/ky; the 11 ky position provides the time marker for calculating the rate). Since there is in the present oceans a more or less linear correspondence between the dissolution indices of surface samples and water depth, the dissolution index scale has been converted to a water depth scale and both are indicated on the same y-axis in Figure 6 (indices at left, depths at right). At each depth level (A, 1.6 - 2.3 km; B, 2.3 - 3 km; C, 3-4 km; D, 4-4.5 km) there is a change of indices with depth-in-core. The difference in amplitude, between different water depth levels, may or may not be real. If it is, saturation values are seen to change most at intermediate levels, that is, where the influence of North Atlantic Deep Water is strongest.

Interpretation of the Preservation Spike. To interpret the preservation signal, we can invoke a single cause for an increase in preservation, combined with a relaxation to initial conditions after the maximum. Alternatively, we can call on two causes: one to increase preservation, another to decrease it. Extraction of CO_2 from the ocean will increase saturation with carbonate, and hence preservation. Either Shackleton's hypothesis of rainforest buildup or Broecker's hypothesis of shelf-carbon buildup can provide for CO_2 extraction and thus yield a single-cause explanation, provided carbon is extracted during early deglaciation. It makes little difference, on the time scales here considered, whether the carbon is extracted from the atmosphere (Shackleton) or from surface water (Broecker).

More than one factor may be involved, however. The data (Figure 6) suggest that the trend of improving preservation is broken rather suddenly, perhaps 12 ky ago, and that the early Holocene is marked by a dissolution pulse (Berger, 1970; Broecker and Broecker, 1974). The coral-reef hypothesis suggests a reason: the deep-sea floor is called on to cover the debt incurred by the continuing buildup of shelf carbonate. Also, there is evidence for a late Holocene dissolution pulse, coinciding with the other late Holocene events cited earlier (bottom-water cooling, benthic $\delta^{13}C$ increase). We suggest that this second pulse reflects, at least in part, invigorated deep mixing for the last several thousand years. Alternatively, or in addition, it reflects a neoglacial shrinking of the biosphere which may have paralleled increased aridity in North Africa and elsewhere, since the Altithermal (see Kellogg, 1980).

We can estimate the amount of carbon buildup (forest or shelf) which is needed to produce the preservation increase during deglaciation, other factors being equal. The amplitude in the preservation signal, in terms of apparent water depth differences, is about one third the depth difference in saturation levels between North Atlantic and North Pacific. The total CO_2 contents of the deep water masses of the two basins differ by about 10%, and this is the chief reason for the difference in saturation. Thus, if we were to extract 3% of the total CO_2 from the deep Pacific, we would drop the saturation levels by an amount comparable to the deglacial effect. The mass of carbon involved is on the order of 60 ACM x 3% = 1.8 ACM. Thus, if carbon buildup and collapse is to produce the preservation signal, it has to be on the order of 2 atmospheric carbon masses, somewhat larger than the present day biosphere. The corresponding $\delta^{13}C$ signal is -25 ‰ x 3% = -0.75 ‰.

Shackleton (1977) likewise suggested a deglacial change by 3% of the ocean's total CO_2, based on his estimate of the amplitude in $\delta^{13}C$ fluctuations (0.7 ‰). As mentioned, he included soil carbon buildup and decay to provide the necessary amplitude. At most one-half of this proposed amplitude is actually seen in the published $\delta^{13}C$ records, and much of it apparently after deglaciation (Figure 5, and original data of Shackleton, 1977). Consequently, it appears that Shackleton's mechanism (or Broecker's) cannot account for more than one-half of the preservation increase, leaving the remainder to be sought in changes of deep circulation or other effects (examples listed in Berger, 1977b).

Box Model Results

Caveat Emptor. From the foregoing it is clear that the relationships between the deep-sea record and the CO_2-record in ice cores are highly complex. The quality of the sedimentary data is insufficient to provide the necessary constraints that would allow us to choose among the hypotheses sketched in Figure 3 (or perhaps to reject all of them). In this situation box models are but a means to guide us in studying the deep-sea record. A simple model will do that about as well as a complex one. Hence, our models are simple - certainly too simple to incorporate the relevant oceanographic parameters, and perhaps too simple to be considered quantitative. This is especially true of the second of the models we here discuss.

Fertility Changes. We have modified Broecker's phosphate-extraction model (Broecker, 1982a) by allowing changes in carbonate saturation and adapting it for continuous signal generation over a 120,000 yr period (Keir and Berger, 1983). The chief ingredients of the model are given in Figure 7. The input function is sea level variation, constructed from the $\delta^{18}O$ record of Core V28-238 (Shackleton and Opdyke, 1973) by a Fourier series fit (see Figure 8A).

The sea level range is arbitrarily set at 150 m, and sea level change is assumed linearly linked to $\delta^{18}O$ change. The minimum flux of carbonate to the sea floor is taken as 2.5 mg cm^{-2} yr^{-1}. The amplitude of fertility variation is taken as 1.6 mg cm^{-2} yr^{-1} (2.5 + 1.6 = 4.1; Figure 8A). This amplitude produces the target pCO_2 change of 90 ppm, as shown in the curve labeled P/2/1.6 (Figure 8B). The label is read as follows: P stands for phosphate-extraction; the number 2 refers to the amount of carbon sequestered on the shelf, in atmospheric carbon masses (2 ACM); the last number refers to the change in I carb, that is, flux of

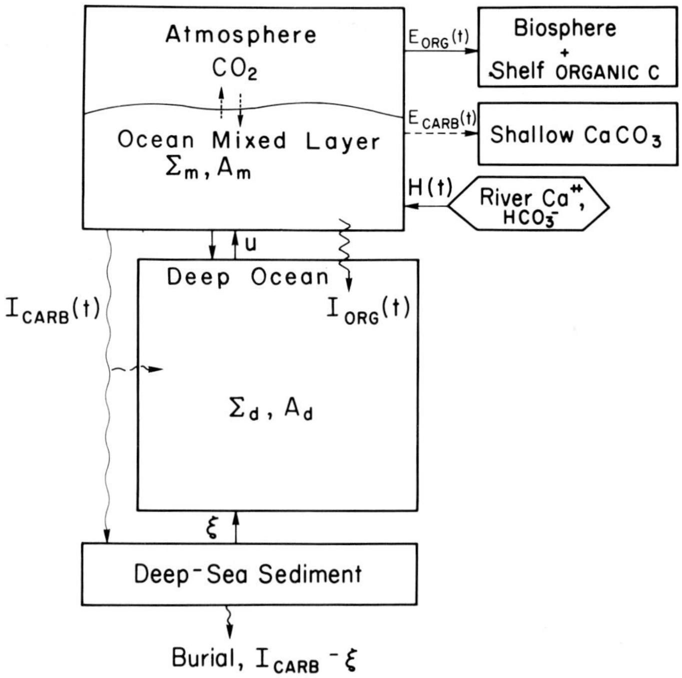

Fig. 7. Box model used for calculating predictions of Broecker's phosphate extraction hypothesis. From Keir and Berger (1983). Σ_m is total CO_2 in mixed layer, A_m is alkalinity in mixed layer , Σ_d and A_d are total CO_2 and alkalinity in the deep water, I_{carb} is particulate carbonate flux, E_{org} and E_{carb} are extraction of organic carbon and carbonate, respectively, and ξ is carbonate dissolution rate.

carbonate to the sea floor (1.6 mg cm^{-2} yr^{-1}). The other two cases shown illustrate the sensitivity of the model to shelf-carbon buildup, and to the amplitude of the change in fertility.

P/2/1 is the Brocker model sensu strictu, since it results from a C/P ratio near 100/1, in the sequestered organic carbon with a mass of 2 ACM. The model yields 60 ppm of CO_2 during deglaciation. By locking up more P per mass of C, or by lowering the efficiency of the dissolved P in the ocean (Weyl hypothesis) we can return to the target plot of P/2/1.6. The case P/0/1.6 describes the situation where the fertility of the ocean is altered without involving shelf-carbon. In effect, this case describes the denitrification mode. Obviously, P/0/1.6 is the most efficient of the three cases, in producing pCO_2 changes.

The $\delta^{13}C$ signal. The phosphate-extraction model P/2/1 produces a $\delta^{13}C$ signal with a change of about 0.7 ‰ for benthic foraminifera. For planktonics, the model predicts no change (Figure 9). The benthic signal, essentially, is the Shackleton (1977) rainforest effect, but in the guise of shelf-carbon buildup and collapse (Niitsuma and Ku, 1977). The planktonic signal is the sum of this effect, and the compensating one from the fertility variation (increasing or decreasing the $\delta^{13}C$ gradient in Figure 4).

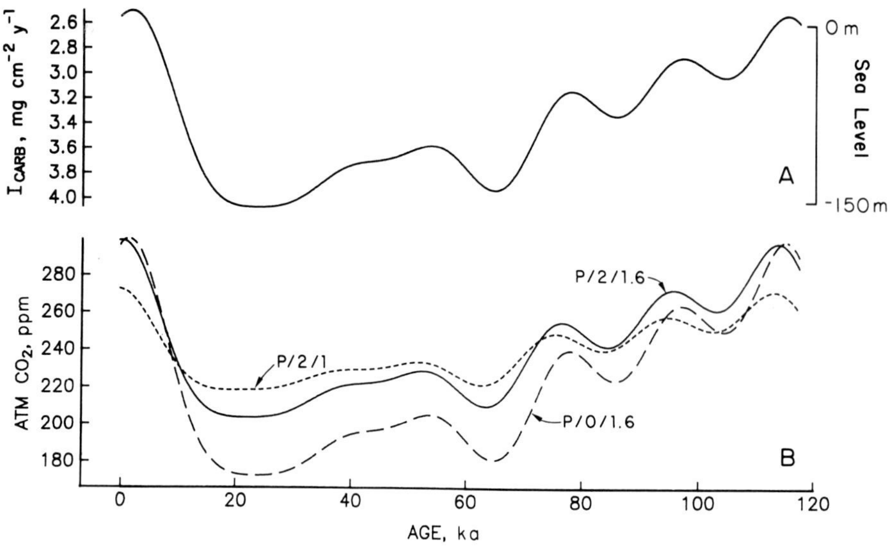

Fig. 8. Output from box model shown in Figure 7. Upper: change in productivity, as seen in carbonate particle flux to the seafloor (I_{carb}, in mg cm^{-2} yr^{-1}). Lower: change in pCO_2, three versions. P/2/1.6 refers to phosphorus model with lockup (and release) of 2 atmospheric carbon masses on the shelf, and a change in I_{carb} of 1.6 mg cm^{-2} yr^{-1}. P/2/1, similar but with a change of 1 mg cm^{-2} yr^{-1}. P/0/1.6, similar but with no carbon reservoir.

Broecker (1982a, b) already drew attention to the compensation phenomenon, identified it in the available data, and used it as chief support for his model. If we accept his arguments, then only P/2/1 will represent his model correctly: a larger fertility amplitude (P/2/1.6) will lead to over-compensation, and a smaller carbon amplitude will produce less fluctuation in the benthics than in the planktonics (Figs. 9B, C).

The difficulty in accepting P/2/1, as hinted earlier, is that in fact the observed change in benthic $\delta^{13}C$ during deglaciation is at most one half of that produced by the P/2/1 model. Thus, the data suggest that the ACM number cannot be higher than 1, hence P/1/*, where * stands for whatever. To retain $\delta^{13}C$ compensation for the planktonics, as in Figure 9A, we must use P/1/0.5. The change in fertility, then, is only of the order of 15%. The effect is to change the scales on Figure 10A while leaving the shape of the curves unaltered. However, instead of producing a deglacial pCO_2 increase near 60 ppm (P/2/1; Figure 8B) the model will then yield only 30 ppm. This is much less than required by the ice-core data.

There is yet another problem. If we grant to Broecker's extraction mechanism all the $\delta^{13}C$ variation of the benthics, then nothing is left for Shackleton's rainforest effect, which is supported by the paleobotanical evidence for forest growth during deglaciation (B. Frenzel, 1981, pers. comm.; F. A. Street, 1982, pers. comm.) Let us, then, share the signal, giving one-half to Broecker and the other half to Shackleton. We now have a model of type P/1/0.25. For this model, the planktonic signal is only partially compensated, so that it fluctuates in parallel with the benthics, but with one-half the amplitude. A partial compensation, in fact, appears closer to the truth than either full- or over-compensation, in our data (Figure 5). By now the pCO_2 yield of the P-extraction model has dropped to 15 ppm; it is no longer of strong interest.

It appears that we must go one step further and entirely uncouple the carbon buildup from the fertility effect. Using denitrification or changes in Redfield ratios of the carbon rain, or changes in carbon/carbonate rain ratios, we can produce models of the type P/0/*, which can yield substantial fluctuations in pCO_2 without undue requirements for $\delta^{13}C$ fluctuation.

The preservation signal. We next ask the P-extraction model to produce a record of carbonate preservation. The driving function (Figure 10A) is a proper derivative of sea level change ($\delta^{18}O$; Figure 8A); it describes the fluctuations in shelf-carbon postulated by the model. Carbonate accumulation (flux to sea floor minus dissolution), and percent preservation follow the shelf-carbon curve rather faithfully, as expected. When carbon is extracted, preservation increases, and when released, it decreases. The computation uses the assumption that dissolution rate is linearly related to the saturation deficit. The zero-extraction case (fertility effect only) also produces a preservation signal. However, its amplitude is quite low (Figure 10C, P/0/1.6). The shapes of the curves for solution indices and fragmentation for Core V28-238 (western equatorial Pacific, Thompson, 1976) are quite similar in shape to the model output. (The amplitude is a different matter: our model ocean has a flat bottom!)

We have argued that P/2/* is unrealistic on the basis of $\delta^{13}C$, and that it should be closer to P/1/* (i.e. deglacial extraction of 1 ACM into the biosphere). We have also argued, again from $\delta^{13}C$, that the fertility change is unlikely to exceed a drop of 20% from glacial to postglacial time. If this be accepted, the model output will have about one-half the amplitude of P/2/1.6 in Figure 10C. The corresponding preservation change on our (flat) model seafloor is from 10% to 16%. Such a change, in the west-equatorial Pacific, corresponds to a depth difference of only 150 to 200 m. Observations suggest a greater range (Figure 6). We speculate that a slowdown of deep mixing rate, by a factor of two, during deglaciation, could produce a preservation signal of the correct magnitude. Preliminary model runs suggest that this may be a viable mechanism. Other effects (such as a changing contribution of carbonate-rich loss from the continents) also may play a role (Berger, 1977b).

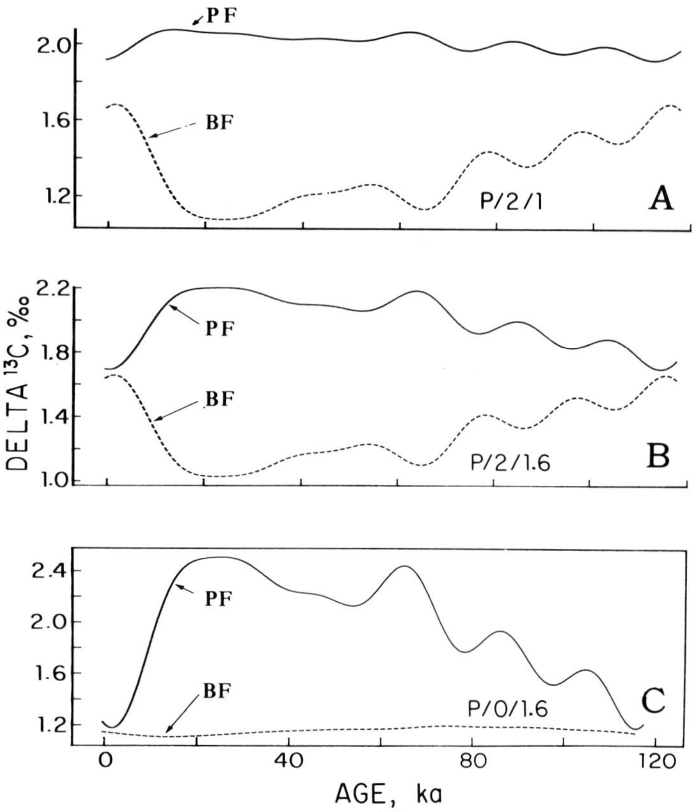

Fig. 9. Model-derived $\delta^{13}C$ stratigraphies for planktonic (solid line) and benthic (broken line) foraminifera (+2.2 ‰ added). Three cases as in Figure 8. A, changes of ocean's $\delta^{13}C$ (from 2 ACM buildup and release), as seen in benthic curve, compensated in the planktonic curve by corresponding changes in fertility (the Broecker 1981 model). B, planktonic signal overcompensated by increased fertility fluctuation. C, fertility effect alone, without carbon reservoir changes.

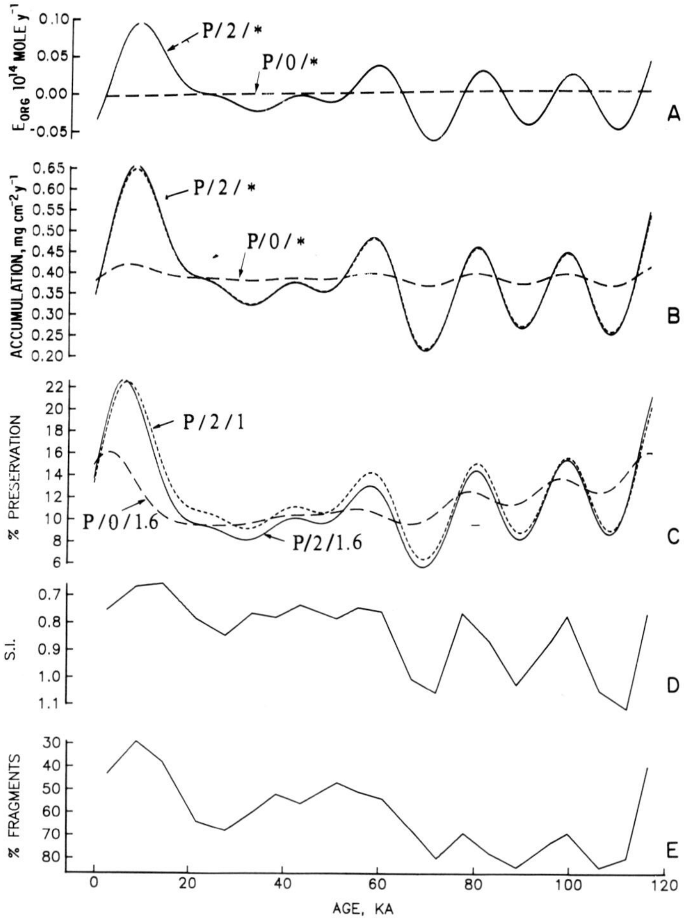

Fig. 10. Effect of shelf-carbon extraction and fertility changes on preservation of carbonate on seafloor. A, B, C, model-derived curves. A shows the rate of deposition of organic carbon on the shelf. The curve is a derivative of the driving function in Figure 8A. B shows the difference between the flux of carbonate to the sea floor, and the dissolution resulting from the fluctuating saturation. The curve follows A because of the effect of carbon extraction on saturation. C gives the proportion of the carbonate flux to the sea floor which is preserved. D and E are data from Thompson (1976). Both show the preservation of foraminifera in the western equatorial Pacific. S.I. (solution index) is based on species percentages (low values = high proportion of delicate species). % Fragments refers to the abundance of broken tests in the samples. From Keir and Berger (1983).

<u>The coral-reef model.</u> When "reef-growth" (that is, shallow-water carbonate deposition) is turned on in the box model of Figure 7 nothing much of interest occurs. There is a rather smallish rise in pCO_2 (10 to 20 ppm depending on reef mass) which results from the drop in alkalinity produced by the additional carbonate precipitation. The alkalinity deficiency never grows very

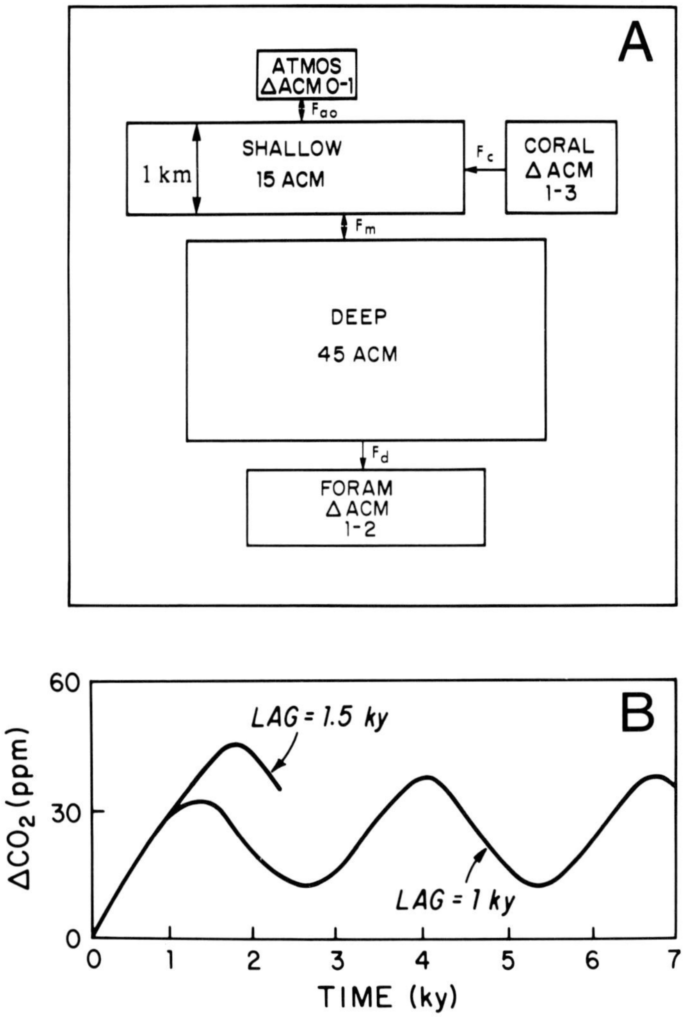

Fig. 11. Box model used for calculating predictions of the coral-reef hypothesis, and pCO_2 output. Time zero refers to start of sea level rise. ΔCO_2 is the pCO_2 addition. From Berger and Killingley (1982).

large, however, being dissipated by mixing with the deep waters of the ocean. This result was already anticipated by Berger and Killingley (1982) who showed that a substantial contribution of atmospheric CO_2 will only be made if mixing of shallow and deep waters is slowed. Their formulation of the coral-reef model is extremely simple (Figure 11). It mainly serves to illustrate that we need to know more about the history of coral reef growth, and about changes in the intensity of ocean mixing during deglaciation.

The boxes are shallow ocean, deep ocean and atmosphere; coral reef and sea floor are source and sink, respectively. The "shallow" layer is 1 km thick, because it was assumed that meltwater would be mixed down to this depth by wind-forcing. The ratio of carbon mass (1:3) is held constant for each mixing step. This fixes the flux F_m. The flux F_{ao} is based on a partitioning of "excess" CO_2 between upper waters and atmosphere in the ratio 80/20. The "excess" CO_2 is delivered by the coral reef growth, which is the forcing function. Growth is linear with time, the rate of buildup is 2 ACM in 5000 years. Dissolution of carbonate on the sea floor is driven by "excess" CO_2, which it removes. At each model step (250 yrs) 20% of the "excess" CO_2 is removed. (The model is not very sensitive to this parameter.) For a mixing lag of between 1000 and 1500 years, we obtain the output shown in Figure 11B. A somewhat greater lag increases the CO_2 available to the atmosphere, because of withholding equilibration with the deep ocean waters. In reality, we suspect that fluctuations will be damped more than shown, although there _is_ a positive feedback involved (more CO_2, warmer climate, less mixing, and _vice versa_). In view of the fact that climatic fluctuations over the last 12,000 years show a quasi-cyclicity of ~2000 years (e.g. Dansgaard _et al_., 1971), such internal feedback cannot be neglected.

The $\delta^{13}C$ signal is virtually unaffected by the shifts in carbonate between shelf and deep sea. The model-generated preservation signal simply follows sea level: poor preservation when sea level rises, good when it falls. In actuality, a better description of the record is "poor preservation during high sea level, good preservation during low sea level". Also, in reality, maximum preservation rather than maximum dissolution is centered on the transgressional phases (Figure 10). The coral-reef effect can explain the sudden end of the deglacial preservation spike, but not its initiation and its high amplitude. It looks as though coral reef buildup (as seen in carbonate dissolution on the sea floor) may occur too late to be of much help in producing the initial rapid rise of pCO_2 seen in the ice cores.

Summary and Conclusions

Where it stands. The phosphate-extraction model, in its classic form (Broecker, 1982a) does not fare well when tested against the data available from the deep-sea record. We suggest that it can by improved by decoupling fertility change from shelf carbon buildup and decay. It is well to remember that there is little or no information on global deglacial shelf carbon mass and its phosphate content. Thus, there is no good reason to prefer P-extraction over other fertility schemes. To assess the efficiency of carbon-decoupled models we note (Figure 8, P/O/1.6) that a one-third decrease in fertility during deglaciation will change the pCO_2 by the necessary amount (~ 100 ppm). In principle it should be possible to document such a decrease from micropaleontology. The $\delta^{13}C$ data are unconvincing, so far.

The coral-reef model depends heavily on a slowdown of mixing during deglaciation. It seems that this is physically reasonable. The model has a problem with timing, as seen in the carbonate preservational record. A similar problem of timing appears to apply to the fertility models:

changes in $\delta^{13}C$, as far as known, occur late during deglaciation or within the Holocene proper. Additional factors may have to be considered to obtain the pCO_2 step seen early during deglaciation. Large-scale warming of meltwater-stabilized surface water comes to mind. Excursions in the $\delta^{18}O$ signal of 1.5 to 2 ‰ have been postulated for early deglaciation, based on deconvolution of the $\delta^{18}O$-record (Berger, 1978). The excursions (if real) presumably are due to both a meltwater contamination effect and a temperature increase. A warming of 5°C (corresponding to a $\delta^{18}O$ of 1 ‰) could deliver about one half of the pCO_2 signal seen in the ice-cores. We may have to call on different mechanisms at different periods during the last 15,000 years, to produce the target signal.

What is needed. The crucial signals from the deep-sea record are $\delta^{13}C$ stratigraphy, preservation, and time-scale ($\delta^{18}O$, ^{14}C). Only the west equatorial Pacific has been studied in the detail necessary to attack the question of choosing between hypotheses which make predictions about these signals. Clearly, since the record is geographically variable, we need similar coverage from other regions (N. S. Atlantic, central Indian ocean, high latitudes). We also need to improve the models. The ocean floor is not flat. There are not one but several ocean basins, which, through exchange of water masses at various depth levels, acquire differing chemistries. These differences change through geologic time. There are major regional differences within ocean basins, between eastern boundary currents and central gyres, and between high and low latitudes. What this amounts to is a need not only for more boxes, but a need to integrate physical oceanography more fully into the geochemical machinery.

Most importantly, we need to remain open-minded about the mix of mechanisms, which produced the pCO_2 signal and their possible succession through time. Our favorite scenario, at present, would run something like this: Influx of meltwater at the beginning of deglaciation (15,000 to 14,000 years ago) tends to produce quasi-stable stratification, which is reinforced by strong warming of the surface layer. Deep mixing is greatly reduced. A fertility-drop ensues. The warming and the decrease in fertility increase the pCO_2. Forests start to grow (13,500 years ago), and after 2000 years the rate of increase in biomass is sufficient to reverse the rise in pCO_2. Temperatures drop. The ocean's deep mixing rate increases and fertility with it, decreasing the atmospheric pCO_2. Glaciers readvance. This is the period of the Younger Dryas (11,000 to 10,000 years ago). The Milankovitch configuration, in the meantime, is that of an interglacial. Forest growth slows or ceases due to the return of quasi-glacial conditions, allowing the pCO_2 to recover. Warming recommences and glaciers start to melt again. The meltwater-temperature-fertility chain is reactivated. As the meltwater effect wanes, coral reefs start growing, keeping the pCO_2 high. The warming reaches a peak in the Hypsithermal (8000-4000 yrs ago); reduced fertility tends to keep the pCO_2 high, but carbonate dissolution on the deep-sea floor, and re-invigoration of deep mixing starts to pull CO_2 down. Neoglaciation (last 2000 years) has set in.

While this scenario may be no closer to the truth than the various others proposed, it does illustrate the potential complexity of the mechanisms involved.

Acknowledgments. The authors are most appreciative of the detailed review of this paper by Taro Takahashi. The work was supported by National Science Foundation Grant OCE82-19553, Office of Naval Research Contract N00014-80-C-0440 and the Electric Power Research Institute.

References

Arrhenius, G., Sediment cores from the East Pacific, Rep. Swed. Deep-Sea Exped. 1947-1948, 5, 1-228, 1952.

Berger, W. H., Planktonic foraminifera: selective solution and the lysocline, Marine Geol., 8, 111-138, 1970.

Berger, W. H., Deep-sea carbonates: Pleistocene dissolution cycles, Jour. Foram. Res., 3, 187-195, 1973.

Berger, W. H., Carbon dioxide excursions and the deep sea record: aspects of the problem, In: N. R. Andersen and A. Malahoff (Eds.), The Fate of Fossil Fuel CO_2 in the Oceans, Plenum Press, New York, 505-542, 1977a.

Berger, W. H., Deep-sea carbonate and the deglaciation preservation spike in pteropods and foraminifera, Nature, 269, 301-304, 1977b.

Berger, W. H., Oxygen-18 stratigraphy in deep-sea sediments: additional evidence for the deglacial meltwater effect, Deep-Sea Res., 25, 473-480, 1978.

Berger, W. H., Paleoceanography: the deep-sea record, In: C. Emiliani (Ed.), The Sea, vol. 7, John Wiley Sons, New York, 1437-1519, 1981.

Berger, W. H., Increase of carbon dioxide in the atmosphere during deglaciation: the coral reef hypothesis, Naturwissenschaften, 69, 87-88, 1982a.

Berger, W. H., On the definition of the Pleistocene-Holocene boundary in deep-sea sediments, Sveriges geologiska undersökning, Avhand. Uppsat., vol. 76, Ser. C 794, 270-280, 1982b.

Berger, W. H., Deglacial CO_2 buildup: constraints on the coral-reef model, Palaeogeogr., Palaeoclimat., Palaeoecol., 40, 235-253, 1982c.

Berger, W. H., and J. S. Killingley, Glacial-Holocene transition in deep-sea carbonates: selective dissolution and the stable isotope signal, Science, 197, 563-566, 1977.

Berger, W. H., and J. S. Killingley, The Worthington Effect and the origin of the Younger Dryas, Jour. Mar. Res., 40 (suppl.), 27-38, 1982.

Berger, W. H., and E. Vincent, Chemostratigraphy and biostratigraphic correlation: exercises in systemic stratigraphy, Proceedings 26th Internat. Geol. Congress, Geology of Oceans Symposium, Paris 1980, Oceanologica Acta Suppl. to vol. 4, 115-127, 1981.

Berger, W. H., and U. von Rad, Cretaceous and Cenozoic sediments from the Atlantic Ocean, In: D. E. Hayes, A. C. Pimm et al., Initial Reports of the Deep Sea Drilling Project, 14, U. S. Gov. Print. Office, Washington, D. C., 787-954, 1972.

Berger, W. H., R. F. Johnson, and J. S. Killingley, "Unmixing" of the deep-sea record and the deglacial meltwater spike, Nature, 269, 661-663, 1977.

Berner, R. A., Early Diagenesis - a Theoretical Approach, Princeton Univ. Press, New Jersey, 241 pp, 1980.

Berner, W., H. Oeschger, and B. Stauffer, Information on the CO_2 cycle from ice core studies, Radiocarbon, 22, 227-235, 1980.

Boyle, E. A., and L. D. Keigwin, Deep circulation of the North Atlantic over the last 200,000 years: geochemical evidence, Science, 218, 784-787, 1982.

Broecker, W. S., Calcite accumulation rates and glacial to interglacial changes in oceanic mixing, In: K. K. Turekian (Ed.), The Late Cenozoic Glacial Ages, Yale Univ. Press, New Haven, 239-265, 1971.

Broecker, W. S., Factors controlling CO_2 content in the oceans and atmosphere, In: G. M. Woodwell and E. V. Pecan (Eds.), Carbon and the Biosphere, AEC Symposium, 30, 32-50, 1973.

Broecker, W. S., Chemical Oceanography, Harcourt, Brace, Jovanovich, New York, 214 pp., 1974.

Broecker, W. S., Glacial to interglacial changes in ocean and atmosphere chemistry, In: A. Berger (Ed.), Climatic Variations and Variability: Facts and Theories, D. Reidel Publ. Co., Holland, 111-121, 1981.

Broecker, W. S., Glacial to interglacial changes in ocean chemistry, In: E. B. Kraus and H. P. Hanson (Eds.), Climatic Variability in the Oceans, Progress in Oceanography, vol. 11, 151-197, 1982a.

Broecker, W. S., Ocean chemistry during glacial time, Geochimica et Cosmochimica Acta, 46, 1689-1705, 1982b.

Broecker, W. S., and S. Broecker, Carbonate dissolution on the western flank of the East Pacific Rise, In: W. W. Hay (Ed.), Studies in Paleo-oceanography, Soc. Econ. Paleont. Mineral. Spec. Pub., 20, 44-57, 1974.

Broecker, W. S., and J. Van Donk, Insolation changes, ice volumes, and the O^{18} record in deep sea cores, Rev. Geophys. Space Phys., 8, 169-198, 1970.

Chen, C., Pleistocene pteropods in pelagic sediments, Nature, 219, 1145-1149, 1968.

CLIMAP Project Members, The surface of the ice-age earth, Science, 191, 1131-1137, 1976.

Dansgaard, W., S. J. Johnsen, H. B. Clausen, and C. C. Langway, Climatic record revealed by the Camp Century Ice Core, In: K. K. Turekian (Ed.), Late Cenozoic Glacial Ages, Yale Univ. Press, New Haven, 37-56, 1971.

Delmas, R. J., J.-M. Ascencio, and M. Legrand, Polar ice evidence that atmospheric CO_2 20,000 yr BP was 50% of present, Nature, 284, 155-157, 1980.

Diester-Haass, L., H. J. Schrader, and J. Thiede, Sedimentological and paleoclimatological investigations of two pelagic ooze cores off Cape Barbas, N.W. Africa, "Meteor" Forschungsergebn. (C), 16, 19-66, 1973.

Douglas, R., and F. Woodruff, Deep-sea benthic foraminifera, In: C. Emiliani (Ed.), The Sea, vol. 7, 1233-1327, 1981.

Duplessy, J.-C., La géochimie des isotopes stables du carbone dans la mer, Centre d'Etudes Nucléaires de Saclay, CEA-N-1565, 1-198, 1972.

Emiliani, C., Pleistocene temperatures, Jour. Geol., 63, 538-578, 1955.

Emiliani, C., Paleotemperature analysis of Caribbean cores P6304-8 and P6304-9 and a generalized temperature curve for the past 425,000 years, Jour. Geol., 74, 109-126, 1966.

Emiliani, C., The cause of the ice ages, Earth Planet. Sci. Lett., 37, 349-352, 1978.

Honjo, S., Material fluxes and modes of sedimentation in the mesapelagic and bathypelagic zones, Jour. of Mar. Res., 38, 53-97, 1980.

Keir, R. S., and W. H. Berger, Atmospheric CO_2 content in the last 120,000 years: the phosphate-extraction model, Jour. Geophys. Res., 88 (C10), 6027-6038, 1983.

Kellogg, W. W., Review of human impact on climate, In: H. Oeschger, B. Messerli, and M. Svilar (Eds.), Das Klima, Analysen und Modelle, Geschichte und Zukunft., Springer-Verlag, Berlin, 18-20, 1980.

Kroopnick, P. M., S. V. Margolis, and C. S. Wong, $\delta^{13}C$ variations in marine carbonate sediments as indicators of the CO_2 balance between the atmosphere and oceans, In: N. R. Andersen and A. Malahoff (Eds.), The Fate of Fossil Fuel CO_2 in the Oceans, Plenum Press, New York, 295-321, 1977.

Kudrass, H.-R., Sedimentation am Kontinentalhang vor Portugal und Marokko im Spätpleistozän und Holozan, "Meteor" Forschungsergebn. (C), 13, 1-63, 1973.

Li, Y. H., T. Takahashi, and W. S. Broecker, Degree of saturation of $CaCO_3$ in the oceans, Jour. Geophys. Res., 74, 5507-5525, 1969.

Luz, B., and N. J. Shackleton, $CaCO_3$ solution in the tropical east Pacific during the past 130,000 years, In: W. V. Sliter, A. W. H. Bé, and W. H. Berger (Eds.), Dissolution of deep-sea carbonates. Cushman Found. Foram. Res. Spec. Pub., 13, 142-150, 1975.

Menzel, D. W., and J. H. Ryther, Nutrients limiting the production of phytoplankton in the Sargasso Sea, with special reference to iron, Deep-Sea Res., 7, 276-281, 1961.

Neftel, A., H. Oeschger, J. Schwander, B. Stauffer, and R. Zumbrunn, New measurements on ice core samples to determine the CO_2 content of the atmosphere during the last 40,000 years, Nature, 295, 220-223, 1982.

Niitsuma, N., and T.-L. Ku, What happened during geomagnetic reversals, Kagaku, 47, 671-678, 1977.

Oba, T., Biostratigraphy and isotopic paleotemperature of some deep-sea cores from the Indian Ocean, Science Rpts. Tohoku University, Sendai, 2nd Ser. (Geology), 41, 129-195, 1969.

Olausson, E., Sediment cores from the Indian Ocean, Repts. Swed. Deep-Sea Exped. 1947-1948, 9, 53-88, 1960.

Peng, T.-H., W. S. Broecker, and W. H. Berger, Rates of benthic mixing in deep-sea sediment as determined by radioactive tracers, Quaternary Res., 11, 141-149, 1979.

Redfield, A. C., B. H. Ketchum, and F. A. Richards, The influence of organisms on the composition of sea-water, In: M. N. Hill (Ed.) The Sea, vol. 2, Interscience, New York, 26-77, 1963.

Ruddiman, W. F., and A. McIntyre, The North Atlantic Ocean during the last deglaciation, Palaeogeogr., Palaeoclimat., Palaeoecol., 35, 145-214, 1981.

Shackleton, N. J., Carbon-13 in Uvigerina: tropical rainforest history and the equatorial Pacific carbonate dissolution cycles. In: N. R. Andersen and A. Malahoff (Eds.), The Fate of Fossil Fuel CO_2 in the Oceans, Plenum Press, New York, 401-427, 1977.

Shackleton, N. J., and N. D. Opdyke, Oxygen isotope and paleomagnetic stratigraphy of equatorial Pacific core V28-238: oxygen isotope temperatures and ice volumes on a 10^5 year and 10^6 year scale, Quaternary Res., 3, 39-55, 1973.

Spencer, D. W., The sediment trap intercomparison experiment - some preliminary data, In: R. F. Anderson and W. Bacon (Eds.), Sediment Trap Intercomparison Experiment, Technical Memo No. 1-81, Woods Hole Oceanographic Institution, 57-104, 1981.

Street, F. A., and A. T. Grove, Environmental and climatic implications of late Quaternary lake-level fluctuations in Africa, Nature, 261, 385-390, 1976.

Thiede, J., Planktonische Foraminiferen in Sedimenten von Ibero-Marokkanischen Kontinentalrand, "Meteor" Forschungsergebn. (C) 8, 15-102, 1971.

Thiede, J., Planktonic foraminifera in hemipelagic sediments: shell preservation off Portugal and Morocco, Geol. Soc. Amer. Bull, 84, 2749-2754, 1973.

Thomas, W. H., and D. L. R. Seibert, Distribution of nitrate, nitrite, phosphate and silicate in the California Current region, 1969, CalCOFl Areas No. 20, VII-X, 2-97, 1974.

Thompson, P. R., Planktonic foraminiferal dissolution and the progress towards a Pleistocene equatorial Pacific transfer function, Jour. Foram. Res., 6, 208-227, 1976.

Thompson, P. R., and T. Saito, Pacific Pleistocene sediments: planktonic foraminifera dissolution cycles and geochronology, Geology, 2, 333-336, 1974.

Vincent, E., and W. H. Berger, Planktonic foraminifera and their use in paleoceanography, In: C. Emiliani (Ed.), The Sea, vol. 7, Wiley-Interscience, New York, 1025-1119, 1981.

Vincent, E., J. S. Killingley, and W. H. Berger, Stable isotope composition of benthic foraminifera from the equatorial Pacific, Nature, 289, 639-643, 1981a.

Vincent, E., J. S. Killingley, and W. H. Berger, Stable isotopes in benthic foraminifera from Ontong-Java Plateau, Box Cores ERDC 112 and 123, Palaeogeogr., Palaeoclimat., Palaeoecol., 33, 221-230, 1981b.

Wefer, G., E. Suess, W. Balzer, G. Liebezeit, P. J. Müller, C. A. Ungerer, and W. Zeuts, Fluxes of biogenic components from sediment trap deployment in circumpolar waters of the Drake Passage, Nature, 299, 145-147, 1982.

Worthington, L. V., Genesis and evolution of water masses, Meteorological Monographs, 8, 63-67, 1968.

ON THE RELATIONSHIP BETWEEN OCEAN CHEMISTRY AND ATMOSPHERIC pCO_2 DURING THE CENOZOIC

Michael L. Bender

Graduate School of Oceanography
University of Rhode Island, Kingston, Rhode Island 02881

Abstract. A simple two-box model of ocean chemistry is used to calculate the changes in atmospheric pCO_2 which would be associated with various related changes in ocean properties. The results show that no major variations of pCO_2 are implied by the observed range of deep ocean temperature, carbonate compensation depth, or seawater calcium concentrations during the Cenozoic. pCO_2 changes associated with the poorly constrained or simply unknown variations in the deep ocean total and preformed nutrient concentrations, and in the ratio of $CaCO_3/CH_2O$ in the raining particulate matter, were probably also small, but the results are equivocal. Atmospheric pCO_2 is a very sensitive function of seawater TCO_2 and TA; as paleoceanographic values of these parameters are not well constrained, it is not yet possible to place useful limits on paleoatmospheric pCO_2 from the sedimentary record.

Introduction

In order to realistically study Cenozoic climates, it is necessary to understand how atmospheric pCO_2 varied. At present, this is known only for the last 20,000 years or so, for which ice core data are available [Oeschger, 1983]. For earlier times, two approaches may be taken to constraining pCO_2. The first is to generate a hypothesis on the way pCO_2 is regulated over long times scales, and use this hypothesis to estimate pCO_2 in the past. pCO_2 regulation has, in fact, received considerable attention. The long term geochemical cycle reflects the balance between weathering (in which CO_2 attacks and dissolves carbonate and silicate minerals on the continents) and reverse weathering (in which the dissolved elements precipitate out as carbonates and silicates, thereby regenerating CO_2). Examples of these reactions are:

$$CaCO_3 + CO_2 + H_2O = Ca^{++} + 2HCO_3^- \qquad (1)$$

$$MgSiO_3 + CO_2 + H_2O = Mg^{++} + 2HCO_3^- + SiO_2 \qquad (2)$$

Weathering reactions go the right, reverse weathering reactions to the left.

Sillen [1967], Holland [1978] and others have suggested that the dissolved and solid components involved in oceanic reverse weathering reactions are at equilibrium, and these equilibria define the pCO_2 (equilibrium model). This conclusion, however, is not widely accepted, in part because reverse weathering reactions involving silicates are kinetically inhibited, and it is not clear what their nature is or where they occur. Because of this problem, Broecker (1971) postulated that feedback mechanisms would drive pCO_2 to a level where weathering and reverse weathering were in balance (kinetic model). High pCO_2 values would cause high weathering rates and low reverse weathering rates, both of which would drive pCO_2 down. Low pCO_2 would cause low weathering rates and high reverse weathering rates, driving pCO_2 up. Three mechanisms have been offered to mediate the feedback: weathering rates increase directly as pCO_2 increases [Broecker, 1971]; photosynthesis rates increase as pCO_2 increases, thereby increasing pCO_2 in soils and groundwater, and speeding up weathering [Holland, 1978]; and temperature increases as pCO_2 rises, speeding up the rates of all chemical reactions including weathering [Walker et al., 1981]. In the kinetic model, there are external forcings on the system. The rate of mountain building is perhaps the most important, in that it regulates the flux to the Earth's surface of new minerals, which form the CO_2 demand.

From this discussion, it is clear that we do not quantitatively understand the way geochemical processes regulate the atmospheric pCO_2, and therefore have no satisfactory basis for estimating its value in the past. We have thus attempted an empirical approach. Regardless of the cause(s), past variations in pCO_2 would have been accompanied by changes in seawater chemistry which would in turn generally be reflected by observable features of the Cenozoic record. The possibility thus arises that the features of sediment geochemistry may be inverted to constrain the atmospheric pCO_2. An attempt to use this approach is described in some detail. As will be seen, it eventually fails because atmospheric pCO_2 is a sensitive function of seawater TCO_2 and TA, parameters not well constrained by the sedimentary record.

While our empirical approach does not constrain pCO_2 to any useful degree, it does give valuable estimates on the magnitude of the pCO_2 variability which would be associated with a change in the calcium carbonate compensation depth, the deep ocean temperature, the $CaCO_3/CH_2O$ ratio in the particle rain, the Redfield ratios, and the total and preformed nutrient concentrations in the deep sea. The calculations are thus valuable for investigating hypotheses about coupled changes in climate and oceanic carbonate chemistry during the Tertiary [e.g., Arthur, 1982] and Quaternary [e.g., Broecker, 1982].

We adopt a two-box model from which atmospheric pCO_2 can be calculated in terms of the carbonate compensation depth (CCD), the degree of calcite saturation (Ω) at the CCD, the seawater $[Ca^{++}]$, the deep ocean $[TCO_2]$, the deep ocean $[PO_4^{3-}]$, the deep ocean temperature, and the flux ratio of $CaCO_3/CH_2O$ to the deep sea. The box model is then used to determine the sensitivity of atmospheric pCO_2 to changes in each of these parameters. Where possible, paleoceanographic values of these parameters are inferred from the sedimentary record and used to constrain the variability in pCO_2.

Effects on Various Parameters on Oceanic pCO_2

Modern Ocean Two Box Model

For our sensitivity tests, we have adopted the following two-box model (Fig. 1), taken largely from Broecker [1982]. The surface ocean and deep ocean temperatures are taken as 21° and $2^\circ C$, respectively; both salinities are $35^\circ/oo$. The surface ocean TCO_2 is taken as 1.961 mM/kg, surface ocean pCO_2 is 300 uatm, and the deep ocean $[PO_4^{3-}]$ is 2.2 uM/kg. Surface ocean TA values are calculated from pCO_2 and TCO_2 to be 2.286 meq/kg. The modern CCD is taken as 4500 m [based on the summary data in Takahashi, 1975], and the degree of calcite saturation (Ω) at the compensation depth is assumed to be 0.90 [adopted from a comparison of CCD as summarized by Takahashi, 1975, and Ω values from Takahashi et al., 1980 and 1981].

The deep ocean TA and TCO_2 values are calculated from the following equations (the subscripts D and S refer to the deep and surface oceans, respectively):

$$TCO_{2D} = TCO_{2S} + 106\ [PO_4^{3-}]_D + 106x[PO_4^{3-}]_D \quad (3)$$

$$TA_D = TA_S + (2x)(106\ [PO_4^{3-}]_D) - 16[PO_4^{3-}]_D \quad (4)$$

x is the molar ratio of $CaCO_3/CH_2O$ in particulate matter falling to the deep ocean. In equation (3), the second and third terms on the right account for addition of CO_2 originating from CH_2O oxidation and $CaCO_3$ dissolution, respectively. The second and third terms on the right of equation (4) refer, respectively, to alkalinity added by $CaCO_3$ dissolution, and titrated by HNO_3 produced by organic N oxidation [Brewer and Goldman, 1976]. The equations assume a C:N:P of 106:16:1.

All carbonate system equations were carried out using the equilibrium constants adopted by Takahashi et al. [1980].

The parameters of the box model for the modern ocean are summarized in Figure 1. This model and the specified parameters constrain x to a value of 0.27.

The pCO_2 of the warm surface ocean is assumed to be in equilibrium with the atmosphere. The pCO_2 of the deep ocean is higher than that of the atmosphere. No gas exchange is allowed between the deep ocean outcrops and the atmosphere, and therefore there is no transfer of CO_2 from the deep ocean to the surface ocean via the atmosphere ["Redfield ocean" of Broecker and Takahashi, 1983]. The contributions of silica and phosphate to the alkalinity of seawater are neglected. The surface ocean is assumed to be completely depleted of nutrients.

In our model, the composition of the deep oceans, together with assumptions about biological cycling, fix the chemistry of the surface ocean. This in turn fixes atmospheric pCO_2. As parameters are varied, the total mass of CO_2 in the ocean-atmosphere system is allowed to vary, and indeed it would be impossible to treat the problem otherwise, given our model.

When considering the following discussion, recall that only pressure-independent terms, such as TCO_2 and TA, are constant within a box. The $CaCO_3$ solubility product increases with pressure and therefore depth within a box, and $[CO_3^=]$ is also pressure dependent. As a result, the degree of saturation decreases with depth within a well mixed box.

Effect of CCD on pCO_2

It is well established that the CCD of the Cenozoic oceans was generally shallower than that of the modern oceans [e.g. Van Andel, 1975]. If $[PO_4^{3-}]_D$ and x have remained constant, this shallower CCD would imply a more acidic ocean and a higher atmospheric pCO_2. How much of a change in pCO_2 is implied? To answer this question, we have calculated pCO_2 as a function of CCD, with $[PO_4^{3-}]_D$, x, Ω_{CCD} and $[TCO_2]_D$ held constant. The approach is as follows. First $[CO_3^=]_{CCD}$ is calculated from the equation

$$[CO_3^=]_{CCD} = 0.9\ K'_{sp}(z)/[Ca^{++}] \quad (z = \text{depth}) \quad (5)$$

$K'_{sp}(z)$ is the apparent solubility product of $CaCO_3$. TA_D (which is, of course, equal to TA_{CCD}) is calculated from $[CO_3^=]_{CCD}$, CCD, and TCO_{2D}.

TCO_{2S} and TA_S are calculated using equations (3) and (4), respectively. Finally, the equilibrium atmospheric pCO_2 is calculated from these latter two values.

In Figure 2, pCO_2 is plotted vs. CCD. The limits of variations in the average CCD during the Cenozoic, 3400 m-4500 m [Van Andel, 1975] are also shown. The change in the CCD alone suggests that, neglecting other factors, pCO_2 has varied between 300-340 uatm during the Cenozoic.

Effect of $[Ca^{++}]$ on pCO_2

If the deep ocean $[Ca^{++}]$ were to change, $[CO_3^{=}]_{CCD}$ would change according to equation (5). We have calculated the effect of such a change on pCO_2 by assuming that the CCD, the deep ocean TCO_2, and the deep ocean-surface ocean differences in TCO_2 and TA all remain constant. From $[CO_3^{=}]_{CCD}$, CCD, and TCO_{2D}, TA_D is calculated. TA_S and TCO_{2S} are calculated from deep-surface differences, and these values are in turn used for calculate pCO_2. Results are given in Figure 3. From this figure, and from indications of the extent of $[Ca^{++}]$ variations in the paleo-oceans, limits can be put on the variability of $[CO_3^{=}]_{CCD}$ and associated variations in pCO_2.

Recent data on the trace element composition of foraminifera [Graham et al., 1982; Delaney, 1983] suggest that the seawater $[Ca^{++}]$ has not varied by more than ± 25% during the past 50 myr. The argument is as follows. The Sr/Ca and Li/Ca ratios have been measured in a large number of

BOX MODEL OF THE MODERN TCO_2 CYCLE

ATMOSPHERE	pCO_2 = 300 UATM
SURFACE OCEAN	TCO_2 = 1.961 MM/KG
	TA = 2.286 MEQ/KG
	$[PO_4^{=}]$ = 0
	$[NO_3]$= 0
PARTICLE RAIN ↓	$CaCO_3/CH_2O$ = 0.27
DEEP OCEAN	TCO_2 = 2.257 MM/KG
	TA = 2.376 MEQ/KG
	$[CA^{++}]$ = 10.3 MM/KG
	CCD = 4500 M
	Ω_{CCD} = 0.9
	$[PO_4^{=}]$ = 2.2 μM/KG
	$[NO_3^{-}]$ = 16 $[PO_4^{=}]$

Fig. 1. A 2-box model representing the CO_2 system in the modern ocean.

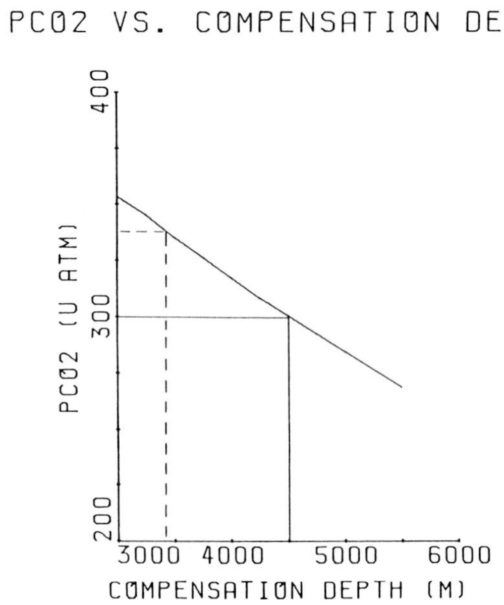

Fig. 2. Results of model calculations for the variation of pCO_2 with CCD. The calculations were done holding constant TCO_{2D}, $CaCO_3/CH_2O$ ratio in the particle rain, $[PO_4^{3-}]_D$, $[Ca^{++}]$ and Ω at the CCD. The dashed line indicates the present CCD, which is also the deepest CCD of the Cenozoic. The solid line indicates the shallowest CCD.

well preserved fossil foraminifera and found to be relatively constant; the same obviously holds for the Li/Sr ratio. The constancy does not reflect diagenesis, which changes the ratios, but is a primary feature of skeletal chemistry. The distribution coefficients of Sr and Li in biogenic calcite are roughly independent of solution composition [Lorens and Bender, 1980; Delaney, 1983], so that foram ratios are direct indicators of the elemental ratios in contemporaneous seawater. Thus there is strong evidence that the relative abundances of Ca:Sr:Li have been nearly constant for most of the Cenozoic.

The marine geochemical cycles of Ca, Sr, and Li are sufficiently decoupled that it seems unlikely that their concentrations would vary sympathetically. Ca is supplied to the oceans by rivers, with a small submarine hydrothermal supplement (3-35% according to Delaney, 1983); it is removed by carbonate sedimentation. The geochemical cycle of Sr differs from that of Ca more than is generally acknowledged. Rivers are its sole source [e.g., Elderfield and Greaves, 1981], with most Sr coming from limestone weathering, but limestone Sr is recycled much more rapidly than Ca [Brass, 1976]. About half the riverine input is removed as marine limestone, the other half being removed in some poorly defined authigenic phase [Turekian, 1964]. The Li cycle is not yet well understood, but is clearly very different from that of Ca and Sr [Delaney, 1983]. Limestones and silicates both contribute significantly to the fluvial flux. The hydrothermal

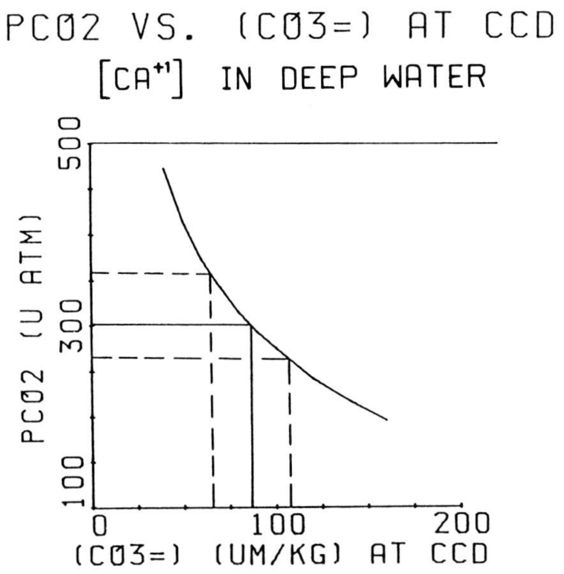

Fig. 3. Results of model calculations for the variation of pCO_2 with $[Ca^{++}]$. The calculations were done holding constant TCO_{2D}, $CaCO_3/CH_2O$ rain ratio, $[PO_4^{3-}]_D$, Ω at the CCD, and CCD. The dashed line indicates the present $[Ca^{++}]$ the solid lines indicate the limits of $[Ca^{++}]$ variation during the Cenozoic.

input may be an order of magnitude greater than that of rivers [Edmond et al., 1979], but hydrothermal fluxes generally are highly uncertain, and Delaney has estimated that for Li as 100%-

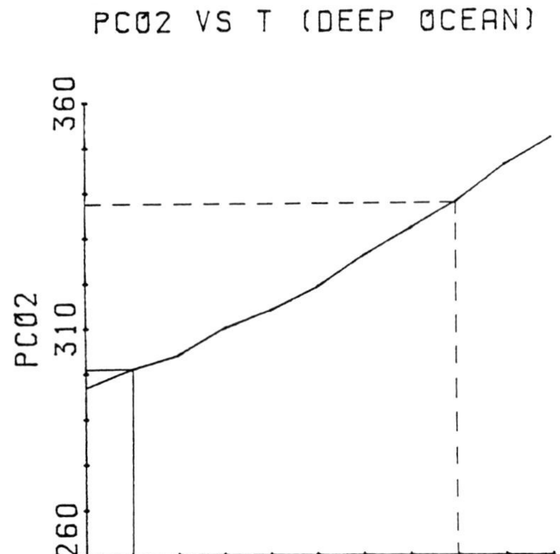

Fig. 4. Results of model calculations for the variation of pCO_2 with deep ocean temperature. The calculations were done holding constant CCD, Ω_{CCD}, $CaCO_3/CH_2O$ rain ratio, $[Ca^{++}]$, $[PO_4^{3-}]_D$, and TCO_{2D}. The dashed line indicates the modern ocean conditions and coldest Cenozoic bottom water temperature; the solid line indicates the warmest Cenozoic bottom water temperature.

1000% of the river flux. It appears that most removal is by aluminosilicate sedimentation and/or low temperature weathering of basalts, although if the hydrothermal flux is small, carbonate sedimentation may be a significant sink.

Not only are the sources and sinks for these elements very different, the large scale controls on removal are also. Calcium is a major ion and its flow through the oceans is tightly balanced by saturation with respect to $CaCO_3$. Sr and Li are trace elements whose flows through the oceans are not governed by any known equilibrium constraints, and are generally modelled as first order kinetic processes.

Because of their widely differing marine geochemistries, we think it is unlikely that the Ca:Sr:Li concentrations in seawater have covaried sympathetically. Rather it seems more likely, although by no means certain, that the constancy of the elemental ratios reflects constancy in their seawater concentrations. We therefore believe that the seawater $[Ca^{++}]$ has varied by ≤25% from its present value. We then assume that, during the Cenozoic, $[CO_3^{=}]_{CCD}$ has been between 75% and 125% of its present value (64 uM/kg - 108 uM/kg). According to Figure 3, this result restricts the variation in pCO_2 caused solely by a change in $[CO_3^{=}]_{CCD}$ to 260-360 uatm.

Effect of Deep Ocean Temperature on pCO_2

As deep ocean temperature increases, pCO_2 increases because a lower alkalinity serves to produce the required degree of $CaCO_3$ saturation. The effect of T_D on pCO_2 was quantified holding constant the deep ocean TCO_2, $[PO_4^{3-}]$, $[Ca^{++}]$, CCD, and Ω_{CCD}. For each value of T_D, the TA_D required to give $\Omega_{CCD} = 0.90$ was calculated. The surface water TA and TCO_2 values were calculated from equations (3) and (4), and used in turn to calculate pCO_2. As T_D rises to the maximum Cenozoic value of about 16°C [Savin et al., 1975; Shackleton and Kennett, 1975], pCO_2 rises to 338 uatm (Fig. 4). Deep ocean temperature therefore has only a small influence on pCO_2.

Effect of $CaCO_3/CH_2O$ in particulate flux on pCO_2.

Assuming constant deep water TA and TCO_2, the box model pCO_2 depends on the ratio of $CaCO_3/CH_2O$ in particles falling to the deep ocean. The reason for this is that the rain ratio reflects the relative amounts of acid (CH_2O) and alkalinity ($CaCO_3$) removed biogenically from surface water.

We calculated the effect of the $CaCO_3/CH_2O$ ratio in the particle flux on pCO_2 using equations (3) and (4). TCO_{2D}, TA_D, and $[PO_4^{3-}]_D$ are held constant at the modern ocean values. x is varied from half to twice its present value. The atmospheric pCO_2 was calculated from TCO_{2S} and TA_S. Results (Figure 5) show that pCO_2 is only slightly sensitive to variations in the $CaCO_3/CH_2O$ ratio. There is no basis for estimating variations in this ratio during the Cenozoic, but we see that pCO2 would have changed significantly as a result of particulate $CaCO_3/CH_2O$ ratio variations only if the average ratio varied by more than a factor of two.

Effect of $[PO_4^{3-}]_D$ on the atmospheric pCO_2

We evaluate the effect of $[PO_4^{3-}]_D$ on pCO_2 from eq. (3) and (4), assuming TA_D, TCO_{2D}, and x are constant (having the values in Figure 1), and varying $[PO_4^{3-}]_D$. pCO_2 increases as $[PO_4^{3-}]_D$ drops, because of decreasing removal of CO_2 from surface water as particulate C_{org}. Model pCO_2 is plotted vs $[PO_4^{3-}]_D$ in Fig. 5. The historical range in $[PO_4^{3-}]_D$ can be evaluated from the difference between the $\delta^{13}C$ of surface and deep water as recorded by the $\delta^{13}C$ of planktonic and benthic foraminifera (Broecker, 1974). As this indicator is used here to constrain both $[PO_4^{3-}]_D$ and (in the next section) TCO_{2D}, we will discuss its development in some detail.

The TCO_2 of deep water is composed of three parts: the surface water component (that TCO_2 which was present when the deep water formed), the part due to $CaCO_3$ dissolution, and the part due to CH_2O oxidation. The deep water $\delta^{13}C$ is the weighted average of the $\delta^{13}C$ of the components of deep water TCO_2:

$$\delta^{13}C_D = \frac{TCO_{2S}\,\delta^{13}C_S}{TCO_{2D}}$$

$$+\frac{TCO_{2(CaCO_3)}\,\delta^{13}C_{CaCO_3} + TCO_{2(CH_2O)}\,\delta^{13}C_{CH_2O}}{TCO_{2D}} \quad (6)$$

The terms in the numerator refer to deep water TCO_2 contributed by surface water downwelling, $CaCO_3$ dissolution, and CH_2O oxidation, respectively. Since $\delta^{13}C_{CaCO_3} \sim 0$ (for example, Williams et al., 1977), and $TCO_2\,(CH_2O) = 106\,[PO_4^{3-}]_D$, we can write:

$$\delta^{13}C_D = \frac{TCO_{2S}}{TCO_{2D}}\,\delta^{13}C_S + \frac{106\,[PO_4^{3-}]_D}{TCO_{2D}}\,\delta^{13}C_{CH_2O} \quad (7)$$

Subtracting from this equation $\delta^{13}C_S$, we get

$$\delta^{13}C_D - \delta^{13}C_S = \delta^{13}C_S$$

$$\cdot\left(\frac{TCO_{2S}}{TCO_{2D}} - 1\right) + \frac{106\,[PO_4^{3-}]_D}{TCO_{2D}}\,\delta^{13}C_{CH_2O} \quad (8)$$

In the modern ocean, $\delta^{13}C_S \sim +2$ [Kroopnick, 1980], and $\delta^{13}C_{CH_2O} \sim -20^{o}/oo$ [Rau et al., 1982]. The first term on the right is much less than the second term and can be neglected, so that

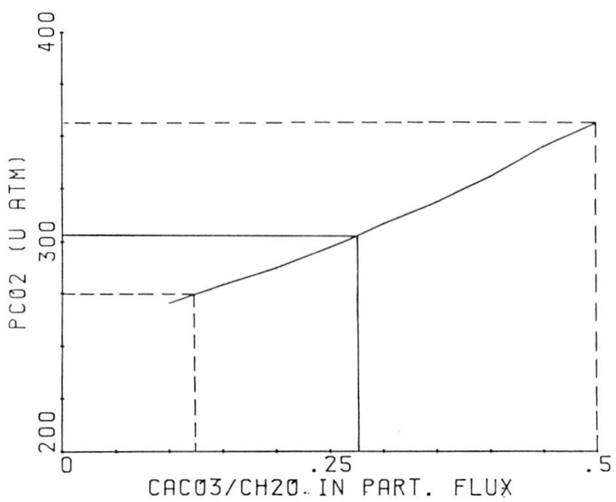

Fig. 5. Results of model calculations for the variation of pCO_2 with the $CaCO_3/CH_2O$ ratio in the particulate rain to the deep ocean. The calculations were done holding constant TCO_{2D}, $[Ca^{++}]$, $[PO_4^{3-}]_D$, Ω at the CCD, and CCD. The dashed line indicates the present ocean values; the solid lines show the variation in pCO_2 resulting from halving or doubling in the $CaCO_3/CH_2O$ rain ratio.

equation (8) simplifies to

$$\delta^{13}C_D - \delta^{13}C_S \approx \frac{106\,[PO_4^{3-}]_D}{TCO_{2D}}\,\delta^{13}C_{CH_2O}$$

$$\approx (-20^{o}/oo)\,\frac{106\,[PO_4^{3-}]_D}{TCO_{2D}} \quad (9)$$

or

$$\delta^{13}C_D - \delta^{13}C_S \approx (-20^{o}/oo)\;TCO_{2(CH_2O)}/TCO_{2D} \quad (10)$$

The $\delta^{13}C$ difference can be evaluated from the $\delta^{13}C$ of benthic and planktonic foraminifera, which reflect the $\delta^{13}C$ of the water in which they grow.

Benthic foraminiferal $\delta^{13}C$ differs from bottom water $\delta^{13}C$ by an amount that, for any one species, is constant to ± 0.3‰ [Belanger et al., 1981; Killingley et al., 1981; Graham et al., 1981]. Using the known fractionation factor, one can get back to bottom water $\delta^{13}C$. The benthic foram $\delta^{13}C$ values vary because of the change in $\delta^{13}C$ of deep water as it ages and gains metabolic CO_2. It is not strictly possible to get the average deep ocean $\delta^{13}C$ from benthic foram $\delta^{13}C$, because the average $\delta^{13}C$ of the box model is not strictly defined in terms of the real, stratified, ocean. Thus, although the benthic $\delta^{13}C$ data can readily be inverted to give deep water $\delta^{13}C$, the interpretation of this data in terms of box models is complicated.

The situation for planktonic foraminifera is

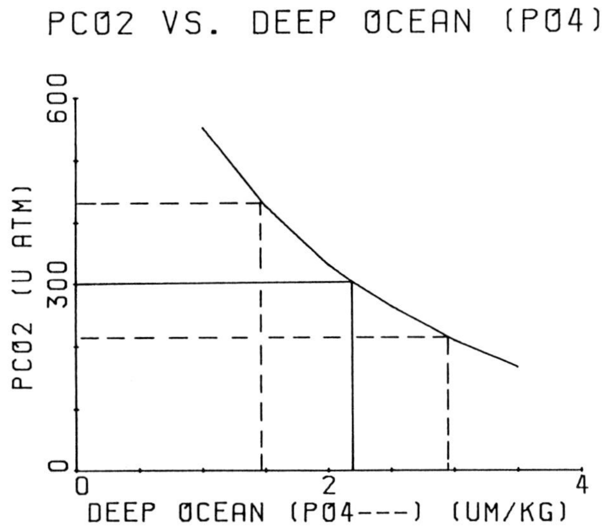

Fig. 6. Results of model calculations for the variation of pCO_2 with $[PO_4^{3-}]_D$. The calculations were done holding constant TCO_{2D}, CCD, Ω_{CCD}, $CaCO_3/CH_2O$ rain ratio, and $[Ca^{++}]$. The dashed line indicates the modern ocean conditions; the solid lines indicate the boundaries on the variability in $[PO_4^{3-}]_D$ during the last 10 myr, assuming constant TCO_{2D}.

even more complex. First, planktonic foraminifera have shorter stratigraphic ranges than benthics, the oldest extent species being Orbulina Universa, which appeared about 15 myr B.P. There is no simple way to determine fractionation factors for extinct species. Second, surface water $\delta^{13}C$ varies [e.g., Kroopnick, 1980]; because of this, we have estimated a mean surface water $\delta^{13}C$ value solely from data on samples from between 35°S and 40°N. Third, as a result of vertical $\delta^{13}C$ variations, temperature variations, and perhaps other unknown factors, $\delta^{13}C$ of planktonic foraminifera vary as a function of latitude [Durazzi, 1981] and test size (e.g., Curry et al., 1981).

Stable isotope studies of Cenozoic planktonic and benthic foraminifera afford a large data base for assessing the $\delta^{13}C$ difference of surface and deep waters during the last 10 myr, for which period we can use the planktonic foram Orbulina Universa to estimate surface water $\delta^{13}C$. We have used planktonic and benthic stable isotope data of Williams et al. [1977], Keigwin [1979, 1982], Bender and Graham [1981], and Loutit [1981] to make that calculation. The results suggest that the surface-deep $\delta^{13}C$ difference has been in the range of +1.4°/oo to 2.5°/oo during the last 10 myr. For comparison, the present $\delta^{13}C$ difference is about 2°/oo. We allow the $[PO_4^{3-}]/TCO_2$ ratio to vary by ±33%, which is slightly more than the $\delta^{13}C$ difference variation. Holding one parameter constant, either PO_4^{3-} or TCO_2 could thus have varied by ±33°/ o.

The effect of a change in $[PO_4^{3-}]_D$ on pCO_2, with TCO_{2D} held constant, can be seen in Fig. 6. A 33% drop in $[PO_4^{3-}]_D$ would cause pCO_2 to rise to 430 uatm, while a 33% increase would cause pCO_2 to fall to 210 uatm.

These calculations can also be used to estimate the effect on pCO_2 of a change in the Redfield ratio (the C/P atom ratio) in organic matter. A drop in the Redfield ratio of 33% (i. e., one-third less CO_2 removed per organic phosphorus atom) would have exactly the same effect as a 33% drop in $[PO_4^{3-}]_D$, and a rise in the Redfield ratio would have the same effect as a comparable rise in $[PO_4^{3-}]_D$. A change in the deep ocean metabolic CO_2 content by this mechanism (as in the case of a change in $[PO_4^{3-}]$) would be reflected by a change in the planktonic-benthonic foram $\delta^{13}C$ difference, according to equation 10, and the pCO_2 change associated with a change in the Redfield ratio can again be constrained to the range of 210-430 uatm.

These calculations also indicate how pCO_2 would change given a change in the deep water preformed $[PO_4^{3-}]$. In our two box model, the entire oceanic PO_4^{3-} pool is assumed to be involved in the biological cycle. In the real ocean, only about 60% of the dissolved PO_4^{3-} is thus involved; the other 40% is preformed, sinking down with new deep water as dissolved PO_4^{3-}, rather than being injected by biological decomposition. How would the atmospheric pCO_2 change if there were a change in the fraction of oceanic PO_4^{3-} which is preformed? Figure 6 again gives an estimate. Changing our model ocean so that 1/3 of the 2.2 u mol/kg deep ocean PO_4^{3-} is preformed would have the same effect as reducing the model $[PO_4^{3-}]$ by 1/3, and would cause pCO_2 to rise to 430 uatm.

Effect of TCO_{2D} on pCO_2

The effect of TCO_{2D} on pCO_2 was calculated by varying TCO_{2D} while holding constant the CCD, $[CO_3^{=}]_{CCD}$, and the surface - deep TA and TCO_2 differences. From TCO_{2D}, CCD, and $[CO_3^{=}]_{CCD}$, TA_D was calculated. From TA_D and TCO_{2D}, and the surface-deep TA and TCO_2 differences, TA_S, TCO_{2S}, and finally pCO_2 were calculated. pCO_2 is plotted vs. TCO_{2D} in Fig. 7.

If TCO_{2D} were 33% lower than the present value, pCO_2 would fall to 100 u atm. If TCO_{2D} were 33% higher than the present value, pCO_2 would rise to 630 u atm. According to the argument in the last section, these values likely represent the limits of pCO_2 variability during the last 10 myr caused by variability in TCO_{2D}. Since the atmospheric pCO_2 can probably be fixed within narrower limits by paleoclimate data alone, we must conclude that it is not now possible to usefully constrain the Cenozoic pCO_2 from the sedimentary record.

From Figures 2-6, it is immediately apparent that changing the TCO_2 of seawater would cause the largest change in the atmospheric pCO_2. Furthermore, such changes would probably leave no imprint in the sedimentary $CaCO_3$ distribution.

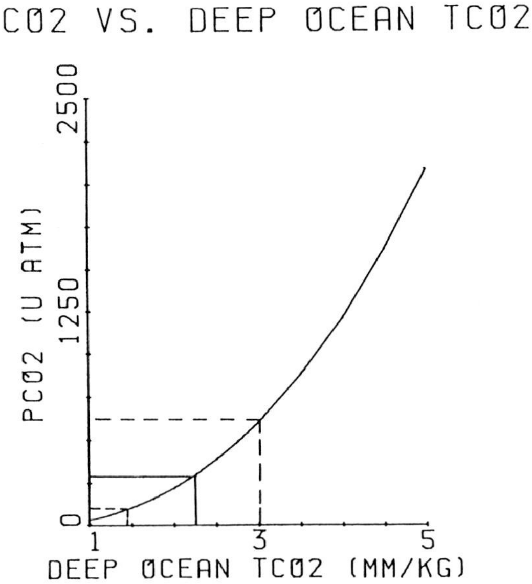

Fig. 7. Results of model calculations for the variation of pCO_2 with TCO_{2D}. The calculations were done holding constant CCD, Ω_{CCD}, $CaCO_3/CH_2O$ rain ratio, $[Ca^{++}]$, and $[PO_4^{3-}]_D$. The dashed line indicates the modern ocean conditions; the solid lines indicate the boundaries on the variability in TCO_{2D} the last 10 myr, assuming constant $[PO_4^{3-}]_D$.

As TCO_{2D} increases from 1 mM/kg to 5 mM/kg, model pCO_2 increases from 31 to 2094 uatm. However, the degree of carbonate saturation throughout the world oceans is virtually unaffected. Calculations not presented here show that model surface water Ω stays in the range 460 ± 40%, and model deep water Ω values would increase along the path of oxygen consumption (i.e., from the deep North Atlantic to the deep North Pacific) in much the same way regardless of TCO_2.

Summary

There is no satisfactory explanation of the long term geochemical control on pCO_2; therefore the pCO_2 of the paleo-atmosphere cannot be predicted *a priori*. However, past variations in pCO_2 can be constrained by modelling the oceanic CO_2 system in terms of variables which either do not greatly affect pCO_2 or are reflected in the sedimentary record. In this paper, we have examined the sensitivity of pCO_2 to compensation depth, seawater $[Ca^{++}]$, $CH_2O/CaCO_3$ particulate flux ratio, deep ocean $[PO_4^{3-}]$, deep ocean TCO_2, and deep ocean temperature. The possible pCO_2 variations which could accompany variations in the CCD, $[Ca^{++}]$, $CaCO_3/CH_2O$ flux ratio, $[PO_4^{3-}]_D$ and T_D would introduce only a small uncertainty in models of Cenozoic climates. Conversely, the changes observed in the CCD and T_D during the Cenozoic do not of themselves imply climatically significant variations of pCO_2. The possible pCO_2 variations which might accompany TCO_2 variations, on the other hand, are so large that Cenozoic pCO_2 cannot be fixed by our approach.

This paper lays out a framework for inverting the sedimentary record to estimate Cenozoic pCO_2. In the last decade, we have acquired the data for estimating most of the parameters required for the inversion, including CCD, $(CO_3^{=})_{CCD}$, paleoceanic (Ca^{++}), deep ocean temperature, and (to a lesser degree) $\delta^{13}C_S - \delta^{13}C_D$. To completely constrain the model, we need to measure $\Delta\delta^{13}C$ much better, along with another parameter recording the total or metabolic CO_2 concentration of deep ocean water. Hopefully the needed results will emerge from future paleoceanographic research.

Acknowledgments. I am grateful to Wallace Broecker for inspiring and encouraging my interest in the chemical history of the oceans, and to the organizers of the Fourth Maurice Ewing Symposium for stimulating me to think about Cenozoic pCO_2 by inviting me to contribute to the Symposium. David Graham of the Woods Hole Oceanographic Institution wrote the computer program used to do the carbonate system calculations, and shared a long collaboration with me on Cenozoic paleoceanography. W. Broecker and Taro Takahashi made very helpful critical reviews of the manuscript. This work was supported by a grant from the National Science Foundation (OCE 79-14594) as part of the CENOP program.

References

Arthur, M. A., The carbon cycle-controls on atmospheric CO_2 and climate in the geologic past. W. H. Bergen et al. (eds.), Climate in Earth History, ISBN 0-309-03329-2, NAS Wash., D.C., 1982.

Belanger, P. E., W. B. Curry and R. K. Mathews, Core-top evaluation of the paleo-oceanographic utility of benthic foraminiferal isotopic ratios. Paleo., Paleo., Paleo., 33, 205-220, 1981.

Bender, M. L., and D. W. Graham, On Late Miocene abyssal hydrography. Marine Micropaleo., 6, 451-464, 1981.

Brass, G. W., The variation of marine $^{87}Sr/^{86}Sr$ ratio during Phanerozoic time: interpretation using a flux model. Geochem. Cosmochim. Acta, 40, 721-730, 1976.

Brewer, P. G., and J. C. Goldman, Alkalinity changes by phytoplankton growth, Limnol. & Oceanog., 21, 108-117, 1976.

Broecker, W. S., A kinetic model for the chemical composition of sea water. Quatern. Res. 1, 188-207, 1971.

Broecker, W., Chemical Oceanography. Harcourt Brace Jovanovich, New York, 1974.

Broecker, W., Ocean chemistry during glacial times. Geochim. Cosmochim. Acta, 46, 1689-1705, 1982.

Curry, W. B., and R. K. Mathews, Paleo-oceanographic utility of oxygen isotopic measurements on planktic foraminifera: Indian Ocean core top evidence. Paleo., Paleo., Paleo., 33, 173-191, 1981.

Delaney, M. L. Foramineral trace elements: uptake, diagenesis, and 100 myr paleochemical history. Ph.D. Thesis, Mass. Inst. of Tech., 253 pp., 1983.

Durazzi, J. T., Stable-isotope studies of planktonic foraminifera in North Atlantic core tops. Paleo., Paleo., Paleo., 33, 157-172, 1981.

Elderfield, H., and M. Greaves, Strontium isotope geochemistry of Icelandic geothermal systems and implications for seawater chemistry. Geochim., Cosmochim., Acta, 45, 2201-2212, 1981.

Graham, D. W., M. L. Bender, D. F. Williams, and L. D. Keigwin, Jr., Strontium-calcium ratios in Cenozoic planktonic foraminifera. Geochim. Cosmochim. Acta, 46, 1281-1292, 1982.

Edmond, J. M., C. Measures, R. E. McDuff, L. H. Chan, R. Collier, B. Grant, L. I. Gordon, and J. B. Corliss, Ridge crest hydrothermal activity and the balances of major and minor elements in the ocean: the Galapagos data. Earth Planet. Sci. Lett., 46, 1-18, 1979.

Graham, D. W., B. Corliss, M. Bender, and L. Keigwin, Jr., Carbon and oxygen isotopic disequilibria of Recent Deep-Sea benthic foraminifera. Marine Micropaleo., 6, 483-497, 1981.

Holland, H. D., The Chemistry of the Atmosphere and Oceans. John Wiley and Sons, New York, 351 pp., 1978.

Keigwin, L. D. Jr., Cenozoic stable isotope stratigraphy, biostratigraphy, and paleoceanography of deep sea sedimentary sequences. Ph. D. Thesis, University of Rhode Island, 1979.

Keigwin, L., Isotopic paleoceanography of the Caribbean and East Pacific: role of Panama uplift in Late Neogene time. Science, 217, 350-353, 1982.

Killingley, J. S., R. F. Johnson and W. H. Berger, Oxygen and carbon isotopes of individual shells of planktonic foraminifera from Ontong-Java Plateau, equatorial Pacific. Paleo., Paleo., Paleo., 33, 193-204, 1981.

Kroopnick, P., The distribution of ^{13}C in the Atlantic Ocean. Earth Planet. Sci. Lett., 49, 469-484, 1980.

Lorens, R., and M. Bender, The imapct of solution chemistry on Mytilus edulis calcite and aragonite. Geochim. Cosmochim. Acta, 44, 1265-1279, 1980.

Loutit, T. S., Miocene stable isotope stratigraphy and paleoceanography. Ph.D. Thesis, University of Rhode Island, 260 pp., 1981.

Oeschger, H., This volume, 1983.

Rau, G. H., R. E. Sweeney and I. R. Kaplan, Plankton ^{13}C:^{12}C ratio changes with latitude: differences between northern and southern oceans. Deep Sea Res., 29, 1035-1039, 1982.

Savin, S. M., R. G. Douglas and F. G. Stehli, Tertiary marine paleotemperatures. Geol. Soc. of Am. Bull., 86, 1499-1510, 1975.

Shackleton, N. J., and J. P. Kennett, Paleotemperature history of the Cenozoic and the initiation of Antactic glaciation: oxygen and carbon isotope analyses in DSDP sites 277, 279, and 281. In: J. P. Kennett, R.E. Houtz, and others, Init. Repts. Deep Sea Drilling Project, 29, U.S. Govt. Printing Office, Washington, DC.,743-755, 1975.

Sillen, L. G., The ocean as a chemical system. Science, 156, 1189-1197, 1967.

Takahashi, T., Carbonate chemistry of sea water and the calcite compensation depth in the oceans. Cushman Foundation for Foraminiferal Research, Spec. Pub., 13, 11-26, 1975.

Takahashi, T., W. S. Broecker, A. E. Bainbridge and R. F. Weiss, Carbonate chemistry of the Atlantic, Pacific, and Indian Oceans: The results of the Geosecs Expeditions, 1972-1978, Tech. Rept. No. 1, CU-1-80. Lamont-Doherty Geological Observatory, Palisades, NY, 1980.

Takahashi, T., W. S. Brocker, and A. E. Bainbridge, The alkalinity and total carbon dioxide concentration in the world oceans, in: B. Bolin, ed., SCOPE Vol. 16, J. Wiley & Sons, NY, 271-286, 1981.

Turekian, K. K., The marine geochemistry of strontium. Geochim. Cosmochim. Acta, 28, 1479-1496, 1964.

Van Andel, T. H., Mesozoic/Cenozoic calcite compensation depth and the global distribution of calcareous sediment. Earth Planet. Sci. Lett., 26, 187-194, 1975.

Walker, J. C. G., P. B. Hays and J. F. Kasting, A negative feedback mechanism for the long-term stabilization of Earth's surface temperature. Jour. Geophys. Res., 86, 9776-9782, 1981.

Williams, D. F., M. A. Sommer, II and M. L. Bender, Carbon isotopic compensation of Recent planktonic foraminifera of the Indian Ocean. Earth Planet. Sci. Lett., 36, 391-403, 1977.

CADMIUM IN BENTHIC FORAMINIFERA
AND ABYSSAL HYDROGRAPHY: EVIDENCE FOR A 41 KYR OBLIQUITY CYCLE

Edward A. Boyle

Department of Earth, Atmospheric and Planetary Sciences,
Massachusetts Institute of Technology

Abstract. Cd/Ca in fossil benthic foraminifera records the variability of bottom water nutrient concentrations. Detailed records of Cd/Ca in benthic foraminifera have been obtained for (1) 4 cores covering the glacial/interglacial transition in the 2500-3500m water depth range at 42°N in the western North Atlantic and (2) a 285 kyr record at 2925 m on the Rio Grande Rise(32°S). Nutrient concentrations in the deep North Atlantic 18,000 years ago were roughly twice as high as at present. This evidence is consistent with a substantial diminution of North Atlantic Deep Water flux relative to southern water sources, but not with a complete cessation of NADW. It is believed that the formation areas of NADW were ice covered during the last glacial maximum, thereby preventing deep water formation by eliminating large heat losses to the atmosphere. The Rio Grande Rise record shows a strong 41 kyr periodicity during the past 285 kyr. These fluctuations are synchronous with changes in the tilt of the earth's axis, with low Cd/Ca (high relative North Atlantic Deep Water flux) corresponding to tilt maxima. This may be interpreted as evidence that ice-cover of the formation areas of NADW is controlled by high-latitude insolation, with increased obliquity corresponding to higher summer insolation and less ice cover.

Introduction

Several lines of evidence indicate that the deep ocean environment has undergone significant fluctuations during the last 200 kyr. Populations of benthic foraminifera have changed markedly (Streeter, 1973; Streeter and Shackleton, 1979; Schnitker, 1974; Corliss, 1979; Corliss, 1982). Duplessy et. al. (1975,1980), Kellogg, Duplessy, and Shackleton (1978);, and Streeter et. al. (1982) provided oxygen isotope evidence that the Norwegian Sea was not a source of bottom water during the most recent glaciation. Curry and Lohmann (1982) and Mix et. al. (1982) have presented carbon isotope evidence which is interpreted as resulting from major changes in North Atlantic Deep Water production on a glacial/interglacial time scale.

Uncertainty exists about the quantitative interpretation of this evidence. Benthic foraminifera do not respond as simply to deep water parameters as was once thought (e.g. Corliss, 1979; Belanger and Streeter, 1980; Bremer and Lohmann, 1982). The oxygen isotope evidence in the Norwegian Sea is clear, but in the open ocean $\delta^{18}O$ changes are subtle and artifacts of species fractionation and of sedimentology are of concern. The interpretation of carbon isotope data has been complicated by the arguments of Shackleton (1977) and Broecker (1982 a,b) that the average oceanic $\delta^{13}C$ can be altered by as much as 1 per mil by a transfer of reduced continental or continental shelf carbon into the ocean. Although this latter effect appears to be less than originally thought (N.J. Shackleton, personal communication; W.H. Berger, personal communication) and the magnitude of this inventory shift can be evaluated at coarse resolution by carbon isotope studies on cores spread throughout the ocean, it is difficult to separate these influences on benthic $\delta^{13}C$ to construct a detailed time history of benthic hydrography.

In view of these difficulties, I have developed a new tracer of deep ocean circulation patterns: the cadmium content of benthic foraminifera tests. This method is described and documented in detail elsewhere (Boyle, 1981; Hester and Boyle, 1982; Boyle and Keigwin, 1982). Briefly, the distribution of cadmium in the ocean is similar to that of phosphorus. Since North Atlantic Deep Water forms from nutrient-depleted near-surface waters of subtropical origin, it produces a low-nutrient 'tongue' in the southerly spreading NADW, variations in the nutrient content of deep Atlantic waters are useful in tracing the thermohaline circulation patterns of the deep ocean. Since cadmium is incorporated into the shells of certain benthic foraminifera in proportion to its concentration in bottom waters overlying the sediment, cadmium can be utilized as a proxy for nutrient distributions.

The use of cadmium as a tracer of past ocean

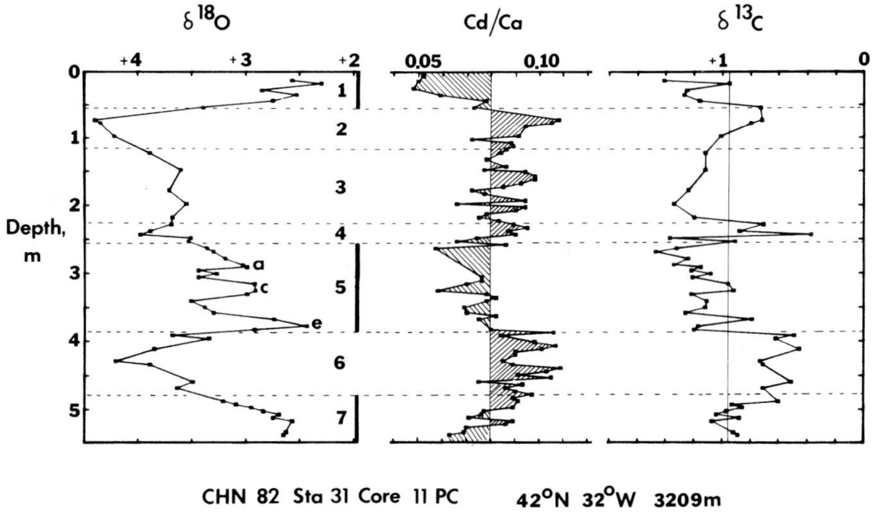

Fig. 1. Cd/Ca, $\delta^{18}O$, and $\delta^{13}C$ data from benthic foraminifera from a high-latitude western North Atlantic sediment core (3209 m water depth). Isotope data relative to PDB (on C. wuellerstorfi) and Cd/Ca data (μmol/mol) reported by Boyle and Keigwin (1982). Cd/Ca data on C. kullenbergi and Uvigerina spp., which show no significant fractionation (Hester and Boyle, 1982). Oxygen Isotope stage numbers given to right of $\delta^{18}O$ data.

circulation is based on two assumptions: (1) the cadmium content of the foraminifera is strictly proportional to that of the bottom water, and (2) the oceanic inventory of cadmium is relatively stable over periods long relative to the periods of oceanic fluctuation.

The first of these assumptions is supported by several types of evidence: core top studies of planktonic and benthic foraminifera (above references) and culturing experiments on living foraminifera utilizing radiotracer Cd (described in a manuscript by Margaret Delaney, the late Allan Bé, and myself submitted to Geochimica Cosmochimica Acta). Although it is still possible that there are residual artifacts not related to water chemistry, these appear to be less severe than those observed in carbon isotope studies of benthic foraminifera.

The second of these assumptions is only approximately true. Studies of the flux of cadmium into the ocean from rivers and estuaries (Boyle, Huested, and Grant, 1982; Shiller, Boyle, and Buesseler, 1982; Grant et. al.; 1982) indicate that the oceanic residence time of cadmium is about 2×10^5 years. As Broecker (1971) has indicated, it is not expected that steady-state mechanisms can produce significant variations on a glacial/interglacial time scale given a residence time of this length. However, it is possible that non-steady-state mechanisms could produce significant inventory changes of short duration (Broecker, 1982 a,b; this volume). The best evidence against this possibility would be detailed studies on cores from throughout the ocean showing that the oceanic inventory is stable. This has not yet been achieved, but studies in progress on an equatorial Pacific ocean core show (Boyle and Keigwin, in preparation) that over the last 225 kyr, glacial oceans have only about 20% more cadmium than interglacial oceans. Although a change of this magnitude is a significant factor in the cadmium variability in deep sea sediment cores, in the Atlantic cores reported here oceanic Cd inventory changes are less important than changes in deep ocean circulation. For simplicity, the signals described here will be discussed as if they were solely due to circulation changes; this assumption is true only to the first approximation, however.

In this paper I will outline some of the principal findings of Cd/Ca studies over the last year. The study of many of these cores is still in progress, and a detailed discussion of Cd/Ca, $\delta^{13}C$, and $\delta^{18}O$ will be provided elsewhere (Boyle and Keigwin, in preparation).

Benthic Hydrography of the Deep Western North Atlantic Over the Past 200,000 Years - Cd/Ca and Isotopic Evidence

Boyle and Keigwin (1982) reported the first joint study of Cd/Ca, $\delta^{18}O$, and $\delta^{13}C$ in benthic foraminifera in a sediment core from the high-latitude western North Atlantic (3209m water depth); these results are reproduced in figure 1. From these data we argued that there is a substantial diminution of the flux of North Atlantic Deep Water (relative to the flux of southern deep and intermediate waters) during periods of extreme glacial climate (oxygen isotope stages 2 and 6). But the Cd/Ca in the benthic foraminifera from this core is never as high as would be expected from complete cessation of North Atlantic Deep Water formation (about 0.14-0.16

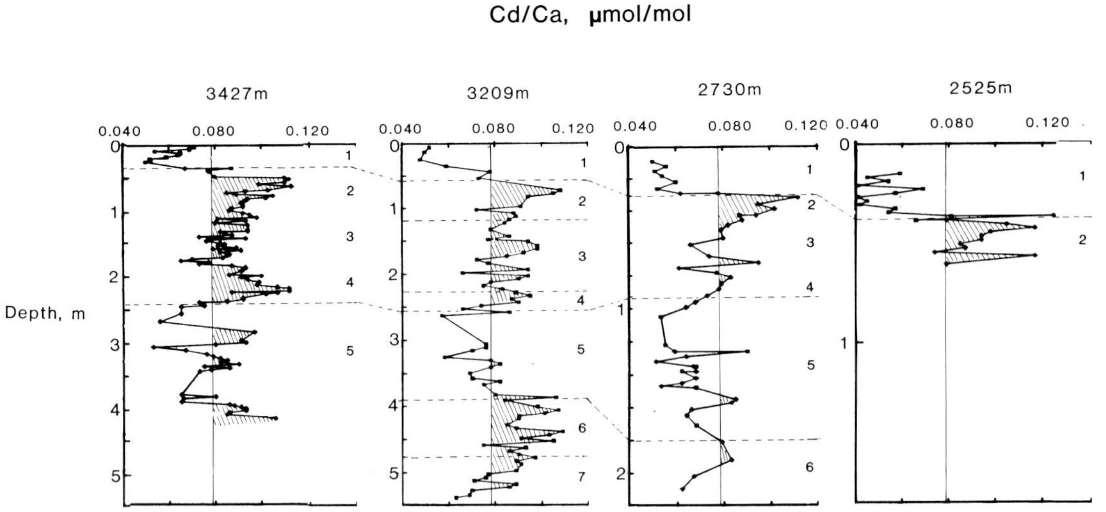

Fig. 2. Cd/Ca records from benthic foraminifera in cores covering a 1000 m depth interval in the North Atlantic at 42 degrees north. Stratigraphy is based on oxygen isotopes (two left-hand cores) and carbonate percentages and Uvigerina percentage data (two right hand cores). Cd/Ca on Uvigerina spp. and C. kullenbergi.

µmole Cd/mole Ca), the maximum observed values being 0.11 µmol/mol. This evidence indicates that there was always a source of nutrient-depleted waters into the deep western North Atlantic.

In order to determine whether the hydrographic changes observed at 3209 m were representative of a larger depth interval, 3 more cores from this region have been studied (figure 2). All of these cores have been analyzed down to the last glacial maximum. It is clear that the magnitude of the glacial/interglacial cadmium change is virtually identical between these cores covering a 1000m depth interval (figure 3). The concurrence of these cores indicates that conclusions based on the 3209 m core were not due to sedimentological or analytical artifacts. Abyssal hydrographic change in the far North Atlantic is uniform over at least that 1000 m depth interval.

As mentioned earlier, it is not yet possible to prove with certainty that all of the cadmium change is due to circulation changes and not to inventory changes. However, as evidence from cores throughout the ocean accumulates, the assumption of constant cadmium inventory will be tested. Once verified, it will be possible to subtract the cadmium (circulation) signal from the carbon isotope signal (circulation + carbon inventory) to determine the exact magnitude of the carbon inventory fluctuations. It is evident that cadmium and carbon isotope data in combination could be more powerful than either separately.

Benthic Hydrography Near The Rio Grande Rise: Evidence For A 41 KYR Cycle

David Johnson (WHOI) collected core AII107 131GGC on the flank of the Rio Grande Rise (31°S,38°W, 2925 m). This core is located in a region of strong chemical gradients between North Atlantic and Southern Ocean waters and can be expected to be particularly sensitive to fluctuations in deep water production rates. Glenn Jones (LDGO) has provided a highly precise and detailed $CaCO_3$ stratigraphy for this core and Bill Curry (WHOI) has provided a benthic oxygen isotope curve (figure 4). Although the sedimentation rate is low, it is sufficient to retain the major features of climate change. The

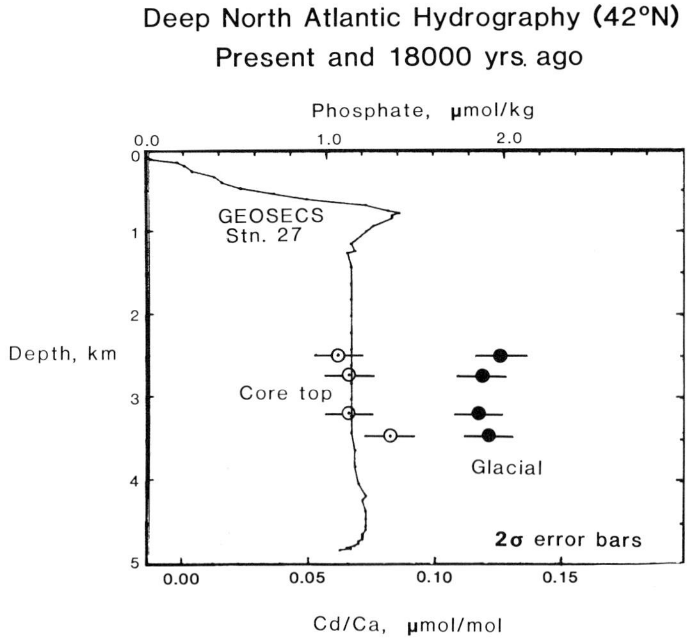

Fig. 3. Benthic hydrography of the western North Atlantic Ocean (42 degrees North) today and 18 kyr ago based on data from the cores shown in Figure 2.

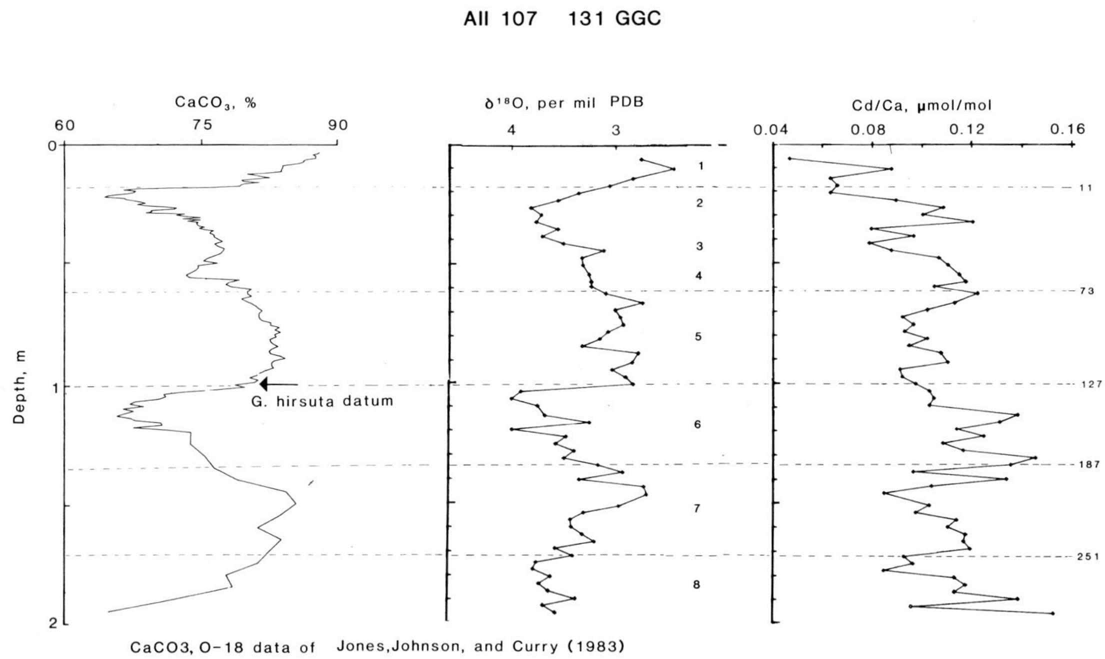

Fig. 4. Carbonate, oxygen isotope (C. wuellerstorfi), and Cd/Ca data (C. wuellerstorfi) from a core in the southwestern Atlantic Ocean. Oxygen Isotope stages indicated to right of $\delta^{18}O$ data; stratigraphic age horizons (kyr) given to right of Cd/Ca.

core was sampled at 3 cm intervals. Benthic foraminifera were handpicked at MIT from 10g samples, and Cd/Ca was analyzed in the species C. wuellerstorfi down the length of the core. This species was chosen because it occurs down the length of the core and its abundance variations are less than other species, thus minimizing artifacts due to biological sediment stirring. This species has not yet been fully calibrated relative to Uvigerina spp. and C. kullenbergi, but preliminary work indicates that any fractionation relative to these species is less than 20%. In any event, the use of a single species downcore eliminates fractionation as a factor in considering fluctuations. Splits of C. wuellerstorfi were provided to W. Curry (WHOI) for carbon and oxygen isotope analysis. The coarse fractions were sent to R. Thunell for micropaleontogical work and the fine fractions were forwarded to G. Jones and D. Ledbetter for mineralogic and grain-size studies. Thunell's planktonic micropaleontology agrees with the isotope stratigraphy, which both differ from the carbonate stratigraphy in the assignment of the 7-8 boundary; the isotopic boundary is used here. The Cd/Ca record from this core is listed in table 1 and plotted in figure 4.

The record in this core differs from that in the North Atlantic cores. There is some indication of a trend towards lower Cd/Ca in the more recent sediments. Although the trend is statistically significant on the basis of the data, I prefer to downplay this feature of the data because trends like this can be generated by mixing intensity variations (e.g. see Shackleton and Opdyke, 1976, for a record which shows $\delta^{18}O$ trends which are not significant: core V28-239). Much of the statistical trend would be eliminated by exclusion of the Holocene, which has not yet been mixed with the 'next glacial period'. The core also shows 'trends' in the oxygen isotope data which are not likely to be significant either. But the fluctuations recorded in this record are certainly significant. Although there is some evidence for the '100 kyr' climate fluctuation, there is a much stronger fluctuation of shorter 'periodicity'. There are about 7 of these fluctuations in a 285,000 year interval (i.e. about a 40 kyr period). It seemed possible that variations in the earth's obliquity (the axis of tilt of the earth's rotation relative to the orbital plane) might be implicated (see Imbrie, Hays, and Shackleton, 1976, for a discussion of orbital parameter changes in climate theories). Therefore this possibility was investigated in more detail. A time scale was constructed by linear interpolation between the oxygen isotope stage boundaries indicated in figure 4. The ages of these boundaries are based on the Shackleton and Opdyke (1973) time scale as slightly modified by dating of the Toba tuff ash layer at the 4-5 boundary (Ninkovitch et. al., 1978). The bottom of the core is clearly still in stage 8, so it was assigned an age of 285 kyr (the 8-9 boundary occurs at about 296 kyr). The linearized record was then subjected to spectral analysis to examine apparent periodicities in a more objective fashion.

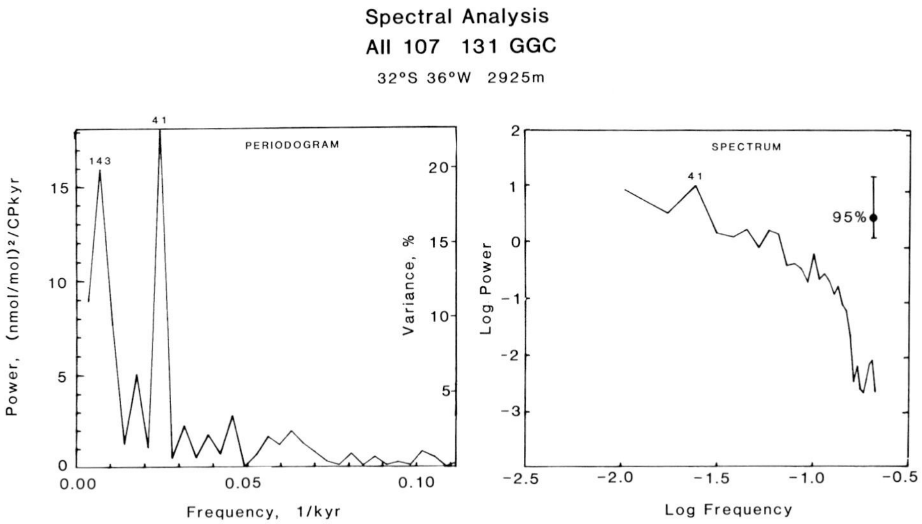

Fig. 5. Spectral analysis of Cd/Ca data in AII107 131GGC. The periodogram was converted into a spectral estimate by averaging 2 adjacent frequency bands.

The record was first detrended by subtraction of the least squares trendline. It was then converted to an equal spaced record by linear interpolation using 128 data points to avoid missing extrema. This procedure doubles the number of data points and introduces some high frequency noise into the spectra. Since the high frequency portion of the spectrum will not be considered here, this noise presents no significant problem. The first and last 10% of the record was tapered by a cosine bell to mimimize leakage. Then the resulting record was analyzed by use of the radix 2 Sande-Tukey Fast Fourier Transform algorithm (Brigham, 1974) on a Tektronix 4052 minicomputer. The periodogram of this record is displayed in figure 5a; the results indicate a very strong 41 kyr periodicity. Mix et. al. (abstract, 1982) report a 41 kyr periodicity in the $\delta^{13}C$ records of North Atlantic core V25-59 and V30-97. In order to assess the statistical significance of this result, adjacent bands were averaged using the Daniel estimator to provide four equivalent degrees of freedom (Koopmans, 1974). The two-sided confidence limits for this estimate are indicated in figure 5b. The periodicity appears to reach the 90% confidence limit relative to white noise.

Although the statistical significance of this result is not overwhelming, it is as strong as many other peaks in spectra from deep-sea sediment cores. In fact the spectrum of obliquity itself over this period would only indicate a slightly more significant result because of the short record length. Mix et. al. (abstract, 1982) report a 41 kyr periodicity in the $\delta^{13}C$ records of North Atlantic core V25-59 and V30-97. Since this cadmium record is one of the strongest indications of a 41 kyr periodicity in abyssal hydrography to date, it is worthwhile to suggest a mechanism that could produce such a periodicity. Fluctuations in the tilt of the earth's axis mainly result in changes in high latitude summer solar insolation. Figure 6 reproduces Vernekar's (1972) calculations for insolation at 70° latitude (the results of Berger (1976) are quite similar). High obliquity produces summer insolation maxima in both hemispheres. The 41 kyr obliquity cycle's influence on high latitude summer insolation is hemisphericly 'synchronous' in the sense that high obliquity produces more summer insolation in both hemispheres.

Ice-cover of NADW formation areas is the hypothesis favored to explain NADW diminishment 18 kyr ago (Streeter and Shackleton, 1979). It is but a small extension of this idea to assume that this ice cover is influenced by high latitude summer insolation. During periods of high obliquity, this modified hypothesis suggests that summer melting of the ice cover would be maximized, so that there would be a higher area of ice-free water during winter. This scenario is thus favorable for NADW production. This fluctuation would then be reflected in NADW chemistry in sensitive areas such as that sampled by this sediment core. The real ocean is capable of vastly more complicated behavior, but this hypothesis provides a 'minimum astonishment' scenario which can be tested by further observation and quantitative models.

One consequence of this hypothesis is that there should be a definite phase relationship between obliquity and NADW flux, with the NADW being maximized either at times of high obliquity, or perhaps slightly later to allow for a phase lag in the system. The estimated coherence of the cadmium record and obliquity is 0.97 (with a 5% probability that this is not significant) and the computed phase angle is 180° (i.e., low cadmium high relative NADW flux) corresponds to higher tilt angle. This coherence is examined in figure 7. The original record is reproduced along with a filtered version of the record. The filter

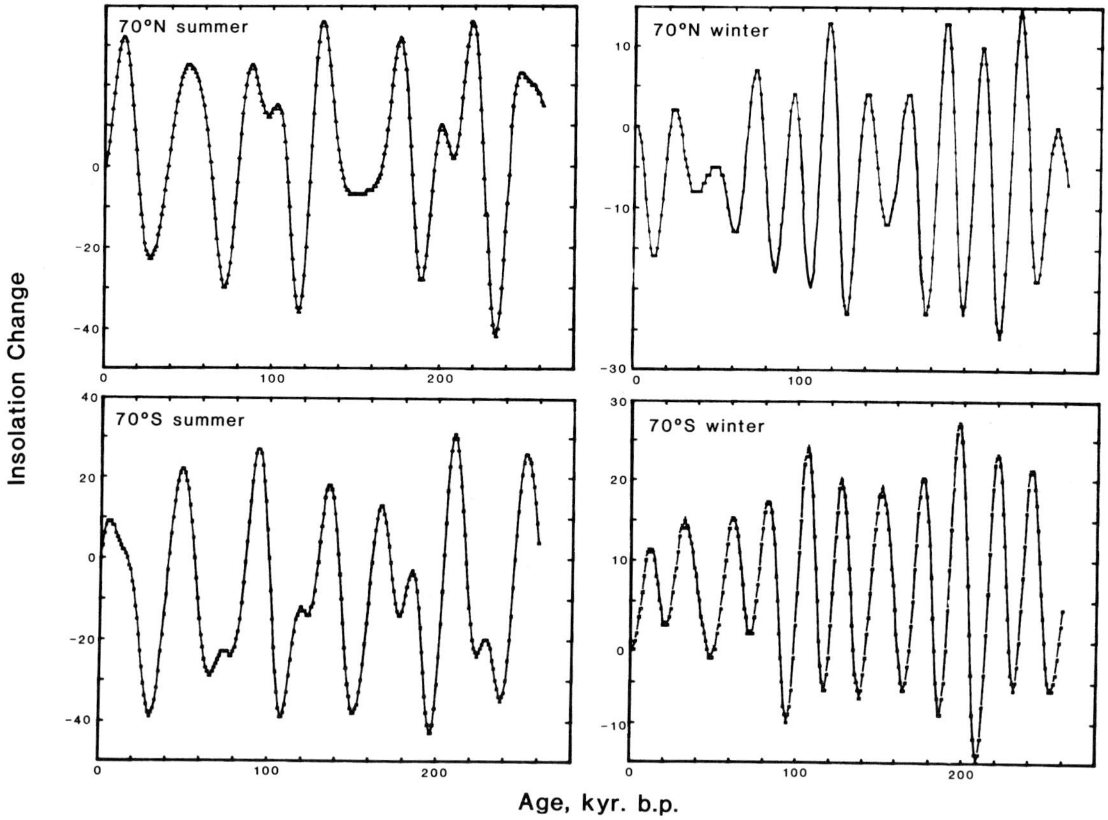

Fig. 6. Insolation calculations for 70 degrees north and south latitude bands, based on the work of Vernekar (1972). Insolation change relative to present given in units of cal/cm^2/day.

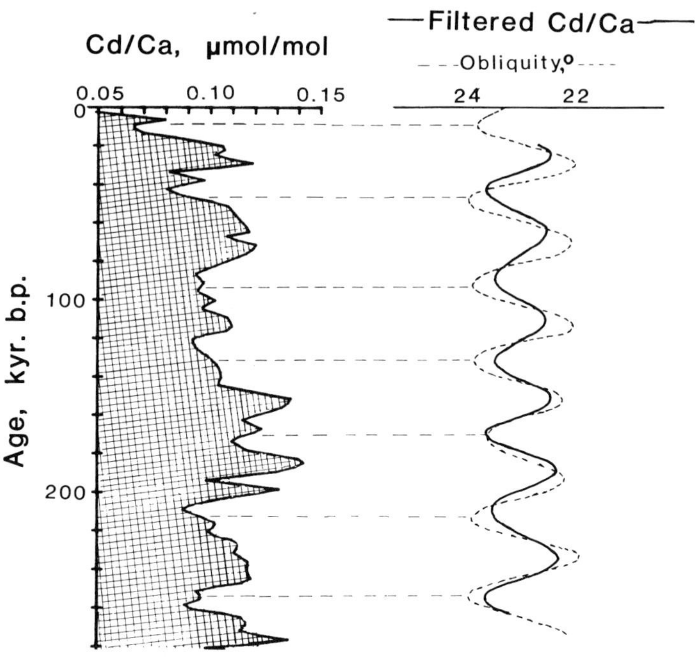

Fig. 7. Comparison of variations in the earth's obliquity and raw and filtered Cd/Ca data from AII107 131GGC.

employed was a frequency-domain bandpass with 50% attenuation at 56 kyr and 29 kyr. It can be seen that within the resolution of this record (probably no better than 5 kyr) the obliquity and Cd/Ca records are in phase over the last 285 kyr. This evidence provides support for the above conceptual model of NADW variability.

It must be emphasized that this core is located in a region of strong hydrographic gradients, and therefore is unusually sensitive to NADW fluctuations. The magnitude of this fluctuation may be relatively small, and it is not clear that the signal should be as strong in other regions. It is also necessary to remark on the low sedimentation rate in the core; it would be desirable to confirm this record in longer cores from higher-sedimentation rate regions of the Rio Grande Rise. Finally, one might wonder why the carbon isotope record from this core does not show a clear 41 kyr signal. This difference between the two tracers can be explained by two factors. (1) Carbon isotope fluctuations are largely due to changes in the inventory of continental reduced carbon (Shackleton, 1977; Broecker, 1982), so that a large non-circulation signal is superimposed on this tracer. (2) In low sedimentation rate cores such as this one, a single sampling of 6 individuals does not represent the mean bioturbated value very well. Since the isotopic measurements on this core were undertaken only for stratigraphic purposes, very few replicates were run. As a result, it is to be expected that the isotopic data suffers from more random error than the cadmium data, where samples were done in triplicate wherever possible (table 1).

Conclusions. (1) Although there is evidence for a substantial diminution of North Atlantic Deep Water (relative to southern source waters), there has been a continuous supply of nutrient depleted waters into the North Atlantic.

(2) Subject to confirmation of the constant cadmium inventory assumption, this evidence indicates that carbon isotope variations in North Atlantic sediment cores can be strongly influenced

TABLE 1. Cd data for core AII107 131GGC. No replicates were excluded.

Depth	Mean Cd/Ca (μmol/mol)	No. of replicates	Std. dev. (n>2) or dev. from mean (n=2)	$\delta^{18}O$ (PDB)	$\delta^{13}C$
6	.049	1	-	2.76	0.96
10	.088	1	-	2.44	1.10
14	.065	2	.001	2.84	0.80
17	.067	2	.006	2.76,3.37	1.07,0.65
20	.065	1	-	3.36	0.67
23	.090	2	.004	3.56	0.55
26	.109	2	.001	3.58,4.06	0.84,0.75
29	.101	2	.012	3.57,3.86	0.71,0.86
32	.120	2	.003	3.66,3.88	1.00,0.98
35	.081	2	.004	3.42,3.70	0.81,0.84
38	.097	3	.017	3.71	0.85
41	.080	2	.007	3.39,3.63	0.93,0.88
44	.088	2	.003	3.32,2.92	0.79
47	.107	3	.021	3.33	0.91
50	.111	2	.009	3.32	0.78
54	.115	2	.007	3.38	0.89,1.00
57	.118	3	.023	3.24	0.94
59	.105	2	.004	3.24	0.87
62	.122	3	.025	3.11,3.10	0.99,1.00
66	.113	3	.013	2.75	0.64
69	.103	3	.021	3.01	0.69
72	.093	2	.004	2.97,2.94	0.66,0.73
75	.097	2	.009	3.08,2.54, 3.29	1.12,0.90 ,0.88
78	.078	2	.004	3.10,3.06	1.14,0.94
81	.081	3	.017	3.16	0.74
84	.084	2	.006	3.33	0.86
87	.087	3	.030	2.79	0.90
91	.091	2	.001	2.85	0.69
94	.094	2	.006	3.36,2.74, 3.00	0.60,0.91 ,0.64
97	.097	3	.031	2.91	0.79
100	.100	2	.001	3.84	0.67
103	.103	3	.015	3.92	0.34
106	.106	3	.015	4.01	0.45
109	.109	2	.013	3.76	0.40
113	.113	2	.016	3.69	0.45
116	.116	3	.014	3.56	0.82
119	.119	2	.001	4.01	0.46
122	.122	3	.014	3.49,3.49	0.69,0.64
125	.125	2	.004	3.62,3.56	0.44,0.73
128	.128	3	.012	3.47,3.35	0.45,0.51
131	.131	2	.025	3.51	0.71
134	.134	2	.005	3.18	0.74
137	.137	1	-	3.03,2.85	0.36,0.85
140	.140	3	.027	3.36	0.59
143	.104	3	.021	2.74	0.98
146	.086	2	.004	2.72	1.14
151	.104	2	.001	3.21,2.74	0.87,0.72

TABLE 1. (continued)

Depth	Mean Cd/Ca (μmol/mol)	No. of replicates	Std. dev. (n>2) or dev. from mean (n=2)	$\delta^{18}O$ (PDB)	$\delta^{13}C$
154	.098	2	.007	0.32	0.58
157	.114	2	.001	3.45	0.45
160	.111	2	.004	3.54,3.35	0.76,0.93
163	.118	3	.014	3.55	0.85
166	.117	3	.017	3.21	0.62
169	.119	2	.005	3.59	0.71
172	.094	2	.006	3.42	0.59
175	.097	2	.000	3.77	0.83
178	.086	2	.005	3.77,3.84	0.77,0.71
181	.113	1	-	3.92,3.52	0.78,0.64
184	.118	2	.027	3.74	0.52
187	.113	2	.027	3.55	0.18
190	.138	1	-	3.39	-0.05
193	.096	1	-	3.70	0.24
196	.152	1	-	3.58	0.38

Pooled relative standard deviation for Cd (134 analyses of 62 samples): 16%. Much of this scatter can be attributed to sampling statistics, because the average sample size was about 6 individuals and biological stirring creates a mix of values at each depth.

by deep ocean circulation changes. A signficant portion of the benthic carbon isotope fluctuations may be due to circulation changes and continental carbon transfers as suggested by Shackleton (1977) and Broecker (1982).

(3) There is evidence for a 41 kyr fluctuation in NADW flow in the South Atlantic Ocean. It is suggested that this fluctuation may be a consequence of the effect of obliquity-induced summer insolation variations on the ice cover of the NADW formation areas.

Acknowledgements. I thank Jim Broda and David Johnson for help and generous provision of core samples. Glenn Jones and Bill Curry generously provided copies of their data on core AII107 131GGC. I thank Doug Luther, Charlie Eriksen, and Carl Wunsch for instruction in spectral analysis. Comments on this work by Lloyd Keigwin, Michael Bender, Wally Broecker, and Bill Curry have been helpful. Core collection and curation at Woods Hole Oceanographic Institution is supported by NSF grant OCE 2025231. This research is supported by NSF grant OCE 8018665.

References

Berger, A. (1976), Obliquity and precession for the last 5,000,000 years, Astron. Astrophys. 51:127-135.

Boyle, E.A. (1981), Cadmium, zinc, copper, and barium in foraminifera tests, Earth Planet. Sci. Lett. 53:11-35.

Boyle, E.A. and L.D. Keigwin (1982), Deep circulation of the North Atlantic during the last 150,000 years: geochemical evidence, Science 218:784-787.

Boyle, E.A., S.S. Huested, and B. Grant (1982) The chemical mass balance of the Amazon Plume II: Copper, nickel, and cadmium, Deep Sea Res., 29:1335-1346.

Belanger, P.E. and S.S. Streeter (1980), Distribution and ecology of benthic foraminifera in the Norwegian-Greenland Sea, Mar. Micropal. 5:401-428.

Bremer, M. L. and G.P. Lohmann (1982), Evidence for primary control of the distribution of certain Atlantic Ocean Benthonic Foraminifera by Degree of Carbonate Saturation, Deep Sea Res., 29: 987-998.

Brigham, E.O. (1974), The Fast Fourier Transform, Prentice-Hall Inc., Englewood Cliffs, N.J.

Broecker, W.S. (1971), Calcite accumulation rates and glacial to interglacial changes in oceanic mixing, in Late Cenezoic Glacial Ages (ed. K.K. Turekian) Yale Univ. Press, New Haven, pp. 239-265.

Broecker, W.S. (1982), Ocean chemistry during glacial time, Geochim. Cosmochim. Acta 46:1689-1705.

Broecker, W.S. (1982), Glacial to interglacial changes in ocean chemistry, Prog. Oceanog. 11:151-197.

Corliss, B.H. (1979), Recent Deep-Sea benthonic foraminifera distributions in the southeast Indian Ocean: inferred bottom-water routes and ecological implications, Mar. Geol. 31:115-138.

Corliss, B.H., (1982) Linkage of North Atlantic and Southern Ocean deep-water circulation during glacial intervals, Nature 298:458-460.

Curry, W. and G.P. Lohman (1982), Carbon Isotopic Changes in Benthic Foraminifera from the Western South Atlantic: Reconstruction of Glacial

Abyssal Circulation Patterns, Quaternary Research, 18:218-235.

Duplessy, J.C., L. Chenouard, and F. Vila (1975), Weyl's theory of Glaciation supported by isotopic study of Norwegian core K11, Science 188:1208-1209.

Duplessy, J.C., J. Moyes, and C. Pujol (1980), Deep water formation in the North Atlantic Ocean during the last ice age, Nature 286:479-482.

Grant, B., Hu Ming-hui, E. Boyle, and J.M. Edmond (1982), Comparision of the trace metal chemistry in the Amazon, Orinoco, and Yangtze plumes, Trans. Am. Geophys. Union 63:48.

Hays, J.D., J. Imbrie, and N.J. Shackleton (1977), Variations in the earth's orbit: a reply, Science 198:529-530.

Hester, K. and E. Boyle (1982), Water chemistry control of cadmium content in recent benthic foraminifera, Nature 298:260-262.

Kellogg, T.B., Duplessy, J.C., and N.J. Shackleton (1978), Planktonic foraminiferal and oxygen isotopic stratigraphy and paleoclimatology of Norwegian Sea deep-sea cores, Boreas 7:61-73.

Koopmans, L.H., The Spectral Analysis of Time Series, Academic Press, New York, pp. 257-280.

Mix, A., R.G. Fairbanks, W. Ruddiman, and A. McIntyre (1982), 1978, $\delta^{18}O$, $\delta^{13}C$, and $CaCO_3$ in Late Quaternary North Atlantic Sediments realted to abyssal circulation, GSA Abstracts with programs 14:567.

Ninkovich, D., N.J. Shackleton, A.A. Abdel-Monem, J.D. Obradovich, and G. Izett, (1978) K-Ar age of the late Pleistocen eruption of Toba, north Sumatra, Nature 276:574-577.

Schnitker, D. (1974), West Atlantic abyssal circulation during the past 120,000 years, Nature 248:385-387.

Shackleton, N.J. and N.D. Opdyke,(1973), Oxygen isotope and paleomagnetic stratigraphy of equatorial Pacific core V28-238, oxygen isotope temperatures and iced volumes on a 10^5 and 10^6 year time scale, Quat. Res. 3:39-55.

Shackleton, N.J. and N.D. Opdyke (1976), Oxygen-isotope and paleomagnetic stratigraphy of Pacific core V28-239 Late Pliecene to Latest Pleistocene, in: Investigation of Late Quaternary Paleoceanography and Paleoclimatology, (eds. R.M. Cline and J.D. Hays) Geol. Soc. Am. Memoir 145:449-464.

Shackleton, N.J. (1977), Carbon-13 in Uvigerina: tropical rainforest history and the equatorial Pacific carbonate dissolution cycles, in The Fate of Fossil Fuel CO_2 (eds. N.R. Andersen and A. Malahoff) Plenum Press, New York, pp. 401-427.

Shiller, A.M., E.A. Boyle, and K. Buesseler (1982), Trace metals in the Mississippi Plume, Trans. Am. Geophys. Union 63:1009.

Streeter, S.S. (1973), Bottom water and benthonic foraminifera in the North Atlantic-Glacial-Interglacial contrasts, Quat. Res. 3:131-141.

Streeter, S.S. and N.J. Shackleton (1979), Paleocirculation of the deep North Atlantic: 150,000 year record in benthic foraminifera and oxygen-18, Science 203:168-171.

Streeter, S.S., P.E. Belanger, T.B. Kellogg, and J.C. Duplessy (1982), Late Pleistocene Paleo-oceanography of the Norwegian-Greenland Sea: Benthic Foraminiferal Evidence, Quat. Res. 18:72-90.

Vernekar, A.D. (1972), Long - Period Global Variations of Incoming Solar Radiation, Met. Monographs, Vol. 12.